Molecular Electronics

IUPAC Secretariat: Bank Court Chambers, 2–3 Pound Way, Templars Square, Cowley, Oxford OX4 3YF, UK

INTERNATIONAL UNION OF PURE AND APPLIED CHEMISTRY

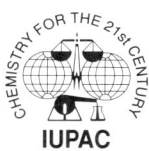

Molecular Electronics

A 'Chemistry for the 21st Century' monograph

EDITED BY

J. JORTNER

School of Chemistry,
Tel Aviv University,
Ramat Aviv, Israel

AND

M. RATNER

Department of Chemistry,
Northwestern University,
Evanston, Illinois, USA

Blackwell
Science

© 1997 International Union of Pure and
Applied Chemistry and published for them by
Blackwell Science Ltd
Editorial Offices:
Osney Mead, Oxford OX2 0EL
25 John Street, London WC1N 2BL
23 Ainslie Place, Edinburgh EH3 6AJ
350 Main Street, Malden
 MA 02148 5018, USA
54 University Street, Carlton
 Victoria 3053, Australia

Other Editorial Offices:
Blackwell Wissenschafts-Verlag GmbH
Kurfürstendamm 57
10707 Berlin, Germany

Blackwell Science KK
MG Kodenmacho Building
7–10 Kodenmacho Nihombashi
Chuo-Ku, Tokyo 104, Japan

First published 1997

Set by Semantic Graphics, Singapore
Printed and bound in Great Britain
by Hartnolls Ltd, Bodmin, Cornwall

The Blackwell Science logo is a
trade mark of Blackwell Science Ltd,
registered at the United Kingdom
Trade Marks Registry

DISTRIBUTORS

Marston Book Services Ltd
PO Box 269
Abingdon
Oxon OX14 4YN
(*Orders*: Tel: 01235 465500
  Fax: 01235 465555)

USA
 Blackwell Science, Inc.
 Commerce Place
 350 Main Street
 Malden, MA 02148 5018
 (*Orders*: Tel: 800 759 6102
   617 388 8250
   Fax: 617 388 8255)

Canada
 Copp Clark Professional
 200 Adelaide St, West, 3rd Floor
 Toronto, Ontario M5H 1W7
 (*Orders*: Tel: 416 597-1616
   800 815-9417
   Fax: 416 597-1617)

Australia
 Blackwell Science Pty Ltd
 54 University Street
 Carlton, Victoria 3053
 (*Orders*: Tel: 3 9347 0300
   Fax: 3 9347 5001)

A catalogue record for this title
is available from the British Library

ISBN 0-632-04284-2

Library of Congress
Cataloging-in-Publication Data

Molecular electronics:
a 'chemistry for the 21st century' monograph/
 edited by J. Jortner and M. Ratner.
   p.   cm.
 Includes bibliographical references
 and index.
 ISBN 0-632-04284-2
 1. Molecular electronics.
 I. Jortner, Joshua.
 II. Ratner, Mark A., 1942–
TK7874.8.M653   1997
621.381—dc21

97–93
CIP

Contents

Contributors

M.G. BAWENDI *Department of Chemistry, Massachusetts Institute of Technology, 77 Massachusetts Avenue, Cambridge, MA 02139, USA* [281]

D.N. BERATAN *Department of Chemistry, University of Pittsburgh, Pittsburgh, PA 15260, USA* [369]

R.R. BIRGE *Department of Chemistry and W.M. Keck Center for Molecular Electronics, Syracuse University, Syracuse, NY 13244, USA* [439]

A.J. BLACK *Department of Chemistry, Harvard University, Cambridge, MA 02138, USA* [343]

L.E. BRUS *Department of Chemistry, Columbia University, 3000 Broadway, New York, NY 10027, USA* [281]

M. BURGHARD *Max-Planck-Institut für Festkörperforschung, Heisenbergstr. 1, D-70569 Stuttgart, Germany* [255]

T.P. BURGIN *Department of Chemistry and Biochemistry, University of South Carolina, Columbia, SC 29208, USA* [191]

R.J. CAVE *Department of Chemistry, Harvey Mudd College, Claremont, CA 91711, USA* [73]

R.I. CUKIER *Department of Chemistry and Center for Fundamental Materials Research, Michigan State University, East Lansing, MI 48824-1322, USA* [119]

C.M. FISCHER *Max-Planck-Institut für Festkörperforschung, Heisenbergstr. 1, D-70569 Stuttgart, Germany* [255]

A. HELLER *Department of Chemical Engineering, The University of Texas at Austin, Austin, TX 78712-1062, USA* [241]

J. JORTNER *School of Chemistry, Tel Aviv University, Ramat Aviv, Israel 69978* [5]

R. KOPELMAN *Department of Chemistry, University of Michigan, Ann Arbor, MI 48109-1055, USA* [393]

A.N. KOROTKOV *Department of Physics, State University of New York, Stony Brook, NY 11794-3800, USA and Nuclear Physics Institute, Moscow State University, Moscow 119899, Russia* [157]

S.J. LANGFORD *School of Chemistry, University of Birmingham, Edgbaston, Birmingham, B15 2TT, UK* [325]

M. MORILLO *Fisica Teorica, Facultad de Fisica, Universidad de Sevilla, Apdo Correos 1065 Sevilla 41080, Spain* [119]

C.J. MULLER *Center for Microelectronic Material and Structures, Yale University, PO Box 208284, New Haven, CT 06520, USA* [191]

R.W. MURRAY *Kenan Laboratories of Chemistry, University of North Carolina, Chapel Hill, NC 27599-3290, USA* [215]

P.F. NEALEY *Department of Chemical Engineering, University of Wisconsin–Madison, 1415 Johnson Dr , Madison, WI 53706, USA* [343]

M.D. NEWTON *Department of Chemistry, Brookhaven National Laboratory, Box 5000, Upton, NY 11973, USA* [73]

D.J. NORRIS *Department of Chemistry and Biochemistry, University of California, San Diego, 9500 Gilman Drive, La Jolla, CA 92093-0340, USA* [281]

J.N. ONUCHIC *Department of Physics, University of California, San Diego, La Jolla, CA 92093, USA* [369]

B. PARSONS *Department of Chemistry and W. M. Keck Center for Molecular Electronics, Syracuse University, Syracuse, NY 13244, USA* [439]

R. RAJAGOPALAN *Applied Materials Inc., 3100 Bowers Avenue, Santa Clara, CA 95054, USA* [241]

M.A. RATNER *Department of Chemistry, Northwestern University, Evanston IL 60208, USA* [5]

F.M. RAYMO *School of Chemistry, University of Birmingham, Edgbaston, Birmingham, B15 2TT, UK* [325]

M.A. REED *Center for Microelectronic Material and Structures, Yale University, PO Box 208284, New Haven, CT 06520, USA* [191]

S.M. RISSER *Department of Physics, Texas Commerce, TX 75429, USA* [369]

S. ROTH *Max-Planck-Institut für Festkörperforschung, Heisenbergstr. 1, D-70569 Stuttgart, Germany* [255]

T. SHIMIDZU *Kansai Research Institute, Kyoto Research Park, 17 Chudoji-minami-machi, Shimogyo-ku, Kyoto 600, Japan* [381]

S.S. SKOURTIS *Department of Chemistry, University of Pittsburgh, Pittsburgh, PA 15260, USA* [369]

Q.W. SONG *Department of Chemistry and W.M. Keck Center for Molecular Electronics, Syracuse University, Syracuse, NY 13244, USA* [439]

J.F. STODDART *School of Chemistry, University of Birmingham, Edgbaston, Birmingham, B15 2TT, UK* [325]

J.R. TALLENT *Department of Chemistry and W.M. Keck Center for Molecular Electronics, Syracuse University, Syracuse, NY 13244, USA* [439]

W. TAN *Department of Chemistry and The Brain Institute, University of Florida, Gainesville, FL 32611, USA* [393]

R.H. TERRILL *Department of Chemistry, University of Illinois, Urbana, IL 61820, USA* [215]

J.M. TOUR *Department of Chemistry and Biochemistry, University of South Carolina, Columbia, SC 29208, USA* [191]

G.M. WHITESIDES *Department of Chemistry, Harvard University, Cambridge, MA 02138, USA* [343]

J.L. WILBUR *Department of Chemistry, Harvard University, Cambridge, MA 02138, USA* [343]

C. ZHOU *Center for Microelectronic Material and Structures, Yale University, PO Box 208284, New Haven, CT 06520, USA* [191]

Introduction

Molecular electronics constitutes a multidisciplinary research area focusing on the potential utilization of molecular scale systems and molecular materials for electronic or optoelectronic applications. The goals of the field are dual. First, it provides molecular materials for a multitude of electronic applications. Second, it attempts to utilize single molecular scale systems, e.g., a molecule, super-molecule, molecular aggregate or cluster, as electronic devices for the processing of optical, electrical, magnetic, chemical or biological signals. The distinction between molecular-based materials for electronics and molecular scale electronic devices [1], pertains both to practical and conceptual aspects. Molecular materials for electronics are already widely realized, and molecular based electronics is a successful and rather well developed field. Organic photoconductors are prevalent active components in xerography, organic liquid-crystal displays are ubiquitous and organic photoresists are widely applied for silicon device fabrication. Molecular materials for more esoteric electronic applications, e.g. organic superconductors, organic transistors, rectifiers, light emitting diodes, hole burning memories and sensors have all been demonstrated.

While molecular materials for electronics already have a respectable past, molecular scale electronic devices have an intriguing future. Nanostructure components of integrated molecular electronics have already been realized in the context of lithographic and chemical preparations of quantum dots and extended quantum dot structures [2]. However, the utilization of a single molecular scale system to process signals is still in the conceptual, speculative and tantalizing stage, and has not yet been realized. The outstanding goal of molecular electronics pertains to the perspective utilization of synthesis, assembly and miniaturization on the molecular level to accomplish a huge density of devices, e.g., molecular wires, rectifiers, optoelectronic triggering switches, transistors and memories. Most of the potential application notions in molecular electronics devolve from more standard electronics. Unique applications in which only molecular systems can be used, e.g., molecular recognition-based sensors, molecular patterned growth or molecular interfaces with biological systems, are still in the embryonic stages, and deserve active exploration.

The conceptual genesis of molecular electronics emerged from Feynman's grand vision that 'there is plenty of room at the bottom' [3] and can be traced to the attempts to replicate some devices from solid state electronics on the molecular level. The complementary source of paradigmatic behaviour stems from biology, where nature has adopted the 'soft' route based on organics towards such functions as memory, information processing, photosensitivity and energy storage. The outstanding challenges of molecular electronics pertain to the advancement of molecular and biophysical microscopic systems, to reproduce the characteristics of microfabricated electronic devices and, even more interestingly, to strive towards unique applications of molecular scale systems. In this sense molecular and cluster systems for electronics applications constitute a bridge between the microfabricated systems (with the silicon devices at their upper limit of their size domain) and of molecular and biological systems (Fig. 1).

1

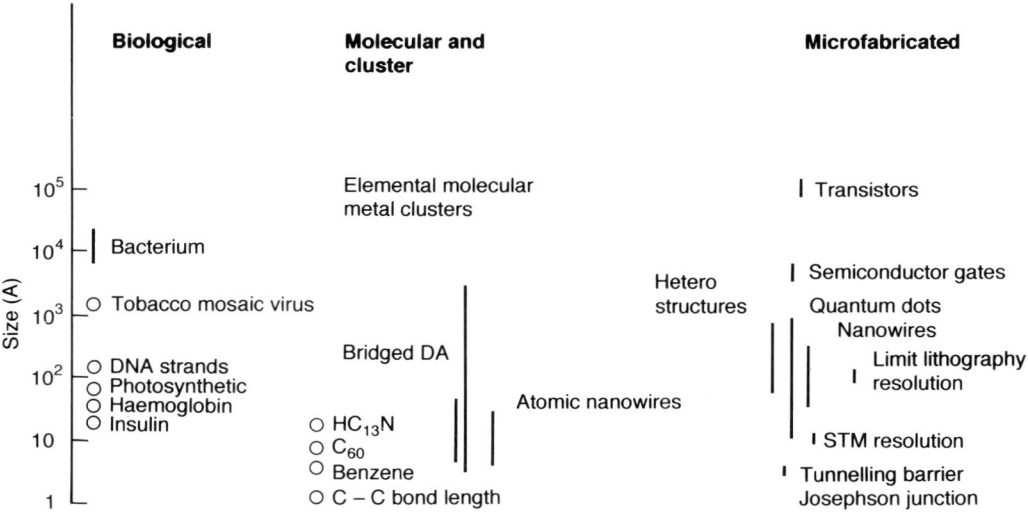

Figure 1

The potential application of molecular size electronic systems involves the intrinsic limitations of quantum mechanics, making microscopic electronics a (fascinating but) distinct field from macroscopic electronics. While in macroscopic electronics the concepts of resistance, capacitance and impedance are well rooted, the microscopic world is different. Here we are concerned with the response properties of a single molecular system [4]. The understanding of how a system's properties evolve from the macroscopic through the mesoscopic to the molecular level represents one of the challenges of the field of molecular based electronics, with cluster chemical physics being instrumental in building these bridges. At the molecular level one in concerned with subnanometre structure interrogated by the direct observation of individual molecules, e.g. via scanning tunnelling or atomic force microscopy [5] with excited-state energetics explored by a single-molecule optical response [6], with nuclear and charge transport dynamics investigated by electron transfer within a single molecule, and with information storage via photoisomerization or by hole burning of a single molecule. With the advent of femtosecond laser sources and their important applications to ultrafast chemical dynamics on the time scale of nuclear motion, the optically triggered response stretches over a time domain of fifteen orders of magnitude, from femtoseconds to seconds.

The exploration of the subnanometre structure, electronic level structure, response and dynamics of individual molecules, small samples containing several molecules, molecular ensembles and clusters will advance the conceptual basis and potential applications of molecular electronics. Necessary conditions for the utilization of molecular devices are high efficiency (yield), error-free operation and connectivity of these systems. Some conceptual intrinsic limitations of molecular based systems pertain to the confinement of elementary excitations, the localization of such excitations, the distinction between coherent and incoherent transport, the connectivity of these systems and the effects of quantum fluctuations [7,8]. The confinement and localization of elementary excitations (e.g. electrons, holes, excitons) within a finite quantum system will impose constraints on the coupling of molecular scale systems to one another and to the outside world. The realization of molecular circuitry constitutes a central conceptual, experimental and technical challenge. In this context, one should inquire what are the current characteristics in circuitry of nanometer dimensions, where ohmic behaviour may be inappropriate, and where electron transfer within the vibronic quasicontinuum of a molecular system and electronic tunnelling within an electronic continuum of a molecular system coupled to a metal will prevail. Another open question pertains to the manifestations of fluctuations, e.g. shot noise or vibrations, which may limit reproducibility.

These potential intrinsic limitations are counterbalanced by some unique features of molecular scale devices. Regarding optical response, time scales for excitation in the femtosecond domain will allow for the preparation of electronically–vibrationally excited wave packets accomplishing selectivity, sensitivity and speed of basic charge and energy transfer processes. Regarding the nature of charge and energy transport, this will involve coherent vs. incoherent motion. Coherent transport may be realized in small scale systems such as atomic wires [8], while in large systems incoherent transfer may be dominant. This brings up an implication of the central concept of cluster size effects: the characteristic dimension at which the molecular behaviour converges to the limiting behaviour of the bulk is property specific and depends on the constituents of the cluster. The exploitation of the discrete electronic level structure of finite systems over a large size domain will result in novel phenomena. Finally, the characteristics of small size and low dimensionality in charge and energy transport provide a rich set of phenomena, opening avenues for the optimization of transport properties in finite molecular scale systems. These are some of the central themes of current research in the field of molecular electronics.

Early work on molecular electronics, with its genesis about three decades ago, was demarked more by vision and enthusiasm than by appropriate experimental system preparation and interrogation methods or a rigorous conceptual framework. The situation is rapidly changing. Nanofabrication, lithography, and molecular self-assembly can be used to prepare structures of well-defined, reproducible nanoscale dimensions. The advent of molecular imaging devices, such as scanning tunnelling microscopy or atomic field microscopy, as well as the utilization of break junctions and other measurement techniques for the characterization of matter on the subnanometre spatial scale (coupled with synthetic methods capable of preparing such structures in an organized and reproducible way), have led to a situation in which the possibility of actual molecular electronics has become a reality. Concurrently, progress in theory pertaining to basic processes, e.g. electron transfer and proton transfer, the coupling between macroscopic electrodes and molecular systems, and the advent of quantized conductance, as well as advances in the elucidation of size effects in clusters, have made central contributions to the conceptual framework of molecular electronics. As this volume demonstrates, the enabling scientific understanding, and the appropriate methodology in conjunction with advanced preparatory, analytical and interrogation techniques, are rapidly falling into place; so that molecular electronics can become a precise scientific and engineering discipline. The relations between spatial structure, electronic level structure and electromagnetic response and dynamics make the research area of molecular electronics rich, fascinating and challenging.

The overall goal of this book is to provide an up-to-date set of authoritative overviews spanning various aspects of molecular electronics and focusing on the scientific and technological perspectives of this research field. The first chapter is meant to serve as an introduction, providing a review of some of the goals, concepts, problems, ideas, experiments and theoretical arsenal of molecular electronics. The general organization of the material in this book is as follows:

1 fundamental processes;

2 molecular circuits;

3 size effects;

4 molecular structure and self assembly;

5 molecular optical devices.

We are grateful to the contributors for their willingness to partake in this endeavour and for their adherence to a time table which allowed us to send the manuscripts to the publisher in July 1996. We are grateful to the International Union of Pure and Applied Chemistry for its endorsement and support of this project within the framework of its publication series in 'Chemistry Towards the 21st Century.' Indeed, molecular electronics is becoming a forefront, high-quality, intellectually and technically advanced multi-disciplinary research area. We thank the editorial office of IUPAC and the staff of Blackwell Science for welcoming and supporting this project. The broad spectrum of the research activity touched on in this book bears witness to the scope and quality of molecular electronics and to the contagious enthusiasm of its practitioners.

Joshua Jortner
Mark Ratner

References

1 Bloor DE. *Mol Cryst Lig Cryst* 1992; **234**: 1.
2 (Chapter 9, this volume.)
3 Feynman RP. In *Engineering and Science*, The Caltech Alumni Magazine, February 1960.
4 Smith FT. *J Chem Phys* 1961; **34**: 793.
5 Binnig G, Rohrer H, Gerber C, Weibel E. *App Phys Lett* 1982; **49**: 57, 1983; **50**: 120.
6 (Chapter 14, this volume.)
7 (Chapter 4, this volume.)
8 (Chapter 5, this volume.)

1 Molecular Electronics: Some Directions

M.A. RATNER* and J. JORTNER†

*Department of Chemistry, Northwestern University, Evanston, IL 60208, USA

†School of Chemistry, Tel Aviv University, Ramat Aviv, Israel 69978

1 Introduction and overview

Early work on molecular electronics [1–12] was demarked more by vision and enthusiasm than by appropriate systems preparation and measurement. More than two decades ago, before the advent of molecular imaging methods, such as scanning tunnelling microscopy (STM) or atomic force microscopy, it was probably easier to speculate on and calculate the possibility of single molecule electronics than to make any appropriate measurements. This situation is rapidly changing: the advent of these imaging microscopies, as well as the utilization of break junctions and other measurement techniques for characterizing matter on a subnanometre space scale, coupled with synthetic methods capable of preparing such structures in an organized and repeatable way, have led to a situation where the possibility of actual molecule-based electronics has become real.

We will use the term molecular electronics in this text to mean the study of response of molecule-based materials, at the level either of molecular-scale electronics (MSE) or molecular materials for electronics (MME) [13]. The scope of the field ranges from tunnelling through individual atomic and molecular wires to the possibility of molecular shift registers and energy storage, from single molecule spectroscopy to molecular hole burning memories, from molecule-based transistors to molecular quantum dots. This includes biological and biomimetic systems as well as entirely synthetic ones, and can be characterized in terms of responses ranging from the femtosecond to seconds and beyond.

Since we use molecular electronics to include both MSE and MME, it might be fair to ask if all measurements on molecular materials are by definition in the area of molecular electronics. The obvious response is a structural one: we mean by molecular electronics, electronic and response properties of molecule-based materials whose characteristic sizes can be described on the nanoscale or smaller scale. Clusters then become part of molecular electronics; indeed, cluster materials are, at present, perhaps better characterized than any other molecular electronic systems, and the generalities of their behaviour can be used as a paradigm for development of concepts, structures and mechanisms for molecular electronics.

In Section 2 of this chapter, we will first describe the different categories of devices that have been envisioned, based on molecular electronics. We will relate them to standard solid state systems, and point out their possible utility in actual devices.

Section 3 is devoted to the basic chemical physics underlying much of molecular electronics, including electron transfer, proton transfer, photo absorption, non-linear optics, descriptions of localization and relaxation, and the coupling of subcomponents.

In Section 4, we will present an example analysis based on molecular wires. We begin with discussion of electron transfer and proceed to an appropriate formulation of

5

electronic flow in circuits with molecular or metallic termini. Applications in STM spectroscopy, and control of molecular currents, will be stressed. Size effects, and the relationship between nanoclusters and single molecule electronics, then form a natural bridge between molecular electronics on the one hand, and cluster science on the other; this is the subject of Section 5. Finally, in Section 6, we consider some of the intrinsic limitations and promise of the field of molecular electronics.

2 Molecular electronic devices

At the level of MSE, there has been essentially no well-characterized experimental study demonstrating particular device characteristics. At the level of MME there are several very important examples, ranging from liquid crystal displays to electrophotography. The nanostructure component of integrated molecular electronics has been developed extensively, in particular in connection with both lithographic [14–18] and chemical [19–23] preparations of quantum dot and extended quantum dot structures (see also Chapter 9). While all of these three areas have been explored extensively, both theoretically and experimentally, with a view toward development of molecular device and molecular response, it is true as of this writing that in the MSE and molecular nanostructure fields no actual devices have yet been demonstrated [24–26].

Nevertheless, the literature has already indicated a number of categories into which molecular devices can be classified. As part of an introduction to the field of molecular electronics, it is perhaps useful to consider the various categories that have been discussed.

2.1 *Molecular wires*

One obvious requirement for construction of an electronic device is a communications channel, to link active elements to each other, to displays, or to other parts of an integrated structure. The silicon structure in current silicon-based electronics provides electronic conduction, and effectively 'wires' different components. In the molecular case, wires have been discussed using organic [27–40], metallic [41–42], and photonic/photoconductive [43–47] motifs.

The simplest and most obvious examples correspond to molecular electronic conductors acting as wires. This subject is treated extensively in Section 4 below, and is perhaps the field in which most experimental work has been concentrated. The essential idea is to use a molecular structure providing electron delocalization, that acts to move charge from one region of an integrated molecular device to another. Suggested wire structures include polyacetylene, polythiophene, polypyrrole and other conductive polymers [48,49], phthalocyanines bonded together either edge-to-edge or in a poker chip arrangement above one another [50–56], DNA [57], polyalkynes [58], redox polymers (see Chapter 6), etc. Organometallic species, such as the Krogmann phase $Cs_2 Pt_2(CN)_4Cl_{0.3}$, as well as specially grown molecular wires based on transition metal chalcogenides, have also been discussed [59].

Since many molecular species are far better photoconductors than ground state conductors, there has also been extensive discussion of photoexcitation for intramolecular transfer either of energy or of electrons (indeed, the most extensive current

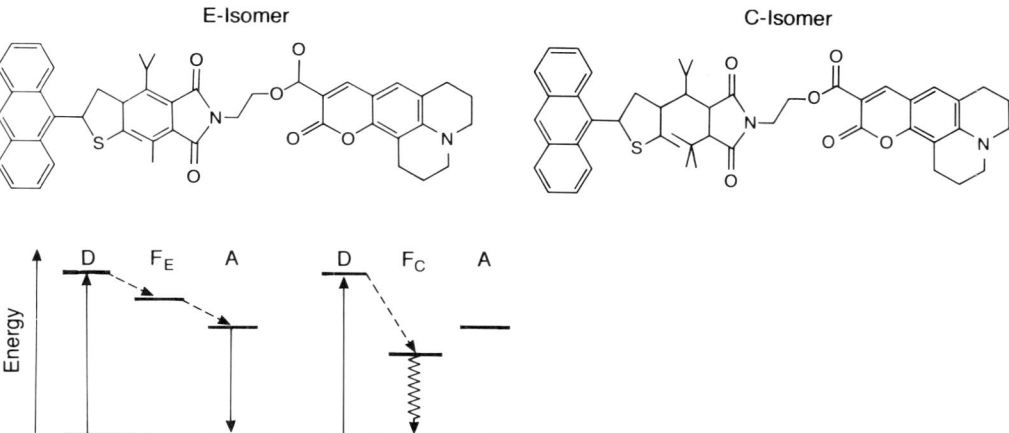

Figure 1.1 A molecular concept for switching, involving intramolecular energy transfer in the molecular system D–F–A. In the E isomer the acceptor level forms the lowest excited state, whereas in the C isomer the lowest state is in F (the Fulgimide). From [60].

application of MME is probably photoconduction in xerography, based on carbazole and other organics).

The usual model for understanding these one-dimensional molecular wires is in terms of a nearest neighbour tunnelling or hopping of an electron from site to site. A very similar construct, involving motion of an exciton from site to site, underlies extensive work in intramolecular energy transfer as an effective wire between photoactive units in an extended structure, such as that in Fig. 1.1 [60]. Note here the similarity between the donor–bridge–acceptor electron-carrying molecular wire and the donor–moderator–acceptor photonic circuit. While there has been much less work on the photonic and photoconductive materials in the MSE field, they enjoy obvious advantages of optoelectronics as opposed to electronics, and we may anticipate more extensive work on them in the future.

2.2 *Switches*

While the wire is essentially a passive component, providing communication between structures, switches act as active components, changing their state in response to an external signal. Many switches have been discussed both in the MME and the MSE fields. The simplest one involve changes in molecular structure in response to an external electrical, optical, chemical, or biological stimulus. Examples would then include molecules such as the photosynthetic reaction centre, that exhibit very large changes in dipole moment upon photoinduced primary charge separation, or molecules that change their structures, and therefore their optical and/or electronic properties, in response to a chemical stimulus. These would include structures such as that in Fig. 1.2, in which the presence of an alkali ion causes the crown ether structure to contract [61,62], and therefore changes the effective optical properties of the bipyridyl. Indeed, metal complexes such as that shown in Fig. 1.3 are extremely sensitive to the outer solvent shell (the conproportionation constant for the binuclear ruthenium complex changes from 10 to 68 400 on changing solvent from water to acetonitrile, probably due

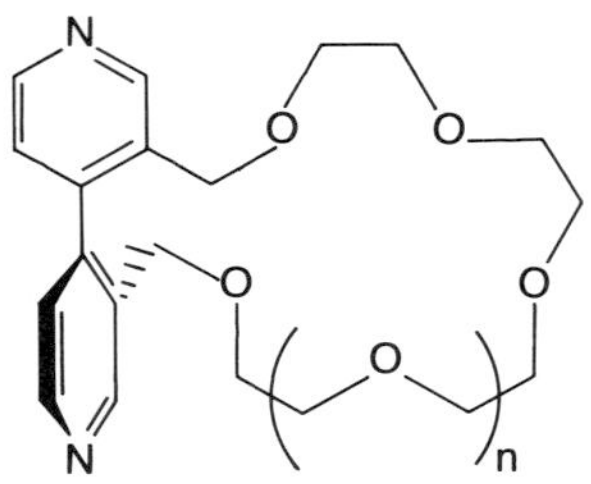

Figure 1.2 A cryptand molecular switch: binding an alkali in the cavity folds the species, to give a rigid structure. From [61].

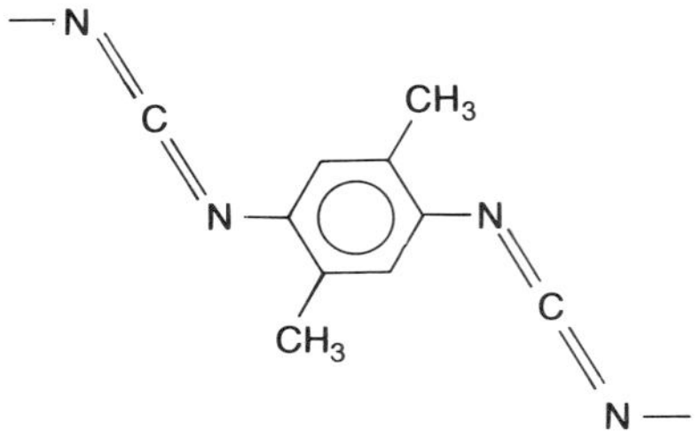

Figure 1.3 When this species acts as a bridging ligand between two $Ru(NH_3)_5$ ends to form a $+3$ mixed valent species, solvent changes cause the conproportionation constant to vary by a factor of 7000. From [62].

to the hydrogen bonding effects of water in the second solvation shell). Therefore, within the general theme of binuclear mixed metal complexes, one could imagine switching due to photons, ions, or solvent as the external stimulus.

Clearly, switches can act as sensors: chemical recognition can, for instance, modify a local site energy, changing conductance and allowing a sensing function [58]. Such switches are discussed extensively in Chapter 10.

Photon-activated switches have been extensively studied; they often involve intramolecular isomerization. For example, Fig. 1.1 shows a fulgide-based photon triad; the optical energy structure changes substantially on photoexcitation, forming the C-isomer from the E-isomer and substantially changing the energy level scheme. This then results in a switch of the intramolecular energy transfer [60].

Photoredox switches of several kinds have also been discussed; e.g. the structure shown in Fig. 1.4 changes from fluorescent to non-fluorescent upon oxidation of the iron; this could correspond to a setting of the molecule, followed by readout using photons [63]. Actual atomic positions have been the basis for several switches; e.g. Fig. 1.5 shows a biologically based switch using an isomerization of an iron ion between two positions in a ligated environment [64]. The motion of the iron will correspond to the switching process, which can be detected optically or electrically. A linear multisite 'photonic wire' can be switched by changing the site energy via a substituent, whose internal electronic state can be changed in response to a chemical or radiative stimulus [65]. Perhaps the most elegant example is atomic switching, based on positioning of a single atom between two larger structures [66]; this has been studied using STM by the IBM group. This is actually atomic-scale electronics rather than molecular-scale electronics, but is indicative of the kinds of manipulation that modern experimental techniques permit.

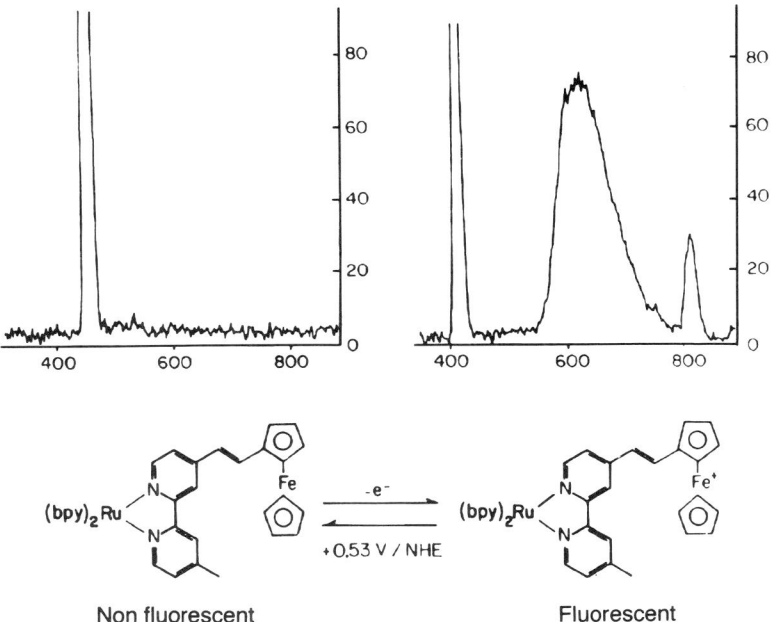

Figure 1.4 A photoredox switch, indicating a very strong change in the fluorescence upon reduction of the iron in ferrocene. From [63].

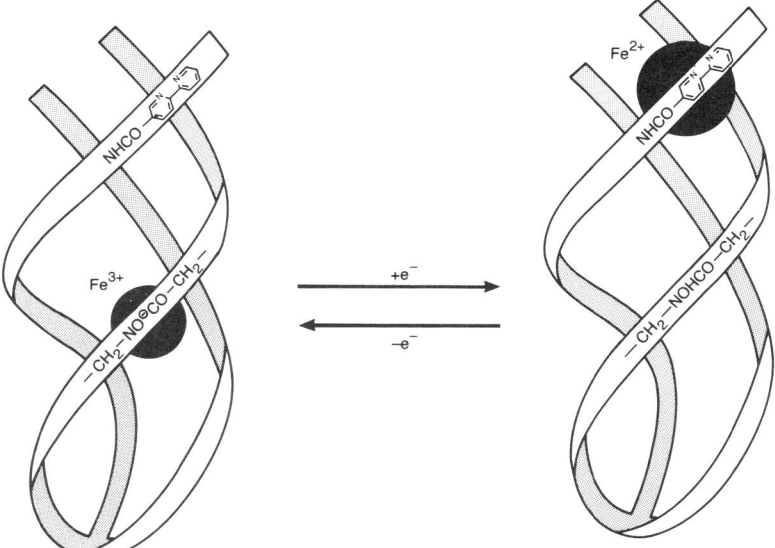

Figure 1.5 Molecular redox switch, showing the interconversion from the Fe^{3+} complex to the Fe^{2+} complex upon exposure to reducing agents. From [64].

The natural prototype for the photon switch is, of course, rhodopsin. In this molecule (Fig. 1.6) the geometry changes, upon optical excitation, from *cis* to *trans* around the double bonded position 11; this transition, which apparently occurs on a subpicosecond time scale and in a coherent fashion [67], forms the molecular basis of vision. This is discussed in Chapter 15.

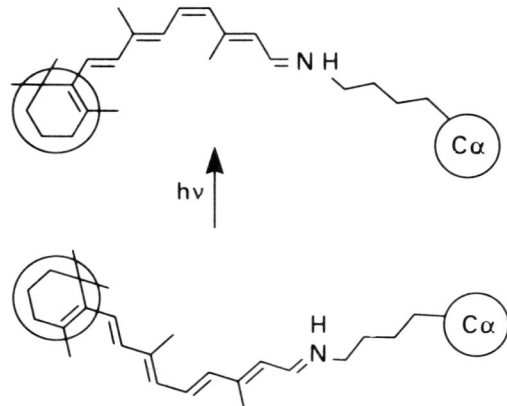

Figure 1.6 The rhodopsin molecule; isomerization around the 11-*cis* bond is a primary step in the vision process. From [327].

Several prototype structures for conformational molecular switches were suggested theoretically by Aviram and co-workers [68,70] and by Hush and co-workers [69]. Some have been investigated experimentally by Tour and collaborators [71]. Potember *et al.* have demonstrated a functional molecular charge transfer switch [72].

2.3   *Memory element*

If molecular electronics is to be patterned after conventional electronics, then clearly memory elements are extremely important; if it is to be patterned after biological systems, memory (albeit of very different type) is also critical. Proposals for nanoelectronic memories range from positioning of protons in hydrogen bonds to photoelectrochemical storage. They include proposals for 'cellular automata' or memories on the basis of localization of electrons, as indicated in the structure of Fig. 1.7 [73].

The most extensively investigated memory is based on quite a different principle, that of photochemical hole burning. The idea (as sketched in Fig. 1.8 [74–76]) is that inhomogeneities in spectral lines are caused by local site differences among molecules that are, in principle, addressable. If the inhomogeneous line of width $\Delta\omega_i$ is broad enough, then different components of it will correspond to specific molecules that can be probed with a very narrow excitation, whose homogeneous line width $\Delta\omega_h$ is much smaller than $\Delta\omega_i$. By photoisomerizing molecules with a very narrow laser line, data can be stored: each addressable position within $\Delta\omega_h$ corresponds to a bit of information, as indicated in Fig. 1.9. The molecule, especially if kept at low temperature, is stable for extremely long time periods. Typical molecules that have been proposed include free base porphyrins, in which the isomerization process is the switching motion of the two

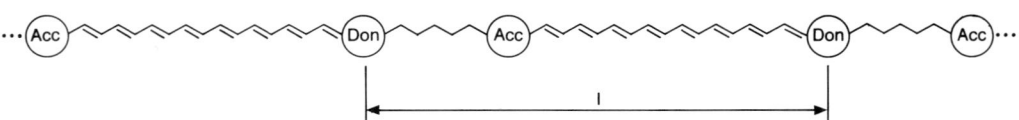

Figure 1.7 Model of a cellular automaton based on the linear arrangement of donor/acceptor polyenes. From [73].

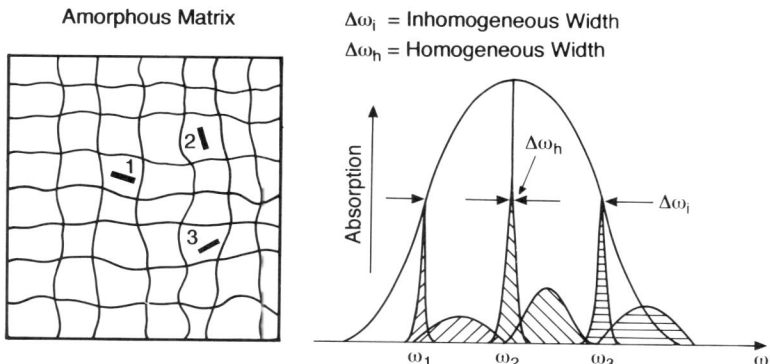

Figure 1.8 Absorption spectra of three chromophores (identical molecules) in an amorphous matrix. The transitions ω_1, ω_2, ω_3 are the zero-phonon transitions to the lowest excited state in different matrix environments. $\Delta\omega_h$ and $\Delta\omega_i$ are, respectively, the widths of the molecular transitions (homogeneous broadening) and the total inhomogeneous band. From [74].

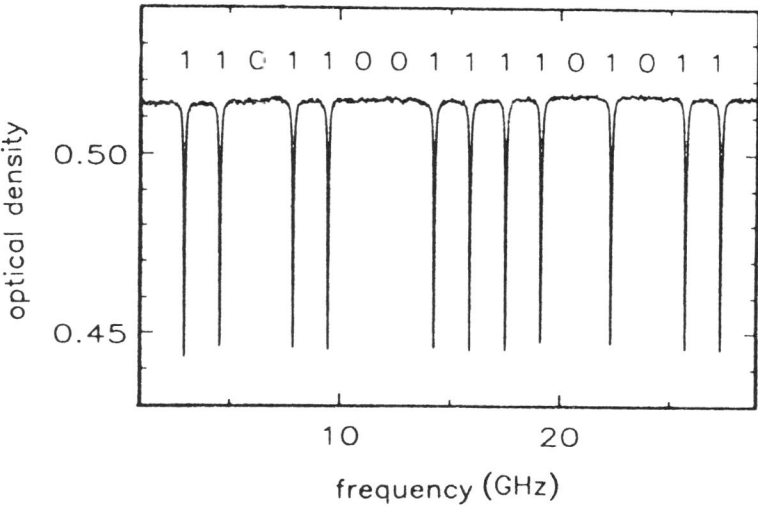

Figure 1.9 A sequence of holes labelled as a bit pattern. The data were obtained for a phthalocyanine chromophore in polymethylmethacrylate at 50 mK. From [74].

protons at the centre of the nitrogen cavity. Impressive demonstrations (multiplexing factors of order 10^5) have been demonstrated for this switching [74].

2.4 Electrodes

In conventional electronics, lead structures are provided by relatively massive silicon islands, or external wires. Molecular electronics has been discussed using both conventional electrodes, as in the ultramicroelectrode applications of electrochemistry or in surface electrochemical structures [11,77,78], as well as using more specialized electrode structures. One attractive specialized structure is provided by polymeric conductors such as polyaniline or doped phthalocyanine, that can conduct both ions and electrons and can therefore act as an electrode either in an electrochemical circuit (battery) of the usual kind or in ultrasmall device applications.

With the relatively easy fabrication of variable size quantum dot structures [19–26; see also Chapter 9], as well as the break junctions and tip structures used in STM [79,80; see also Chapter 5], several results have appeared in the literature measuring transport in molecular wires using nanoscale metallic electrodes [38,81–84]. This is an example of the use of nanostructures in association with molecule-based electronics for integrated device purposes; this subject is discussed in Chapter 5.

2.5   *Energy conversion*

Primary fast events in bacterial photosynthesis consist of photoexcitation, energy transport in the antenna complex, formation of a photoexcited special pair, and very rapid intramolecular transfer from the special pair to the accepting pheophytin [85]. This photosynthetic structure is sketched in Fig. 1.10. Artificial photosynthesis, based on synthesized molecules, has been extensively investigated [11] as a possible mechanism for molecular photoelectronic energy conversion [86,87], using candidate structures such as that in Fig. 1.11. In this so-called extended tetrad structure, photoexcitation in the centre site causes facile, effective separation of the electron and the hole, forming a giant dipole which is stabilized at the ends of the pentad. This is very similar to the structures discussed above in connection with switching. Multimolecular structures, based on a Langmuir–Blodgett film technique [88], are sketched in Fig. 1.12. Here the molecular photodiode is built on a thin, optically transparent gold electrode; photoexcitation at the chromophore S results in separation of the electrons to the acceptor site near the metal, and holes to the donor site. In this sense, the molecular structure bears an analogy to the P/N junction in semiconductor photodevices.

More obviously molecular behaviour is shown in systems that mimic the allosteric process — that is, systems whose dimensions change upon receipt of a stimulus. One example is a simple process of ionostriction, in which binding of an ion causes molecular reduction in size. An example of this was cited in our discussion of switching (Fig. 1.2). Other applications have also been discussed, based on substantial changes in molecular geometry; indeed, heat shrink tubing is, in this sense, an application of molecular electronics.

2.6   *Sensors*

Since much of the information in the biological, atmospheric and condensed-phase worlds is carried by molecules or ions, it makes sense that molecular electronics should concentrate on sensing applications. Important examples include the biosensors based on immobilized enzymes linked to an electrode through molecular conductors, as discussed in Chapter 7. Similarly, ionic conductor systems can be used as sensors for any electroactive molecular ionic species; standard solid state examples include the yttrium-stabilized zirconium sensor for oxygen, but applications involving polymeric electrolytes have been extensively discussed [89]. Also, the selective binding of ions or solvents, as discussed in Sections 2.2 and 2.5 above, can be used as sensors for those ions or solvents [58,90; N. Lewis, personal communication]. Finally, specially constructed molecular cavities, e.g. those based on modified cyclodextrins, can be used to

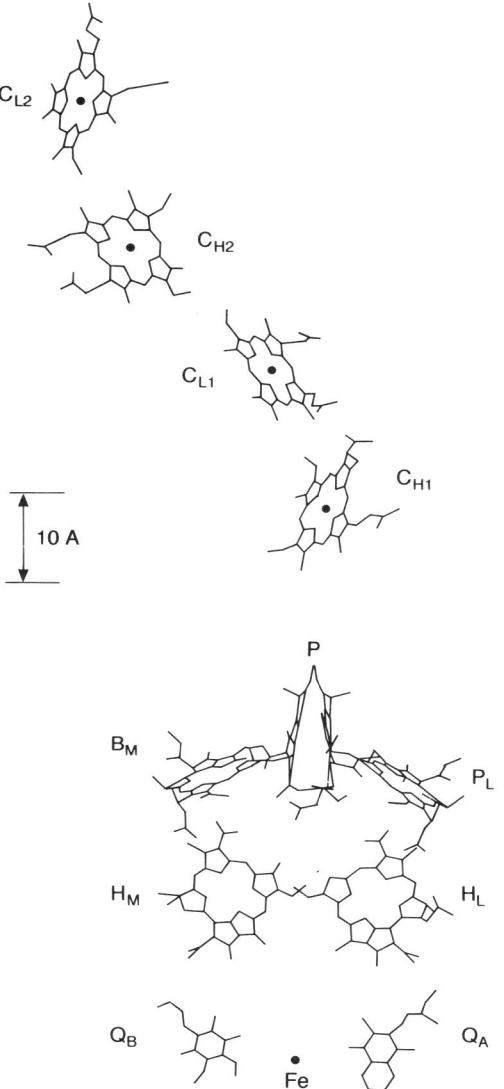

Figure 1.10 A schematic diagram of the reactive components in the reaction centre (this one from *Rps. viridis*). The bacteriochlorophyll dimer, P, is initially excited, and transfers an electron first to the bacteriopheophytin H_L, and subsequently to the quinone Q_A. The electron transfer from P to H_L is essentially irreversible, and occurs on a time scale of roughly 3 ps. From [85].

bind specifically particular target species; this 'molecular recognition' can then be monitored by, in this case, surface acoustic wave detection (Fig. 1.13) [91,92].

2.7 *Optics and optical switches*

Optical electronics, in particular non-linear optical applications to signal processing and switching, is really a separate field of its own, and lies outside the scope of the current text; molecular species are very promising as non-linear optical components in modulators, switches and limiters for optical situations. Optical applications

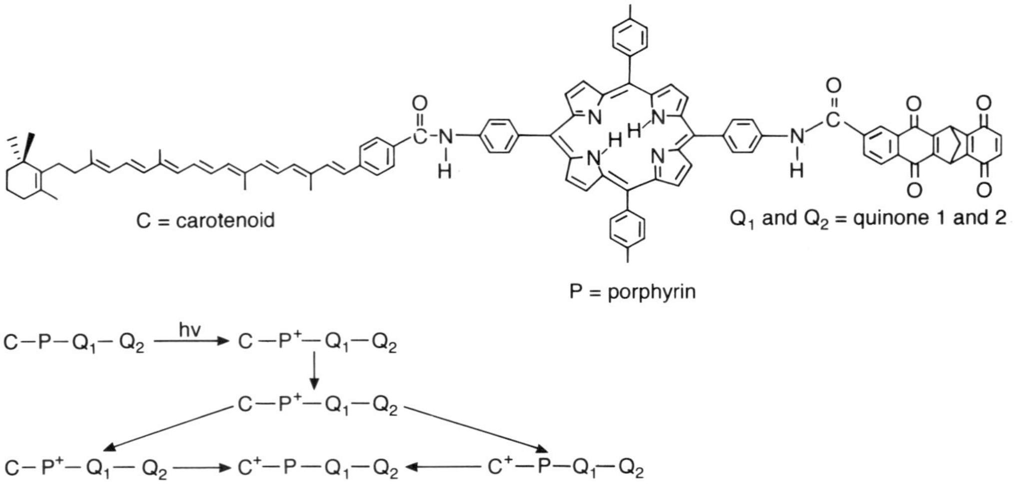

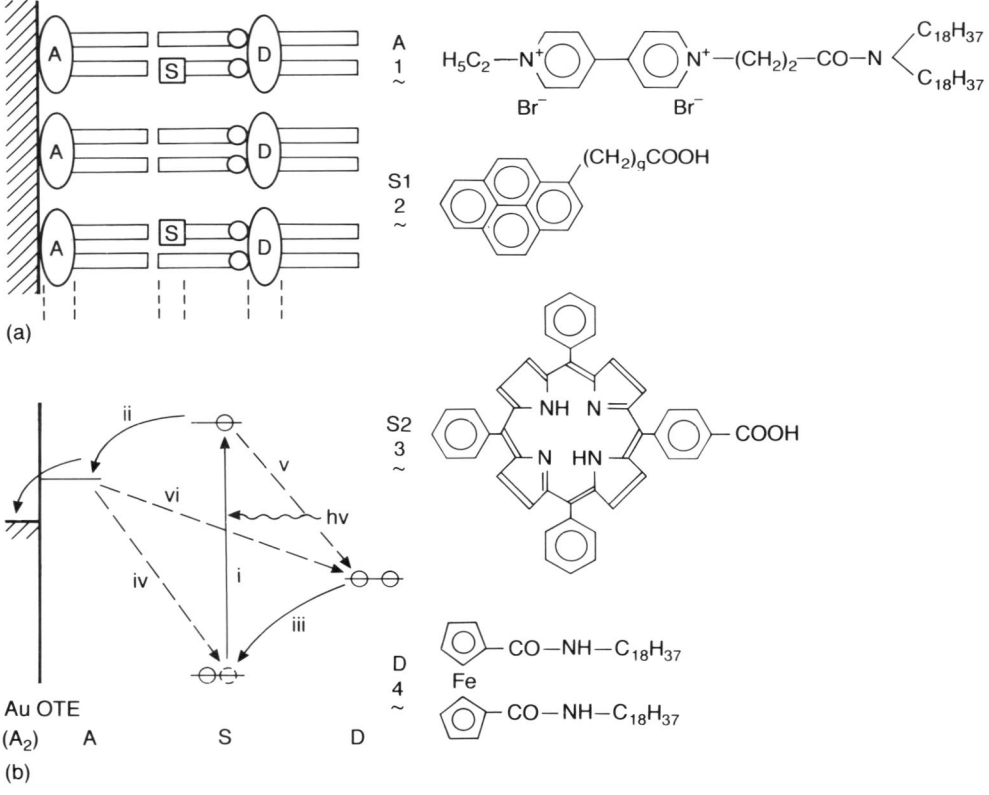

Figure 1.11 The pathways for preparation of a long-lived charge-separated excited state intermediate from the radiation of a tetrad molecule, a carotenoid porphyrin diquinone. The initial excitation occurs in the porphyrin segment, and the charge separation, eventually, goes to the two ends of the molecule. From [11].

Figure 1.12 A molecule photodiode, based on a heterogeneous Langmuir–Blodgett film. Both the structural and energy diagrams are shown (a) and (b). The acceptor/spacer and donor species used in the construction are shown on the right hand side of the figure. From [88].

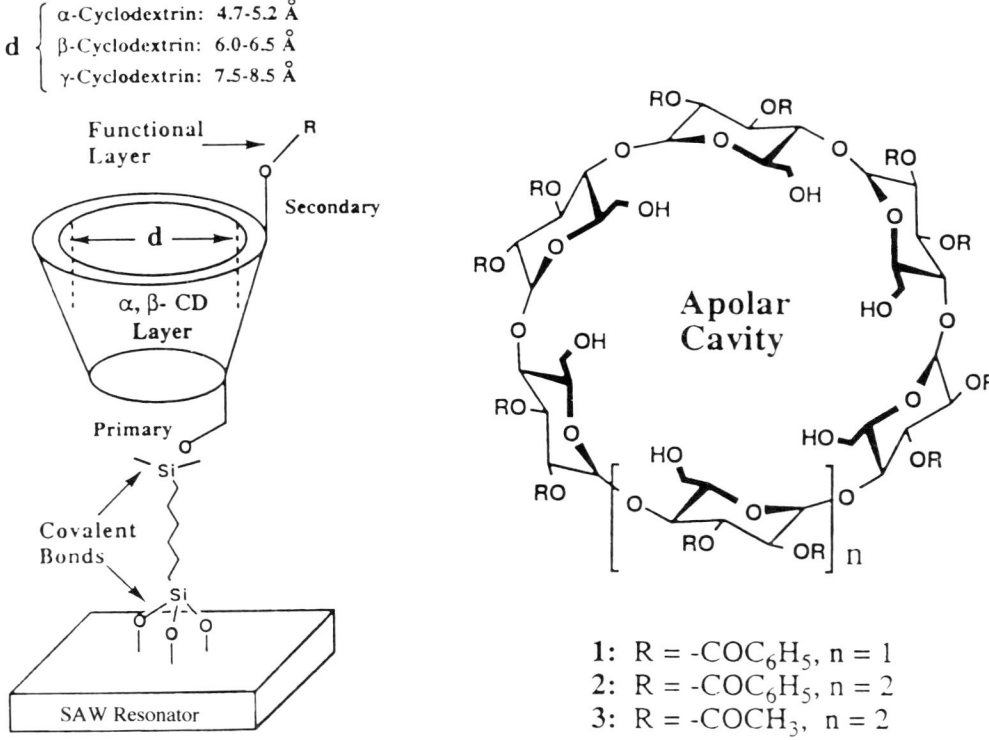

α-Cyclodextrin: 4.7-5.2 Å
β-Cyclodextrin: 6.0-6.5 Å
γ-Cyclodextrin: 7.5-8.5 Å

1: R = -COC$_6$H$_5$, n = 1
2: R = -COC$_6$H$_5$, n = 2
3: R = -COCH$_3$, n = 2

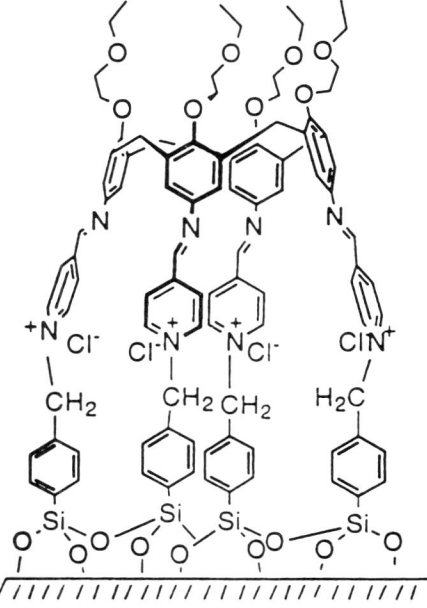

Figure 1.13 A self-assembled monolayer chemical sensor, based on functionalized cyclodextrins on a surface acoustic wave resonator. The acceptance of molecules into the cup-shaped cyclodextrin receptor is controlled by chemical substituents at the top of the cyclodextrin, indicated by R in the first scheme. From D. Li (personal communication).

combined with electronics have already been discussed in Sections 2.1, 2.2, 2.3 and 2.5 above. Chapter 14 examines recent fascinating and important advances in near-field optics.

2.8 *Displays*

Many molecular structures form classic liquid crystals, and the application of liquid crystals to display technology is one of the most important current applications of molecule-based electronics. Photoluminescent displays, based on the light-emitting properties of molecules, have moved forward very rapidly in the past few years [93–96]. In particular, the *p*-paraphenylenevinylene conductive polymer has been used in some very effective, high-quantum-yield electroluminescent structures; see Fig. 1.14. In a sense, this device application is precisely the inverse of the photo-electric charge separation mechanism: here electron and hole enter from the two electrodes, and are combined at the molecular chromophore to give a luminescent signal.

2.9 *Electrochemical devices*

Although electrochemistry is, in principle, separable from molecular electronics because it requires ion flow, nevertheless electrochemical devices are of major interest in connection with molecular electronics. Systems such as smart windows [97], molecular and ionic sensors based on current flow, and separation structures based on artificial ion channels all involve electronic applications at the molecular level. An elegant structure is shown in Fig. 1.15; it is essentially a synthetic ion channel and can be used for ion separations [98]. A different strategy is based on molecular lining of a very narrow pore in a ceramic plate; when the absorbed molecules are negatively charged, the cation/anion separation through the narrow pore is highly effective [99].

2.10 *Heterostructure and quantum well devices*

Molecular polymerization chemistry can be used to prepare heterostructured slab materials with a large number of applications. Because of the molecular nature of the heterostructure components, molecular functions (binding, recognition, switching) are available that cannot be found in traditional semiconductor heterostructures [100]. This is discussed in Chapter 13.

2.11 *Information processing*

There have been interesting and extended proposals for the use of molecular structures as optical shift registers [101] and cellular automata [73]. Graphic and interferometric applications have also been discussed. Bacteriorhodopsin has, in fact, been used in several demonstration systems for optical information processing, including spatial light modulation, interferometry, holographic pattern recognition, and optical data storage [102–104].

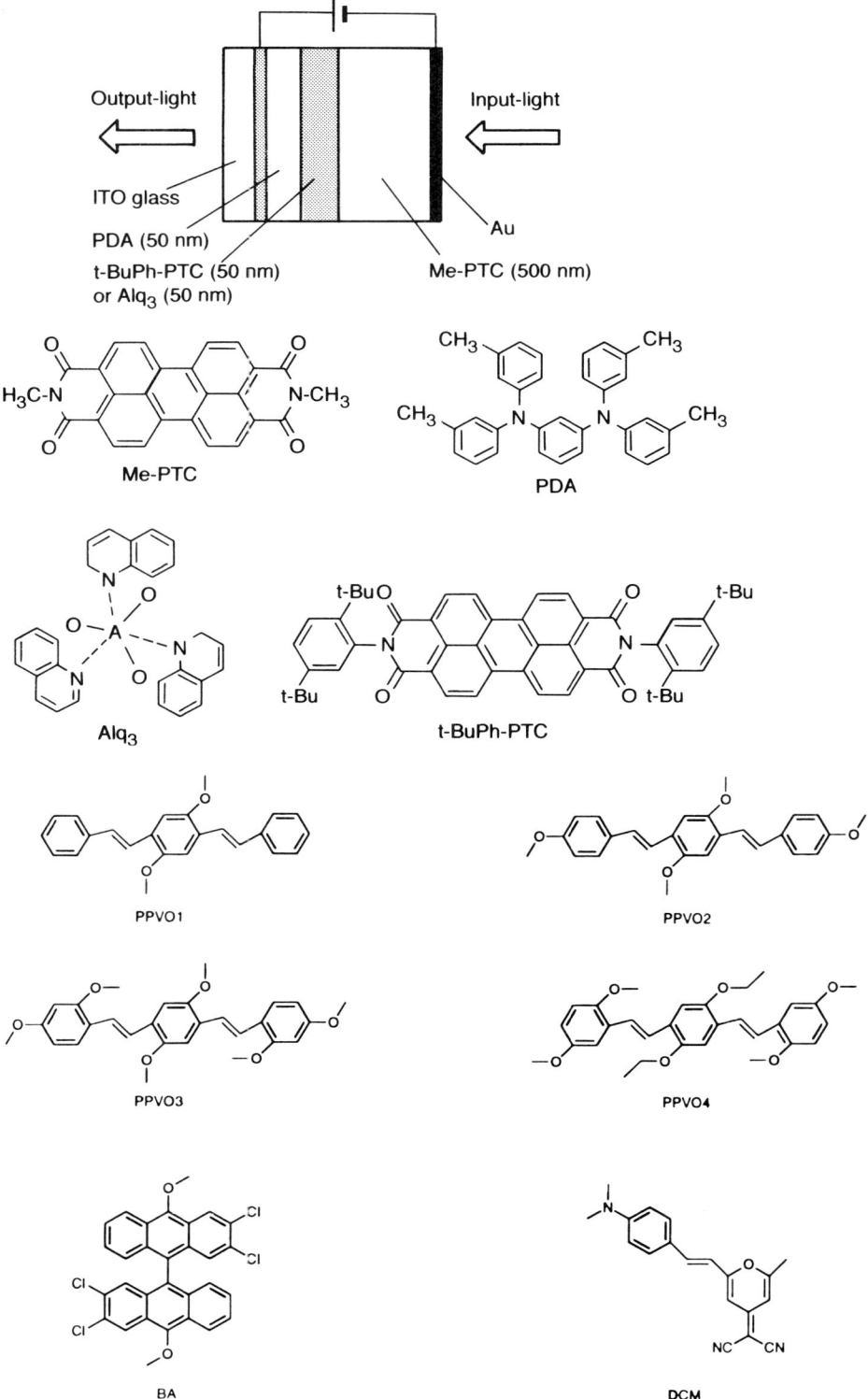

Figure 1.14 Some examples of organic light emitting diodes, based on hole and electron injection in a multilayer structure. Different substituted emissive components are shown, including polycyclics or aluminium complexes. Various polyparaphenylenevinylenes or PDA or Me PTC are used as conducting elements in a schematic illustration of the device configuration. From [315].

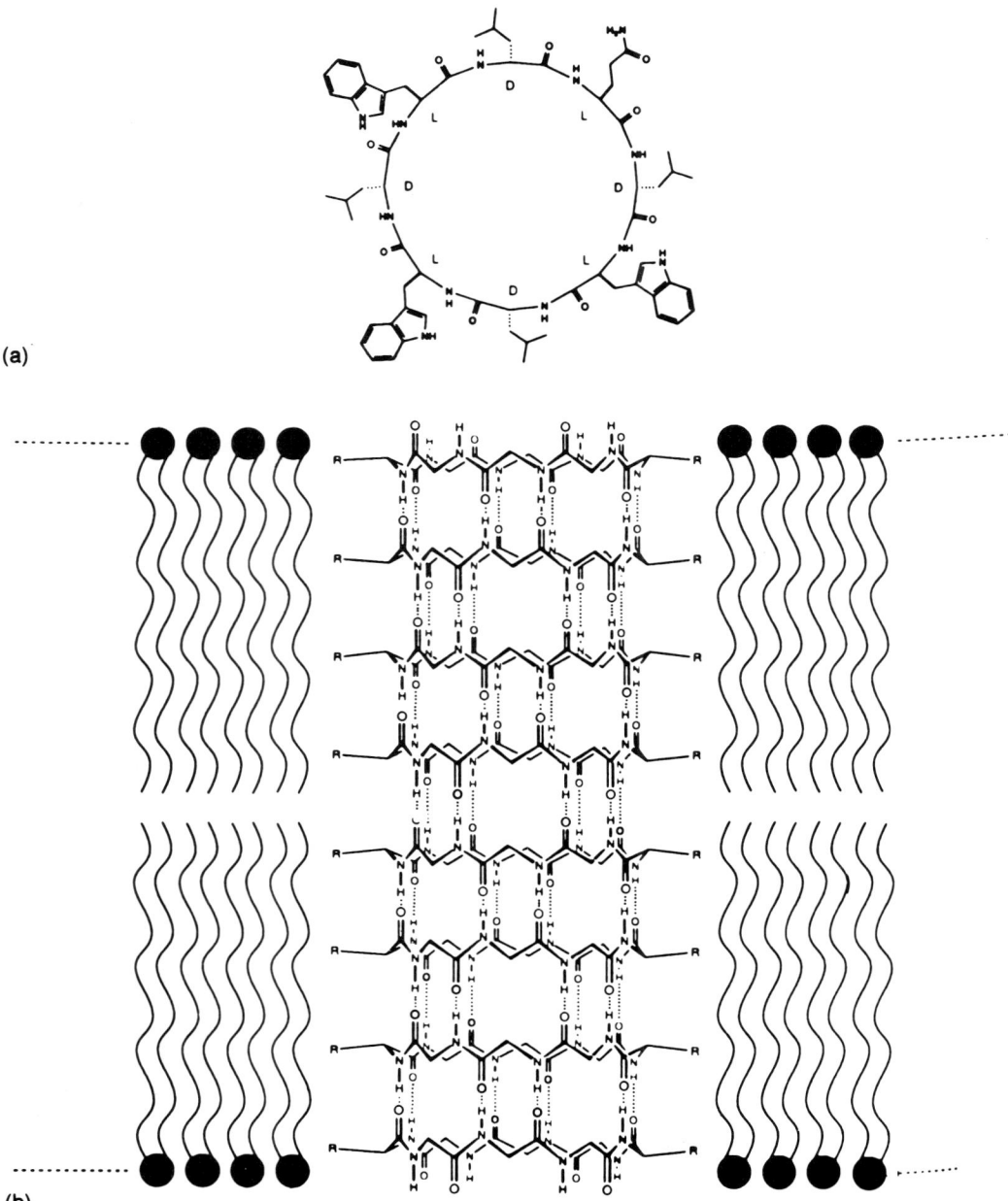

Figure 1.15 (a) The chemical structure of a synthetic ion channel, based on assembled D and L amino acids. (b) The peptide subunits are shown in a self-assembled tubular configuration embedded in a lipid bilayer membrane. For clarity most side chains are omitted; a top view of the cylindrical molecule is shown in (a). From [98].

2.12 *Diodes and rectifiers*

Although a molecular rectifier based on a single molecule structure with continuum electrodes was discussed theoretically more than 20 years ago [105], and extensive synthetic efforts toward preparing such a molecular rectifier have been ongoing in a number of laboratories [106,107], the first definitive results were published in 1993 [108] and 1994 [109]. Figures 1.16 and 1.17 show the I/V characteristics of two reported molecular rectification systems; neither of them is actually a true single molecule layer, but both demonstrate substantial asymmetry in the I/V characteristics, and amount to molecular rectification. Demonstration of effective rectification by a single molecule remains an unfulfilled goal. Chapter 8 provides a discussion of this topic.

2.13 *Intrinsically novel molecular devices*

Much of the area of molecular electronics has, self-consciously, been devoted to mimicking standard semiconductor electronics. We have already referred to several instances (molecular shape and volume changes, sensing) involving molecular functions that cannot be mimicked by semiconductors. Other important applications will be based on the use of molecular recognition and molecular signal transduction [58,90–92; N. Lewis, personal communication]. It may well be, indeed, that it is these unique aspects of molecular as opposed to solid-state, electronics that will be most novel, and most important in applications.

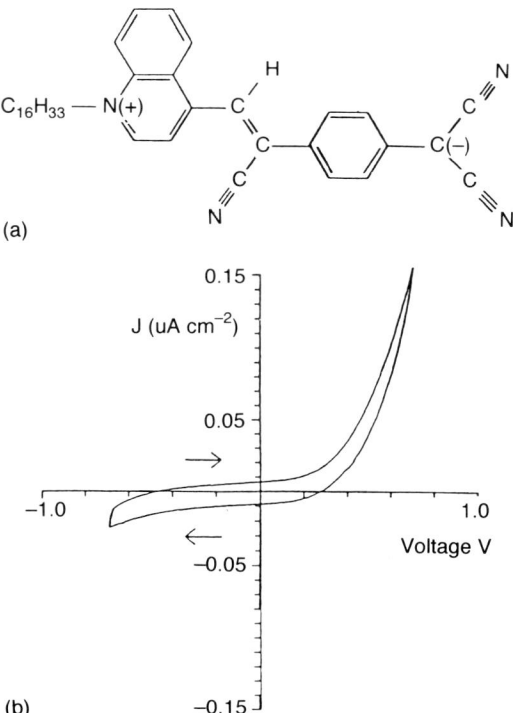

(a)

(b)

Figure 1.16 (a) A zwitterionic molecule used in Langmuir–Blodgett films for molecular rectification. (b) shows the actual I/V characteristic, showing a rectification-like asymmetry in the current response. From [108].

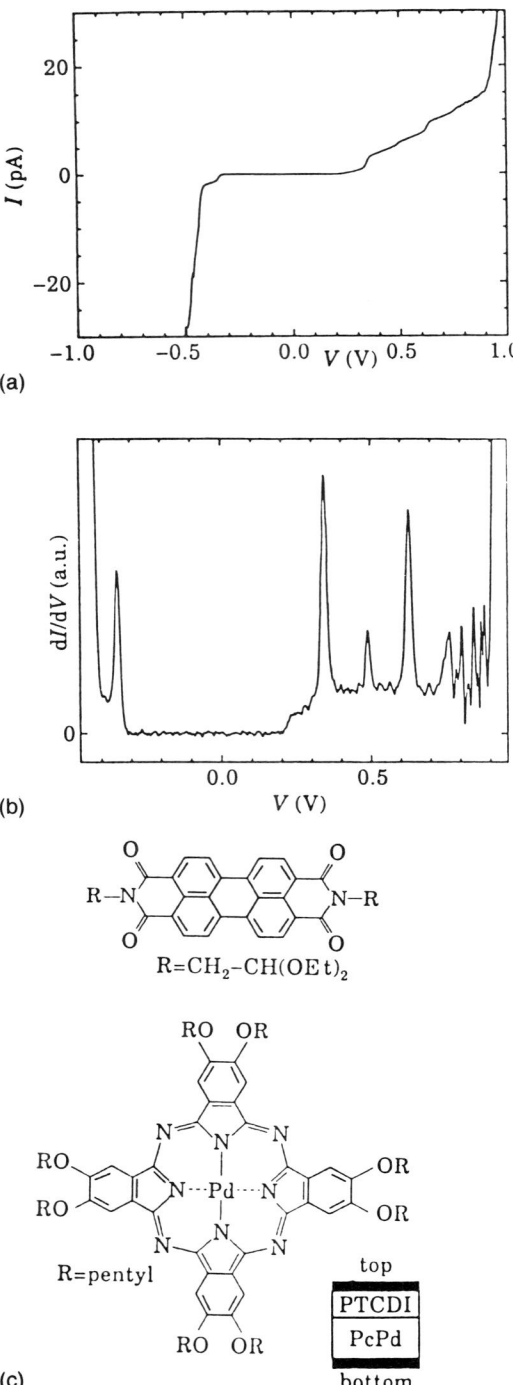

(a)

(b)

(c)

R=CH₂–CH(OEt)₂

R=pentyl

top

PTCDI
PcPd

bottom

Figure 1.17 (a) and (b) The asymmetric current voltage characteristic of a Langmuir–Blodgett film, based on (c) a bilayer structure of a phthalocyanine and a quinone. The derivative structure has jumps in the conductance, which may be evidence for either Coulomb blockade or eigenvalue staircase behaviour. From [109].

3 Basic processes

3.1 *Electron transfer*

3.1.1 GENERALITIES

Energy storage and disposal via electron transfer (ET) in molecular, supermolecular, and biophysical systems [110,111] is expected to provide a central conceptual and technical basis for molecular electronics, pertaining both to molecular devices and to MMEs. ET in molecular systems is relevant both for molecular materials and devices, while ET in supermolecules may be pertinent for devices. ET in supermolecules falls into two general categories.

1 Supermolecules [112–115] consisting of an electron donor (D) and an electron acceptor (A), linked by a non-rigid or a rigid molecular bridge (B).

2 Biophysical systems such as the photosynthetic reaction centres (RC) of bacteria and plants [85], where the primary process (the conversion of solar energy into chemical energy) proceeds via a sequence of well-organized, highly efficient, directional and specific ET processes between prosthetic groups embedded in the protein medium.

ET processes fall into two major types: (i) ET in ground electronic states, and (ii) ET in electronically–vibrationally excited states. ET in ground electronic states is important for electronic transport in MMEs, as well as for charge transfer in molecular devices. The latter can be induced by charge injection in external fields, by electrochemical processes, or triggered by the formation of a precursor ion. ET in electronically-vibrationally excited states is triggered by optical excitation and may encompass also a variety of other intramolecular and intermolecular radiationless processes, e.g. electronic–vibrational relaxation, vibrational energy redistribution, or medium-induced vibrational relaxation in conjunction with ET. These excited-state ET processes are significant for time-resolved triggering of charge transport and separation in molecular devices, e.g. molecular wires, switches, and rectifiers. These may consist of synthetic supermolecules or of components of biophysical systems, which may be realized in artificial photosynthesis. From the mechanistic point of view, photophysical and photobiological ET from an electronically excited donor proceeds via superexchange mediated or direct processes $DBA \xrightarrow{h\nu} D^*BA \xrightarrow{ET} D^+BA^-$ or $DA \xrightarrow{h\nu} D^*A \xrightarrow{ET} D^+A^-$.

The control of ET in DBA or DA systems in solution [110], in a solid [111,116], in a protein [85,111] or within an 'isolated' solvent-free supermolecule [117] can be accomplished by the following.

1 Structural control. 'Molecular engineering' of the D, A and B subunits determines the molecular energetics and the direct D–A or superexchange D–B–A electronic coupling.

2 Intramolecular dynamic control of the nuclear equilibrium configurational changes (i.e. nuclear distortions) accompanying ET [111].

3 Medium control of 'conventional' ET in a solvent or in a cluster. The function of the medium on $DBA \rightarrow D^+BA^-$ ET is [117]: (i) the energetic stabilization of the ionic states; (ii) the coupling of the electronic states with the medium nuclear motion, which originates from short-range and long-range interactions in polar solvents, short-range interactions with C–H group dipoles in non-polar hydrocarbons and with polar amino acid residues in protein.

4 Dynamic medium control of ET [118–122] involving: (i) the medium acting as a heat bath — the relaxation of the medium polar modes is often fast on the time scale of the electronic ET processes, which then constitute the rate-determining step; (ii) dynamic solvation effects of DBA or/and D^+BA^- determine the ET dynamics when condition (i) is violated and solvent-controlled ET may be exhibited; (iii) specific dynamic control of pathways by solvent motions ('gating'); (iv) very slow solvent relaxations such as in glassy matrices, that lead to reduced solvent reorganization energies.

The structural, intramolecular, solvent, and dynamic control of ET will allow for the design of molecular systems where ET is: (i) ultrafast (on the time scale of ~1 ps to ~100 fs), overwhelming any energy waste processes; (ii) highly efficient, eliminating any back reactions; (iii) stable with respect to the predictable variation of molecular and medium properties; (iv) practically invariant with respect to temperature changes.

3.1.2 BASIC ET THEORY

The non-adiabatic ET rate is given by [85,111,116–119,123]

$$k = (2\pi/h)V^2F, \tag{1.1}$$

where V is the electronic coupling and F is the thermally averaged nuclear vibrational Franck–Condon factor. This microscopic description rests on the following description and conditions.

1 ET is described as a radiationless transition.

2 The Born–Oppenheimer separability of electronic and nuclear motion applies, allowing description of the system in terms of diabatic potential surfaces (Fig. 1.18).

3 The electronic coupling is sufficiently weak to warrant the description of the radiationless transition in the non-adiabatic limit.

4 Microscopic ET rates are insensitive to medium dynamics. This state of affairs is realized under one of the following conditions: (i) the common situation of fast medium vibrational dynamics, which allows for the separation of time scales with the microscopic ET rate constants constituting the rate-determining step [118]; (ii) the microscopic ET rates weakly depend on the distribution in the initial D*BA vibronic manifold [124].

The electronic coupling in the DBA system $V = V_{DA} + V_{super}$ consists of a sum of a direct D–A exchange contribution $V_{DA} = (\phi^0_{DA}|\hat{H}|\phi^0_{D^+A^-})$ between the electronic states of DA and D^+A^-, and a superexchange [111,124] $V_{super} \simeq V_{DB}V_{BA}/\delta E_B$, where δE_B is the vertical energy difference between the potential energy surfaces (PES) of DBA and D^+B^-A. For superexchange between closed shell D, B and A systems both direct and superexchange interactions require the evaluation of individual pair V_{DA}, V_{DB} or V_{BA} couplings, while for intramolecular superexchange in D–B–A supermolecules molecular electronic structure calculations have been utilized [125–128]. This subject is extensively discussed in Chapter 2. In both cases many electron computations, transcending the naive one-electron picture, have to be invoked for quantitative study. The accumulated information concerning the distance dependence of both direct and superexchange interactions is that both interactions are expected to exhibit an exponential distance dependence [85,111–116,123,133,134].

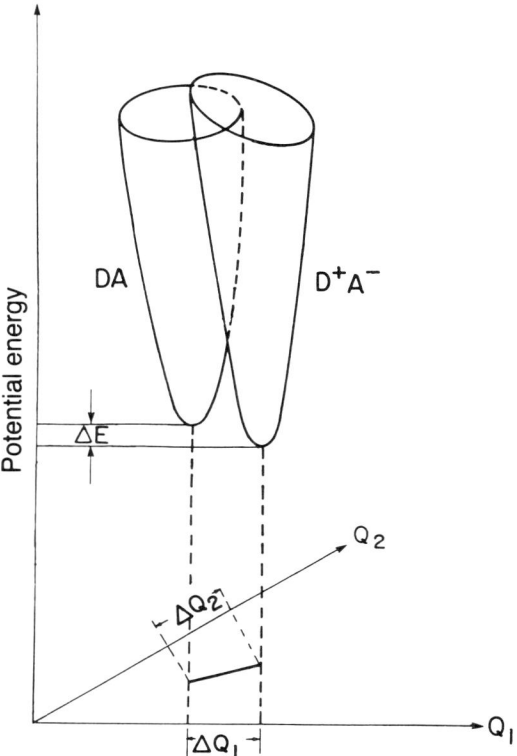

Figure 1.18 Diabatic nuclear potential energy surfaces for $DA \rightarrow D^+A^-$ ET. The nuclear coordinates incorporate the medium and the intramolecular coordinates. The displacements of the equilibrium configurations are marked as ΔQ_1, ΔQ_2.

$$V = a \exp(-\beta R_{DA}), \qquad\qquad (1.2)$$

where R_{DA} is the (either edge-to-edge or centre-to-centre) D–A distance. The distance dependence of intramolecular superexchange interactions in synthetic supermolecules and in the photosynthetic reaction centre (Fig. 1.19) was inferred from the analysis of either ET rates or optical charge transfer spectra. The general trend confirms the exponential relation (Eqn 1.2), quantified by β. The exponent β is system specific, so that the exponential distance dependence of V [and of $k \propto V^2 \propto \exp(-2\beta R_{DA})$] is not universal. On the basis of these exponential relations one cannot infer, *a priori*, whether the electronic coupling is direct or superexchange mediated, and further theoretical input is required. The minimization of β will be desirable for molecular wires. It has been proposed (as follows from simple perturbation theory) that β can be minimized by reducing the gap between the state with the electron localized on D or A, and that with the electron vertically excited on B [129]. Indeed, if this gap vanishes, resonant tunnelling is expected, as seen in inelastic electron tunnelling spectroscopy [130] and in recent Langmuir–Blodgett measurements [108,109], as well as inferred from simple model computations [131–139]. In these truly resonant situations, the exponential decay of Eqn 1.2 is not expected. In conductive polymers such as $(CH)_x$, essentially resonant transfers (vibronically mediated) occur, and Eqn 1.2 is not exhibited.

The nuclear Franck–Condon factor F in Eqn 1.1 contains the thermal average of the overlap integrals between the initial and final vibrational states of the system, under the

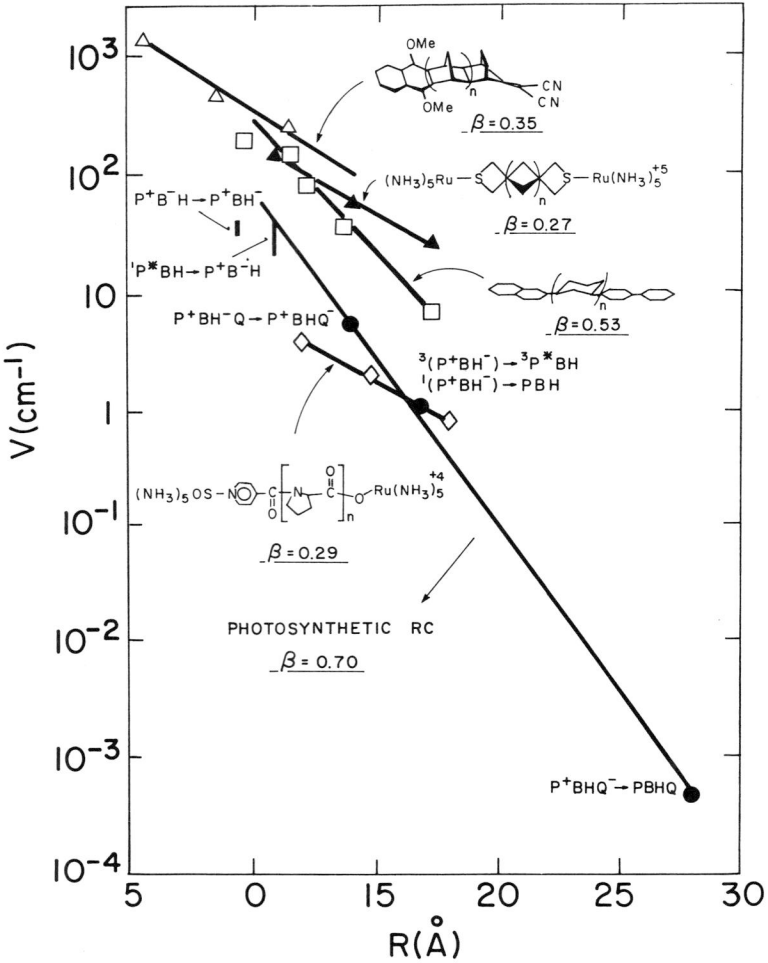

Figure 1.19 The distance dependence of V for synthetic supermolecules obtained from kinetic and spectroscopic data [112–115,125] and for the bacterial photosynthetic RC obtained from analysis of kinetic data. From [146–149]. The system-specific β values of Eqn 1.2 (in Å^{-1}) are presented on the figure.

restriction of energy conservation. For two harmonic potential energy surfaces characterized by identical reduced masses $\{\mu_i\}$, frequencies $\{\omega_i\}$, normal modes $\{Q_i\}$ and nuclear displacement $\{\Delta Q_i\}$ of the equilibrium coordinates (Fig. 1.18), F takes the form of a Fourier transform [123]

$$F = (2\pi)^{-1} \exp(-G) \int_{-\infty}^{\infty} dt \, \exp(i\Delta Et)\exp[G_+(t) + G_-(t)] , \tag{1.3}$$

where ΔE is the energy gap, and

$$G_\pm(t) = \sum_i (\Delta_i^2/2) \left\{ \frac{\bar{v}_i + 1}{v_i} \right\}^{1/2} \exp(\pm i\omega_i t) \tag{1.4}$$

$$G = G_+(0) + G_-(0) \tag{1.5}$$

$$\bar{v}_i = [\exp(h\omega_i/k_B T) - 1]^{-1} \tag{1.6}$$

$$\Delta_i = (\mu_i \omega_i / h)^{1/2} \Delta Q_i. \tag{1.7}$$

$\bar{v}_i$ denotes the thermal average and Δ_i is the reduced displacement of the ith mode, while $\exp(-G)$ is the Debye–Waller factor. Equation 1.3 was advanced for multiphonon optical transitions, for radiationless transitions in general and for ET in particular.

The vibrational modes incorporated in Eqn 1.3 include:

1 Low-frequency medium modes, usually approximated by a single frequency ω_m with an effective Δ_m. 'Intelligent guesses' for glasses result in $\omega_m \approx 10$–100 cm^{-1}. It seems that for a polar solvent a phonon picture is not applicable, while for a protein medium [140,141] $\omega_m \approx 100 \text{ cm}^{-1}$. The medium reorganization energy is $\lambda_m = S_m h \omega_m$, where $S_m = \Delta_m^2/2$. For polar solvents $\lambda_m \approx 4000$–8000 cm^{-1} while for the protein RC $\lambda_m \approx 800$–3000 cm^{-1} [140,141].

2 Intramolecular modes in the range $\omega \approx 100$–3000 cm^{-1}. A complete treatment of the ET requires the incorporation of all the coupled (quantum) intramolecular modes, providing the basis for intramolecular ET dynamics [117]. Often, but not exclusively, the intramolecular modes can be approximated by a single molecular frequency $\omega_c \approx 1500 \text{ cm}^{-1}$ and a dimensionless shift $S_c = \Delta_c^2/2 \approx 0.5$–$1.0$ [140,141].

Energy gap laws constitute a major generalization of microscopic relaxation phenomena, which originate from the dependence of F (Eqn 1.3) on ΔE (or ΔG). The quantum nature of the intramolecular vibrational modes is usually prevalent at all the relevant temperatures, as $k_B T \ll h \omega_c$. When a single mode can be handled in terms of the high temperature limit ($k_B T > h \omega_m$), Eqn 1.3 reduces to [123]

$$F = (4\pi \lambda_m k_B T)^{-1/2} \exp(-S_c) \sum_{n=0}^{\infty} \frac{S_c^n}{n!} \exp\left[-\frac{(\Delta E + \lambda_m + n h \omega_c)^2}{4\lambda_m k_B T} \right] \tag{1.8}$$

providing a useful contribution for the calculation of nuclear contributions to the ET rate over a broad temperature domain. In the limit $S_c = 0$, Eqn 1.8 reduces to the classical Marcus relation [111], $F_{cl}(S_c = 0) = (4\pi \lambda_m k_B T)^{-1/2} \exp[-(\Delta E + \lambda_m)^2/4\lambda_m k_B T]$ with a Gaussian activation energy. The (free) energy dependence of F (Fig. 1.20) provides a demonstration of the classical Marcus relation in the normal region ($-\Delta E \leqslant \lambda_m$) and is the activationless domain ($\Delta E = -\lambda_m$), while in the inverted region ($-\Delta E \geqslant \lambda_m$) marked deviations from the classical relation are exhibited [i.e. $F > F(S_c = 0)$] due to the vibrational excitation of the intramolecular quantum modes of D and A, accompanying ET. Quantum effects in the inverted region include a marked enhancement of the ET rate at constant ΔG (Fig. 1.20) and in a surprisingly weak temperature dependence of k for strongly exoergic reactions [140,141]. The contribution of high-frequency intramolecular modes to ET dynamics opens up the possibility of mode-specific ET. Provided that vibrational relaxation of a high-frequency mode is slow on the time scale of ET (involving an equilibrated manifold of all the other medium and intramolecular modes), non-adiabatic ET from distinct photoselected vibrational states will be state specific, as demonstrated by Spears [142,143] for the ET process $[Co(Cp)_2] [Co(CO)_4] \rightarrow [Co(Cp)_2] [Co(CO)_4]^-$ from vibrationally excited high-frequency CO modes ($\omega \approx 2000 \text{ cm}^{-1}$).

The energy gap dependence of F (Fig. 1.20) implies an optimization principle for activationless ET. The F factor is maximized ($F = F_{MAX}$) for $-\Delta E = \lambda_m + n h \omega_c$, with the

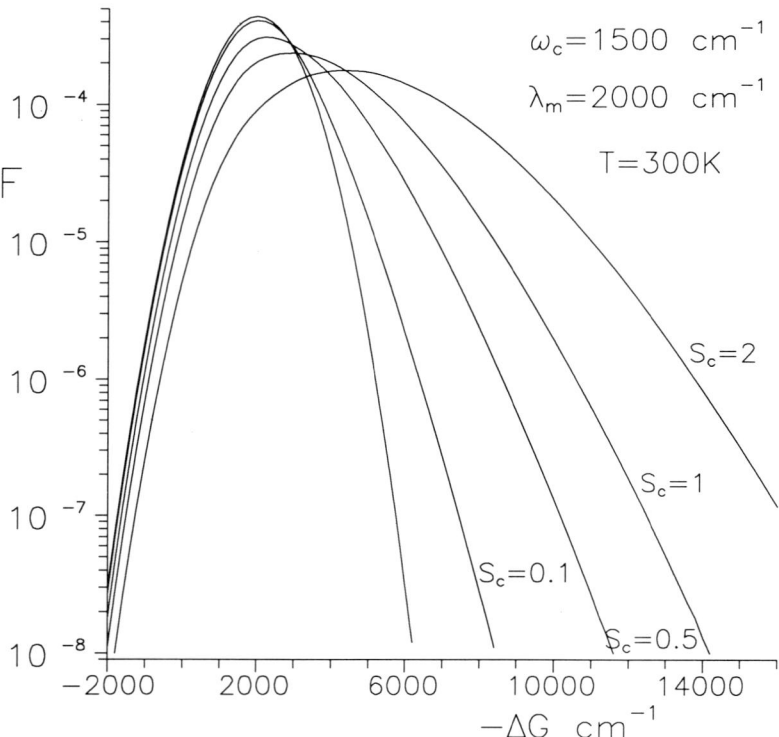

Figure 1.20 The energy gap dependence of the nuclear Franck–Condon factor, which incorporates the role of high frequency intramolecular mode(s).

dominating contribution to the sum in Eqn 1.8 originating from the nth term, so that $F_{MAX} \approx (4\pi\lambda_m k_B T)^{-1/2}(S_c^n/n!)\exp(-S_c)$. This situation corresponds to activationless ET (i.e. for the lowest intersection of the multidimensional nuclear potential surfaces occurring at the minimum of the initial DA state) with a weakly temperature dependent $k \propto T^{-1/2}$ non-Arrhenius-type rate. For typical values of $S_c \approx 1$ the activationless ET is realized with $F_{MAX} \approx (4\pi\lambda_m k_B T)^{-1/2} \exp(-S_c)$. The maximization of the Franck–Condon factor involves the optimization of the nuclear contribution to the ET rate.

In the non-adiabatic limit the optimal activationless rate is $k = (2\pi V^2/h)F_{MAX}$. For characteristic values $\lambda_m = 4000 \text{ cm}^{-1}$ for ET in polar solvents or $\lambda_m \approx 1000 \text{ cm}^{-1}$ in protein RC together with $S_c \approx 1$, we estimate (at room temperature) the activationless ET rate

$$k \text{ (s}^{-1}) \approx (1.4\text{--}2.7) \times 10^8 \text{ } (V/\text{cm}^{-1})^2 \tag{1.9}$$

For typical V values (Fig. 1.19), we estimate $k \approx 2 \times 10^{10} \text{ s}^{-1}$ for $V = 10 \text{ cm}^{-1}$ and $V \approx 2 \times 10^{12} \text{ s}^{-1}$ for $V \approx 100 \text{ cm}^{-1}$. The upper limit of the maximized activationless ET rate is determined by the breakdown of the non-adiabatic formalism, which requires the extension of the non-adiabatic theory.

3.1.3 EXTENSIONS OF NON-ADIABATIC ET THEORY

Adiabatic ET. In Section 3.1.2, we have spelled out the conditions for the applicability of non-adiabatic ET theory. Condition 3 on p. 22 for weak electronic coupling implies that the Landau–Zener parameter is small, i.e. $\gamma \approx 2V^2/\hbar\omega_m(\lambda_m\hbar\omega_m)^{1/2} < 1$, where ω_m is an effective medium frequency [144,145]. Taking characteristic values of $\hbar\omega_m = 100 \text{ cm}^{-1}$ both for a protein medium and a polar solvent, non-adiabatic ET prevails for $V \leqslant 200 \text{ cm}^{-1}$ in a polar solvent ($\lambda_m \approx 4000 \text{ cm}^{-1}$) and $V \leqslant 100 \text{ cm}^{-1}$ for a protein medium ($\lambda_m \approx 1000 \text{ cm}^{-1}$). In Fig. 1.21, we have marked the upper limit for the V values, which still correspond to non-adiabatic ET. When the electronic coupling is sufficiently strong, i.e. $\gamma > 1$, the adiabatic limit for ET applies. The ET rate is then given by the Holstein formula

$$k = (\omega_m/2\pi)\exp(-E_a/k_BT). \tag{1.10}$$

For activationless ET, the adiabatic rate constant is $k \approx \omega_m/2\pi$.

Solvent-controlled ET. The breakdown of assumption 4 (p. 22) implies that solvent relaxation, rather than the microscopic electronic processes, constitutes the rate-determining step for ET. For ET in a system solely characterized by coupling to the medium ($S_c = 0$), which corresponds to the normal Marcus region ($\Delta E \leqslant -\lambda_m$), the realization of solvent-controlled ET is determined by the magnitude of the solvent adiabaticity parameter [118] $\kappa = 4\pi V^2\langle\tau\rangle/\hbar\lambda_m$, where $\langle\tau\rangle$ is the longitudinal dielectric relaxation time τ_L, e.g. $\tau_L = 200 \text{ fs}$ for water, $\tau_L = 190 \text{ fs}$ for acetonitrile and $\tau_L = 1.2 \text{ ps}$ for methyl acetate at room temperature [118]. For the membrane protein medium of the RC [146–151], molecular dynamics simulations give [152] $\langle\tau\rangle = 100 \text{ fs}$ over the temperature domain 10–300 K. The medium-controlled ET rate is given by $k = k^{NA}/(1 + \kappa)$, where the non-adiabatic rate k^{NA} is given by Eqn 1.1. In the limit $\kappa \gg 1$, $k \propto k^{NA}/\kappa \propto \langle\tau\rangle^{-1}$, being independent of V (Fig. 1.21).

Is this formalism relevant for activationless ET? Medium-controlled ET will be manifested only provided that the microscopic rates are sensitive to the details of the distribution of the initial states. For activationless ET both model calculations in the classical limit and numerical computations reveal that the microscopic ET rates k_i are quite insensitive to the initial vibrational state [124]. This weak excess energy dependence of k_i is compatible with the weak temperature dependence of activationless ET.

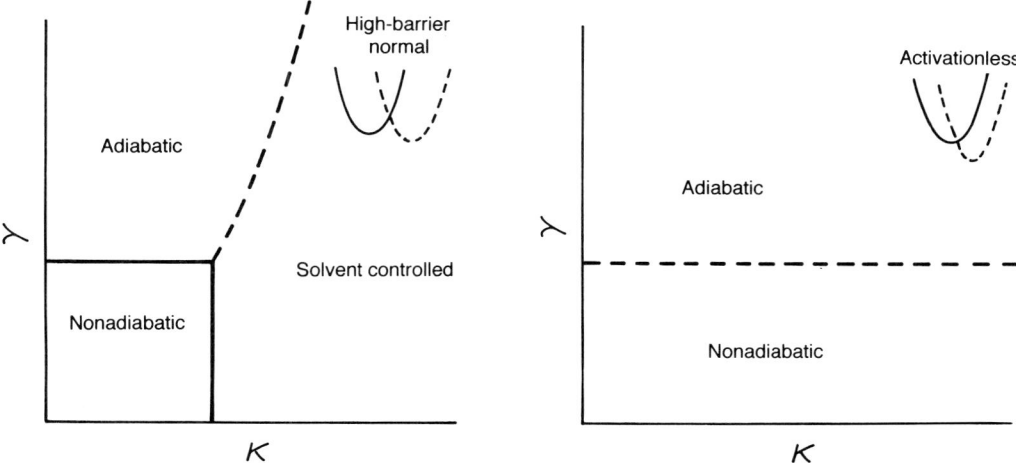

Figure 1.21 Domains for electron transfer (ET) for the normal limit and for the activationless case.

Accordingly, activationless ET is invariant with respect to medium relaxation dynamics, with the rate being independent of κ (Fig. 1.21). This analysis provides a possible explanation for the recent experimental observations of some ET rates, which substantially exceed the τ_L^{-1} limit predicted for solvent-controlled ET. Kobayashi et al. [153] reported ultrafast ET rates in (Nile blue$^+$) (TMPD) with $k \approx (100\ \text{fs})^{-1}$ which correspond to $k \approx 50/\tau_L$, while Heitele et al. [154] observed for ET in porphyrin–quinone cyclophanes independence of $k \approx (1\ \text{ps})^{-1}$ on ($\kappa = 0.1$–400), i.e. over a broad region. ET in both systems seems to correspond to activationless ET (or inverted region [124]). The primary ET reactions in the photosynthetic RC are nearly activationless [146–149], so that the medium dynamics (with $\langle \tau \rangle \approx 100$ fs) [152] are not manifest explicitly in the rates. Thus activationless ET can be appreciably faster than previously expected, and is not limited by solvent dynamics.

An upper limit for activationless ET rates. For activationless ET the non-adiabatic limit is expected to prevail, being characterized by $k \propto V^2$ with increasing V, until the adiabatic limit for ET will be achieved constituting an upper limit for the rate. An approximate estimate for the upper limit for the activationless ET rate can be inferred from Eqn 1.10 in conjunction with the Landau–Zener parameter $\gamma \leqslant 1$ [144,145], which, for a typical value of $\omega_M \approx 30$–100 cm^{-1}, results in $(\omega_m/2\pi)^{-1} \sim 2$–7 ps. The fastest room temperature ET rates recorded to date in synthetic supermolecules $k \approx (100\ \text{fs})^{-1}$ [152] and in the RC [with the rates for primary charge separation $^1\text{P*BH} \xrightarrow{k_1} \text{P}^+\text{B}^-\text{H} \xrightarrow{k_2} \text{P}^+\text{BH}^-$ being $k_1 \approx (3000\ \text{fs})^{-1}$ and $k_2 \approx (1000\ \text{fs})^{-1}$ at $T = 300$ K] [155–158] are roughly comparable to the theoretical upper limit.

The pertinent time scales for ET can be inferred by drawing the analogy between these processes and radiationless transitions. The relevant physical parameters are the (pure) electronic coupling V, the (average) Franck–Condon factors (FC), the medium-induced vibronic density of states ρ, the medium-induced vibrational relaxation rate $\Gamma = h/\tau_{VR}$ (where τ_{VR} is the vibrational relaxation time) and the characteristic medium frequency ω_m. The condition for strong interstate coupling, i.e. $V(\text{FC})^{1/2}\rho \gg 1$, implies that the time scale for non-adiabatic ET $\tau_{ET} = [(2\pi/h)V^2(\text{FC})\rho]^{-1}$ is longer than the mixing lifetime $\tau_{MIX} = [V(\text{FC})^{1/2}/h]^{-1}$. In turn, for conventional non-adiabatic ET the separation of time scales implies that $\tau_{VR} \ll \tau_{ET}$. Finally, τ_{VR} cannot be faster than the period of vibrational motion, i.e. $\tau_{VR} \geqslant h/\omega_m$. The concept of the medium-induced vibrational relaxation time τ_{VR} of the internal vibrational states of the D and A centres has to be extended to incorporate also the solvent relaxation time $\langle \tau \rangle$ induced by the charge distribution (Eqn 1.3). The description of solvent relaxation should also include short-time, short-range inertial effects. Thus, for conventional non-adiabatic ET occurring from a thermally equilibrated manifold,

$$\tau_{ET} \gg \tau_{MIX}; \quad \tau_{ET} \gg \tau_{VR} \geqslant h/\omega_m; \quad \tau_{ET} \gg \langle \tau \rangle. \tag{1.11}$$

The lower limit for the time scale of ET dynamics is provided by the time scale of nuclear motion h/ω_m, which constitutes the adiabatic limit for activationless ET according to Eqn 1.10.

For ultrafast ET, when τ_{ET} is comparable to or smaller than τ_{VR} and/or $\langle \tau \rangle$, non-adiabatic level depletion occurs from a non-equilibrated vibronic (and solvent) manifold. The ET dynamics is determined by the excess vibrational energy dependence of the microscopic interstate non-adiabatic ET rates (Sections 3.1.1 and 3.1.3), and by vibrational and solvent intrastate vibrational energy relaxation rates. With the advent of femtosecond lasers [159], the dynamics of coherent vibrational wave packets of D*A (or D*BA) states becomes amenable to experimental interrogation [159–161]. The dynamics involves the interplay between intrastate vibrational relaxation, interstate electron transfer, and dephasing of the vibrational wavepacket. In addition to the level depletion time scales τ_{ET} and τ_{VR}, the pure vibrational dephasing time scale $\tau_p \sim 100$–1000 fs in solution [157] and in biophysical systems [158] enters. Exploration of the response of molecular electronic systems to femtosecond laser excitation will provide a new approach for the exploration of ET dynamics on the time scale of nuclear motion, attaining selectivity, sensitivity, and high yield of basic ET processes.

3.1.4 COHERENT AND INCOHERENT TRANSFER

To this point we have been concerned with two-centre ET. When multisite electron transfers are considered, or when a bridge structure intervenes between the donor and the acceptor, the question of coherent or hopping transfer through the bridge comes up. In the special case of resonant transfer (D, B, A degenerate), considerations similar to those relevant for the small polaron problem (Holstein) suggest that a criterion can be found based on the comparison of the scattering length and the lattice spacing [162]. For a linear chain, the band energy is

$$E_k = 2V \cos(ka), \tag{1.12}$$

where a is the lattice constant, V the intersite coupling and k the pseudomomentum. Introducing the scattering time τ_{scatt}, the electron velocity

$$v_e = (1/h)\partial E_k/\partial k = -(2Va/h)\sin(ka) \tag{1.13}$$

and the mean free path λ

$$\lambda = v_e \tau_{scatt}, \tag{1.14}$$

we find for the linear chain

$$\lambda = -(2\tau_{scatt} Va/h)\sin(ka). \tag{1.15}$$

The condition for incoherent behaviour is then given by

$$\lambda \leqslant a \quad \text{or} \tag{1.16}$$

$$2|V|\tau_{scatt}/h \leqslant 1. \tag{1.17a}$$

Thus incoherent ET occurs for small $|V|$ and/or short τ_{scatt}. For increased $|V|$ and/or long τ_{scatt} the system becomes delocalized. Coherent transfer occurs when

$$2|V|\tau_{scatt}/h > 1. \tag{1.17b}$$

Of course, the band description of the electronic structure disregards the important effects of disorder, where diagonal disorder, i.e. energetic spread of the site energies, results in Anderson–Mott localization [163]. In reduced dimensionality solid-state structures, weak localization effects occur even without full Anderson localization [164]; analogous behaviours might be anticipated in molecular wires.

For the (far more interesting) case of the ET in which the D, A sites are not degenerate with the bridge (wire), the conditions for incoherent vs coherent transfer are less clear. Still, it is not unreasonable to expect that the onset of incoherent behaviour will occur when

$$\tau_{\text{thermalization}} \sim \tau_{\text{tunnelling}}, \tag{1.18}$$

where the thermalization time reflects the trapping of the charge carrier via vibrational coupling. The tunnelling time issue has been widely investigated [165]. Accordingly, we tentatively associate τ_{tunnel} with the Landauer–Buttiker time

$$\tau_{\text{LB}} = \int_{x_1}^{x_2} [m/2(V_0(x) - E)]^{1/2} \, dx; \tag{1.19}$$

here x_1, x_2, m, V_0 and E are, respectively, the two turning points, electron mass, potential and injection energy. This suggests that incoherent (bridge-localized) behaviour will occur with small gaps between donor and bridge levels, while for longer bridges high temperatures and strong bridge electron/vibration coupling will prevail. These are not found in most of the standard ET systems, but will be of real significance in assuring true molecular wire behaviour.

3.2   Proton transfer

The proton is about 2000 times heavier than the electron, but substantially lighter than all other atoms of interest in chemistry. Therefore, proton transfer reactions (like strongly coupled electron transfer reactions) can often be viewed in a kind of adiabatic picture involving a single potential surface on which the light particle (the proton) responds to the dynamics of the other particles instantaneously, and the rate of transfer of the proton (like that of the electron) depends essentially on the nuclear dynamics of the remainder of the molecular structure [166,167].

Unlike electrons, however, protons can be studied crystallographically, characterized by vibrational spectroscopy, and replaced in a facile way by long-lived deuterium. While electrons can be localized, and indeed electron localization is an important aspect of modern solid-state science, localization of protons within molecular or solid-state structures is the common situation. On the other hand, in symmetric molecular structures protons often have a choice of localization sites; by localizing them in different sites, one can (in principle, and perhaps in practice) store information in the form of dipoles.

Information storage using proton location is of interest in the solid state, where hydrogen-bonded ferroelectric memories were investigated decades ago. It has also been proposed in the molecular electronics area, in particular the hemiquinone structures suggested by Aviram and co-workers [168] and by Todd and co-workers [169]. In these hemiquinones, protons move as a pair between, for example, paraquinone and a parahydroquinone (Fig. 1.22). The actual rate of transfer will clearly be strongly geometry dependent, but it has not been well studied experimentally.

Much closer to realization are the hole burning memories based on the location of the two interior protons in the porphyrin structures of Fig. 1.23. This was already discussed in Section 1: these hole burning memories have been demonstrated [74–76]. The actual

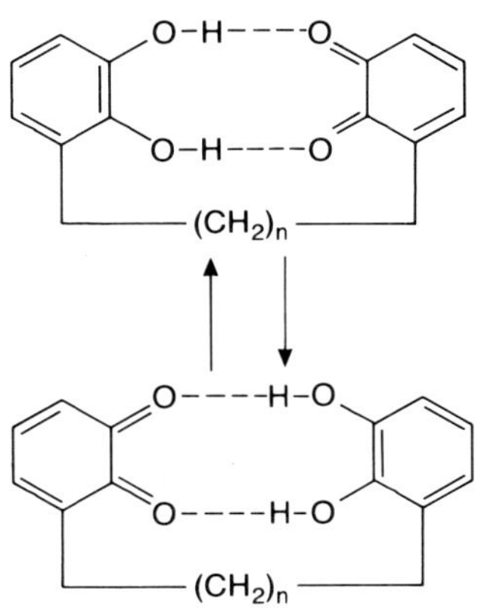

Figure 1.22 A 'hemiquinone' molecule, proposed as an information storage system. On switching of the two protons from the right to the left, the dipole moment would change, and this change in valence tautomer is measurable. From [168].

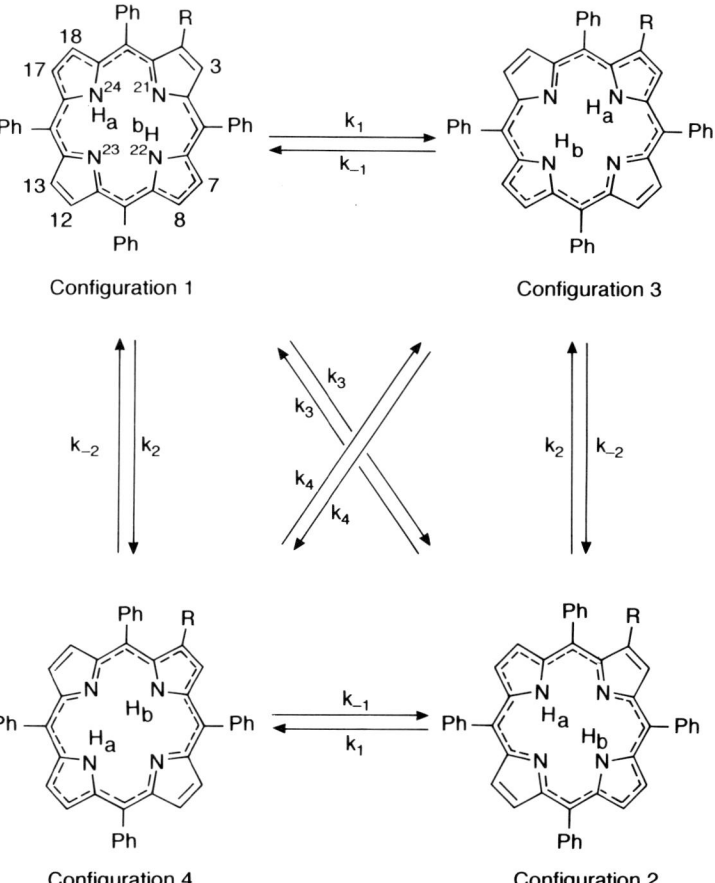

Figure 1.23 Intramolecular proton transfer among the four nuclear configurations in two-substituted 5,10,15,20-tetraphenylporphyrin. From [317].

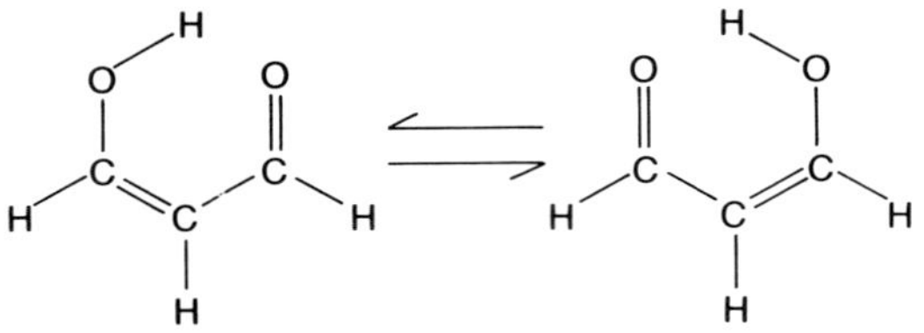

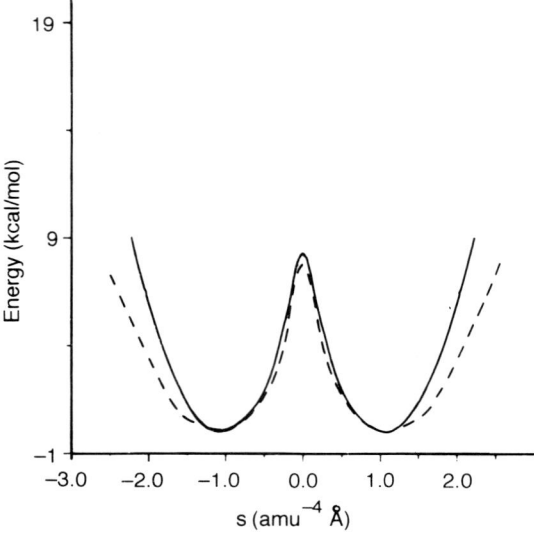

Figures 1.24 The malonaldehyde system, a prototype for single proton transfer reactions. The motion of the proton between the two oxygens is conditioned on the entire reorganization of the bonding network in the molecule.

localization corresponds to the two protons moving, as indicated in Fig. 1.23, among the four interior nitrogen atoms of the macrocycle. This has been studied extensively [74,170], and it appears to be a correlated, double pair motion of the protons — note that it changes the electrical moments of the molecule, although the molecule remains non-dipolar (as long as the symmetry is not broken).

Extensive theoretical work has been done on simple proton transfer reactions involving one proton [167,171–173]. For example, Fig. 1.24 shows the prototype malonaldehyde species, which has two resonant forms, corresponding to the proton localized on right or left oxygen. These tautomeric species, at low temperatures, correspond to two equivalent minima in a double well potential as sketched in Fig. 1.25. The motion of the proton from one side of the double well to another can be caused by thermal excitation, by collisions, or by photoprocesses. If one functionalizes the malonaldehyde (e.g. by putting a methyl group on one side) the two minima become inequivalent. The transfer of the proton from one side to the other can be controlled by photoexcitation; extensive study of this process, both experimental [174] and theoretical [175], has appeared recently. The issue of proton transfer is discussed extensively in Chapter 3 of this volume.

Figure 1.25 The malonaldehyde system. The calculated intrinsic reaction path from malonaldehyde is shown by the dashed line. From [318].

3.3 *Photoabsorption*

In all molecular electronics applications, the system must interact with external stimuli and probes. The stimuli are generally in the form of a chemical species (electron, proton, molecule, or ion for recognition), or wave-like excitation (pressure wave, electromagnetic excitation). Photoabsorption or photoexcitation is perhaps the most general way to couple a molecular electronics system to its environment. Molecular electronics processes such as energy transduction (including photosynthesis and vision), optoelectronics, photodetection (including luminescence and quenching detection), photoinduced electron transfer, optical switching and electrophotography all depend on the photoabsorption process. Elegant quantum size effects in small clusters (Section 5), in particular semiconductor nanoclusters, will also clearly depend upon the photoabsorption processes; these are discussed in Chapter 9.

The essential chemical physics of photoabsorption will vary depending on the nature of the molecular system being studied. For straightforward absorption, the golden rule yields the absorption rate in the form [176]

$$W_{g \to x} = 2\frac{\pi}{h}|\langle x|\mu|g\rangle|^2\delta(E_x - E_g \pm h\omega) \tag{1.20}$$

Here x and g label the excited and ground states, respectively; their relative energies are E_x and E_g. The radiation is assumed to be monochromatic with a frequency ω; the dipole moment operator μ couples ground and excited states. The states involved, assumed to be of crude Born–Oppenheimer type, are products of vibrational and electronic parts — the Franck–Condon factors then are simple vibrational overlaps. If damping or decay processes are important in the excited state, the δ function of Eqn 1.20 is replaced by

$$\frac{\Gamma}{\Gamma^2 + (E_x - E_g \pm h\omega)^2}$$

with the damping parameter Γ describing dephasing and lifetime effects in the excited state.

Depending on the nature of the molecular systems involved, the ground and excited states are written differently. For a simple single-site, two-level model as sketched in Fig. 1.26, the electronic states are the two levels. For two-site situations, such as donor/acceptor species, bichromophoric molecules, or bridge-linked photoinduced electron transfer systems such as those in Fig. 1.27, an important analysis was given by Hush [177], in terms of the two levels in Fig. 1.28. Making Mulliken-like assumptions concerning the orthogonality of the D^+A^- and the ground DA states, the Hush relationship is then [176,177]

$$V(\text{cm}^{-1}) = \frac{0.0206}{R_c}(\varepsilon_{MAX}\nu_{MAX}\Delta\nu_{1/2})^{1/2}. \tag{1.21}$$

Here V is the electronic coupling that mixes the DA and D^+A^- states, and R_c is the centre-to-centre distance in angstroms separating them; ε_{MAX}, ν_{MAX} and $\Delta\delta\nu_{1/2}$ are, respectively, the molar extinction coefficient, the position of the maximum absorption

Energy

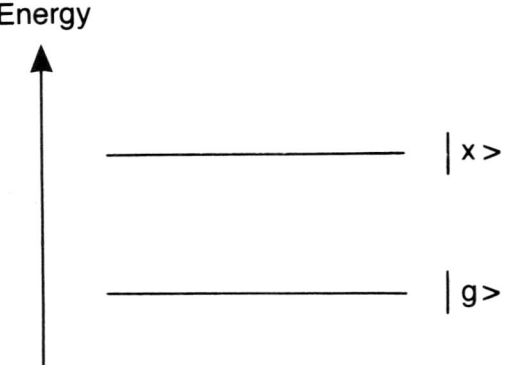

$|x\rangle$

$|g\rangle$

Figure 1.26 A simple two-level model for photoabsorption.

frequency for the charge transfer absorption, and the full width at half height of that absorption (the latter two in cm^{-1}). This relationship suggests that the coupling vanishes when the absorption vanishes or the distance becomes very large, and that the square of the matrix element is proportional to the extinction coefficient. This relationship has been used very broadly for a number of purposes, including estimation of distances or of mixing matrix elements on the basis of optical transition strengths [145,177–179].

In fact, the situation is more complicated. Especially when coupling between donors and acceptors is weak, a third process enters [180], involving the locally excited donor acceptor state DA* (the highest level shown in Fig. 1.28). The spectroscopic states then are admixtures of the ground state, charge transfer state, and the locally excited state (DA, DA*, D^+A^-). Understanding of both photoabsorption and subsequent radiative decay is then substantially modified from the simple two-level model given by Hush. A number of applications, including the effect of solvent dependence, based on this three-level model have been given recently [180].

As molecular electronic systems become larger, the nature of the interaction can change. For two-site donor/acceptor systems with very long bridges, the Förster transfer mechanism may be dominant; this allows for very long range energy transfer. As the molecular structure becomes larger, and can no longer be treated as a two-site system, one approaches a band-type description or an excitonic description, depending on how strongly coupled the local excitations may be. Design of systems to optimize photoexcitation processes ranges from simple energy-level tuning, such as used in selection of dye chromophores, to more sophisticated notions such as those involved in artificial photosynthetic molecules [86,87] like that in Fig. 1.11. Here the central chromophore is responsible for photoabsorption, and should have a large oscillator strength and a relatively slow radiative decay. The coupling to the terminal sites, which results in the separation of charge to form both a giant dipole and a long-lived polar state, then depends on tuning the intramolecular relaxation channel from the initially excited state to the charge-separated product. Among the most sensitive aspects of artificial photosynthesis are this control of subsequent intramolecular relaxation following initial excitation, and of the initial absorption cross-section.

Finally, in truly extended systems such as nanoclusters of silicon or cadmium selenide, the photoexcitations involve both delocalized states over the entire cluster, with remnant excitonic couplings (between the excited state electron and the ground state hole subjected to exciton confinement discussed in Section 5.4), and localized (trapped) electrons or holes; these latter features are especially characteristic of smaller clusters, and become weaker in larger ones. Calculational work from several laboratories has shown that the quantum size effect (determined by the actual extent of the cluster) and the excitonic coupling combine to provide the general shift in absorption wavelength (bluer for smaller crystals) seen in these materials [181–183]. This is significant for the design of optoelectronic systems with selected frequency dependence.

3.4 *Molecular non-linear optics*

Molecular electronics includes not only linear interaction of molecular systems with photons (absorption or emission of single photon), but also higher order interactions. If, for example, the molecular system is exposed to a radiation field, the overall polarization of the molecular species can be expanded in the field strength, giving

$$p_i = \mu_i + \sum_{j \geq i} \alpha_{ij} E_j + \sum_{k \geq j \geq i} \beta_{ijk} E_j E_k + \sum_{l \geq k \geq j \geq i} \gamma_{ijkl} E_j E_k E_l + \dots . \qquad (1.22)$$

Here p_i is the polarization of a molecule along its i axis; the dipole moment in the absence of the radiation field is given by μ_i. The parameters α, β and γ are, respectively, the polarizability and the first and second hyperpolarizabilities of the molecule, while E_j is the electromagnetic field along the jth molecular axis. Because the atomic unit of the electromagnetic field corresponds to 5.1×10^9 V cm^{-1}, even strong laser fields are small compared with atomic fields, and therefore one expects the subsequent terms in this equation to be small compared with the first non-vanishing term. In general, polarizabilities and hyperpolarizabilities are frequency dependent.

Non-linear optical response of molecules generally refers to the β and γ terms, corresponding to first and second hyperpolarizabilities. These can be useful for optoelectronic applications such as frequency doubling, frequency modulation, intensity-dependent refractive index, optical Kerr effect, and light intensity modulation by light. This is an extensive field, and really lies more in the area of molecular optics than molecular electronics. The field has generated a very substantial literature of its own [184–186].

3.5 *Coupling of subcomponents*

The power of electronics arises when different components can be linked, so that an overall device can be prepared that has more than one function or logical step. The obvious example is the construction of a computer using simple logic gates. In analysing the prospects for molecular electronics, therefore, it is important to consider how the different subcomponents might be linked. This is one aspect of the addressing problem: even if particular molecular electronic components, such as switches, gates, rectifiers, or energy transducers could be produced, coupling these to one another and to the outside world remains a problem.

It is perhaps useful to distinguish the subcomponent coupling problem in photonic situations from that in electronic or ionic situations.

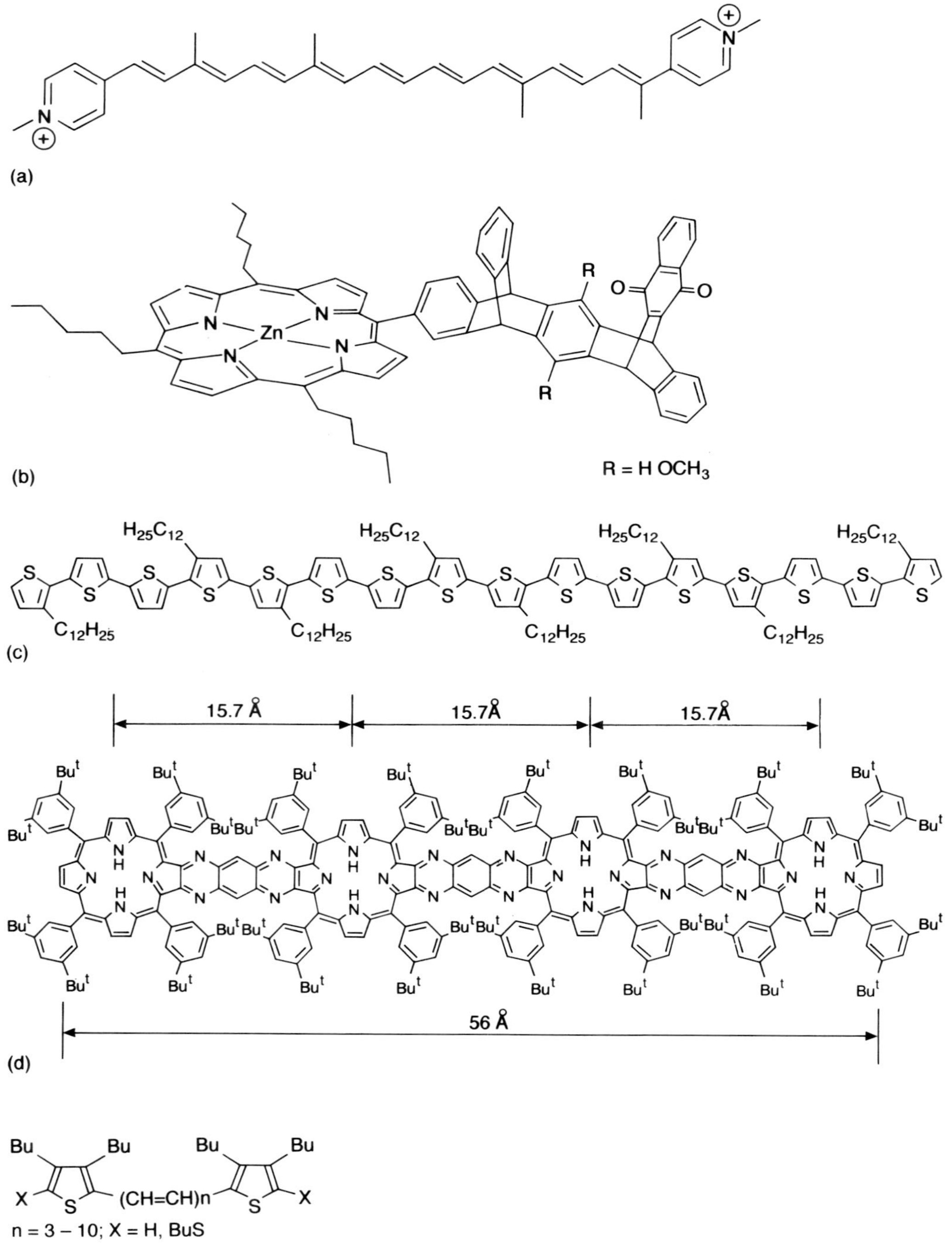

(a)

(b)

R = H OCH₃

(c)

15.7 Å   15.7 Å   15.7 Å

56 Å

(d)

Bu   Bu   Bu   Bu

X⟍ₛ⟋—(CH=CH)n—ₛ⟋X

n = 3 – 10; X = H, BuS

(e)

Figure 1.27 A series of donor/bridge acceptor compounds used to study long range intramolecular electron transfer. In most of these structures, the electron is not thought to be localized on the bridge; rather, the bridge orbitals act to facilitate electron transfer by the so-called superexchange mechanism, that involves coherent transfer without electron localization on the bridge. (a) The so-called caroviologen wire of Lehn [210]. (b) A rigid fused porphyrin/quinone structure [87]. (c) The extended structure of a long polythiophene bridge [320]. (d) An oligoporphyrin [55]. (e) A simple conjugated polyacetylene-like structure [321]. (f) The staffane structures [322]. The polycyclic compounds pioneered by Verhoeven *et al.* [323] are shown in (g), while the important stereoisomerically controlled molecules based on decalin are shown in (h) [324].

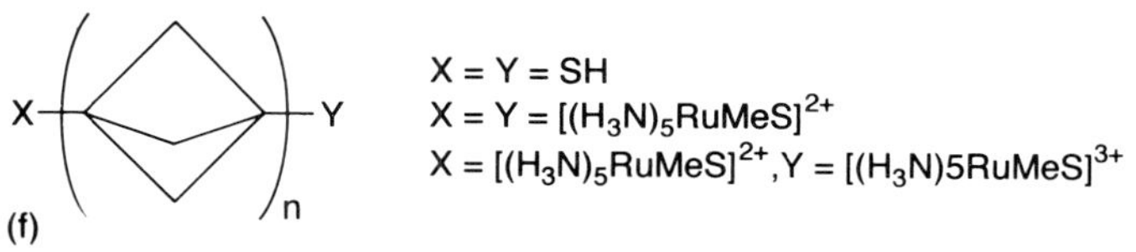

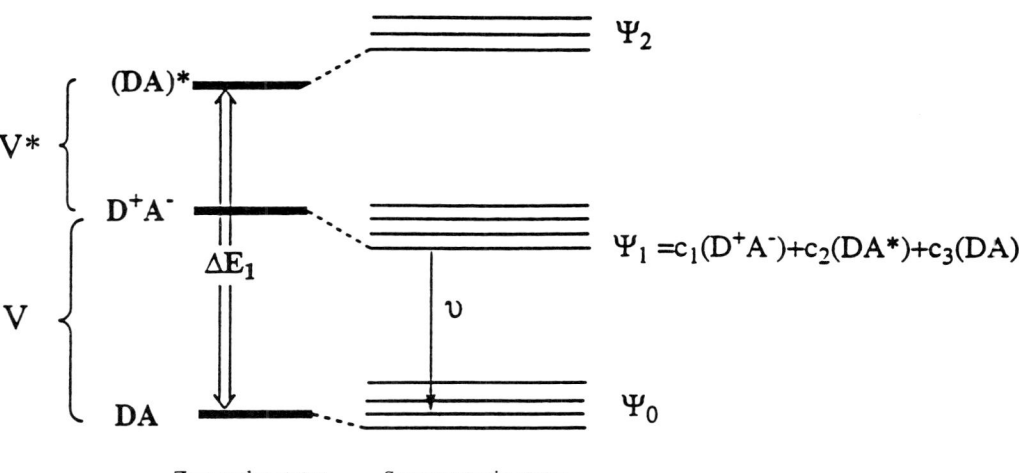

(f)

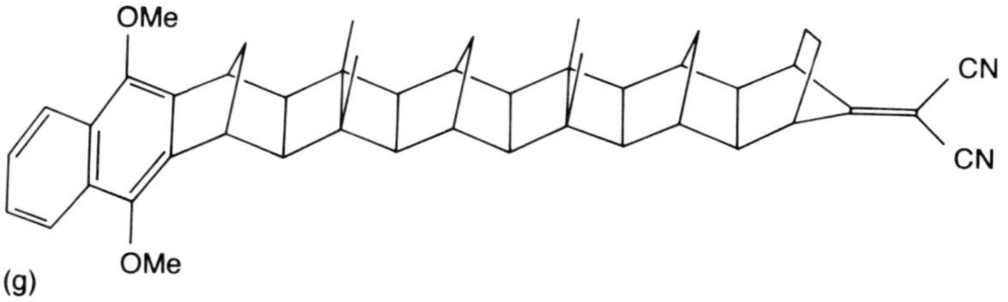

$X = Y = SH$
$X = Y = [(H_3N)_5RuMeS]^{2+}$
$X = [(H_3N)_5RuMeS]^{2+}, Y = [(H_3N)5RuMeS]^{3+}$

(g)

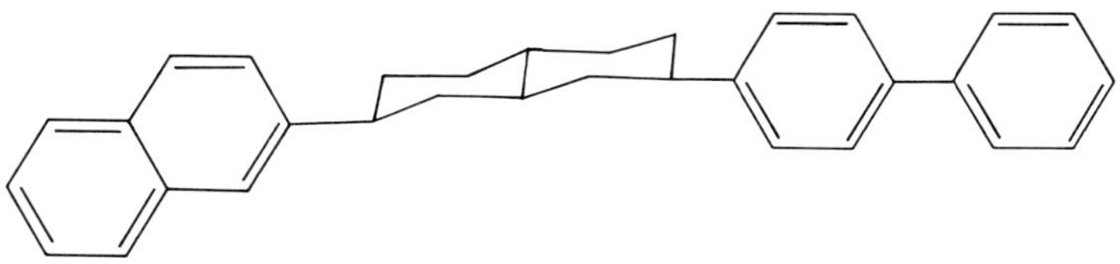

(h)

Figure 1.27 (*Continued*)

$$\Psi_2$$

$(DA)^*$

V^*

D^+A^-

ΔE_1

V

$\Psi_1 = c_1(D^+A^-) + c_2(DA^*) + c_3(DA)$

υ

DA Ψ_0

Zero order states   Spectroscopic states

Figure 1.28 The three-state model for understanding photoinduced electron transfer; the zero order states consist of the neutral donor/acceptor system, the charge transfer donor/acceptor system, and the photoexcited donor/acceptor system; all three of these mix to form the spectroscopic states.

For electronic couplings, corresponding to electron or hole charge transport in the device, the question arises as to how mesoscopic- or nanoscopic-scale components relate to macroscopic ones. In macroscopic design (including contemporary computer design), junctions are assumed ohmic, so that electron flow simply follows Ohm's law, with interfaces providing an extra resistance component. At the molecular level, ohmic behaviour is not expected in general, because of the back-scattering processes, quantum coherence effects, Schottky barriers, image and dipolar coulomb barriers, tunnelling at interfaces, coulomb charging energies, and the sparseness of molecular energy levels [129–137]. Conditions under which the microscopic Hamiltonian interaction between subcomponents leads to the ohmic regime remain under investigation [187–189]. (Similar design problems actually occur in semiconductor design, especially in regimes in which ballistic transport may occur [164,190].)

In the particular case where one is interested in the coupling of an electrode (considered as a macroscopic metallic band structure) to a molecule, the standard model was put forth by Newns and Anderson [191,192], and is discussed extensively in Section 4 below. In the case of molecule-to-molecule coupling, the interaction is purely Hamiltonian. Under these conditions, if the density of final states is sufficiently large, we have an electron transfer process of the usual kind, and expect adiabatic and non-adiabatic limits (in the former, the rate of charge transfer is simply proportional to the density of states on the final molecular acceptor, and in the latter it also includes a proportionality to the square of the coupling matrix element; cf. Section 3.1). When the density of final states is not this large, one expects more specific quantum effects, including level-matching resonances and back-scattering behaviours. Control, both spatial and temporal, of these relaxation/charge flow problems can provide a rich set of phenomena — e.g. Wasielewski and co-workers [193] have demonstrated a molecular switch based on the use of two sequential intramolecular photoabsorptions. Section 4 treats more specifically the problem of molecular electronic wires — that is, molecular structures intended to provide electric linkages between subcomponents [194].

Molecular photonic wires have been discussed both experimentally and theoretically [43,44]. The essential problem is to provide a molecular structure that effectively links donor and acceptor, so that photoexcitation of the donor is effectively transferred to luminescence behaviour on the acceptor site. Theoretically, there are two standard limits for the energy transfer rate from donor to acceptor. In the model, donor and acceptor species are considered as independent two-level structures, and the efficiency of energy transfer is determined by the long range electrostatic interaction. Restricting the latter to the dipolar term, one can write, analytically, the rate constant for transfer as [195]

$$k_{\text{Förster}} = \frac{3}{2} \kappa^2 \tau_D^{-1} (R_c/R)^6.$$ (1.23)

Here κ, τ_D and R are, respectively, a geometric factor depending on the relative angular orientation of the two chromophores, the radiative lifetime of the free donor, and the donor–acceptor distance. The R^6 dependence, arising from the square of the R^{-3} dipole–dipole interaction, is characteristic of the Förster mechanism. The characteristic distance R_c is determined by the spectral overlap:

$$R_c^6 = \frac{9000 \ln 10 \phi_D}{128\pi^3 n^4 N} \int_0^\infty \frac{dv}{v^4} f_D(v)\varepsilon_A(v),$$ (1.24)

with ϕ_D the donor emission quantum yield. The quantities f_D and ε_A are, respectively, the donor emission and acceptor absorption lineshapes. Thus the Förster mechanism decays with distance as R^6 and depends on the spectral overlap integral (last term in Eqn 1.24), and on the relative orientation. Förster transfer is useful in many regimes, but its requirement for spectral overlap and appropriate angular projection puts severe constraints on it as a mechanism for efficient long-range energy transfer. Nevertheless, it is apparently dominant for molecules with large transition dipole moments separated by long distances.

For situations in which the molecular electronic wavefunctions of D,A overlap, exchange coupling can exceed the Förster dipole interaction. For exchange coupling (often referred to as the Dexter coupling) [196], the energy transfer rate constant becomes

$$k_{\mathrm{Dexter}} = \frac{2\pi}{h} K J e^{-2R/L}.$$

(1.25)

Here J is the unweighted spectral overlap

$$J = \int f_{\mathrm{D}}(\nu) \varepsilon_{\mathrm{A}}(\nu)\, d\nu,$$

(1.26)

while K is an energy parameter and L is an effective average orbital radius for D and A. While the exchange mechanism falls off more quickly (exponentially, compared to R^6) than the Förster one, it still can dominate for weakly dipolar systems at shorter separations.

The Förster and Dexter results arise from the use of golden rule arguments, and assume that the D–A coupling is relatively weak. In particular, they assume that vibrational relaxation is fast compared with transfer. Clearly, if a synthetic molecular electronic structure were prepared with strong enough electronic coupling between D and A, these assumptions might fail, and more efficient transfer (a better wire) might be seen. In any case, by changing the coupling chemically, such a wire might be turned on or off. Recent theoretical studies [197] have suggested that excitations might move by effectively separate hole and electron transfer, i.e. via charge transfer states. This novel suggestion merits further study, as it might be more effective than Förster or Dexter transport in certain situations.

A particularly lovely example of photonic wires is given by Lindsey and his collaborators, who have developed molecular photonic wires based on fused porphyrinic structures [44]. Figure 1.29 shows one structure, with a dipyrromethene dye as the optical input (absorber), and a free base porphyrin as the output (fluorescer). The

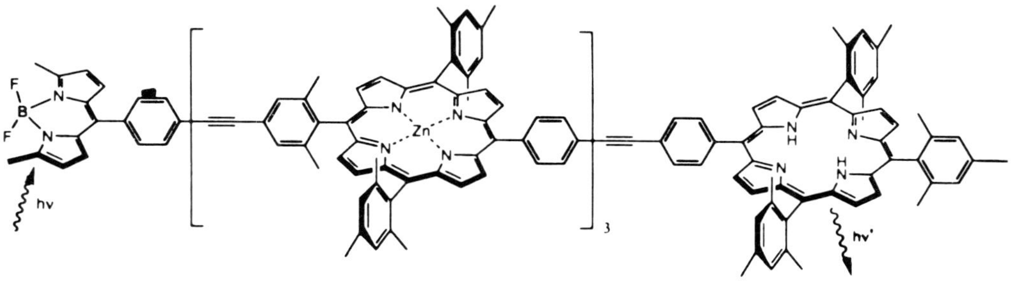

Figure 1.29 The photonic molecular wire, based on peripherally bridged zinc porphyrins, with biphenyl alkyne bridges. The input light is absorbed by the boron methene dye, while the free base porphyrin provides the output of radiation. From [44].

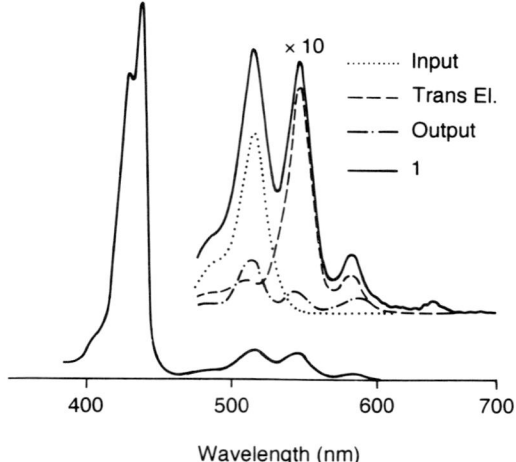

Figure 1.30 The absorption spectrum of the photon wire structure of Fig. 1.29. The overall structure is the solid line, while the dotted or dashed lines show the absorption spectra of the various subcomponents. From [44].

structure as drawn has an extended length of roughly 90 Å; it is soluble and stable, so that it can be prepared and manipulated in a modular fashion.

Figure 1.30 shows the absorption spectrum of the full donor/wire/acceptor structure of Fig. 1.29; note that the absorption in the range between 450 and 600 nm is essentially a superposition of the three component parts. At 485 nm (where the experimental studies were made), roughly 62% of the light is absorbed by the input dye.

Under illumination at 485 nm, 92% of the fluorescence arises from the free base porphyrin (acceptor). The fluorescence quenching of the input dye corresponds to 96%

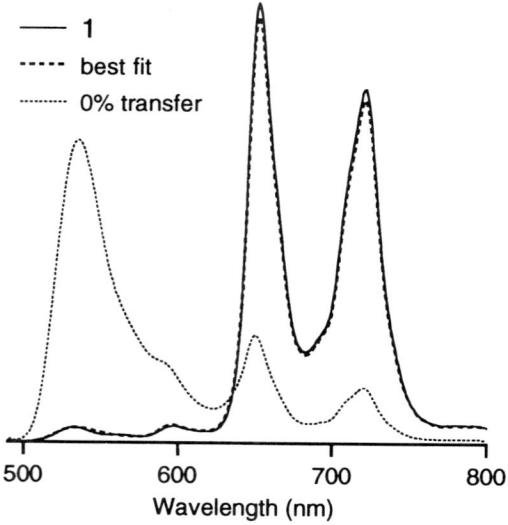

Figure 1.31 The fluorescence emission spectrum of the structure in Fig. 1.29 at room temperature. A simulated spectrum is shown assuming both zero per cent transfer of energy, and a best fit to step-wise energy transfer. The best fit description is given in the text; the assignments of the emission maxima are the dipyrromethene dye (534 nm), zinc porphyrins (597, 646 nm) and the free-base porphyrin (650, 720 nm). From [44].

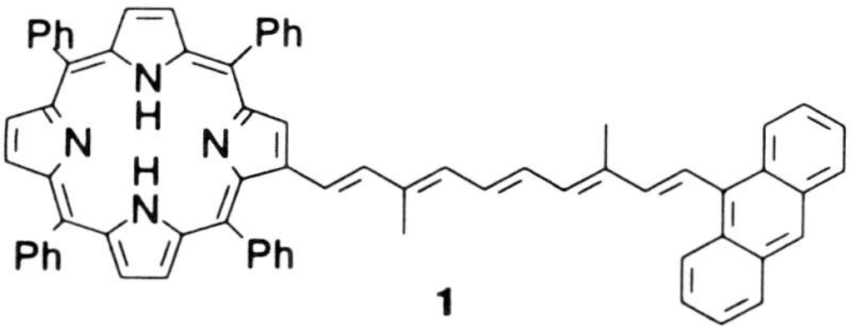

Figure 1.32 An extended spacer structure between an anthracene and a porphyrin, also used as an energy transfer molecular wire. From Wolf *et al.* [325].

energy transfer to the neighbouring site porphyrin. To fit the fluorescence submission spectrum after excitation at 485 nm, Wagner and Lindsey suggested step-wise energy transfer among the five pi-type components of the assembled donor/wire/acceptor of Fig. 1.29. Figure 1.31 shows the observed emission spectrum and the best fit, assuming energy transfer efficiencies, in a step-wise fashion from input to output, of 95%, 93%, 93%, and 93%. This gives an overall signal transmission efficiency of roughly 76%; note that the fit is excellent. If no energy transfer occurs (dotted line in Fig. 1.31), one simply observes fluorescence from the dye.

The Förster energy transfer efficiencies for this same structure are, respectively, 93%, 28%, 28%, and 77%; this gives an overall transmission efficiency of less than 6%. The observable efficiency is 13 times higher, indicating that energy transfer is somehow mediated by the diarylethyne linkage. If one chooses to ignore the linkage altogether, and simply to calculate the direct Förster transfer from the input to the output over 90 Å, the efficiency is 0.1%. As Lindsey and Wagner suggest [44], this certainly implies that a very efficient mechanism of energy transfer, probably involving direct π interactions down the conjugated chain, is active in these systems.

It is striking that, in comparison with the massive amount of theoretical and experimental work on bridge-assisted electron transfer, the sort of bridge-assisted, highly efficient photon wire behaviour seen in these structures has not been investigated extensively, and direct theoretical models have not been adduced. This will certainly be one of the future directions for molecular electronics. As Lindsey and Wagner point out [44], questions such as 'how far can excited state energy transfer be transmitted?' and 'can this form of photonic single transmission be gated?' are critical and challenging. The important point, currently, is that synthetic structures such as that in Fig. 1.29 (and similar structures investigated by other groups [43], such as Fig. 1.32) are available, and measurements of their transfer efficiencies present interesting challenges to our understanding.

4 Charge transfer circuits

4.1 *An example analysis: molecular wires*

Molecular wires — that is, molecular structures that act to interconnect other components, either by electronic charge motion or by energy transfer — are obviously crucial components of any molecular electronic circuitry. They are also closely related to the extremely important phenomenon of electron transfer, discussed extensively in Section 2.1 above, and are under intense experimental investigation in a number of candidate structures and candidate circuits. As such, they are a perfect prototype for discussion of molecule-based electronics.

The fundamental viewpoint for understanding charge transfer in molecular wires begins with considerations of molecular electron transfer, as already discussed. One simplification occurs because these are intramolecular electron transfer situations so that diffusion and assembly of components is no longer a dynamical problem. From the point of view of simple electron transfer theory, this would suggest that (if the wire coupling is relatively inefficient) non-adiabatic electron transfer models (Section 3.1) should be adaptable to describe the situation.

An additional complexity enters here, however, that requires a different analysis in each case. If the reservoirs to which the molecular wire is affixed are vibronic continua, as is the usual case in intramolecular electron transfer kinetics, then the whole process will be sensitive to the effective state densities in the vibronic continuum, and therefore can be controlled by Franck–Condon factors, and by the constraints they impose on energy deposition in the inverted regime (Section 3.1). If, on the other hand, the transfer is between electronic continua (such as will occur in transfer between metallic electrodes, or in STM measurements), Franck–Condon constraints are replaced by considerations of state density in the electrodes themselves [130–137,188,189,198–200]. For simple metallic electrodes, these state densities are slowly varying with energy; one then does not expect an inverted regime, and the actual dynamics/energetics of the electrode become less important.

If the passage of energy or charge through a molecular wire is considered from the point of view of chemical kinetics, it is important to know what the initial and final states are. For intramolecular electron transfer, these kinetic descriptions are clear only if there is a substantial localization on the donor site before reaction, and on the acceptor site afterwards. When bridges become quasi-resonant, the initial and final states are no longer clearly defined and, therefore, neither is the chemical rate constant. This situation would occur for structures in which the generalized bridge, or wire, has eigenstates that are nearly resonant with either donor or acceptor states; under these conditions the rate constant for donor to acceptor transfer is poorly defined experimentally and theoretically. This is not true in the conductance problem. In Section 4.3 we discuss these crucial differences, and the resulting experimental and theoretical challenges in analysing molecular wires.

4.2   *Classification of circuits*

A molecular wire is essentially a structure that links two active end structures. In this section, we will be considering electrical charge transport, and therefore the molecular wires should transport charge (photon molecular wires were considered in Sections 3.3 and 3.5 above). For simplification purposes, we represent the wire-coupled device structure as the triad DBA, and Table 1.1 lists limiting situations. The terminal structures D and A, which are normally thought of as donor and acceptor, could be either molecules (characterized by discrete molecular electronic structures) or electrodes, macroscopic or mesoscopic (and characterized, respectively, by continuous and semicontinuous state densities).

The distinction between structure B acting as a spacer or bridge on the one hand, or

Table 1.1 D–B–A charge transfer circuits.

Case	D	B	A	Examples (Refs)
1	Molecule	Space/bridge	Molecule	[324]
2	Molecule	Wire	Molecule	[210]
3	Molecule	Wire	Electrode	[11]
4	Molecule	Space/bridge	Electrode	[11]
5	Electrode	Space/bridge	Electrode	Chapter 6
6	Electrode	Wire	Electrode	[37,326]

as a molecular wire on the other [27–39], depends upon the decay of the charge transfer efficiency between D and A as a function of their separation, and on the nature of the electron transport through the B structure. If the electron is localized on the B structure, it is useful to distinguish this as a wire situation. If the electron is never actually localized on the bridge structure (although the bridge may, and generally does, facilitate electron transfer via a superexchange process) [201–208], then this structure acts as a spacer. In bridge-assisted electron transfer reactions, the so-called 'chemical mechanism' [209] is precisely the one that we are calling wire behaviour — that is, when the electron is localized on the bridging structure B. Other wire-like behaviours can occur due to resonances between bridge and donor states (see Section 4.3 below).

Cases 1 and 2 in Table 1.1 characterize intramolecular electron transfer reactions. These have been extensively discussed in the literature, including in Chapter 2 of this volume. Characteristic structures such as those of Fig. 1.27 generally lie in this category. When the energy levels in the B structure come close (within thermal energies) to the occupied levels on the A or D structures, the electron can be localized within the wire. Examples include the caroviologen molecular wires of Lehn and collaborators [210], as shown in Fig. 1.27a, conductive polymers such as polyacetylene in which the charge is carried by dressed electrons, the long polyene structures of Wasielewski [211] (Fig. 1.33) or (apparently at least at ambient temperature) the photosynthetic reaction centre. In this case, after about three picoseconds, following photoexcitation of the special pair in the native reaction centre (which we can generalize as the D element in

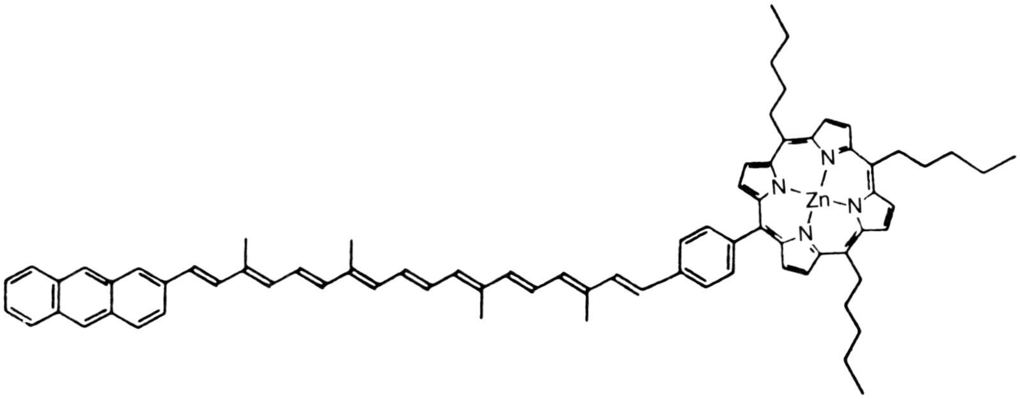

Figure 1.33 An extended D–B–A molecule, for which electron localization on the alkene bridge may occur at high enough T. From [211].

Table 1.1), the electron is localized on the bridging bacteriochlorophyll (the B structure). Subsequently, the electron is localized at 17 Å (centre-to-centre distance) away from the centre on the bacteriopheophytin centre. The primary charge separation in the native bacterial photosynthetic reaction centre involves the wire behaviour, while in modified reaction centres (made by site mutagenesis of amino acid residues or chemical substitution of the prosthetic groups) the spacer behaviour can prevail [157; M.E. Michel-Beyerle, personal communication].

Case 3 of Table 1.1 includes the electrode biosensors discussed in Chapter 7. Here one deals with charge transfer between a molecular structure (the redox enzyme itself) and the continuous structure (electrode) modulated by a generalized molecular wire [11]. In many cases, the wire consists of localized electron hopping sites, such as ferrocenes or other π-type structures. The wire itself is a structural entity, very often a soft structural entity such as a siloxane. The electronic motion between the redox enzyme and the electrode is thought to occur by electron hopping, incoherently, among the sites on the molecular wire structure.

In STM experiments, involving molecular adlayers, both cases 4 and 5 of Table 1.1 can occur [59,80,84,212–216]. If characteristic gap energies E_g in Fig. 1.34 are large compared with thermal energies, neither an electron nor hole is ever actually localized in the polymer bridge, but rather transfer occurs between the continuum levels of the electrodes modulated in a superexchange-type fashion by the molecular energy levels; this is considered more extensively below, and in several recent theoretical discussions [126–129,131–139,217–223]. As in intramolecular electron transfer (Section 3.1), one expects the distance dependence of these coherent electron transfer processes to be roughly exponential, essentially because overlaps are exponential and the virtual mixing process is thought of as sequential, from the D structure to the various components of the B structure to the A structure. Proofs of the exponential behaviour can be given in various limits: in the simple case of barrier tunnelling, the rate process is exponential in distance, following Eqn 1.2 or Eqn 1.27, with the parameter β scaling as the square root of the barrier height for tunnelling [224]. In the case of a linear chain, it is easy to prove (for instance by iteration of the Lippman–Schwinger equation [202]) that the decay constant is given by

$$k(r) = k_0 \exp(-2\beta[R - R_0]) \tag{1.27}$$

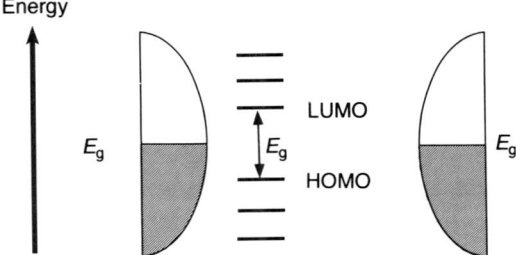

Figure 1.34 Schematic energy level diagram, showing the molecular orbitals of an adsorbed molecule, with the gap including the Fermi level of the electrodes (continuum structures to the left and right). With this energetic configuration, there is no significant electron transfer between molecule and electrode.

$$\beta = \frac{-1}{a} \ln(2t/E_\mathrm{g}).\tag{1.28}$$

Here R_0, k_0, a, t, and E_g are, respectively, a reference separation and the corresponding rate, the length between electronic basis function in the bridge structure, the mixing matrix elements between those near-neighbour bridge structures, and the energy gap of Fig. 1.34. Finally, one can use statistical arguments to show that in large systems [225], such as proteins, exponential behaviour should also occur, with the β coefficient being given by statistical distributions of pathways (as described in Chapter 12) [226].

Case 5 of Table 1.1 has been seen in STM experiments, break-junction studies, inelastic electron tunnelling spectroscopy experiments [227,228], and charge transfer through thin assembled Langmuir–Blodgett films [229; see also 108,109]; these are described in Chapters 5 and 8 of this volume. Molecular wire behaviour can be seen when scattering events due to electron/vibrational coupling, and relatively small values of the gaps in Fig. 1.34, result in localization of the electrons on the bridging structures [230–233]. The interdigitated array electrode structures studied by Murray [77] and described in Chapter 6 of this volume are examples of case 6.

4.3 *Rate constants and conductances*

Rate constants are generally defined using a first-order kinetic scheme, such as that in Eqn 1.29

$$\mathrm{d}\{D\}/\mathrm{d}t = -k\{D\}.\tag{1.29}$$

Here D is the donor state, and k is the rate constant. When this scheme does not hold at all times, and simple exponential rate behaviour is not observed, one can define an instantaneous rate constant in time; this is often called a rate coefficient [221,234–238]. Simple exponential decay may not be observed under several conditions: for example, when photoexcited electron transfer is measured from an initial state that has not yet reached thermal equilibrium, it is not surprising that non-exponential decay is observed; elegant examples have been given in the recent literature by Barbara, Hupp, Scherer, and co-workers in binuclear metal complexes [145,160,161,239], and this situation has been well characterized theoretically, using a generalized golden rule expression, by Coalson, Evans, and Nitzan [236–238]. Again, in the photosynthetic reaction centre, non-exponential decay has been clearly observed [155–158]. In this case, non-exponentiality could arise either because of non-equilibration in the initial site (the photoexcited special pair) or because of inhomogeneous structures caused by differing energy levels or geometries in the membrane active site [234]. Electron transfer reactions in inhomogeneous media such as glasses will, for the same reason, be expected to be non-exponential. Inhomogeneity in, e.g. the electron transfer integrals or matrix elements that enter into non-adiabatic electron transfer expressions can, once again, give non-exponential decay [234].

The conductance, the rough analogue of the rate constant in the situation where charge is transferred between two electrodes with continuous distributions of energy levels, always begins not from a molecular donor level, as in Eqn 1.1, but rather from an electrode structure to which a voltage is applied as a function of time. The assumed linear relationship between voltage and current, as in Eqn 1.30,

$$I = gV, \tag{1.30}$$

defines a conductance; here I, g, and V are, respectively, current, conductance, and applied voltage. There is an important conceptual difference between the conductance and the rate constant: the initial state, and the final state, for conductance measurements are always well defined, and correspond to electrodes. However, in the DBA electron transfer systems, one can approach a degenerate case for which the population on the initial component A is poorly defined. For example, suppose that A is a molecule with discrete energy levels, as is B. Assume now that one of the empty energy levels of B becomes essentially degenerate with the highest occupied level of D. Under these conditions these states will mix efficiently, and one can no longer define an initial population in the D state; measurement of a rate constant for transfer from D then becomes poorly defined. Such behaviour does not occur in the conductance: for example, the so-called Sharvin limit [240,241] in STM corresponds to the case where the continuous electrode and the tip are brought into near contact with one another; say with only one atom between. Even if this atom contains levels that are resonant with the Fermi levels of the two continua, nevertheless the conductance remains well defined, even in the presence of resonances. This distinguishes conduction from rate constant, in the case of resonance transfer.

Remarkable examples of quantized conductance have been seen in atomic wires, using either STM or break junctions [199]. The atomic wires can be as simple as one or two Xe atoms [242,243], or as elaborate as a multichannel, multiatom metal nanowire [79,244,245]. Clear examples are shown in Chapter 5: the observed conductance in the latter case is often, but not always, an integral multiple of the conductance quantum $(2e^2/h = (12.9 \text{ k}\Omega)^{-1})$. For longer wires (or for the Xe case), the conductance is usually explained using the Landauer expression

$$g = \frac{2e^2}{h} \sum_i T_i,$$

with T_i the squared scattering matrix element in channel i. If $T_i = 1$ for all i, then the conductance g is indeed an integral multiple of $2e^2/h$. This is observed for short (less than several nm) metal wires. Longer metal wires lead to heat dissipation and instability. For wires of Xe or S, the scattering matrix is substantially less than unity, and both calculated and observed conductances are small — roughly 10^7 and 10^{10} Ω for Xe and Xe_2 wires, respectively [242,243]. In those cases, the actual calculations were given by Lang [242], based on a jellium model of the electrodes and a self-consistent density functional study of the wire.

4.4   Conductance in model systems

A special and intriguing case, that generalizes the concepts of intramolecular electron transfer (cases 1 and 2 of Table 1.1) to the situation of STM [212–214] (cases 5 and 6 of Table 1.1), is sketched in Fig. 1.35. Here an STM tip and a macroscopic electrode form the terminals of a circuit, with a molecular species in between [84,215,216]. This

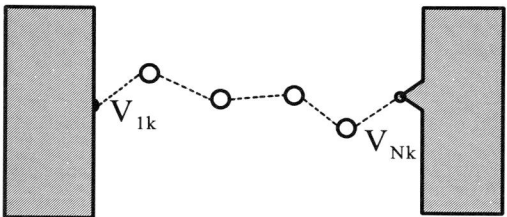

Figure 1.35 The schematic structure of a metallic support–molecular wire–STM tip electrode structure, investigated using scattering theory models to understand the molecular wire behaviour itself.

molecular species is to act as wire, connecting the two electrodes, and we are interested in a transport of charge through this wire as a function of the chemical, and physical, parameters of the system.

STM spectroscopy or break-junction measurements [79] offer the possibility to observe the tunnelling current as a function of the parameters of the system, in particular the electronic structure parameters of the molecular wire itself, and the applied voltage through the electrodes.

In the simplest case, when the mixing is sufficiently strong that no transport limitation is imposed by the molecule, the Tersoff–Hamann limit [246], generally used for interpreting STM imaging, is obtained:

$$g = W\rho(E_F, \vec{r}_{tip}). \tag{1.31}$$

Here the right hand side is simply proportional to the density of bare surface states calculated at the Fermi surface at the position of the scanning tip [247]. For STM imaging of surfaces, and in some cases for STM imaging of absorbed species, this Tersoff–Hamann limit is appropriate [248]. For the general case of molecular situations on surfaces, however, the molecule itself can attenuate the current, coulombic effects and barrier tunnelling can occur, and a scattering approach is more useful.

A second set of limiting cases, those described as cases 1–4 of Table 1.1, have already been discussed.

The presence of the macroscopic electrodes in Fig. 1.35 changes the nature of the electronic states within the molecular bridge itself. The simplest way to understand this is in terms of a model developed by Newns [191] and Anderson [192], and illustrated schematically in Fig. 1.36. The metal is considered as a one-dimensional tight binding

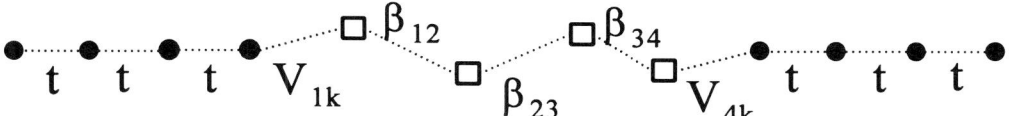

Figure 1.36 The Newns–Anderson model for chemisorption, with an active molecular wire between. The figure shows a one-dimensional tight binding band in the metal (dots) and the site orbitals themselves (boxes). When cyclic boundary conditions are assumed for the metal, the terminal sites of the molecular wire are coupled with the same strength to all of the levels within the wire. The parameters t, β_{ij}, V_{1k} and V_{Nk} are, respectively, the one-fourth of the metal band width, the tunnelling integrals within the molecule, and the tunnelling integrals between the terminal atom or the molecule and that of the metal.

band (essentially a Huckel metal), that couples with a set of site orbitals that represent the molecule. Newns and Anderson [191,192] showed that an equivalent representation is obtained by considering the eigenstates (band states) of the metal electrodes; they then couple randomly to the terminal sites of the molecular bridge, as indicated in Fig. 1.36. The molecule's interaction with continua introduces self-energy and damping effects, that will substantially change the nature of the transport spectra.

A direct approach to the problem of electron tunnelling in both STM imaging and molecular conductance problems has been developed in an important series of papers by Joachim and co-workers [81,129,139,198,199; C. Joachim, personal communication]. They compute the metallic electrode wave-functions directly, using full periodic boundary conditions in the two directions perpendicular to the STM junction; the molecular species within the STM is represented explicitly, using an extended Hückel model Hamiltonian (electron repulsions are ignored). Important results include both direct simulation pictures of STM images and a direct comparison of calculated and measured conductance through a single C_{60} molecule; the number obtained, 54 MΩ, is less than 0.1% of the quantum of conductance $2e/h$, implying extensive resistive scattering in the molecular junction. Datta's group [188,189] has developed and applied similarly appropriate and useful tight-binding-based methods, and have, for benzenedithiol spacer molecules, obtained GΩ-sized molecular resistances quite comparable to those deduced experimentally [38,82].

In a simple scattering theory approach [131–137,249], the current through a molecular bridge is given by

$$j = e\nu = \frac{2\pi e}{h} \sum_{if}^{\text{reservoirs}} f(E_i)[1 - f(E_f + eW)] |t_{if}|^2 \delta(E_i - E_f). \tag{1.32}$$

Here j, e, ν, and W are, respectively, the current, the electronic charge, the rate of electron transfer, and the applied voltage. The Fermi functions $f(E)$ denote the population of the states; the two Fermi factors then say that the electron must come from an occupied site, and flow to an empty one. The last term conserves energy between initial and final states. The t-matrix element t_{if} measures the scattering cross-section from the initial state i in the reservoir on the left to the final state f in the reservoir on the right. Note that we have not linearized in voltage, and therefore highly non-linear current/voltage relationships can occur.

The scattering approach has several significant advantages; essentially one can prove that 'the conductance is a sum of the transmission probabilities of electrons over all energies, weighted by the required Fermion occupation numbers [250–252]. Thus the current in the situation of interest, for an average of many channels or for a single junction, arises from the same scattering formalism.

The molecule structure enters only in the scattering matrix. This is formally given by [131–137,249]

$$t = V + VG^0 t = V + VGV + \dots \tag{1.33}$$

with the Green's function itself given by

$$G^0(Z) = (Z - H_0)^{-1} \tag{1.34}$$

$$G(Z) = (Z - H)^{-1}. \tag{1.35}$$

V is the direct coupling between the initial and final states, which we can ignore for the molecular bridge situation. We can then write the t-matrix between state A in the metal on the left and state B in the metal on the right as

$$|t_{AB}|^2 = |V_{A1}|^2 |V_{NB}|^2 |G_{N1}|^2. \tag{1.36}$$

Here we have assumed that the interaction elements between site 1 on the molecular chain and the many states on the left electrode are the same for all those electrode states (this is part of the Newns and Anderson model), and similarly the end site (labelled N) in the bridge couples equally to all bands on the right hand side. The Green's function element, the last term in Eqn 1.36, describes the mixing to the molecular interactions between the first and the end sites in the molecule.

The coupling of the molecular levels to the continuum levels results in a self-energy effect, which enters only on the end orbitals of the molecular structure as indicated by Eqns 1.37–1.39:

$$H_{11} \rightarrow H_{11} - \Sigma_1(Z) \tag{1.37}$$

$$H_{NN} \rightarrow H_{NN} - \Sigma_N(Z) \tag{1.38}$$

$$H_{xy} = H_{xy}^{\text{molecular}}; \text{ except } H_{11}, H_{NN} \tag{1.39}$$

Self-energy contains both real and imaginary parts, that are proportional to the couplings V_{1p} and V_{Np}, respectively, where p labels a site in the continuum level. That is, formally the self-energy is given by

$$\Sigma_k = \lim_{s \to 0} \sum_p V_{kp}^2 / (E - E_p^0 + is). \tag{1.40}$$

This rather heavy formalism is necessary to take into account the continuum aspects of the electrodes; similar considerations do not actually enter into electron transfer, for exactly the same kinds of reasons that the rate constant for electron transfer differs from the conductance in an STM experiment.

The scattering formalism just discussed has been used by several workers [131–137,188,189,249]; C. Joachim, personal communication]. Using the simplest possible model, in which the molecule itself is represented by a tight binding model (essentially a Hückel Hamiltonian), several fascinating results can be obtained both analytically and numerically.

1 The distance dependence of the conductance differs depending on whether or not molecular energy levels of the isolated bridge are near the Fermi surface of the electrodes. That is, in the small voltage limit, the conductance is a function of N (Fig. 1.37) and can be exponentially decaying (corresponding to superexchange behaviour, or large gaps in Fig. 1.34), decreasing more slowly, as a power law, when the molecular levels are near the Fermi surface, and exhibiting essentially no decay (resonance transfer, with oscillations) when degeneracies occur between molecular energy levels and the Fermi level.

2 If the conductance is examined as a function of the Fermi level position, one observes resonance-type behaviour, but the current never diverges (Fig. 1.38). This generalizes the McConnell relationship of superexchange [201,253], which is relevant

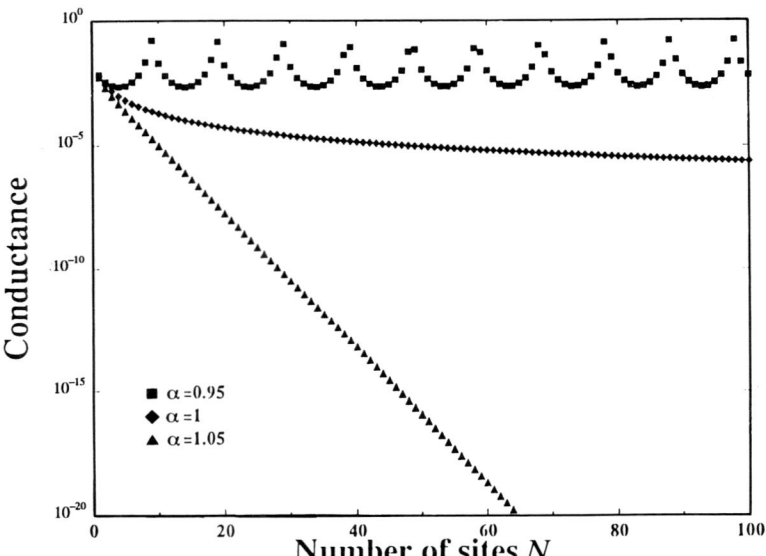

Figure 1.37 The conductance as a function of the number of sites, for the models of Figs. 1.35 and 1.36. The α parameter measures the energy misfit, and is given by $\alpha = E_b/2V$. E_b is the site energy within the molecular wire while V is the site–site coupling (Huckel β) within the wire. The Fermi energy $E_f = 0$. When the Fermi surface of the electrodes is far from an occupied molecular orbital, exponential decay ($\alpha = 1.05$) results. As the Fermi surface touches a molecular orbital, power law decay results ($\alpha = 1$). Finally, when the Fermi levels overlap within the band of molecular orbitals, long-range transfer without significant decay, but marked by oscillations is shown ($\alpha = 0.95$). From [133].

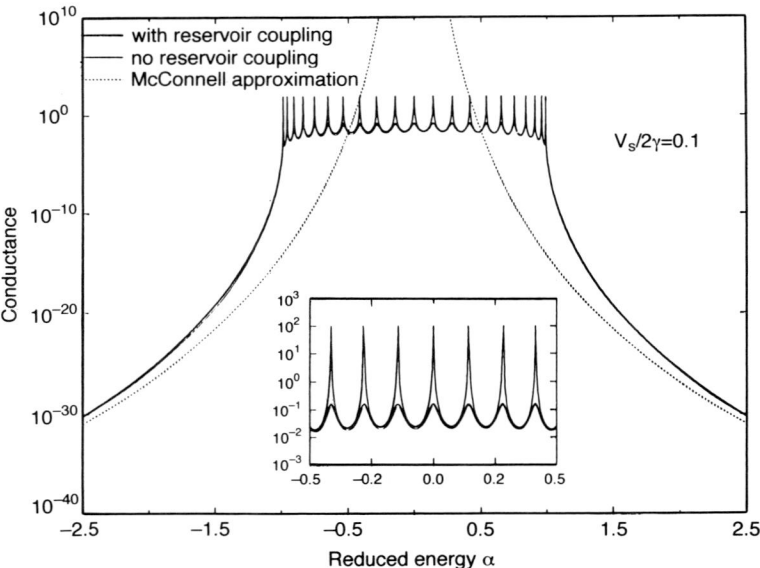

Figure 1.38 Conductance as a function of reduced energy α, for the molecular wire structure of Fig. 1.37. This particular structure has 21 sites on the bridge. Note the resonances that occur because of molecular orbitals crossing the Fermi level; the McConnell approximation is the simple formula arising from site-to-site superexchange without accounting for the continuum due to the electrodes, and yields a divergence when the donor energy is degenerate with the site energies within the wire. From [133].

for electron transfer (that is, in the absence of continua), and results in divergences when resonances occur between the bridge and the donor/acceptor.

3 If the molecular chain is disordered — that is, if the energies of the different sites in the molecular chain are allowed to vary — then only a statistical definition of the conductance follows [131–134,254,270]. Figure 1.39 demonstrates that the conductance is sensitive to diagonal disorder: for resonant type transfers, increased disorder removes the resonances, and reduces the overall conductance. (In semiconductor nanostructures, similar so-called 'weak localization' processes also reduce conductance [164].) Conversely, for situations in which the conductance is relatively small (off resonance), the increase of disorder can actually increase the conductivity, due to the adventitious crossings of disordered eigenstates with the Fermi surface.

4 The coupling of the substrate and adsorbate electronic states can radically alter the space-dependent scattering, modifying substantially the STM image. For example [129,139], one can compare the images of benzene adsorbed at different sites on the Pt(111) face, obtained experimentally and computed using a scattering formalism. Effects of the electronic structure modifications are notable.

5 One striking aspect of nanoscale electronics is the Coulomb blockade and the Coulomb staircase (see Chapter 4). These arise from the fact that there is a capacitance effect associated with passing electrical charge over a small interface between two conductive particles [255–259]. It is perhaps more intuitive to think of this in terms of Fig. 1.40, which shows an isolated quantum dot (small particle) suspended between two electrodes, appearing as the plates of a capacitor [260–264]. To move electrical charge

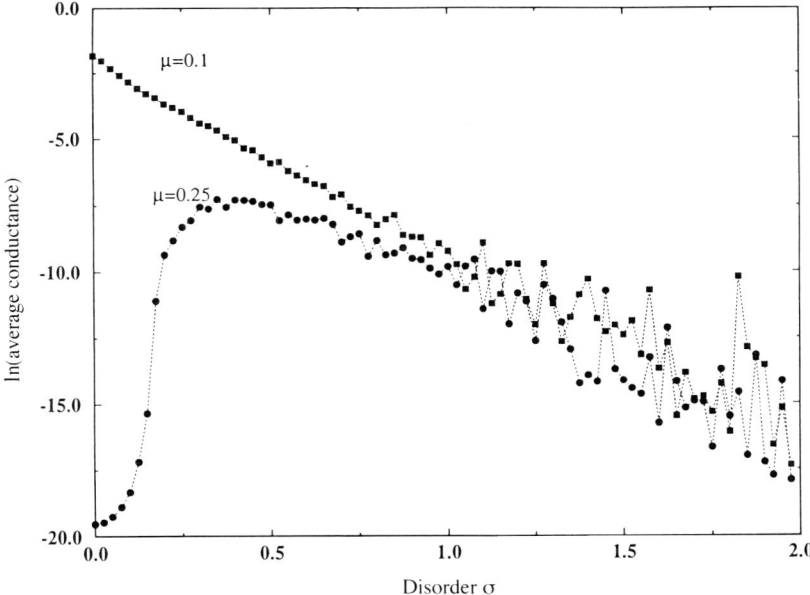

Figure 1.39 Conductance for disordered molecular wire [as that in Fig. 1.37, but with different, random site energies (Huckel α)]. As the disorder (spacing among the atomic energy levels) is increased, the logarithm of the average conductance decreases when the molecular wire is resonant with the donor electrode, but actually increases for cases (like the lowest line of Fig. 1.37) in which initially there was no resonance. The scattering then causes adventitious resonances, that (for small disorders) actually increase the conductance. From [270].

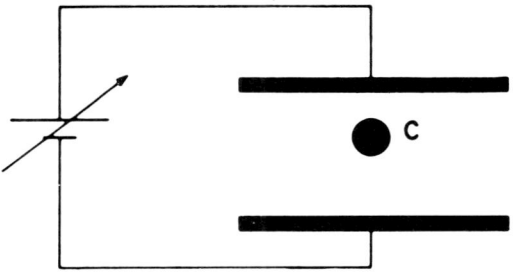

Figure 1.40 A model for the mesoscopic Coulomb blockade and Coulomb staircase phenomena, seen in a single quantum dot. The quantum dot particle is suspended between electrodes, and characterized by a capacitance C. From [260].

onto the capacitor requires overcoming Coulombic repulsion (Coulomb self-energy) on the capacitive particle. This means that no current will flow until the applied potential is large enough to overcome this self-repulsion — this is the Coulomb blockade. Similarly, there will be single electron steps as a function of applied voltage, as the number of electrons on the quantum dot increases or decreases by one. This is shown schematically in the current/voltage plot of Fig. 1.41 and the conductance/voltage plot of Fig. 1.42.

These simple Coulomb staircases occur when the only effect that the electrons feel upon entering the dot particle is the Coulombic repulsion. If, however, the dot particle is small enough, one also expects it to have characteristic eigenvalues. One then has both an eigenvalue staircase and a Coulomb blockade, and expects to see jumps in the current as a function of voltage whenever one of the many body affinity states of the dot (that is, the eigenstates that include both the quantization due to the size of the particle and the Coulomb repulsion) crosses the Fermi surface [131,132]. Such irregular spectra are well known in an inelastic electron tunnelling spectroscopy [227,228], but should also characterize STM-type measurements involving molecular wires, where the quantized nature of the states in the wire becomes relevant.

Work by Roth and co-workers [109], shown in Fig. 1.17, demonstrates both apparent molecular rectification and a staircase-like conductance structure in Langmuir–Blodgett films made of a layer of phthalocyanine and a layer of quinone. Similar behaviour has been seen in conductance via gold nanodots [265,266]. Both a rectification and a

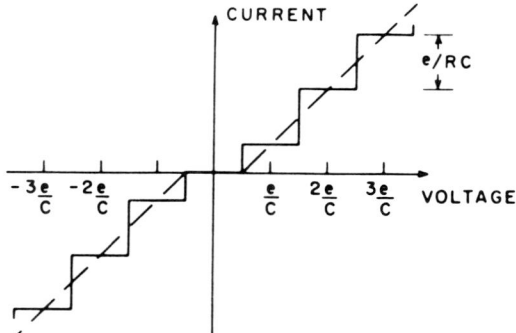

Figure 1.41 The expected current is a function of voltage for the single dot structure of Fig. 1.40. Notice that there is no current until the voltage reaches a value of $e/2C$: this is the Coulomb blockade. Following that, there is a series of steps in the current, spaced by e/C and with height e/RC: this is the Coulomb staircase. From [260].

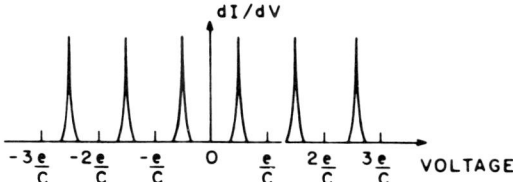

Figure 1.42 The derivative of the current trace of Fig. 1.41, showing a series of evenly spaced conductance peaks. From [260].

staircase-like structure are predicted from the scattering formalism we have discussed, but (in apparent agreement with experiment) the staircase is very different from the Coulomb blockade in mesoscopic structures, as exhibited in Fig. 1.42. In particular, both the amplitudes and the spacings are no longer regular; this may well be a signature of the molecular eigenvalue staircase [131,132].

Given the elegant advances in the synthesis of controlled nanostructure arrays using chemical means, these blockade and staircase effects might well act to control the nature of electron flow, and therefore as sensitive electrometers or detectors. While this has been speculated about considerably in the mesoscopic literature [255–264], chemically prepared structures, such as the gold nanospheres functionalized with thiols on the surface discussed in Chapter 11 of this volume, might well provide an even more controlled and even more accurate set of molecular electronic devices.

4.5 *Incoherence effects and vibrational coupling*

The scattering theory just described ignores all interactions between the electrons and the vibrations, as well as all dephasing effects. In actual systems, one might anticipate, rather, that the electron will be thermalized and proceed through the wire-like structure as a defect (the so-called soliton defect if the wire has a degenerate ground state such as polyacetylene [48,49]). There have been some suggestions in the literature that this is in fact true [230–233], but (to our knowledge) actual characterization of either intramolecular electron transfer or STM occurring by electron defect motion has not been convincingly demonstrated. Nevertheless, the scattering formalism that we have just discussed is clearly inadequate for dealing with the vibrations. A variation of similar scattering arguments to include vibrations has been recently given [221], and one can anticipate both a great deal more experiment in this regard and substantially more sophisticated theory.

The Coulomb blockade phenomenon in standard mesoscopic physics is washed out by fluctuations at very low temperatures (generally a few kelvin at most) [255–264]. With molecular electronic systems, in which the characteristic sizes may be substantially smaller, the Coulomb staircase intervals should be appropriately larger (it scales roughly as inverse radius) and the molecular eigenvalue ladder should also result in substantial structure in the conductance [131,132,255,265,266]. So long as measurements are made with macroscopic electrode cross-sections against the molecular wire structure, one would expect averaging over the different manifestations of the molecular wire, and over the fluctuations within the molecular wire. Averaging over these fluctuations will result in a current of ohmic-type, from which a diffusion rate or an electron transfer rate can be defined. Murray [267; R.W. Murray, personal communication], Natan [268], and others [189] have made measurements on systems involving distributions of functionalized gold nanospheres; they observe generally ohmic behaviour, which can then be correlated with average electron hopping times and dynamics.

When the electrodes present a smaller, nanoscopic cross-section to the molecular situation (while still remaining macroscopic themselves, so that fluctuations [269] within the electrode are unimportant — see Landauer for important caveats along this score) [200,250–252], one would expect to see the fluctuations within the quantum system. This is the situation with ordinary, and especially with high-impedance, STM experiments, in which observed currents to date come tantalizingly close to demonstrating molecular wire behaviour (nanoamp currents through a single molecular strand) [33,38,81,83,198,199,265,266]. Some examples of this were given for the model molecular wires in Fig. 1.39: notice that only the *average* of the conductance then becomes defined, and that, over a finite samples, these averages retain some noise. Investigating the log normal distributions expected within this disorder [270] is one fascinating possibility for understanding the limitations of a quantum molecular wire.

5 Size effects

5.1 *Clusters*

The conceptual framework and the emerging practice of molecular electronics encompass both molecular materials and molecular-scale devices. In this broad context, the borderline between molecular and condensed matter chemical and physical properties is of intrinsic significance. These pertain to the bridging between a single molecule and bulk matter, with respect to the electronic-level structure, the response to (linear and non-linear) optical excitations and transport of elementary excitations and dynamics. The bridge between molecular, surface and condensed matter chemical physics is provided by clusters, i.e. finite aggregates containing $2–10^9$ particles [271–273]. The genesis of the development of cluster science can be traced to the work of Mie in 1908 [274] on the optical properties of small particles, while the work of Kubo in 1962 [275] pioneered the exploration of the electronic structure of small metallic particles, addressing the important issue of the non-metal 'transition' in finite systems, which can be realized when the thermal energy exceeds the energy gap. Subsequently alternative mechanisms for metal–non-metal transitions in clusters, e.g. the closure of an electronic band gap [276,277], were advanced. During the 1970s a wealth of information emerged on electronic transport and optical properties of metal clusters embedded in insulators, e.g. granular metals [278] and metal–rare gas mixtures [279], where intercluster interactions can be described in terms of percolation theory and tunnelling effects [280]. Concurrently, studies of electronic, optical and catalytic properties of clusters supported on surfaces emerged [281]. An additional dimension to the exploration of clusters originated in 1977 with Friedel's [271] emphasis of the properties of 'clean and isolated clusters', which rested on the generation of cluster beams [282,283], on structural concepts for close packing of hard spheres [284,285] in conjunction with electron diffraction studies [286], on spectroscopic and dynamics studies of van der Waals molecules [287,288], and on studies of excess electron localization in supercritical polar fluids [289]. An arsenal of experimental, computational, and theoretical methods [290–293] is currently applied to the study of clusters. While the exploration of isolated clusters in supersonic jets substantially contributes to the conceptual framework, studies of systems based on clusters, e.g. embedded clusters in the bulk [294], supported clusters on surfaces [295], packed quantum dot solids [296] and cluster ensembled structures [297], will provide potential materials understanding and devices for molecular electronics.

5.2 *Cluster size effects*

A key concept for the quantification of the unique characteristics of atomic and molecular clusters pertains to size effects [298–300]. These involve the evolution of structural, thermodynamic, electronic, energetic, electromagnetic, dynamic, and chemical features of finite systems with increasing cluster size (Fig. 1.43). Cluster size effects fall into two distinct domains.

1 Specific size effects. In the 'small cluster' size domain an irregular size dependence of the relevant cluster properties $\chi(n)$ (where n is the number of units) is exhibited. This irregular pattern is manifested most dramatically in the existence of 'magic numbers' in $\chi(n)$ vs n, which reflect shell closure effects. Typical examples involve the structural closed shells of Mackay icosahedra [284,285] in clusters of rare gas atoms [286] and of spherical large molecules [301], the enhanced energetic stability and increased ionization potentials for electronic closed shells in metal clusters [302], and the expected increased stability of the Fermion closed shell structure in $(^3\text{He})_N$ clusters [303].

2 Smooth size effects for 'large' clusters. In this size domain a quantitative description (Fig. 1.43) was advanced for the 'transition' of energetic, electronic, spectroscopic, electrodynamic, and dynamic attributes of clusters to the infinite bulk system in terms of cluster size equations (CSEs) [298–300],

$$\chi(n) = \chi(\infty) + An^{-\beta}, \tag{1.41}$$

where A is the constant and β ($\beta \geq 0$) is a positive exponent.

CSEs, which are quantified by Eqn 1.41, can be traced to two distinct physical origins: cluster packing and excluded volume effects. These two categories will now be considered.

5.2.1 SIZE EFFECTS ORIGINATING FROM CLUSTER PACKING

These are the consequences of the large surface/volume fraction of clusters. For sufficiently large clusters, the fraction of surface atoms is given by $F = \alpha n^{-1/3}$. A straightforward utilization of this result pertains to the description of extensive variables Y, e.g. the internal energy, entropy, or magnetization, of the cluster. Viewing the cluster of n units as a composite system consisting of surface and volume subsystems, the value of the extensive variable $Y(n)$ is obtained from a subsystem additivity rule

$$Y(n) = n(1 - F)y_v + nFy_s, \tag{1.42}$$

where y_v and y_s are the corresponding variables (per unit) for the bulk and for the surface, respectively. The total value of the variable per unit $y(n) = Y(n)/n$ is

$$y(n) = y_v + \alpha(y_s - y_v)n^{-1/3}. \tag{1.43}$$

Equation 1.43 constitutes a simple application of the liquid drop model of nuclear physics.

5.2.2 SIZE EFFECTS ORIGINATING FROM EXCLUDED VOLUME CONTRIBUTIONS

A multitude of energetic and spectroscopic size effects can be described in terms of the infinite system observable, corrected for the excluded volume contribution. The CSE (Eqn 1.41) then assumes the general form

$$\chi(n) = \chi(\infty) + C(n), \tag{1.44}$$

where

$$C(n) = An^{-\beta} \tag{1.45}$$

is the excluded volume correction term. $C(n)$ accounts for the modification of the bulk value for the observable in the cluster, due to the excluded volume outside it (i.e. the range R_c to ∞). R_c is the cluster radius, which is related to the radius R_0 of a single constituent by $R_c = R_0 n^{1/3}$. Accordingly, the excluded volume correction can be expressed in the alternative form

$$C(R_c) = A(R_c/R_0)^{-3\beta}. \tag{1.46}$$

The CSE (Eqn 1.44) is better than it appears at first sight for the quantification of physical observables. All the short-range contributions to $\chi(n)$, which are usually difficult to evaluate, are incorporated in $\chi(\infty)$, which is often taken from experiment. What is explicitly evaluated is the correction term $C(n)$ arising from the excluded volume contributions. These are determined by long-range effects, which are amenable to reliable calculation. We shall now apply excluded volume effects to the energetics and spectroscopy of clusters.

5.2.3 EXCLUDED VOLUME EFFECTS ON CLUSTER IONIZATION POTENTIALS

The CSE was successfully applied to a multitude of energetic size effects [298–300]. For the size dependence of the cluster ionization potentials (IP) and electron affinities (EA) the excluded volume contribution originates from the charging energy. The application of the continuum dielectric model to $C(n)$ results in

$$IP(Z; R_c) = IP(\infty) + \frac{(2Z+1)e^2}{2R_c}f(\varepsilon_0, \varepsilon_\infty) \tag{1.47a}$$

$$EA(Z; R_c) = EA(\infty) + \frac{(2Z-1)e^2}{2R_c}f(\varepsilon_0, \varepsilon_\infty) \tag{1.47b}$$

where Z is the cluster initial charge and ε_0 and ε_∞ are the static and optical (high frequency) dielectric constants, respectively. The function $f(\varepsilon_0, \varepsilon_\infty)$ depends on the nature of the ionization or electron attachment, being

$$f(\varepsilon_0, \varepsilon_\infty) = (1 - 1/\varepsilon_0) \tag{1.48a}$$

for the adiabatic ionization potential and electron affinity, and

$$f(\varepsilon_s, \varepsilon_\infty) = (1 - 2/\varepsilon_0 + 1/\varepsilon_\infty) \qquad (1.48b)$$

for the vertical ionization potential and electron affinity. IP(∞) and EA(∞) are the ionization potential and the electron affinity of the corresponding macroscopic system. Accordingly $C(n)$ (Eqn 1.45) is characterized by $\beta = 1/3$. This situation is realized for the vertical ionization potentials of rare gas clusters [298], the vertical ionization potentials of microscopically solvated anion clusters [304,305] and the vertical binding energies of interior excess electron states in large $(NH_3)_n^-$ clusters [306]. The excluded volume contribution of an impurity in a herocluster depends on the location of the impurity. The distinction between surface and interior impurity states is of interest in the context of anion solvation in water clusters, e.g. $X^-(H_2O)_n$ (X = F, Cl, Br, I) [305] and of rare gas heteroclusters [300]. A theory of site-specific cluster ionization potentials was developed [307].

The ionization potentials and electron affinities of metal clusters are of importance for molecular electronics, being given from the charging model by $\varepsilon \rightarrow \infty$, where IP($\infty$) is the work function (or band energy) of the (macroscopic) metal and EA(∞) is its bulk band energy. These relations account well for the smooth cluster size dependence of the electron affinities of sufficiently large metal clusters (Fig. 1.44), which faithfully follow the artist's view of size effects (Fig. 1.43).

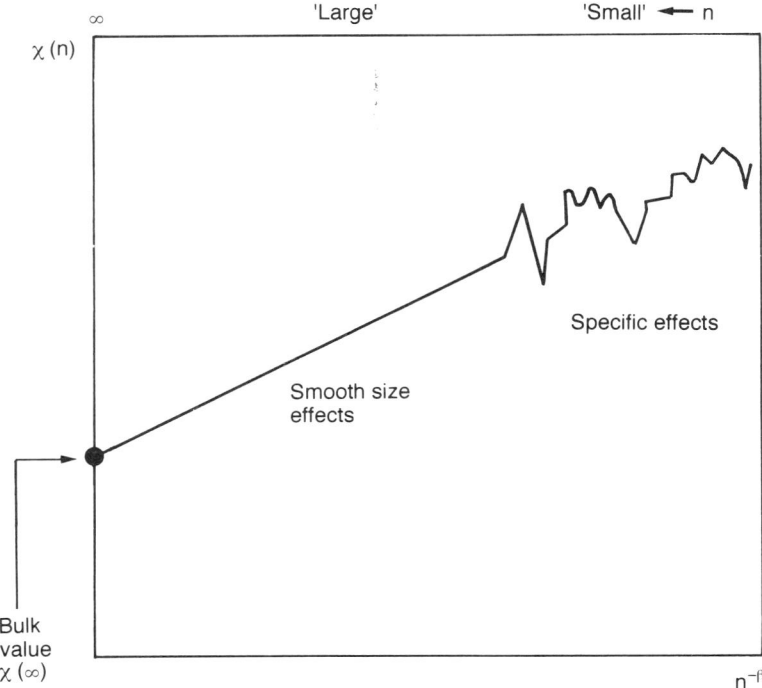

Figure 1.43 The cluster size dependence of a cluster property $\chi(n)$ on the number n of the cluster constituents. The data are plotted vs $n^{-\beta}$ where $\beta \leqslant 0$. 'Small' clusters reveal specific size effects, while 'large' clusters are expected to exhibit for many properties a smooth size dependence of $\chi(n)$, which converges for $n \rightarrow \infty$ to the bulk value $\chi(\infty)$.

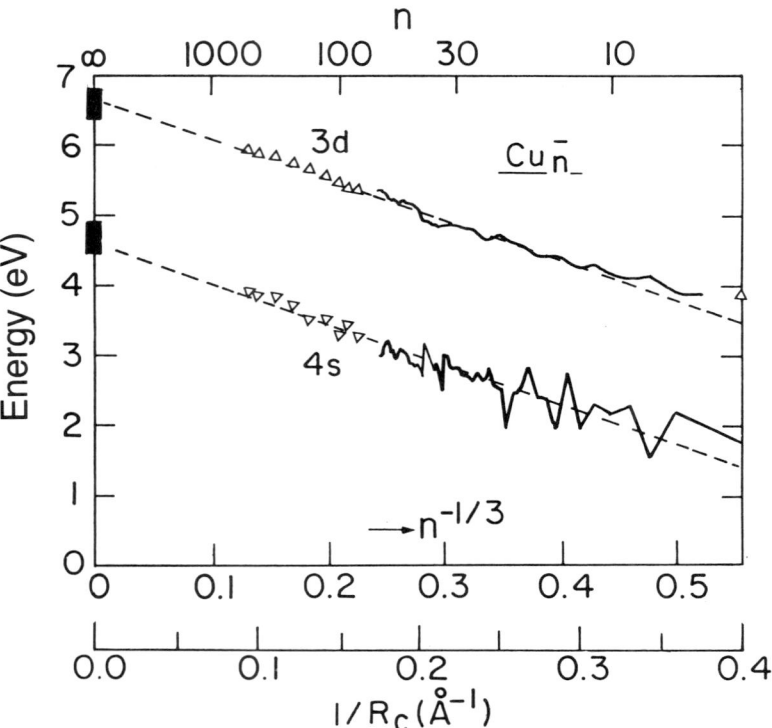

Figure 1.44 The dependence of the energies of the ns and nd bands of Cu_n^- ($n = 1$–411) clusters [328]. The electron affinities of the negative clusters reveal the CSE behaviour for $n > 60$, converging to the corresponding bulk energies (marked by solid bars).

5.3   *A digression on molecular charging effects and Coulomb blockade in finite systems*

Cluster size effects can provide a link between electrical properties of mesoscopic systems [256,257] and of clusters, with the issue of Coulomb blockade and Coulomb staircase [256–259] providing a pertinent case. In a double-junction mesoscopic device, consisting of an island and two metallic electrodes, which is subjected to an external voltage V, excess charges can accumulate on the island, giving rise to a Coulomb blockade (Fig. 1.41). This is manifested by the delay in the voltage for the entry of an extra electron into the bridge. The voltage delay scales linearly with (semiclassically) the junction capacitance, or (microscopically) the electron–electron repulsion in the island. It will be interesting to provide a connection between the Coulomb blockade in the mesoscopic metal–island–metal system and a corresponding metal–cluster–metal system. We shall provide qualitative arguments for the cluster size dependence of the Coulomb blockade caused by the presence of charge in the cluster, in the cluster size domain of a smooth dependence of the ionization potential and electron affinity on the cluster size (Section 5.2.3). The addition of a single electron to a cluster of charge Z results in the cluster charging energy

$$W_c(R_c) = EA(Z+1, R_c) - EA(Z, R_c) \tag{1.49}$$

which, according to Eqn 1.47 is

$$W_c(R_c) = \delta E + f(\varepsilon_0, \varepsilon_\infty)e^2/R_c, \tag{1.50}$$

where $\delta E = EA(Z+1, \infty) - EA(Z, \infty)$, corresponding to the charging energy of the bulk material. In analogy to the macroscopic case, the charging energy of capacitor C_a with the charge Q is $W = Q^2/2C_a$, and this energy is supplied by the voltage source with the applied voltage $V = Q/C_a$. For the charging of the cluster we take $Q = e$. Following the arguments for molecular wires [131,132] the charging energy of the cluster is taken to be given roughly by Eqn 1.49. One then gets $W_c(R_c) = e^2/C_a$ and the linear relationship $V_{MIN} = W_c(R_c)/e$ between $W_c(R_c)$ and the minimal voltage V_{MIN} required to inject an electron into the cluster. For $V < V_{MIN}$ no electron injection takes place, while for $V > V_{MIN}$ electron injection prevails. For sufficiently large clusters (i.e. R_c considerably exceeding the constituent radius and being smaller than the size of the mesoscopic system) the minimal voltage to overcome the Coulomb blockade obeys a CSE of the form $V_{MIN} = [\delta E + AR_c^{-1}]/e$, decreasing linearly with R_c^{-1}. For a three-dimensional cluster $V_{MIN} \approx [\delta E + \bar{A}n^{-1/3}]/e$ while for a two-dimensional cluster (Section 5.2), i.e. cluster island, $V_{MIN} \approx [\delta E + \bar{\bar{A}}n^{-1/2}]/e$. This analysis provides a heuristic description of the connection between the Coulomb blockade in the mesoscopic system and in (large) clusters. With a further decrease of the cluster size the molecular limit is approached, where specific effects on the electronic level structure will be exhibited. Double-junction molecular wire devices have to be treated by molecular orbital arguments and the application of the Hubbard model [131,132]. Simple scaling arguments in the molecular limit [131,132] can be applied, asserting that for a particle in a box of radius (or length) R_c the energy level separation scales as R_c^{-2}, but since the number of electrons scales as R_c^3, then for the molecular staircase system $V_{MIN} \propto R_c^{-1}$. The observation of the Coulomb blockade is facilitated for mesoscopic systems, quantum dots, and also for large clusters. The observation of the eigenvalue staircase [83] requires smaller, molecular-sized objects.

5.4 Confinement

The interaction of the boundary of a cluster with an electron or hole will result in their localization (confinement) within the cluster, provided that this interaction is repulsive. Such a situation prevails for semiconductor [308–312] or rare gas (e.g. Ne and Ar) [298,312] clusters. For the case of a single charge carrier, cluster boundary scattering effects on either coherent or incoherent motion (Section 3.1.4) will be exhibited. Cluster boundary interactions are manifested by the confinement of large radius Wannier–Mott excitons (i.e. electron–hole pairs) in such clusters. In this context one encounters an interesting quantum mechanical problem: the interplay between the attractive two-body electron–hole interaction and the one-body interaction exerted by the cluster boundary [308–312]. Consider an exciton [313] characterized by the binding energy

$$B = \mu e^4/2\varepsilon^2 h^2 N^2 \tag{1.51}$$

and a principal quantum number N and a radius

$$r_N = N^2 \varepsilon h^2/\mu e^2, \tag{1.52}$$

where ε is the (optical) dielectric constant and μ the effective mass (i.e. $\mu^{-1} = m_e^{-1} + m_h^{-1}$

with m_e and m_h corresponding to the effective mass of the electron and hole, respectively).

Two limiting situations can be realized [308–312] for such an exciton with increasing cluster size (characterized by the cluster radius R_c).

1 Strong confinement, $R_c \ll r_N$. This situation corresponds to the individual confinement of the electron and the hole within the potential well, with negligible spatial correlation. The energy of the lowest excited state was inferred from an electrostatic model

$$E_1 = \pi^2 h^2/\mu e^2 - 1.8e^2/\varepsilon R_c \tag{1.53}$$

with the first term corresponding to quantum localization of both electron and hole within a spherical box and the second term representing the interaction between them.

2 Weak confinement $R_c \gg r_N$. Now the characterization of the exciton as a pseudoparticle is preserved, being confined in the cluster. The energy E_N of the Nth hydrogenic exciton state is [308–312]

$$E_N(R_c) = E_N(\infty) + \pi^2 h^2/[2(m_e + m_h)R_c^2], \tag{1.54}$$

where the exciton energy in the infinite system is

$$E_N(\infty) = E_G - B/N^2 \tag{1.55}$$

with E_G being the band gap. The second term in Eqn 1.54 represents the confinement of the pseudoparticle in a box. Equation 1.54 constitutes [298] a CSE of the interesting form $E_N(n) = E_N(\infty) - \alpha n^{-2/3}$.

Exciton confinement brings up a unique feature of cluster boundary effects. The spectroscopic manifestations of exciton confinement were established [298,312]. Of considerable interest are the effects of confinement on the exciton dynamics, e.g. transport and thermal ionization. In the realm of molecular electronics one should explore ways to overcome confinement of electrons, holes, and excitons via appropriate molecular circuitry.

5.5   *The unified description of cluster size effects*

The description of 'smooth' cluster size effects, which originate either from cluster packing or from excluded volume contributions, results in the quantification of the gradual 'transition' from the large finite cluster to the bulk systems. The CSE (Eqn 1.41) provides a unified description for the energetic, quantum, electronic, spectroscopic, and electrodynamic size effects [298,300]. Table 1.2 provides an overview of the parameters of the CSEs for several physical attributes, specified in terms of the β exponent. Table 1.2 also shows the validity conditions for the applicability range of these CSEs, specified in terms of the minimal number n_- of units for the onset of bulk properties (B). From the data of Table 1.2, i.e. distinct CSEs (with different values of β and n_-) for different physical properties, it is apparent that the cluster size effects are not universal.

The CSEs provide the quantitative answer to one of the central questions in the area of cluster chemical physics [298,300]: what is the minimal cluster size for which the cluster properties become size invariant and do not differ in any significant way from those of a macroscopic sample of the same material? According to the CSE, the relative

Table 1.2 Estimates of 'critical' cluster sizes for the attainment of bulk properties.

Observable	β	n_{*}	$|A/\chi(\infty)|$	n_c†
Cohesive energy of alkali clusters	1/3	5	1	10^6
Ionization potentials of rare gas clusters	1/3	20	0.1	10^3
Ionization potentials and electron affinities of metal clusters	1/3	60	0.7	3×10^5
Photoionization of $I^-(H_2O)_2$	1/3	125	1	10^6
Photoionization of $(NH_3)_n^-$	1/3	150	3	3×10^7
Collective vibrational compression modes of Ar_n	1/3	20	5	10^8
Confined Wannier excitons in rare gas clusters	2/3	700	1	10^3
Dispersive spectral shift, $M \cdot A_n$ cluster	1	10	7	7×10^2
Dispersive spectral width, $M \cdot A_n$ cluster	3	10	1	7

* Rough estimates for the minimal cluster size ($n > n\_$) for the applicability of the CSE.
† Relative deviation of physical property from bulk value $\leqslant 1\%$.

deviation from the bulk value for a specific physical property $\chi(n)$ will be

$$|\chi(n) - \chi(\infty)|/|\chi(\infty)| = |A/\chi(\infty)|n^{-\beta}. \tag{1.56}$$

Defining somewhat arbitrarily the realization of bulk properties for $n \geqslant n_c$, where the relative deviation is $\leqslant 1\%$, one estimates the number of units, n_c, in the 'critical' cluster size to be

$$n_c = 100^{1/\beta}|A/\chi(\infty)|^{1/\beta}. \tag{1.57}$$

Table 1.2 provides a compilation of the values of $|A/\chi(\infty)|$. The catalogue of n_c values (Fig. 1.45) reveals a variation of n_c over eight orders of magnitude for various physical

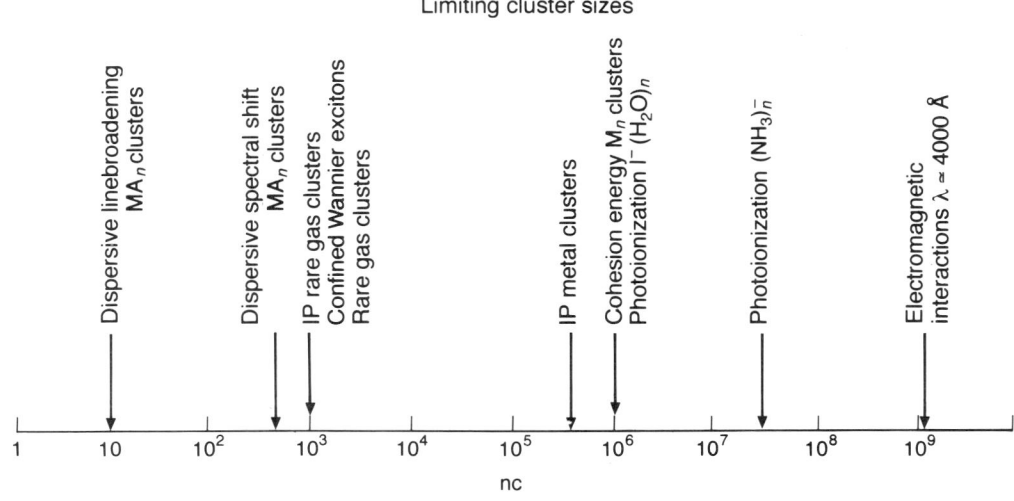

Figure 1.45 An overview of the 'critical' cluster size (n_c) for the attainment of bulk properties for energetic, spectroscopic, and electromagnetic observables.

properties. The largest value of n_c corresponds to electromagnetic interactions (e.g. the Einstein coefficients for spontaneous and stimulated emission) which are characterized by the dimensionless size parameter $(2\pi R_0/\lambda)n^{1/3}$, where λ is the wavelength of light [298]. On the other extreme, dispersive interactions result in the lowest values of n_c. From the foregoing analysis we conclude that the CSEs provide a unified (but not universal) description for the merging between the properties of microscopic large finite systems and those of the macroscopic bulk material. The non-universality of cluster size effects implies a wide variance of different properties (e.g. optical and electrical) of a finite system from the corresponding bulk values.

5.6   *Dimensionality scaling of cluster size equations*

The preceding analysis of cluster size effects mostly focused on the usual physical situation when both the finite cluster and the corresponding reference bulk material are three-dimensional ($D = 3$). Cluster chemical physics in low dimensionality [68–70,314] is of interest, pertaining to the following physical situations.

1 Lower geometrical cluster dimensionalities. These may involve finite 'cluster wires' converging to an infinite wire ($D = 1$), or planar clusters converging to an infinite plane ($D = 2$) for supported 'islands' on a substrate.

2 Fractal clusters. These are characterized by Hausdorf dimensionality D, which satisfies two conditions: (i) the correlation length for the (self-similar) fractal structure is smaller than the cluster radius; (ii) with increasing cluster size, the cluster converges to the macroscopic fractal structure. Examples of fractal clusters include clusters of porous materials.

3 Dimensionality scaling due to cluster packing. The fraction of the surface atoms in a cluster of Hausdorf dimensionality D is $F = \bar{a}n^{-1/D}$, where $\bar{a}$ is a numerical constant. These heuristic relations constitute the generalization of the liquid drop model for D dimensions. A more careful examination of the dispersion relations for elementary excitations in fractal clusters is still required.

4 Dimensionality scaling of excluded volume effects. We consider now the situation when the cluster and the excluded volume are characterized by a common dimensionality, D ($\neq 3$). The extension of the CSEs just requires the replacements of the excluded volume corrections $C(R_c)$ (Eqn 1.45) by contributions over a D-dimensional space. The packing of the constituents is given by setting $R_c \propto n^{1/D}$. The excluded volume contribution for the ionization potentials is now $C_{(D)}(R_c) \propto R_c^{D-4}$ and $C_{(D)}(n) \propto n^{(D-4)/D}$.

Such CSEs, originating from dimensionality scaling, will be of interest for the ionization, spectroscopic response, and dynamics of clusters of lower ($D = 1$ and $D = 2$) dimensionality and of fractal clusters. The dimensional feature of cluster behaviour is particularly relevant to molecular systems, since molecules are generally not spherical (excluding esoteric cases such as fullerenes and cage compounds). Indeed, quantization effects perpendicular to the transport direction are crucial both in mesoscopic devices [163] and molecular wires.

6   Perspectives

As the other chapters in this book make clear, both molecular electronics and molecule-based electronics are research areas being pursued with state-of-the-art experimental techniques, theoretical ideas and applications. In particular, the advent of novel synthetic methodologies has permitted molecular assemblies and nanostructured molecular arrays to be prepared, whose structural aspects can be designed rather than obtained by chance. Local probes, including scanning microscopies, near-field optics, and imaging permit investigation and analysis of these structures, again on nanometer scale.

Most of the application notions in molecular electronics devolve from more standard electronics: these range from molecular wires, molecular rectifiers, molecular transistors, molecular switches, molecular field-effect transistors (FETs) and molecular amplifiers to molecular-based machines and memories. Unique applications in which only molecules could be used, such as molecular recognition-based sensors, molecular patterned growth, or molecular interfaces with biological systems, have been pursued less actively.

As this volume demonstrates, the enabling scientific understanding, appropriate methodology, and analytical techniques are rapidly falling into place, so that molecular electronics can become a precise scientific and engineering discipline.

In this chapter we focused on molecular wires, which are perhaps the most obvious component of the molecular electronics tool kit. Even within this area, however, a great deal remains to be learned. As mentioned and cited in Section 4, examples of situations in which ignorance is profound, and importance is great, include the following.

1 What are the limiting current behaviours in circuitry of nanometric dimensions? Ideas such as conductance are originally based on ohmic considerations, in which charge carriers attain terminal velocities, via collisions, in being transported through conductive structures. For very short transport distances, these considerations may be inappropriate. In current computer design, ohmic behaviour is nearly always assumed in designing chips for particular tasks [190]. With structure sizes at the nanometer level, it is not clear that ohmic behaviour may be obtained: one could imagine quantum transport, ballistic electron transport, or electron dynamics occurring in connection with the Coulomb blockade. The actual measurement of such small currents is now possible, and the first experiments characterizing current flow through molecular wires are beginning to appear. The understanding of these quantum currents will require consideration not only of the wire itself, but also of the leads and structures in which that wire is embedded. The comparison with molecular electron transfer, in which the vibronic states and the solvent provide a vibronic continuum in which memory is lost and irreversibility occurs, is suggestive: the continuum of electronic states in the electrodes, or leads, can indeed provide irreversibility and memory loss. In contrast to the Franck–Condon factors which occur in electron transfer, however, electrons can enter empty states of the electrode band structure at any energy without energy penalty; that is, there is no gap law for high energy injection. This is useful in designing circuitry based on resonant tunnelling transport of electrons.

2 What is the role of fluctuations? In classical statistical mechanics, ensembles are used to describe observable behaviours. Characteristically, then, the level of fluctuation scales as the inverse square root of the number of systems sampled. In molecular wires, such ensemble averaging does not occur: the current measured consists of quantum transport of electrons through quantum states of the wire between (continuum) states of the electrodes. One then wonders what role fluctuations may play. That is, will the observed current always be the same for a given applied voltage, or will statistical fluctuations such as shot noise or vibrations result in currents whose mean is substantially smaller than the fluctuation? The considerations here are interesting and complex: the classical ensemble average is probably not the correct way to proceed, since the measurement is not made on an ensemble. The quantum dynamics, on the other hand, also implies a statistical consideration in terms of preparation of the initial state, or density. This subject has only begun to be explored theoretically, and (obviously) no fluctuation spectrum of the conductance through a single molecular wire has yet been reported.

3 How will optical excitation change behaviours? Optically induced molecular currents in molecular wires present a different set of considerations, just as photoinduced electron transfer behaves quite differently from ground state electron transfer. Here the initial state is prepared by short time interaction of the molecular wire structure with an arbitrary pulse; this could produce initial states for transfer that do not occur in ordinary (or even molecular wire) ground state circuitry. Sensors based on optically induced molecular wire transfer have not yet been considered, but they could clearly offer substantial advantages with respect to sensitivity and speed.

4 How do localized or coherent behaviours evolve? The precise nature of the current in molecular wires has not been clarified theoretically, and (in the absence of definitive experiments) it is not clear what the thermal, temporal, spatial, or frictional aspects of this current would be. In very short molecular wires (of the order of a few repeat sites), or for very rigid ones such as carbon nanotubes, coherent transfer might well occur; under these conditions, transfer through molecular resonances has been shown to produce sharp spikes in the conductance, and substantial increases in current. These observations are reminiscent of those seen in inelastic tunnelling spectroscopy, in which, again, transport occurs when molecular resonances (eigenstates) cross the Fermi surface. For very long wires, on the other hand, the transport is not ballistic: under these conditions, organic wires (if they are to conduct at all) must do so by an incoherent mechanism, in which the electron becomes localized along the wire and drifts, under the influence of a field, with a large effective mass. The diffusive regime has been suggested theoretically, and may have been seen experimentally in electron transfer reactions; again, in the absence of direct experimental observation of molecular wire behaviour, no definitive statements can be made. Nevertheless, we would expect that for wire lengths that substantially exceed those required for stabilization of a polaronic or excitonic defect, the incoherent, drift transport should be observed. This is just what happens in conductive polymers.

Localization of the carrier can occur through static trapping by defects, or through the dynamic trapping afforded by the electron/vibration interaction (forming a polaron).

An enticing possibility arises when considering the material of which nanowires are made. As the section above on size effects makes clear, the characteristic dimension at which molecular behaviour attains the limiting behaviour of the bulk solid will vary depending both on the property observed and on the constituents of the cluster. For example, one would expect Coulomb staircase structures to remain spaced by characteristic eigenvalue spacings if the appropriate quantum dot is made either very small (so that affinity levels are not solely fixed by Coulomb repulsion) or of strongly localized, very narrow band, molecular materials; this is in sharp contrast to mesoscopic quantum dots made of broad-band semiconductors or metals, in which the step size in the Coulomb staircase scales as inverse radius, and becomes exceedingly small (less than 20 K) for characteristic mesoscopic dimensions.

Similarly, it is intriguing that very thin molecular wires probably carry charge, over long distances, by a defect mechanism based on soliton or polaron transport — that is, the carriers should be characterized by large effective mass, slow diffusion mobility, and a small deBroglie wavelength. In atomic wires, on the other hand, wavelengths will remain large because of relatively weak electron–phonon scattering (at least if the metallic order is relatively good). Under these conditions, surface scattering effects should be important, and the material can become semiconductive when the transverse dimension becomes smaller than the characteristic wavelength of the transporting electron [163]. This then provides the intriguing possibility that at a very small diameter, molecular wires might actually be more efficient than metallic wires; their relatively high effective mass and slow mobility is more than made up for by their long life time and diffusive behaviour — in this case, perhaps the path is narrow enough so that the tortoise really is more mobile than the hare.

The area of molecular electronics seems to have come of age as a scientific and engineering discipline — its future should be striking and rich.

Acknowledgements. M.R. is very pleased to thank the Chemistry Division of the ONR and NSF for support. J.J. is pleased to thank the Volkswagen Foundation for support of his work on electron transfer.

7 References

1 Carter FL, ed. *Molecular Electronic Devices.* New York: Dekker, 1982.
2 Carter FL, Siatkowski RE, Wohltgen H, eds. *Molecular Electronic Devices.* Amsterdam: Elsevier, 1988.
3 Aviram A, ed. *Molecular Electronics — Science and Technology.* New York: Engineering Foundation, 1989.
4 Aviram A, ed. *Molecular Electronics — Science and Technology.* Washington, DC: AIP, 1992.
5 Metzger RM, Day P, Papavassiliou GC, eds. *Lower-Dimensional Systems and Molecular Electronics.* New York: Plenum, 1991.
6 Borissov M, ed. *Molecular Electronics.* Singapore: World, 1987.
7 Birge RR, ed. *Molecular and Biomolecular Electronics.* Washington, DC: ACS, 1994.
8 Hong FT, ed. *Molecular Electronics.* New York: Plenum, 1988.
9 Petty MC, Bryce MR, Bloor D, eds. *Introduction to Molecular Electronics,* New York: Oxford University Press, 1995.
10 Miller JR. *Adv Mater* 1990; **378**: 695, 601.
11 Mirkin CA, Ratner MA. *Annu Rev Phys Chem* 1992; **43**: 719.
12 Lazarev PI, ed. *Molecular Electronics.* Dordrecht: Kluwer, 1991.
13 Bloor DM. *Mol Cryst Liq Cryst* 1992; **234**: 1.
14 Hultheen JC, van Duyne RP. *J Vac Sci Technol* 1995; **A13**, 1553.
15 Pease RFW. *J Vac Sci Technol* 1992; **B10**: 278.
16 Smith HI, Schattenburg ML. *IBM J Res Dev* 1993; **37**: 319.
17 Dagata JA, Marrian CRK. *Technology of Proximal Probe Lithography.* Washington, DC: SPIE, Bellingham, 1993, V.10.
18 Foss CA, Hornyak GL, Stockert JA, Martin CR. *J Phys Chem* 1994; **98**: 2963.
19 Weller H. *Angew Chem Int Ed Engl* 1993; **32**: 41.
20 Sacra A, Norris DJ, Murray CB, Bawendi MG. *J Chem Phys* 1995; **103**: 5236.
21 Colvin VL, Cunningham KL, Alivisatos AP. *J Chem Phys* 1994; **101**: 7122.

22 Alivisatos AP. *Science* 1995; **271**: 933.

23 Bawendi MG, Steigerwald ML, Brus LE. *Annu Rev Phys Chem* 1990; **41**: 477.

24 Barnier F, Yassar A, Hajlaoui R *et al*. *J Am Chem Soc* 1993; **115**: 8716.

25 Dodabalapur A, Katz HE, Torsi L, Haddon RC. *Science* 1995; **269**: 1560.

26 Haddon RC. *J Am Chem Soc* 1996; **118**: 3041.

27 Wu CG, Bein TR. *Science* 1994; **264**: 157.

28 Kugimiya S, Lazrak T, Blanchard-Desce M, Lehn JM. *J Chem Soc Chem Commun* 1991; **17**: 1179.

29 Lehn JM. *Angew Chem Int Ed Engl* 1990; **29**: 1304.

30 Schumann JS, Pearson DL, Tour JM. *Angew Chem Int Ed Engl* 1995; **67**: 199.

31 Jörgensen M, Bechgaard K. *J Org Chem* 1994; **59**: 5877.

32 Crossley MJ, Burn PJ. *J Chem Soc Chem Commun* 1991; 1569.

33 Bauerle P, Fischer T, Bidlingmeier B, Stabel A, Rabe JP. *Angew Chem Int Ed Engl* 1995; **34**: 303.

34 Ribou AC, Launay JP, Takahashi K, Nihara T, Tarutani S, Spangler CW. *Inorg Chem* 1994; **33**: 1325.

35 Wu CG, Bein T. *Science* 1994; **264**: 1757.

36 Tolbert LM. *Acc Chem Res* 1992; **25**: 561.

37 Sailor MJ, Curtis C. *Adv Mater* 1994; **6**: 688.

38 Dorogi M, Gomez J, Osifchin R, Andres RP, Reifenberger R. *Phys Rev B* 1995; **52**: 9071.

39 Graber KC, Freeman RG, Hommer MB, Natan MJ. *Anal Chem* 1995; **67**: 735.

40 Baumgarten M, Müller U, Müllen K. In: Aviram A, ed. *Molecular Electronics — Science and Technology*. Washington, DC: AIP, 1992: 68.

41 Pascual JI, Mendez J, Gomez-Herrero J, Bard AM, Garcia J, Binh VT. *Phys Rev Lett* 1993; **71**: 1852.

42 Krans JM, Muller CJ, Yanson IK, Govaert TCM, Mesper R, Van Ruitenbeek JM. *Phys Rev B* 1993; **48**: 1472.

43 Wolf HC. In: Aviram A, ed. *Molecular Electronics — Science and Technology*. Washington, DC: AIP, 1992: 237.

44 Wagner RW, Lindsey JS. *J Am Chem Soc* 1994; **116**: 9759.

45 Itoh Y, Webber SE, Rogers MA. *Macromolecules* 1989; **22**: 2766.

46 Serroni S, Denti G, Campagna S, Ciano M, Balzani V. *Chem Comm* 1991; 944.

47 Prathapan S, Johnson TE, Lindsey JS. *J Am Chem Soc* 1993; **115**: 7519.

48 Salaneck WR, Lundstrom I, Ranby B, eds. *Conjugated Polymers and Related Compounds*. London: Oxford University Press, 1993.

49 Bredas JL, Silbey R, eds. *Conjugated Polymers*. Dordrecht: Kluwer, 1991.

50 Donovan KJ, Scott K, Sudiwala RV *et al*. *Thin Sol Films* 1993; **232**: 110.

51 Schramm CJ, Scaringe RP, Stojakovic DR, Hoffman BM, Ibers JA, Marks TJ. *J Am Chem Soc* 1980; **102**: 6702.

52 Hale PD, Ratner MA. *J Chem Phys* 1985; **83**: 5277.

53 Thompson JA, Murceta K, Miller DC *et al*. *Inorg Chem* 1993; **32**: 3546.

54 Hiejma T, Yakushi K. *J Chem Phys* 1995; **103**: 3950.

55 Crossley MJ, Burn PL. *J Chem Soc Chem Comm* 1991; 1569.

56 Lu TX, Reimers JR, Crossley MJ, Hush NS. *J Phys Chem* 1994; **98**: 11878.

57 Turro NJ. *Pure Appl Chem* 1995; **67**: 199.

58 Zhou Q, Swager TM. *J Am Chem Soc* 1995; **117**: 12593.

59 Jerome D, Caron LG, eds. *Low-Dimensional Conductors and Superconductors*. New York: Plenum, 1987.

60 Walz J, Ulrich K, Port H, Wolf HC, Wonner J, Effenberger F. *Chem Phys Lett* 1993; **213**: 322.

61 Ward MD. *Chem Soc Rev* 1995; 21.

62 Naklicki ML, Crutchley RJ. *J Am Chem Soc* 1994; **116**: 6045.

63 Schaumburg K. In: Göpel W, Ziegler C, eds. *Nanostructures Based on Molecular Materials*. Weinheim: VCH, 1992: 156.

64 Zelikovitch L, Libman J, Shanzer A. *Nature* 1995; **374**: 790.

65 Lindsey JM. *J Am Chem Soc* 1996; **118**: 3996.

66 Eigler DM, Lutz CP, Rudge WE. *Nature* 1991; **352**: 600.

67 Peteanu LV, Schoenlein RW, Wang Q, Mathies RA, Shank CV. *Proc Natl Acad Sci* 1993; **90**: 11762.

68 Aviram A. *J Am Chem Soc* 1988; **110**: 5687.

69 Hush NS, Wong AT, Bacskay GB, Reimers JR. *J Am Chem Soc* 1990; **117**: 4192.

70 Farazdel A, Dupuis M, Clementi E, Aviram A. *J Am Chem Soc* 1990; **112**: 4206.

71 Tour JM, Wu R, Schumm JS. In: Aviram A, ed. *Molecular Electronics — Science and Technology*. Washington, DC: AIP, 1992: 77.

72 Potember RS, Poehler TO, Cowan DO. *Appl Phys Lett* 1979; **34**: 405.

73 Roth S. In: Göpel W, Ziegler Ch, eds. *Nanostructures Based on Molecular Materials*. Weinheim: VCH, 1992: 65.

74 Haarer D. *Mol Cryst Liq Cryst* 1993; **236**: 1.

75 Volker S. *Annu Rev Phys Chem* 1989; **40**: 499.

76 Moerner WE. *Persistent Spectral Hole Burning Science and Applications*. Berlin: Springer, 1988.

77 Chidsey CED, Murray RW. *Science* 1986; **231**: 25.

78 Wrighton MS. *Science* 1986; **231**: 32.

79 Muller CJ, VanRuitenbeek JM, deJogh CJ. *Physica* 1992; **C191**: 485.

80 Chen CJ. *Introduction to Scanning Tunneling Microscopy*. New York: Oxford University Press, 1993.

81 Joachim C, Gimzewski JK, Schlitter RD, Chavy C. *Phys Rev Lett* 1995; **74**: 2102.

82 Miller TG, McElfresh MW, Reifenberger R. *Phys Rev B* 1993; **48**: 7499.

83 Andres RP, Bein T, Dorogi M *et al. Science* 1996; **272**: 1323.

84 Bumm LA, Arnold JJ, Cygan MT *et al. Science* 1996; **271**: 1705.

85 Jortner J, Pullman B, eds. *Perspectives in Photosynthesis*. Dordrecht: Kluwer, 1990.

86 Gust D, Moore TA, Moore AM. *Acc Chem Res* 1993; **26**: 198.

87 Wasielewski MR. *Chem Rev* 1992; **92**: 435.

88 Fujihira M. In: Göpel W, Zeigler Ch, eds. *Nanostructures Based on Molecular Materials*. Weinheim: VCH, 1992: 27.

89 *Electrochim. Acta* 1995; **40**: (13/14) 67.

90 Herr BR, Mirkin CA. *J Am Chem Soc* 1994; **116**: 1157.

91 Yang Y, McBranch D, Swanson B, Li D. *Materials Research Symposium*. In press.

92 Moore LW, Springer KW, Shi J-X *et al. Adv Mater* In press.

93 Bradley DDC, Tsutsui T, eds. *Organic Electroluminescence*. Cambridge: Cambridge University Press, 1995.

94 Burroughes JH, Bradley DDC, Brown AR *et al. Nature* 1990; **347**: 539.

95 Braun D, Heeger AJ. *Thin Solid Films* 1992; **216**: 96.

96 Cornil J, dos Santos DA, Beljonne D, Bredas JL. *J Phys Chem* 1995; **99**: 5604.

97 Gustafsson JC, Inganas O, Andersson AM. *Synth Met* 1994; **62**: 17.

98 Ghadiri MR, Granja JR, Buehler LK. *Nature* 1994; **369**: 301.

99 Martin CR. *Science* 1994; **266**: 1761.

100 Shimidzu T. In: Aviram A, ed. *Molecular Electronics — Science and Technology*. Washington, DC: AIP, 1992: 129.

101 Hopfield JJ, Onuchic JN, Beratan DN. *Science* 1988; **241**: 817.

102 Hampp N, Rhoma R, Bräuchle C *et al.* In: Aviram A, ed. *Molecular Electronics — Science and Technology*. Washington, DC: AIP, 1992: 181.

103 Hong FT, Hong FH, Needlemann RB, Ni B, Chang M. In: Aviram A, ed. *Molecular Electronics — Science and Technology*. Washington, DC: AIP, 1992: 204.

104 Birge RA. In: Birge RA, ed. *Molecular and Biomolecular Electronics*. Washington, DC: ACS, 1994: 1.

105 Aviram A, Ratner MA. *Chem Phys Lett* 1974; **29**: 277.

106 Metzger RM. In: Birge RA, ed. *Molecular and Biomolecular Electronics.* Washington, DC: ACS, 1994: 81.

107 Metzger RM. In: Metzger RM, Day P, Papavassiliou, eds. *Lower-Dimensional Systems and Molecular Electronics.* New York: Plenum, 1991: 691.

108 Martin AS, Sambles JR, Ashwell GJ. *Phys Rev Lett* 1993; **70**: 218.

109 Fischer CM, Burghard M, Roth S, V Klitzing K. *Europhys Lett* 1994; **28**: 129.

110 Marcus RA. *Annu Rev Phys Chem* 1964; **15**: 155.

111 Marcus RA, Sutin N. *Biochim Biophys Acta* 1985; **811**: 265.

112 Closs GL, Calcaterra LT, Green NJ, Penfield KW, Miller JR. *J Phys Chem* 1986; **90**: 3673.

113 Stein CA, Lewis NA, Seitz GJ. *J Am Chem Soc* 1982; **104**: 2596.

114 Isied SS, Vassilian A, Wishart JF *et al. J Am Chem Soc* 1988; **109**: 635.

115 Penfield KW, Miller JR, Paddon-Row MN *et al. J Am Chem Soc* 1987; **109**: 5061.

116 Mikkelsen KV, Ratner MA. *Chem Rev* 1987; **87**: 112.

117 Bixon M, Jortner J. *J Phys Chem* 1993; **97**: 13061.

118 Rips I, Jortner J. *J Chem Phys* 1987; **87**: 6513.

119 Ratner MA. *Naval Res Rev* 1994; **XLVI**: 49.

120 Hoffman BM, Ratner MA. *Inorg Chim Acta* 1996; **243**: 233.

121 Gaines GL, O'Neil MP, Svec WA, Niemczyk MP, Wasieleswki MR. *J Am Chem Soc* 1991; **113**: 719.

122 Sutin N, Brunschwig B, Johnson MK *et al.*, eds. *Electron Transfer in Biology and the Solid State.* Washington, DC: ACS 1990.

123 Jortner J. *J Chem Phys* 1976; **64**: 4860.

124 Bixon M, Jortner J. *Chem Phys* 1993; **176**: 467.

125 Siddarth P, Marcus RA. *J Phys Chem* 1990; **94**: 2985.

126 Liang C, Newton MD. *J Phys Chem* 1993; **97**: 13083.

127 Mikkelsen KV, Ratner MA. *J Phys Chem* 1989; **93**: 1759.

128 Shepard MJ, Paddon-Row MN, Jordan KD. *Chem Phys* 1993; **176**: 289.

129 Joachim C, Vinussa JF. *Europhys Lett* 1996; **33**: 635.

130 Hipps KW, Mazur U. *J Phys Chem* 1994; **98**: 5824.

131 Mujica V, Kemp M, Roitberg A, Ratner MA. *J Chem Phys* 1996; **104**: 7296.

132 Averin DV, Korotkov AN, Likharev KK. *Phys Rev B* 1991; **44**: 6199.

133 Mujica V, Kemp M, Ratner MA. *J Chem Phys* 1994; **101**: 6849.

134 Mujica V, Kemp M, Ratner MA. *J Chem Phys* 1994; **101**: 6856.

135 Sumetskii M. *Phys Rev B* 1993; **48**: 4586.

136 Sumetskii M. *J Phys Cond Matter* 1991; **3**: 2651.

137 Sumetskii M. *Sov Phys JETP* 1985; **62**: 355.

138 Landauer R. *IBM J Res Dev* 1957; **1**: 223.

139 Sautet P, Joachim Ch. *Chem Phys Lett* 1991; **185**: 23.

140 Bixon M, Jortner J. *J Phys Chem* 1991; **95**: 1941.

141 Todd MD, Nitzan A, Ratner M, Hupp JT. *J Photochem Photobiol* 1994; **A82**: 87.

142 Spears KG, Wen X, Arrivo S. *J Phys Chem* 1994; **98**: 9693.

143 Spears KG. *J Phys Chem* 1995; **99**: 2469.

144 Newton MD, Sutin N. *Annu Rev Phys Chem* 1984; **35**: 437.

145 Barbara P, Meyer TJ, Ratner MA. *J Phys Chem* 1996; **100**: 13148.

146 Michel-Beyerle ME, ed. *Reaction Centers of Photosynthetic Bacteria.* Berlin: Springer, 1990.

147 Breton J, Vermeglio A, eds. *The Photosynthetic Reaction Center, Structure, Spectroscopy and Dynamics,* Vols I, II. New York: Plenum, 1987, 1952.

148 Michel-Beyerle ME, ed. *The Reaction Center of Photosynthetic Bacteria: Structure and Dynamics.* Berlin: Springer, 1995.

149 Michel-Beyerle ME, Small GJ, eds. *Chem Phys* 1995; **197**: 223–474.

150 Bixon M, Jortner J, Michel-Beyerle ME. *Biophys Biochim Acta* 1991; **1056**: 30.

151 Bixon M, Jortner J, Michel-Beyerle ME. *Chem Phys* 1995; **197**: 389.

152 Schulten K, Tesch M. *Chem Phys* 1991; **158**: 421.

153 Kobayashi T, Takagi Y, Kondori H, Kemnits K, Yoshihara K. *Chem Phys Lett* 1991; **180**: 416.

154 Heitele H, Pöllinger F, Haberle T, Michel-Beyerle ME, Staab H. *J Phys Chem* 1994; **98**: 7402.

155 Holzapfel W, Finkele U, Kaiser W *et al. Proc Natl Acad Sci USA* 1990; **87**: 5168.

156 Chan CK, DiMango TM, Chen KXQ, Norris J, Fleming GR. *Proc Natl Acad Sci USA* 1991; **88**: 11202.

157 Dressler K, Umlauf E, Schmidt S *et al. Chem Phys Lett* 1991; **193**: 270.

158 Vos MH, Rappaport F, Malbry JC, Breton J, Martin JL. *Nature* 1993; **363**: 320.

159 Wynne K, Galli C, Hochstrasser RM. *J Chem Phys* 1994; **100**: 4797.

160 Reid PJ, Silva C, Barbara PF, Karki L, Hupp JT. *J Phys Chem* 1995; **99**: 2609.

161 Arnett DC, Vohringer P, Scherer NF. *J Am Chem Soc* 1995; **117**: 12262.

162 Holstein T. *Ann Phys (NY)* 1959; **8**: 343.

163 Mott NF. *The Metal Insulator Transition.* London: Taylor & Francis, 1972.

164 Beenakker CWJ, van Houten H. *Solid State Phys* 1995; **44**: 1.

165 Buttiker M, Landauer R. *Phys Rev Lett* 1982; **49**: 1739.

166 Bountis T, ed. *Proton Transfer in Hydrogen-Bonded Systems.* New York: Plenum, 1992.

167 Staib A, Borgis D, Hynes JT. *J Chem Phys* 1995; **102**: 2487.

168 Aviram A, Seiden PE, Ratner MA. In: Carter F, ed. *Molecular Electronic Devices.* New York: Dekker, 1981.

169 Todd MD, Todd RH, Mikkelsen KV. *J Mol Struct* 1994; **314**: 39.

170 Stockli A, Meier BH, Kreig R, Meyer R, Ernst RR. *J Chem Phys* 1990; **93**: 1502.

171 Rösch N, Ratner MA. *J Chem Phys* 1974; **61**: 3344.

172 Hammes-Schiffer S, Tully JC. *J Chem Phys* 1994; **101**: 4657.

173 Morillo M, Cukier R. *J Chem Phys* 1993; **98**: 4548.

174 Barbara PF, Jarzeba W. *Acc Chem Res* 1985; **21**: 195.

175 Fleming GR. *Chemical Applications of Ultrafast Spectroscopy.* Oxford: Clarendon Press, 1986.

176 Schatz GC, Ratner MA. *Quantum Mechanics in Chemistry.* Englewood Cliffs, NJ: Prentice-Hall, 1994.

177 Hush NS. *Prog Inorg Chem* 1967; **8**: 391.

178 Nelsen SF, Adamus J, Wolff J. *J Am Chem Soc* 1994; **116**: 1589.

179 Hupp JT, Neyhart GA, Meyer TJ. *J Am Chem Soc* 1986; **108**: 5349.

180 Bixon M, Jortner J, Verhoeven JW. *J Am Chem Soc* 1994; **116**: 7349.

181 Ramakrishna MV, Friesner RA. *Israel J Chem* 1993; **33**: 3.

182 Hill NA, Whaley KB. *J Chem Phys* 1994; **100**: 2831.

183 Liu H-J, Hupp JT, Ratner MA. *J Phys Chem* 1996; **100** 12204.

184 Prasad PN, Williams DJ. *Introduction to Nonlinear Optical Effects in Molecules and Polymers.* New York: Wiley, 1991.

185 Messir J, Kajar F, Prasad P, Ulrich D, eds. *Nonlinear Optical Effects in Organic Polymers.* Dordrecht: Kluwer, 1989.

186 Ratner MA, ed. *Int J Quant Chem* Spec. No. on nonlinear optics 1992; **43**: 516.

187 Landauer R. In: Kramer B, Bergmann G, Bruynseraede Y, eds. *Localization, Interaction and Transport Phenomena.* New York: Springer, 1985: 38–50.

188 Datta S. *Electronic Transport in Mesoscopic Systems.* Cambridge: Cambridge University Press, 1995.

189 Samanta MP, Tian W, Datta S, Henderson JI, Kubiak CP. *Phys Rev B* 1996; **53**: 7626.

190 Lundstrom M. *Fundamentals of Carrier Transport.* Reading: Addison-Wesley, 1990.

191 Newns DM. *Phys Rev* 1969; **178**: 1123.

192 Anderson PW. *Phys Rev* 1961; **124**: 41.

193 O'Neil MP, Niemczyk MP, Svec WA, Gosztola P, Gaines GL, Wasielewski MR. *Science* 1992; **257**: 63.

194 Liu ZF, Hashimoto K, Fujishima A. In: Aviram A, ed. *Molecular Electronics — Science and Technology.* Washington, DC: AIP, 1992: 218.

195 Michl J, Bonacic-Koutecky V. *Electronic Aspects of Organic Photochemistry.* New York: Wiley, 1990.

196 Dexter DL. *J Chem Phys* 1953; **21**: 836.

197 Scholes GD, Harcourt RH. *J Chem Phys* 1996; **104**: 5054.

198 Joachim C, Gimzewski JK. *Europhys Lett* 1995; **30**: 409.

199 Joachim C, Müllen K, Roth S, Ratner MA, eds. *Atomic and Molecular Wires*. Dordrecht: Kluwer, in press.

200 Landauer R. *Z Phys* 1987; **368**: 217.

201 McConnell HM. *J Chem Phys* 1961; **35**: 508.

202 Ratner MA. *J Phys Chem* 1990; **94**: 4877.

203 Broo A, Larsson S. *J Phys Chem* 1991; **95**: 4925.

204 Kim K, Jordan KD, Paddon-Row MN. *J Phys Chem* 1994; **98**: 11053.

205 Curtiss LA, Naleway CA, Miller JR. *Chem Phys* 1993; **176**: 387.

206 Mikkelsen KV, Kifkov L, Nar H, Farver O. *Proc Natl Acad Sci* 1993; **90**: 5443.

207 Liu Y-P, Newton MD. *J Phys Chem* 1995; **99**: 11382.

208 Newton MD. *Chem Rev* 1991; **91**: 676.

209 Creutz C. *Prog Inorg Chem* 1983; **30**: 1.

210 Lehn J-M. *Angew Chem Int Ed Engl* 1988; **27**: 90.

211 Wasielewski MR, Johnson DG, Svec WA, Kersey KM, Cragg DE, Minsek DW. In: Norris JR, Meisel D, eds. *Photochemical Energy Conversion*. New York: Elsevier, 1989: 135.

212 Rabe J. In: Petty MC, Bryce MR, Bloor D, eds. *Introduction to Molecular Electronics*. New York: Oxford University Press, 1995: 261.

213 Behm RJ, Garcia N, Rohrer H, eds. *Scanning Tunneling Microscopy and Related Methods*. Dordrecht: Kluwer, 1990.

214 Fujita T, Nakai H, Nakasuji H. *J Chem Phys* 1996; **104**: 2410.

215 Arca M, Mirkin MV, Bard AJ. *J Phys Chem* 1995; **99**: 5040.

216 Basame S, White HS. *J Phys Chem* 1995; **99**: 16430.

217 Ou-Yang H, Marcus RA, Källebring B. *J Chem Phys* 1994; **100**: 7814.

218 Ren J, Whangbo MH, Bengel H, Maganov SG. *J Phys Chem* 1993; **97**: 4767.

219 Mikkelsen KV, Pedersen SV, Lund H, Swanstrom P. *J Phys Chem* 1991; **95**: 8892.

220 Reimers JR, Hush NS. *J Photochem Photobiol* 1994; **A82**: 31.

221 Petrov EG, Tolokh IS, Demidenko AA, Gorbach VV. *Chem Phys* 1995; **193**: 237.

222 Baratoff A, Persson BNJ. *J Vac Sci Tech* 1988; **A6**: 331.

223 Mujica V, Doyen G. *Int J Quant Chem* 1993; **527**: 687.

224 Miller JR, Beitz JV. *J Chem Phys* 1981; **74**: 6746.

225 Kemp M, Roitberg A, Mujica V, Ratner MA. *Science*, submitted.

226 Onuchic JM, Beratan DN, Winkler JR, Gray HB. *Annu Rev Phys Chem* 1992; **21**: 349.

227 Hipps KW, Mazur U. *J Phys Chem* 1994; **98**: 5824.

228 Mazur U, Hipps KW. *J Phys Chem* 1994; **98**: 8169.

229 Hamm S, Wachtel H. *J Chem Phys* 1995; **103**: 10689.

230 Skourtis S, Mukamel S. *Chem Phys* 1995; **197**: 367.

231 Felts AK, Pollard WT, Friesner RA. *J Phys Chem* 1995; **99**: 2929.

232 Davis W, Wasielewski M, Ratner MA, Kosloff R. In preparation.

233 Hey R, Schreiber M. *J Chem Phys* 1995; **103**: 10726.

234 Wang Z, Pearlstein RM, Jia Y, Fleming GR, Norris J. *Chem Phys* 1993; **176**: 421.

235 Cho M, Silbey R. *J Chem Phys* 1995; **103**: 595.

236 Coalson RD, Evans DG, Nitzan A. *J Chem Phys* 1994; **101**: 436.

237 Neria E, Nitzan A. *Chem Phys* 1994; **183**: 351.

238 Neria E, Nitzan A. *J Chem Phys* 1993; **99**: 1109.

239 Scherer NF, Jonas D, Fleming GR. *J Chem Phys* 1993; **99**: 153.

240 Sharvin YV. *Sol Phys JETP* 1965; **48**: 984.

241 Landauer R. *Phys Lett* 1981; **85a**: 91.

242 Lang ND. *Phys Rev B* 1995; **52**: 5335.

243 Yazdani A, Eigler DM, Lang ND. *Science* 1996; **272**: 1921.

244 Olesen L, Laegsgaard E, Stensgaard I *et al*. *Phys Rev Lett* 1994; **72**: 2251.

245 Pomerantz M, Aviram A, McCorkle RA, Roland PA, Schrott AG. In: Aviram A, ed. *Molecular Electronics — Science and Technology.* Washington, DC: AIP, 1992: 39.

246 Tersoff J, Hamann DR. *Phys Rev B* 1985; **31**: 805.

247 Rabe J. In: Petty MC, Bryce MR, Bloor D, eds. *Introduction to Molecular Electronics.* New York: Oxford University Press, 1995: 28.

248 Benkataraman V, Flynn GW, Wilbur JL, Folkers JP, Whitesides GM. *J Phys Chem* 1995; **99**: 8684.

249 Bratkovsky AM, Sutton AP, Todorov TN. *Phys Rev B* 1995; **52**: 5036.

250 Meirav V, Kastner MA, Wind SJ. *Phys Rev Lett* 1990; **65**: 771.

251 Landauer R. In: Kramer B, Bergman O, Bruynseraede Y, eds. *Localization, Interaction and Transport Phenomena.* New York: Springer, 1985: 38.

252 Joachim Ch, Sauter P. In: Behm RJ, Garcia N, Rohrer H, eds. *Scanning Tunneling Microscopy and Related Methods Methods.* Dordrecht: Kluwer Academic, 1990.

253 Kurnikov IV, Beratan DN. *J Chem Phys,* in press.

254 Mujica V, Kemp M, Roitberg A, Ratner MA. In: Ludena EV, Vashishta P, Bishop RF, eds. *Condensed Matter Theories II.* Commack: Nova Science, 1996, in press.

255 Su B, Goldman VJ, Cunningham JE. *Phys Rev B* 1990; **46**: 7644.

256 Grabert H, Devoret MH, eds. *Single Charge Tunneling, Coulomb Blockade Phenomena in Nanostructures.* New York: Plenum, 1992.

257 Altschuler BL, Lee PA, Webb RA, eds. *Mesoscopic Phenomena in Solids.* Amsterdam: Elsevier, 1991.

258 Kirk WP, Reed MA, eds. *Nanostructures and Mesoscopic Systems.* New York: Academic, 1991.

259 Korotkov AN, Averin DV, Likharev KK. *Physica B* 1990; **1265**: 927.

260 Ruggiero ST, Barner JBB. *Z Phys B* 1991; **85**: 333.

261 Grabert H. *Z Phys B* 1991; **85**: 319.

262 Kastner MA. *Phys Today* 1993; **Jan**: 24.

263 Reed M. *Sci Am* 1993; **Jan**: 118.

264 Ben-Jacob E, Gefen Y. *Phys Lett* 1985; **108A**: 287.

265 Nejoh H, Ueda M, Aono M. *Jap J Appl Phys* 1993; **32**: 1480.

266 Andres RP, Bein T, Dorogi M *et al. Science* 1996; **272**: 1323.

267 Carter MT, Rowe GK, Richardson JN, Tender GM, Terrill RH, Murray RW. *J Am Chem Soc* 1995; **117**: 2896.

268 Grabar KC, Freeman RG, Hommer MB, Natan MJ. *Anal Chem* 1995; **34**: 735.

269 Kouwenhoven L. *Science* 1996; **271**: 1689.

270 Kemp M, Roitberg A, Mujica V, Wanta T, Ratner MA. *J Phys Chem* 1996; **100**: 8349.

271 Friedel J. *J Phys (Paris)* 1977; **C2**: 38.1.

272 Jortner J. *Ber Bunsenges Phys Chem* 1984; **88**: 188.

273 Bjørnholm S. *Contemp Phys* 1990; **31**: 309.

274 Mie G. *Ann Phys* 1908; **25**: 377.

275 Kubo R. *J Phys Soc Japan* 1962; **17**: 975.

276 Rademann K, Keiser B, Even U, Hensel F. *Phys Rev Lett* 1987; **59**: 2319.

277 Brechignac C, Boyer M, Chauzac P, Delaretaz G, Labastie P, Wöste L. *Phys Rev Lett* 1988; **60**: 275.

278 Webman I, Cohen MH, Jortner J. *Phys Rev B* 1978; **17**: 4555.

279 Schwentner N, Koch EE, Jortner J. *Electronic Excitations in Condensed Rare Gases.* Berlin: Springer Verlag, 1985.

280 Zallen R. *The Physics of Amorphous Solids.* New York: John Wiley, 1983.

281 Donohoe AJ, Robins JL. *J Cryst Growth* 1972; **17**: 70.

282 Becker EW, Klingelhöfer R, Lohse P. *Z Naturforsch* 1962; **17a**: 462.

283 Hagena OF, Obert W. *J Chem Phys* 1962; **56**: 1793.

284 Mackay AL. *Acta Crystallogr* 1962; **15**: 916.

285 Hoare MR. *Adv Chem Phys* 1969; **40**: 49.

286 Farges J, DeFeraudy MF, Raoult B, Torchet G. *J Chem Phys* 1983; **78**: 5067.

287 Smalley RE, Levy DH, Wharton L. *J Chem Phys* 1974; **64**: 3266.

288 Beswick JA, Jortner J. *Chem Phys Lett* 1977; **49**: 13.

289 Gaathon A, Jortner J. *Can J Chem* 1977; **55**: 1801.

290 Proceedings of the Seventh International Symposium on Small Particles and Inorganic Clusters, Kobe, Japan, 1994. *Surface Rev Letters* 1996; **3**: 1–1222.

291 Jena P, Khanna SN, Rao BK, eds. *Physics and Chemistry of Finite Systems: From Clusters to Crystals*. Dordrecht: Kluwer, 1992.

292 Berry SR, Burdett J, Castleman AW, eds. Small Particles and Inorganic Clusters. *Z Phys D, Atoms, Molecules and Clusters* 1993; **26**: 1.

293 Echt O, Recknagel E, eds. Proceedings of the Fifth International Symposium on Small Particles and Inorganic Clusters, Konstanz, 1991. *Z Phys D, Atoms, Molecules and Clusters* 1990; **19**: 20.

294 Habrich W, Fedrigo S, Butlet J, Lindsay DM. *Z Phys D* 1991; **19**: 157.

295 Nishitani R, Kasuya A, Kubota S, Nishina Y. *Z Phys D* 1991; **19**: 333.

296 Cagan CR, Murray CB, Nirmal M, Bawendi MG. *Phys Rev Lett* 1996; **76**: 1517.

297 Hakkinen H, Manninen M. *Phys Rev Lett* 1996; **76**: 1599.

298 Jortner J. *Z Phys D* 1992; **24**: 247.

299 Jortner J, Ben-Horin N. *J Chem Phys* 1993; **98**: 9346.

300 Jortner J. *J Chim Phys* 1995; **92**: 205.

301 Martin TP, Näher U, Schaber H, Zimmerman U. *Phys Rev Lett* 1993; **70**: 3079.

302 de Heer W, Knight W, Chou M, Cohen ML. *Solid State Phys* 1986; **40**: 93.

303 Stringari B, Treiner J. *J Chem Phys* 1987; **87**: 5021.

304 Combariza JE, Kestner NR, Jortner J. *Chem Phys Lett* 1994; **221**: 156.

305 Markovich G, Pollack S, Giniger R, Cheshnovsky O. *Z Phys D* 1993; **26**: 98.

306 Lee GH, Arnold ST, Eaton JG *et al*. *Z Phys D* 1991; **20**: 9.

307 Cheshnovsky O, Giniger R, Markovich G, Makov G, Nitzan A, Jortner J. *J Chem Phys* 1995; **92**: 397.

308 Brus LE. *J Chem Phys* 1984; **80**: 4403.

309 Brus LE. *J Chem Phys* 1986; **90**: 2555.

310 Efros Al L, Efros AL. *Sov Phys Semiconductors* 1982; **16**: 772.

311 Ramakrishna MV, Frieser RA. *Phys Rev Lett* 1991; **67**: 629.

312 Möller T. *Z Phys D* 1991; **20**: 1.

313 Schwentner N, Koch E, Jortner J. *Electronic Excitations in Condensed Rare Gases*. Heidelberg: Springer, 1985.

314 Jortner J. *Z Phys Chem* 1994; **184**: 283.

315 Katsume T, Hiramoto M, Yokoyama M. *Appl Phys Lett* 1994; **64**: 2546.

316 Frederiksen P, Bjornholm T, Madsen HG, Bechgaard K. *J Mater Chem* 1994; **4**: 676.

317 Crossley MJ. *J Am Chem Soc* 1987; **109**: 2336.

318 Bosch E, Moreno M, Luch JM, Bertran J. *J Chem Phys* 1990; **93**: 5685.

319 Wasielewski MR, Niemczyk MR, Johnson DG, Svec WA, Minsek DW. *Tetrahedron* 1989; **45**: 4785.

320 Guay J, Diaz A, Wu R, Tour JM. *J Am Chem Soc* 1993; **115**: 1869.

321 Spangler CW, Me MQ. *Polymer Preprints* 1994; **35**: 316.

322 Kaszynski P, Friedli AC, Michl J. *J Am Chem Soc* 1992; **14**: 601.

323 Verhoeven JW, Wegewijs B, Kroon J, Rettschnick RPH, Paddon-Row MN, Oliver AM. *J Photochem Photobiol* 1994; **83**: 161.

324 Closs GL, Miller JR. *Science* 1988; **240**: 440.

325 Holl N, Port H, Wolf HC, Strobel H, Effenberger F. *Chem Phys* 1993; **176**: 215.

326 Klein JD, Herrick RD, Palmer D, Sailor MJ, Brumlik CJ, Martin CR. *Chem Mater* 1993; **5**: 902.

327 Gilbert A, Baggott J. *Essentials of Molecular Photochemistry*. London: Blackwell, 1991.

328 Cheshnovsky O, Taylor KJ, Conceicao J, Smalley RE. *Phy Rev Lett* 1990; **64**: 1785.

2 Molecular Control of Electron and Hole Transfer Processes: Electronic Structure Theory and Applications

M.D. NEWTON* and R.J. CAVE†

*Department of Chemistry, Brookhaven National Laboratory, Box 5000, Upton, NY 11973, USA,

†Department of Chemistry, Harvey Mudd College, Claremont, CA 91711, USA

1 Introduction

Recent decades have seen remarkable advances in microscopic understanding of electron transfer (ET) processess in widely ranging contexts, including solid-state, liquid solution, and complex biological assemblies [1–6]. This understanding is reflected in theoretical models of rapidly increasing sophistication [7–14], which relate the dynamical and kinetic behaviour of ET processes to the underlying structural, energetic, and electronic properties of the reactive systems. Typically, one identifies, and treats quantum mechanically, local molecular donor (D) and acceptor (A) sites, and then formulates the manner in which the effective coupling facilitating the ET process is mediated by the energetic and electronic features of the intervening medium [the 'bridge' (B)] as well as the surrounding environment (Fig. 2.1). The energetics and dynamics associated with activation are treated with either classical or quantum mechanical models. The theoretical models play an important dual role, on the one hand leading (in conjuction with modern computing power) to realistic computational implementation, and on the other, allowing analysis of the results of such calculations (as well as those from experiment) in terms of compact predictive models grounded in

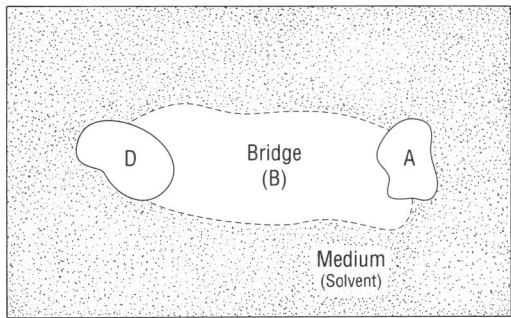

$$\psi_i \iff \{ DBA \} \quad \text{(initial state)}$$

$$\psi_f \iff \{ D^+BA^- \} \quad \text{(final state)}$$

Figure 2.1 Generic electron transfer (ET) system, DBA/D$^+$BA$^-$, comprising local donor (D) and acceptor (A) sites, the intervening bridge (B), and the surrounding medium (or solvent). In the two-state approximation (TSA), the kinetics may be modelled in terms of initial (ψ_i) and final (ψ_f) state wavefunctions, in which the transferring charge is localized primarily on the D and A sites, respectively.

simple concepts of chemical structure and bonding. The power of the current armament of theoretical tools for confronting the challenges posed by ET dynamics is underscored by their generic applicability, e.g. to thermal, optical, and photoinitiated processes, both homogeneous and interfacial (e.g. at electrodes).

The continuing challenge, of course, is to convert the rapidly accumulating mechanistic information about ET kinetics (often representable in terms of simple rate constants) into precise tools for fine-tuned control of the kinetics and design of molecular-based systems which meet specified ET characteristics. Progress toward these latter objectives is yielding increasingly productive contact with the world of micro-electronic devices — i.e. molecular electronics [15,16], the guiding focus of the current volume. For some time, the literature has offered inspiring examples of the fruitful application of orbital or other quantum chemical concepts in formulating idealized models for devices such as rectifiers, switches and registers, and in general articulating the concept of 'molecular wires' [17–27]. The close relationship between scanning tunnelling microscopy (STM) and chemical electron transfer processes has been noted in a number of recent papers [20,24,28]. In the biological arena, one of nature's premier 'devices', photosynthesis, has stimulated intensive theoretical and computational studies in recent years [29–32].

With the above background in mind, the primary goal of this chapter is to report recent advances in the modelling, calculation, and analysis of electronic coupling in complex molecular aggregates, thereby allowing an assessment of current progress toward the goal of molecular-level control and design. The control of electron transfer kinetics (i.e. enhancing desired processes, while inhibiting others) involves, of course, system energetics (especially activation and reorganization energies) as well as electronic coupling, which is most directly relevant only after the system has reached the appropriate point (or region) along the 'reaction coordinate'. Nevertheless, to focus the discussion in this chapter, we will consider such energetics, and the associated molecular and solvent coordinates which control them, only to the extent that they bear on the analysis of the electronic coupling.

In the following sections we will first discuss the formulation of basic ET models, including the definition of initial and final states, the role of orbitals and one-particle models in a many-electron context, the utility of various effective Hamiltonians, and the role of vibronic [10,33] as well as purely electronic effects. With these theoretical tools in hand, we will then examine very recent applications to complex molecular systems using the techniques of computational quantum chemistry, followed by detailed analysis of the numerical results. We will then conclude with some comments regarding the current 'state of the art' and remaining challenges.

2   Theoretical foundation: preliminaries

Before launching into a detailed analysis of the electronic aspects of long-range D–A coupling, it is appropriate to establish a kinetic context and introduce some concepts and distinctions crucial to the subsequent discussion. We are interested in D–A coupling primarily as a controlling factor in ET kinetics, although it also plays a central role in a number of related processes, including photoelectron and electron transmission spectroscopy [34], magnetic exchange [32,35], and energy transfer [36,37].

2.1 Kinetic context

A convenient point of departure is provided by the standard non-adiabatic transition state theory (TST) rate constant expression [2,3].

$$k_{ET}^{TST} = (2\pi/h)(T_{if})^2(\text{FCWD}), \tag{2.1}$$

where the 'transfer integral' T_{if} is the *effective* electronic Hamiltonian matrix element coupling the initial (ψ_i) and final (ψ_f) states, which differ, respectively, by having an electron localized primarily at the D and A sites, as illustrated schematically in Fig. 2.1. The Franck–Condon weighted density of states (FCWD) reflects the influence of all of the nuclear (inertial) modes of the system, generally represented in terms of effective normal coordinates (Q_{v_i} and Q_{w_f}, respectively, for the initial and final states) and the associated quantum-mechanical Franck–Condon factors. For sufficiently high temperature (where $h\nu \ll k_B T$ for all Q), FCWD takes on a limiting form proportional to the classical Arrhenius activation factor,

$$\text{FCWD} \propto \exp(-E_a/k_B T) \tag{2.2}$$

where the effective activation energy (actually a free energy) for the case of linear coupling to the medium is given in terms of the 'reorganization energy' (E_r) and the free energy change (ΔG^0) for the process (see Fig. 2.2),

$$E_a = (E_r + \Delta G^0)^2/4E_r. \tag{2.3}$$

At lower temperatures, where quantal effects become appreciable for the high-frequency modes, Eqn 2.1 may be recast as a superposition of vibronic state-to-state processes [10],

$$k_{ET}^{TST} = (2\pi/h) \sum_{v_i w_f} (P_{v_i})(V_{v_i w_f})^2(\text{FCWD}')_{v_i w_f} \tag{2.4a}$$

where P_{v_i} is the normalized distribution of initial vibronic states (typically in terms of

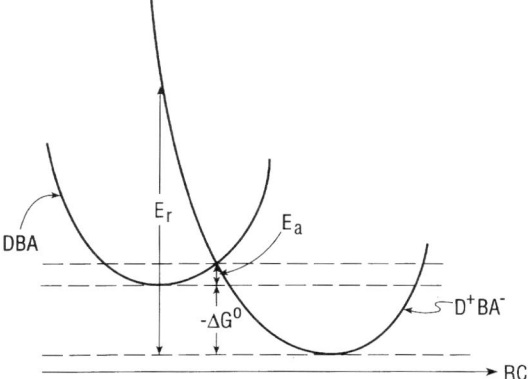

Figure 2.2 Effective energy profiles along the reaction coordinate (RC) for the initial and final diabatic states, indicating the reorganization energy (E_r), activation energy (E_a), and reaction driving force ($-\Delta G^0$). In a linear system, with parabolic profiles of equal curvature, the sum of the vertical energy gap at the equilibrium configuration for the initial state (DBA) and $-\Delta G^0$ is equal to E_r [2,3].

Boltzmann factors), where the vibronic factor $(V_{v_i w_f})^2$ is given by

$$(V_{v_i w_f})^2 = (T_{if})^2 (S_{v_i w_f})^2, \tag{2.4b}$$

and $(S_{v_i w_f})^2$ is a vibrational Franck–Condon factor (i.e. the square of the corresponding vibrational overlap integral). The implicit relationship between the quantities $(\text{FCWD})_{v_i w_f}$, and FCWD, as obtained from comparison of Eqns 2.1 and 2.4, is given by

$$\text{FCWD} = \sum_{v_i w_f} P_{v_i} (S_{v_i w_f})^2 (\text{FCWD}')_{v_i w_f}. \tag{2.5}$$

In Eqns 2.1, 2.4b, and 2.5 we have employed the Condon approximation [38], factoring T_{if} out of the full vibronic matrix element, with the understanding that T_{if} is to be evaluated for values of the nuclear coordinates pertinent to the configuration or range of configurations of the system in which the primary electronic transition occurs. The validity of the Condon factorization depends, among other things, on the extent to which T_{if} varies with the coordinates Q, a topic which we return to when discussing computational results in Section 5. The coordinates of interest in this connection include the reaction coordinate (RC, as in Fig. 2.2), as well as others such as conformational modes of the DBA system (Fig. 2.1). The influence of fluctuations in these coordinates (and hence in the magnitude of T_{if}) on the overall kinetics depends in detail on the relationship between the timescale for such fluctuations and the timescales of the other dynamical processes [7,10,39].

The non-adiabatic expressions for k_{ET} (Eqns 2.1 and 2.4a) will be valid provided that the rate-determining step is the primary 'electron hop' (whose probability is controlled by T_{if}) in contrast to possible alternative dynamical bottlenecks associated with the various inertial degrees of freedom [7,10]. Within the TST regime at high temperature, the Landau–Zener (LZ) model [40a,b] in the case of a harmonic system (see Fig. 2.2) shows that the non-adiabatic limit is valid when the following inequality is obeyed [2]:

$$(T_{if})^2 (\pi^{3/2})/(h\omega_{eff})(k_B T E_r)^{1/2} \ll 1 \tag{2.6}$$

where v_{eff} is the effective frequency associated with vibrational motion along the reaction coordinate. As the coupling T_{if} increases beyond the regime defined by inequality 2.6, not only must one depart from the non-adiabatic limit, but the TST framework itself may become invalid as inertial contributions (e.g. solvent dynamics) begin to play a role in the rate-determining process [7,10], leading even to the possibility that a simple rate constant may not be adequate to account for the kinetics [7]. Furthermore, the simple expression for E_a (Eqn 2.3), based on the harmonic weak coupling (small T_{if}) limit (Fig. 2.2), must be modified to reflect the consequences of avoided crossing on the height and shape of the barrier [2,3,14].

At lower temperature, where nuclear quantal effects may be significant, analogues of the preceding dynamical analysis may be carried out at the vibronic level (i.e. with $V_{v_i w_f}$ (Eqn 2.4b) replacing T_{if}) and with E_r limited to the contribution to reorganization energy from the low-frequency modes [10].

Aside from the initial and final states discussed above, additional complications may arise due to the presence of low-lying intermediate states (such as those associated with the intervening bridge (Fig. 2.1)) and may be treated either in a high-temperature or low-temperature framework [41,42].

In Fig. 2.3 we illustrate the diversity of ET processes of chemical interest. These include:

1 thermally activated ET proceeding through the crossing region (Fig. 2.3a);

2 optical ET (often designated as 'intervalence transfer' in the case of binuclear mixed valence transition metal complexes), occurring vertically from the equilibrium configuration of the initial state (Fig. 2.3a); and

3 photoinitiated ET (following initial photoexcitation, typically to a locally excited non-charge transfer state), involving charge separation (CS), followed by subsequent photochemistry (not shown) occurring in competition with charge recombination (CR) back to the ground state (Fig. 2.3b).

2.2   *What particles (and how many) are actually transferred in 'electron transfer' processes?*

In a process of the type (Fig. 2.1)

$$DBA \rightarrow D^+BA^- \tag{2.7}$$

a single electron is nominally transferred from D to A (analogous processes involving multiple electron transfer are also possible). Analysis of the D–A coupling, based on a detailed examination of the electronic manifolds of B, however, reveals that in general the process 2.7 is more properly viewed as a superposition of a number of charge transfer processes ('pathways') occurring in parallel, including transfer of an electron in one direction and transfer of a 'hole' (an electron-deficient site) in the opposite direction, as well as more complex processes involving both electrons and holes [13,14,43,44]. It is essential to recognize the *physical* (as opposed to *semantic*) basis of these distinctions. Superposed on the net transfer of charge between D and A, one may also expect local response (i.e. polarization relaxation) of the system to the oxidation and reduction occurring, respectively, at D and A [14,45]. We note that 'system' in this connection may be broadly understood to include the 'solvent' as well as the DBA 'solute'.

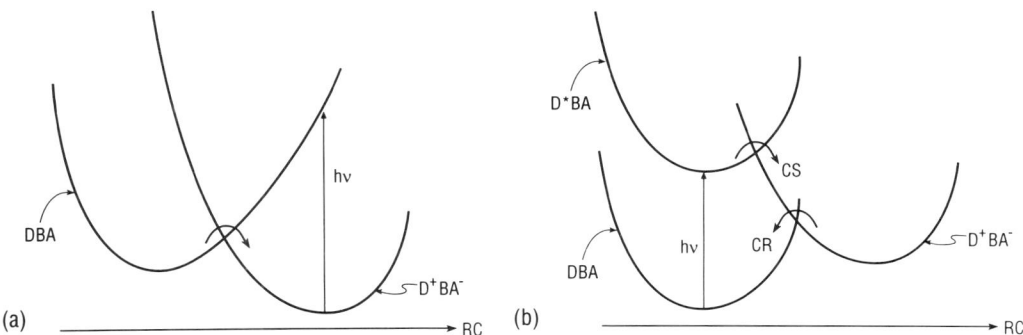

Figure 2.3 (a) Schematic representation of optical and thermal electron transfer (ET), corresponding, respectively, to the vertical transition with excitation energy *h*v and passage through the transition state (or crossing) region. In experimental studies, the thermodynamically stable state, which is the final state in the thermal ET process, generally serves as the initial state in the corresponding optical process. (b) Sequence of photoinitiated electron transfer: charge separation (CS) from a locally excited state, followed by charge recombination (CR) back to the ground state. The CS, CR notation is generally limited to cases where the D and A sites are initially charge neutral (as drawn).

With the foregoing multiparticle perspective in mind, it is of great interest to determine from combined analysis and calculation the extent to which the many-electron quantity T_{if} may be cast as an effective one-particle (i.e. orbital) matrix element, T_{DA}, where DA in this context denotes the 'donor' and 'acceptor' orbitals, confined primarily to the D and A sites, respectively. In particular, it has been found in a number of cases [14] that the many-electron (relaxation) effects may be captured by electronic Franck–Condon-type overlap integrals, S_{if}^{el}, with magnitude $\gtrsim 0.9$:

$$T_{if} = S_{if}^{el} T_{DA} \tag{2.8}$$

where T_{DA} may be viewed as the matrix element of an effective one-particle transition operator. In addition to 'solute' (i.e. DBA) contributions to S_{if}^{el}, similar effects due to solvent have been reported [9].

It still remains to assess the contributions of electron and hole processes to T_{DA} in various cases of interest, as discussed in Sections 3 and 5. For simplicity we shall use the expression 'electron transfer' generically in the remainder of the chapter, noting the electron/hole distinction wherever appropriate.

2.3   *The nature of 'electron tunnelling'*

The D–A coupling embodied in T_{if} (or T_{DA}) is often described as an effective 'electron tunnelling' process. To focus the discussion we will consider specifically the dependence of T_{if} on the D–A separation distance (r_{DA}) and compare the predictions of two quite distinct models, each of which yields exponential decay:

$$T_{if}(r_{DA}) \propto \exp(-(\beta/2)r_{DA}) \tag{2.9}$$

(the decay coefficient is defined so that T_{if}^2 in Eqn 2.1 decays with coefficient β). The phenomenological Gamow (WKB) model [46] is based on tunnelling through an electronically homogeneous medium represented by an effective one-dimensional rectangular barrier of height Δ^G and width r_{DA}:

$$\beta^G = 2(2m_e \Delta^G)^{1/2}/\hbar \tag{2.10}$$

where m_e is the mass (actual or effective) of the tunnelling electron.

Most modern treatments of bridge-mediated electronic coupling do not involve tunnelling in the literal sense of the Gamow approach and employ electronically inhomogeneous superexchange models [47,48], which take detailed account of the electronic structure of the DBA system (Fig. 2.1). In implementing these models one has the choice of how to subdivide the bridge into subunits, B_j. The optimal choice involves various tradeoffs, including chemical transferability (favouring small subunits), compactness of representation when it comes to synthesizing the overall coupling, T_{if} (favouring larger subunits), and theoretical considerations such as the applicability of perturbation theory (here the tradeoff is complex, but often favours larger subunits [49,50]). As a simple example for illustration, we adopt the McConnell model [47] based on a bridge consisting of a homologous sequence of units, each of which is represented by a single orbital with energy Δ relative to the common energy of the D and A orbitals, and with nearest-neighbour (NN) bridge orbitals coupled by the 'hopping integral' t. From this model, for $|t/\Delta| \ll 1$ we obtain

$$\beta^{SE} = -2(\ln|t/\Delta|)/(\Delta r) \tag{2.11}$$

where Δr is the width of one subunit.

In comparison with the result from the homogeneous model (Eqn 2.10), β^{SE} reflects the molecular and electronic structure of B, not only through the hopping integral t but also Δ and Δr, since the values of all the parameters depend on the specific choice of subunit (we also note, in contrast to Eqn 2.10, that in Eqn 2.11 the electron mass is implicit in t and Δ). Another important distinction lies in the role of gap Δ in Eqns 2.10 and 2.11. In the latter case, Δ (for a particular bridge subunit) is in principle (and in some cases in practice) a spectroscopic observable, whereas the phenomenological Δ^G, which represents the full bridge in Eqn 2.10, is not in general directly accessible as an observable. Effective Δ^G values may, of course, be inferred from Eqn 2.10 on the basis of experimental β values, but attempts to use experimentally determined energy gaps (e.g. from polarography) frequently lead to exaggerated β magnitudes [48]. Equation 2.10 is more usefully applied to the direct or 'through-space' D–A coupling when no bridge is present. In this case, Δ^G may be identified with the relevant ionization potential of D [28].

In Section 3 we introduce generalized superexchange models in which bridge units may have several orbitals (both occupied and unoccupied). The consequences of departing from the weak perturbation limit ($|t/\Delta| \ll 1$) are also explored, including the approach to the opposite limit, in which intermediate charge transfer states involving the bridge approach resonance with D and A levels. The results of detailed quantum mechanical calculations are analysed (Section 5) in terms of these models, and the extent to which complex DBA systems display exponential distance dependence is assessed.

2.4   *Influence of medium on T_{if}*

While medium effects will not be dealt with here in great detail, it is worthwhile summarizing the distinct types of influence which they might exert on T_{if} magnitudes (over and above the modest many-electron Franck–Condon effect noted above). By medium we refer to the environment of the DBA complex (Fig. 2.1). Typically, it will be a polar solvent. Solvent molecules may, of course, constitute part of the 'bridge' (as, e.g., in the case of solvent-separated ion pairs [51]), which does not necessarily comprise a complete sequence of covalent linkages between D and A.

The influence of a polar medium is exerted primarily by its electrostatic field, either long range (as representable by a continuum or molecular-level model) or short range (e.g. via specific hydrogen bonding). The field (which varies with fluctuations of the medium) can modulate the tails of the D and A orbitals either directly, by controlling the degree of radial localization of the orbitals, or indirectly, by modifying the energy gaps (Δ) which control superexchange coupling. In practive, such effects may be quite small [52]. In addition to these electrostatic effects, the electronic manifold of the medium in the immediate vicinity of the bridge may provide superexchange pathways which substantially affect the overall magnitude of T_{if}, thus in effect serving to expand the size of the bridge. The operational choice of the 'bridge' is guided by the objective of including all sites for which direct orbital participation has an appreciable effect on T_{if}.

In some cases, of course, local D and A groups may be defined so as to be in contact (as in contact ion pairs), thus eliminating the need for a 'bridge' altogether.

3   Theoretical models for T_{if} in the two-state approximation

We now turn to the task of formulating models for the initial and final states in an ET process and the effective operator which couples them in the transfer matrix element T_{if} (Eqn 2.1). At the outset, we shall confine our attention to the so-called two-state approximation (TSA), in which the dynamics of ET is assumed to be adequately accounted for by the two-component space (denoted as the D–A space) spanned by states in which the transferring electron is primarily confined to the D and A sites [53,54]. (When one or both of the two states are degenerate or near degenerate, a somewhat more elaborate 'two-level' model must be adopted [14].) In general, the validity of the TSA rests on the requirement that the gap separating the intermediate ET states involving the bridge from those of the D–A space is large relative to the strength of coupling of D and A to the bridge. The large gap condition may be relaxed if the latter coupling is small relative to the band width of the intermediate bridge states [53]. While the basis set for representing the D–A space is to a large extent arbitrary, convenience generally dictates the choice of either the appropriate adiabatic states (i.e. which diagonalize the electronic Hamiltonian of the system), denoted ψ_1 and ψ_2, or the 'diabatic' states ψ_i and ψ_f, related to ψ_1 and ψ_2 by a unitary transformation,

$$\psi_1 = \cos\theta\ \psi_i + \sin\theta\ \psi_f \tag{2.12a}$$

$$\psi_2 = -\sin\theta\ \psi_i + \cos\theta\ \psi_f, \tag{2.12b}$$

defined operationally so as to correspond optimally to the actual (non-stationary) states presumed to be involved in the dynamical process underlying the ET kinetics.

For ET initiated in the ground state, ψ_1 and ψ_2 will generally be the two lowest energy adiabatic states, whereas higher energy states will be involved in photoinitiated ET (Fig. 2.3). We emphasize that the definition of ψ_i and ψ_f is never unique, and the utility of a given prescription must ultimately rest on the results of detailed applications, which we consider below. Clearly, in an ET process ψ_i and ψ_f are designed to correspond to charge-localized valence bond structures. Such structures, aside from their chemically intuitive appeal, also have the advantage that their coupling is dominated by the electronic Hamiltonian, with only minor non-Born Oppenheimer coupling (i.e. from the nuclear momentum operator) expected [2].

In thermally activated ET we are interested in the electronic states at the transistion state (TS). When the system is at equilibrium in either the initial or final state (where D and A are well out of resonance), the diabatic states can be taken as essentially the same as their adiabatic counterparts. When the system with weakly coupled D and A is suddenly carried into the TS by a fluctuation, we adopt the picture that the system remains in the (now non-stationary) ψ_i state until it dynamically 'tunnels' to ψ_f. In the TS configuration, where D and A are essentially in resonance, the adiabatic states would depart sharply from ψ_i and ψ_f (due to delocalization associated with the resonant D and A sites). However, in the kinetic model based on TST, the TS does not last long enough for the possibility of adiabatic (stationary) states to be relevant, and ψ_f is assumed to decay irreversibly to its equilibrium configuration [2]. The required resonance of D and A is a statement of the Franck–Condon control of thermally activated ET [2,3,14]; i.e. at the TS

$$\langle \psi_i H^{el} \psi_i \rangle = \langle \psi_f H^{el} \psi_f \rangle \tag{2.13}$$

and this electronic energy matching will be central to most of the models for T_{if} given below. Of course, vibronic effects allow some departure from strict electronic energy matching in thermal processes. In optical ET, the photon energy balances the electronic mismatch associated with vertical excitation from equilibrium. Here, as noted above, the distinction between adiabatic and diabatic states is likely to be minor (for a possible exception, see [55]) and a different formulation of the effective coupling, T_{if}, is required [14], as discussed in Section 4.

Returning to the case of thermal ET, we now define the effective coupling T_{if} as [14]

$$T_{if} = (E_1 - E_2)/2 = (\langle \psi_1 H^{el} \psi_1 \rangle - \langle \psi_2 H^{el} \psi_2 \rangle)/2 \tag{2.14}$$

where E_1 (E_2) is the eigenvalue associated with ψ_1 (ψ_2) in the TS. The diabatic energy matching at the TS (Eqn 2.13) then implies (via Eqn 2.12, where we now have 50–50 mixing or $\cos \theta = \sin \theta = 1/\sqrt{2}$),

$$T_{if} = \langle \psi_i H^{el} \psi_f \rangle. \tag{2.15}$$

The sign of T_{if} is a physical observable, but is contingent on the phase conventions entailed in the definitions of ψ_i and ψ_f (and hence via Eqn 2.12, also ψ_1 and ψ_2), as discussed in Section 5 [14,43].

We now consider various approximate methods of evaluating T_{if}, using either the adiabatic (Eqn 2.14) or diabatic (Eqn 2.15) representation.

3.1 *Superexchange via perturbation theory*

The diabatic states ψ_i and ψ_f as defined in the TSA have the charge involved in the ET process localized predominantly on the D and A sites, respectively, but with tails extending into the bridge region. It is the overlap of these tails which is usually crucial in determining the magnitude of T_{if}. This role of the electronic manifolds of the bridge in mediating the long-range coupling of D and A sites is given precise form in superexchange theories, which were introduced in Section 2.3 and which we now examine further in the remainder of the chapter. A detailed understanding of superexchange coupling is achieved by introducing an auxiliary (zero-order) basis of diabatic states (taken generally as orthonormal), which by construction have the transferring charge *entirely* localized on the various sites of the system (D, A, and the 'units' of B, as discussed in Section 2). We denote these states as

$$\psi_D^0 \equiv \psi_i^0 \tag{2.16a}$$

$$\psi_A^0 \equiv \psi_f^0 \tag{2.16b}$$

$$\psi_{B_j}^0 \ (j = 1, n) \tag{2.16c}$$

where n is the total number of intermediate bridge states (particular choices of bridge units, m in number, with $m \leq n$, are presented below). These states, which we denote collectively as $\{\psi_k^0\}$, may be considered as eigenfunctions of some suitable 0th order Hamiltonian with zero-order energies E_k^0:

$$H^{\mathrm{el}} = (H^0)^{\mathrm{el}} + V^{\mathrm{el}} \tag{2.17a}$$

$$(H^0)^{\mathrm{el}}\psi_k^0 = E_k^0\psi_k^0 \tag{2.17b}$$

and with vanishing first-order energies (i.e. $V_{kk}^{\mathrm{el}} \equiv \langle\psi_k^0 V^{\mathrm{el}}\psi_k^0\rangle = 0$).

Equation 2.15, which involves H^{el} and the 'full' diabatic states ψ_i and ψ_f (i.e. 'dressed' with their long-range tails), may now be recast in the equivalent form,

$$T_{\mathrm{if}} = \langle\psi_D^0|T^{\mathrm{eff}}|\psi_A^0\rangle \tag{2.18}$$

which implicitly defines the effective transition operator, T^{eff}.

3.1.1   RESULTS SECOND ORDER IN COUPLING TO THE BRIDGE

We first adopt a 'dynamical' point of view and evaluate T_{if} as given by Eqn 2.18, with T^{eff} taken as the 'transition operator' from scattering theory [12].

$$T(E) = V^{\mathrm{el}} + V^{\mathrm{el}}G(E)V^{\mathrm{el}} \tag{2.19a}$$

where $G(E)$ is the electronic Green function

$$G(E) = (E - H^{\mathrm{el}} + i\varepsilon)^{-1} \tag{2.19b}$$

(in the remainder of the chapter we suppress the el superscript).

The Green function depends on the continuous variable E, the 'tunnelling energy' [12,54], and a positive infinitesimal constant ε, for which the limit $\varepsilon \to 0$ is ultimately taken. In the general case, which could accommodate vibronic effects and the influence of low-lying bridge states beyond the simple TSA defined above, the full Green function (including self-energy contributions) should be employed, and ε may play a crucial role [56]. In the TSA, however, where the D and A sites are relatively weakly coupled to the bridge, one need only consider the bridge Green function, $G_B(E)$, which operates exclusively on the space defined by the $\psi_{B_j}^0$ (see Eqn 2.16c),

$$G_B(E) = (E - H_B + i\varepsilon)^{-1}, \tag{2.20}$$

where H_B may be obtained from H using the projection operator for the B space ($\{\psi_{B_j}^0\}$) and where ε may be effectively set to zero. Switching to a matrix representation, we approximate Eqn 2.19a as

$$T_{\mathrm{if}}(E) = V_{\mathrm{if}} + \sum_{j,k}^n V_{\mathrm{i}j}(G_B(E))_{jk}V_{k\mathrm{f}} \tag{2.21}$$

where $V_{\mathrm{if}} \equiv \langle\psi_i^0 V\psi_f^0\rangle$, etc., and the double sum is over the bridge states $\{\psi_{B_j}^0\}$. The optimal choice of the tunnelling energy has received detailed attention in the literature [54]. When the $\{\psi_{B_j}^0\}$ are well separated energetically from ψ_D^0 and ψ_A^0, it is adequate to set $E = E_{D/A}^0$, the mean value of E_D^0 and E_A^0 (in the general case, when the D and A sites of the DBA system are not symmetry-equivalent, the condition given by Eqn 2.13 does not necessarily correspond to $E_D^0 = E_A^0$ [14,32,57]; see also the results from partitioning theory [11] given below).

Equation 2.21 displays T_{if} as a direct or 'through-space' term and a bridge-mediated superexchange term (often denoted 'through-bond', although formal covalent bonds linking D and A via the bridge are not required [43,48]) which, consistent with the TSA, is lowest (i.e. first) order in D–B (V_{ij}) and D–A (V_{kf}) coupling (i.e. overall second order in coupling to the bridge).

Switching now to a stationary point of view, we note that Eqn 2.21 can be reinterpreted as

$$T_{if} = \langle \psi_D^0 H^{eff}(E) \psi_A^0 \rangle \tag{2.22}$$

where H^{eff} is an effective energy-dependent Hamiltonian which is defined in the ψ_D^0, ψ_A^0 space and which may be formulated by application of the Lowdin partitioning theory [11]. As above, an approximate result (good to second order in coupling to the bridge) is obtained by replacing E with the zero-order value, $E_{D/A}^0$ [53].

Yet another second-order expression for T_{if} is obtained by adopting the adiabatic representation (Eqn 2.14 and evaluating E_1 and E_2 using second-order Rayleigh–Schrodinger perturbation theory (RSPT) with the $\{\psi_k^0\}$ basis, taking the $\psi_{B_j}^0$ as eigenfunctions of H_B (i.e. where V in Eqn 2.17a includes only interactions between the bridge and the D/A sites):

$$T_{if} = \sum_j (V_{ij}V_{jf})/\Delta_j^0 \tag{2.23}$$

where $\Delta_j^0 \equiv E_{D/A}^0 - E_{B_j}^0$ (note that these energy gaps are in general *negative*). This result, derived by McConnell [47], is equivalent to those given by Eqns 2.21 and 2.22, if the same $\psi_{B_j}^0$ basis is employed (i.e. bridge eigenfunctions). In the terminology introduced earlier, Eqn 2.23 represents the superexchange result when the entire bridge is one 'unit' containing all n zero-order states.

It is to be emphasized that aside from the limitation of a second-order treatment of the coupling of D and A to the bridge, all of the formulations of T_{if} in this section can be implemented in a non-perturbative manner [i.e. the $(G_B(E))_{jk}$ elements appearing in Eqn 2.21, or their counterparts in Eqns 2.22 and 2.23, may be treated exactly, either explicitly [58,59], or implicitly [60,61]]. Equations 2.21–2.23 provide upper limits for the magnitude of T_{if} as defined by Eqn 2.14 [53]. We now apply perturbation theory to the treatment of the bridge and generate higher order superexchange expressions.

3.1.2 HIGHER ORDER SUPEREXCHANGE MODELS

Further elaboration of Eqn 2.19a and 2.21 may be achieved by invoking the Dyson equation [56,57],

$$G(E) = G^0(E) + G^0(E)VG(E) \tag{2.24}$$

where

$$G^0(E) = (E - H^0 + i\varepsilon)^{-1}. \tag{2.25}$$

Equation 2.19a together with Eqn 2.24 yields an iterative version of the Lippmann–

Schwinger equation [12]

$$T(E) = V + VG^0(E)T. \tag{2.26}$$

As shown by Ratner [12], successive iteration yields

$$T(E) = V + VG^0(E)V + \ldots + V(G^0(E)V)^{p-1} + \ldots \tag{2.27}$$

where the pth term is pth order in the coupling V. Adopting the matrix representation in the basis $\{\psi_k^0\}$, including only the bridge component of $G^0(E)$ (cf. Eqns 2.19b and 2.20) with $E = E_{D/A}^{(0)}$ and ε set to zero,

$$(G_B^0)_{jk} = \delta_{jk}/\Delta_j^0 \ (j, k = 1, n) \tag{2.28}$$

and truncating Eqn 2.27 at the value of p which yields convergence to the desired tolerance, allows T_{if} to be displayed as the superposition (with constructive or destructive interference) of all superexchange 'pathways' up to order p, subject to the constraint entailed in the use of G_B — i.e. the restriction of first-order dependence on V_{ij} and V_{kf} for $p > 1$, leaving up to $(p - 2)$th order dependence on intra-bridge coupling elements, V_{jk}. Smooth convergence is expected when the ratios V_{jk}/Δ_k^0 are sufficiently small in magnitude compared with unity.

For example, truncating at $p = 3$ yields

$$T_{if} = V_{if} + \sum_j^n V_{ij}V_{jf}/\Delta_j^0 + \sum_{j,k}^n V_{ij}V_{jk}V_{kf}/\Delta_j^0\Delta_k^0. \tag{2.29}$$

Thus the pth order superexchange pathways include all sojourns from D to A involving stops at $p - 2$ virtual intermediate bridge states, including multiple visits ('retracings' or 'back scattering' [49,50,62]) to a given state.

The previous results may be extended to accommodate the case of non-orthogonal $\psi_{B_j}^0$, with overlap matrix elements $S_{jk} \equiv \langle \psi_{B_j} | \psi_{B_k} \rangle$. The corresponding generalization of Eqn 2.21 has been reported in [61]. In general one must also take account of non-orthogonality between the D,A orbitals and those of the bridge (see also [12]).

The higher order results obtained from the 'scattering approach' (e.g. Eqn 2.29) can also be obtained by extending the second-order RSPT result based on the adiabatic splitting (i.e. Eqns 2.14 and 2.23) to the higher order required when the $\{\psi_{B_j}^0\}$ are no longer eigenfunctions of H_B [14,47]. For the case of a linear homologous sequence of n bridge 'units', each with a single state, $\psi_{B_j}^0$, and coupled only by NN interactions, where

$$V_{jk} = t\delta_{jk\pm1} \ (j, k = 1, n) \tag{2.30a}$$

$$V_{i1} = V_{nf} = T \tag{2.30b}$$

$$\Delta_j^0 = \Delta \tag{2.30c}$$

we obtain the celebrated exponential n dependence predicted originally by McConnell [47] (introduced earlier in Section 2.3),

$$T_{if} = (T^2/\Delta)(t/\Delta)^{n-1}. \tag{2.31}$$

This result of McConnell is actually somewhat restricted [14] since it assumes that the two matrix elements of V in Eqn 2.30b have the same sign, and likewise for the elements in Eqn 2.30a. The more general situation, and the possibility of sign alternation with n, are discussed in later sections. Higher order expressions for T_{if} based on the partitioning method (Eqn 2.22) have also been reported [11].

3.2 *Variational models*

T_{if} may be evaluated entirely in terms of variationally determined wavefunctions [14,43], either from the adiabatic splitting (Eqn 2.14), where E_1 and E_2 are evaluated separately, or from Eqn 2.15, where charge-localized wavefunctions (ψ_i and ψ_f) are obtained directly. An interesting feature of this latter approach is that in general the ψ_i and ψ_f so obtained are not orthogonal and thus require use of the more general expression [14]

$$T_{if} = H_{if} - (S_{if})(H_{ff})/(1 - S_{if}^2) \tag{2.32a}$$

$$T_{fi} = H_{fi} - (S_{fi})(H_{ii})/(1 - S_{if}^2) \tag{2.32b}$$

where $S_{if} = \langle \psi_i | \psi_f \rangle$ and where the T matrix is Hermitian ($T_{if} = T_{fi}$) due to the assumed energy degeneracy (Eqn 2.13). Analogous situations in which $S_{if} \neq 0$ arise in some cases where ψ_i and ψ_f are obtained directly using perturbation theory [14].

3.3 *Extensions and embellishments of the McConnell model*

Before proceeding to the analysis of cases involving units with multiple states and simultaneous electron and hole transfer, we pursue in more detail the nature of the McConnell-type models, extending Eqn 2.31 to the case of a heterogeneous bridge, but retaining the NN coupling model with one state per unit:

$$T_{if} = (T_{i1}T_{nf}/\Delta_1^0) \prod_{j=1}^{n-1} (t_{jj+1}/\Delta_{j+1}^0) \tag{2.33}$$

where the coupling elements and energy gaps are straightforward generalizations of Eqn 2.30, and with $n \geqslant 1$.

It is not possible to reach simple conclusions about sign patterns without taking account of the many-electron nature of the states and the details of the orbital interactions [14], a topic which we defer to Section 3.4. However, it is of interest to enquire about the utility of the simple form of Eqn (33) in treating actual bridges of chemical interest. It is known, for example, that non-nearest-neighbour interactions are very important when superexchange is analysed in terms of local (i.e. two-centre) bonding or anti-bonding orbitals [37,49,63,64]. Furthermore, the presence of side chains [62] or the occurrence of 'retracing' or 'back scattering' [62,63] would also be expected to complicate the coupling, thus perhaps greatly reducing the utility of the concept of a pathway of the type exemplified by Eqn 2.33, with its attractive feature of factorization into contributions from each bridge unit. Nevertheless, computational applications of the pathway concept, which exploit the simple form of Eqn 2.33, seem capable of giving useful insight into the nature of long-range D–A coupling in complex molecular assemblies [65], thus indicating the utility of the concept at least in an *effective* sense.

Some progress in rationalizing this situation has been achieved by formulating 'renormalized' t and Δ parameters, which mimic the effects of some of the complications noted above, while allowing the form of Eqn 2.33 to be maintained [58,62,63,66–69]. Nevertheless, the 'intersection' of different pathways, which becomes increasingly important as the degree of connectivity increases (e.g. as in polycyclic bridges), leads to interference effects which require a departure from the 'single effective pathway' picture (Eqn 2.33) [58], as exemplified by the superposition displayed in Eqn 2.29. When it becomes essential to include multiple states on each site, the form of Eqn 2.33 may still be retained in the sense that scalar multiplication is replaced by matrix multiplication, where the scalar T (t) factors become linear (rectangular) arrays and the Δ become diagonal square arrays [11,43].

Finally, we note that an exact implementation of Eqn 2.23 for the homologous linear bridge with NN coupling gives an extension of Eqn 2.31 not limited by the $t/\Delta \ll 1$ constraint [53]:

$$T_{if} = 2(-1)^n (T^2/t) f_n(\alpha) \tag{2.34}$$

where

$$f_n(\alpha) = g(\alpha)/[(\alpha + g(\alpha))^{n+1} - (\alpha - g(\alpha))^{n+1}] \tag{2.35a}$$

$$g(\alpha) = (\alpha^2 - 1)^{1/2} \tag{2.35b}$$

and

$$\alpha = |\Delta/2t|. \tag{2.35c}$$

Equation 2.34 is seen to yield the limiting Eqn 2.31 when $|\alpha| \gg 1$. Remarkably, Eqn 2.35 yields nearly exact exponential n dependence essentially down to $|\alpha| = 1$, while oscillations of T_{if} are observed for $|\alpha| < 1$ [53,70]. A more detailed analysis which included self-energy terms in the bridge Green function (Eqn 2.20), observed a narrow region with $1/n$ falloff near $|\alpha| = 1$ and eliminated singularities in the $|\alpha| < 1$ region [20].

3.4   *Many-particle perspectives*

The many-electron wavefunctions pertaining to a DBA system are conveniently represented in terms of one-electron orbitals. We have already noted (Section 2.2) that the overall many-electron coupling element T_{if} may often be represented to good approximation as an effective one-particle quantity (i.e. T_{DA} in Eqn 2.8), in which the particle is exchanged between ortibals localized on D and A. We now take a more detailed look at the coupling by decomposing it into 'hole' and 'electron' contributions associated, respectively, with the occupied and unoccupied orbitals of B. To illustrate this situation we adopt an independent-particle model for the many-electron Hamiltonian,

$$H = \sum_i h(r_i) \tag{2.36}$$

and employ single-determinant orbital wavefunctions to represent ψ_D^0, ψ_A^0, and $\{\psi_{B_j}^0\}$. Relative to the reference occupied bridge manifold in ψ_D^0 and ψ_A^0, the $\{\psi_{B_j}^0\}$ involve (in a virtual sense) either removal of electrons from the occupied manifold or addition of electrons to the unoccupied manifold. Not surprisingly the picture of hole and electron contributions depends on the choice of bridge units.

3.4.1 CASE OF THE ENTIRE BRIDGE AS ONE UNIT

In this case, which corresponds to the second-order expression given by Eqn 2.23, each bridge eigenfunction $\psi_{B_j}^0$ has one hole or one excess electron. Thus, if we define the occupied and unoccupied orbital manifolds as

$$\phi_j^h \ (j = 1, n_{occ}) \tag{2.37a}$$

$$\phi_j^e \ (j = 1, n_{unocc}) \tag{2.37b}$$

the many-electron matrix elements V_{ij} may be expressed in terms of the orbital matrix elements,

$$T_{xj}^h = \int \phi_x^0 h \phi_j^h \tag{2.38a}$$

$$T_{xj}^e = \int \phi_x^0 h \phi_j^e \tag{2.38b}$$

where $x \equiv D$ or A and ϕ_D^0 and ϕ_A^0 are the localized D and A orbitals. Taking due account of the antisymmetry of a single determinant (i.e. the permutational symmetry) with respect to interchange of electrons [14,71], we may now re-express Eqn 2.23 in the following orbital form [14],

$$T_{DA} = T_{DA}^h + T_{DA}^e \tag{2.39}$$

where

$$T_{DA}^y = \sum_j^{n_y} T_{Dj}^y T_{jA}^y / \Delta_j^y \tag{2.40}$$

and where $y =$ h or e, $n_y = n_{occ}$ or n_{unocc}, and Δ_j^y is the *orbital* energy gap defined analogously to the many-electron *state* energy gap introduced in Eqn 2.23; note that Δ_j^h is *positive*:

$$\Delta_j^h = \varepsilon_{D/A}^0 - \varepsilon_j^h \tag{2.41a}$$

$$\Delta_j^e = \varepsilon_{D/A}^0 - \varepsilon_j^e \tag{2.41b}$$

$$\varepsilon_{D/A}^0 = \varepsilon_D^0 = \varepsilon_A^0. \tag{2.41c}$$

Thus a decomposition based on the one-electron eigenfunctions of the bridge yields a simple additive superposition of hole and electron pathways, although the question as to whether the interference is constructive or destructive is not immediately obvious [14]. An indirect numerical method for achieving the h/e partitioning (Eqn 2.30) when the explicit orbital contributions are not available has been reported recently [44].

3.4.2 ALTERNATIVE LOCAL ORBITAL MODEL

To get more insight into the nature of the h/e interference, we now adopt an alternative partitioning according to which each of m subunits of B has one occupied (ϕ_j^h) and one empty (ϕ_j^e) orbital (i.e. $n = 2m$; see comment after Eqn 2.16). A perturbative treatment of the type used in Section 3.1.2 yields the following pure hole and electron contributions in the local orbital framework [14]:

$$T_{DA}^y = (T_{D1}^y T_{nA}^y / \Delta_1^y) \prod_{j=1}^{m-1} (t_{jj+1}^y / \Delta_{j+1}^y) \ (y = h, e) \tag{2.42}$$

where t_{jj+1} is an NN orbital coupling element within the bridge. For a typical case of interest, namely where the ϕ_j^h and ϕ_j^e are, respectively, local bonding and antibonding sigma orbitals, the phases of the orbitals can be arranged so that all the T^y and t^y elements on the right hand side of Eqn 2.42 for the $y = h$ case are normally expected to be *negative*, whereas the two T^y factors in Eqn 2.42 with $y = e$ have *opposite* signs, while the t^e's are *positive* [14]. The important consequence is that: (i) both T_{DA}^h and T_{DA}^e alternate in sign with m [i.e. the parity rule, with negative sign (the 'normal' case) for m even]; and (ii) T_{DA}^h and T_{DA}^e interfere *constructively*. However, this partitioning based on local orbitals is *not* the same as that based on bridge eigenfunctions (Eqn 2.39). While it may be assumed that h (see Eqn 2.36) does not couple the orbitals on a given unit, coupling is expected between an NN pair, ϕ_j^h and ϕ_{j+1}^e: i.e. the orbital eigenfunctions employed in Section 3.4.1 would involve some mixing of such pairs, and these contributions must be added to the pure hole and electron terms displayed in Eqn 2.42, as illustrated in the next section.

3.4.3   INCLUSION OF HYBRID PATHWAYS

Continuing with the local orbital model of Section 3.4.2, we consider all possible pathways of the 'forward' type (i.e. where an electron moves toward A or a hole moves toward D, with no 'retracings'). To enhance the multiparticle perspective we proceed in the context of explicit single-determinant wavefunctions.

A rich diversity of NN pathways is obtained if the occupations of both orbital sets ($\{\phi_j^e\}$ and $\{\phi_j^h\}$) are allowed to vary, subject to the above assumptions and restrictions. The different types of pathways are illustrated schematically for the case $m = 2$ in Fig. 2.4 [43]. It is of particular interest to identify the minimal number of particles (electrons (e) or holes (h)) necessary to characterize each pathway. Towards this end we define a reference (or 'vacuum') configuration and indicate explicitly only those occupation changes relative to it. The resulting virtual transitions are then of four types: e or h transfer (i.e. the passage of an e or h either to or from one of the bridge orbitals ϕ_j^e or ϕ_j^h, respectively), and the creation (+ eh) or destruction (–eh) of eh pairs. Figure 2.4(a) displays pure e transfer and corresponds to a one-particle process, whereas the hybrid pathways (Fig. 2.4c,d) involve up to three particles (both e and h) relative to the vacuum level. While the process in Fig. 2.4(b) is pure hole transfer (i.e. involving only holes on the bridge sites), it is seen to be of the three-particle type. Since one expects a close correspondence between e and h pathways, the asymmetry of Fig. 2.4(a) (one-particle) vs Fig. 2.4(b) (three-particle) may be surprising, and in fact, the expected isomorphism may be recovered by simply adopting another vacuum level for hole transfer (Fig. 2.5). Thus Figs 2.4(a) and 2.5 bear the expected mirror image one-particle relationship.

While Fig. 2.4(a–d) displays the four basic types of pathways possible for $m = 2$, there are actually a total of six pathways. The other two are variants of Fig. 2.4(c) and 4(d)

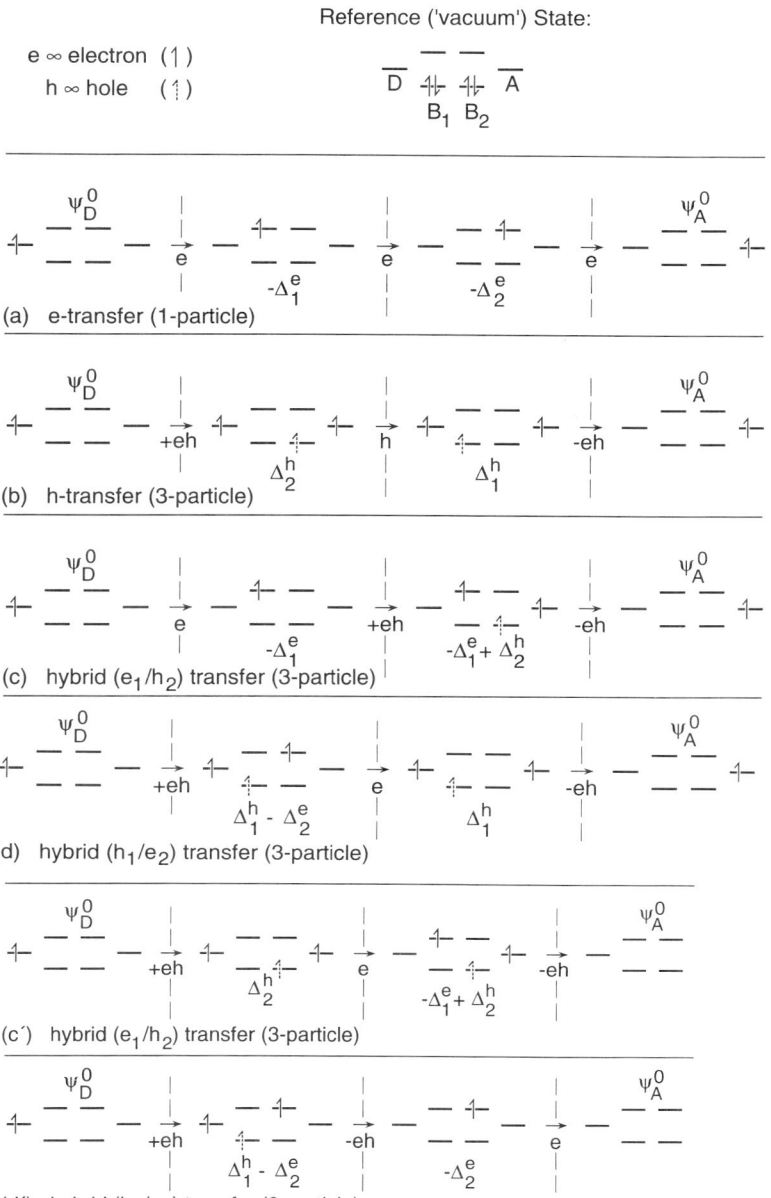

Figure 2.4 Schematic depiction of nearest neighbour (NN) superexchange coupling via two bridge units (B_1 and B_2), each of which has an occupied (ϕ^h) and an unoccupied (ϕ^e) orbital. Orbital occupations relative to the defined reference (or 'vacuum') state are indicated. The excitation energies of the virtual intermediate states relative to the degenerate initial (ψ_D^0) and final (ψ_A^0) state are expressed in terms of the orbital energy differences defined in Eqn 2.41. The three primitive steps ($\rightarrow$) in each 'pathway' correspond to electron (e) or hole (h) transfer, or the creation (+ eh) or destruction (– eh) of an electron/hole pair. Relative to the vacuum, the various states require the specification of at most three particles — the added electron and, in cases b–d, the eh pair. Pathway c' is obtained by interchanging the order of the e and + eh steps in pathway c, and pathway d' is obtained by interchanging the e and – eh steps in pathway d. The set of six processes defines all pathways coupling ψ_D^0 and ψ_A^0 in terms of NN forward-directed (i.e. D to A) steps. In Scheme I of [43], also reprinted with permission, the sign of the 'e$_1$h$_2$' contribution to T_{if} (from pathways c and c') is incorrect due to a typographical error and is corrected in the present work. (Figure 1 of [43], reprinted with permission. Copyright (1993) American Chemical Society.)

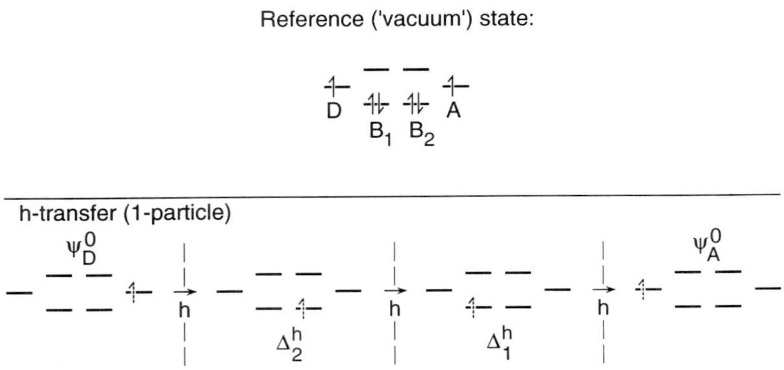

Reference ('vacuum') state:

h-transfer (1-particle)

Figure 2.5 Alternative representation of hole transfer requiring the specification of only a single particle (h). See Fig. 2.4 caption for definition of symbols. (Figure 2 of [43], reprinted with permission. Copyright (1993) American Chemical Society.)

obtained, respectively, by interchanging the first and second virtual transfers in Fig. 2.4(c) and Fig. 2.4(c′) and the second and third transfers in Fig. 2.4(d) and Fig. 2.4(d′). Figures 2.4 and 2.5 include the energy gaps associated with the intermediate states ψ_j^0 [the state and orbital gaps are the same, aside from the sign difference in the case of holes (see Eqn 2.41), in view of the independent-particle model adopted here].

The various contributions to T_{DA} are assembled in Scheme 2.1, where once again proper account is taken of the permutation symmetry of the single-determinant wavefunctions ([43]).

3.4.4   SCHEME 2.1

$$T_{DA} \text{ contribution}$$

(a)	'e'	$(T_{D1}^e)(t_{12}^{ee})(T_{2A}^e)/(\Delta_1^e \Delta_2^e)$
(b)	'h'	$(T_{D1}^h)(t_{12}^{hh})(T_{2A}^h)/(\Delta_1^h \Delta_2^h)$
(c + c′)	'e$_1$h$_2$'	$-(T_{D1}^e)(t_{12}^{eh})(T_{2A}^h)(-1/\Delta_1^e + 1/\Delta_2^h)/(-\Delta_1^e + \Delta_2^h)$
(d + d′)	'h$_1$e$_2$'	$-(T_{D1}^h)(t_{12}^{he})(T_{2A}^e)(1/\Delta_1^h - 1/\Delta_2^e)/(\Delta_1^h - \Delta_2^e)$

Summation of the contributions from all six pathways yields the following compact expression:

$$T_{DA} = \sum_{x,y = e,h} (T_{D1}^x)(t_{12}^{xy})(T_{2A}^y)/(\Delta_1^x)(\Delta_2^y). \tag{2.43}$$

Thus the initial evaluation of T_{DA} in terms of explicit energy gaps for the *six* intermediate *many-particle* states ψ_j^0 (as displayed in Fig. 2.4 and Scheme 2.1) has been converted to an equivalent expression based on *four* intermediate *one-particle* states in the orbital space, a result which is similar in form to the conventional third-order (for $m = 2$) perturbative result (cf. Eqn 2.29). The two diagonal terms in Eqn 2.43, hh and ee, correspond to the pure hole and electron processes shown in Figs 2.4(b) (or 2.5) and 2.4(a), respectively. The two cross-terms, eh and he, have a less obvious origin, but are clearly seen to be vestiges of the four hybrid processes (Figs 2.4c,c′,d,d′), an insight which underscores the value of the many-particle perspective.

Generalizing to arbitrary m, we find that the two-orbital/site NN model yields 2^m

different pathways, whose relative contributions to T_{DA} (both signs and magnitudes) are controlled by the joint action of the orbital parameters T, t and Δ. Clearly, the relative importance of the pure electron and hole pathways is expected to diminish with m [43]. Specific computational results for the case of alkyl spacer units will be presented in Section 5, where the one-particle operator h (Eqn 2.36) is identified with the Fock one-electron Hamiltonian.

In the special case where each ϕ_j^h, ϕ_j^e pair may be taken as linear combinations of a common pair of hybrid atomic orbitals, more insight may be obtained by recasting Eqn 2.43, generalized for arbitrary m, as

$$T_{DA} = (\bar{T}_{D1}\bar{T}_{mA})(1/\Delta_1^h - 1/\Delta_1^e)\prod_{j=1}^{m-1}\eta_{jj+1} \tag{2.44a}$$

where

$$\eta_{jj+1} = (\bar{t}_{jj+1})(1/\Delta_{j+1}^h - 1/\Delta_{j+1}^e) \tag{2.44b}$$

and where the effective coupling elements $\bar{T}$ and $\bar{t}$ depend on the details of ϕ_j^h and ϕ_j^e. For example, if each ϕ_j^h, ϕ_j^e pair is, respectively, a symmetric and antisymmetric combination of symmetry-equivalent hybrid atomic orbitals, then we may write

$$\bar{T}_{xy} = T_{xy}^h = T_{xy}^e \ (xy \equiv D1 \text{ or } mA) \tag{2.45a}$$

and

$$\bar{t}_{jj+1} = t_{jj+1}^h = -t_{jj+1}^e . \tag{2.45b}$$

The factors η_{jj+1} associated with each unit of the bridge convey the *constructive* nature of the h and e interference since Δ_j^h and Δ_j^e have opposite signs (see Eqn 2.41) [13,14]. By analogy with Eqn 2.11, $-2\ln|\eta_{jj+1}|$ may be considered a local decay coefficient, a quantity dealt with in Section 5.2.

The examples given here in Section 3.4, in spite of the use of a grossly oversimplified NN model, nevertheless serve to show how the role of hole and electron processes in 'electron transfer' is somewhat contingent on the choice of the states (or orbitals) used in analysing the problem (e.g. bridge orbital eigenfunctions vs local orbitals).

4　More general models for T_{if}

The discussion of coupling so far has been confined to the TSA for situations of resonant ET in the TS, using models limited for the most part to perturbation theory. In this section we introduce more general formulations of diabatic states and electronic coupling (T_{if}), allowing us to relax the above constraints, while remaining within an electronic framework. Consideration of specific vibronic effects may, of course, be required in some cases [33,72].

4.1　The Mulliken–Hush model

Remaining for the moment at the level of the TSA, we invoke the Mulliken–Hush (MH) model for non-resonant coupling [73,74]. Although introduced initially in the weak perturbation limit ($|T_{if}| \ll \Delta_{if}$, where Δ_{if} is the diabatic vertical gap for optical ET, displayed as $h\nu$ in Fig. 2.3a), it is fundamentally a non-perturbative model, exact within the TSA subject only to the assumption that the off-diagonal dipole moment matrix element between ψ_i and ψ_f, μ_{if}, is negligible [75,76]. Thus if the adiabatic (a) and diabatic (d) Hamiltonian and dipole matrices are defined, respectively, as

$$\mathbf{H}_a \equiv \begin{pmatrix} 0 & 0 \\ 0 & \Delta E_{12} \end{pmatrix} \tag{2.46a}$$

$$\boldsymbol{\mu}_a \equiv \begin{pmatrix} 0 & \boldsymbol{\mu}_{tr} \\ \boldsymbol{\mu}_{tr} & \Delta\boldsymbol{\mu}_{12} \end{pmatrix} \tag{2.46b}$$

$$\mathbf{H}_d \equiv \begin{pmatrix} 0 & T_{if} \\ T_{if} & \Delta_{if} \end{pmatrix} \tag{2.47a}$$

$$\boldsymbol{\mu}_d \equiv \begin{pmatrix} 0 & 0 \\ 0 & \Delta\boldsymbol{\mu}_{if} \end{pmatrix} \tag{2.47b}$$

where the assumption that $\mu_{if} = 0$ is imposed in Eqn. 2.47b, and where ΔE_{12} and μ_{tr} are, respectively, the energy eigenvalue difference and the adiabatic transition dipole moment, exact solution of the two-state secular equation yields the following relationship when μ_{tr} and $\Delta\mu_{12}$ are parallel [77]

$$T_{if} = (|\boldsymbol{\mu}_{tr}|/|\Delta\boldsymbol{\mu}_{if}|)\Delta E_{12}. \tag{2.48}$$

It is customary to evaluate $|\Delta\boldsymbol{\mu}_{if}|$ from the relationship

$$|\Delta\boldsymbol{\mu}_{if}| = (r_{DA})e \tag{2.49}$$

where r_{DA} is the assumed separation distance between the centroids of the D and A orbitals (based, for example, on crystal structure data), and where e is the magnitude of the electronic charge.

Thus in optical ET, Eqn 2.48 provides an estimate of T_{if} in terms of the vertical excitation energy (ΔE_{12}), the transition moment $\boldsymbol{\mu}_{tr}$ and the assumed r_{DA}. A frequently used expression equivalent to Eqn 2.48 is given by [76]:

$$T_{if} = 2.06 \times 10^{-2}(\nu_{max}\varepsilon_{max}\Delta\nu_{1/2})^{1/2}/r_{DA} \tag{2.50}$$

where $|\boldsymbol{\mu}_{tr}|$ has been expressed in terms of the optical parameters ν_{max} (cm^{-1}), bandwidth $\Delta\nu_{1/2}$ (cm^{-1}), and molar absorptivity ε_{max} (cm^{-1} mol^{-1}), and where r_{DA} is in Å, and T_{if} in cm^{-1}.

In fact, the exact two-state solution also yields an expression for $\Delta\boldsymbol{\mu}_{if}$ entirely in terms of the elements of $\boldsymbol{\mu}_a$ (Eqn 2.46b) [77]:

$$|\Delta\boldsymbol{\mu}_{if}| = [(\Delta\mu_{12})^2 + 4(\boldsymbol{\mu}_{tr})^2]^{1/2}. \tag{2.51}$$

Thus Eqn 2.48, with $|\Delta\boldsymbol{\mu}_{if}|$ replaced by the right hand side of Eqn 2.51, allows T_{if} to be specified exclusively in terms of adiabatic observables [77]: i.e. ΔE_{12} and $\boldsymbol{\mu}_{tr}$ (obtainable from spectral energy and intensity) and $\Delta\boldsymbol{\mu}_{12}$ (obtainable from Stark spectroscopy). Of course, these quantities may also be obtained from quantum calculations. In the limit of resonant ET (i.e. $\Delta_{if} = 0$) we obtain the expected result (Eqn 2.14)

$$T_{if} = \Delta E_{12}/2 \tag{2.52}$$

since in this case (where $\cos\theta = \sin\theta = 1/\sqrt{2}$ in Eqn 2.12), $\boldsymbol{\mu}_{tr} = \Delta\boldsymbol{\mu}_{if}/2$.

In the case of bridge-mediated coupling, the MH model may be generalized using a perturbative superexchange model. As an example, for a single bridge state $\psi_{B_1}^0$ we obtain [14]:

$$T_{if} = V_{D1} V_{1A}/(\Delta_1^0)_{eff} \tag{2.53}$$

where the effective energy gap is given by

$$(\Delta_1^0)_{eff} = 2(\Delta_{i1}^0 \Delta_{f1}^0)/(\Delta_{i1}^0 + \Delta_{f1}^0), \tag{2.54}$$

based on the assumption $\mu_{11} = (\mu_{ii} + \mu_{ff})/2$ and with $\Delta_{x1}^0 \equiv E_x^0 - E_1^0$, $x = $ i, f. The latter quantities are vertical gaps based, respectively, on the initial and final states and are generally evaluated at the equilibrium value of the reaction coordinate for the initial state (approximated here as ψ_i^0), in contrast to the gaps in Section 3, which pertain to the TS. The variation of gap with reaction coordinate yields a corresponding variation in superexchange coupling T_{if}, thus allowing a measure of the breakdown of the Condon approximation. We distinguish the two cases of interest with superscripts eq and TS, and consider for simplicity a thermoneutral process ($\Delta G^0 = 0$), with ψ_i^0, ψ_f^0, and $\psi_{B_1}^0$ all having parabolic free energy profiles of equal curvature and with the minimum of $\psi_{B_1}^0$ and the TS at the same point along the reaction coordinate. We find [78]:

$$T_{if}^{eq}/T_{if}^{TS} = 1 - [(E_r/2)/(|\Delta_{i1}^0| - E_r/2)]^2 \tag{2.55}$$

where E_r is the reorganization energy (displayed in Fig. 2.2). The variation of T_{if} with position along the reaction coordinate is actually rather modest (e.g. the ratio is >0.9 for $|\Delta_{i1}^0| > 2E_r$), thus supporting the use of the Condon approximation.

Much of the above analysis rested on the assumption that all dipole vectors are parallel. In general one must adopt a reference direction (e.g. as given by $\Delta\mu_{12}$ in Eqn 2.46b) and then use the projections of the other dipole matrix elements along this direction [77].

4.2 *Generalized Mulliken–Hush model*

We now consider more general models, designed to be broadly applicable to:
1 ground and excited state processes (either thermal or optical);
2 multistate systems (e.g. involving both CS and CR processes; see Fig. 2.3);
3 general coupling situations (i.e. beyond the weak perturbation limit);
4 computational implementation which (i) allows flexible inclusion of electron correlation; (ii) applies to arbitrary geometries (avoiding searches for crossings or seams, thus providing for general tests of the Condon approximation).

We obtain a model satisfying these criteria, denoted as the generalized Mulliken Hush (GMH) model [77], by exploiting the full consequences of the MH model, recognizing that one of its crucial assumptions, namely that transition moments connecting diabatic states localized at different sites are zero ($\mu_{if} = 0$), provides a *general* method for defining diabatic states in terms of purely adiabatic quantities. Specifically, we take the transformation that diagonalizes the adiabatic dipole moment matrix as the transformation to the Mulliken–Hush diabatic states. When one applies this to the adiabatic (diagonal) Hamiltonian, the diabatic Hamiltonian is obtained (previous examples of the use of diagonalization of the dipole moment matrix to define diabatic states in a pair-wise fashion can be found in [79–82].

When the multistate GMH transformation as defined here yields more than one diabatic state localized on the same site, we impose the additional condition that the block of the Hamiltonian associated with a single site be diagonal, thus yielding states diabatic in the GMH sense with respect to intersite coupling, but 'locally adiabatic' within each site or local region [77]. The GMH analysis employs the component of each dipole vector in the direction defined by the dipole difference for the initial and final adiabatic states (two-state case) or by the average of such differences when several ET processes are considered for a given system.

As already noted above in connection with the two-state MH model (now recognized as a special case of GMH), we emphasize that the GMH method can be implemented solely in terms of experimental quantities. Observed excitation energies, transition dipoles from intensity measurements, and adiabatic dipole moment differences from Stark measurements yield direct experimental estimates of diabatic coupling elements according to the above procedure. At the multistate level, the relative sign relations among the transition moments must be known, and may be taken from calculations if not available from analyses of the experimental data. (These phase relations are 'observables', as distinct from phase 'conventions'; e.g. for a three-state system two distinct cases arise: an even or odd number of positive transition dipole moments.) Finally, we mention that as in the two-state case (Section 4.1), the diagonal GMH diabatic dipole moment matrix directly yields estimates of the centroids of the different states of the system of interest and the associated r_{DA} values, quantities of great utility in interpreting charge transfer data obtained from Stark spectroscopy or quantum calculations [77,83].

One of the central physical assumptions underlying the GMH model — the diagonal form of the diabatic dipole moment matrix — has been given quantitative support by considering alternative formulations of diabatic states which take detailed account of the eigenvectors obtained from multistate quantum chemical calculations [84]. These techniques are in the spirit of the so-called least-motion block diagonal formulations [85,86].

5   Computational applications

The theoretical models discussed in Sections 3 and 4 have played an important role in quantum chemical applications to a wide variety of chemical systems. The power of current-day computational capability makes molecular systems of considerable complexity (including assemblies with up to ~100 atoms and several hundred electrons) accessible to detailed electronic structural treatment, using either *ab initio* or semi-empirical techniques [14]. These techniques range from simple pathway approaches [57,66] to one-electron path integral methods [87] [which permit molecular level treatment of the medium, but which include only 'electron' (and not 'hole') contributions to T_{if}], to many-electron models of the independent-particle [50,61], self-consistent field (SCF) [34,36,43,45,64], or configuration interaction (CI) type. The CI methods, as well as related perturbative many-body techniques, allow the inclusion of electron correlation [64,77,88,89,107] and are generally essential for an even-handed treatment of a manifold of different electronic states [77]. Density functional theory (DFT) techniques may also be applied to ET systems [89].

Up to the present, most electronic structure calculations applied to ET processes have been restricted to isolated DBA systems. Aside from T_{if}, these calculations permit the evaluation of the contribution to reorganization energy from discrete vibrational modes of DBA [90]. Some recent treatments have included the effect of a solvent reaction field, using cavities of realistic shape which take detailed account of the molecular structure of the solute [52] (i.e. the DBA system). In the cases studied with these realistic models, solvent does not have a significant influence on T_{if} magnitudes, although the available data base is small.

A technical problem arises in cases of DBA systems with negative charge (either with or without solvent present), since calculations often do not yield stability with respect to loss of an electron. Yet even for such 'unbound' systems, SCF techniques typically yield wavefunctions from which reasonable T_{if} values may be obtained, provided that the orbital basis does not contain very diffuse functions [34].

In the remainder of this section we focus on various DBA systems with electronically saturated bridges, and present a selective set of results and conclusions from detailed quantum calculations chosen to elicit the key factors which control D–A coupling, both in terms of trends in T_{if} values and the analysis of these trends made possible by the theoretical concepts associated with the superexchange model developed in Section 3. Thus we examine the dependence of T_{if} on a number of chemical factors which characterize the DBA systems, including (i) charge type, (ii) symmetry of DB and BA interaction, (iii) conformation of D and A relative to B, (iv) internal conformation of B, (v) covalent vs H-bonded or non-bonded interactions within DBA, (vi) energy gaps (Δ), and (vii) role of surrounding solvent.

The role of the superexchange model is two-fold: first to provide illuminating decompositions of T_{if} into compact sets of elementary constituents; and second, to offer a basis for fashioning these constituents into transferable parameters which may be used, with suitable tuning, to design new systems of desired specifications from reference systems of known properties.

A word is in order concerning the correspondence between estimates of T_{if} obtained from theoretical calculations and from experiement. The course of action advocated in the previous paragraph is very timely precisely because the diverse body of results available at present indicates that the degree of correspondence is indeed quite good, certainly at the semi-quantitative level (some specific comments are offered below). Precise quantitative comparisons remain difficult, not only because of remaining deficiencies in the electronic structure models (and associated treatment of vibronic and medium effects), but also because of the difficulties inherent in extracting T_{if} values from experimental kinetic data.

5.1 *Orbital models*

The results reported below are based on SCF orbital wavefunctions, carried out either at the single- (Sections 5.2 and 5.3) or multi- (Section 5.4) configuration level. The results are all of the *ab initio* type, except for Section 5.4, which includes some CI results based on the semi-empirical INDO/S method [91].

The single-configuration SCF model has been applied to T_{if} evaluation at three distinct levels, distinguished by the degree of state-specific electronic relaxation included [14,43]:

1 fully relaxed SCF (charge-localized ψ_i, ψ_f);
2 symmetrically delocalized SCF [ψ_1 and ψ_2 (Eqn 2.12), with mean-field relaxation]
3 frozen orbitals for the *n*-electron DBA system (based on the $(n \pm 1)$-electron SCF parent).

These three levels lead, respectively, to evaluation of T_{if} as

1 direct coupling (Eqn 2.32);
2 adiabatic splitting (Eqn 2.14);
3 adiabatic splitting (Eqn 2.14) via Koopmans' theorem (KT) [92] (i.e. orbital energy splitting).

The performance of the three levels has been thoroughly discussed in the literature [14,34,43,45,49,64,88]. The methods are most easily applied to symmetrical DBA systems, since the symmetry helps to specify the reaction coordinate at the TS, where most applications have been made (in this situation, the fully relaxed SCF wavefunction is in general spatially symmetry-broken). Treatment of non-symmetrical systems is also feasible [36,93,94]. The frozen-orbital/Koopmans' theorem (KT) approach is attractive for purposes of analysis, since initial (ψ_i) and final (ψ_f) states may be characterized by a common set of orbitals, conveniently chosen to correspond to the desired assignment of bridge subunits, as discussed in Section 5.3. Fortunately, the KT model yields results which are generally in reasonable agreement with the two SCF models, at least for purposes of semi-quantitative evaluation and analysis. The one-particle model based on KT may be considered as a particular example of the generic orbital model for T_{ij} derived in Section 3.4 (Eqn 2.43) in the context of a set of many-electron configurations.

With respect to the delocalized, spatially restricted SCF level (for symmetrical DBA systems), the charge-localized model may be viewed as introducing electron correlation (associated with the charge-state-specific relaxation). More elaborate incorporation of electron correlation has also been investigated (using both many-body-perturbation theory and MCSCF models) and found to make appreciable contributions in some cases, although generally not having a major effect on T_{if} magnitudes [64,88,89].

The approaches outlined above have been implemented at the *ab initio* level with a variety of orbital basis sets, including minimal (e.g. STO-3G), split-valence (e.g. 3-21G), and extended (e.g. 6-31G*) types [34,43,49,63,64,95]. The 3-21G basis is generally found adequate, and even the minimal STO-3G set is able to give a semi-quantitative account of many trends.

5.2   *Trends in T_{if} for saturated bridges*

We now consider the results of calculations for families of DBA systems with saturated organic bridges, which illustrate how T_{if} varies with D–A separation, molecular conformation, nature of the coupling between the units of the system, and degree of solvation. While many other bridge types are of interest, an examination of electronically saturated bridges is especially revealing, showing how prior naive pictures of such

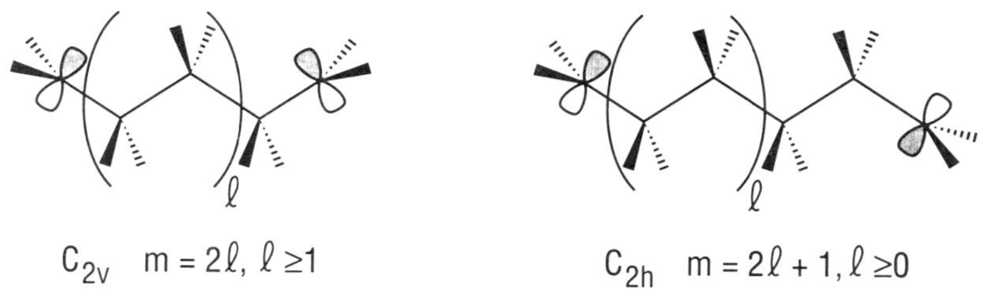

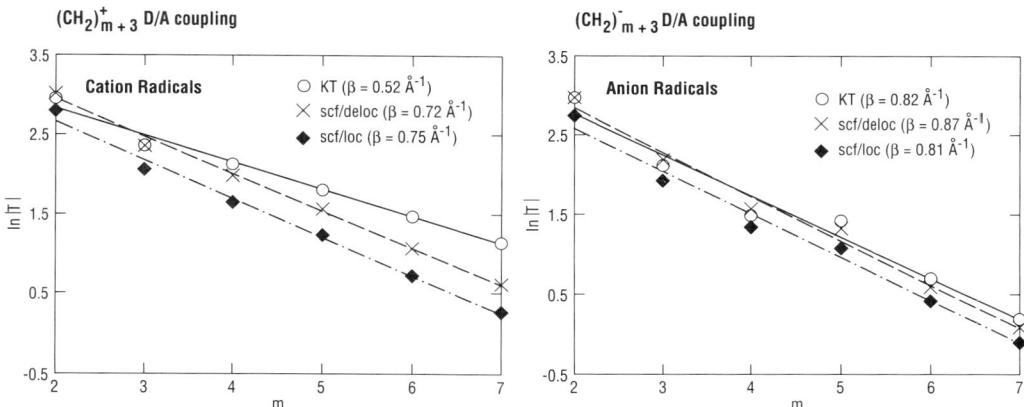

$$C_{2v} \quad m = 2\ell, \ \ell \geq 1 \qquad\qquad C_{2h} \quad m = 2\ell + 1, \ell \geq 0$$

$$\mathbf{1}\pi(m)$$

Figure 2.6 π-type D–A orbitals (the non-bonding orbitals of the terminal CH_2 groups) linked by a *trans*-alkane bridge [$(CH_2)_{m+1}$] possessing m covalent CC bonds (the covalent bonds connecting the D and A groups to the bridge contribute little to T_{if} [34,43,64]). The even and odd m members correspond, respectively, to C_{2v} and C_{2h} point group symmetry.

bridges as merely inert insulating 'spacers' have given way to a detailed appreciation of their effectiveness in facilitating D–A coupling.

5.2.1   DECAY COEFFICIENTS (β) FOR HOMOLOGOUS BRIDGES

Coupling through *trans*-staggered alkane bridges in $(CH_2)_{m+3}^{\pm}$ radical cations and anions, $\mathbf{1}\pi(m)$, as in Fig. 2.6, is displayed in Fig. 2.7 at three different levels (see Section 5.1): KT energy splittings (Eqn 2.14), energy splittings (Eqn 2.14) for SCF delocalized wavefunctions (SCF/deloc), and direct interaction (Eqn 2.32) of charge-localized wavefunctions (SCF/loc). The calculations are all based on symmetrical (C_{2v} or C_{2h}) molecular structures, which may be taken as representative of the transition state for activated ET [43]. In each DBA system, D and A are the terminal CH_2 groups,

Figure 2.7 Plots of $\ln|T_{if}|$ for $(CH_2)_{m+3}$ radical cations and anions, $\mathbf{1}\pi(m)$, presented as a function of the number of bonds (m) in the $(CH_2)_{m+1}$ bridge linking donor and acceptor CH_2 groups (see Fig. 2.6). Decay coefficients (β) for radical cation and anion systems are given at three levels, using a 3-21G orbital basis: energy splitting at the KT level (Eqn 2.14); direct SCF energy splitting (Eqn 2.14); and direct evaluation of T_{if} (Eqn 2.32) using charge-localized SCF wavefunctions (see [43]). The β values (Å^{-1}) are based on linear least-squares fits of $\ln|T_{if}|$ vs m, for $m = 2$–7, with $\Delta r = 1.27$ Å (see Eqn 2.9).

and m denotes the number of CC bonds within the $(CH_2)_{m+1}$ bridge (the sigma bonds linking D and A to B have a minor influence on the overall D–A coupling [43]). Consistent with the discussion in connection with Eqn 2.8, the T_{if} values based on the most fully relaxed wavefunctions (SCF/loc) have the smallest magnitudes. It is seen that

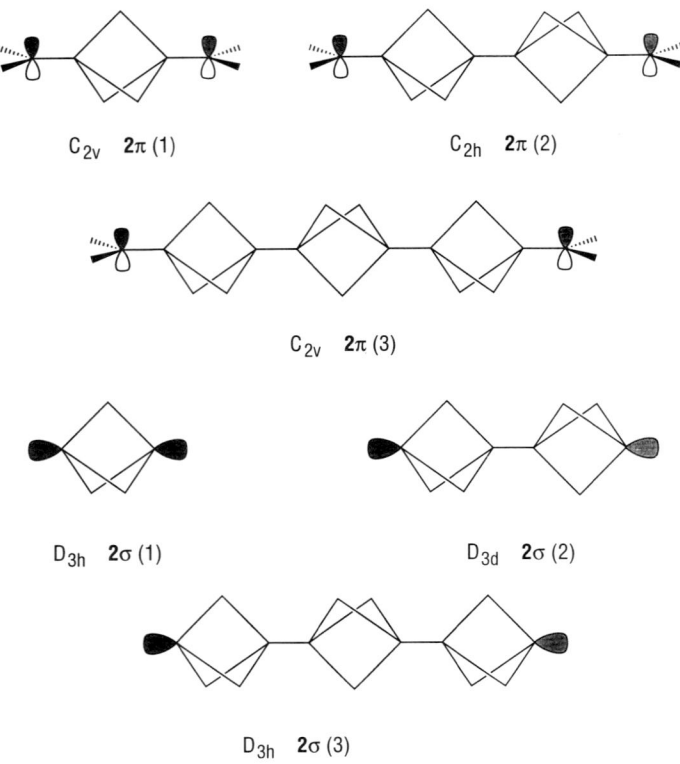

C_{2v} **2π** (1) C_{2h} **2π** (2)

C_{2v} **2π** (3)

D_{3h} **2σ** (1) D_{3d} **2σ** (2)

D_{3h} **2σ** (3)

(a)   bicyclo [1.1.1] pentane spacers

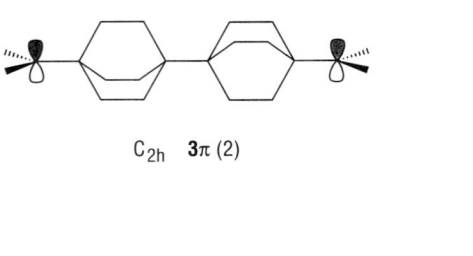

C_{2h} **3π** (2)

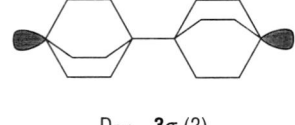

D_{3d} **3σ** (2)

(b)   bicyclo [2.2.2] octane spacers

Figure 2.8 (a) π-type [**2π**(m)] and σ-type [**2σ**(m)] D–A orbitals linked by bicyclo [1.1.1] pentyl bridge units ($m = 1$–3), with adjacent units related by a staggered conformation. The odd and even m members correspond, respectively, to C_{2v} and C_{2h} point group symmetry for the **2π**(m) series, and D_{3h} and D_{3d} symmetry for **2σ**(m) series. (b) Analogues of **2π**(2) and **2σ**(2), with bicyclo [1,1,1] pentane units replaced by bicyclo [2.2.2] octane units.

the overall variation of T_{if} (in the displayed $m = 2-7$ range) is roughly exponential (linear regression coefficients, r, for $\ln|T_{if}|$ fits to m are >0.98 [43]), even though local fluctuations are evident (see also [63] and [95]). In studies of longer alkane chains (with a variety of D and A groups) a tendency toward flattening of the falloff with increasing chain length has been observed for radical cations [63]. The two SCF models (loc and deloc) give similar results, with β slightly greater for radical anions, while the 'one-electron' (KT) model gives an appreciably smaller β value for the radical cations. Calculated βs in the range 0.7–0.9 Å^{-1} are compatible with estimates (~0.7–1.0 Å) inferred from experimental kinetic data (see Table V of [43]). As discussed below, the quasi-exponential falloff by no means implies that the simple McConnell type model (Eqn 2.23) is valid. (Here and in the remainder of the chapter we suppress the SE superscript for β introduced in Eqn 2.11.)

Turning now to more complex bridges comprising bicyclo [1.1.1] pentane units ('staffanes' [43,63]), depicted in Fig. 2.8(a), we display T_{if} and β values for 1–3-membered bridges in Table 2.1 (at the KT level). For the π-type D–A systems [$2\pi(m)$, where m denotes the number of bicyclo units], the decay characteristics are similar to those displayed by the $1\pi(m)$ systems. For the analogous σ-type D–A systems [$2\sigma(m)$], the limited data suggest a greater difference in falloff for cation vs anion systems [note also the different sign pattern, in comparison with that for $2\pi(m)$]. The $2\sigma(1)$ cases have been omitted, since the coupling for them is dominated by direct through-space D–A coupling [43,49]. Throughout this chapter, as in [43] and [49], the coupling is presented as $- T_{if}$, for which a positive sign corresponds to 'normal' coupling (see also footnote † in Table 2.1).

Finally, we emphasize that the type of falloff exemplified by the data of Fig. 2.7 and Table 2.1 is not limited to DBA systems fully linked by covalent bonds [34,43,95]. Figure 2.9 displays 1–3-membered bridges [$4\sigma(m)$] comprising methane molecules in

Table 2.1 Coupling through linked staffane bridges [$2\pi(m)$ and $2\sigma(m)$].*

D–A type	m	Radical cations†‡		Radical anions†‡	
		$- T_{if}^{KT}$	β_{mm+1}	$- T_{if}^{KT}$	β_{mm+1}
π	1	20.6	0.67	19.0	0.85
	2	6.52	0.66	4.35	0.84
	3	2.14	—	1.00	—
σ	2	– 38.9	0.56	– 20.0	1.24
	3	15.6	—	2.63	—

* See Fig. 2.8(a). Based on energy splitting (Eqn 2.14) at the KT level, using the 3-21G orbital basis [43]. The $m = 1$ σ case is not included since it is dominated by through-space coupling.
† T_{if} given in mhartree (1 mhartree = 0.027 ev $\sim k_B T$ at room temperature). The sign convention [43,49] is based on the phases of the D and A orbitals in the two MOs of primarily D–A character and assigns a positive (negative) sign to the quantity $- T_{if}$ when the occupation of the 'in-phase' ('out-of-phase') MO is energetically favoured [the positive sign for $- T_{if}$ corresponds to the 'normal' situation in which in-phase ('bonding') orbital interactions are characterized by a negative transfer integral]. For $2\pi(m)$, the in-phase MOs are taken as those transforming as $a_1(C_{2v})$ or $b_u(C_{2h})$, while for $2\sigma(m)$, in-phase corresponds to $a_1(C_{2v})$ or $a_g(C_{2h})$.
‡ Local β (Å^{-1}) defined analogously to Eqn 2.9: $\beta_{mm+1} = - 2 \ln|T_{if}(m + 1)/T_{if}(m)|/\Delta r$, where $\Delta r = 3.37$ Å is the mean axial 'length' of a staffane unit (see [43]).

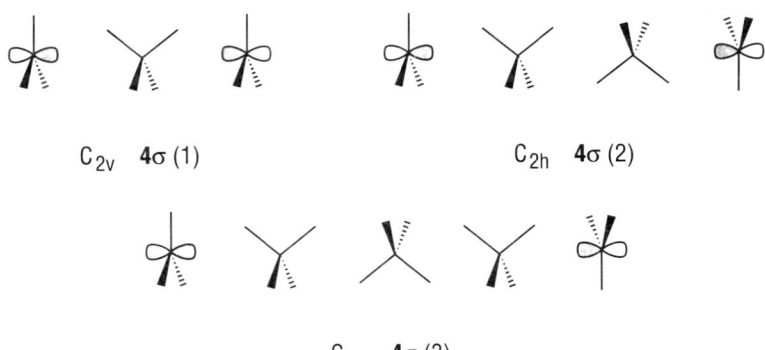

C_{2v} **4σ** (1) C_{2h} **4σ** (2)

C_{2v} **4σ** (3)

Figure 2.9 Coupling of D–A orbitals (the non-bonding orbitals of the terminal CH_3 groups) in a relative σ orientation, mediated by a non-bonded sequence of m CH_4 spacer units. The odd and even m members correspond, respectively, to C_{2v} and C_{2h} point group symmetry.

van der Waals contact (NN C···C separations are 3.4 Å [43]), with σ-type D–A coupling represented by CH_3 groups in van der Waals contact with B. Table 2.2 shows falloff for coupling in the radical anions quite similar to that found for the covalent systems, while the falloff is somewhat faster for the radical cations. Estimates of β ~ 1.2 Å^{-1} for both electron and hole transfer through non-covalently linked glassy media have been obtained from experimental kinetic data [48,96], with some correlation observed between β and the estimated effective energy gap (Δ) [96].

Table 2.2 shows similar decay through non-bonded radical cation systems in which CH_4 is replaced by H_2O (the results for the corresponding radical anions are omitted, since they were found quite sensitive to orbital basis set; this may be a manifestation of the unbound nature of the model radical anion calculations, as noted in Section 5.1).

5.2.2   CONFORMATIONAL DEPENDENCE

Since T_{if} is now known to be dominated by non-nearest neighbour interactions, when cast in terms of local bonds (and anti-bonds) [34,43,49,63,64,97], sensitivity to molecular conformation is expected [43,49]. Table 2.3 displays results [49] for confor-

Table 2.2 Distance dependence of coupling through non-covalently linked bridge systems.*
(Table 3 of [110], reprinted with kind permission from Elsevier Science.)

		$\beta^*_{m\,m+1}$	
Bridge unit	m	Radical cation	Radical anion
$CH_4[4\sigma(m)]$†	1	0.94	0.84
	2	1.01	0.86
H_2O‡	1	0.96	
	2	0.94	

* See footnote ‡ of Table 2.1; Δr = 3.4 and 2.8 Å, respectively, for CH_4 and H_2O.
† [43].
‡ Analogous to the **4σ**(m) structures (Fig. 2.9), but with CH_4's replaced by H_2O's. The orientation of the H_2O's corresponds to the CH_2 moieties of the CH_4 units in **4σ**(m) whose planes are perpendicular to the D–A axis. Van der waals C···O contacts of 3.1 Å were assigned.

Table 2.3 Conformational dependence of T_{if} for $(CH_2)_6^{\pm}$ radical ions [**1**π(3)].* (Table 2 of [110], reprinted with kind permission from Elsevier Science.)

Conformation of D and A orbitals relative to bridge†	Conformation of C_4 bridge	$-T_{if}$	
		Radical cation	Radical anion
In-plane	*trans* (C_{2h})	− 11.0	− ˙8.4
	gauche (C_2)	− 2.4	− 4.5
	cis (C_{2v})	− 0.3	− 4.3
Perpendicular to plane	*trans* (C_{2h})	− 1.1	+ 0.3
	gauche (C_2)	+ 3.6	+ 2.0
	cis (C_{2v})	+ 7.9	+ 9.9

* [49]; sign convention for T_{if} (mhartree) as defined in footnote † of Table 2.1. For the $(CH_2)_6^{\pm}$ systems, 'in phase' is taken as transforming according to a_g, a, and a_1 for, respectively, C_{2h}, C_2, and C_{2v} symmetry with 'in-plane' D–A conformers, and a_u, a, and b_1 for the corresponding 'perpendicular' conformers.
† Angle of the D(A) π orbitals relative to plane formed by the carbon atom of the terminal D(A) methylene group and the closest two carbon atoms of the bridge.

mational variants of **1**π(3), both with regard to D–A conformation relative to B and the internal conformation of B. Clearly the effects are quite pronounced, involving both sign and magnitude. Indeed these results underscore the importance of considering the sign as well as magnitude in any attempt to gain a comprehensive understanding of the nature of D–A coupling (the signs, albeit bona-fide observables [14,43,49], are nevertheless contingent on the underlying sign conventions, as noted in the table footnotes). It is interesting to note that for both D/A–B conformations, the *trans→gauche→cis* modification of B always results in a monotonic increase in the algebraic value of T_{if}. It should be emphasized that the 'in-plane' and 'perpendicular' orientations of D–A orbitals relative to B lead to coupling mechanisms of essentially different qualitative types [49]: in the former case, coupling is dominated by the CC framework (even though CH bonds make appreciable contributions), whereas in the latter case, coupling to the CH bonds becomes either an important (*gauche* B) or the exclusive (*trans* or *cis* B) mechanism.

Table 2.4 displays the effect of staggered vs eclipsed conformation of the two bicyclo units in **2**π(2), **2**σ(2), **3**π(2), and **3**σ(2). The staggered conformers are displayed in Fig. 2.8. One might expect the bicyclo [2.2.2] systems to be more sensitive than the

Table 2.4 Conformational dependence of $|T_{if}|$ for polycyclic bridge units.*

DBA	$|T_{if}^{ecl}/T_{if}^{stagg}|$	
	Radical cation	Radical anion
3π(2)	1.6	⩾10†
2π(2), **2**σ(2), **3**σ(2)	1.0–1.1	1.0–1.2

* Based on systems with two bridge units [**2**π(2), **2**σ(2), **3**π(2), and **3**σ(2)] [43]. Staggered (stagg) and eclipsed (ecl) conformers correspond, respectively, to the C_{2h} and D_{3d} $m = 2$ structures as drawn (Fig. 2.8), and to the C_{2v} and D_{3h} variants obtained by rotating one bridge unit 60° relative to the other.
† Inequality is due to sensitivity of calculated T_{if} value for the staggered conformer [43].

bicyclo [1.1.1] systems, because in the former case, the relevant pairs of adjacent bonds involve closer contacts due to the larger bond angles at the bridgeheads [43]. Such behaviour is indeed found for the case of σ-type coupling, but *not* π-type.

5.2.3   DEPENDENCE OF T_{if} ON THE NATURE OF THE BONDING IN DBA

We have already seen above that similar decay of T_{if} with D–A separation is observed whether the D, A and bridge units are covalently bonded or in non-bonded contact, a result which is not surprising in view of the dominant role of non-nearest neighbour interactions noted above. In Table 2.5 we compare results for covalently linked systems and hydrogen-bonded (H-bond) systems. The H-bond is thought to play an important role in many cases of long-range D–A coupling, especially in protein-mediated coupling [65].

Each of the DBA systems in Table 2.5 possesses two three-bond linkages, either fully covalently linked, as for the boat conformer of cyclohexane, **5π** (see below), or hydrogen bonded, as in the carboxylic acid dimer, **6π** (see below). As expected from Section 5.2.2, the results are quite dependent on the conformation of D and A relative to B. However, it is clear that the strength of coupling through the H-bonds is comparable to that through a sequence of CC bonds, even though r_{DA} separation is greater in the former case. The carboxyl group, of course, represents an exception to our focus on fully saturated bridges. However, the π-bond of the carboxyl does not appear crucial here,

Table 2.5 Comparison of coupling through three-bond links:* covalent bonds vs hydrogen bonds.†

DBA	D–A conformation‡	$- T_{if}$ (mhartree)§	
		Radical cation	Radical anion
Covalent links	in plane	− 0.7	− 9.1
(r_{DA} = 5.8 Å)	out of plane	+ 2.6	+ 4.3
5π			
H-bonded links‖	in plane	− 5.7	− 5.3
(r_{DA} = 6.8 Å)	out of plane	+ 0.9	+ 1.1
6π			

* For the case of the hydrogen bond, the 'three-bond' viewpoint implicitly treats the OH···O unit as an effective single bond. The single bonds linking the terminal CH_2 groups to the bridge are not counted in the total number of linker bonds since they are very weakly coupled to the D and A orbitals (i.e. the 2pπ-type orbitals) [43].

† Results based on the 3-21G orbital basis set, at the KT level (Eqn 2.14).

‡ The 'conformation' of the D(A) orbital is defined by the dihedral angles between the plane perpendicular to the D(A) CH_2 group and the planes defined by the D(A) carbon atom and the closest two atoms in each of the two linkers it is attached to. By this criterion, **5π** and **6π** as drawn correspond, respectively, to dihedral angles 60° and 90°, and are designated as 'out of plane.' The 'in-plane' conformers, obtained by 90° rotations of the D and A groups, correspond, respectively, to dihedral angles of 30° and 0°.

§ Sign convention as described in footnote † of Table 2.1. In C_{2v} symmetry (**5π**), 'in phase' is taken to correspond to the a_1 (as drawn) or b_1 ('in-plane' conformer) representations, whereas for C_{2h} symmetry (**6π**), the corresponding representations are a_u (as drawn) or a_g.

‖ The hydrogen-bonded O···O separation is 2.6 Å, based on the structures of benzoic acid dimers [107].

since the H-bond-mediated coupling is actually much more favourable for the 'in-plane' orientation of the D–A orbitals, which involves only overlap with the sigma framework of the bridge.

5π

5.2.4 INFLUENCE OF SOLVENT

The selection of the 'bridge' in any model for T_{if} is guided by the assumption that it will give adequate account of all important pathways for superexchange coupling. Relative to a given DBA system in isolation, one might expect (see Section 2) an influence from the surrounding environment (taken here as solvent), either through modulation of the importance of already existing pathways through B by perturbing the relevant energy gaps Δ (see Section 3), or by providing additional pathways by virtue of its own electronic manifolds. As an example, we return to the *trans*-staggered alkane spacer [$1\pi(m)$] and compare in Table 2.6 results for β obtained: (i) *in vacuo*; (ii) with an

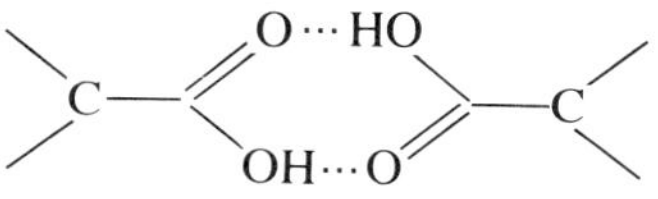

6π

Table 2.6 Effect of peripheral aqueous solvent on coupling through *trans*-alkyl bridges [$1\pi(m)$].* (Table 6 of [110], reprinted with kind permission from Elsevier Science.)

Nature of hydration	Decay coefficient, β (Å^{-1})†	
	Radical cations	Radical anions
Solvent-free‡	0.85	0.95
Dielectric continuum§	0.85	0.95
Specific hydration‖ [$(CH_2)_{m+3}(H_2O)_{m+1}$]$^\pm$	0.81	0.95

* Based on SCF energy splittings (Eqn 2.14), with the 6-31G** orbital basis, and including m in the range 2–6 (see [52]).

† Equation 2.9, based on a linear least-squares fit of $\ln|T_{if}|$ as a function of r_{DA}. The β values are somewhat larger than those obtained with the 3-21G basis ('SCF/deloc' results displayed in Fig. 2.7).

‡ *In vacuo* results from [52].

§ Based on dielectric cavity model described in [52].

‖ An H_2O interacts with each of $m+1$ CH_2 triads for each $(CH_2)_{m+3}$ species, oriented so as to allow the H_2O lone pairs to overlap optimally with the carbon framework (as indicated in **7**), with van der Waals $C\cdots O$ contacts of 3.1 Å.

aqueous medium represented by a realistic molecular cavity in a dielectic continuum [52]; and (iii) with explicit addition of water molecules (see Structure **7**, see below) in the quantum chemical calculations (Y.P. Liu and M.D. Newton, unpublished). Solvent is seen to exert no influence on β via the long-range interaction provided by the dielectric continuum model [52]. Even with the enlarged superexchange model entailed in the inclusion of water molecules, the only effect is a very modest reduction in β for radical cations. Qualitatively, this result is not surprising, since the relatively high-lying lone pair orbitals of H_2O are expected to participate (at least to some extent) in hole-type superexchange (most important for coupling in radical cations), whereas the unoccupied water orbitals are not expected to be effective in promoting corresponding 'electron'-type pathways. Aside from β values, the magnitudes of T_{if} are essentially unaffected by inclusion of a polarized dielectric [52], and even with the explicit waters the effects are minor (<10%).

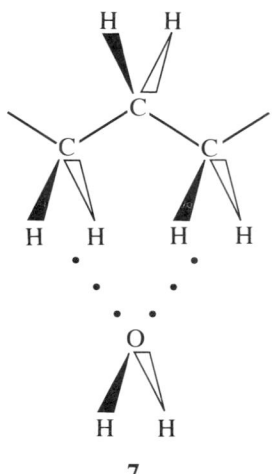

7

5.3   *Analysis of superexchange coupling in terms of natural bond orbitals*

Considerable insight into the nature of superexchange coupling through saturated bridges has been obtained in an additive perturbative framework by adopting two-centre 'bridge units' containing localized 'bond' and 'anti-bond' molecular orbitals (LMOs). These represent an effective compromise [43] between even more localized (one-centre) atomic orbitals (which interact too strongly to make perturbation theory applicable) and larger molecular fragments (amenable to a perturbative treatment and of demonstrated utility [11,50], but less obviously likely to provide a basis for transferable parameters). A convenient (though by no means unique) definition of LMOs is the natural bond orbital (NBO) scheme of Weinhold and co-workers [98]. Detailed studies [49,63,97] have shown that NBOs yield a workable perturbative scheme (e.g. along the lines of Eqn 2.29), in which overall T_{if} magnitudes at the one-electron KT level may often be recovered additively at a level of sufficient quantitative accuracy to make a detailed NBO-based mechanistic analysis of great value. These treatments are all within the 'second-order' constraint of the TSA (Sections 2 and 3), which precludes intermediate return 'visits' to the D site.

Table 2.7 Gaps ($|\Delta_j|$) for *trans*-staggered alkane bridge systems [$1\pi(m)$] in a local (NBO) basis.†‡

Bond type	Δ_j
CC	10.9–12.2
CH	8.2–8.4
CC*	8.7–10.1
CH*	11.4–11.7

† The indicated quantities represent the magnitude (in eV) of the lowest energy gap (Δ_j) between D–A orbitals and the highest lying occupied NBO of the indicated bond type (CC or CH) for the radical cation systems, and the lowest lying unoccupied NBO of the given anti-bond type (CC* or CH*) for the radical anion systems.

‡ The orbital gaps are based on the diagonal elements of the Fock matrix in the NBO basis, corresponding to α-spin (for the radical cations) and β-spin (for the radical anion) manifolds of the neutral diradical parent (treated at the UHF level with the 3-21G basis) [43].

5.3.1   NBO GAPS AND HOPPING INTEGRALS

We display in Tables 2.7 and 2.8 typical values, respectively, of gaps (Δ_j) and hopping integrals (T, t) in the NBO basis obtained from the Hartree–Fock MOs of the neutral parent (i.e. the frozen orbitals which define the KT model). For radical anions, the $|t/\Delta|$ ratios are all well below unity, whereas for radical cations, $|t/\Delta|$ is found to approach 0.5

Table 2.8 Hopping integrals for *trans*-staggered alkane bridge systems [$1\pi(m)$] in a local (NBO) basis.†

		NBO pairs		
		1,2‡	1,3‡	1,4‡
Coupling within bridge§‖				
C–C	C–C	145-148	29–30	10–11
C–C	C–C*	25-29	48–50	8–9
C–C*	C–C*	6	29–30	40–42
D–A and CC or CC* (radical cations)				
D–A	C–C	53	17–18	4
D–A	C–C*	96	41	2
D–A and CC or CC‖ (radical anions)				
D–A	C–C	79	23	7
D–A	C–C*	76	41	2

† The listed quantities are the magnitudes (in mhartrees) of the hopping integrals for the indicated NBO pairs, based on the Fock matrix [43] (see footnote ‡ of Table 2.7).

‡ 1,2 (nearest neighbour), 1,3 (next nearest neighbour), and 1,4 denote successively longer range NBO interactions (the NBO pairs involved in 1,2 interactions share a common C atom).

§ The values given for coupling between bridge NBOs apply to both radical cation and anion species.

‖ While the NBO Fock matrix is not in general diagonal with respect to a CC/CC* (or CH/CH*) pair sharing a common set of atoms, the magnitude of such coupling is small (≤ 0.01 hartree) [43].

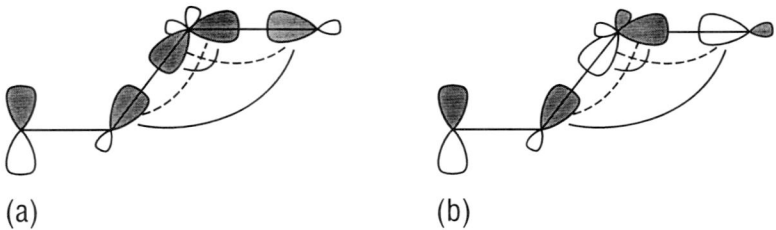

(a)   (b)

Figure 2.10 Schematic orbital diagram (roughly according to scale, based on calculated results), depicting the nearest neighbour interaction of a pair of (a) bonding (CC) and (b) anti-bonding (CC*) natural bond orbitals (NBOs). (Fig. 5 of [49], reprinted with permission. Copyright (1992) American Chemical Society.)

in some cases. Thus for these cations, smooth convergence certainly cannot be taken for granted, even though it is often found in practice when D and A groups are modelled by CH_2 [43,49,63,97] or other [63] groups. Chemical variation of the D–A groups may, of course, substantially reduce the NBO gaps (Δ). In such cases [34,63], model one-electron Hamiltonians in an NBO basis are still of value even if perturbation theory must be abandoned.

The sharp disparity in magnitudes of t_{12} for CC and CC* NBOs (Table 2.8) may be qualitatively understood in terms of the different nodal structure of the respective NBOs [49]: primarily constructive (destructive) interference pertains when adjacent CC (CC*) bonds interact, as shown schematically in Fig. 2.10.

5.3.2   NEAREST NEIGHBOUR VS NON-NEAREST NEIGHBOUR COUPLING

The order-of-magnitude difference in NN t values does not lead to as much disparity in hole vs electron coupling as one might expect, because of the relatively minor role which NN interactions play in overall coupling [34,49,97]. A qualitative sampling of this situation is offered in Fig. 2.11, which displays the most important individual pathway (i.e. the contribution to the perturbation sum of greatest magnitude) for 10 different DBA systems, characterized by π- and σ-type D–A groups coupled to various acyclic, cyclic and bicyclic bridges (obtained for radical anions at the 3-12G/KT level) [49]. These pathways all involve the carbon framework, but in only one case (the first entry on the left) does an NN McConnell path appear, while the other cases involve jumps over at least one link in a covalent sequence (see also the similar analyses in [34] and [64]). For the π-type D–A cases, coupling between D–A groups and the sigma bonds linking them to the bridge is very weak, a consequence of the π–σ orthogonality [49,63,64]. In larger bridges [63,64], pathways involving jumps over one to three NN covalent links are typically found to be the most important.

It is to be emphasized that the 'most important' pathways are generally not dominant (and in fact may have sign opposite to that of the overall T_{if} [49]), since T_{if} is generally the superposition of a very large number of competing terms, especially when the extended orbital basis sets desirable for adequate variational flexibility are employed [49,64]. In this sense, the NBO units are too 'fine-grained', and a 'coarser grained' selection of bridge units would be necessary to obtain a very compact representation of T_{if} (i.e. a small set of effective pathways).

Radical anions

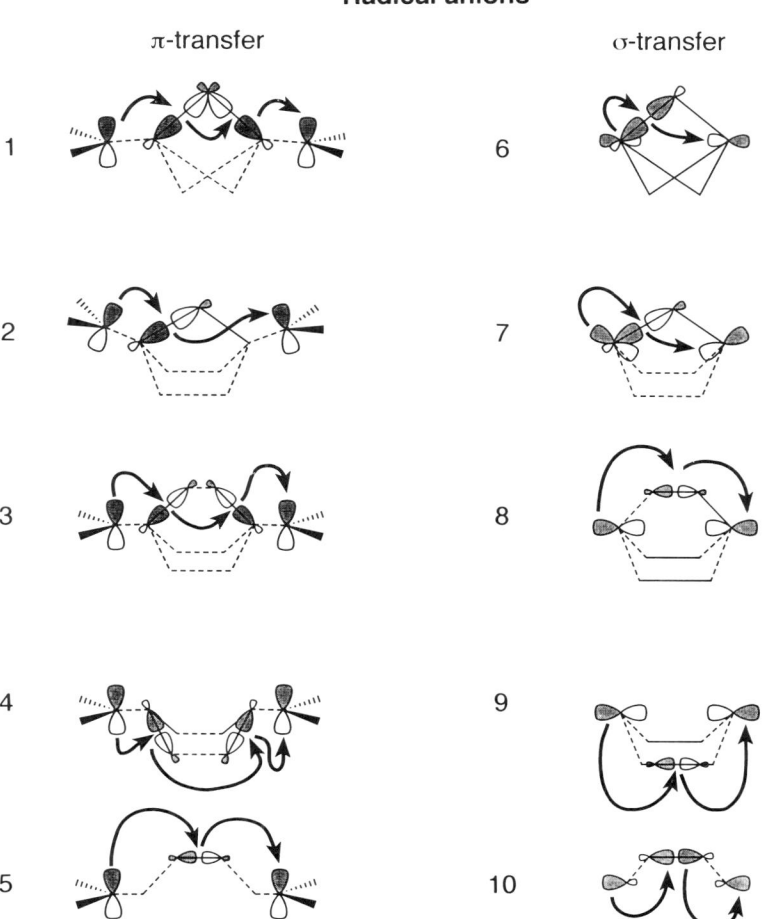

Figure 2.11 Orbital diagrams indicating the most important pathways for electron transfer in radical anion systems. Results are obtained from perturbation analysis based on NBOs obtained at the KT/3-21G level. The orbital lobes are drawn approximately to scale on the basis of the calculated NBOs. The bonds (or anti-bonds) involved in each pathway are denoted by solid lines. Solid lines are also used to denote pathways symmetry-equivalent to those explicitly depicted by orbitals. Other bonds are either not shown (CH bonds, except for those on the terminal CH_2 groups in the case of π-transfer) or indicated by dashed lines. Each primitive step in a given pathway is denoted by an arrow. (After Fig. 3 of [49], reprinted with permission. Copyright (1992) American Chemical Society.)

5.3.3   *TRANS*-STAGGERED BRIDGES [$1\pi(m)$]

A decomposition of T_{if} into hole, electron, and hybrid contributions for radical cations and anions of $1\pi(m)$ is presented in Table 2.9. For simplicity, we consider only CC and CC* contributions obtained at the STO-3G/KT level [43]. Comparable results have been obtained with inclusion of CH and CH* NBOs and a larger (i.e. 3-21G) basis set, although quantitative differences have been noted [64]. While pure hole pathways are always more important than pure electron pathways for coupling in radical cations, and vice versa for radical anions, we see that the hybrid pathways tend to increase in relative importance as the bridge is lengthened.

Table 2.9 Through-space (TS), hole (h), electron (e), and hybrid (h/e) contributions (%) to T_{if} from CC and CC* NBOs† for odd-membered *trans*-staggered alkane bridges [$1\pi(m)$]. (Table IV of [43], reprinted with permission. Copyright (1993) American Chemical Society.)

m	TS‡	h§	e§	h/e§	Mc‖
Radical cations					
1	– 7	65	42	0	65
3	– 2	65	12	25	12
5	0	58	2	40	2
7	0	53	0	47	1
Radical anions					
1	– 4	36	68	0	68
3	– 4	20	53	31	0
5	– 2	8	38	56	0
7	– 1	3	25	73	0

† The perturbative estimate of T_{if} based on the CC and CC* NBOs of the spacer group (terminal CC and CC* bonds linking the D–A CH_2 groups to the spacer are omitted); obtained with an STO-3G basis.

‡ Through-space contribution (based on Fock matrix element between D and A NBOs). Since the TS term differs in sign from the other contributions, it is assigned a negative per cent contribution.

§ h, e, and h/e denote contributions based, respectively, on pathways of the hole, electron and hybrid (both hole and electron) type, as discussed in Section 3.4. Together with the TS term, these contributions account for 100% of the CC and CC* contributions to T_{if}.

‖ Mc denotes the contributions from the nearest neighbour pathway of the hole type (radical cations) or electron type (radical anions), as given by Eqn 2.31.

An attempt to incorporate the complexity of the pathways for π-type D–A coupling mediated through a *trans*-staggered alkane bridge in terms of an effective or renormalized McConnell model has been reported for radical cations in [63]. With a model restricted to CC bonds and pathways of the 'forward' type (i.e. no 'retracings', even though these in fact make appreciable contributions to T_{if}), numerical coefficients were determined such that the scaled McConnell model reproduced the full perturbative result within the model adopted. The scaling coefficients displayed in Table 2.10 reflect in a compact manner the growth in number of *actual* pathways which contribute to the *effective* McConnell pathway. The scale factor is given approximately by

$$f_m \sim (1.6)^{m-2} \tag{2.56}$$

and makes a contribution of -0.7 Å^{-1} to the decay coefficient β (Eqn 2.9), offsetting the value of $+1.4$ Å^{-1} due to the *actual* (i.e. unscaled) McConnell result.

Further systematic work of this type should be of help in bridging the current gap between effective pathway models [65] and those based explicitly on detailed orbital calculations.

Considering now the effect of varying the D–A group, keeping the bridge constant, we summarize in Table 2.11 results at the 3-21G/KT level [63], which display the correlation between the D–A energy (either the absolute $E_{D/A}^0$, or the mean gap relative to B) and the decay coefficient (β) for transfer in radical cations (assumed to be primarily of the hole type). The overall trend is basically monotonic in the sense expected from the

Table 2.10 Renormalized McConnell-type (NN) model for coupling through *trans*-staggered alkane bridges ($T_{if} \propto f_m(t_{12}/\Delta)^{m-1}$).*

m	f_m
2	1.00
4	2.44
6	6.27
8	16.10

* Based on results reported in Table 7 of [63], at the 3-21G/KT and 6-31+G/KT levels for radical cations with bridges the same as for $1\pi(m)$, m even, but with terminal vinyl D, A groups (t_{12} refers to NN coupling between C–C σ NBOs and was assigned a typical value of – 4.0 eV, while the orbital energy gap Δ was taken as 10 eV). The 'renormalization' coefficients f_m were evaluated so as to reproduce the numerical results from perturbative calculations based on all possible 'forward' pathways involving the C–C NBOs (i.e. all 'forward' pathways of the pure 'hole' type).

simple McConnell-type model (Eqn 2.11), assuming a common effective t and terminal hopping integral T (see Eqn 2.31) for all D–A cases.

5.3.4   A SIMPLE TEST OF ADDITIVITY

In Table 2.12 we examine the possibility of simple additivity relations for radical anions with σ- or π-type coupling through bridges with a variable number of *cis* 3-bond linkers [49]. Only the through-bond component is considered (obtained by subtracting the NBO through-space term from the total KT/3-21G result for T_{if}). The through-space term makes a substantial, and essentially (within 10%) constant contribution to T_{if} for the three σ cases [43,49]. The contrast between patterns for σ and π coupling is striking: for σ coupling, the through-bond terms are simply proportional to the number of linkers, whereas no such pattern is displayed for π coupling (even taking due account of direction cosines governing the orientation of the D–A π-orbitals relative to each linker). This situation is at least superficially compatible with the 'most important' pathways shown in Fig. 2.11: i.e. they are all of the same type for σ coupling. However,

Table 2.11 Correlation of decay coefficient (β) with D–A energy level ($E^0_{D/A}$) and gap (Δ) for *trans*-staggered $[X(CH_2)_{m-1}X]^+$ radical cations.*

X	$E^0_{D/A}$† (Δ)‡	β§
H_2N-	– 9.8 (11.2)	0.50
HS-	– 10.1 (11.7)	0.43
$H_2C{=}CH$-	– 10.1 (10.8)	0.41
$HC{\equiv}C$-	– 10.9 (10.3)	0.20
HO-	– 12.6 (9.0)	0.08

* Adapted from data given in [63]; energies in eV, β in Å^{-1} (based on $\Delta r = 1.27$ Å), m is the number of CC sigma bonds (represented by NBOs based on the KT level with a 3-21G orbital basis.)
† Limiting values for $m = 14$ ($E^0_{D/A}$ depends only weakly on m).
‡ Mean value of Δ^h_j over all internal CC bonds of bridge, based on $m = 6$ results.
§ Limiting results for large m ($\beta_{12,14}$; see definition of local β given in footnote ‡ of Table 2.1). The local β's diminish in magnitude by up to a factor of two as m ranges from 2 to 16.

Table 2.12 Additivity relationships among T_{if} values (mhartree) for radical anion coupling through *cis* three-bond bridge linkers.*

Number of linkers	D–A symmetry type	
	σ	π
1	− 19.8 (− 19.8)†	− 4.3 (− 4.3)
2	− 37.9 (− 39.6)	+ 4.1 (− 2.1)
3	− 53.2 (− 59.4)	− 6.0 (− 6.5)

* The relevant DBA systems are schematically depicted in the last three entries of each column in Fig. 2.11. The sign conventions for the $-T_{if}$ quantities are as given in footnote † of Table 2.1, with 'in phase' corresponding to a_1 (C_{2v}) or a_{1g} (C_{2h}). Results are taken from Table II of [49] and are NBO through-bond contributions for the radical anions treated at the 3-21G/KT level.
† The quantities in parentheses are the expected values based on simple additive contributions from each linker, where each linker is assumed to follow an effective McConnell relationship proportional to T^2 (see Eqn 2.31), and T is weighted by an appropriate direction cosine in the case of π-type D–A groups.

as emphasized above, these pathways are not dominant. In an analogous study of norbornyl-based bridges [99], specific interference between linkers has been elucidated in terms of NBO-based pathways, and in general such effects are expected to disrupt simple additivity schemes for treating multiple linkers or pathways (see also [100,101]).

5.4   Applications of the GMH

The final set of applications deals with systems involving optical or excited state ET. These systems allow many of the general features of the GMH (Section 4) to be exploited.

5.4.1   ET IN THE FOUR-STATE $Zn_2(OH_2)^+$ MANIFOLD

The $Zn_2(OH_2)^+$ cluster (see Fig. 2.12) provides a model system which allows several features of D–A coupling of generic interest to be probed in detail. The four lowest energy electronic states (all spin doublets) are dominated by the four valence bond structures:

$$Zn(s^2)/Zn^+(s) \tag{2.57a}$$

$$Zn^+(s)/Zn(s^2) \tag{2.57b}$$

$$Zn(sp\sigma)/Zn^+(s) \tag{2.57c}$$

$$Zn^+(s)/Zn(sp\sigma) \tag{2.57d}$$

where s and pσ refer to the valence 4s and 4pσ Zn orbitals. Complexing an axial water molecule to one Zn atom, on the side opposite to the distant Zn, provides an additional degree of freedom (r_{ZnO}), which allows the homonuclear symmetry to be broken in a controlled fashion. The relative energies of the four low-lying adiabatic states (obtained by state-average *ab initio* CASSCF calculations [84]) are displayed in Fig. 2.12 for three

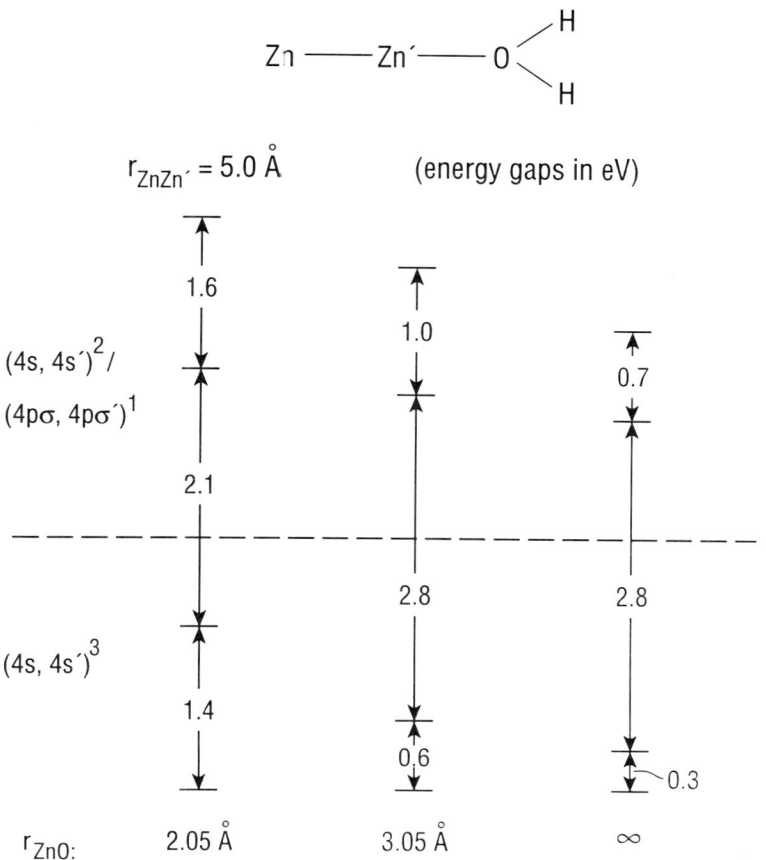

Figure 2.12 Adiabatic energy gaps for the lowest four states of the $Zn_2(OH_2)^+$ complex, based on state-averaged CASSCF calculations using a 12-orbital/three-electron active space to allow correlation of the electrons in the full valence manifold of the Zn_2 moiety, and with an augmented split-valence orbital basis for Zn and an SCF minimal basis for H_2O [77]. The lower and higher energy pairs of states at each of the three ZnO distances correspond roughly to the respective valence electronic configurations $(4s, 4s')^3$ and $(4s, 4s)^2(4p\sigma, 4p\sigma')^1$, where the prime denotes the Zn atom nearest the water molecule.

different values of r_{ZnO}, with r_{ZnZn} fixed at 5.0 Å. It is not immediately clear how to extract from the four-state adiabatic picture the effective coupling elements which govern the four possible ET processes among the diabatic states represented schematically in Eqn 2.57: these processes comprise thermal ground state ET (2.57a→2.57b); excited state ET (2.57c→2.57d), and two optical processes (2.57a→2.57d and 2.57b→2.57c). For completeness we note that the other possible optical processes (2.57a→2.57c and 2.57b→2.57d) involve local excitations of weak intensity because of the singlet→triplet character of the local transitions (corresponding to the designation of the excited states as 'trip-doublets' [102]). In the following we denote orbitals on the hydrated Zn with a prime.

Analysis of the four-state system defined above by use of the GMH model (Section 4.2) in conjunction with the full adiabatic dipole matrix and the state energies yields straightforwardly the results given in Table 2.13. The strength of coupling and the decay coefficients (β) are seen to display a sharp dependence on the character of the

Table 2.13 Values of $|T_{if}|$ (mhartree) and decay coefficients, β (Å^{-1}) for four electron transfer (ET) processes in $Zn_2(OH_2)^+$.*

ET process†	r_{ZnO} (Å)		
	2.05	3.05	∞
$|T_{if}|$ for r_{ZnZn} = 5.0 Å			
s/s′	10.5	8.0	7.3
p/p′	13.0	14.7	13.9
s/p′	51.7	38.5	24.4
p/s′	22.5	22.1	24.4
β based on range r_{ZnZn} = 5.0–9.0 Å‡			
s/s′	2.36	2.47	2.55
p/p′	1.10	1.26	1.55
s/p′	1.00	1.40	1.78
p/s′	1.81	1.83	1.78

* See structure in Fig. 2.12 and description of calculations in caption.
† The four ET processes are labelled in terms of the dominant valence orbitals involved in each case (see configurations 2.57 a–d); results are based on the diabatic states determined by the GMH analysis [84].
‡ Linear regression coefficient $r \geqslant 0.99$ in least squares fits of $\ln|T_{if}|^2$ vs r_{ZnZn}.

orbital pairs nominally involved in the four different processes (see above) and on the closeness of the perturbing water molecule (only in the limit of large r_{ZnO} do the sp′ and ps′ processes become equivalent). The variations with r_{ZnO} (i.e. departure from the limit given by the Condon approximation) are complex, displaying examples of non-monotonic dependence. Since the $Zn_2(OH_2)^+$ complex lacks a 'bridge', the coupling is direct or through-space, and indeed the β magnitudes are comparable to those found for coupling through vacuum [14,28,95] and correlate well with the ionization potentials (IP) of the corresponding neutral parents (i.e. as given by Eqn 2.10, with Δ^G replaced by IP). Attempts to approximate the results of the four-state analysis (Table 2.13) by a conventional pairwise approach (i.e. successive applications of the two-state GMH to pairs of adiabatic states) yield increasingly large departures from the four-state results as r_{ZnZn} decreases; e.g. for $r_{ZnZn} < 6$ Å, departures of up to 50% in T_{if} magnitudes are found. Furthermore, in the limit of weak hydration (i.e. large r_{ZnO}), treatment at the level of the TSA is simply incapable of accounting for the two distinct optical processes (sp′ and ps′).

5.4.2 SYMMETRY AND SOLVENT EFFECTS

We now apply the GMH approach to the three-state problem defined by the ground (DBA), locally excited (D*BA), and charge transfer (D^+BA^-) states of the two systems depicted in Fig. 2.13 (**8 and 9**), where D and A are, respectively, anthracene and olefin moieties, fused to bridges with frameworks comprising units based on norbornane and cyclobutane. Due to the orbital symmetries pertaining to an assumed C_s plane, the charge separation, CS ($D*BA \rightarrow D^+BA^-$), and charge recombination, CR ($D^+BA^- \rightarrow$ DBA), processes are, respectively, symmetry-forbidden and symmetry-allowed [103]. In this study, the adiabatic dipole matrix and state energies were obtained from INDO/S

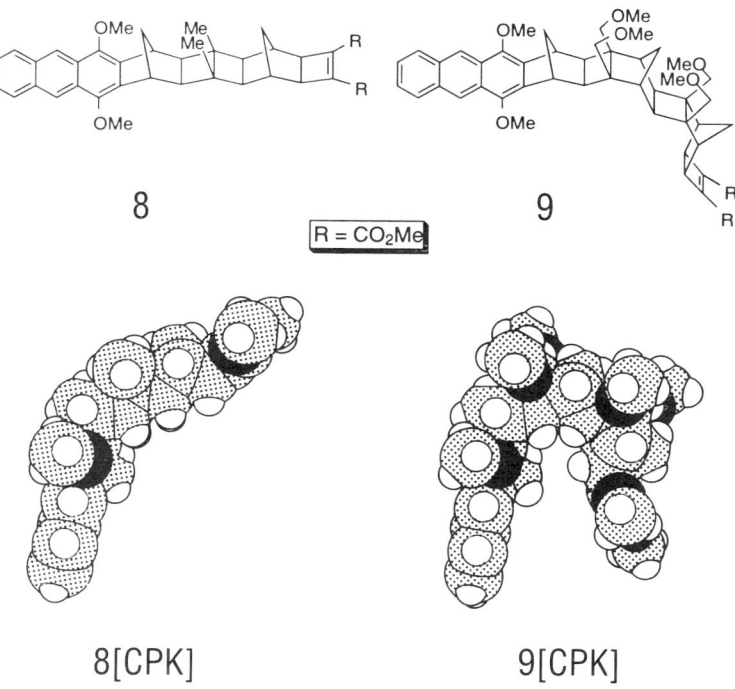

Figure 2.13 DBA systems **8** and **9**, involving anthracene donor and substituted olefin acceptor groups linked, respectively, by double relays of seven and nine CC bonds [103]. Space-filling profiles are depicted by CPK structures. (Reproduced from [103] with permission. Copyright (1995) American Chemical Society.)

CI calculations [91], designed to provide a 'balanced' treatment of the D*BA and D$^+$BA$^-$ excited states by including in the CI all single excitations from the highest occupied MO (homo) of the ground state. In order to mimic the influence of solvent, the DBA structures were first optimized (at the MM2 force field level [104]), starting with a small sample of assumed interaction geometries between a given solvent molecule (S) and **8** on **9**, obtained by placing S at various points (and with different orientations) adjacent to B on the underside of DBA (Fig. 2.13). The salient results are displayed in Table 2.14 for the cases S = n-pentane, S = acetonitrile (MeCN), S = benzonitrile (ϕCN), and also for the case where S is absent. In all cases, the results for a given DBA/S pair correspond to a Boltzmann-weighted rms-average of T_{if} over the set of solvent structures sampled.

As expected on symmetry grounds, $|T_{if}^{CS}| \ll |T_{if}^{CR}|$ is generally found (the symmetry-breaking influence of solvent yields non-zero values for T_{if}^{CS}). The primary mechanistic question is the precise role of solvent in enhancing the coupling as a function of solvent and bridge type. We see that solvent has only a modest (CS) or essentially no (CR) effect on coupling in the system with the 'straight' bridge (**8**), whereas there is a strong solvent effect for both CS and CR in the system with the bent ('C-clamp') bridge (**9**), especially for the solvent with the lowest lying empty orbitals (ϕCN). It is also found (in results not shown) that removal of the central portion of the bridge (tying off the dangling orbitals so created with H atoms) causes a drastic reduction of coupling strength in **8**, while a similar treatment of **9** has very little effect [103].

Table 2.14 Solvent effect on $|T_{if}|$ (cm^{-1})* for systems **8** and **9**. (Table 7 of [110], reprinted with kind permission from Elsevier Science.)

Solvent	8		9	
	CS†‡(exp)§	CR‖	CS†‡(exp)§	CR‖
Solvent-free	4 —	68	0.1 —	3
n-pentane	4 —	70	16 —	60
MeCN	6 (15)	62	7 (21)	16
φCN	19 (12)	72	46 (65)	231
	'Indirect' solvent effect (dominant DBA coupling)		'Direct' solvent effect (dominant DSA coupling)	

* Units of cm^{-1} (8066 cm^{-1} = 1 eV) are used here (in contrast to the use elsewhere of mhartree) to accommodate the very small magnitude of some of the T_{if} values. Results based on GMH diabatic states obtained from INDO/S CI calculations [101].
† CS≡charge separation (see Fig. 2.3b): D*BA→D$^+$BA$^-$.
‡ Symmetry-forbidden for C$_s$ point group symmetry (see Fig. 2.13).
§ Experimental values [103].
‖ CR≡charge recombination (see Fig. 2.3b): D$^+$BA$^-$→DBA.

We are thus led to the following picture. In **8**, the superexchange coupling is dominated by the bridge (B), and this B-mediated coupling is indirectly affected to a modest degree by the solvent, especially in the case of CS, due to solvent-induced symmetry breaking of the zero-order C$_s$ wavefunctions (the solvent perturbs the molecular structure of the DBA moiety very little from C$_s$ geometry). In contrast, a solvent molecule within the C-clamp cavity in **9** creates a dominant solvent-mediated superexchange pathway (DSA). The accessible orbitals of φCN make it especially effective in this role, but even the saturated pentane molecule proves effective. These results underscore the point made above — namely, that covalent bonding is not a requirement for effective superexchange coupling (Section 5.2.1). Finally, we note the good agreement between the calculated T_{if} values and the values available from analysis of experimental kinetic data [105].

5.4.3 EFFECTIVE ET DISTANCES

As a final illustration of the power of the GMH model, we evaluate the effective ET distances (r_{DA}, based on Eqns 2.49 and 2.51) associated with the processes discussed in Section 5.4.1 and 5.4.2, and also for two mixed-valence metal ion complexes, comparing the results with estimates (r_{DA}^0) based on molecular structure. The results are summarized in Table 2.15. For Zn$_2$(OH$_2$)$^+$, the r_{DA} are all appreciably shorter than the reference value, $r_{DA}^0 = r_{ZnZn}$. The choice of r_{DA}^0 in the case of the larger D and A groups in **8** and **9** is not so obvious. It turns out that the GMH values for both the CS and CR processes are close to the respective separation distances of the anthracene and double-bond midpoints. In contrast, for the two binuclear Ru complexes the r_{DA} values (obtained entirely in terms of experimental dipole and energy quantities [75,76,109]) are greatly reduced relative to r_{RuRu} values, reflecting the role of strong ligand-field mixing in determining the centroids of the effective D and A orbitals.

Table 2.15 Effective separation of D–A sites, r_{DA} (Å).

ET process	From molecular geometry (r_{DA}^0)	From GMH analysis*
$Zn_2(H_2O) + $†		
s/s′	5.00	4.67
s/p′	5.00	3.60
p/s′	5.00	3.88
p/p′	5.00	4.40
8		
CS‡	11.6§	12.1‖
CR¶	11.6§	11.8‖
9		
CS‡	7.1§	7.2‖
CR¶	7.1§	7.0‖
$Ru^{2+}LRu^{3+}$		
IT** { L = pz	6.8††	1.4‡‡
IT** { L = bpy	11.3††	5.2‡‡

* See Section 4 and [77].

† From four-state GMH analysis, using results of CASSCF calculations (Section 5.4 and [84]), with $r_{ZnO} = 3.05$ Å ($r_{DA}^0 = r_{ZnZn}$).

‡ Charge separation process (see Fig. 2.3b).

§ Distance separating the midpoints of the anthracene (D) and olefinic C=C bond (Å).

‖ Based on three-state GMH analysis, using results from INDO/S CI calculations [103].

¶ Charge recombination process (see Fig. 2.3b).

** Intervalence transfer (IT), the optical ET process depicted in Fig. 2.3a; pz and bpy denote the pyrazine and 4,4-bipyridine ligands. Each Ru also has five peripheral NH_3 ligands.

†† $r_{DA}^0 = r_{RuRu}$.

‡‡ From two-state GMH analysis using experimental adiabatic data (transition energy [76], dipole moment shift [75,109], and transition dipole [76]).

6 Concluding remarks

Quantitative evaluation and analysis of long-range electron coupling involving many classes of donor–acceptor pairs and mediated by a wide range of intervening bridge types has become a major activity in theoretical chemistry. The techniques of computational molecular quantum mechanics in conjunction with orbital-based theories of electronic coupling, including particularly the powerful hierarchy of superexchange-models, have achieved major success in identifying in precise form the fundamental molecular principles controlling the coupling pertinent to long-range electron (and hole) transfer. Fine-tuning of system characteristics — e.g. by variation of chemical composition of the primary sites (DBA) or the surrounding medium, or applying a variable external field — provides the basis for rational system design which is crucial to the emerging field of molecular electronics.

In the present chapter we have discussed general principles and theories (Sections 2–4) and then offered a number of specific examples of applications, mostly illustrating the properties of the important class of bridging materials comprising electronically saturated species. While the decomposition of bridges into local bonds and antibonds or similar 'fine-grained' components will continue to serve as a rich source of insight into coupling through complex chemical spacers, practical considerations for important cases of electron transfer in very large, extended systems pose a major challenge for the future — namely, to synthesize from the fine-grained ingredients, the more coarse-grained elements needed in the formulation of effective pathways suitable for manageable treatment of the very large systems. Electron transfer in protein systems is already playing an active role in this endeavour, driven by the goal of elucidating biochemical mechanisms, and major attention is being focused on electron transfer in a number of other extended systems, [1,15]. The rapid growth in computer power, and the development of sophisticated numerical techniques to exploit this power, will play an essential role by permitting exact solutions for suitable reference systems characterized by very large-scale model molecular Hamiltonians. Efforts toward design of materials with specified performance characteristics should also be greatly facilitated by development of tools for sensitivity analysis, permitting an assessment of whether charge transport is strongly controlled by a relatively small set of localized regions in an extended system, as opposed to more highly distributed mechanisms.

Acknowledgements. We wish to acknowledge helpful discussions with Professor M.B. Zimmt and Dr Y-P. Liu. This research was carried out at Brookhaven National Laboratory under contract DE-AC02-76CH00016 with the US Department of Energy and supported by its Division of Chemical Sciences, Office of Basic Energy Sciences.

7   References

 1 Mikkelsen KV, Ratner MA. *Chem Rev* 1987; **87**: 113.
 2 Marcus RA, Sutin N. *Biochim Biophys Acta* 1985; **811**: 265.
 3 Newton MD, Sutin N. *Annu Rev Phys Chem* 1984; **35**: 437.
 4 *Chem Rev* 1992; **92**: No. 3.
 5 Bertrand P. *Structure and Bonding* 1991; **75**: 1.
 6 Moser CC, Keske JM, Warncke K, Farid RS, Dutton PL. *Nature* 1992; **355**: 796.
 7 Sumi H, Marcus RA. *J Chem Phys* 1986; **84**: 4894.
 8 Bader JS, Kuharski RA, Chandler D. *J Chem Phys* 1990; **93**: 230.
 9 Kim HJ, Hynes JT. *J Chem Phys* 1992; **96**: 5088.
10 Jortner J, Bixon M. *J Chem Phys* 1988; **88**: 167.
11 Larsson S. *J Am Chem Soc* 1981; **103**: 4034.
12 Ratner MA. *J Phys Chem* 1990; **94**: 4877.
13 Onuchic JN, Beratan DN. *J Chem Phys* 1990; **92**: 722.
14 Newton MD. *Chem Rev* 1991; **91**: 767.
15 Mirkin CA, Ratner, MA. *Annu Rev Phys Chem* 1992; **43**: 719.
16 Aviram A, ed. In: *Molecular Electronics — Science and Technology.* New York: Am Inst Phys, 1992.
17 Aviram A, Ratner MA. *Chem Phys Lett* 1974; **29**: 277.
18 Aviram A. *J Am Chem Soc* 1988; **110**: 5687.
19 Kemp M, Mujica V, Ratner MA. *J Chem Phys* 1994; **101**: 5172.
20 Mujica V, Kemp M, Ratner MA. *J Chem Phys* 1994; **101**: 6856.
21 Petrov EG, Tolokh IS, Demidenko AA, Gorbach VV. *Chem Phys* 1995; **193**: 237.
22 Joachim C. *New J Chem* 1991; **15**: 223.
23 Waldeck DH, Beratan DN. *Science* 1993; **261**: 576.

24 Farazdel A, Dupuis M. *Phys Rev* 1991; **B44**: 3909.

25 Hush NS, Wong AT, Bacskay BG, Reimers JR. *J Am Chem Soc* 1990; **112**: 4192.

26 Reimers JR, Hush NS. *Chem Phys* 1993; **176**: 407.

27 Broo A, Zerner MC. *Chem Phys* 1995; **196**: 423.

28 Cave RJ, Baxter DV, Goddard III WA, Baldeschwieler JD. *J Chem Phys* 1987; **87**: 926.

29 Warshel A, Creighton S, Parson WW. *J Phys Chem* 1988; **92**: 2696.

30 Scherer POJ, Fischer SF. *Chem Phys* 1989; **131**: 115.

31 Marchi M, Gehlen JN, Chandler D, Newton MD. *J Am Chem Soc* 1993; **115**: 4178.

32 Marcus RA. *Chem Phys Lett* 1987; **133**: 471.

33 Reimers JR, Hush NS. *Chem Phys* 1990; **146**: 105.

34 Jordan KD, Paddon-Row MN. *Chem Rev* 1992; **92**: 395.

35 Michel-Beyerle ME, Bixon M, Jortner J. *Chem Phys Lett* 1988; **151**: 188.

36 Koga N, Sameshima K, Morokuma K. *J Phys Chem* 1993; **97**: 13117.

37 Scholes GD, Ghiggino KP, Oliver AM, Paddon-Row MN. *J Phys Chem* 1993; **97**: 11871.

38 Onuchic JN, Beratan DN, Hopfield JJ. *J Phys Chem* 1986; **90**: 3707.

39 Tang J. *J Chem Phys* 1993; **98**: 6263.

40a Landau L. *Phys Z Sowjet* 1932; **2**: 46.

40b Zener C. *Proc Roy Soc London* 1932; Ser A: 696.

41 Kuznetsov AM, Ulstrup J. *J Chem Phys* 1981; **75**: 2047.

42 Beratan DN, Onuchic JN, Hopfield JJ. *J Chem Phys* 1987; **86**: 4488.

43 Liang C, Newton MD. *J Phys Chem* 1993; **97**: 3199.

44 Stuchebrukhov AA. *Chem Phys Lett* 1994; **225**: 55.

45 Newton MD, Ohta K, Zhong E. *J Phys Chem* 1991; **95**: 2317.

46 Gamow G. *Z Phys* 1928; **51**: 204.

47 McConnell HM. *J Chem Phys* 1961; **35**: 508.

48 Miller JR, Beitz JV. *J Chem Phys* 1981; **74**: 6746.

49 Liang C, Newton MD. *J Phys Chem* 1992; **96**: 2855.

50 Siddarth P, Marcus RA. *J Phys Chem* 1992; **96**: 3213.

51 Arnold BR, Noukakis D, Farid S, Goodman JL, Gould IR. *J Am Chem Soc* 1995; **117**: 4399.

52 Liu Y-P, Newton MD. *J Phys Chem* 1995; **99**: 12382.

53 Evenson JW, Karplus M. *J Chem Phys* 1992; **96**: 5272.

54 Skourtis SS, Beratan DN, Onuchic JN. *Chem Phys* 1993; **176**: 501.

55 Wynne K, Galli C, Hochstrasser RM. *J Chem Phys* 1994; **100**: 4797.

56 Todd MD, Nitzan A, Ratner MA. *J Phys Chem* 1993; **97**: 29.

57 Beratan DN, Onuchic JN. *Adv Chem Ser* 1991; **228**: 71.

58 Onuchic JN, de Andrade PCP. *J Chem Phys* 1991; **95**: 1131.

59 Gruschus JM, Kuki A. *J Phys Chem* 1993; **97**: 5581.

60 Okada A, Kakitani T, Inoue J. *J Phys Chem* 1995; **99**: 2946.

61 Stuchebrukhov AA, Marcus RA. *J Phys Chem* 1995; **99**: 7581.

62 Regan JJ, Risser SM, Beratan DN, Onuchic JN. *J Phys Chem* 1993; **97**: 13083.

63 Shephard MJ, Paddon-Row MN, Jordan KD. *Chem Phys* 1993; **176**: 289.

64 Curtis LA, Naleway CA, Miller JR. *Chem Phys* 1993; **176**: 387.

65 Beratan DN, Betts JN, Onuchic JN. *J Phys Chem* 1992; **96**: 2852.

66 Skourtis SS, Regan JJ, Onuchic JN. *J Phys Chem* 1994; **98**: 3379.

67 Arnobio A, da Gama S. *J Theor Biol* 1990; **142**: 251.

68 Goldman C. *Phys Rev A* 1991; **43**: 4500.

69 Lopez-Castillo J-M, Filali-Mouhim A, Jay-Gerin J-P. *J Phys Chem* 1993; **97**: 9266.

70 Evenson JW, Karplus M. *Science* 1993; **262**: 1247.

71 Richardson DE, Taube H. *J Am Chem Soc* 1983; **105**: 40.

72 Bixon M, Jortner J, Verhoeven JW. *J Am Chem Soc* 1994; **116**: 7349.

73 Mulliken RS. *J Am Chem Soc* 1952; **64**: 811.

74 Hush NS. *Electrochim Acta* 1968; **13**: 1005.

75 Reimers JR, Hush NS. *J Phys Chem* 1991; **95**: 9773.

76 Creutz C, Newton MD, Sutin N. *J Photochem Photobiol A: Chem* 1994; **82**: 47.

77 Cave RJ, Newton MD. *Chem Phys Lett* 1996; **249**: 15.

78 Sutin N. *Adv Chem Ser* 1991; **228**: 25.

79 Werner H-J, Meyer W. *J Chem Phys* 1981; **74**: 5802.

80 Macias A, Riera A. *J Phys* 1978; **B11**: L489.

81 Kato S, Amatatsu Y. *J Chem Phys* 1990; **92**: 7241.

82 Kim HJ, Bianco R, Gertner BJ, Hynes JT. *J Phys Chem* 1993; **97**: 1723.

83 Shin Y-GK, Brunschwig BS, Creutz C, Sutin N. *J Am Chem Soc* 1995; **117**: 8668.

84 Cave RJ, Newton MD. *J Chem Phys* 1997; in press.

85 Pacher T, Cederbaum LS, Köppel H. *J Chem Phys* 1988; **89**: 7367.

86 Domcke W, Woywood C, Sengle M. *Chem Phys Lett* 1994; **226**: 257.

87 Marchi M, Chandler D. *J Chem Phys* 1991; **95**: 889.

88 Braga M, Larsson S. *Chem Phys Lett* 1993; **213**: 217.

89 Kim K, Jordan KD, Paddon-Row MN. *J Phys Chem* 1994; **98**: 11053.

90 Newton MD. *J Phys Chem* 1991; **95**: 30.

91 Zerner MC, Loew GH, Kirchner RF, Mueller-Westerhoff UT. *J Am Chem Soc* 1980; **102**: 589.

92 Koopmans T. *Physica* 1993; **1**: 104.

93 Braga M, Larsson S. *J Phys Chem* 1993; **97**: 8929.

94 Curtiss LA, Naleway CA, Miller JR. *J Phys Chem* 1995; **99**: 1182.

95 Curtiss LA, Naleway CA, Miller JR. *J Phys Chem* 1993; **97**: 4050.

96 Krongauz VV. *J Phys Chem* 1992; **96**: 2609.

97 Naleway CA, Curtiss LA, Miller JR. *J Phys Chem* 1991; **95**: 8434.

98 Reed AE, Curtiss LA, Weinhold F. *Chem Rev* 1988; **88**: 899.

99 Shephard MJ, Paddon-Row MN, Jordan KD. *J Am Chem Soc* 1994; **116**: 5328.

100 Beratan DN. *J Am Chem Soc* 1986; **108**: 4321.

101 Onuchic JN, Beratan DN. *J Am Chem Soc* 1987; **109**: 6771.

102 Reimers JR, Hush NS. *Inorg Chim Acta* 1994; **226**: 33.

103 Cave RJ, Newton MD, Kumar K, Zimmt MB. *J Phys Chem* 1995; **99**: 17501.

104 Allinger NL. *J Amer Chem Soc* 1977; **99**: 8127.

105 Kumar K, Sin Z, Waldeck DH, Zimmt MB. *J Am Chem Soc* 1996; **118**: 243.

106 Sanz JF, Malrieu JP. *J Phys Chem* 1993; **97**: 99.

107 Colapietro M, Domenicano A. *Acta Cryst* 1978; **B34**: 3277; 1982; **B38**: 1953.

108 Ringnalda MN, Langlois J, Greeley BH, Russo TV, Müller RP, Marten B *et al.* PS-GVB v2.0. Schrödinger, 1994.

109 Oh DH, Sano M, Boxer SG. *J Am Chem Soc* 1991; **113**: 6880.

110 Newton MD. *J Electroanal Chem* 1996; in press.

3 External Field Control of Tunnelling in Dissipative Two-state Systems

R.I. CUKIER* and M. MORILLO*†

*Department of Chemistry and Center for Fundamental Materials Research, Michigan State University, East Lansing, MI 48824-1322, USA

†Fisica Teorica, Facultad de Fisica, Universidad de Sevilla, Apdo Correos 1065 Sevilla 41080, Spain

1 Introduction

Solid-state devices have provided the desired stability, storage and switchability required for the realization of computational systems. Limitations on their further miniaturization have stimulated the field of molecular electronics [1–7], where molecules and their assemblies are to become the circuit elements. One molecular paradigm that has often been featured in this arena is that of molecules that can tautomerize or otherwise exhibit two-state behaviour. Advances in supramolecular chemistry coupled with potentialities inferred from biological electron transfer processes have played a large role in guiding these investigations. For example, the precise molecular geometry of the photosynthetic reaction centre permits a photoinitiated chain of electron transfer reactions to occur with the result of separating charge with high efficiency; and this must and does occur repetitively and reproducibly [8]. In this sense, nature has control over the charge separation. However, for the purposes of storage and switching, the system must be connected to the external world in a way that permits the *external* control of the state of the system. This is the sense in which we will use the term 'control'. In fact, in 1974 Aviram and Ratner [9] suggested just this strategy in the context of molecular electronics when they proposed that a molecular rectifier be based on biasing an electron transfer molecule with an external electric field.

Our objective is an investigation of the prospect of controlling tunnelling by the application of an external field. Tunnelling is responsible for the transfer of a 'particle' from one spatially localized region to another by the quantum mechanical process of wavefunction penetration of a potential energy barrier [10]. Examples of tunnelling species include protons, hydrogen and heavier atoms, chemical groups such as methyl group, electrons, and defects and impurities in solids [11–18]. Throughout this work we will assume that distinct species can be identified by their spatial localization. Naturally, if the barrier separating these localized states is so low that it does not make sense to consider localized particles as a zero-order description of the system, we would not be able to contemplate storing information by virtue of the system's state. As is well known [10], the probability of tunnelling increases as the particle's mass decreases, the height and width of the barrier of the potential decreases, and maximizes for a symmetric potential profile. At the most basic level, an external field can control tunnelling by biasing the potential profile. For example, if the potential profile experienced by the tunnel particle were symmetric in the absence of an external field, then its application could provide an asymmetry that would greatly reduce the possibility of tunnelling. Conversely, an asymmetric potential profile in the absence of an external field could be symmetrized by its application and greatly enhance the tunnelling. As a tunnelling probability is strongly dependent on the degree of asymmetry in the potential, external fields can have dramatic effects on the tunnelling.

In order to externally control tunnelling, the tunnel particle must be connected to the external field. As it is often the case that tunnel systems exhibit a different charge distribution in their different localized states, an appropriate coupling would be via an electric field. In simplest form (for a charge-neutral tunnel system), the coupling then is between the dipole of the particle and the electric field. Other couplings can be addressed. For example, if the tunnel particle is best described as a hydrogen atom, the coupling would be to an external pressure field. In view of the flexibility of electric fields with regard to intensity and frequency, and the prevalence of tunnel systems that exhibit significant charge displacement on transfer, electric fields should be viewed as prime candidates for controlling tunnelling. We will, in fact, see that a number of possibilities for the control of tunnelling can occur when fields whose strength and frequency can be manipulated precisely are used.

In general, a discussion of tunnelling should be carried out from the perspective of the characteristics of the potential energy surface that the particle experiences. However, as was appreciated from the outset, a reduction from a spatially dependent description to one in terms of quantum states is very profitable and appropriate. In particular, if the barrier characteristics and particle mass are appropriate, a reduction to a two-state system is possible. Qualitatively, all that is required for this reduction is a set of energy levels for the particle that consist of two closely spaced levels, the tunnel doublet, followed by higher excited states that are not thermally accessible to the particle. In addition, with an external field, we have to assume that excitations from the tunnel pair to higher state will not occur. This pattern of energy levels is typically the case for the transfer of electrons [19] and can occur for suitably chosen proton, hydrogen and small-group transfers [20], as well as for defects and impurity tunnelling in solids [18]. We shall assume the correctness of a reduction to a two-level system throughout this chapter.

A tunnel system is typically not isolated from its surroundings. In addition to the coupling of the tunnel system to, e.g. an electric field, the influence of the tunnel system's coupling to the surroundings must be investigated. There are two distinct sources of coupling to the surroundings: external and internal. By external we shall mean that the applied field also has a stochastic component. We view the use of a fluctuating external field as a potential method of controlling the tunnelling characteristics in much the same way as a deterministic external field. The simplicity that follows from using an external fluctuating field is that the reaction of the system itself on the field (both the deterministic and stochastic parts) can be neglected. Thus, the external field affects the tunnelling system, but not vice versa. A fluctuating external field has the potential to convert an oscillating tunnel system population to a decaying one, and lead to a rate process rather than the usual periodic behaviour of an isolated tunnel system. The other coupling to the surroundings we will consider, and refer to as internal, is that created by the fluctuations of the medium in which the tunnel system is typically embedded. Internal coupling is characteristic of most tunnel systems as they are typically embedded in a medium. The complication that arises is from the effect that the tunnel system has on the medium in addition to the effect that the medium has on the tunnel system. In most studies incorporating the effect of the medium, the coupling is modelled as that of a charge to medium dipoles (the dielectric medium approximation that dominates electron transfer theory) or displacement field of a hydrogen atom, e.g. to the medium's acoustic phonons. Again, the role of the medium fluctuations will be to potentially lead to a rate process for the population decay of the tunnel system. In both cases of external and internal fluctuations, our efforts will be devoted to understanding the role of the fluctuations on the possibilities for controlling tunnelling. In what follows, we shall refer to the fluctuating field as a *noise*, for both internal and external fluctuations.

There are a number of parameters that enter the control of tunnelling expressions presented herein. The tunnel system is characterized by $\hbar\omega_0$, essentially the separation between the energy levels of what would be a degenerate system in the absence of the interaction responsible for tunnelling; its value is the province of quantum chemical calculations. The external field is characterized by an intensity b and, for a periodic field, its frequency Ω. Note that b is, e.g. the dot product of the transition dipole of the tunnelling species and the electric field intensity. For some fixed dipole value, the electric field intensity can be used to control the value of b. The coupling to the surroundings for external noise will be characterized by a strength (bandwidth) Δ and a correlation time τ_c. For external noise, these latter parameters can be specified at will. When the coupling is internal and the medium can be treated classically Δ and τ_c are simply related to the thermal properties of the medium. τ_c is a correlation time for, e.g. dielectric relaxation in the medium and $\Delta \approx \sqrt{E_r k_B T/\hbar}$, with T the temperature, is a thermal bandwidth. The reorganization energy E_r is a measure of the strength of the coupling of the medium to the tunnel system. Also, the medium's interaction with the tunnel system introduces a solvation free energy ΔG^0 as the difference in system–medium (free) energy when the system is localized in its two possible states. This is a thermodynamic, equilibrium property. When the medium must be treated quantum mechanically in the sense that the characteristic medium frequencies are not small compared with the temperature, the coupling between the medium and tunnel system is characterized by a spectral density. Then, more particulars of the medium enter the theory. Nevertheless, the control of tunnelling expressions we will develop will be qualitatively the same as the simpler classical medium expression.

Our exposition will start with systems isolated from their surroundings and then follow a path of increasing complexity that arises from the external and internal couplings to the surroundings. Throughout we will always focus on the equations of motion of the tunnel system so that the entire dynamic process can be discussed for different initial conditions. This contrasts with the approach of directly obtaining an approximation to the population evolution by, e.g. a golden rule calculation. This latter approach is directed toward obtaining a rate constant for the rate of population decay. While this is often the regime that emerges in the presence of coupling to a medium, we shall address examples where the approach to an equilibrium state from some prepared initial state is not monotonous. When the external field is time dependent an equation of motion method should be used, as golden rule calculations rely on short time approximations that may be violated in general.

Section 2 shows that a deterministic external field can control the coherent oscillations of an otherwise isolated tunnel system. Two methods of control are considered. First, a method where the external field is periodic and there is a specific connection between the amplitude and frequency of the field. This will be referred to as Floquet control, as the idea arises from the suggestion that the clock-like tunnelling of an isolated system can be stopped by the application of such a field [21,22]. Second, a method where an external field is applied that may be constant or time varying with no *a priori* connection between amplitude and magnitude of the field. We will show that Floquet fields of appropriate strength can indeed suppress tunnelling indefinitely. We will also show that sufficiently strong fields of arbitrary frequency can also strongly suppress tunnelling, the more so as the field frequency goes to zero. The role of an external noise source on these results is addressed in Section 3. By adding a stochastic term to the system–external field coupling term, we obtain stochastic equations of motion for the elements of the relevant system operators. These are solved analytically for appropriate parameter regimes and also numerically by stochastic trajectory simulations. We will find a number of regimes of behaviour. For strong coupling where $\Delta/\omega_0 \gg 1$ the population evolution of an initially localized state for a constant in time external field is an exponential decay. When the coupling to the medium is not so strong, $\Delta\tau_c \ll 1$, the equations of motion for the system averaged over the noise describe an oscillatory decay process for the population. We will find that tunnelling can be controlled with the application of an external field though naturally the evolution of population has an entirely different character than that of an isolated system. Section 4 considers the role of the medium fluctuations (internal noise) in the possibility of tunnel control when the medium can be treated classically. We describe the system plus medium with the spin Boson Hamiltonian, where the system–medium interaction is bilinear in the system and medium operators, augmented by the coupling to the external field. An analysis of the system dynamics leads to equations of motion that are similar in appearance to those of Section 3 where now the noise term is specified by the statistical properties of the medium. Therefore, similar regimes to those of Section 3 emerge. The Arrhenius form of the rate constant familiar from charge transfer theory in the absence of an external field is obtained. A constant external field acts simply as another source of reaction free energy ΔG^0. As this appears in an exponential dependence, the effect on a rate constant is dramatic. When the external field is time dependent, there is no rate constant, strictly speaking, but we are able to derive useful expressions for an external field period-averaged rate constant. This rate constant shows that control of tunnelling with periodic external fields can also be accomplished in the presence of internal noise. Finally, in Section 5, we consider control of tunnelling when the medium should be treated quantum mechanically. A more elaborate formalism is used here, a projection operator method [23,24], as low temperature requires a special treatment. In particular, the tunnel splitting ω_0 is no longer suitable as the natural expansion parameter. The Hamiltonian should first be rearranged into a form suitable for carrying out perturbation theory valid at low temperature. The result of our calculations is a set of equations of motion for the system density matrix elements with coefficients specified in terms of the spectral density that arises in spin Boson Hamiltonian theories [13,25,26], and the effects arising from the coupling to the external field. We are able to recast our general expressions in a way that relates them to the simpler classical medium expressions obtained in Section 4. While the parameters change, the same principles for the control of tunnelling as in the classical medium case emerge here. Thus, we are able to provide a rather complete picture of how to control tunnelling in two-level systems.

The exposition herein reflects mainly our own work. Some of the material presented in Sections 2–4 has appeared previously [22,27]. Here we have tried to present these ideas in a more unified way, starting with the simpler case of an isolated system and concluding with the complex situation involving the quantum aspects of both the tunnelling object and the medium degrees of freedom (Section 5). Throughout, we have tried to provide analytic expressions that illustrate the principles that can be found for the control of tunnelling with external fields. This requires the use of fields of simple form, either constant or sinusoidal. At the expense of some numerical analysis, our results can also accommodate the use of fields of arbitrary shape.

We conclude this introduction by presenting the following common notation for the states of the two-level system. The ground and excited state of the system Hamiltonian are denoted by $|1\rangle$ and $|2\rangle$, respectively. These are the spectroscopic, stationary states of the two-level system. Since our focus will be on tunnelling between spatially localized states, we also introduce the orthogonal left $|L\rangle$ and right $|R\rangle$ localized states that are related to the $|1\rangle$ and $|2\rangle$ eigenstates according to

$$|L\rangle = (1/\sqrt{2})[|1\rangle + |2\rangle] \text{ and } |R\rangle = (1/\sqrt{2})[|1\rangle - |2\rangle] \,. \tag{3.1}$$

These localized states are not eigenfunctions of the two-level system Hamiltonian. If the system wavefunction is initially, e.g. $|R\rangle$, its time evolution will exhibit coherent oscillations at the tunnel frequency between these localized states. Because the system has only two states, it is convenient to use a spin representation. In terms of the above basis sets, the Pauli spin operators and the unit operator are defined as

$$s_x = |1\rangle\langle2| + |2\rangle\langle1| = |L\rangle\langle L| - |R\rangle\langle R|$$

$$is_y = |1\rangle\langle2| - |2\rangle\langle1| = |R\rangle\langle L| - |L\rangle\langle R| \tag{3.2}$$

$$s_z = |1\rangle\langle1| - |2\rangle\langle2| = |R\rangle\langle L| + |L\rangle\langle R|$$

$$\hat{1} = |1\rangle\langle1| + |2\rangle\langle2| = |L\rangle\langle L| + |R\rangle\langle R|.$$

In the $|1\rangle$, $|2\rangle$ basis, the Pauli operators have the following matrix representation

$$s_x = \begin{pmatrix} 0 & 1 \\ 1 & 0 \end{pmatrix}, \quad is_y = \begin{pmatrix} 0 & 1 \\ -1 & 0 \end{pmatrix}, \quad s_z = \begin{pmatrix} 1 & 0 \\ 0 & -1 \end{pmatrix}. \tag{3.3}$$

2   Tunnelling in a two-state system subject to an external field

The Hamiltonian appropriate to a symmetric two-level system (TLS) in the presence of an external field that couples to the transition dipole of the tunnelling object is

$$H(t) = \frac{\hbar\omega_0}{2} s_z + 2b(t)s_x. \tag{3.4}$$

Here, $|2V| = \hbar\omega_0$ is the splitting of the tunnel doublet in the absence of coupling to a medium (solvent) and the coupling energy to the external field is denoted as $b(t)$ (the factor of 2 is included for notational convenience). The spin operators, s_x, s_y, and s_z, are the conventional Pauli matrices, as defined in Eqn. 3.3. The above Hamiltonian is written on the basis of the tunnel doublet eigenstates $|1\rangle$ and $|2\rangle$ of the unperturbed system. The coupling V is responsible for the tunnelling between the left-localized $|L\rangle$ and right-localized $|R\rangle$ states, which are defined as the symmetric and antisymmetric linear combinations of the eigenstates as given in Eqn. 3.1. The problem can be formulated in the localized basis as well; it is purely a matter of convenience as to which basis is used. The origin of the $b(t)$ term is the coupling of an external electric field $\mathbf{E}(t)$ to the transition dipole moment $\mathbf{m}$ of the tunnelling object. It has the form [28–30]

$$2b(t) = \mathbf{m} \cdot \mathbf{E}(t), \tag{3.5}$$

where $\mathbf{m} = |\mathbf{m}|\mathbf{i}$ with $\mathbf{i}$ a unit vector pointing in the direction of the transition dipole moment, and $|\mathbf{m}| = |\langle 1|e\mathbf{r}|2\rangle|$ is its magnitude.

When the external perturbation is periodic in time, e.g.

$$b(t) = b \cos \Omega t, \tag{3.6}$$

an isolated TLS can be analysed using Floquet theory [31–34]. As shown by e.g. Shirley [32], Floquet states can be obtained from the eigenvalue–eigenfunction equation

$$\left[H(t) - i\hbar \frac{\partial}{\partial t} \right] \mu_\alpha(t) = \varepsilon_\alpha \mu_\alpha(t), \tag{3.7}$$

where $\mu_\alpha(t + T) = \mu_\alpha(t)$ with $T = 2\pi/\Omega$. This demonstrates that the Floquet eigenfunctions are periodic in the external frequency. The $\varepsilon = \varepsilon(b, \Omega)$ are known as the quasi-energies. Note that the Schrödinger equation is obtained by setting the right-hand side of Eqn. 3.7 to zero, A solution of the Schrödinger equation is a linear combination of these Floquet eigenfunctions and is not a periodic function of time, except under special circumstances that we now describe. Consider the following linear combinations of these Floquet states:

$$|L_F(t)\rangle = \frac{1}{\sqrt{2}} [|\mu_1(t)\rangle + |\mu_2(t)\rangle], \; |R_F(t)\rangle$$

$$= \frac{1}{\sqrt{2}} [|\mu_1(t)\rangle - |\mu_2(t)\rangle]. \tag{3.8}$$

We expand the wavefunction $\Psi(t)$ of the Hamiltonian $H(t)$, using this basis, as

$$\Psi(t) = a_1(t)|L_F(t)\rangle + a_2(t)|R_F(t)\rangle, \tag{3.9}$$

where the $a_\alpha(t)$s ($\alpha = 1, 2$) are the expansion coefficients. As an initial condition we choose a left localized state; thus $a_1(0) = 1$ and $a_2(0) = 0$. Inserting this $\Psi(t)$ in Schrödinger's equation yields the equations of motion for the coefficients

$$-i\hbar \frac{\partial a_1}{\partial t} = -\left(\frac{\varepsilon_1 + \varepsilon_2}{2} \right) a_1 - \left(\frac{\varepsilon_1 - \varepsilon_2}{2} \right) a_2; \; -i\hbar \frac{\partial a_2}{\partial t} = -\left(\frac{\varepsilon_1 + \varepsilon_2}{2} \right) a_2 - \left(\frac{\varepsilon_1 - \varepsilon_2}{2} \right) a_1. \tag{3.10}$$

If, for some b and Ω values of the Hamiltonian given in Eqns. 3.4 and 3.6, the quasi-energies are degenerate, i.e. $\varepsilon_1 = \varepsilon_2$, then $|a_1(t)| = |a_1(0)|$. Therefore, $a_1(t)$ would be constant up to an irrelevant phase factor and $a_2(t)$ would be zero for all times. Thus, if the system is initially in state $|L_F(0)\rangle$, and if $\varepsilon_1 = \varepsilon_2$, then the system would remain in the state $|L_F(t)\rangle$ for all time.

Since $|L_F(t)\rangle = |L_F(t + T)\rangle$, if the initial state is chosen such that $\Psi(0) = |L_F(0)\rangle$, then, for multiples of the period T, the system will again be in state $|L_F(0)\rangle$. Now, if we can guarantee that, during the time interval $0 < t < T$ the system remains close to state $|L_F(0)\rangle$, then it will stay essentially in this state forever. Grossmann and Hanggi [21] point out that, for external fields of interest here, the two Floquet eigenstates are similar in shape and have the same parity as the doublet eigenstates of the Hamiltonian in the absence of the external field. Therefore, $|L_F(0)\rangle$ corresponds to a left-localized state. Let us now discuss conditions that will guarantee this outcome based on the system's equations of motion. The Heisenberg representation of the equation of motion for the operator O is

$$i\hbar\dot{O}(t) = [O(t), H(t)]. \tag{3.11}$$

With the Hamiltonian of Eqn. (3.4), we obtain

$$\dot{s}_x(t) = -\omega_0 s_y(t)$$

$$\dot{s}_y(t) = \omega_0 s_x(t) - \frac{4b(t)}{\hbar} s_z(t) \tag{3.12}$$

$$\dot{s}_z(t) = \frac{4b(t)}{\hbar} s_y(t).$$

From the above, the expectation values $x(t) = \langle L|s_x(t)|L\rangle$, $y(t) = \langle L|s_y(t)|L\rangle$, and $z(t) = \langle L|s_z(t)|L\rangle$, obtained with the use of Eqn. 3.4, satisfy the set of equations

$$\dot{x}(t) = -\omega_0 y(t)$$

$$\dot{y}(t) = \omega_0 x(t) - \frac{4b(t)}{\hbar} z(t) \tag{3.13}$$

$$\dot{z}(t) = \frac{4b(t)}{\hbar} y(t).$$

The connection between the probability of being on the left side at time t, $P_L(t)$, and $x(t)$ is obtained via

$$P_L(t) = |\langle L|L(t)\rangle|^2 = \langle L|L(t)\rangle\langle L(t)|L\rangle = \langle L|\frac{1}{2}(\hat{1} + s_x(t))|L\rangle = \frac{1}{2}(1 + x(t)). \tag{3.14}$$

The conditions for the degeneracy of the quasi-energies can be obtained from Shirley's [32] and Aravind and Hirschfelder's [34] work. The conditions are implicit but, for one limiting case, can be given a simple analytic expression. Shirley showed that if

$$\omega_0 \ll \max[\Omega, (b\Omega/\hbar)^{1/2}] \tag{3.15a}$$

the difference of Floquet quasi-energies can be expressed as

$$\varepsilon_1 - \varepsilon_2 = \hbar\Omega - \hbar\omega_0 J_0(4b/\hbar\Omega), \tag{3.15b}$$

where J_0 is the Bessel function of order zero. This equality is modulo Ω. The condition $\varepsilon_1 = \varepsilon_2$, obtained from the first zero of the Bessel function, then yields the important relation

$$4b/\hbar\Omega = 2.4048\ldots \tag{3.16}$$

As the tunnel term is small compared with the external driving term, the solution of Eqn. 3.13 will be dominated by the external term and, with the given initial condition of the population being in the left well, the probability $P_L(t)$ will stay close to its initial value during one tunnel period. Note that the condition $\omega_0 \ll \Omega$ indicates that there are many Floquet periods within one bare tunnel period $2\pi/\omega_0$. Thus, the solution of the equations of motion with the conditions stipulated above, and b and Ω connected via Eqn. 3.16, should lead to localization of the tunnel object in the initial state within a Floquet period and this localization should persist indefinitely, according to the Floquet periodicity condition. Even though the relation Eqn. 3.15a is approximate, we should expect to see localization for a very long time.

To check this analytical result, we have integrated the equations of motion given in Eqn. 3.13 by use of a Runge–Kutta scheme [27]. For example, choosing the parameters $\Omega/\omega_0 = 52.80$ and $4b/\hbar\omega_0 = 126.973$, which satisfy the conditions of Eqns. 3.15 and 3.16, we obtain the results displayed in Fig. 3.1. We have extended this plot to more than 100 bare tunnel periods without evidence that the suppression of tunnelling is lost. Thus, Floquet suppression of tunnelling is possible for long times compared with the bare tunnel time.

Not all Floquet fields provide localization for long times compared with the tunnel period. All the Floquet periodicity conditions ($\varepsilon_1 = \varepsilon_2$) asserts is that what happens in one Floquet period T will be repeated indefinitely. There are external fields providing Floquet periodicity that do not satisfy the conditions of Eqn. 3.15. For example, the Floquet condition for a field with frequency equal to the tunnel frequency implies a field strength that is not large enough to guarantee complete localization within one Floquet period [21,22]. This is illustrated in Fig. 3.2 where $\omega_0 = \Omega$ and $4b/\hbar\omega_0 = 2.0924$. The Floquet periodicity is evident, but there is no localization.

On the other hand, there are fields with strengths of sufficient size that, even if the

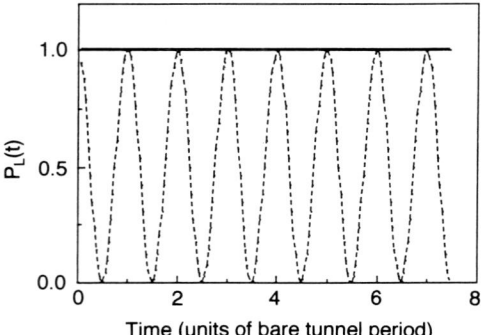

Figure 3.1 Time evolution of $P_L(t)$ in units of the bare tunnel period. Parameters used are for Floquet localization (see Eqn. 3.16 with $\Omega/\omega_0 = 52.8$ and $4b/\hbar\omega_0 = 126.973$. Solid line, Floquet; dotted line, bare tunnelling.

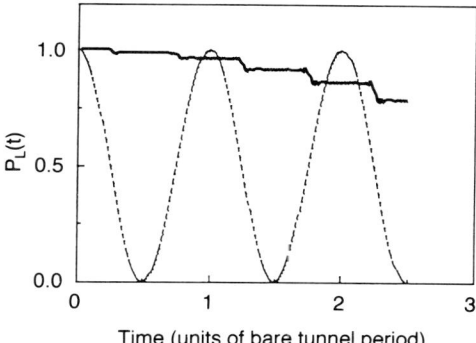

Figure 3.2 Time evolution of $P_L(t)$ in units of the bare tunnel period. Solid line, Floquet; dotted line, bare tunnelling. Parameters used are for an external field frequency equal to the tunnel frequency $\Omega = \omega_0$, and the amplitude b is the same as for Fig. 3.1.

Floquet periodicity and localization conditions are violated, there can be approximate localization for some time. Thus, another strategy for controlling tunnelling can be based on the use of external fields that are strong enough to produce localization. It may be that the external field is still periodic but does not yield simple expressions for the quasi-energies. Then, investigating the degree of localization is best done numerically. In Fig. 3.2, we consider such a case with the field strength $4b/\hbar\omega_0 = 126.973$, and the frequency chosen as $\Omega = \omega_0$. Clearly, tunnelling is suppressed, but not for many inverse bare tunnel periods. A better way to suppress tunnelling, when the Floquet periodicity and localization conditions are not met, is with the use of a large, constant external field. This follows from inspection of Eqn. 3.13, which indicates that a static external field should maximize the influence of the external field term relative to the tunnel contribution. In Fig. 3.3, we use the same b value as for Fig. 3.2 but set $\Omega = 0$, a static field. Tunnelling is suppressed, and maximally so, for this choice of b. Note that the data of Fig. 3.3 extend for 150 bare tunnel periods, and the probability has still remained at one. Thus, the conclusion is reached that strong static fields are more efficient at tunnel suppression than oscillating fields not satisfying the Floquet conditions.

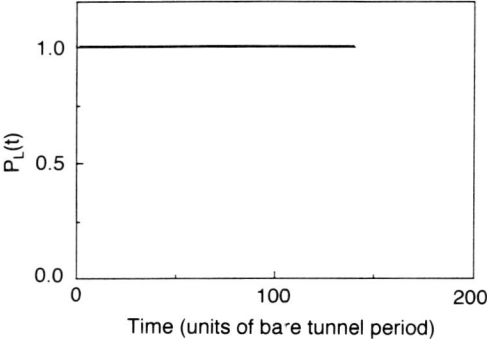

Figure 3.3 Time evolution of $P_L(t)$ in units of the bare tunnel period. Solid line, Floquet. The bare tunnelling is not displayed, for clarity of presentation. Parameters used are for an external field frequency equal to zero, and the amplitude b is the same as for Fig. 3.1.

3   The effect of external noise on the control of tunnelling

In the previous section we analysed the influence of a deterministic external field on the tunnelling dynamics of a two-level system. Here, we will consider the implications to tunnelling when the external field also has a fluctuating component. In the simplest case, the statistical properties of the imposed field can be prescribed independently of the characteristics of the tunnel system. That is, both the systematic and fluctuating components of the external field influence the tunnel system but the tunnel system itself does not affect the properties of the external field. External fluctuating fields can be imposed electronically, and they have found application in stochastically driven barrier passage problems [35]. Macroscopic quantum devices, such as SQUIDs [36], atoms or molecules driven by laser fields [37], and inversion molecules in the gas phase, such as ammonia [38], are appropriate candidates for systems dominated by external fields. In this case, the tunnel system can be described by an externally modulated TLS through the introduction of a phenomenological Hamiltonian, in contrast to the situation with internal fluctuations. When there are internal fluctuations, arising from the interaction between the tunnel system and the medium, the Hamiltonian describing the tunnel system, the medium and their coupling must be analysed. This subject will be discussed in Sections 4 and 5. Regardless of the origin of the noise, one of its effects is to provide the fluctuations required to destroy the quantum coherence of the tunnel process and, in suitable cases, lead to a rate process for the decay of the population. An important feature of an external field is that its statistical properties are controllable in a way not possible with the internal fluctuations that arise from the medium. Again of interest is the use of Floquet fields to, for example, provide periodicity and localization. But now the influence of the noise on the averaged dynamics must be taken into account. In particular, we may question the extent to which Floquet properties of the isolated system carry over to the noise-modulated situation.

To analyse the effect to tunneling originating from a random component of the external field, we add a random term $F(t)$ to the Hamiltonian of Eqn. 3.4:

$$H = \frac{\hbar\omega_0}{2} s_z + [2b(t) + F(t)]s_x. \tag{3.17}$$

Using Eqn. 3.11 for each realization of the noise results in the following stochastic equations of motion for the time-dependent coefficients $x(t)$, $y(t)$, $z(t)$ defined above (Eqn. 3.13):

$$\dot{x}(t) = -\omega_0 y(t)$$
$$\dot{y}(t) = \omega_0 x(t) - [f(t) + 4b(t)/\hbar]z(t) \tag{3.18}$$
$$\dot{z}(t) = +[f(t) + 4b(t)/\hbar]y(t),$$

where the noise

$$f(t) = 2F(t)/\hbar \tag{3.19}$$

will be characterized by an Ornstein–Uhlenbeck (OU) process [39]:

$$\langle f(t)\rangle = 0 \tag{3.20}$$

$$\langle f(t)f(s)\rangle = \Delta^2 e^{-|t-s|/\tau_c}.$$

The quantities Δ and τ_c are, respectively, the strength and decay constant of the noise [40]. When the noise is external to the system, and therefore not necessarily thermal in character, Δ and τ_c can be varied in a controlled manner, and are not restricted by the physical properties of the material. The noise-averaged population in the left-localized state $P_L(t)$ will be evaluated from

$$P_L(t) = \frac{1}{2}[1 + \langle x(t) \rangle] \tag{3.21}$$

where the $\langle \rangle$ indicates an average over realizations of the OU noise.

The stochastic equations of motion, Eqn. 3.18, can be solved numerically by generating trajectories for the different realizations of the noise. Then the desired averages can be obtained by averaging over a sufficiently large number of stochastic trajectories. The numerical method is discussed in [22]. Also, for certain parameter regimes, analytic approximations for the averaged behaviour of the system can be found. We will first discuss several analytic approximations, and then compare them with the numerically obtained results.

To this end it is useful to introduce several dimensionless parameters formed from ω_0, τ_c and Δ that will allow us to identify several limiting regimes. The two regimes where analytic approximations can be developed correspond to strong and weak coupling limits. In the strong coupling limit $\Delta/\omega_0 \gg 1$. The noise is so strong, compared with the tunnel frequency, that the system is always characterized by exponential relaxation of the population regardless of the correlation time of the noise. In the weak coupling limit, $\Delta\tau_c \ll 1$, equations of motion for the averages of x, y, and z can be obtained, which may be solved for the time course of $P_L(t)$. In this regime, the coherent oscillations of the isolated tunnel system are still destroyed, but the population evolution can display an oscillatory behavior for sufficiently weak noise.

3.1. *Strong coupling limit*

We now analyse the system of stochastic equations presented in Eqn. 3.18, where $f(t)$, the OU process, is a zero average Gaussian random process with correlation function given in Eqn. 3.20. In particular, we are interested in the quantity $P_L(t)$, defined in Eqn. 3.21, that is the noise-averaged decay probability of an initially left-localized system. To analyse the strong coupling regime where $\Delta/\omega_0 \gg 1$, we define the quantities [27,41]

$$U_{\pm} = \frac{1}{\sqrt{2}}(-z \pm iy)\exp\left(\mp i\int^t [f(t') + 4b(t')/\hbar]dt'\right). \tag{3.22}$$

Writing $y(t)$ in terms of U_+ and U_-, and using this result in Eqn. 3.18, yields

$$\dot{x}(t) = -\omega_0 \frac{i}{\sqrt{2}}\left[U_+ \exp\left(+i\int^t [f(t') + 4b(t')/\hbar]dt'\right)\right.$$
$$\left. - U_- \exp\left(-i\int^t [f(t') + 4b(t')/\hbar]dt'\right)\right]. \tag{3.23}$$

Taking the time derivative of Eqn. 3.22, using Eqns 3.18 and solving formally for $U_+(t)$

and $U_-(t)$, we can rewrite Eqn. 3.23 as

$$\dot{x}(t) = -\frac{\omega_0^2}{4} \int_0^t x(t') \left\{ \exp\left(+i \int_{t'}^t [f(t'') + 4b(t'')/\hbar] dt'' \right) \right.$$

$$\left. + \exp\left(-i \int_{t'}^t [f(t'') + 4b(t'')/\hbar] dt'' \right) \right\} dt'. \tag{3.24}$$

Now average over the realizations of the noise and assume the factorization property

$$\left\langle x(t') \exp\left(i \int_{t'}^t f(t'') dt'' \right) \right\rangle \simeq \langle x(t') \rangle \left\langle \exp\left(i \int_{t'}^t f(t'') dt'' \right) \right\rangle \tag{3.25}$$

to write

$$\langle \dot{x}(t) \rangle = -\frac{\omega_0^2}{2} \int_0^t \left\langle \exp\left(i \int_{t'}^t f(t'') dt'' \right) \right\rangle \cos\left(\int_{t'}^t [4b(t'')/\hbar] dt'' \right) \langle x(t') \rangle dt'. \tag{3.26}$$

The factorization property is consistent with the standard assumption that there is a time-scale separation between the (slow) population decay and the fast microscopic fluctuations responsible for the decay. The Gaussian property of the noise permits evaluation of the average as [39]

$$\left\langle \exp\left(i \int_{t'}^t f(t'') dt'' \right) \right\rangle = \exp\left(-\int_{t'}^t \int_{t'}^s \langle f(s) f(s') \rangle ds ds' \right). \tag{3.27}$$

Using this result and Eqn 3.20 in Eqn 3.26 yields

$$\langle \dot{x}(t) \rangle = -\frac{\omega_0^2}{2} \int_0^t e^{-(\Delta\tau_c)^2 \left[\frac{u}{\tau_c} + e^{-u/\tau_c} - 1 \right]} \cos\left[\int_{t-u}^t dt'' \frac{4b(t'')}{\hbar} \right] \langle x(t-u) \rangle du. \tag{3.28}$$

Because the system is initially left localized, and we are only considering the initial evolution of this state, the average decay rate is characterized by a one-way (irreversible) expression. When $\Delta/\omega_0 \gg 1$, the kernel defined by Eqn 3.28 decays sufficiently fast at a local time, exponential relaxation for Eqn 3.28 results. Note that the kernel, excluding the external field part, has the form of a stochastically modulated oscillator familiar from Kubo-type dephasing theory [42,43]. As in that theory, it is important to distinguish between short or long correlation times, respectively, $\omega_0\tau_c \ll 1$ and $\omega_0\tau_c \sim 1$. For $\omega_0\tau_c \sim 1$, where $\Delta\tau_c \gg 1$ we are likely to be in a slow modulation regime whereby an Arrhenius-type rate constant is obtained. If $b(t)$ is sinusoidal, as in Eqn 3.6, and its frequency, Ω, is smaller than the kernel's decay rate, then we obtain the expression

$$\langle \dot{x}(t) \rangle = -\Gamma(t) \langle x(t) \rangle \tag{3.29}$$

with

$$\Gamma(t) = \sqrt{\frac{\pi}{2}} \frac{\omega_0^2}{2\Delta} \exp\left[-\left(\frac{4b}{\hbar\Delta} \right)^2 \frac{\cos^2 \Omega t}{2} \right]. \tag{3.30}$$

In this regime all information about the correlation time τ_c is lost since the fluctuation speed of the external field is slow. Equation 3.30 is the tunnelling version of the strong coupling, slow modulation limit of a stochastically modulated oscillator. In this case, the relaxation may be viewed as controlled in part by a Boltzmann factor for finding, in an ensemble of two-level systems, those that have been symmeterized by the fluctuations.

For $\omega_0 \tau_c \ll 1$, where it is more likely that the condition $\Delta \tau_c \ll 1$ is obeyed, a fast modulation limit is obtained with $\Gamma(t)$ given by

$$\Gamma(t) = \frac{1}{2} \frac{\omega_0^2}{\Delta^2 \tau_c + \left(\dfrac{4b}{\hbar \Delta}\right)^2 \dfrac{\cos^2 \Omega t}{\tau_c}}. \tag{3.31}$$

Here, the fluctuation time is short and it will affect the population transfer rate. It is also possible to obtain an approximate expression for the case of intermediate modulation, $\Delta \tau_c \sim 1$ [27].

3.2   *Weak coupling limit*

In this regime, characterized by $\Delta \tau_c \ll 1$, the coherent oscillations of the isolated tunnel system are still destroyed, but the population evolution can display an oscillatory behaviour for sufficiently weak noise. The perturbative approach used above for the strong coupling limit is inadequate to deal with this case. We then resort to a new perturbation expansion in the parameter $\Delta \tau_c$ that yields equations of motion for the averages of x, y and z. To carry out the solution of Eqns 3.18, we separate their deterministic and stochastic parts and write them in matrix form as

$$\dot{\mathbf{u}}(t) = \mathbf{A}(t) \cdot \mathbf{u}(t) + f(t) \mathbf{C} \cdot \mathbf{u}(t) \tag{3.32}$$

where $\mathbf{u}(t) = (x(t), y(t), z(t))$ and the matrices $\mathbf{A}$ and $\mathbf{C}$ are:

$$\mathbf{A}(t) = \begin{pmatrix} 0 & -\omega_0 & 0 \\ \omega_0 & 0 & -\dfrac{4b}{\hbar} \cos \Omega t \\ 0 & \dfrac{4b}{\hbar} \cos \Omega t & 0 \end{pmatrix}, \tag{3.33}$$

and

$$\mathbf{C} = \begin{pmatrix} 0 & 0 & 0 \\ 0 & 0 & -1 \\ 0 & 1 & 0 \end{pmatrix}. \tag{3.34}$$

In order to focus on the effects of the noise, it is convenient to transform to an interaction representation

$$\mathbf{r}(t) = T_+ \exp\left[-\int^t \mathbf{A}(s) \mathrm{d}s\right] \cdot \mathbf{u}(t), \tag{3.35}$$

where T_+ denotes time ordering. The formal solution for the average of Eqn 3.35 can be written as

$$\langle \mathbf{r}(t) \rangle = \left\langle T_+ \exp\left[\int^t \mathbf{L}(s) \mathrm{d}s\right]\right\rangle \mathbf{r}(0) = \left\langle T_+ \exp\left[\int^t \mathrm{d}t_1 \int^t \mathrm{d}t_2 \langle \mathbf{L}(t_1) \cdot \mathbf{L}(t_2) \rangle\right]\right\rangle \mathbf{r}(0) \tag{3.36}$$

where $\mathbf{L}(t) = f(t)\mathbf{C}(t)$ and

$$\mathbf{C}(t) = T_+ \left[\exp - \int^t \mathbf{A}(s)\mathrm{d}s \right] \cdot \mathbf{C} \cdot T_- \left[\exp + \int^t \mathbf{A}(s)\mathrm{d}s \right]. \tag{3.37}$$

The second equality in Eqn 3.36 follows from the Gaussian property of the noise [39,40,43,44]. Furthermore, $\langle \mathbf{L} \rangle = 0$, and

$$\langle \mathbf{L}(t) \cdot \mathbf{L}(s) \rangle = \Delta^2 e^{-|t-s|/\tau_c} \mathbf{C}(t) \cdot \mathbf{C}(s). \tag{3.38}$$

In order to progress with these formal expressions, we now invoke the weak coupling limit $\Delta\tau_c \ll 1$, which permits the elimination of the time ordering in Eqn 3.36 [39]. Then, by time differentiation of Eqn 3.36, without the time ordering, the following equations of motion for the averages are obtained

$$\langle \dot{\mathbf{u}}(t) \rangle = \mathbf{A}(t) \cdot \langle \mathbf{u}(t) \rangle + \int_0^t \langle f(t)f(t_2) \rangle \cdot \mathbf{C} \cdot \left[T_- \exp \int_{t_2}^t \mathbf{A}(s)\mathrm{d}s \right] \cdot \mathbf{C}$$

$$\cdot \left[T_+ \exp \left(- \int_{t_2}^t \mathbf{A}(s)\mathrm{d}s \right) \right] \mathrm{d}t_2 \cdot \langle \mathbf{u}(t) \rangle. \tag{3.39}$$

Local in time equations of motion for the averages can be obtained by noting that for $\omega_0\tau_c \ll 1$, the kernel $\langle f(t)f(t_2) \rangle$ decays rapidly relative to the average system evolution. Also, assuming that the oscillation of the external field is slow on this time scale, $\sin \Omega t < 1$, we obtain the time-local equations of motion

$$\langle \dot{x}(t) \rangle = -\omega_0\langle y(t) \rangle$$

$$\langle \dot{y}(t) \rangle = \omega_0\langle x(t) \rangle - \frac{4b}{\hbar} \cos \Omega t \langle z(t) \rangle - \Delta^2\tau_c\langle y(t) \rangle \tag{3.40}$$

$$\langle \dot{z}(t) \rangle = \frac{4b}{\hbar} \cos \Omega t \langle y(t) \rangle - \Delta^2\tau_c\langle z(t) \rangle.$$

In the absence of the systematic part of the external field ($b = 0$), the evolution of $\langle x(t) \rangle$ reduces to that of a damped harmonic oscillator [41]. With the field present, no such reduction is possible and the solution of the averaged equations of motion in Eqn 3.40 must be carried out numerically.

3.3   Numerical solutions and their comparison with the above limiting results

The analytic expressions derived above can be compared with the numerical solutions obtained by the stochastic trajectory solution method. We first discuss the results in the strong coupling limit, where Δ/ω_0 is large. All the plots we will discuss are for the noise-averaged population in the left-localized state, $P_L(t)$, as defined in Eqn 3.21, with $P_L(0) = 1.0$. Figure 3.4 shows a plot with $\Delta/\omega_0 = 5.0$, $\omega_0\tau_c = 1$ and a constant external field of amplitude $4b/\hbar\omega_0 = 2.0924$. Here, $\Delta\tau_c = 5.0$, which corresponds to a slow modulation regime [40] where the relaxation kernel in Eqn 3.28 can be approximated as a Gaussian decay in time. Clearly, the Arrhenius-like expression for the rate, given by Eqn 3.30, describes this regime. Note that the systematic part of the external field plays the role of an activation energy. Larger field amplitudes further suppress the tunnelling

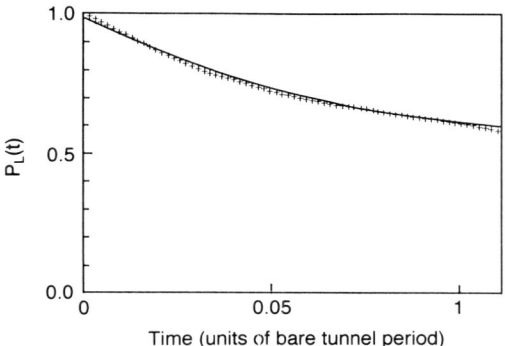

Figure 3.4 Time evolution of $P_L(t)$ in units of the bare tunnel period for $\omega_0\tau_c = 1.0$, $\Delta/\omega_0 = 5.0$ with a constant systematic external field of amplitude $4b/\hbar\omega_0 = 2.0924$. The + line corresponds to the numerical solution of the stochastic equations, and the solid line to the approximate expression of Eqn 3.30 (the slow modulatiuon limit).

in accord with the prediction of Eqn 3.30. If the field also has a time variation that is not too rapid ($\sin \Omega\tau_c$ small) then results similar to those of Fig. 3.4 are also obtained. Thus, it is clear that for the strong coupling and slow modulation regimes, the Arrhenius formula of Eqn 3.30 is an excellent description of the population evolution.

As we showed in Section 2, a Floquet field can alter the tunnelling period so that it is no longer determined by the tunnel splitting frequency ω_0 but by the Floquet frequency Ω. In order to suppress the tunnelling in one Floquet period, the amplitude of the systematic external field must be sufficiently large, and the frequency must be correspondingly adjusted. Because the strength of the deterministic part of the external field is so strong, even in the presence of noise the population should still remain essentially localized for an extremely long time. This insensitivity to the presence of noise reflects the strength of the external field. Certainly, once noise is present, Floquet localization cannot last indefinitely, although it may remain for experimentally relevant times. This can be understood by reference to Eqn 3.18 where, for a large external field relative to the noise strength, the noise terms can be neglected for some length of time, and then the averaged equations obey Floquet-type equations of motion.

We now investigate the average dynamics when the Floquet field amplitude is not so large as to produce localization in one period. The plot in Fig. 3.5 corresponds to a Floquet condition external field, with and without the noise. The external field amplitude $4b = 2.0924\hbar\omega_0$ is the Floquet amplitude for the field frequency when $\Omega = \omega_0$. This solution does not follow from the Bessel condition of Eqn 3.16 as it violated the condition of Eqn 3.15a; rather, it is obtained from the parameter values given in [21]. Without noise, the system exhibits Floquet periodicity. The presence of noise destroys this periodicity and produces a decay similar to that of Fig. 3.4.

Turning now to the weak coupling limit, defined by $\Delta\tau_c \ll 1$, we present results for $\omega_0\tau_c = 0.1$ and $\Delta/\omega_0 = 1.0$ so that $\Delta\tau_c = 0.1$, and various values of the other parameters. In Figs 3.6 and 3.7 plots for no systematic and a constant systematic field are shown. Figure 3.6 exhibits the behaviour of an under-damped harmonic oscillator. As noted after Eqn 3.40, the averaged population obeys such an equation of motion in the absence of the systematic external field. With this field present, the damped oscillations have a superimposed monotonic decay. The equations of motion given in Eqn 3.40

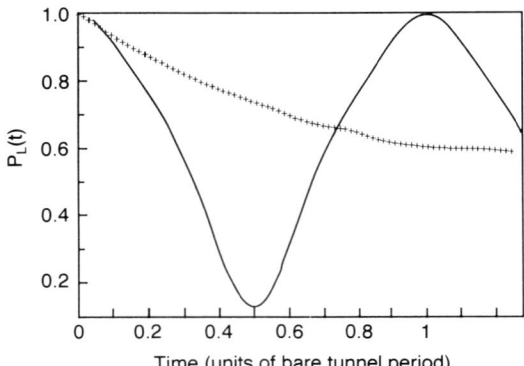

Figure 3.5 Time evolution of $P_L(t)$ in units of the bare tunnel period. Here, $\omega_0\tau_c = 1.0$, $\Delta/\omega_0 = 5.0$ with a systematic external field of amplitude $4b/\hbar\omega_0 = 2.0924$ and frequency $\Omega = \omega_0$. For these values of the amplitude and frequency the quasi-energies are degenerate and produce the Floquet periodicity condition. The + line corresponds to the numerical solution of the stochastic equations, and the solid line to the system evolution in the absence of the noise.

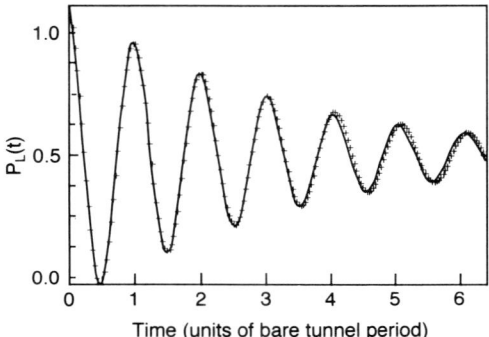

Figure 3.6 Time evolution of $P_L(t)$ in units of the bare tunnel period for $\omega_0\tau_c = 0.1$, $\Delta/\omega_0 = 1.0$ with no systematic external field. The + line corresponds to the numerical solution of the stochastic equations, and the solid line to the solution of the approximate averaged equations given in Eqn 3.40. These data are for the weak coupling limit, $\Delta\tau_c \ll 1$.

were solved numerically, and these solutions adequately match those obtained by numerical solution of the stochastic equations. The combination of noise and asymmetry of the two-level system, induced by the non-zero systematic external field, produces the decay. If the noise was absent, the asymmetric tunnel system would exhibit fast, small amplitude oscillations close to the initial population value. The weak noise always pushes the system towards equilibrium, but with the systematic field present, there is a decaying component to the evolution. Figure 3.8 contains the plots for a Floquet field with the same parameters as for Fig. 3.5. The characteristic Floquet periodicity is still in evidence, but the noise is producing a slow approach to equilibrium. Even though there is a systematic external field, because it is varying through positive and negative values, there now is no monotonic decay superimposed on the time evolution, as for Fig. 3.7.

Floquet localization theory is usually applied to non-dissipative systems [45]. The same quasi-energy concept can, however, be used for dissipative systems. In particular, the quasi-energy degeneracy condition for obtaining a time periodic wavefunction will

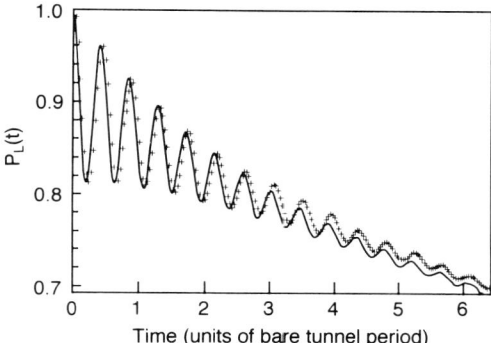

Figure 3.7 Time evolution of $P_L(t)$ in units of the bare tunnel period for $\omega_0\tau_c = 0.1$, $\Delta/\omega_0 = 1.0$ with a constant systematic external field of amplitude $4b/\hbar\omega_0 = 2.0924$. The + line corresponds to the numerical solution of the stochastic equations, and the solid line to the solution of the approximate averaged equations given in Eqn 3.40. These data are for the weak coupling limit, $\Delta\tau_c \ll 1$.

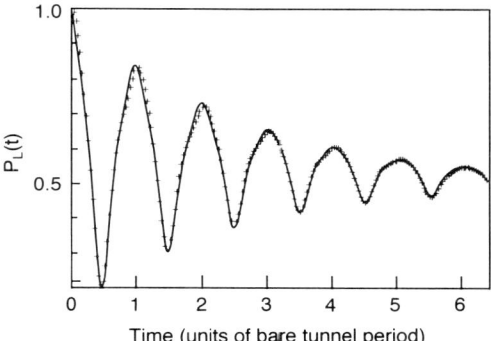

Figure 3.8 Time evolution of $P_L(t)$ in units of the bare tunnel period for $\omega_0\tau_c = 0.1$, $\Delta/\omega_0 = 1.0$ with a systematic external field of amplitude $4b/\hbar\omega_0 = 2.0924$ and frequency $\Omega = \omega_0$. For these values of amplitude and frequency the quasi-energies are degenerate and produce the Floquet periodicity condition. The + line corresponds to the numerical solution of the stochastic equations, and the solid line to the solution of the averaged equations given in Eqn 3.40.

still hold, but there will be superimposed on this periodicity a time decay from the dissipation. The data in Fig. 3.8 indeed display this behaviour. If, however, the external field is oscillatory but does not satisfy the Floquet condition, then the oscillations of the population are related to the tunnel period, not the Floquet period. This feature may be understood from the form of Eqn 3.40. The Floquet fields are of weak amplitude relative to the tunnel splitting. But, when the Floquet condition is satisfied, Floquet theory enforces a periodicity on the solutions of the Floquet period. When the same amplitude field is used, but the frequency is moved away from the Floquet frequency, the oscillations in the solutions of Eqn 3.40 are dominated by the tunnel frequency.

Finally, in Fig. 3.9, we show the results of the numerical solution of the stochastic equations for the conditions $\omega_0\tau_c = 1$ and $\Delta/\omega_0 = 1$. Here, we are neither in a strong nor in a weak coupling limit, so there is no analytic expression for the evolution of the average population. Shown are results for a constant external field whose amplitude is the same as that in Fig. 3.7, and a Floquet field with parameters as in Fig. 3.8. For the constant systematic external field, the initial decays are similar but, subsequently, the

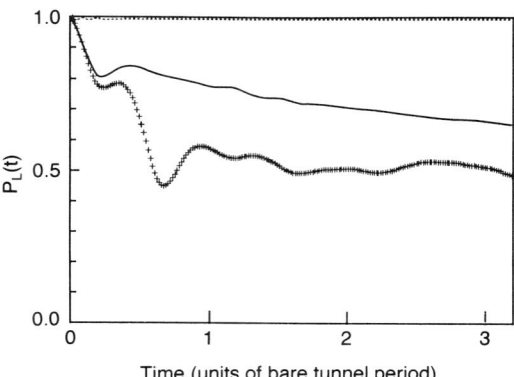

Figure 3.9 Time evolution of $P_L(t)$ in units of the bare tunnel period for $\omega_0 \tau_c = 1.0$, $\Delta/\omega_0 = 1.0$ with a systematic external field of frequency zero and amplitude $4b/\hbar\omega_0 = 2.0924$ (solid line), frequency zero and amplitude 9.0924 (dashed line) and frequency $\Omega = \omega_0$ and amplitude 2.0924 (+ line). The plots are generated by numerical solution of the stochastic equations.

Fig. 3.9 data exhibit a faster and less oscillatory decay. We speculate that the smoothing of the decay is due to the time convolution, as opposed to local in time, equations of motion for averaged quantities that would result from an analysis of the time-ordered equations given in Eqn 3.39. For the Floquet field, the approach to equilibrium is very rapid. In contrast with Fig. 3.8, the Floquet oscillations are strongly suppressed for the same reason as noted in the discussion of the constant field case. We note that constant systematic external fields produce slower decays than do time-varying fields. Also included in Fig. 3.9 is the plot for a rather strong constant systematic external field. Once again, strong fields localize the tunnel system.

4   The control of tunnelling reactions with external fields

Most tunnelling processes are strongly affected by the coupling of the tunnel system to the medium in which it is immersed [11–18]. Thus, it is not sufficient to consider the noise process to be unaffected by the tunnel system, as is the case for the external noise case treated in Section 3. Typically, the medium's coupling to the tunnel process is strong and, consequently, both permits the transfer by tunnelling (by equalizing the energy of the transferring particle in its initial and final states) and renders the process incoherent so that the transfer probability obeys a rate law. An intensively investigated case of tunnelling in a medium is that of hydrogen atom transfer in rigidly bound hydrogen-bonded systems such as dicarboxylic acids [20]. Here, the tautomeric states are sufficiently separated spatially that the barrier to transfer is high. The hydrogen tunnel between two states that are localized around the minima of a double-well potential. Then an appropriate model is a two-level tunnel system coupled to a medium. (Actually, in some dicarboxylic acids thermal excitation to higher tunnel pairs is possible, but this possibility will not be considered here [46–48].) Another important case is electron transfer in condensed phases [16,49,50]. Again, an appropriate model is a two-level system coupled to a medium, as the excited electronic states are much higher in energy than could be accessed thermally.

To analyse these medium effects on both Floquet and ordinary external field modifications of tunnelling, we need a model for the tunnel system and its coupling to the medium. In recent years, medium effects on tunnelling reactions have been discussed by using as a model a two-level system linearly coupled to an infinite number of bath excitations, represented as Bosons (such as acoustic modes). This is the standard spin Boson Hamiltonian [13,25,26], whose origin from a double-well potential perspective has been carefully discussed by, for example, Leggett *et al.* [25]. Our objective is to analyse the effect of an external field on the dynamics arising from this Hamiltonian, and investigate again the possibility of controlling tunnelling. In this section we will restrict attention to a high-temperature bath where the medium can be treated classically, and one might expect that a stochastic description of the medium would be appropriate. Not surprisingly, there is a close connection with the equations of motion obtained in the previous section.

A Hamiltonian appropriate to a two-level system linearly coupled to an infinite number of bosonic excitations with the addition of an external field that couples to the transition dipole of the tunnelling object is [13,51]

$$H = \frac{\hbar\omega_0}{2} s_z + \sum_j \left[\frac{p_j^2}{2} + \frac{1}{2} \omega_j^2 \left(q_j - \frac{\gamma_j}{\omega_j^2} s_x \right)^2 \right] + 2b(t)s_x. \tag{3.41}$$

The bath oscillators are characterized by the coordinates and momenta, q_js and p_js, respectively, and the masses of the oscillators are set to one for convenience. The TLS–medium couplings are characterized by the γ_js and the other terms are as defined in Eqn 3.4. If the initial and final transfer states are different chemically there will be a difference in the equilibrium solvation free energies of these states, a ΔG^0 contribution. Rather than introduce this quantity as a separate s_x term in the above Hamiltonian, we will incorporate it in the $b(t)$ term, for convenience. It is worth noting that since both these terms are proportional to s_x, a constant external field may be viewed as another source of free energy though, obviously, their physical origins are distinct.

Using the Heisenberg representation of the equations of motion given in Eqn 3.11 for the total system, we obtain for the Hamiltonian of Eqn 3.41 the following operator equations of motion for the spin variables:

$$\dot{s}_x(t) = -\omega_0 s_y(t)$$

$$\dot{s}_y(t) = \omega_0 s_x(t) + \frac{2}{\hbar} \sum_j \gamma_j q_j(t) s_z(t) - \frac{4b(t)}{\hbar} s_z(t) \tag{3.42a}$$

$$\dot{s}_z(t) = -\frac{2}{\hbar} \sum_j \gamma_j q_j(t) s_y(t) + \frac{4b(t)}{\hbar} s_y(t)$$

and for the medium coordinate operators:

$$q_j(t) = \left[\left(q_j - \frac{\gamma_j}{\omega_j^2} s_x \right) \cos \omega_j t + \frac{p_j}{\omega_j} \sin \omega_j t \right] + \frac{\gamma_j}{\omega_j^2} s_x(t) - \int_0^t \mathrm{d}s \frac{\gamma_j}{\omega_j^2} \cos \omega_j(t-s)\dot{s}_x(s). \tag{3.42b}$$

Note that $q_j = q_j(0)$ and $p_j = p_j(0)$ are diagonal operators in spin space but the operator $q_j(t)$ is not. On the right-hand side of Eqn 3.42b the first term corresponds to the time evolution of the oscillator coordinate due solely to the influence of the harmonic force for a fixed value of the spin variable s_x. The next two terms give the time evolution of the oscillator position due to the coupling with the spin variables. Note that the external field does not influence the oscillator position in an explicit fashion. If it did, it would correspond to the situation of external fields that modify the properties of the medium in the absence of the tunnel system. This would, indeed, happen for sufficiently strong fields, but we do not consider this effect here [52].

We are interested in the evolution of a system that is initially prepared in the $|L\rangle$ state. In the absence of the medium and the external field, the system undergoes a coherent oscillation at frequency ω_0, corresponding to the tunnelling between the left and right states ($|L\rangle$ and $|R\rangle$) of the unperturbed system. With the medium and external field present, the coupled equations, Eqns 3.42, cannot be solved exactly, as the spin and medium degrees of freedom are non-linearly coupled. Two approximations are required to obtain a tractable system of equations. As motivation for the first approximation, we expect that, in order to define left and right states, there must be sufficiently long residence time in the left and right wells of the double-well system underlying the TLS; then, the transit time between these states must be slow, compared with the microscopic time scale of the bath fluctuations. The time scale of left–right transfer is gauged by the tunnel splitting ω_0, and therefore $\dot{s}_x(t)$ is slow, compared with the microscopic time scale. We will use this assumption to eliminate the last term on the right-hand side of Eqn 3.42b. In order to do this, we need estimates for the sizes of the three terms on the right-hand side of Eqn 3.42b, and this requires a discussion of the spin–medium interaction.

As shown below (cf. Eqn 3.50) the spin–medium interaction can be characterized by the correlation function $K(t)$ that in the high-temperature limit can be written as [53]

$$K(t) = (4k_B T/\hbar^2) \sum_j \frac{\gamma_j^2}{\omega_j^2} \cos \omega_j t. \tag{3.43}$$

The condition under which the classical approximation is correct is $\beta \hbar \omega_d < 1$, where ω_d is a typical cutoff frequency of the medium oscillators. A useful model for $K(t)$, corresponding to a Debye spectrum, is characterized by a strength parameter Δ and a correlation time τ_c [42,43,54]. Thus,

$$K(t) = \Delta^2 e^{-t/\tau_c}, \tag{3.44}$$

where the noise strength Δ is given by [55]

$$\Delta^2 = 2E_r k_B T/\hbar^2 \tag{3.45a}$$

and E_r is the medium's reorganization energy. It is straightforward to show that E_r is given in terms of the couplings and frequencies of the oscillators by the expression

$$E_r = 2 \sum_j (\gamma_j/\omega_j)^2 \tag{3.45b}$$

and corresponds to the following energetic difference. E_r is the difference of the medium potential energy after the tunnelling system has transferred to the right-hand state with the medium still in the equilibrium configuration for the left-hand state and the equilibrium potential energy of the medium with the tunnelling system in the right-hand state. Note that we have used the same symbols here for the strength and correlation time as in Section 3 in order to emphasize the formal similarity between the two situations, although the physical origins of the two noises are distinct.

With these facts in hand, inserting Eqn 3.42b in Eqn 3.42a leads to the following estimates for the contributions of the three terms on the right-hand side of Eqn 3.42b. The first is order $\hbar\Delta$; it scales with the strength of the noise, since it involves the thermal average of the oscillator coordinate. The second is of order $(\hbar\Delta)^2/k_B T$, since it is proportional to E_r (cf. Eqn 3.45) and the third is of order $(\hbar\omega_0/k_B T)(\hbar\Delta^2\tau_c)$. This latter estimate is obtained by noting the time integral will introduce the factor of τ_c and the time derivative of s_z introduces the ω_0. Comparison of the three terms shows that to drop the third term requires $\omega_0\tau_c \ll 1$.

The second approximation we will make is to set $s_x(t) = s_x(0)$ in Eqn (3.42b). Note that this is consistent with the first approximation of slow motion of the population. If this second approximation is not made, it will turn out that the resulting equations of motion are not consistent, in the sense that there will be complex terms that violate the reality condition of quantum expectation values. Thus, the two approximations are closely tied together. With them, the spin equations of motion are:

$$\dot{s}_x(t) = -\omega_0 s_y(t)$$

$$\dot{s}_y(t) = +\omega_0 s_x(t) + \frac{\mathscr{F}(t)}{\hbar}s_z(t) - \frac{4b(t)}{\hbar}s_z(t) \tag{3.46}$$

$$\dot{s}_z(t) = -\frac{\mathscr{F}(t)}{\hbar}s_y(t) + \frac{4b(t)}{\hbar}s_y(t)$$

with the operator $\mathscr{F}(t)$ defined as

$$\mathscr{F}(t) = 2\sum_j \gamma_j q_j(t) = 2\left[\sum_j\left[\gamma_j\left(q_j - \frac{\gamma_j s_x}{\omega_j^2}\right)\cos\omega_j t + \frac{p_j\gamma_j}{\omega_j}\sin\omega_j t\right] + \sum_j\left(\frac{\gamma_j}{\omega_j}\right)^2 s_x\right]. \tag{3.47}$$

This simplification of the bath dynamics amounts to a time-dependent self-consistent field approximation whereby the bath responds to the expectation value of the spin variable [56]. The LL matrix element (with respect to the spin states) of $\mathscr{F}(t)$ will still be denoted by $\mathscr{F}(t)$, as its LR and RL elements are zero. The statistical properties of $\mathscr{F}(t)$ will be specified once we choose an initial ensemble. Suppose that initially the tunnelling system population is in the left side and the medium is in equilibrium *with respect to this state*. Thus,

$$\rho_{in} = \frac{1}{C}e^{-\beta H_0}|L\rangle\langle L| \tag{3.48}$$

where

$$H_0 = \sum_j \frac{p_j^2}{2} + \frac{\omega_j^2}{2}\left(q_j - \frac{\gamma_j}{\omega_j^2}s_x\right)^2. \tag{3.49}$$

Then the statistical properties of $\mathscr{F}(t)$ are those of a Gaussian process whose average and two-point correlation function are as follows:

$$\langle \mathscr{F}(t) \rangle = E_r \tag{3.50a}$$

and

$$\langle \delta \mathscr{F}(t) \delta \mathscr{F}(0) \rangle = \hbar^2 K(t), \tag{3.50b}$$

where $\delta \mathscr{F}(t) = \mathscr{F}(t) - \langle \mathscr{F}(t) \rangle$ and the $\langle \rangle$ denotes averaging over the initial ensemble defined by Eqns 3.48 and 3.49. The use of the conditional initial condition is responsible for the 'noise' being given by a Gaussian process. Other initial conditions would not be as convenient.

If we take matrix elements of the spin-operator equations, we will obtain algebraic equations that can be solved when averaged over the Gaussian process. The population of the left state is obtained from the matrix element $\langle L|s_x(t)|L \rangle$. Taking the LL matrix elements of Eqn 3.46 leads, once again, to equations that are given entirely in terms of the LL elements, which we again define as $x(t) = \langle L|s_x(t)|L \rangle$, and similarly for the other matrix elements of the Pauli matrices. We then obtain equations of motion of the same form as in Eqn 3.18 with the key difference that the stochastic properties of the noise are given by Eqns 3.50. Note that the noise average is not zero here, but is related to the system–bath coupling through the reorganization energy E_r, and the correlation function of the noise depends on E_r and the medium relaxation time τ_c.

For most of these reactions, the reorganization energy and medium relaxation time have numerical values that correspond to the strong coupling $\Delta/\omega_0 \gg 1$ and slow modulation $\Delta\tau_c \gg 1$ regime. For example, even for those proton transfer reactions where the coupling to the medium is typically via acoustic phonons [57] and, therefore, weak compared with electrostatic interactions, estimates of Δ (cf. Eqn 3.45) for standard temperature are at least 100 cm^{-1}, while the tunnel splittings are around 1–10 cm^{-1} and the correlation times satisfy $\tau_c > 1$ ps. For electron transfer reactions and for proton transfer reactions where the coupling is electrostatic (the electron or proton charge interacting with the medium dipoles) medium reorganization energies are considerably larger than for acoustic phonon coupling, so the corresponding Δs are larger [49]. Thus, except for low temperatures or small E_rs, where the classical description is not correct, we may analyse the equations of motion of Eqn 3.46 with techniques appropriate to the strong coupling, slow modulation regime. In particular, the method introduced in Section 3.1 can be applied here as the only formal difference is that now $\langle \mathscr{F} \rangle$ is not zero. The result is

$$\langle \dot{x}(t) \rangle = -\frac{\omega_0^2}{2} \int_0^t e^{-(\Delta u)^2/2} \cos\left[\frac{E_r}{\hbar} u - \int_{t-u}^t dt'' \frac{4b(t'')}{\hbar}\right] \langle x(t-u) \rangle du, \tag{3.51}$$

where the E_r term arises from the non-zero value of $\langle \mathscr{F} \rangle$ and recall that the solvation free energy term is included in the definition of $b(t)$, i.e. $b(t) \rightarrow b(t) - \Delta G^0/4$. The same restriction to a one-way rate expression as in Section 3.1 applies to this result. Note that the use of the initial ensemble of Eqn 3.48 with the medium displaced to be in equilibrium with the left-localized state has to be modified if we want to describe the reverse (right to left state) process. Then we need an initial ensemble with the medium oscillators centred around the right-localized state, The resulting noise would have the same correlation function as in Eqn 3.50b but its average value would be $-E_r$ (cf. Eqn 3.50a). As a result the rate equation for the reverse process would be obtained by changing E_r to $-E_r$ in Eqn 3.51. Thus, the classical medium relaxation equation that accounts for the forward and backward processes, obtained by adding the respective rate kernels, is

$$\langle \dot{x}(t) \rangle = -\int_0^t k(t,u) \langle x(t-u) \rangle du$$

$$= -\omega_0^2 \int_0^t e^{-(\Delta u)^2/2} \cos\left(\frac{E_r}{\hbar} u\right) \cos\left[\int_{t-u}^t dt'' \frac{4b(t'')}{\hbar}\right] \langle x(t-u) \rangle du. \tag{3.52}$$

A more general relaxation equation appropriate to a quantum mechanical treatment of the medium is presented in Section 5 (see Eqn 3.90).

The analysis of Eqn 3.51 is simplified by first assuming a constant field (or one whose time variation is slow compared with the decay time of the integrand). Then, for a rapidly decaying rate kernel, the rate constant expression is

$$k = \frac{\omega_0^2}{2} \int_0^\infty \exp\left(-\frac{\Delta^2 t^2}{2}\right) \cos\left[\left(\frac{E_r - 4b}{\hbar}\right) t\right] dt = \frac{\omega_0^2}{2} \frac{\sqrt{\pi/2}}{\Delta} e^{-\frac{(E_r - 4b)^2}{2(\Delta \hbar)^2}}. \tag{3.53}$$

This rate constant expression reduces to the standard Arrhenius form of charge transfer theory in the absence of the external field. In that case b is just $-\Delta G^0/4$ and the exponent contains the activation energy familiar from high-temperature charge transfer kinetics. The part of b arising from the coupling to the external field $\mathbf{E}$ depends on the angle θ between $\mathbf{E}$ and the dipole moment $\mathbf{m}$. If the orientations of the tunnel systems are distributed according to some distribution function, then we must average the rate constant over this distribution of orientations with respect to the fixed external field direction. In a liquid, or a disordered solid, this orientation distribution will be isotropic. Two interesting cases can be singled out. First, consider a case where $|E_r + \Delta G^0| \ll |b|$, with b proportional to the external field strength. Then the orientationally averaged rate $\bar{k}_{or}$ is

$$\bar{k}_{or} = \frac{\omega_0^2}{2} \int_{-1}^1 \frac{\sqrt{2\pi}}{(2\Delta/\hbar)} e^{-\frac{4m^2 E^2 \cos^2 \theta}{2\Delta^2}} d(\cos \theta) \simeq (\pi/8) \frac{\hbar \omega_0^2}{mE} \tag{3.54}$$

for $mE > \hbar\Delta$. Since for zero external field, $k(E=0) = \omega_0^2 [\pi/2]^{1/2}/2\Delta$, the ratio of rates is

$$\frac{\bar{k}_{or}(E)}{k(0)} = \frac{\sqrt{\pi}}{2} \frac{\sqrt{E_r k_B T}}{mE}. \tag{3.55}$$

For a sufficiently large electric field, the average rate constant is reduced, with respect to its value at zero field. The electric field then suppresses the tunnelling.

Second, when $|E_r + \Delta G^0|$ is not small, we can still write the average rate as

$$\bar{k}_{or} = \frac{\omega_0^2}{2} \frac{\sqrt{2\pi}}{4mE(\Delta/\hbar)} \int_{|E_r + \Delta G^0| - 2mE}^{|E_r + \Delta G^0| + 2mE} e^{-y^2/2\Delta^2} dy \simeq \frac{\pi \hbar \omega_0^2}{16mE}. \tag{3.56}$$

The integration above is accurate for $|E_r + \Delta G^0| = 2mE$; that is, external field values that compensate the reorganization plus solvation free energy. Also, since, as noted above, we are now assuming that $|E_r + \Delta G^0|$ is not small, and, in particular, $|E_r + \Delta G^0|$ will be larger compared with $k_B T$, the upper limit of integration in Eqn 3.56 can be extended to infinity. Therefore,

$$\frac{\bar{k}_{or}(E)}{k(0)} = \frac{\sqrt{2\pi}}{8} \frac{\hbar \Delta}{mE} e^{\frac{|E_r + \Delta G^0|}{4k_B T}}, \tag{3.57}$$

that shows that the rate is exponentially enhanced with respect to its zero field value. This is a consequence of the external field acting as a reaction free energy. Depending on the value of tunnel system–external field angle θ, the tunnelling can be enhanced or suppressed by the external field. On average, the enhancement dominates and leads to the overall rate enhancement.

We now turn to an analysis of Eqn 3.52 when the time variation of the external field is significant within the time scale of the kernel's decay rate. We consider the time-dependent quantity

$$\Gamma(t) = \int_0^\infty k(t, \tau) d\tau. \tag{3.58a}$$

$\Gamma(t)$ is not a rate constant as it is time dependent. However, for a sinusoidal field whose frequency is large compared with its amplitude, Γ will be approximately constant. For an external field varying as $b \cos(\Omega t)$, a useful quantity is the average of $\Gamma(t)$ over the period of the external field:

$$\bar{\Gamma} \equiv \frac{\Omega}{2\pi} \int_0^{2\pi/\Omega} \Gamma(s) ds. \tag{3.58b}$$

A simple analytic expression for $\bar{\Gamma}$ can be obtained by use of the following expansion [58],

$$e^{i(4b/\hbar\Omega)\sin \Omega t} = \sum_{p=-\infty}^{p=\infty} J_p(4b/\hbar\Omega)e^{ip\Omega t}, \tag{3.59}$$

in Eqn 3.52. The contribution of the fast oscillations arising from products of the Bessel functions of different orders vanishes and $\bar{\Gamma}$ has the form

$$\bar{\Gamma} = \frac{\hbar\sqrt{\pi\omega_0^2}}{4\sqrt{E_r k_B T}} \sum_{p=-\infty}^{p=\infty} J_p^2 (4b/\hbar\Omega) \left[\exp\left(-\frac{(E_r + \Delta G^0 + p\hbar\Omega)^2}{4E_r k_B T} \right) \right.$$
$$\left. + \exp\left(-\frac{(-E_r + \Delta G^0 + p\hbar\Omega)^2}{4E_r k_B T} \right) \right]. \tag{3.60}$$

Note that this rate constant is the sum of the rate constants for the forward and backward (left to right plus right to left) transitions. The structure of Eqn 3.60 is a weighted sum of effective activation energy terms (the terms in the exponent). An interesting possibility for control of tunnelling is to consider parameters where the terms for $p > 0$ are small, because of small values arising from the effective activation energies, while for the $p = 0$ term we use the Floquet condition of Eqns 3.15 and 3.16

whereby $J_0(4b/\hbar\Omega) = 0$. The resulting rate constant would then be very small. This result is consistent with the assertion that if a field is strong enough to cause localization in the absence of noise, it will be strong enough to do so in the presence of noise; in particular, the noise arising from the medium fluctuations. On the other hand, Eqn 3.60 shows that a time-dependent external field could greatly enhance a rate constant by compensating an unfavourable $|E_r + \Delta G^0|$ value with an appropriate set of $p\hbar\Omega$ values. Another result that is apparent from Eqn 3.60 is that for $4b/\hbar\Omega \ll 1$ only the zero-order Bessel function contributes and its value is close to unity. Then $\bar{\Gamma}$ reduces to the rate constant in the absence of the external field. It is also the case that in this limit $\Gamma(t)$ becomes independent of the external field and equal to $\bar{\Gamma}$. This rate constant can be either small or large, depending on the value of $|E_r + \Delta G^0|$.

In summary, the use of external fields to control tunnelling, even in the presence of the 'noise' arising from the coupling to the medium, is a theoretically sound prospect. Rates can be either enhanced or suppressed with appropriate parameter choices. As detailed above, analytic expressions can be readily found for constant and sinusoidal external fields. The use of pulsed or shaped fields can also be contemplated but then the rate expression in Eqn 3.52 is best analysed by numerical quadrature.

5 Control of tunnelling at low temperatures

The previous section considered the control of tunnelling when the temperature is sufficiently high compared with the medium mode frequencies, $\beta\hbar\omega/2 \ll 1$, that a classical description of the medium is appropriate. Here, we will consider cases where the quantum treatment of the nuclear degrees of freedom is required. This circumstance arises in two ways. First, for so-called outer sphere reactions, where the medium consists of, e.g. the dipoles of a solvent, as the temperature is decreased a substantial number of medium modes will have frequencies ω that violate the condition $\beta\hbar\omega/2 \ll 1$. Second, many electron transfer reactions involve coupling to the internal vibrational modes of the reacting species. These modes typically have frequencies such that $\beta\hbar\omega/2 \gg 1$ even at room temperature; thus, room temperature can be a low temperature in this sense. The medium here is formed from the vibrational modes that couple to the change transfer. Farazdel *et al.* [59], in fact, considered the possibility of controlling tunnelling by application of a constant external field in an isolated molecule with high-frequency vibrational modes.

It will be convenient to deal with the equations of motion for the density matrix instead of the Heisenberg equations of motion for the operators. The strategy will be to transform the Hamiltonian to a form suitable for a perturbation theory that is still valid when $\bar{V}$, the Franck–Condon renormalized tunnel splitting, is no longer a suitable expansion parameter. This is the case, for example, when $\bar{V}$ is large compared with the temperature or with the equilibrium solvation free energy difference of the initial and final states [13,23,24,60,61]. A projection operator method will be used to obtain an equation of motion for the system's density matrix to second order in the perturbation that we will refer to as a kinetic equation. Previously, we carried out a projection operator analysis along these lines in the absence of an external field [23,24]. When there is a time-dependent external field, a similar methodology can be developed as we now discuss. Applying the unitary transformation

$$U = \exp\left(\frac{i}{\hbar} s_x \sum_j \frac{\gamma_j p_j}{\omega_j^2} \right) \qquad (3.61)$$

to the Hamiltonian of Eqn 3.41, and dividing the interaction term in the transformed Hamiltonian into its thermally averaged part and the fluctuations about it leads to

$$H = H_{\text{medium}} + H_S + H_F(t) + H_{\text{in}} \qquad (3.62a)$$

where

$$H_{\text{medium}} = \sum_j \left(\frac{p_j^2}{2} + \omega_j^2 \frac{q_j^2}{2} \right),$$

$$H_S = \bar{V} s_z,$$

$$H_F(t) = 2b(t)s_x, \qquad (3.62b)$$

$$H_{\text{in}} = \frac{V}{2}(\Phi^+ s_z + \Phi^- i s_y).$$

The interaction part of the Hamiltonian involves the operators $\Phi^{\pm}$

$$\Phi^+ = \Pi_- + \Pi_+ - 2\langle \Pi \rangle,$$
$$\Phi^- = \Pi_- - \Pi_+ \qquad (3.63)$$

that are related to shifting operators $\Pi_{\pm}$ by

$$\Pi_{\pm} = \exp\left(\pm \frac{2i}{\hbar} \sum_j \frac{\gamma_j}{\omega_j^2} p_j \right) \qquad (3.64)$$

and their equilibrium averages taken with $\rho_b = \exp(-\beta H_{\text{medium}})/tr_b \exp(-\beta H_{\text{medium}})$ [10], the density matrix describing the medium at equilibrium

$$\langle \Pi \rangle \equiv \langle \Pi_{\pm} \rangle = tr_b \rho_b \Pi_{\pm} = \exp\left(-\frac{1}{\hbar} \sum_j \frac{\gamma_j^2}{\omega_j^3} \coth(\beta \hbar \omega_j/2) \right), \qquad (3.65)$$

with $\beta = 1/k_B T$ and tr_b denoting a trace over the medium degrees of freedom. The quantity $\bar{V}$ in Eqn 3.62 is the Franck–Condon renormalized tunnel splitting

$$\bar{V} = V \langle \Pi \rangle. \qquad (3.66)$$

The Hamiltonian of Eqn 3.62 has been separated into several parts containing the system variable and its interaction with the external field, the medium variables, and a system–medium interaction H_{in}. This latter term will be considered to be small. Note that this interaction part, whose thermal average is zero, can be small even when $\bar{V}$ is not.

The evolution of the overall density matrix $\rho(t)$ is given by the Liouville equation

$$i\hbar \frac{\partial \rho}{\partial t} = [H(t), \rho(t)]. \qquad (3.67)$$

We want a kinetic equation for $\sigma(t) = tr_b \rho(t)$, the reduced density matrix of the system.

To carry out this reduction we use a projection operator method [62]. A suitable projection operator P is defined by $PO = \rho_b tr_b O$. Applying this projection operator to Eqn 3.67 yields the exact but formal expression, with $Q = 1 - P$,

$$\frac{\partial \sigma(t)}{\partial t} = tr_b L(t)\rho_b\sigma(t) + \int_0^t d\tau tr_b L(t)(Te^{\int_\tau^t ds QL(s)Q})QL(t)\rho_b\sigma(t) + IVT \tag{3.68}$$

where IVT denotes the contribution to the dynamics from the initial value of the density matrix. The choice of an initial ensemble $\rho(0) = \rho_b\sigma(0)$ makes the IVT zero. In Eqn 3.68,

$$L(t) = L_0(t) + L_1$$

$$L_0(t)\ldots \equiv -\frac{i}{\hbar}[H_S + H_F(t) + H_{medium}, \ldots] \equiv L_S + L_F(t) + L_b \tag{3.69}$$

$$L_1 \ldots \equiv -\frac{i}{\hbar}[H_{in}, \ldots].$$

As the Hamiltonian has an explicit time dependence arising from the external field, we have had to introduce a time ordering in Eqn 3.68 to deal with the non-commuting time-dependent operators [39]. We use the notation

$$Te^{\int_\tau^t ds QL(s)Q}O \equiv \left(T_+ e^{-\frac{i}{\hbar}\int_\tau^t ds QH(s)Q}\right)O\left(T_- e^{+\frac{i}{\hbar}\int_\tau^t ds QH(s)Q}\right) \tag{3.70}$$

where $T_+ (T_-)$ denotes time ordering with the largest time to the left (right).

Working to second order in H_{in} and taking into account the following useful relations valid for any operator A

$$PL_0(t)QA = 0,$$

$$QL_0(t)QA = L_0(t)QA,$$

$$PL_b PA = 0, \tag{3.71}$$

$$PL_1 QA = PL_1 A$$

leads to

$$\frac{\partial \sigma(t)}{\partial t} = -\frac{i}{\hbar}[H_S + H_F(t), \sigma(t)] - \frac{1}{\hbar^2}\int_0^t d\tau tr_b [H_{in}, [\widetilde{\widetilde{H}}_{in}(t,\tau), \rho_b \widetilde{\widetilde{\sigma}}(t, \tau)]] \tag{3.72}$$

with the definitions

$$\widetilde{\widetilde{H}}_{in}(t,\tau) = \left(T_+ e^{-\frac{i}{\hbar}\int_\tau^t ds[H_S + H_F(s) + H_b]}\right)H_{in}\left(T_- e^{+\frac{i}{\hbar}\int_\tau^t ds[H_S + H_F(s) + H_b]}\right) \tag{3.73}$$

and

$$\widetilde{\widetilde{\sigma}}(t,\tau) = \left(T_+ e^{-\frac{i}{\hbar}\int_\tau^t ds[H_S + H_F(s)]}\right)\sigma(\tau)\left(T_- e^{+\frac{i}{\hbar}\int_\tau^t ds[H_S + H_F(s)]}\right). \tag{3.74}$$

Use has been made of $[H_S, H_b] = [H_F(t), H_b] = 0$, to simplify Eqn 3.74.

The difficulty in analysing the above equation of motion centres on the time-ordering requirement arising from the non-commutativity of the time-dependent external field operator and the system operator $[H_F(t), H_S] \neq 0$. Specializing to the specific form of our Hamiltonian will permit progress, as evaluating this commutator with the use of Eqn 3.62b shows that it is proportional to the small quantity $\bar{V}b(t)$. The modification of the evolution in Eqns 3.73 and 3.74 from assuming commutativity is therefore negligible. Then Eqn 3.73 becomes

$$\widetilde{\widetilde{H}}_{in}(t, \tau) = \left(e^{-\frac{i}{\hbar}(t-\tau)H_S}\right)\left(e^{-\frac{i}{\hbar}\int_\tau^t ds H_F(s)}\right)\left(e^{-\frac{i}{\hbar}(t-\tau)H_b}\right)$$
$$\times H_{in}\left(e^{+\frac{i}{\hbar}(t-\tau)H_b}\right)\left(e^{+\frac{i}{\hbar}\int_\tau^t ds H_F(s)}\right)\left(e^{+\frac{i}{\hbar}(t-\tau)H_S}\right) \tag{3.75}$$

and similarly for Eqn 3.74. With the form of H_{in} in Eqn 5.62b we have

$$\left(e^{-\frac{i}{\hbar}(t-\tau)H_b}\right)H_{in}\left(e^{+\frac{i}{\hbar}(t-\tau)H_b}\right) = \frac{V}{2}\left(\Phi^+(\tau-t)s_z + \Phi^-(\tau-t)is_y\right) \tag{3.76}$$

where we have explicitly written the time dependence of the medium operators $\Phi^\pm$ in the Heisenberg representation. Note that the argument of the operators in Eqn 3.76 is $(\tau - t)$ according to the conventional relation between the operators in the Schrödinger and Heisenberg representations [10]. The action of the field propagator on the spin operators in Eqn 3.75

$$\left(e^{-\frac{i}{\hbar}\lambda(t,\tau)s_x}\right)s_\alpha\left(e^{+\frac{i}{\hbar}\lambda(t,\tau)s_x}\right) = s_\alpha(\lambda(t,\tau)), \tag{3.77a}$$

where we have defined

$$\lambda(t, \tau) = \int_\tau^t 2b(s)ds, \tag{3.77b}$$

is readily found by generating second-order differential equations in λ for the spin operators $s_\alpha(\lambda)$. This yields

$$s_z(\lambda(t, \tau)) = \cos 2\lambda(t, \tau)s_z - \sin 2\lambda(t,\tau)s_y$$
$$s_y(\lambda(t, \tau)) = \cos 2\lambda(t, \tau)s_y + \sin 2\lambda(t,\tau)s_z. \tag{3.78}$$

At this point we may obtain a more explicit expression for Eqn 3.72 by taking its matrix elements in the basis of eigenstates of H_S, the system Hamiltonian. We focus attention on the relaxation term, as the oscillating term's evaluation is straightforward. By expanding the double commutator and using the cyclic permutation property of the trace over the medium operators to rearrange terms we find the kinetic equation

$$\frac{d\sigma_{ij}(t)}{dt} = \left(-\frac{i}{\hbar}[(H_S + H_F(t)), \sigma(t)]\right)_{ij} + \sum_{kl}\int_0^t R_{ijkl}(t, \tau)\widetilde{\widetilde{\sigma}}_{kl}(t, \tau)d\tau \tag{3.79a}$$

with

$$R_{ijkl}(t, \tau) = \Gamma^+_{ljik}(t, \tau) + \Gamma^-_{ljik}(t, \tau) - \delta_{lj}\sum_r \Gamma^+_{irrk}(t, \tau) - \delta_{ik}\sum_r \Gamma^-_{lrrj}(t, \tau)$$

$$\Gamma^{+}_{ljik}(t,\tau) = \frac{1}{\hbar^2}\,\mathrm{e}^{-i\omega_{ik}(t-\tau)}\langle\widetilde{H}_{\mathrm{in}}(t-\tau)_{lj}\,\bar{H}_{\mathrm{in}}(\lambda(t,\tau))_{ik}\rangle \tag{3.79b}$$

$$\Gamma^{-}_{ljik}(t,\tau) = \frac{1}{\hbar^2}\,\mathrm{e}^{-i\omega_{lj}(t-\tau)}\langle\bar{H}_{\mathrm{in}}(\lambda(t,\tau))_{lj}\,\widetilde{H}_{\mathrm{in}}(t-\tau)_{ik}\rangle$$

and we have defined $\bar{H}_{\mathrm{in}}$ and $\widetilde{H}_{\mathrm{in}}$ as

$$\bar{H}_{\mathrm{in}}(\lambda(t,\tau))_{lj} = \frac{V}{2}\,(\Phi^{+}s_z(\lambda(t,\tau))_{lj} + \Phi^{-}is_y(\lambda(t,\tau))_{lj})$$

$$\tag{3.80}$$

$$\widetilde{H}_{\mathrm{in}}(t-\tau)_{lj} = \frac{V}{2}\,(\Phi^{+}(t-\tau)(s_z)_{lj} + \Phi^{-}(t-\tau)(is_y)_{lj}).$$

Recall that the brackets denote the averages over the equilibrium medium density matrix as given in Eqn 3.65. These equilibrium time correlation functions are non-stationary and therefore depend explicitly on t and τ (not just their difference) because of the time dependence of the external field. $\widetilde{H}_{\mathrm{in}}(t-\tau)$ represents the conventional Heisenberg operator at time $t-\tau$ obtained from the corresponding Schrödinger operator with the action of the medium propagator, while $\bar{H}_{\mathrm{in}}(\lambda(t,\tau))$ denotes the operator obatined from the action of the external field propagator (cf. Eqn 3.77).

The relaxation part of Eqn 3.79 is expressed in terms of the quantity $\widetilde{\widetilde{\sigma}}(t,\tau)$ that we now need to express in terms of $\sigma(\tau)$ in order to obtain a closed equation for σ. These reduced density matrices are connected via Eqn 3.74. Thus, the relaxation equation in Eqn 3.79 can be written as

$$\frac{\mathrm{d}\sigma_{ij}(t)}{\mathrm{d}t} = \left(-\frac{i}{\hbar}\,[(H_{\mathrm{S}}+H_{\mathrm{F}}(t)),\sigma(t)]\right)_{ij} + \sum_{kl}\int_0^t \widetilde{R}_{ijkl}(t,\tau)\sigma_{kl}(\tau)\mathrm{d}\tau, \tag{3.81}$$

where the tilde on $\widetilde{R}_{ijkl}$ symbolizes the redefined relaxation tensor obtained from the conversion of the $\widetilde{\widetilde{\sigma}}(t,\tau)$ matrix elements to the $\sigma(\tau)$ matrix elements. We stress that Eqn 3.79 and 3.81 are just different expressions of the same kinetic equation.

The kinetic equation as written in Eqn 3.81 is not of the convolution structure because of the presence of the time-dependent external field. If the external field were constant (or absent) then Eqn 3.81 would have a convolution structure and could be expressed in Laplace transform space with the transformed relaxation kernel $\hat{R}(\mu)$. (The $\hat{}$ denotes the Laplace transform of the function and we use μ for the Laplace transform argument.) The conventional time-local kinetic equation is obtained by replacing $\hat{R}(\mu)$ with $\hat{R}(\mu=0)$. This replacement is correct when $\hat{R}(\mu)$ is an analytic function of μ in the neighbourhood of $\mu=0$. A case where this replacement is not valid is for very low temperatures and a medium whose mode spectrum (the ω_js of Eqn 3.62b) is ohmic. Here, $\hat{R}(\mu)$ has a branch point at $\mu=0$. The consequences of this feature have been discussed by Leggett *et al.* in detail [25]. Even in the ohmic case, when the temperature is not so low, time-local equations give a satisfactory account of the dynamics [26]. In the following we will assume the adequacy of this replacement, as it describes the kinetics in most regimes.

A connection with a local-in-time, Redfield-like theory [63], including the effect of an external field, can now be made. Equation 3.74 shows that another way of writing

$\widetilde{\widetilde{\sigma}}(t, \tau)$ is as $\sigma^0(t)$, where this latter notation expresses the feature that $\sigma^0(t)$ is the zero-order solution of Eqn 3.79a with the condition that at the initial time $t = \tau$, the density matrix was $\sigma(\tau)$. Therefore, in Eqn 3.79a, $\widetilde{\widetilde{\sigma}}(t, \tau)$ can be replaced by $\sigma^0(t)$. Since Eqn 3.79 has been obtained by second-order perturbation theory, it is correct, to the order of the calculation, to replace $\sigma^0(t)$ by $\sigma(t)$. Therefore, we can write the following Redfield-like kinetic equation

$$\frac{d\sigma_{ij}(t)}{dt} = \left(-\frac{i}{\hbar} [(H_S + H_F(t)), \sigma(t)] \right)_{ij} + \sum_{kl} \left[\int_0^\infty R_{ijkl}(t, \tau) d\tau \right] \sigma_{kl}(t). \tag{3.82}$$

We stress that even if the kinetic equation in Eqn 3.81 is approximated by its time local version, it still differs from Eqn 3.82 as the relaxation tensors are different. The difference arises from the presence of the external field. In the absence of the external field, the time evolution connecting $\widetilde{\widetilde{\sigma}}(t, \tau)$ and $\sigma(\tau)$ is slow as it originates from the small quantity $\bar{V}$. Thus, without an external field, and assuming the existence of the time integral in Eqn 3.82 (or that the $\hat{R}_{ijkl}(\mu = 0)$ exist), Eqns 3.81 and 3.82 are equivalent. When the external field is present, the interaction representation evolution in Eqn 3.74 includes its potentially significant affect on the system dynamics, and leads to differences in the forms of the kinetic equations.

Before presenting specific results for our two-level system model, we note that the derivation of the kinetic equation in Eqn 3.79 is not restricted to this particular model. As long as the interaction Hamiltonian can be written as a product of system and medium operators, the structure in Eqn 3.79 will be obtained. In particular, the kinetic equation applies to externally driven systems with several tunnel doublets. Of interest there would be the possibility of externally induced transitions between different tunnel doublets and their affect on the population transfer from one localized state to the other.

The analysis of the population dynamics implied by the above kinetic equation requires solutions of the coupled equations of motions for the density matrix elements. The equations of motion have coefficients given by the R_{ijkl}, the elements of the relaxation tensor R. These elements can be expressed in terms of the medium, system–medium and system–external field coupling parameters. This calculation is carried out in the appendix. We first focus on the structure of the equations of motion and then list the various independent elements of R. Since a two-level system's density matrix is characterized by three independent quantities, we may choose the following linear combinations as our variables:

$$p(t) = \sigma_{12}(t) + \sigma_{21}(t)$$
$$d_-(t) = i(\sigma_{21}(t) - \sigma_{12}(t)) \tag{3.83}$$
$$d_+(t) = \sigma_{11}(t) - \sigma_{22}(t).$$

In the basis of eigenstates of the system Hamiltonian, $p(t)$ is the difference in population between the left- and right-localized states, the quantity of interest for tunnelling dynamics. It is related to the average left-state population $P_L(t)$ used in the previous sections, according to $p(t) = 2P_L(t) - 1$.

We now analyse the general kinetic equation, Eqn 3.79. To do so it is convenient to relate the matrix elements of $\widetilde{\widetilde{\sigma}}(t, \tau) = \sigma^0(t)$ to those of $\sigma(\tau)$ according to the relation of Eqn 3.74, which is simply a rotation. Thus,

$$
\begin{pmatrix} p^0(t) \\ d_+^0(t) \\ d_-^0(t) \end{pmatrix} = \begin{pmatrix} 1 & 0 & 0 \\ 0 & \cos 2\lambda(t, \tau) & -\sin 2\lambda(t, \tau) \\ 0 & \sin 2\lambda(t, \tau) & \cos 2\lambda(t, \tau) \end{pmatrix} \begin{pmatrix} p(\tau) \\ d_+(\tau) \\ d_-(\tau) \end{pmatrix}.
\tag{3.84}
$$

With this connection, and Eqn 3.79, we eventually obtain the following equations of motion

$$
\dot{p}(t) = 2(\bar{V}/\hbar)d_-(t) + \int_0^t [g(t, \tau) + c(t, \tau)] p(\tau)\mathrm{d}\tau + \int_0^t [\alpha(t, \tau) + \beta(t, \tau)]\mathrm{d}\tau
$$

$$
\dot{d}_-(t) = -2(\bar{V}/\hbar)p(t) + 2(2b(t)/\hbar)d_+(t)
$$

$$
+ \int_0^t \{[g(t, \tau) - c(t, \tau)]\cos(2\lambda(t, \tau)) + i[\alpha(t, \tau) - \beta(t, \tau)]\sin(2\lambda(t, \tau))\}d_-(\tau)\mathrm{d}\tau
$$

$$
+ \int_0^t \{i[\beta(t, \tau) - \alpha(t, \tau)]\cos(2\lambda(t, \tau)) + [g(t, \tau) - c(t, \tau)]\sin(2\lambda(t, \tau))\}d_+(\tau)\mathrm{d}\tau \tag{3.85}
$$

$$
\dot{d}_+(t) = -2(b(t)/\hbar)d_-(t) + \int_0^t \{2c(t, \tau)\cos(2\lambda(t, \tau)) + 2i\gamma(t, \tau)\sin(2\lambda(t, \tau))\}d_+(\tau)\mathrm{d}\tau
$$

$$
+ \int_0^t \{2i\gamma(t, \tau)\cos(2\lambda(t, \tau)) - 2c(t, \tau)\sin(2\lambda(t, \tau))\}d_-(\tau)\mathrm{d}\tau.
$$

The coefficients in these equations are expressed in terms of the elements of the relaxation tensor R as follows:

$$
\alpha = R_{1211}, \quad \beta = R_{2111}, \quad g = R_{1212}, \quad c = R_{1111}, \quad \gamma = R_{1112}. \tag{3.86}
$$

For our model Hamiltonian there are only five independent elements. Their explicit expressions are presented in the appendix. Along with the definition of $\lambda(t, \tau)$ in Eqn 3.77b they completely specify the equations of motion. Two combinations that will be relevant to approximate solutions of these equations are $\alpha + \beta$ and $c + g$. Their explicit forms, obtained with the use of Eqn 3.A6, p. 154, are

$$
g(t, \tau) + c(t, \tau) = -2(\bar{V}/\hbar)^2[\mathrm{e}^{W(t-\tau)} + \mathrm{e}^{W(\tau-t)} - 2]\cos 2\lambda(t, \tau) \tag{3.87}
$$

and

$$
\alpha(t, \tau) + \beta(t, \tau) = 2(\bar{V}/\hbar)^2[\mathrm{e}^{W(\tau-t)} - \mathrm{e}^{W(t-\tau)}]i \sin 2\lambda(t, \tau), \tag{3.88}
$$

where $W(t)$ is defined in Eqn 3.A5. Note that Eqn 3.A5 relates $W(t)$ to the spectral density of the system–medium interaction.

Clearly, Eqns 3.85 are very complex. We first analyse them for a constant external field. For an initial condition, $p(0) = 1$, $d_+(0) = d_-(0) = 0$, corresponding to the system localized on the left side, an accurate approximate equation of motion can be developed. Since the relaxation terms in the second and third equations in Eqn 3.85 are explicitly of order $\bar{V}^2$, and considering that $d_\pm(t)$ vanish initially, their effect on the evolution of $p(t)$ may be neglected except perhaps for very long times. Then, the equations of motion in Eqn 3.85 simplify to

$$\dot{p}(t) = 2(\bar{V}/\hbar)d_-(t) + \int_0^t [g(t-\tau) + c(t-\tau)]p(\tau)d\tau + \int_0^t [\alpha(t-\tau) + \beta(t-\tau)]d\tau$$

$$\dot{d}_-(t) = -2(\bar{V}/\hbar)p(t) + 4(b/\hbar)d_+(t) + O(\bar{V}^2) \tag{3.89}$$

$$\dot{d}_+(t) = -4(b/\hbar)d_-(t) + O(\bar{V}^2).$$

These equations have a convolution structure that will permit Laplace transformation. Formally solving the second and third equations for $d_-(t)$ and inserting them into the first equation provides a closed equation for $p(t)$ whose Laplace transformed solution is a generalization [24] of the so-called non-interacting blip approximation [25]. The advantage of this convolution formulation is its validity at temperatures smaller than $\bar{V}$. The tunnel system's interaction with the medium at such low temperatures may not be sufficiently strong as to completely destroy the coherence of the tunnelling event. There is no guarantee of a rate process, as would be implied by a time-local kinetic equation. As the temperature is raised, $\bar{V}$ decreases rapidly and leads to incoherent tunnelling as described by a rate process.

Turning now to time-dependent external fields, the non-convolution structure of equations of motion precludes a simple analysis in terms of Laplace transforms for arbitrary time variations. For general initial conditions and arbitrarily low temperature only a numerical solution of Eqn 3.85 is possible. We can proceed analytically if we focus on the physically interesting initial condition used above and for temperatures high enough that the renormalized tunnel splitting $\bar{V}$ is a negligible quantity, even though $V = \bar{V}/\langle\Pi\rangle = \bar{V}e^{W(0)/2}$ is itself small but not negligible. Even at these temperatures the medium dynamics may still have to be treated quantum mechanically as $\beta\hbar\omega/2$ may not be small. Under these conditions, terms proportional to $\bar{V}$ and $\bar{V}^2$ are neglected while those proportional to V^2 are kept in Eqns 3.87–3.89. In this way the evolution of $p(t)$ is decoupled from that of $d_\pm(t)$ and is given by the first order in time equation

$$\dot{p}(t) = -\int_0^t [k(t,\tau)]p(\tau)d\tau + \int_0^t [\alpha(t,\tau) + \beta(t,\tau)]d\tau \tag{3.90}$$

where the rate kernel $k(t,\tau)$ is defined as

$$k(t,\tau) \equiv -(g(t,\tau) + c(t,\tau)) = 2(\bar{V}/\hbar)^2 [e^{W(t-\tau)} + e^{W(\tau-t)} - 2]\cos 2\lambda(t,\tau)$$

$$\approx 2(V/\hbar)^2 [e^{W(t-\tau)-W(0)} + e^{W(\tau-t)-W(0)}]\cos 2\lambda(t,\tau)$$

$$= (2V/\hbar)^2 e^{\Psi(t-\tau)-\Psi(0)} \cos G(t-\tau)\cos 2\lambda(t,\tau). \tag{3.91}$$

The inhomogeneous term $\alpha + \beta$ of Eqn 3.88 can be rewritten, with the definitions in Eqn 3.A6, as

$$\alpha(t,\tau) + \beta(t,\tau) = -(2V/\hbar)^2 e^{\Psi(t-\tau)-\Psi(0)} \sin G(t-\tau)\sin 2\lambda(t,\tau). \tag{3.92}$$

The rate kernel $k(t,\tau)$ is the generalization to the quantum regime of the kernel appearing in Eqn 3.52. It is the sum of the forward and backward (left to right and right to left) rate kernels. At high temperature $\beta\hbar\omega/2 \ll 1$ and for slow modulation the rate kernel obtained from Eqn 3.91 coincides with the kernel in Eqn 3.52. This reduction is obtained by using the short time expansions of $\Psi(t)$ and $G(t)$, and recognizing the definitions of $\bar{V}$ given in Eqns 3.65 and 3.66 and E_r given in Eqn 3.45b.

For a constant external field, the inhomogeneous term in Eqn 3.90 accounts for the decay of $p(t)$ to its correct long-time equilibrium value (within the limits of the approximations used to derive Eqn 3.90). For a constant external field, we may Laplace transform Eqn 3.90 and extract the equilibrium solution p_{eq} as

$$p_{eq} = p(t \to \infty) = \lim_{\mu \to 0} \mu \hat{p}(\mu) = \frac{\hat{\alpha}(0) + \hat{\beta}(0)}{\hat{k}(0)}. \tag{3.93}$$

Following the methodology of Leggett *et al.* [25] (see their Appendix E) we find from Eqns 3.91 and 3.92 that Eqn 3.93 becomes

$$p_{eq} = -\tanh(2b/k_B T). \tag{3.94}$$

This gives the correct Boltzmann population distribution of the left and right states (for $\bar{V}$ small compared to b). Actually, when the coupling is strong, it is not straightforward to decide what the system's quantum mechanical equilibrium distribution should be, as the separation between system and medium cannot be made in a simple fashion. Indeed, the equilibrium form of the reduced density matrix for the system is not known explicitly when the system and medium are quantum mechanical.

We now analyse the effects arising from a quantum medium in the presence of a time-dependent external field on the quantity

$$\Gamma(t) = \int_0^\infty k(t, \tau) d\tau. \tag{3.95}$$

Again, we do not call $\Gamma(t)$ a rate constant as it is a time-dependent quantity. As we saw in the classical medium case, however, for a sinusoidal external field we can, to a good approximation, obtain a constant value. In preparation for this analysis, let us define $\hat{k}_0(\mu)$ as the Laplace-transformed rate kernel in the absence of an external field:

$$\hat{k}_0(\mu) = \int_0^\infty k(t) e^{-\mu t} = (2V/\hbar)^2 \int_0^\infty e^{\Psi(t) - \Psi(0)} \cos G(t) e^{-\mu t}. \tag{3.96}$$

We will assume the existence of the limit $\hat{k}_0(\mu \to 0) \equiv \hat{k}_0$. Let us use, as an external field,

$$,b(t) = b_c + b_0 \cos \Omega t,$$

so that there is a constant and an oscillatory part. Using in Eqns 3.91 and 3.95 this external field expression, the expansion of Eqn 3.59, and again averaging over one period of the external field (see Eqn 3.58b), leads to the result

$$\bar{\Gamma} = Re \sum_{m = -\infty}^{m = +\infty} J_m^2 \left(\frac{4b_0}{\hbar \Omega}\right) \hat{k}_0(im\Omega - ib_c). \tag{3.97}$$

Thus, knowledge of $\hat{k}_0(\mu)$, the field-free Laplace-transformed rate kernel, will provide the rate constant in the presence of just a constant field according to

$$\bar{\Gamma} = Re \; \hat{k}_0(- ib_c). \tag{3.98}$$

When a sinusoidal external field is present, and the parameters that enter Eqn 3.97 are chosen such that the first term dominates, then

$$\bar{\Gamma} = Re \; J_0^2 \left(\frac{4b_0}{\hbar\Omega} \right) \hat{k}_0(- ib_c), \tag{3.99}$$

and, just as for the classical medium case, if a Floquet condition is met by a suitable choice of b_0 and Ω the rate constant will be close to zero.

The analysis carried out in this section provides the generalization to a quantum medium of the results of Section 4. The new feature that is a consequence of a quantum medium is the dependence on the spectral density of the system–medium interaction, in contrast to a classical medium treatment where it is the reorganization energy (a moment of the spectral density) that enters as a parameter. The same is true of tunnelling in the absence of an external field.

6 Concluding remarks

The theme that we have elaborated upon here, from a variety of perspectives, is that tunnelling can be controlled by the application of an external field. The general projection operator approach presented in Section 5 shows that for all but very low temperatures, a rate regime, or a regime with a time-dependent rate, does exist. The rate constant (or cycle-averaged rate constant) depends on the external field, and the medium and tunnel system parameters, in a sufficiently simple fashion (see Eqn 3.97) that various strategies for suppressing tunnelling can be readily contemplated. The difference in rate expressions for low and high temperatures relative to the medium's characteristic frequency is that, in the former case, it is the medium spectral density that enters the rate expression, while in the latter only the medium reorganization energy is relevant.

When the coupling between system and bath is not so strong, the tunnelling dynamics can exhibit a number of regimes with an oscillatory decay to equilibrium, as we focused on in Sections 2 and 3. By employing methods that are directed to obtaining equations of motion for the density matrix of the tunnel system, the behaviour of the system evolution beyond rate constant regimes can be investigated.

An important feature of all the methods of analysis that we used is that the external field does not have to be weak. It is incorporated in the zero-order Hamiltonian. The time ordering implied by the non-commutativity of the system and external field operators would preclude progress; however, the commutator is proportional to the (small) renormalized tunnel splitting $\bar{V}$, and this permits a sufficient disentanglement of the propagators to lead to useful equations of motion for the system degrees of freedom. Note that the relaxation terms of the equations of motion are functions of the external field, and this feature leads to the dramatic variation of the rate constants with external field.

Finally, let us remark on the difference between our methodology and Redfield theory as applied to nuclear magnetic relaxation (NMR). While we have discussed the differences regarding time-local versus time non-local expressions in Section 5, there are actually more fundamental distinctions. In the typical application of Redfield theory to NMR, the systematic terms oscillate at the Larmor frequency. The relaxation tensor is independent of the Larmor frequency, and depends on the system–medium coupling strength. In the two-level tunnel system, both the systematic and the relaxation terms are proportional to the tunnel splitting, which is an intrinsically small quantity. In NMR, a distinction between secular (non-oscillatory) and non-secular (oscillating at frequencies related to the Larmor frequency that are large compared with the relaxation terms) is made [63]. The non-secular terms lead to negligible effects on the relaxation of the spin system. In contrast, for the two-level tunnel system, the tunnel splitting is very small and, as a consequence, a distinction between secular and non-secular terms cannot be used to simplify the theory. In fact, as shown in Section 5, the α and β terms defined by Eqns 3.86 are non-secular in the NMR sense, and are responsible here for obtaining the correct equilibrium state of the spin system (see Eqn 3.93).

7   Appendix

In this appendix we evaluate the elements of the relaxation tensor R. Equations 3.79b, 3.80 and 3.63–3.66 show that the basic quantities of interest that must be evaluated are the $\Gamma^{\pm}_{ijkl}$. They are obtained from Eqns 3.79b and 3.80. As the ω_{ik} in our model are proportional to $\bar{V}$, these frequencies are very small and may be neglected in the expressions for $\Gamma^{\pm}_{ijkl}$. With this approximation we find

$$\Gamma^{+}_{1111} = \Gamma^{+}_{2222} = -\Gamma^{+}_{1122} = -\Gamma^{+}_{2211} = \langle \Phi^{+}(t-\tau)\Phi^{+}\rangle \cos 2\lambda(t,\tau)$$

$$\Gamma^{+}_{1212} = \Gamma^{+}_{2121} = -\Gamma^{+}_{2112} = -\Gamma^{+}_{1221} = \langle \Phi^{-}(t-\tau)\Phi^{-}\rangle \cos 2\lambda(t,\tau)$$

$$\Gamma^{+}_{1211} = \Gamma^{+}_{2122} = -\Gamma^{+}_{2111} = -\Gamma^{+}_{1222} = \langle \Phi^{-}(t-\tau)\Phi^{-}\rangle i \sin 2\lambda(t,\tau) \tag{3.A1a}$$

$$\Gamma^{+}_{1112} = \Gamma^{+}_{2221} = -\Gamma^{+}_{1121} = -\Gamma^{+}_{2212} = \langle \Phi^{+}(t-\tau)\Phi^{+}\rangle i \sin 2\lambda(t,\tau)$$

and

$$\Gamma^{-}_{1111} = \Gamma^{-}_{2222} = -\Gamma^{-}_{1122} = -\Gamma^{-}_{2211} = \langle \Phi^{+}\Phi^{+}(t-\tau)\rangle \cos 2\lambda(t,\tau)$$

$$\Gamma^{-}_{1212} = \Gamma^{-}_{2121} = -\Gamma^{-}_{2112} = -\Gamma^{-}_{1221} = \langle \Phi^{-}\Phi^{-}(t-\tau)\rangle \cos 2\lambda(t,\tau)$$

$$\Gamma^{-}_{1211} = \Gamma^{-}_{2122} = -\Gamma^{-}_{2111} = -\Gamma^{-}_{1222} = \langle \Phi^{+}\Phi^{+}(t-\tau)\rangle i \sin 2\lambda(t,\tau) \tag{3.A1b}$$

$$\Gamma^{-}_{1112} = \Gamma^{-}_{2221} = -\Gamma^{-}_{1121} = -\Gamma^{-}_{2212} = \langle \Phi^{-}\Phi^{-}(t-\tau)\rangle i \sin 2\lambda(t,\tau).$$

In obtaining these equations we have used the results

$$\langle \Phi^{+}\Phi^{-}(t-\tau)\rangle = \langle \Phi^{-}\Phi^{-}(t-\tau)\rangle = \langle \Phi^{+}(t-\tau)\Phi^{-}\rangle = \langle \Phi^{-}(t-\tau)\Phi^{+}\rangle = 0. \tag{3.A2}$$

The non-zero correlation functions are

$$\langle \Phi^{+}(t-\tau)\Phi^{+}\rangle = 4e^{-W(0)}[\cosh W(t-\tau) - 1]$$

$$\langle \Phi^{+}\Phi^{+}(t-\tau)\rangle = 4e^{-W(0)}[\cosh W(\tau-t) - 1]$$

$$\langle \Phi^{-}(t-\tau)\Phi^{-}\rangle = -4e^{-W(0)} \sinh W(t-\tau) \tag{3.A3}$$

$$\langle \Phi^{-}\Phi^{-}(t-\tau)\rangle = -4e^{-W(0)} \sinh W(\tau-t)$$

where

$$W(t) = \sum_j (2\gamma_j^2/\hbar\omega_j^3)[\coth(\beta\hbar\omega_j/2)\cos\omega_j t - i\sin\omega_j t] \equiv \Psi(t) - iG(t). \qquad (3.A4)$$

Note that $W(t=0)/2$ is the exponent in the Franck–Condon factor renormalizing the tunnel splitting (see Eqns 3.65 and 3.66). It arises from the overlap of the medium wavefunctions centered around the left and right local states of the tunnelling system. The quantum equilibrium time correlation functions in Eqns 3.A2 and 3.A3 are evaluated by expressing the time evolution of the momentum operator $p(t)$ appearing in Eqn 3.64 in terms of its initial values $q(0)$, $p(0)$ according to the medium harmonic oscillator dynamics of Eqn 3.62b. The resulting Gaussian average over these initial-time coordinate and momentum operators can then be obtained by standard techniques [10]. It is conventional to express $W(t)$ as an integral over a spectral density $J(\omega)$ of the medium modes [13,16,25] as

$$W(t) = \int_{\omega_{min}}^{\omega_{max}} (J(\omega)/\omega^2)[\coth(\beta\hbar\omega/2)\cos\omega t - i\sin\omega t]d\omega \qquad (3.A5a)$$

with

$$J(\omega) \equiv \sum_j (2\gamma_j^2/\hbar\omega_j)\delta(\omega - \omega_j). \qquad (3.A5b)$$

Note that the spectral density reflects the nature of the coupling between the system and the medium.

With these averages and Eqn 3.A1, the elements of the R tensor can be obtained from Eqn 3.79b. They are:

$$R_{1111} = R_{1221} = R_{2222} = R_{2112} = -R_{1122} = -R_{2211}$$
$$= -(\bar{V}/\hbar)^2 \cos 2\lambda(t, \tau)[\sinh W(\tau - t) + \sinh W(t - \tau)]$$

$$R_{1212} = R_{2121} = -2(\bar{V}/\hbar)^2 \cos 2\lambda(t, \tau)[\cosh W(\tau - t) + \cosh W(t - \tau) - 2] + R_{1111}$$

$$R_{1211} = R_{2122} = -(\bar{V}/\hbar)^2 i \sin 2\lambda(t, \tau)\{[\sinh W(t - \tau) - \sinh W(\tau - t)] - 2[\cosh W(\tau - t) - 1]\}$$

$$R_{2111} = R_{1222} = -(\bar{V}/\hbar)^2 i \sin 2\lambda(t, \tau)\{[\sinh W(t - \tau) - \sinh W(\tau - t)] + 2[\cosh W(\tau - t) - 1]\}$$

$$R_{1112} = R_{2221} = -R_{1121} = -R_{2212} = (\bar{V}/\hbar)^2 i \sin 2\lambda(t, \tau)[\sinh W(t - \tau) + \sinh W(\tau - t)].$$

$$(3.A6)$$

These are the explicit expressions for the five independent quantities characterizing R that are used in Eqn 3.86.

Acknowledgements. Support by NATO (R.I.C. and M.M.), the Center for Fundamental Materials Research at Michigan State University (R.I.C. and M.M.), the DGICYT of Spain and the Junta de Andalucia (M.M.) and the National Science Foundation (R.I.C.) for our research in this area is gratefully acknowledged.

8   References

1 Carter FL, ed. *Molecular Electronic Devices*. New York: Marcel Dekker, 1982.
2 Carter FL, ed. *Molecular Electronic Devices II*. New York: Marcel Dekker, 1987.

3 Carter FL, Siatkowski RE, Wohltjen H, eds. *Molecular Electronic Devices.* Amsterdam: North-Holland, 1988.

4 Hong FT, ed. *Molecular Electronics Biosensors and Biocomputers.* New York: Plenum Press, 1989.

5 Aviram A, ed. *Molecular Electronics — Science and Technology.* New York: AIP, 1992.

6 Ashwell GJ, ed. *Molecular Electronics.* New York: John Wiley, 1992.

7 Sienicki K, ed. *Molecular Electronics and Molecular Electronic Devices.* Boca Raton: CRC Press, 1994.

8 Breton J, Vermiglio A, eds. *The Photosynthetic Bacterial Reaction Center — Structure and Dynamics.* New York: Plenum, 1988.

9 Aviram A, Ratner MA. *Chem Phys Letts* 1974; **29**: 277.

10 Messiah A. *Quantum Mechanics.* Amsterdam: North-Holland, 1966: vol I.

11 Chance B, De Vault DC, Frauenfelder H, Marcus RA, Schreiffer JR, Sutin N, eds. *Tunneling in Biological Systems.* New York: Academic Press, 1979.

12 DeVault D. *Quantum Mechanical Tunneling in Biological Systems.* London: Cambridge University Press, 1984.

13 Fain B. *Theory of Rate Processes in Condensed Media.* Berlin: Springer, 1980.

14 Levich VG. In: Henderson H, Yost W, eds. *Physical Chemistry — An Advanced Treatise.* New York: Academic, 1970: 985.

15 Marcus R, Sutin N. *Biochim Biophys Acta* 1985; **811**: 265.

16 Ulstrup J. *Charge Transfer Processes in Condensed Media.* Berlin: Springer, 1979.

17 Bottker H, Bryksin VV. *Hopping Conduction in Solids.* Berlin: Akademic, 1985.

18 Stoneham AM. *Theory of Defects in Solids.* Oxford: Clarendon, 1975.

19 Newton MD. *Chem Rev* 1991; **91**: 767.

20 Benderskii VA, Makarov DE, Wight CA. *Adv Chem Phys* 1994; **88**: 1.

21 Grossmann F, Hanggi P. *Europhys Lett* 1992; **18**: 571.

22 Cukier RI, Morillo M. *Chem Phys* 1994; **183**: 375.

23 Morillo M, Cukier RI, Tij M. *Physica A* 1991; **179**: 411.

24 Morillo M, Tij M. *Physica A* 1991; **179**: 428.

25 Leggett AJ, Chakravarty S, Dorsey AT *et al. Rev Mod Phys* 1987; **59**: 1.

26 Silbey R, Harris RA. *J Phys Chem* 1989; **93**: 7062.

27 Morillo M, Cukier RI. *J Chem Phys* 1993; **98**: 4548.

28 Eckert M, Zundel G. *J Phys Chem* 1987; **91**: 5170.

29 Weidemann EG, Zundel G. *Z Naturforsch* 1970; **25A**: 627.

30 Janoschek R, Weidemann EG, Pfeiffer H, Zundel G. *J Am Chem Soc* 1972; **94**: 2387.

31 *Adv Chem Phys* 1989; **73**.

32 Shirley JH. *Phys Rev B* 1965; **138**: 979.

33 Sambe H. *Phys Rev A* 1973; **7**: 2203.

34 Aravind PK, Hirschfelder JO. *J Phys Chem* 1984; **88**: 4788.

35 Moss F. In: Weiss GH, ed. *Some Problems in Statistical Physics.* Philadelphia: SIAM, 1992.

36 Tesche CD. *Ann NY Acad Sci* 1986; **480**: 236.

37 Chu F. *Adv Chem Phys* 1986; **73**: 739.

38 Oka T. *Adv Mol Phys* 1973; **9**: 127.

39 van Kampen NG. *Stochastic Processes in Physics and Chemistry.* Amsterdam: North-Holland, 1981.

40 Toda M, Kubo R, Hashitsume N. *Statistical Physics II, Non-equilibrium Statistical Mechanics.* Berlin: Springer, 1991: Ch 2.

41 Korst NN, Nikitin EE. *Theor Exp Chem* 1965; **1**: 5.

42 Kubo R. In: Haar D, ed. *Fluctuations, Relaxation and Resonance in Magnetic Systems.* New York: Plenum, 1962: 23.

43 Kubo R. *Adv Chem Phys* 1969; **15**: 101.

44 Kubo R. *J Phys Soc Japan* 1954; **9**: 935.

45 Friedmann PP. *Adv Chem Phys* 1986; **73**: 197.

46 Cukier RI, Morillo M. *J Chem Phys* 1990; **93**: 2364.

47 Stockli A, Meier BH, Kreis R, Meyer R, Ernst RR. *J Chem Phys* 1990; **93**: 1502.

48 Meyer R, Ernst RR. *J Chem Phys* 1990; **93**: 5518.

49 Marcus RA, Sutin N. *Biochim Biophys Acta* 1985; **811**: 265.

50 Marcus RA. *Rev Mod Phys* 1993; **65**: 599.

51 Caldeira AO, Leggett AJ. *Am Phys* 1983; **149**: 374.

52 Kielich S. In: Davies M, ed. *Dielectric and Related Molecular Processes.* London: Chemical Society, 1972: 192.

53 Zwanzig R. *J Stat Phys* 1973; **9**: 215.

54 Zusman LD. *Soc Phys JETP* 1976; **42**: 794.

55 Cukier RI, Morillo M. *J Chem Phys* 1989; **91**: 857.

56 Kosloff R. *J Phys Chem* 1988; **92**: 2087.

57 Skinner JL, Trommsdorff HP. *J Chem Phys* 1988; **89**: 897.

58 Dakhnovskii Y. *J Chem Phys* 1994; **100**: 6492.

59 Farazdel A, Dupuis M, Clementi E, Aviram A. *J Am Chem Soc* 1990; **112**: 4206.

60 Petzinger KG. *Phys Rev B* 1982; **26**: 6530.

61 Romero-Rochin V, Oppenheim I. *Physica A* 1989; **155**: 52.

62 Zwanzig R. *J Chem Phys* 1960; **33**: 1338.

63 Redfield AG. *Adv Mag Res* 1965; **1**: 1.

4 Coulomb Blockade and Digital Single-electron Devices

A.N. KOROTKOV

Department of Physics, State University of New York, Stony Brook, NY 11794-3800, USA and Nuclear Physics Institute, Moscow State University, Moscow 119899, Russia

1 Introduction

The continuous trend in scaling down component size leads integrated electronics into the domain of so-called 'mesoscopics', the dimension area between the microscopic and macroscopic worlds. The Coulomb blockade and other single-electron charging effects (for reviews see [1–5]) are among the most prominent phenomena of mesoscopic physics. Although these effects are caused by single electrons that make a connection with the microscopic world, they are usually well described in terms of macroscopic 'electrical engineering' quantities like capacitances and resistances. Single-electron effects will play an important role in almost any electronic device with dimensions less than ~30 nm. Moreover, they can be used as the new physical basis for the operation of nanoscale digital circuits.

The main component of single-electronics is a tunnel junction with a very small capacitance. These junctions can be implemented using a variety of materials: metal–insulator–metal structures, GaAs quantum dots, silicon structures, large molecules with conducting cores, etc. If the size (and, hence, the effective electrical capacitance C) of the junction is sufficiently small, then the tunnelling of only one electron may produce a noticeable change e/C of the voltage across the junction. The discreteness of this change, which is a consequence of the electrical charge discreteness, leads to a number of effects which constitute the field of single-electronics. The most widely known effect is the Coulomb blockade. This is the suppression of tunnelling at voltages $|V| < e/2C$, because in this case the tunnelling would increase the electrostatic energy of the capacitor: $C(V \pm e/C)^2/2 > CV^2/2$. Besides the 'Coulomb blockade' and 'single-electronics', the other key words of the field are 'correlated tunnelling' and 'single-charge tunnelling'.

The typical capacitance in most present-day experiments is of the order of 10^{-16} F (the simplest well-established technology uses metal junctions with an area about 50×50 nm^2). It corresponds to the voltage scale e/C of the order of 1 mV that is already sufficiently large to allow experimental studies. To avoid the smearing of single-electron effects by thermal fluctuations, the thermal energy $k_B T$ should be much less than the typical one-electron charging energy,

$$k_B T \ll e^2/2C. \tag{4.1}$$

For $C \approx 10^{-16}$ F this condition limits the temperature to $T \lesssim 1$ K. As a consequence, at the present stage of development the practical use of single-electron devices is limited to scientific experimentation and fundamental metrology. To achieve real industrial impact the operation temperature must be increased up to 300 K (or at least to 77 K); that requires a dramatic decrease of the typical size of the components. There have already been a considerable number of experiments in which the room temperature or liquid nitrogen temperature operation of simple single-electron 'devices' was reported, but these systems are not sufficiently reproducible yet.

Capacitances as small as 10^{-19} F can be achieved in principle in the 'molecular electronic devices' using conducting clusters of atoms (with diameter of 1 nm or less) embedded in a molecular matrix. In this case the energy scale $e^2/2C$ would be about 1 eV, and room temperature operation could be ensured. Notice that at the size scale below a few nm, single-electronics gradually enters the areas of chemistry and atomic physics. Here, the discreteness of energy levels in conducting 'islands' becomes an important factor, the electric capacitance becomes a not well-defined quantity, and the Coulomb blockade energy gradually converts into the energy of ionization and electron affinity. However, the ideas of single-electronics are applicable even at this size scale, opening the possibility of information processing in molecular-level devices.

2   Brief history

The influence of single-electron effects on the conductance of thin granular metallic film was understood by Gorter [6] in 1951. During the next 20 years important contributions to this field came from Neugebauer and Webb [7], Giaver and Zeller [8], and Lambe and Jaklevic [9]. The first quantitative theory for a two-junction system was developed in 1975 by Kulik and Shekhter [10].

In its present-day meaning single-electronics was launched in the mid-1980s when the detailed theory of correlated tunnelling (now known as the 'orthodox' theory) developed by Averin and Likharev [11,12] was almost immediately supported by experiments with single-electron transistors [13,14]. Since that time there has been constantly growing interest both in theoretical and experimental single-electronics, so that the total number of publications at the present time is well above a thousand.

Single-electron effects have been studied experimentally using various materials and technologies. The oldest and most developed technique is the fabrication of small tunnel junctions by overlapping narrow strips of metal films using electron beam patterning and double-angle evaporation [13,15–27]. Among the most significant achievements towards possible applications are single-electron transistors with charge sensitivity better than 10^{-4} $e/\mathrm{Hz}^{1/2}$ (at 10 Hz) [28], low-temperature absolute thermometers with ~1% accuracy [29], the prototype of the dc current standard with relative accuracy about 10^{-8} [30], and the single-electron trap ('memory cell') with a retention time of more than 12 h [31].

Since 1990 [32] single-electron charging effects have also been studied in tunnelling through small islands ('quantum dots') of 2D electron gas in GaAs-based heterostructures. It was predicted [33,34] that in these structures the coexistence of energy and charge quantizations should play a much more important role than in metal islands of comparable size. This fact was soon confirmed experimentally [35,36] (structures exhibiting both types of quantization are now sometimes called 'artificial atoms' [5]). The study of single-electron effects in the systems of quantum dots has so far been of mainly scientific interest [37–41], and the technology is still far from practical applications. An exception is the experiment [42] in which the side-gated constrictions in δ-doped GaAs were used to demonstrate the operation of the few-electron 'memory cell' at 4 K. The Coulomb barrier in this case appears due to tunnelling between randomly positioned small conducting islands in the constriction.

There is a considerable number of experiments in which single electrons tunnel from the tip of the scanning tunnelling microscope to the substrate via a small conducting particle [43–54]. The effect may survive up to room temperature for sufficiently small metal particles [47,48], and it can be even stronger when tunnelling via single molecules is studied [49–54]. Scanning tunnelling microscope can be also used for the fabrication of the single-electron circuits [55].

Quite promising results were obtained recently in experiments with silicon-based structures [56–60]. The memory effects and the operation of single-electron transistors were reported even at room temperature. With the vast experience accumulated in silicon technologies, silicon-based devices could be a real way to integrated single-electronics.

Ultradense integrated circuits are the most fascinating goal of single-electronics. There have been many theoretical suggestions on this topic, e.g. digital circuits based on single-electron transistors [12,61–63], the logic which uses single electrons to represent logic bits [64–69] (including various 'wireless' logics [70,71]) and background charge-independent devices [72]. The practical realization of integrated single-electron circuits remains a questionable issue because very serious technological problems need to be solved to reach this goal. Nevertheless, the rapid progress in experimental single-electronics during the last few years combined with the rapid improvement of nanotechnology makes room temperature single-electronics a candidate for the next generation of ultradense digital devices.

3 'Orthodox' theory of single-electronics

The main object of single-electronics is a small tunnel junction. The simplest approach, described below, works very well for metallic junctions. Several specific features of semiconductor and molecular-level systems will be considered in Section 6.

The tunnel junction consists of two electrodes separated by an insulating layer, and naturally has some electric capacitance C depending on the geometry (in the simple case of the planar capacitor $C = \varepsilon\varepsilon_0 S/d$ where S is the area, d is the insulator thickness, and ε is its dielectric constant). In contrast to the usual capacitor, in the tunnel junction electrons can pass through a sufficiently thin barrier (typically several nanometres). Let us assume the linear $I–V$ curve (in the absence of single-electron effects), $I = V/R$, that is the typical case for metallic systems.

'Orthodox' single-electronics [1] deals with junctions having sufficiently large resistances,

$$R \gg R_Q = \pi\hbar/2e^2 \approx 6.4 \text{ k}\Omega. \tag{4.2}$$

To understand the physical meaning of this condition, note that the rate of tunnelling in a junction biased by some voltage V is $\Gamma = V/eR$ so that the typical time between tunnelling events is $1/\Gamma = eR/V$. The 'duration' of a tunnelling event due to the uncertainty principle is $\hbar/eV$ (there are several other definitions of the tunnelling time — see, e.g. [72], however, they are not relevant to this problem). Hence, Eqn 4.2 simply means that the tunnelling events do not overlap, and we can speak about the separate tunnelling of single electrons.

In the case of small capacitance C the voltage $V = V_b$ before the tunnelling event is considerably different from the voltage $V_a = V_b - e/C$ after the event. Hence, it is not clear which value should be used for the calculation of the tunnelling rate. The simple guess is that we can take the average value as the effective voltage

$$V_{eff} = \frac{V_b + V_a}{2} = V_b - \frac{e}{2C}. \tag{4.3}$$

This guess coincides with the result of the orthodox theory. In fact, the effective voltage should be related to the change W of the electrostatic (free) energy of the system (energy gain due to tunnelling). In the case of a single capacitor charged initially by $Q = CV$ this change will be

$$W = \frac{Q^2}{2C} - \frac{(Q-e)^2}{2C} = e\left(V - \frac{e}{2C}\right) = eV_{eff}, \tag{4.4}$$

which coincides with Eqn 4.3. This derivation is still valid if we consider the tunnel junction as a part of the complex circuit. The only difference is that we need to use the effective (total) capacitance of the junction calculated taking account of the rest of the circuit.

For a given energy gain W the tunnelling rate in orthodox theory is calculated using the formula

$$\Gamma = \frac{W}{e^2 R(1 - \exp(-W/k_B T))} \tag{4.5}$$

where T is the temperature. In case of zero temperature this expression transforms to $\Gamma = V_{eff}/eR$ for positive $W = eV_{eff}$ and $\Gamma = 0$ for negative W.

Absence of tunnelling for $W < 0$ is natural because the processes which increase the free energy are forbidden by the second law of thermodynamics. Using Eqn 4.4 we see that $W < 0$ if the voltage across the junction is less than the threshold value $V_t = e/2C_{eff}$ that corresponds to the charge $Q = e/2$. This is the condition of the Coulomb blockade of tunnelling (see Fig. 4.1). For finite temperature the tunnelling inside the blockade region is possible but it is strongly suppressed as long as $k_B T \ll e^2/2C_{eff}$.

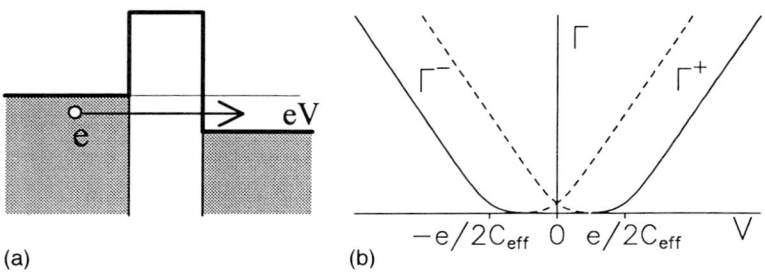

(a)   (b)

Figure 4.1 (a) Schematic energy diagram of tunnel junction. (b) The tunneling rates Γ^+ and Γ^- for both directions as functions of the voltage V across the junction. The Coulomb blockade suppresses the tunnelling at $|V| < e/2C_{eff}$ (solid lines). The cusps of the curves are rounded due to finite temperature. The dashed lines show Γ^+ and Γ^- for the case without single-electron effects ($e/C_{eff} = 0$).

Electron transport in an arbitrary single-electron circuit consisting of tunnel junctions, capacitors, and voltage sources is described by the orthodox theory as a sequence of jumps of single electrons. For any given charge state of the system one should calculated the tunnelling rates for all junctions. In which particular junction and at exactly what moment the next tunnelling will occur, is a matter of chance with the probabilities determined by the corresponding rates. After the jump the charge state changes, and one should calculate all rates anew. These rates determine the probability distribution for the next jump, and so on. This scheme may be used to implement a Monte Carlo algorithm [73] for the simulation of the electron transport. Another approach [1,10] is to solve the kinetic equation

$$\frac{d}{dt}p(k) = \sum_{m \neq k} p(m)\Gamma(m \rightarrow k) - p(k) \sum_{m \neq k} \Gamma(k \rightarrow m); \quad \sum_{k} p(k) = 1, \tag{4.6}$$

which describes the evolution of the probability distribution $p(k)$ among all possible charge configurations.

The orthodox theory can also treat systems with ohmic resistances, if they are considerably larger than the quantum unit R_Q. According to the fluctuation–dissipation theorem the spectral density of the quantum fluctuations of current through the ohmic resistance R_0 is $S_I(\omega) = 2\hbar\omega/R_0$ (at zero temperature). Corresponding r.m.s. fluctuations of the charge can be estimated as $\Delta q \sim (S_I(\omega)\Delta\omega)^{1/2}/\omega \sim (\hbar/R_0)^{1/2}$ for $\Delta\omega \sim \omega$. Hence, inequality $R_0 \gg R_Q$ allows one to neglect the quantum fluctuations of the charge, $\Delta q \ll e$. Such an ohmic resistor is considered in orthodox theory as an open circuit when the effective capacitances are calculated. (One can say that the charge transferred through R_0 'during' the tunnelling event is negligible, $(e/C)/R_0 \times \hbar/(e^2/C) \ll e$.) The charge transfer through the resistor during the time between tunnelling events leads to the gradual change in time of the tunnelling rates (this change is stochastic at finite temperature due to Nyquist noise).

Despite its simplicity, the orthodox theory of single-electronics is sufficient to describe most experimental results quantitatively. Among the most important developments beyond the orthodox theory are the account of arbitrary electrodynamic environment [74,75] (in particular, arbitrary ohmic resistances), the theory of simultaneous tunnelling (cotunnelling) in several junctions [77,78], and the account of energy quantization [33,34,79–83].

4 Single-electron transistor

The most thoroughly studied single-electron device is the single-electron transistor (SET), also called the SET-transistors, [11,12] which is the simplest circuit in terms of fabrication. Its basic part consists of two tunnel junctions in series (Fig. 4.2a). Using this example, let us illustrate the use of the orthodox theory.

The voltage drops $V_{1,2}(n)$ across the junctions are functions of the number n of excess electrons on the central island

$$V_j(n) = V\frac{C_1 C_2}{C_j C_\Sigma} + (-1)^j \frac{Q_0 - ne}{C_\Sigma}, \tag{4.7}$$

where V is the bias voltage, $C_\Sigma = C_1 + C_2$ is the total capacitance of the central island,

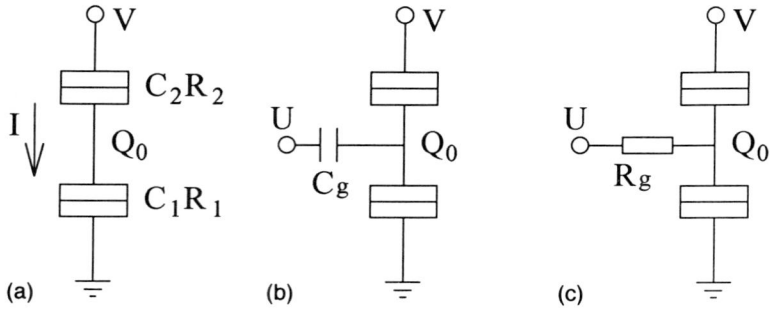

Figure 4.2 Single-electron transistor (SET): (a) the basic part consisting of two tunnel junctions in series; (b) capacitively coupled SET (C-SET); and (c) resistively coupled SET (R-SET). The current through the SET depends on the subelectron fraction of the charge Q_0.

and Q_0 is its initial (background) charge. The energy gain $W_j^{\pm}(n)$ due to tunnelling ($\pm$ denotes two different directions of tunnelling) can be calculated as

$$W_j^{\pm}(n) = e(\pm V_j(n) - e/2C_\Sigma) \tag{4.8}$$

because the effective capacitance is C_Σ for any tunnelling. The next step is the calculation of the rates $\Gamma_j^{\pm}(n)$ using Eqn 4.5. The stationary (denoted st) solution of the kinetic equation 4.6 for the probabilities $p(n)$ of different charge states n is as follows

$$p_{st}(n) \times (\Gamma_1^+(n) + \Gamma_2^-(n)) = p_{st}(n+1) \times (\Gamma_1^-(n+1) + \Gamma_2^+(n+1)), \quad \sum p_{st}(n) = 1, \tag{4.9}$$

and the average (dc) current I can be calculated as

$$I = \sum_n p_{st}(n)(\Gamma_1^+(n) - \Gamma_1^-(n)). \tag{4.10}$$

Figure 4.3 shows several dc I–V curves of the symmetric double-junction system

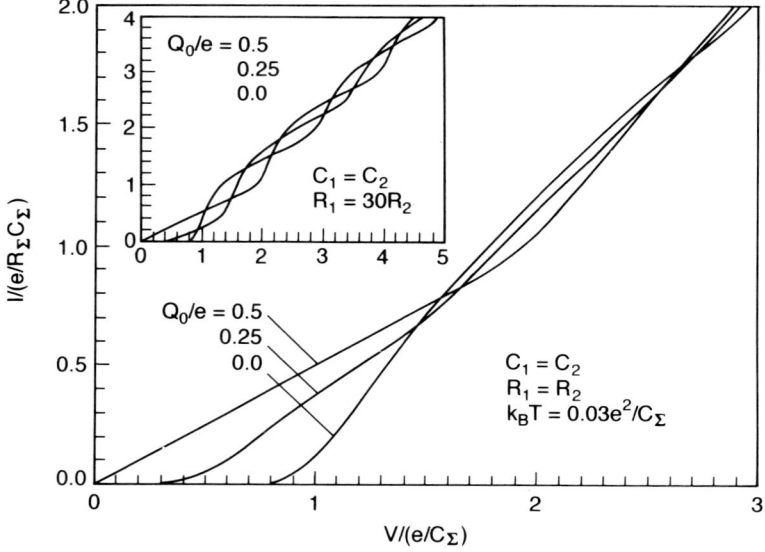

Figure 4.3 The typical I–V curves for the symmetrical single-electron transistor (SET) and the SET with different resistances of junctions (inset) calculated using Eqn 4.5–4.10 for different Q_0.

$(C_1 = C_2, R_1 = R_2)$ calculated in this way. The Coulomb blockade suppresses the current when the voltage is not sufficient to provide the energy for single-electron charging of the central island (notice also the rounding of the curve cusps due to finite temperature). The threshold voltage V_t depends on the background charge, and its maximal value is e/C_Σ. The same value determines the $I–V$ curve offset at large voltages, $I = (V - e/C_\Sigma)/R_\Sigma$. The Coulomb blockade completely disappears ($V_t = 0$) for half-integer background charge, $Q_0 = (k + 1/2)e$, because the states with effective charge $e/2$ and $- e/2$ have equal energies. The current is a periodic function of Q_0 (Fig. 4.4) because the addition of the integer electron charge is compensated by the tunnelling of one electron in or out of the central island. This periodic dependence is usually called Coulomb oscillations. Very high (subelectron) sensitivity to the charge of the central island is the basis of the SET operation. Controlling Q_0 by a capacitively coupled gate (C-SET; Fig. 4.2b) or via a coupling resistor (R-SET; Fig. 4.2c), controls the flow of electrons tunnelling through the SET.

R-SET is quite difficult to implement because the coupling resistance R_g should be much larger than R_Q to prevent quantum fluctuations of Q_0; simultaneously the resistor size should be small so that its stray capacitance does not significantly increase C_Σ. Experimental demonstration of the R-SET is still an unsolved problem despite significant progress in this direction [23,84]. Similar difficulty has not so far allowed the experimental study of the so-called RC-SET [85] which would be very useful in digital circuits because of its multistable characteristics.

In contrast, C-SET has been demonstrated repeatedly by many scientific groups using different materials and technologies (see Section 2). In some laboratories it is a routine device which is used to measure very small charge variations, e.g. in other single-electron circuits. The gate voltage U (see Fig. 4.2b) induces the effective charge into the central island, $Q_0 \rightarrow Q_0 + C_g U$ (C_g is the gate capacitance); hence, Fig. 4.4 can be considered as a control curve of the C-SET. The gate voltage period is equal to

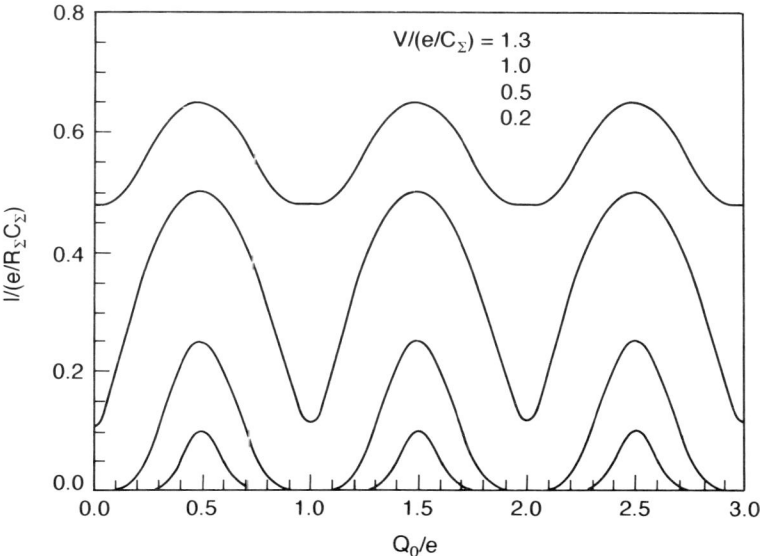

Figure 4.4 The typical theoretical dependence of the current through the symmetrical single-electron transistor on the induced charge Q_0 for different bias voltages V.

$\Delta U = e/C_g$. If C_g is comparable to the junction capacitance, its contribution to the total capacitance should also be taken into account. To calculate characteristics of C-SET it is sufficient to use Eqns 4.7–4.10 with the substitution $C_1 \rightarrow C_1 + \alpha C_g$, $C_2 \rightarrow C_2 + (1 - \alpha)C_g$, $Q_0 \rightarrow Q_0 + C_g U - \alpha C_g V$, where α is an arbitrary number (usually $\alpha = 0$ or $\alpha = 1$ is used).

If the resistances of the two junctions of the C-SET are considerably different, then the *I–V* curve shows substantial periodic oscillations with period $\Delta V = e/C_1$ (for $R_1 \gg R_2$), called the Coulomb staircase. The cusps of the *I–V* curves shown in the inset of Fig. 4.3 correspond to condition $W_2 = 0$ in Eqn 4.8. Each period of the staircase corresponds to an additional electron on the central island. The Coulomb staircase is typical for experiments using the scanning tunnelling microscope because the tunnel junction between its tip and the conducting particle is typically much smaller than the junction between particle and substrate. In contrast, the Coulomb staircase in C-SETs made of metal films is usually very weak, because this technology is able to produce junctions of the same parameters (in this case the staircase may be the evidence of a bad sample).

The theory of the SET is well confirmed experimentally. Figure 4.5 shows the example of the layout, *I–V* curve, and the dependence of the current on the gate voltage for the C-SET made of metal films [26]. Unusually high operation temperature (up to 30 K) was achieved in this experiment by the use of film anodization (recently the same group increased the temperature up to 100 K [25] using the three-angle evaporation technique) typically metal film SETs operate at $T < 4$ K. Simple orthodox theory described above is usually sufficient for good quantitative agreement with experimental data for metallic C-SETs. Some additional factors should typically be taken into account for semiconductor C-SETs and double-junction structures with a molecule or small cluster as the central island (see Section 6). Figure 4.6 shows the SEM image and the current–gate voltage dependence for the recently demonstrated Si-based SET [60] operable at temperatures over 80 K. The current in this device was actually carried by the tunnelling holes, and the energy level spacing was comparable to the Coulomb blockade energy; this is why the authors of [60] call this device a single hole quantum dot transistor. Notice that the Coulomb oscillations in Fig. 4.6 are not exactly periodic and they are superimposed on the monotonic dependence on the gate voltage. We will discuss the reason for this difference from the behaviour of metallic SETs in Section 6.

The C-SET can be used as a highly sensitive electrometer. The charge sensitivity of the SET is limited by its noise. In experiments the spectral density of this noise usually has $1/f$ dependence [20,24,28,85] that can be explained as the random capture of electrons by impurities in the tunnel barriers and/or the substrate in the vicinity of the SET. The noise decreases with the improvement of the technology. The lower limit is determined by the intrinsic thermal/shot noise of the SET [87–90] which was recently measured [91] at relatively high frequency where the contribution of $1/f$ noise is small. The ultimate sensitivity [87–89] limited by the intrinsic noise of the C-SET is given by $\delta Q_0 \approx 2.7 e (k_B T C_\Sigma/e^2)^{1/2}(RC_\Sigma \Delta f)^{1/2}$ where Δf is the bandwidth. For typical parameters of present-day experiments, $C_\Sigma \sim 3 \times 10^{-16}$ F, $T \sim 0.1$ K, $R \sim 10^5\ \Omega$, the ultimate sensitivity is about $2 \times 10^{-6} e/\text{Hz}^{1/2}$, while the best experimental sensitivity recorded so far is $7 \times 10^{-5} e/\text{Hz}^{1/2}$ at 10 Hz [28].

Fitting of the experimental results obtained at $T < 0.1$ K usually shows that the real

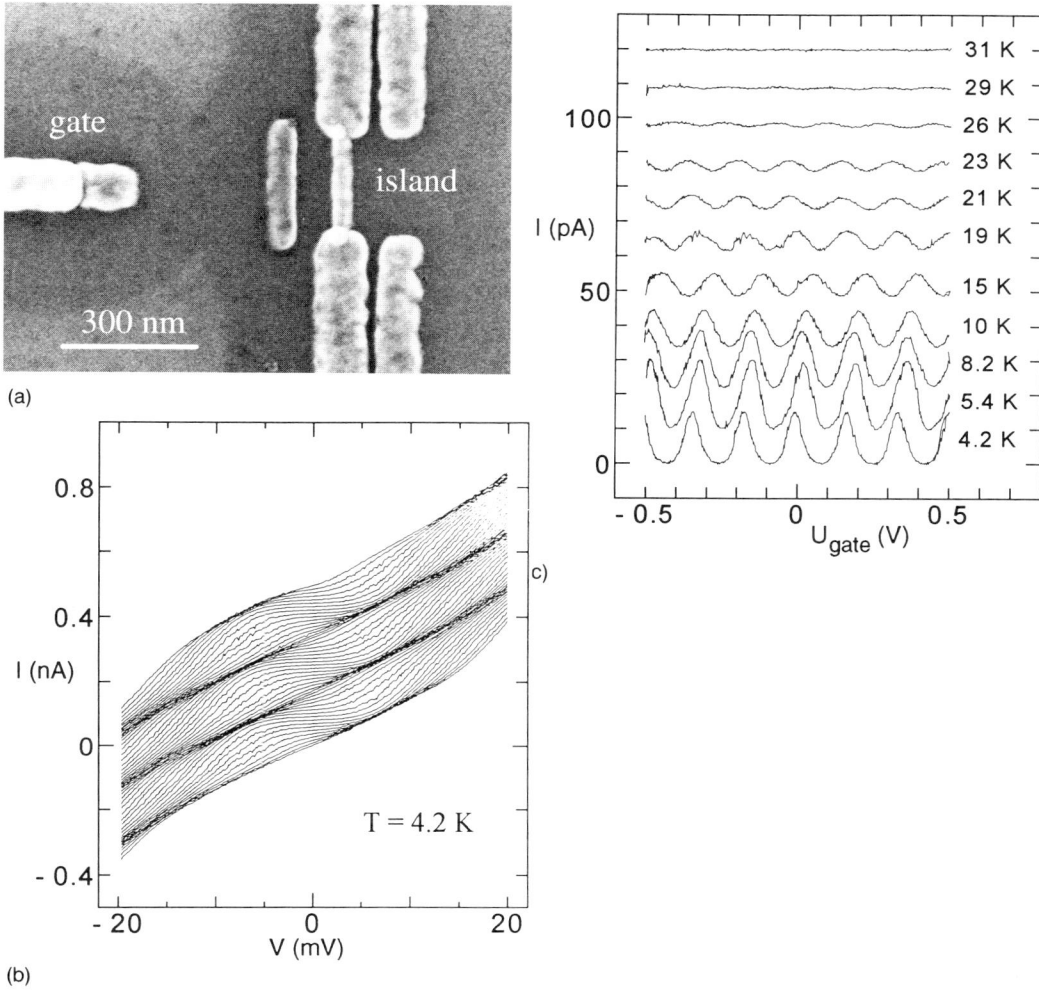

Figure 4.5 Experimental realization [26] of the capacitively coupled single-electron transistor using narrow metal films: (a) layout; (b) I–V curves for different gate voltages; and (c) the dependence of the current on the gate voltage at different temperatures (courtesy of Y. Nakamura). The curves in (b) and (c) are shifted vertically for clarity.

temperature is higher than the temperature of the cryostat. This can be explained as heating due to the transport current [24,92,93] and imperfect microwave isolation of the SET from the environment.

The orthodox theory should be modified for calculation of the small current well below the Coulomb blockade threshold at low temperatures. In this case the single-electron tunnelling is blocked, and the current is due to 'simultaneous' tunnelling (cotunnelling) of two electrons through both junctions [77,78]. Because of the quantum nature of the process involving the whole electrical circuit, cotunnelling is also called macroscopic quantum tunnelling of charge (q-MQT). The rate of such a process is proportional to the product $(R_Q/R_1)(R_Q/R_2)$, and, hence, is relatively small for $R_i \gg R_Q$.

5 Digital single-electron devices

The most important potential application of single-electronics is integrated digital

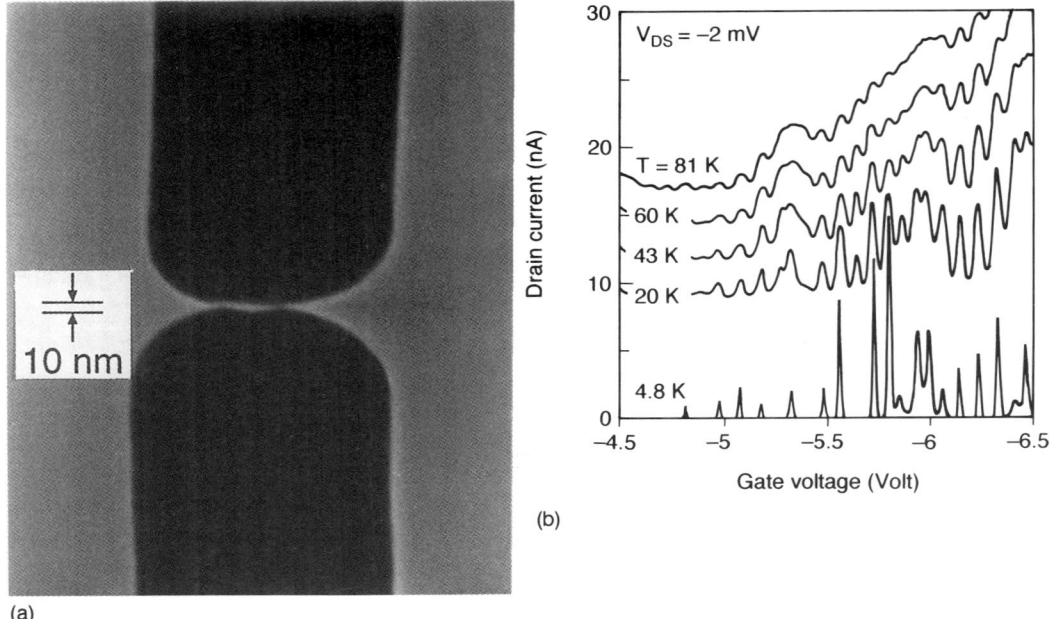

(a)

(b)

Figure 4.6 Si-based single-electron transistor [60]: (a) electron micrograph of the structure and (b) the dependence of the current on the gate voltage (courtesy of E. Leobandung). The current is actually due to the hole tunnelling, so the structure is named single hole quantum dot transistor.

electronics, which could substitute conventional semiconductor transistor technology at the size scale below 30 nm. There have been many theoretical suggestions on this subject; we will discuss several of them.

5.1   *Logic/memory using SET transistors*

Conceptually the simplest way to realize digital single-electronics is to use SETs instead of field-effect transistors (FETs) in circuits resembling conventional electronics. It is possible to use C-SETs or R-SETs. Because R-SET is still too difficult for fabrication, let us limit the discussion to C-SET circuits.

For C-SET the dc input current is zero, hence the power amplification is formally infinite. However, the voltage gain K_V is not large [12] (in contrast to semiconductor MOSFETs),

$$K_V \lesssim C_g/\min(C_1, C_2). \tag{4.11}$$

The condition $K_V > 1$ which is necessary for the operation of logic devices requires the gate capacitance C_g to be larger than the junction capacitance.

The buffer/inverter can be realized by one SET in series with a load resistor R_L. Notice that the fabrication of such a resistor is not a big problem, in contrast to the resistor for R-SET, because there is no limitation on its stray capacitance. However, to reduce the number of technological steps, it is more reasonable to use a tunnel junction instead of the load resistor [94]. Calculations show that for good operation of buffer/inverter, R_L should be at least 10 times larger than the junction resistance [62]. Hence, the additional power dissipation in the load resistor will be much larger than in the SET.

To reduce the power consumption complementary circuits can be used [12,61–3]. It is important that in contrast to CMOS technology in which n-MOS and p-MOS transistors are physically different, both complementary SETs can be physically identical because of the periodic dependence of the current on the gate voltage. To achieve the complementary action, the operating point of one transistor should be on the rising branch of this dependence while for the other transistor it should be on the falling branch. This can be done with the use of additional capacitors [61] or different background charges in complementary transistors [62] (Fig. 4.7a). However, even without any special effort, complementary action occurs automatically in the simplest case of two symmetrical transistors with zero background charges [62]. It is interesting that in terms of the maximal operation temperature this simplest case is very close to the optimal one.

The maximum temperature at which the complementary inverter still amplifies the signal is equal to $0.026e^2/Ck_B$ [62] where C is the capacitance of one tunnel junction.

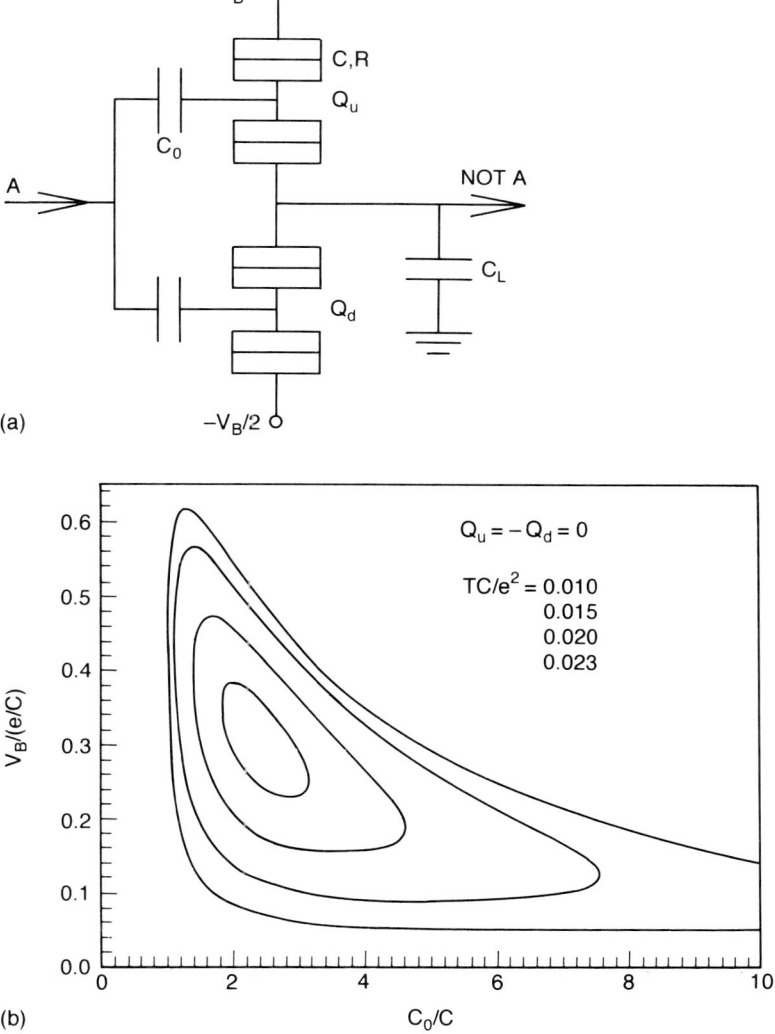

Figure 4.7 (a) The complementary inverter made of two single-electron transistors and (b) its parameter window for different temperatures [62].

Notice that the same maximum temperature can be obtained for a resistively loaded transistor if R_L is very large. The maximum temperature is achieved when the gate capacitance C_g is about twice the junction capacitance. The optimal C_g (corresponding to largest parameter margins) increases when the temperature decreases, so that $C_g/C \approx 3$ seems to be more or less the best choice for experimental realization. To have reasonable parameter margins, the temperature should be approximately half the maximal temperature, $T \sim 0.01 e^2/Ck_B$. In this case the margins for bias voltage and C_g are sufficiently wide (see Fig. 4.7b), and the critical margin is that for the fluctuations of background charges (about $0.1e$).

The operation point which optimizes the maximal temperature of the complementary inverter corresponds to relatively larger power consumption, i.e. about $2 \times 10^{-3} e^2/RC^2$ per SET. However, in the 'power-saving' mode for a price of slight reduction of the operation temperature, the power consumption can be reduced down to $10^{-4} e^2/RC^2$ per transistor [62].

The switching time of the complementary inverter is close to $3RC_L$ where C_L is the load capacitance (see Fig. 4.7a). Relatively large load capacitance, $C_L \gtrsim 300C$, should be used in order to make the fluctuations of the output voltage due to the shot noise in the transistor negligible.

Two inverters connected in a circle constitute the bistable flip-flop which can be used as a static memory cell. Almost all results of analysis of the inverter are directly applicable to the flip-flop. Slightly lower temperature and slightly narrower parameter margins are required for the operation of the logic gates based on SETs [63]. A possible structure of the NOR gate [63] is shown in Fig. 4.8(a). Notice that in contrast to the SET inverter which is similar to the circuit used in conventional digital electronics, design of the SET NOR gate differs from the conventional one. (The direct reproduction of the design is impossible because of different characteristics of SET and MOSFET transistors.) The operation of a SET NOR gate is illustrated in Fig. 4.8(b). One can see that the threshold lines are close to the perfect (square) shape. Inversion of the bias voltage transforms a NOR gate into a NAND gate with the similar characteristics. NOR and NAND gates accompanied by the NOT gate (inverter) are more than sufficient for performing arbitrary logic functions. However, special design for some other gates, e.g. SET XOR gate [63], can help to make the logic more efficient. The single-electron transformer [97] can be also useful in the SET logic.

For a technology with a minimal feature size of 2 nm one can expect capacitances of the tunnel junctions as low as 3×10^{-19} F. This corresponds to $e^2/Ck_B = 6 \times 10^3$ K; hence, the maximum temperature at which the SET still amplifies the signal is close to 150 K. This would allow the operation of the SET logic at liquid nitrogen temperatures. (We see that room temperature operation requires the fabrication technology at the subnanometer level.) For estimation of the typical switching time let us take $R \approx 300$ kΩ and $C_L \approx 10^3 C \approx 3 \times 10^{-16}$ F; then this time is about 1 ns. The power consumption per transistor is quite small, about 2×10^{-8} W for the parameters above in a typical operation point (in a 'power-saving operation point' it is about 10^{-9} W). However, because the density is very large, power dissipation is a serious problem. For example, at 10^{11} transistors per cm^2 even in the power saving mode the total power is of the order of 100 W cm^{-2}. Probably, an even more difficult problem of the logic/memory based on SETs is the need to keep fluctuations of background charge within

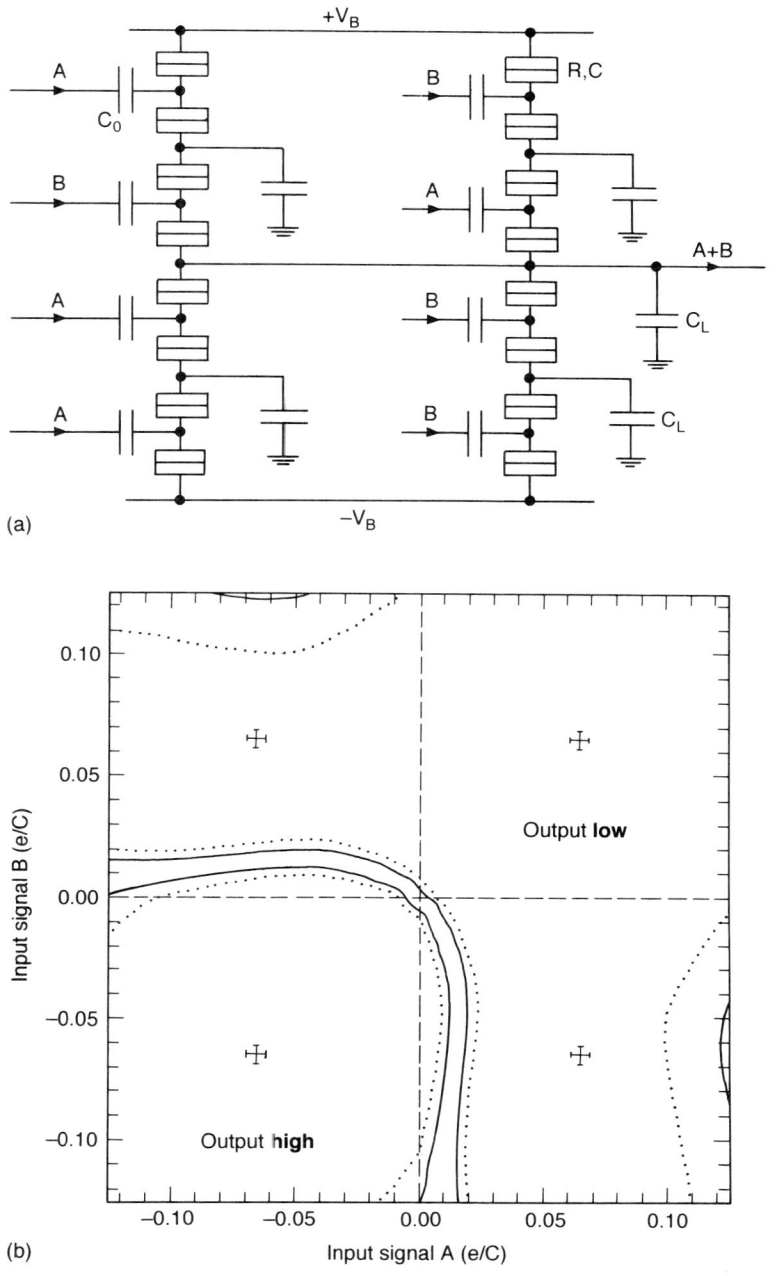

Figure 4.8 (a) The NOR gate made of single-electron transistors and (b) its typical output characteristics on the plane of input signal amplitudes [62]. The solid lines in (b) show the 'active' region where the output cannot be definitely interpreted, and the areas between solid and dashed lines correspond to the noise margins.

margins of the order of $0.1e$. This is a common problem for any integrated single-electronics; we discuss the possible solutions in Section 5.4.

It should be emphasized that a single SET logic device can be relatively easily fabricated using present-day technology. Multilayer technology which allows relatively large gate capacitances and solves the problem of connections between circuit elements has already been developed [28,96]. It is expected that the first SET logic devices will be demonstrated within a few years (they will probably operate at $T < 1$ K and require individual adjustment of background charge at each island).

5.2 *SEL logic and single-electron trap*

In the logic/memory based on SETs the logic unity and zero are represented by different dc voltage levels, similar to conventional digital electronics. Another possibility is to represent bits by single electrons [64–69], so that one extra electron in a conducting island would correspond to logical unity, while the absence of an extra electron would correspond to logical zero. The circuits based on this truly single-electron approach are called single-electron logic (SEL) [64–6]. The apparent advantage of this idea is a low power dissipation because in a static state there is no current, and the logical processing of one bit of information requires only few tunnelling events.

In the initially proposed SEL logic [64] single electrons propagate together with information along the relatively long arrays of tunnel junctions and ohmic resistors. In this scheme the proper dc biasing is a difficult problem because the bias should be distributed in a specific way among the large number of cells. To resolve this problem it was suggested [65] to separate the propagation of electrons and information: electrons tunnel across the elementary cell which is a short biased array of junctions, while the information propagates from one cell to another perpendicular to the motion of electrons.

Figure 4.9(a) shows the basic cell of the SEL family considered in Refs [64–6]. Notice that it is similar to the complementary SET inverter, however the important difference is that the capacitance of the middle island of an SEL cell is of the order of the junction capacitance (in contrast to large C_L in the SET inverter). Inputs X and Y determine the charge state of the middle island. For example, if the lower branch of the cell is 'closed' by the signal Y, and the signal X opens the upper branch of the cell, then one extra electron tunnels through the upper branch to the middle island. This creates the logical 'unity'. Parameters are chosen in a way that the next electron cannot come because of the increased potential of the island. The extra electron can be removed (creating logical 'zero') from the middle island by closing the upper branch and opening the lower one. The charge of the middle electrode, being the output of the cell, is used to affect the charge state of the next cell.

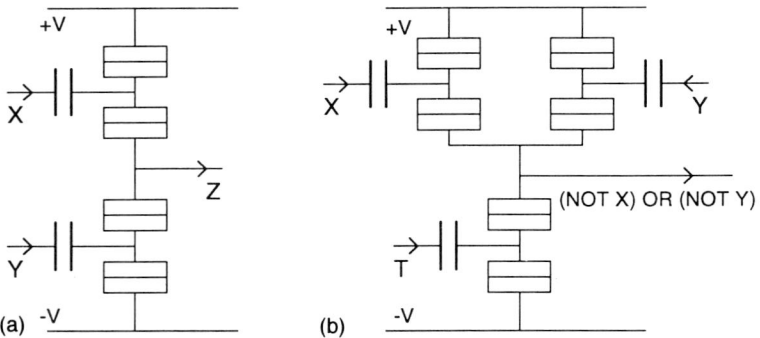

Figure 4.9 (a) The basic cell of the single-electron logic and (b) the SEL NOR gate [65,66,67]

Figure 4.9(b) shows the SEL logical gate NOR. Signals X and Y are logical inputs. The middle island becomes charged by an extra electron when one of the upper branches opens. Clock signal T discharges the middle island at the end of the clock signal (upper branches should be closed at this time).

In contrast to the SET circuits, there is a strong back action from the output to the input in SEL circuits. Numerical simulations have proved [64–6] that the proper choice of parameters provides unidirectionality of the signal propagation. However, because of the back action, the parameter margins are considerably narrower than in the SET logic case.

Another problem of SEL logic is that the information coded by a single electron can be destroyed by a single erroneous event due to cotunnelling or thermoactivated tunnelling. A possible solution would be the use of multijunction arrays as branches of SEL circuits, however this possibility has not yet been studied quantitatively.

The problems mentioned above make SEL logic circuits much more difficult to implement than SET logic at least at the present stage. However, it is hoped that the problems will eventually be solved by the search for optimal design and improvement of the technology. This hope is strongly supported by a successful experimental demonstration [18,21,31] of the 'memory cell' in which a logical bit is represented by a single electron on the conducting island. This circuit, which is usually called a 'single-electron trap', consists of several (typically, five to seven) tunnel junctions in series with a capacitor (Fig. 4.10a). Experimentally (Fig. 4.10b), this is an array of metal junctions which ends with a relatively large island so that its capacitance to the ground C_S is comparable with the junction capacitance C. The number of electrons on the island can be changed by application of the bias voltage U (Fig. 4.10a). Several charge

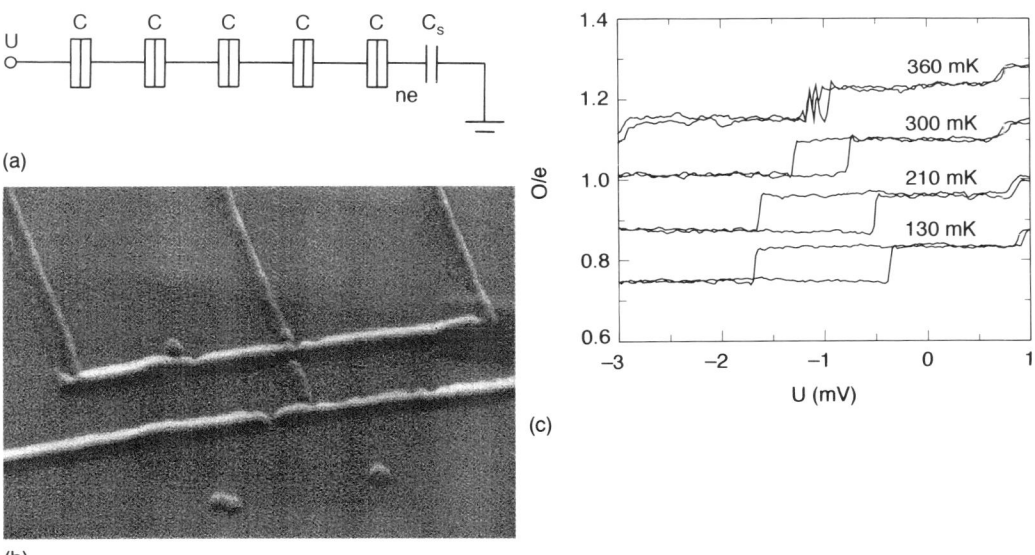

(a)

(b)

(c)

Figure 4.10 The single-electron trap: (a) schematic drawing; (b) electron micrograph of the structure; and (c) the hysteretic dependence of the trapped charge (multiplied by the coupling coefficient) on the voltage $U = V_{trap}$ [31] (courtesy of P. Dresselhaus). The charge is measured by the single-electron transistor (upper part of the layout). The height of each loop in (c) corresponds to one extra electron in the trap.

states can be stable for the same U because of the Coulomb barrier created by the array of junctions. In the case of zero background charges the tunnelling is blocked when $|V| < V_t = Ne/2C_{eff}$, $C_{eff} = C + ((N-1)/C + 1/C_S)^{-1}$ where V is the voltage across the array consisting of N junctions. One additional electron in the edge island changes V by $\Delta V = e/(C_S + C/N)$. Hence, as many as $m = 1 + int(2V_t/\Delta V)$ different states can be within the Coulomb blockade range. Two stable states ($m = 2$) which differ by one electron on the edge island represent logical unity and zero in a single-electron trap (Fig. 4.10c).

Similar to the SEL logic circuits, erroneous switching of the single-electron trap is due to thermoactivated processes and cotunnelling. Both processes are suppressed with an increase of the number of junctions in the array. An error rate less than one switching per 12 h was demonstrated at a temperature of 50 mK in the seven-junction array made of aluminium tunnel junctions [31]. The charge state of the island was monitored with the help of a nearby single-electron transistor (Fig. 4.10b). Theoretical consideration shows that in principle an error rate below 10^{-17} s^{-1} can be achieved in a similar trap [97].

If the capacitance C_S of the storage island is relatively large so that e/C_S is considerably smaller than the Coulomb blockade threshold V_t, then there are many, $m \approx 2V_t/(e/C_S)$, stable states within the blockade range. Representation of the logical bit by a single electron is ineffective in this case, however the bit can be stored as several electrons on the island. For example, $q = +me/2$ can correspond to unity, and $q = -me/2$ corresponds to zero. The power dissipation during the writing process is larger than in the single-electron case, however for $m \sim 10$–30 it is still extremely small. The advantage of the multielectron storage is that single erroneous tunnelling events do not destroy the information, and hence, the simple refreshing of information can be used to avoid errors (in the single-electron case refreshing is possible only with the use of redundancy).

The multielectron storage based on the Coulomb blockade was demonstrated [41] using a side-gated constriction in a δ-doped layer of GaAs. The arrays of tunnel junctions appeared naturally in the constriction due to disorder. Several tens of electrons were used to represent a bit. The operation was confirmed up to liquid helium temperature (4.2 K), and the storage time was as long as several hours.

Single-electron memory effects at room temperature were reported in silicon-based structures [56]. The current through the narrow ultrathin poly-Si film showed hysteresis as a function of gate voltage. This effect was ascribed to the trapping of single electrons in small naturally formed grains of poly-Si. The use of disorder for the creation of extremely small islands (far beyond the limits of modern lithography) offers the possibility of high temperature operation. However, such a technique obviously has a problem with the reproducibility of sample characteristics because of the random nature of the island creation.

Let us also mention very promising recent experiments with the room temperature traping of single electrons on a small floating gate above a narrow FET channel [98,99,100].

5.3   *Wireless single-electron logic*

Both SET circuits and SEL logic considered above require wires for the power supply and connections between circuit elements. Although the necessity of wires is not a principal problem, it is obviously inconvenient at the few-nanometer size scale. In the Wireless single-electron logic (WISE) proposed in [70] the power is supplied by an alternating external electric field, and the capacitive coupling between neighbouring cells is due to their close location. The 'device' consists of many conducting islands, and the logical functions are determined by their specific arrangement (Fig. 4.11). Small 'puddles' of 2D electron gas, small metallic droplets on an insulating substrate, or conducting clusters in a dielectric matrix are possible implementations of the islands. The basic cell of the logic is a short chain of closely located islands so that electrons can tunnel between neighbouring islands. There is no tunnelling between different chains because of the larger separation.

Application of in-plane electric field E creates the voltage between the islands. When E exceeds the Coulomb blockade threshold E_t, the tunnelling occurs somewhere inside the chain, producing an electron–hole pair. The electric field drags the components of the pair apart towards the opposite edges of the chain, creating the polarized state. If now the field E is decreased, the pair is eventually destroyed, however it will occur at the field E_a considerably smaller than E_t. Stability of both polarized and non-polarized states for E between E_a and E_t allows us to use these states as logical zero and unity.

The polarization change can propagate along a line of closely located chains (Fig. 4.11a). Suppose that all chains are not polarized initially, and E is slightly less than E_t. This is a metastable state. If one chain becomes polarized, the field of extra electron (hole) on the edge island increases the potential difference between neighbouring islands of the next chain (Fig. 4.11a). This makes tunnelling energetically favourable and leads to polarization of the next chain. This in turn polarizes the next chain, and so on. The unidirectional propagation (in Fig. 4.11 from left to right) is a consequence of the asymmetry of the circuit.

The natural fan-out of the signal into two lines can be realized if both edge islands of a chain are used to trigger the next chains (Fig. 4.11b).

A 'bi-controlled' chain (fifth from the right in Fig. 4.11c) which can be triggered by the polarization of either of two neighbouring input chains can be used as the basic part of the logical gate OR. The logical AND can be designed similar to the OR gate, but with slightly greater distance between the 'bi-controlled' chain and the neighbouring input chains, in order to decrease their influence. Another possibility is to make the island of the 'bi-controlled' chain slightly smaller in order to increase the Coulomb blockade energy.

Because of the asymmetry between logical zero and unity the design of the inverter is relatively complex. The circuit shown in Fig. 4.11(d) implements the logical function (NOT A).AND.B if the signal from input A comes before the signal from input B. This circuit can be used as NOT A, if logical unity always comes from input B and it comes later than signal A.

According to numerical simulations, the correct operation of the circuits shown in Fig. 4.11 requires that the magnitude of external filed E lies within a 5% margin [70]. This number also gives a crude estimate of the margins for other parameters (fluctuations of radius, spacing, etc.)

These logical gates, together with propagation lines and fan-out circuits, are sufficient for computing. In the simplest mode of operation, all chains inside a device initially have zero polarization and external field is zero. Then external field increases up to a

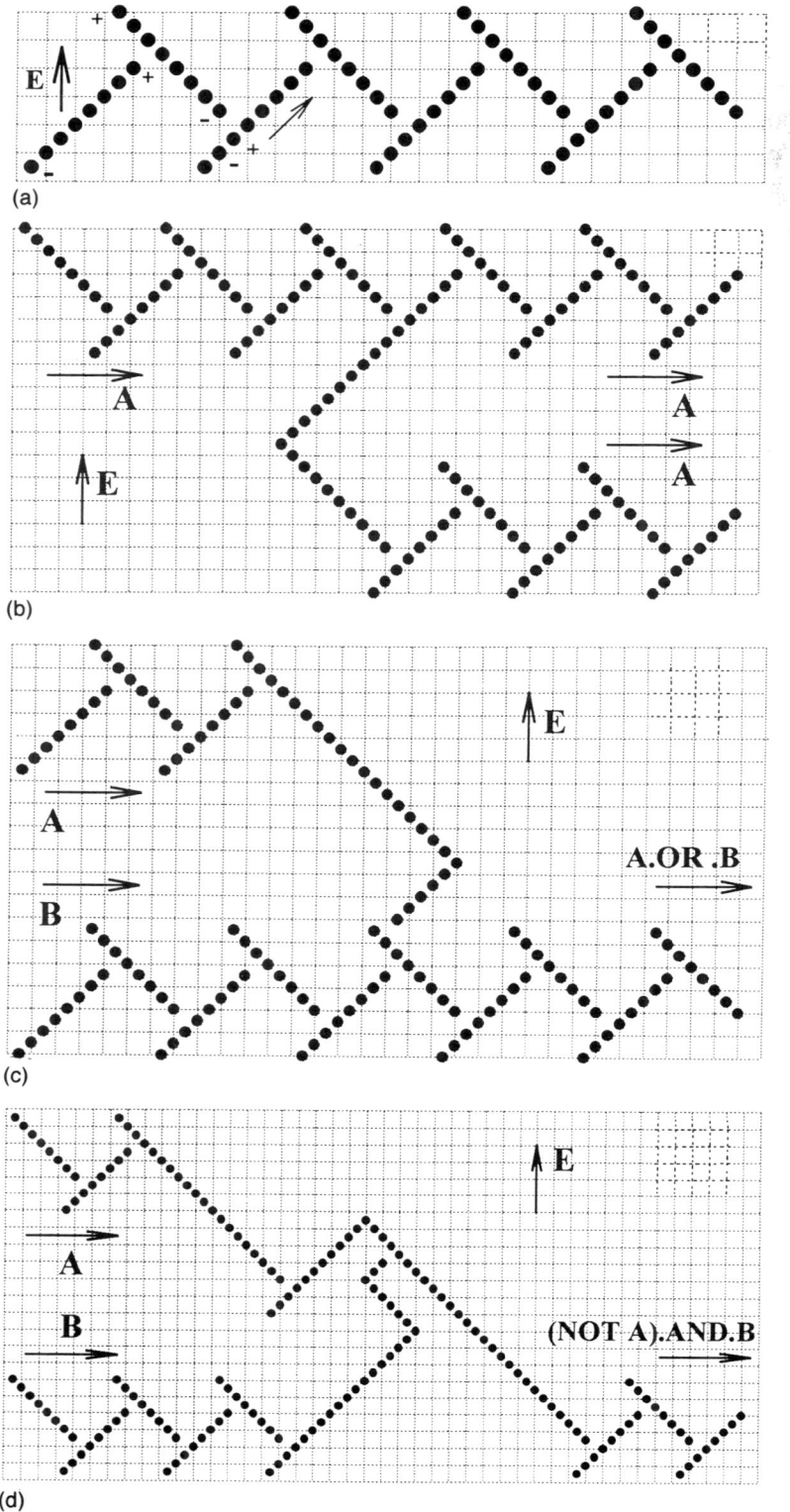

Figure 4.11 'Wireless' single-electron logic [70] based on tunnelling between small conducting islands and biased by electric field E. (a) The propagation line; (b) the circuit for fan-out; (c) the logical gate OR (gate AND has a similar design); and (d) the gate (NOT A).AND.B which can be used as an inverter.

value for which all gates operate correctly, and cells start to switch in accordance with the input information flowing from the edges of the device. The result of the computation is the final polarization of output cells which can be read out, for example, by SETs. This simplest mode of operation can obviously be improved by the use of periodic changes of the external field ('clock cycles'). Properly chosen levels of the field can reset some cells but preserve the information in other cells.

Note that the wireless single-electron logic proposed in [70] somewhat resembles the earlier proposed Ground state computing devices [101,102]. In both ideas the bistable polarization of the basic cell as well as only the nearest neighbour coupling are used. The main difference is the absence of the power supply in ground state computing, so that the only driving force is the fixed polarization of the cells at the 'edge' of the device. The small total energy gain (proportional to the number of 'edge' cells) should be distributed evenly between all 'bulk' cells to ensure their deterministic sequential switching. Hence, an integrated ground state computing device cannot operate in the mode of sequential switching of cells. In order to reach the ground state, a significant part of the device should be involved in the macroscopic quantum process [77,78] ('simultaneous' switching of many cells), and this transition would require practically infinite time because of the exponential dependence on the number of cells. In contrast to the ground state computing, the principle of operation of the wireless single-electron logic allows traditional computing by the sequential switching of cells in the device of arbitrary large integration scale.

There is no static power dissipation in wireless single-electron logic. Typically the switching of a cell requires energy of the order of only e^2/C where C is a typical capacitance. However, this dissipation can be further reduced. In a recent suggestion called single-electron parametron [71], the robust signal propagation along the shift register can cost even less, ultimately much less than $k_B T$ per switching of a cell. The possibility of logic devices with the energy dissipation below thermal limit was proven long ago [103,104]. Single-electron parametron seems to be the first realization of such a device based on the classical dynamics of the discrete internal degree of freedom.

This idea is shown in Fig. 4.12 (it represents the simplest, though not the best mode

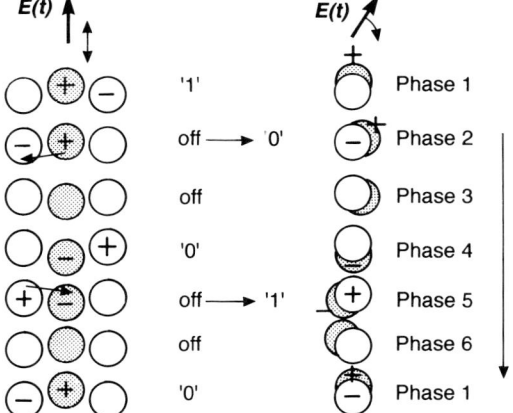

Figure 4.12 Shift register of the single-electron parametron [71]. Information propagation is caused by the rotating electric field $E(t)$.

of operation). The basic cell is a chain consisting of three islands. Rotating electric field changes the polarization of the chain four times per period: islands are neutral (state 'off') when the field is perpendicular to the plane of the chain, and the chain is necessarily polarized (state '0' or '1') when the field is in plane. When the neutral state becomes the polarized one, it can evolve into two different states: the electron from the central island can jump on either of two outer islands. The result is determined by the polarization of the neighbouring (previous) chain which has became polarized earlier because of the change in the chain orientation along the propagation line. Notice that the next chain does not influence the decision because it is in a neutral phase at this time. The resulting polarization will in turn determine the polarization of the next chain when it will enter the polarized phase. For the circuit shown in Fig. 4.12 the signal propagation speed is six steps per period of field rotation, and the transmission rate is two bits per period, so on average each bit requires three chains.

Power dissipation less than $k_B T$ per switching of a chain is achieved at low rotation frequency, $\omega \ll (k_B T)^2 / e^3 R E d$, where R is the tunnel resistance and Ed is the voltage between islands induced by the in-plane component of the field. In this case the switching consists of the large number of electron jumps back and forth. In the adiabatic limit the energy $k_B T \ln 2$ is first taken from the thermostat (when neutral and polarized states have equal energies, the entropy is reduced by one bit) and then this energy is returned to the thermostat. In the first approximation the total power dissipation per switching is proportional to the switching speed.

In comparison with the wireless logic of [70], single-electron parametron also offers greater parameter margins [71]. Numerical simulations for a particular 'layout' show the margin to be about 20% for the amplitude of the rotating electric field.

5.4   *The problem of background charge*

Fluctuating background charge is a very serious, possibly the most serious, problem of integrated single-electronics. Single-electron devices are so sensitive to the induced charge that a single charged impurity in the close vicinity of a device can significantly influence its operation. In the case of a single circuit, background charges can be adjusted individually with the help of additional gates. There is obviously no such possibility for integrated circuits. What could be a solution of this problem?

First, the problem may turn out to be not so serious after all. There is some experimental evidence [9,14] that even in rather dirty systems the background charge tends to relax to zero. Theoretically this could be understood, for example, as being due to the attraction of the charged impurities to conducting surfaces by the image charge force. In general, it can be hoped that a narrow statistical distribution of background charges might occur naturally in some materials.

Second, it might be that the problem can be solved with the use of extremely pure materials. For example, considering molecular electronic devices in which all circuit elements are reproducible on the atomic level, there may be an extremely low concentration of impurities.

Third, instead of capacitively coupled single-electron devices, we can try to use resistively coupled circuits. For example, R-SET is not influenced by background charges at all. However, there are problems with this. The R-SET is obviously much more difficult to make than C-SET, and also the R-SET as a voltage amplifier requires significantly lower temperatures [105] because of the Nyquist noise in the coupling resistor.

Finally, one more possibility is to come up with some capacitively coupled devices which would work in the environment of fluctuating background charges. One example of such a 'Q_0-independent device' was suggested recently [72]. The idea is to use C-SET in a mode when the ramping input signal drives the SET through several periods of its control characteristic. In this case the output signal will oscillate (Fig. 4.4), and for any initial Q_0 the amplitude of oscillation is equal to the maximal swing of the control characteristic. Such transistors can be used in a very high density memory (10^{11} bits per cm^2 or even more) to read out the stored information [72] (Fig. 4.13). Suppose that similar to traditional non-volatile semiconductor memories [106], the digital bits are stored in a form of electric charge Q on the floating gate located in the vicinity of the SET. In the case of a very small gate (of the order of 10 nm) this charge is just a few (10–20) electrons. The charge can be changed, for example, by its injection/extraction through the dielectric layer via Fowler–Nordheim tunnelling [106] (the graded barrier would considerably improve the operation [72]). The cell is selected by the simultaneous application of the voltages of different polarity to word and bit lines (small voltage difference between two bit lines is used for SET biasing — see Fig. 4.13). To read out the stored information we try to write the logical unity in. If unity has already been stored, the charge on the gate does not change, and the SET remains in the initial state. However, if logical zero has been stored on the gate, then its charge will gradually increase up to the level corresponding to logical unity. During this increase the current through the SET oscillates; this can be registered by an FET sense amplifier (one FET may serve about 100 memory cells). The previously stored information is destroyed during read-out, hence it should be restored later. Notice that voltage amplification by

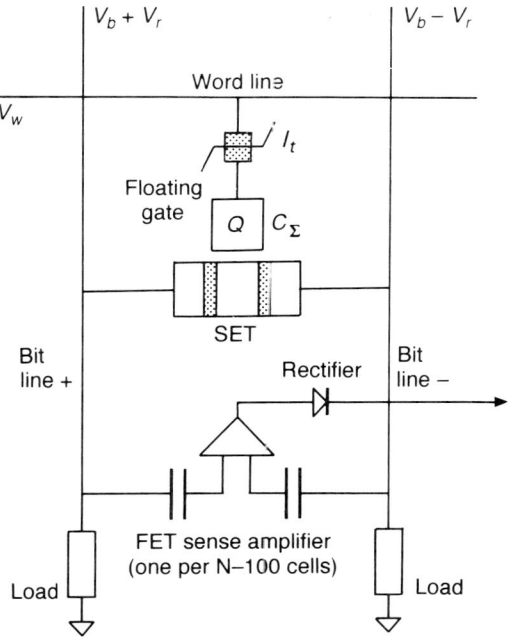

Figure 4.13 Ultradense hybrid SET/FET memory operating in Q_0-independent mode [69].

the SET is not required in this mode of operation, and this fact significantly increases (by a factor of approximately five) the maximal operation temperature.

Estimates show [72] that the density of 10^{11} bits per cm^2 and the room temperature operation of such a memory is feasible for ~4 nm minimum feature size technology. Estimated read/write time is about 3 ns and is limited both by the time of the floating gate charging and by the intrinsic noise [87] of the SET.

Single-electron devices operating in Q_0-independent mode seem to be the most radical solution of the problem of background charge fluctuations. However, although this idea can be used in memory devices, it can hardly be applied to logic circuits.

6 Single-electronics in semiconductors, clusters of metal atoms, and molecular systems

Single-electron effects become stronger with a decrease of the typical size. Also, they necessarily acquire new features. Eventually the field of single-electronics transforms into the field of atomic physics and chemistry; however, the basic ideas of single-electron devices are applicable even at this level. They can be used for information processing in hypothetical molecular electronics devices.

The orthodox theory works well for metallic systems down to approximately the 1-nm size scale. At this scale the level discreteness in small metal particles (clusters of atoms) starts to play an important role. Also, increasing Coulomb energy becomes comparable to the height of the tunnel barriers leading to highly non-linear I–V curves. In semiconductors these effects are important at considerably larger size scales and they are typical in experiments with quantum dots. Obviously, these effects should also be taken into account when tunnelling via molecules is studied. This is why in this section we consider together the features of single-electron circuits based on semiconductor quantum dots, clusters of metal atoms, and single molecules.

6.1 *The level discreteness*

The orthodox theory of single-electronics assumes the continuous energy spectrum of all electrodes. It should be somewhat modified [33,34,79–83] to take into account the level discreteness. As an example consider the SET with a discrete spectrum of electrons in the central island. The complete description of the charge state now includes not only the total number of electrons on the island, but also the occupation of individual levels. In one-electron approximation (neglecting the collective excitations [107]) the electron addition energy depends on two integer parameters k and n:

$$E_{k,n} = \varepsilon_k + e(ne + Q_0 + e/2)/C_\Sigma, \tag{4.12}$$

where k is the level number (ε_k is the energy spectrum) and n is the total number of excess electrons on the island. Note that the contribution from the background charge can be included into the definition of ε_k. The finite bias voltage can be taken into account in the same way as for the usual SET (Section 4). The tunnelling rates should be calculated for each level individually. The rate of electron tunnelling to/from the empty/occupied kth level via jth junction is given by expression

$$\Gamma = \Gamma_j \frac{1}{1 + \exp(-W/k_B T)}, \quad W = \pm(-1)^j \varepsilon_k \pm eV_j(n) - e^2/2C_\Sigma, \tag{4.13}$$

where $\pm$ stands for the direction of tunnelling, $V_j(n)$ is the voltage drop across the junction given by Eqn 4.7, and Γ_j depends on the matrix element of tunnelling and electron density in the external electrode (Γ_j can also depend on n and k). For the calculation of the average current and other characteristics Eqn 4.13 should be supplemented by some model describing the energy relaxation of the electrons on the island. Equations 4.5 and 4.8 of the orthodox theory can be obtained by summing Eqn 4.13 over all energy levels in the case of negligible level spacing, Fermi distribution of electrons on the island, and constant Γ_j (then $R_j = \delta/e^2\Gamma_j$ where δ is the average level spacing).

The $I\text{--}V$ curve of a SET with level discreteness contains the step-like features (Fig. 4.14a) which appear when the discrete level in the island crosses the Fermi level in the external electrode. The position of the step along the voltage axis corresponds to $W = 0$ in Eqn 4.13 and depends on two integer parameters k and n (in contrast to only one parameter n in usual SET) as well as on the junction number j. In the general case the arrangement of steps can be quite complicated, however in typical cases the simple classification is possible. For example, Fig. 4.14(a) shows the $I\text{--}V$ curve in the case when the level spacing δ (equidistant two-fold degenerate spectrum is assumed) is considerably less than the Coulomb energy e^2/C_Σ, and the barrier transparencies are significantly different. The level discreteness produces the fine structure superimposed on the Coulomb staircase. Notice that the level spacing contributes to the period of the Coulomb staircase, hence there is no pure periodicity in the case of realistic non-equidistant spectrum ε_k. The slow energy relaxation of the electrons on the island leads to some smoothing of the Coulomb staircase [79–81].

The discrete levels also modify the dependence of the current on the induced charge Q_0. In this case it can have the multipeak shape (compare Figs 4.14b and 4.4). The slight asymmetry of the peaks in Fig. 4.14(b) is due to small difference between two tunnel barriers. Perfect periodicity is absent if the spectrum ε_k is not equidistant because it influences the position of the peaks. The level spacing δ also contributes to the average period making it larger than e.

In the orthodox theory the total number of conduction electrons on the island is large, so that it is possible to extract any number of them. This leads to some sort of electron–'hole' symmetry. In the case of quantum dots or molecular-scale devices it is possible to have just a few conduction electrons on the island, so that there is obviously no such symmetry. Even complete asymmetry, when initially there are no conduction electrons on the island and they appear only due to transport, is quite typical (in semiconductors the same situation is also possible for holes). In this case the relative importance of the charge quantization and energy discreteness depends not only on the ratio $(e^2/C_\Sigma)/\delta$ but also on the ratio Γ_e/Γ_c of emitter and collector barrier transparencies so that the actual parameter is $\alpha = (e/C_\Sigma)/\delta \times \Gamma_e/(\Gamma_e + \Gamma_c)$ [32,33]. For example, even if $(e^2/C_\Sigma) > \delta$ but the collector barrier is much lower so that $\alpha \ll 1$, electrons do not accumulate on the central island, and steps on the $I\text{--}V$ curve reflect only the spectrum ε_k. The Coulomb staircase is noticeable only when $\alpha \gtrsim 1$.

Equations 4.12 and 4.13 are based on the classical expression $E_{int} = (me)^2/2C_\Sigma$ for the

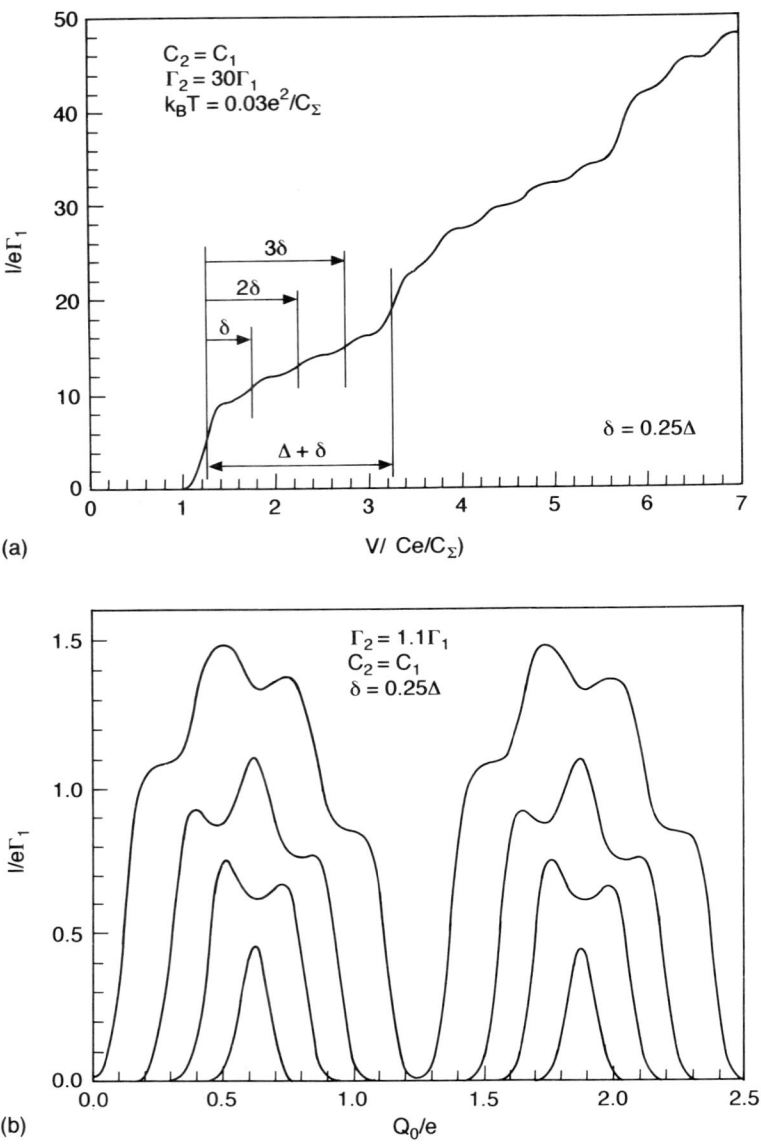

Figure 4.14 (a) The typical I–V curve and (b) the typical I–Q_0 dependence calculated for the single-electron transistor (SET) with discrete spectrum of the central island. Fine structure in (a) is due to the level spacing δ while the Coulomb staircase is determined mainly by the Coulomb energy $\Delta = e^2/C_\Sigma$. The I–Q_0 dependence can have the multipeak shape in contrast to the usual SET (see Fig. 4.4).

interaction energy of m electrons on the island. In the few-electron case (small m) the absence of electrostatic self-interaction of an individual electron makes this simple expression considerably inaccurate, and the better approximation is $E_{int} = m(m - 1)e^2/2C_\Sigma$ [79–81]. The accuracy of this approximation is confirmed by exact calculation [108] of the interaction energy of a few (up to 30) electrons on the sphere even in the extremely 'quantum' case when electrostatic energy is much smaller than the Fermi energy.

The separation of the electrostatic and one-electron energy in Eqn 4.12 is definitely only a simple approximation, and in the exact theory the many-body problem should be solved. This problem is simplified in the case when only low-temperature low-voltage conductance is studied; then the transport is determined by the ground states of the configurations with m and $m + 1$ electrons on the island. The finite voltage case requires also the calculation of excitations. There is some progress in this direction (see e.g. [107,109]). However, the exact calculation is difficult not only because of the mathematical complexity of the problem, but also because the result is very sensitive to the geometry of the island which is usually not known accurately. The most widely used approximation is still Eqn 4.12, and it explains surprisingly well the experimental data (in some experiments the slow variation of the capacitance should also be taken into account — see Section 6.2).

The theory of single-electron transport in systems with discrete levels [33,34,79–83] was confirmed experimentally both in metal and semiconductor structures. Let us discuss the difference of the typical parameters of these structures. First, let us estimate the energy level discreteness in a spherical cluster of aluminium atoms with diameter $d = 1$ nm (it would contain only about 30 atoms). In the free electron gas approximation the average spacing δ (per spin) between levels is given by expression

$$\delta = \frac{1}{g(\varepsilon_F)v} = \frac{2\hbar^2\pi^2}{vm(3\pi^2\rho)^{1/3}}, \tag{4.14}$$

where v is the volume, m is the effective electron mass, and ρ is the electron concentration. For the value $\rho = 1.8 \times 10^{23}$ cm^{-3} we obtain $\delta \approx 0.15$ eV. Estimating the typical single-electron Coulomb energy, $\Delta = e^2/C_\Sigma$, let us take $C_\Sigma = \beta 2\pi\varepsilon\varepsilon_0 d$ with $\varepsilon \approx 5$ and the geometrical factor $\beta \approx 3$; then $E_c \approx 0.2$ eV. We see that in metallic systems the level discreteness becomes comparable to the Coulomb energy roughly at the 1-nm size scale, and the influence of energy quantization is negligible when the typical size is larger than a few nanometres. That is why the level discreteness is so difficult to observe in metallic single-electron devices.

The interplay between two effects in the metallic system was demonstrated experimentally for the first time only recently [110] using transport through a very small aluminium particle with volume about 130 nm^3. The corresponding spacing was $\delta \approx 0.7$ meV while the charging energy was $\Delta \approx 12$ meV (the geometry was close to the plane capacitor that increased C_Σ in comparison with the estimate above). Because of the relatively small energy scale, the level discreteness showed up on the I–V curve only at temperatures below 2 K (most measurements were done at $T = 0.3$ K); at higher temperatures only the Coulomb staircase was observed. Note that the step-like features for the aluminium electrodes in the normal state were transformed in this experiment into the peak-features [110] for superconducting electrodes because their shape directly corresponds to the density of states in electrodes.

The step-like features due to the level discreteness superimposed on the Coulomb staircase were also observed in the experiment [54] with metal clusters $Pt_{309}Phen_{36}O_{30}$. The level spacing δ was up to 50 mV while the single-electron charging energy was up to 500 mV, and the discreteness was clearly observed at 4.2 K (the measurements at higher temperature were not reported in the paper). It is remarkable that the conducting particle used in this experiment can be described by the chemical formula (hence, formally this is a single molecule), and the orthodox theory (modified for the account of discreteness) is still very well applicable to this system. The experiments confirm that in metal systems the level discreteness is a small effect in comparison with single-electron charging effects when the size scale is larger than roughly 1 nm.

In semiconductor systems the level discreteness becomes important at a considerably larger size scale. This is caused by typically much lower electron concentration and lower effective electron mass (see Eqn 4.14). For example, in Si-based systems with doping level $\rho \sim 10^{21}$ cm^{-3}, δ would become comparable to the Coulomb energy at $d \sim 5$ nm. For much lower doping concentration the interplay between the level discreteness and the Coulomb effects was reported [60] at $d \sim 20$ nm. The irregular position of the Coulomb oscillation peaks in Fig. 4.6 [60] can be ascribed to the irregular energy difference between neighbouring discrete levels. Fluctuations of the peak height can be caused by the different tunnelling matrix elements for different levels.

A more dramatic increase of the level spacing occurs in semiconductor systems with a two-dimensional electron gas. In this case δ does not depend on the electron concentration, $\delta = \pi \hbar^2 / 2mS$, where S is the island area. Let us estimate the electrical capacitance of the conducting island of 2D gas as $C_\Sigma = \varepsilon\varepsilon_0 S/a$. Here a is the effective distance from a conducting electrode in the plane capacitor geometry which is a good approximation when the 'vertical' transport via a quantum dot is studied. In the case of 'lateral' transport (when conducting electrodes are in the same plane) this expression can be used with $a \sim 0.2d$ proportional to the diameter d of the dot. The ratio $\delta/(e^2/C_\Sigma)$ is equal to $\pi \hbar^2 \varepsilon\varepsilon_0 / 2ma = a_B/2a$ where $a_B = 4\pi\varepsilon\varepsilon_0\hbar^2/me^2$ is the Bohr radius in the given material. This is a natural result since by definition the Bohr radius corresponds to the length scale at which Coulomb and quantum energies coincide. In GaAs the Bohr radius is as large as 10 nm. This is why both the level discreteness and the single-electron effects are important [33,34,82,83,111] in experiments with electron transport through GaAs-based quantum dots [35–37,41,113–118] when the size scale a is comparable to 10 nm. Stressing the analogy with atomic physics in which the Bohr radius determines the size of the electron orbit, semiconductor quantum dots are sometimes called 'artificial atoms' [5].

The relative importance of the two effects is quite different in experiments with vertical and lateral transport via quantum dots. In the vertical geometry a is close to the barrier width (in fact, the finite well width and the existence of two barriers should be taken into account [33,34]). The typical barrier width is 3–10 nm. Hence, the ratio $\delta/(e/C_\Sigma)$ is typically of the order of unity, and even can be larger than unity. That is why the level discreteness is always important in vertical transport via GaAs quantum dots and can have a major effect. In the first experimental study [112] of such a transport there was no sign of the single-electron charging effect, and the steps on the I–V curve were determined purely by the energy spectrum ε_k. This was because the collector barrier was much more transparent than the emitter barrier leading to $\alpha \ll 1$ (see above). Similar experiments [35] with the increased thickness of the collector barrier showed the Coulomb staircase with a fine structure due to ε_k. It was possible to change the major effect simply by applying the different polarity of the voltage [34] because that interchanged the emitter and collector. The interplay between two effects in vertical tunnelling via a quantum dot was also reported by several other groups (see, e.g. [113–115]).

In experiments with lateral transport via a quantum dot, the single-electron charging energy is typically considerably larger than the level spacing. The estimate above gives for GaAs dot $\delta/(e^2/C_\Sigma) \sim 25$ nm/d, so that for the typical dot diameter $d \sim 0.5$ μm this ratio is about 0.05 (this ratio somewhat increases if we take into account the capacitance increase due to coupling to electrodes). Note that the application of a strong magnetic field changes the energy spectrum and can considerably increase the level spacing δ.

There are many experiments demonstrating the coexistence of the energy and charge quantizations in lateral transport via a quantum dot [36,37,41,116–119]. Figure 4.15(a) shows the experimental I–V curve with the Coulomb staircase and the fine structure due to the level discreteness [116]. The inset shows the layout of the metal gates which form the quantum dot in the two-dimensional electron gas beneath them. The dependence of the current on the voltage of the central gate C is shown in Fig. 4.15(b). This gate does not much affect the tunnel barriers but changes the induced charge in the dot. The multipeak shape of the dependence is a consequence of the level discreteness (Fig. 4.14b).

The theory described above can also be applied to tunnelling through single molecules. Experimental I–V curves in such systems [49–53] typically have a region of Coulomb blockade and cusps or steps resembling the Coulomb staircase, and these features are usually discussed in terms of single-electron transport. If the molecule contains a relatively large cluster of metal atoms [54], good agreement even with simple orthodox theory can be expected. However, if the cluster consists of just a few atoms or there is no metal cluster at all, the theory of single-electronics should be used with some

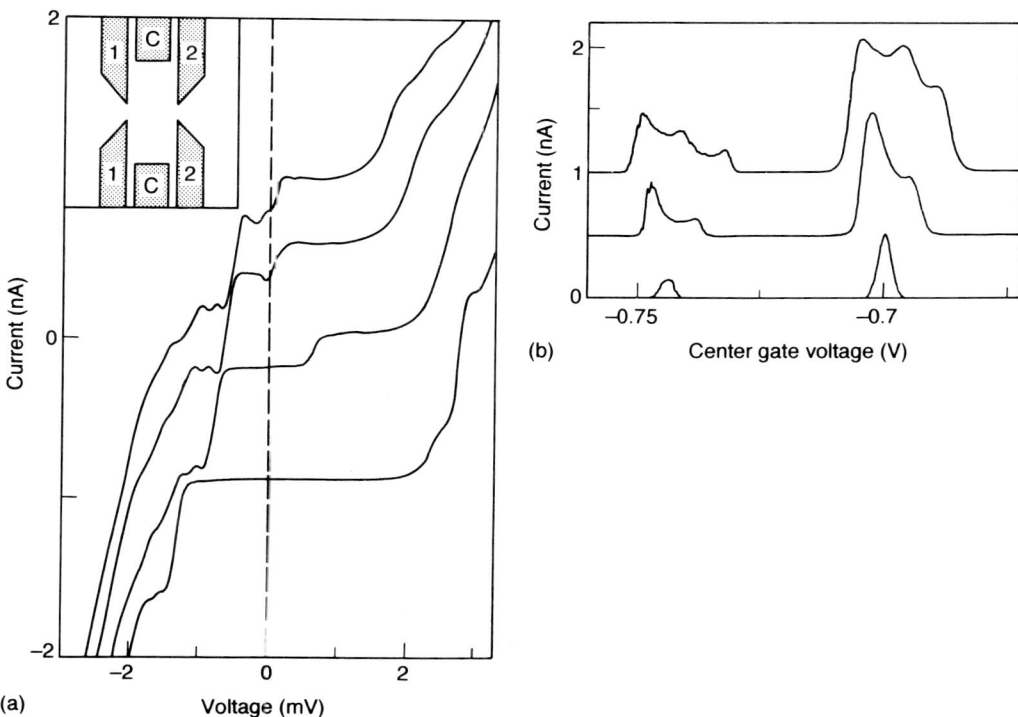

Figure 4.15 (a) The experimental I–V curve and (b) the dependence of the current on the gate voltage [116] for the C-SET based on the GaAs quantum dot (courtesy of A.T. Johnson). Notice the fine structure on the I–V curve and the multipeak shape of curves in (b) due to the level discreteness. Also notice the non-linearity of the I–V curve because of the barrier suppression.

caution. First of all, the level discreteness is not a small correction in this case but a major factor. Typically the separation of the Coulomb energy and one-electron spectrum assumed in Eqn 4.12 should fail, and the excitation spectrum should considerably depend on the charge number n. The calculation of capacitances could be used only for crude estimates and typically there should be no symmetry between addition and removal of electrons (electron affinity and ionization energy can be quite different) leading to highly asymmetric I–V curves. In contrast to orthodox theory, it can be impossible to add or remove more than 2–3 electrons to/from a molecule without its mechanical breakdown or chemical transformation. Thus, the experimental results in the single-molecule systems can considerably differ from the predictions of the standard theory of single-electronics.

On the other hand, it is surprising that such a macroscopic quantity as the capacitance can be sometimes used even at the microscopic size scale. In the model considered in [108] only a few conducting electrons are sufficient to establish a well-defined capacitance (the formal definition fluctuates only slightly with the number of electrons). As an interesting example note that the first three ionization energies of the single aluminium atom (5.97, 18.8, and 28.5 eV [120]) correspond to the dimensionless sequence $1 : 3.1 : 4.8$ which is very close to the orthodox sequence $1 : 3 : 5$.

6.2   *Barrier dependence on the voltage*

In the previous subsection the single-electron charging energy e^2/C of the 1-nm aluminium grain was estimated as 0.2 eV. This number is comparable to the energy height H of the tunnel barrier which depends on the material and is typically between 0.3 eV (thermally grown aluminium oxide) and 3 eV (vacuum barrier). This would lead to highly non-linear I–V curves of single-electron devices.

If the barrier has low transparency then the typical voltage of the I–V curve non-linearity is even much less than H/e and is comparable to $\hbar/e\tau$ where τ is the traversal time of tunnelling [in case of the rectangular barrier $\tau = l/(2H/m)^{1/2}$ where l is the barrier width]. For example, if $H = 1$ eV and $l = 2$ nm, then $\hbar/e\tau = 0.2$ eV.

Hence, the finite height of the barrier becomes an important factor [1,121,122] in metallic single-electron devices typically at the size scale of 1 nm (in some materials it appears considerably earlier [86]). The suppression of the tunnel barrier by the applied voltage is always a strong effect in experiments with lateral transport via semiconductor quantum dots because of typically low barrier height. For example, Fig. 4.15(a) shows experimental I–V curves [116] in which the voltage scale of the exponential non-linearity of the I–V curve is comparable with the period of the Coulomb staircase. Note that in semiconductor devices barrier suppression is typically important even when the relatively large size scale does not allow us to resolve individual levels. Finite barrier height is obviously also important in the single-molecule systems because of large typical voltages.

The effect can be taken into account within orthodox theory (neglecting for simplicity the level discreteness) by introduction of the non-linear 'seed' I–V curve $I_0(V)$ of the tunnel junction [1]. The tunnelling rates in this case are given by the general expression

$$\Gamma = \frac{I_0(W/e)}{e(1 - \exp(-W/k_B T))} \tag{4.15}$$

instead of Eqn 4.5 (W is the energy gain due to tunnelling). A more accurate approximation [123,124] takes into account the change of the image charge potential due to Coulomb blockade and gives the additional factor $\exp(e^2\tau/12C\hbar)$.

When the non-linearity of $I_0(V)$ is relatively small at the single-electron voltage scale $V \sim e/C$ (it implies $\hbar/\tau \gg e^2/C$), the I–V curve of the SET preserves the usual Coulomb features. However, the current grows exponentially with voltage, so that it becomes impossible to measure experimentally the offset voltage, and the Coulomb staircase becomes smoother [116,121] (see Fig. 4.15a). In case of strong non-linearity (which has not yet been achieved experimentally) the Coulomb staircase should completely disappear and give place to new periodic features with different period [122].

Let us mention one more effect which is important in semiconductor single-electron devices. In contrast to metallic systems, the geometrical size of a semiconductor conducting island can depend on the number of electrons on the island and on the gate voltage. Hence, the capacitance is not constant, leading to non-periodicity of the Coulomb staircase and non-periodic dependence on the gate voltage in an SET. The change of geometric size also leads to change of width of a tunnel barrier while the barrier height can be directly affected by the gate voltage. Sufficiently large gate voltage can either completely deplete the conducting island or remove the tunnel barrier depending on the polarity. As a consequence, on the large scale of the gate voltage semiconductor SETs usually behave like FETs (see Figs 4.6 and 4.15b): starting from the state with negligible current, one can finish with the perfectly open transport channel. Single-electron quasi-periodic dependence on the gate voltage (Coulomb oscillations) is observed in a relatively narrow range of gate voltage when the conducting island has already appeared but the tunnel resistance of the barrier is still larger than the quantum unit R_Q.

7   Conclusion

We have considered only some of the issues related to the field of single-electronics. For example, we did not mention single-electron effects in superconducting systems [1], including the possibility of measuring experimentally the parity of the total number of electrons in superconducting islands [19,125]. Another interesting subject is single-electron oscillations with frequency determined by the dc current, $f = I/e$ [1,23,74,84, 126,127]. This relation can be inverted: the magnitude of the dc current can be accurately controlled by the frequency of applied ac bias [15–17] that is used in the single-electron turnstile [16] and pump [17]. We have also not discussed the problem of cotunnelling [77,78], the effect of the electromagnetic environment [75,76], photon-assisted tunnelling [128–130], coherent effects [131,132], and many other issues.

Single-electronics has been a rapidly growing field during the last 10 years, and this growth continues. It is already clear that single-electronics is interesting not only from the purely scientific point of view, but can also be used in applications. The simplest application is the use of the SET for various purposes as a very sensitive electrometer capable of measuring subelectron charges. Another application is the standard of dc current [30] based on the single-electron pump. It is quite possible that arrays of small tunnel junctions will be used as low-temperature thermometers [22] other applications are being developed.

The most important potential application is ultradense (up to 10^{12} cells per cm^2)

integrated digital electronics which was the main topic of the present review. The question of whether the propects are real is however still uncertain. The main problem is the need for new technology capable of dealing with objects of the order of 1 nm. This length scale is imposed by the requirement of room-temperature operation. It is likely that such technology should use conducting clusters of atoms embedded in the molecular matrix; hence we speak about molecular electronics devices. Another major obstacle to integrated single-electronics is the random distribution of the background charge. If technology does not provide a solution to this problem, only circuits operating in Q_0-independent mode [72] will be practical, and this will considerably limit the variety of possible devices.

Despite the problems, ultradense integrated single-electron circuits will hopefully eventually be realized and will be able to substitute CMOS technology to continue the exponential growth of computer performance. The rapid progress in experimental single-electronics, in particular the recent demonstration of devices operating at the temperature of liquid nitrogen and even at room temperature, strongly supports this hope.

Acknowledgements. The author thanks D.V. Averin and K.K. Likharev for the numerous discussions and critical reading of the manuscript. The author is also grateful to Y. Nakamura, E. Leobandung, P.D. Dresselhaus and A.T. Johnson for providing figures with experimental results. The work was supported in part by ONR grant No. N00014-93-1-0880 and AFOSR grant No. 91-0445.

8 References

1 Averin DV, Likharev KK. In: Altshuler BL, Lee PA, Webb RA, eds. *Mesoscopic Phenomena in Solids.* Amsterdam: Elsevier, 1991: 173.
2 Likharev KK. *IBM J Res Dev* 1988; **32**: 144.
3 Grabert H, Devoret MH, eds. *Single Charge Tunnelling.* New York: Plenum, 1992.
4 Likharev KK, Claeson T. *Sci Am* 1992; **266**: 50.
5 Kastner MA. *Phys Today* 1993; **46** (1): 24.
6 Gorter CJ. *Physica* 1951; **15**: 777.
7 Neugebauer CA, Webb MB. *J Appl Phys* 1962; **33**: 74.
8 Giaver I, Zeller HR. *Phys Rev* 1969; **181**: 789.
9 Lambe J, Jaklevic RC. *Phys Rev Lett* 1969; **20**: 1504.
10 Kulik IO, Shekhter RI. *Sov Phys JETP* 1975; **41**: 308.
11 Averin DV, Likharev KK. *J Low Temp Phys* 1986; **62**: 345.
12 Likharev KK. *IEEE Trans Magn* 1987; **23**: 1142.
13 Fulton TA, Dolan GC. *Phys Rev Lett* 1987; **59**: 109.
14 Kuzmin LS, Likharev KK. *JETP Lett* 1987; **45**: 496.
15 Delsing P, Likharev KK, Kuzmin LS, Claeson T. *Phys Rev Lett* 1989; **63**: 1861.
16 Geerligs LJ, Anderegg VF, Holweg PAM *et al. Phys Rev Lett* 1990; **64**: 2691.
17 Pothier H, Lafarge P, Orfila RF, Urbina C, Esteve D, Devoret MH. *Physica B* 1991; **169**: 573.
18 Fulton TA, Gammel PL, Dunklleberger LN. *Phys Rev Lett* 1991; **67**: 3148.
19 Tuominen MT, Hergenrother JM, Tighe TS, Tinkham M. *Phys Rev Lett* 1992; **69**: 1997.
20 Zimmerli G, Eilies TM, Kautz RL, Martinis JM. *Appl Phys Lett* 1992; **61**: 237.
21 LaFarge P, Joyez P, Pothier H *et al. CR Acad Sci Paris Ser II* 1992; **314**: 883.
22 Pekola JP, Hirvi KP, Kauppinen JP, Paalanen MA. *Phys Rev Lett* 1994; **73**: 2903.

23 Kuzmin LS, Pashkin Yu A. *Physica B* 1994; **194–196**: 1713.
24 Verbrugh SM, Benhamadi ML, Visscher EH, Mooij JE. *J Appl Phys* 1995; **78**: 2830.
25 Nakamura Y, Chen CD, Tsai JS. *Jpn Appl Phys* 1996; **35**: L1465.
26 Nakamura Y, Klein DL, Tsai JS. *Appl Phys Lett* 1996; **68**: 275.
27 Wahlgren P, Delsing P, Haviland DB. *Phys Rev B* 1995; **52**: 2293.
28 Visscher EN, Verbrugh SM, Lindeman J, Hadley P, Mooij JE. *Appl Phys Lett* 1995; **66**: 305.
29 Hirvi KP, Kauppunen JP, Korotkov AN, Paalanen MA, Pekola JP. *Appl Phys Lett* 1995; **67**: 2096.
30 Keller MW, Martinis JM, Zimmerman NM, Steinbach AH. *Appl Phys Lett* 1996; **69**: 1804.
31 Dresselhaus PD, Ji L, Han S, Lukens JE, Likharev KK. *Phys Rev Lett* 1994; **72**: 3226.
32 Meirav U, Kastner MA, Wind SJ. *Phys Rev Lett* 1990; **65**: 771.
33 Korotkov AN, Averin DV, Likharev KK. *Physica B* 1990; **165**: 927.
34 Averin DV, Korotkov AN, Likharev KK. *Phys Rev B* 1991; **44**: 6199.
35 Bo Su, Goldman VJ, Cunningham JE. *Science* 1992; **255**: 313.
36 McEuen PL, Foxman EB, Meirav U *et al*. *Phys Rev Lett* 1991; **66**: 1926.
37 van der Vaart NC, Godijn SF, Nazarov YV *et al*. *Phys Rev Lett* 1995; **74**: 4702.
38 Duruöz CI, Clarke RM, Marcus CM, Harris Jr JS. *Phys Rev Lett* 1995; **74**: 3237.
39 Waugh FR, Berry MH, Mar DJ, Westervelt RM, Campman KL, Gossard AC. *Phys Rev Lett* 1995; **75**: 705.
40 Kouwenhoven LP, Jauhar S, Orenstein J *et al*. *Phys Rev Lett* 1994; **73**: 3443.
41 Klein O, Chamon C de C, Tang D *et al*. *Phys Rev Lett* 1995; **74**: 785.
42 Nakazato K, Blaikie RJ, Ahmed H. *J Appl Phys* 1994; **75**: 5123.
43 van Bentum PJM, Smokers RTM, van Kempen H. *Phys Rev Lett* 1988; **60**: 2543.
44 Wilkins R, Ben-Jacob E, Jacklevic RC. *Phys Rev Lett* 1989; **63**: 801.
45 Wan J-C, McGreer KA, Anand N, Nowak E, Goldman AM. *Phys Rev B* 1990; **42**: 5604.
46 Hanna AE, Tinkham M. *Phys Rev B* 1991; **44**: 5919.
47 Schönenberger C, van Houten H, Donkersloot HC. *Europhys Lett* 1992; **20**: 249.
48 Dorogi M, Gomes J, Osifchin R, Andres RP, Reifenberger R. *Phys Rev B* 1995; **52**: 9071.
49 Nejoh H. *Nature* 1991; **353**: 640.
50 Nejoh H, Ueda M, Aono M. *Jpn J Appl Phys (Part 1)* 1993; **32**: 1480.
51 Soldatov ES, Khanin VV, Trifonov AS *et al*. *Phys Rev Lett* 1996 (submitted).
52 Fischer CM, Burghard M, Roth S, von Klitzing K. *Europhys Lett* 1994; **28**: 129.
53 Zubilov AA, Gubin SP, Korotkov AN *et al*. *Tech Phys Lett* 1994; **20**: 195.
54 Dubois JGA, Gerritsen JW, Shafranjuk SE *et al*. *Europhys Lett* 1996; **33**: 279.
55 Matsumoto K, Ishii M, Segawa K, Oka Y, Vartanian BJ, Harris JS. *Appl Phys Lett* 1996; **68**: 34.
56 Yano K, Ishii T, Hashimoto T, Kobayashi T, Murai F, Seki K. *IEEE Trans Electron Dev* 1994; **41**: 1628.
57 Takahashi Y, Nagase M, Namatsu H *et al*. *Electron Lett* 1995; **31** (2): 136.
58 Fujiwara A, Takahashi Y, Mirase K, Tabe M. *Appl Phys Lett* 1995; **67**: 2957.
59 Matsuoka H, Kimura S. *Appl Phys Lett* 1995; **66**: 613.
60 Leobandung E, Guo L, Wang Y, Chou SY. *Appl Phys Lett* 1995; **67**: 938, 2339.
61 Tucker JR. *J Appl Phys* 1992; **72**: 43339.
62 Korotkov AN, Chen RH, Likharev KK. *J Appl Phys* 1995; **78**: 2520.
63 Chen RH, Korotkov AN, Likharev KK. *Appl Phys Lett* 1996; **68**: 1954.
64 Likharev KK, Semenov VK. In: *Extended Abstracts of International Superconductive Electronics Conference*, Tokyo, 1987: 182.
65 Likharev KK, Polonsky SF, Vyshenskii SV. Reprint, 1990.
66 Averin DV, Likharev KK. In: Grabert H, Devoret MH, eds. *Single Charge Tunneling*. New York: Plenum, 1992: 311.
67 Nazarov YuV, Vyshenskii SV. In: Koch H, Lubbig H, eds. *Single-Electron Tunneling and Mesoscopic Devices*. Berlin: Springer, 1992: 61.
68 Nakazato K, White JD. *IEDM* 1992; **1992**: 487.

69 Ancona MG. *J Appl Phys* 1996; **79**: 526.

70 Korotkov AN. *Appl Phys Lett* 1995; **67**: 2412.

71 Likharev KK, Korotkov AN. *Science* 1996; **273**: 763.

72 Likharev KK, Korotkov AN. In: *Proceedings of ISDRS'95 (Charlottesville, Virginia)*.

73 Hauge EH, Støvneng JA. *Rev Mod Phys* 1989; **61**: 917.

74 Bakhvalov NS, Kazacha GS, Likharev KK, Serduykova SI. *IEEE Trans Magn* 1989; **25**: 1436.

75 Nazarov YuV. *JETP Lett* 1989; **49**: 126.

76 Ingold G-L, Nazarov YuV. In: Grabert H, Devoret MH, eds. *Single Charge Tunneling*. New York: Plenum, 1992: 21.

77 Averin DV, Odintsov AA. *Phys Lett A* 1989; **140**: 251.

78 Averin DV, Nazarov YuV. In: Grabert H, Devoret MH, eds. *Single Charge Tunneling*. New York: Plenum, 1992: 217.

79 Averin DV, Korotkov AN. *Sov Phys JETP* 1990; **70**: 937.

80 Averin DV, Korotkov AN. *J Low Temp Phys* 1990; **80**: 173.

81 Averin DV, Korotkov AN. In: Lazarev PI, ed. *Molecular Electronics*. Dordrecht: Kluwer, 1991: 9.

82 Beenakker CWJ. *Phys Rev B* 1991; **44**: 1646.

83 van Houten H, Beenakker CWJ, Staring AAM. In: Grabert H, Devoret MH, eds. *Single Charge Tunneling*. New York: Plenum, 1992.

84 Kuzmin LS, Haviland DB. *Phys Rev Lett* 1991; **67**: 2890.

85 Korotkov AN. *Phys Rev B* 1994; **49**: 16518.

86 Kuzmin LS, Pashkin YuA, Tavkhelidze AN. *Appl Phys Lett* 1996; **68**: 2902.

87 Korotkov AN, Averin DV, Likharev KK, Vasenko SV. In: Koch H, Lubbig H, eds. *Single-electron Tunneling and Mesoscopic Devices*. Berlin: Springer, 1992: 45.

88 Korotkov AN, PhD thesis, Moscow State University, 1991.

89 Korotkov AN. *Phys Rev B* 1994; **49**: 10381.

90 Hershfield S, Davies JH, Hyldgaard P, Stanton CJ, Wilkins JW. *Phys Rev B* 1993; **47**: 1967.

91 Birk H, de Jong MJM, Schönenberger C. *Phys Rev Lett.* 1995; **75**: 1610.

92 Kautz RL, Zimmerli G, Martinis JM. *J Appl Phys* 1993; **73**: 2386.

93 Korotkov AN, Samuelsen MR, Vasenko SA. *J Appl Phys* 1994; **76**: 3623.

94 Fukui H, Fujishima M, Hoh K. In: *Extended Abstracts, ICSSD '94*. Yokohama, 1994: 331.

95 Averin DV, Korotkov AN, Nazarov YuV. *Phys Rev Lett* 1991; **66**: 2818.

96 Zimmerli G, Kautz RL, Martinis JM. *Appl Phys Lett* 1992; **61**: 2616.

97 Fonseca LRC, Korotkov AN, Likharev KK, Odintsov AA. *J Appl Phys* 1995; **78**: 3238.

98 Tiwari S, Rana F, Hanai H, Hartstein A, Crabbe EF, Chan K. *Appl Phys Lett* 1996; **68**: 1377.

99 Guo L, Leobandung E, Chou SY. *Appl Phys Lett* 1997; **70**: 850.

100 Nakajima A, Futatsugi T, Kasemura K, Fukano T, Yokoyama N. *IEDM* 1996; **952**.

101 Lent CS, Tougaw PD, Porod W, Bernstein GH. *Nanotechnology* 1993; **4**: 49.

102 Bandyopadhyay S, Das B, Miller AE. *Nanotechnology* 1994; **5**: 113.

103 Bennett C. *IBM J Res Devel* 1973; **17**: 525.

104 Likharev KK. *Int J Theor Phys* 1982; **21**: 311.

105 Korotkov AN. In preparation.

106 Hu C, ed. *Nonvolatile Semiconductor Memories*. New York: IEEE, 1991.

107 Weinmann D, Häusler W, Kramer B. *Phys Rev Lett* 1995; **74**: 984.

108 Belkhir L. *Phys Rev B* 1994; **50**: 8885.

109 Jovanovic D, Leburton J-P. *Phys Rev B* 1994; **49**: 7474.

110 Ralph DC, Black CT, Tinkham M. *Phys Rev Lett* 1995; **74**: 3241.

111 Groshev A. *Phys Rev B* 1990; **42**: 5895.

112 Reed MA, Randall JN, Aggarval RJ, Matyi RJ, Moore TM, Wetsel AE. *Phys Rev Lett* 1988; **60**: 535.

113 Guéret P, Blank N, Germann R, Rothiizen H. *Phys Rev Lett* 1992; **68**: 1896.

114 Tarucha S, Austing DG, Honda T, von der Hage RJ, Kouwenhoven LP. *Phys Rev Lett* 1996; **77**: 3613.

115 Tewordt M, Law VJ, Nicholls JT *et al. Solid State Electr* 1994; **37**: 793.

116 Johnson AT, Kouwenhoven LP, de Jong W, van der Vaart NC, Harmans CJPM, Foxon CT. *Phys Rev Lett* 1992; **69**: 1592.

117 Foxman EB, McEuen PL, Meirav U *et al. Phys Rev B* 1993; **47**: 10020.

118 Weis J, Haug RJ, von Klitzing K, Ploog K. *Semicond Sci Technol* 1994; **9**: 1890.

119 Kouwenhoven LP, McEuen PL. In: Timp G, ed. *Nano-Science and Technology.* To be published.

120 Landolt-Börnstein. *Numerical Data*, Vol I/1. Berlin: Springer, 1950: 211.

121 Korotkov AN, Nazarov YuV. *Physica B* 1991; **173**: 217.

122 Lutwyche MT, Wada Y. *J Appl Phys* 1994; **75**: 3654.

123 Korotkov AN. *Phys Rev B* 1994; **49**: 11508.

124 Averin DV. *Phys Rev B* 1994; **50**: 8934.

125 Averin DV, Nazarov YuV. *Phys Rev Lett* 1992; **69**: 1993.

126 Korotkov AN, Averin DV, Likharev KK. *Phys Rev B* 1994; **49**: 1915.

127 Korotkov AN. *Phys Rev B* 1994; **50**: 17674.

128 Likharev KK, Devyatov IA. *Physica B* 1994; **194–196**: 1341.

129 Bruder C, Schoeller H. *Phys Rev Lett* 1994; **72**: 1076.

130 Kouwenhoven LP, Jauhar S, Orenstein J *et al. Phys Rev Lett* 1994; **73**: 3443.

131 Yacoby A, Heiblum M, Mahalu D, Shtrikman H. *Phys Rev Lett* 1995; **74**: 4047.

132 Büttiker M, Stafford CA. *Phys Rev Lett* 1996; **76**: 495.

5 Mesoscopic Phenomena Studied with Mechanically Controllable Break Junctions at Room Temperature

C. ZHOU,* C.J. MULLER,* M.A. REED,* T.P. BURGIN† and J.M. TOUR†

*Center for Microelectronic Material and Structures, Yale University, PO Box 208284, New Haven, CT 06520, USA

†Department of Chemistry and Biochemistry, University of South Carolina, Columbia, SC 29208, USA

1 Introduction

The scaling-down of electronic device size in integrated circuits has resulted in tremendous effort in research on mesoscopic systems, of which molecular electronics is at the very forefront. The core of molecular electronics at this stage is to be able to probe a few or a single molecule and measure the electric and electro-optic properties. While there are many approaches toward this challenging problem, the mechanically controllable break junction (MCB) is thought to be one of the promising tools.

The MCB was recently developed by Muller *et al.* [1,2], following the pioneering work done by Morland [3]. As shown in Fig. 5.1, a conventional MCB consists of breaking a thin metal wire of the material to be used as electrodes, typically 100 μm in diameter, held fixed to a bending beam by two droplets of epoxy. Generally a notch is

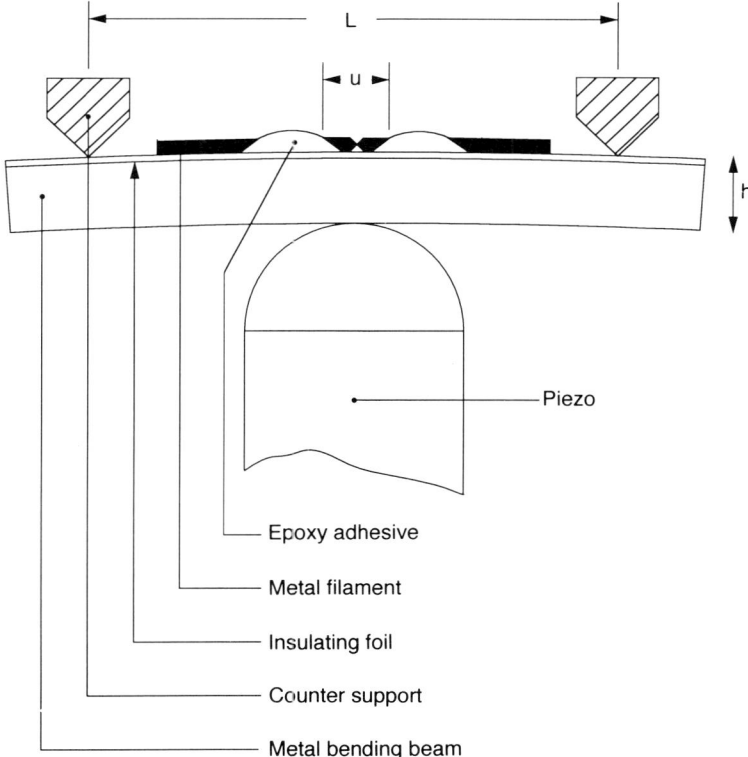

Figure 5.1 A schematic drawing of the sample mounting in a three-point bending configuration showing the unglued section u, and the distance L, between the counter supports.

cut in the wire between the two anchors, so that by bending the beam the surface of the beam elongates and the wire is broken at the notch, creating two electrodes with clean surfaces formed *in situ*. After the break, the distance of the two facing ends can be adjusted with atomic resolution by fine tuning the deflection of the bending beam, which is controlled by the voltage applied to the piezo element. With this technique, both atomic scale point contacts with adjustable contact size and vacuum tunnel junctions with adjustable barrier width have been studied in many material systems.

One approach to study materials at the atomic scale is using scanning tunnelling microscopy (STM). The necessity of using a feedback system to create and maintain the tip–substrate separation mechanically stable severely limits their application for long-term or variable environment measurements. Unlike STM, even without a feedback system, MCB is intrinsically stable. This high stability is achieved by a huge reduction factor defined as the ratio between the length variations of the piezo element and the unglued section. Within the range one can bend the beam, the extension of the beam surface distributes uniformly, and the reduction factor can be estimated as [1]:

$$\alpha = \frac{L^2}{4uh} \tag{5.1}$$

where L is the distance between the two counter supports, u is the distance between the two unglued anchors, and h is the thickness of the bending beam, as shown in Fig. 5.1. Figure 5.2 shows a scanning electron microscope (SEM) micrograph of a conventional MCB and an enlarged view of the centre part. For a typical MCB, L is 1 inch, the beam thickness is 500 µm, and u is around 300 µm. Thus we estimate the reduction factor is around 400–1000. Measurements done at liquid helium temperature show that this kind of MCB can achieve a stability of 100 fm [1].

In addition, MCB has the advantage that the two electrodes are made of the same material, which can help to create a symmetrical electrical system if we deposit molecules with symmetrical structure between the two electrodes. Finally, MCB can be conveniently loaded into a cryogenic environment, which can help to clarify all kinds of

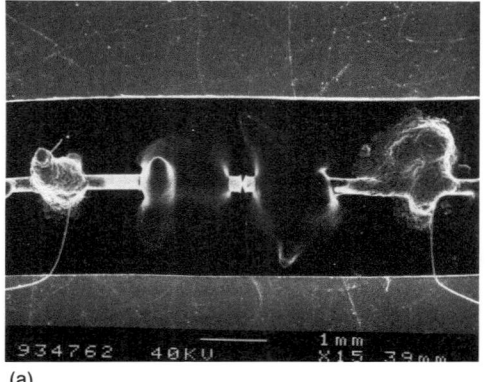
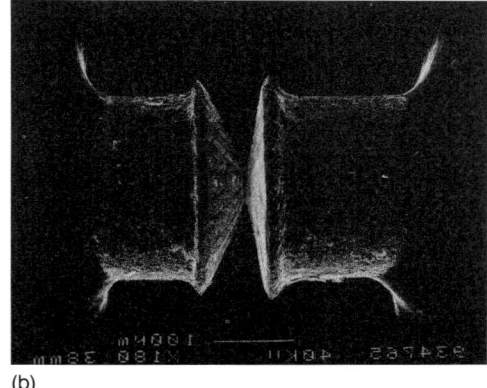

(a)   (b)

Figure 5.2 (a) Scanning electron microscope (SEM) micrograph of a conventional break junction. The wide beam is the substrate covered with a thin layer of insulator, on which the metal wire is mounted with two droplets of epoxy. (b) Enlarged view of the centre part of the metal wire. The notch on the metal is clearly visible; distance between unglued sections u is 300 µm.

interesting phenomena such as Coulomb blockade, quantum confinement effects and spin effects.

MCB has been widely used to study the quantum point contact at liquid helium temperature [4,5], superconductive vacuum tunnelling junctions [6], and Kondo scattering [7]. For the first time we have carried out quantum point contact and vacuum tunnelling experiments at room tempeature. Conductance quantization is clearly observed at room temperature with quantum point contacts realized with MCBs, which will be covered in Section 2. Results from the deposition of benzene-1,4-dithiol in MCBs will be covered in Section 3. In Section 4, we report our effort to improve the MCBs by fabricating MCB in silicon with the assistance of advanced e-beam lithography and silicon micromachining technology. The microfabricated MCBs are 100 times smaller than the conventional ones and are found to be at least two orders more stable than the conventional MCBs.

2   Quantum point contact realized with conventional MCB at room temperature

2.1   *Introduction*

It has been pointed out [8] for a long time that the conductance of an ideally smooth and narrow constriction between two conducting reservoirs assumes only quantized values, in multiples of $G_0 = 2e^2/h$ when the constriction size is comparable with the Fermi wavelength (λ_F) of the electrons. In recent years, considerable interest has been focused on controllable two-dimensional electron gas (2DEG) point contacts after the initial discovery by van Wees *et al.* [9] and Wharam *et al.* [10]. As a natural step, a number of people have been trying to produce small controllable metallic contacts in order to investigate whether these systems (with a much higher Fermi energy compared with the 2DEG system) also exhibit the conductance quantization effect. Both STM [11–15] and MCB [4,16–18] have been employed to study atomic-scale metallic contact. Transitions between plateaus while adjusting the contact size with atomic resolution are generally observed. Nevertheless the results are complex and their interpretation is not straightforward.

The basic theory of conductance quantization in a point contact system has been known for a long time [8,19]. A long and perfectly clean cylindrical conductor between two reservoirs can be viewed as an electron waveguide, as shown in Fig. 5.3(a). We define k as the wave vector in the direction of propagation and use the cylindrical coordinates r, ϕ, z. While electrons are confined by the lateral boundary and thus take quantized modes when the radius is of the order of Fermi wavelength, we can approximate the longitudinal modes as free electron propagation. As a result, the electron wavefunction of the one-dimensional conductor can be described as

$$J_m\left(\gamma_{mn}\frac{r}{R}\right)\exp(im\phi)\exp(ikz),$$

where γ_{mn} is the nth root of the Bessel function $J_m(x)$.

Assuming the energy of the subband (m,n) is $E_{m,n}$, the dispersion relation is:

$$E_{mn}(k) = E_{m,n} + \frac{\hbar^2 k^2}{2m}. \tag{5.2}$$

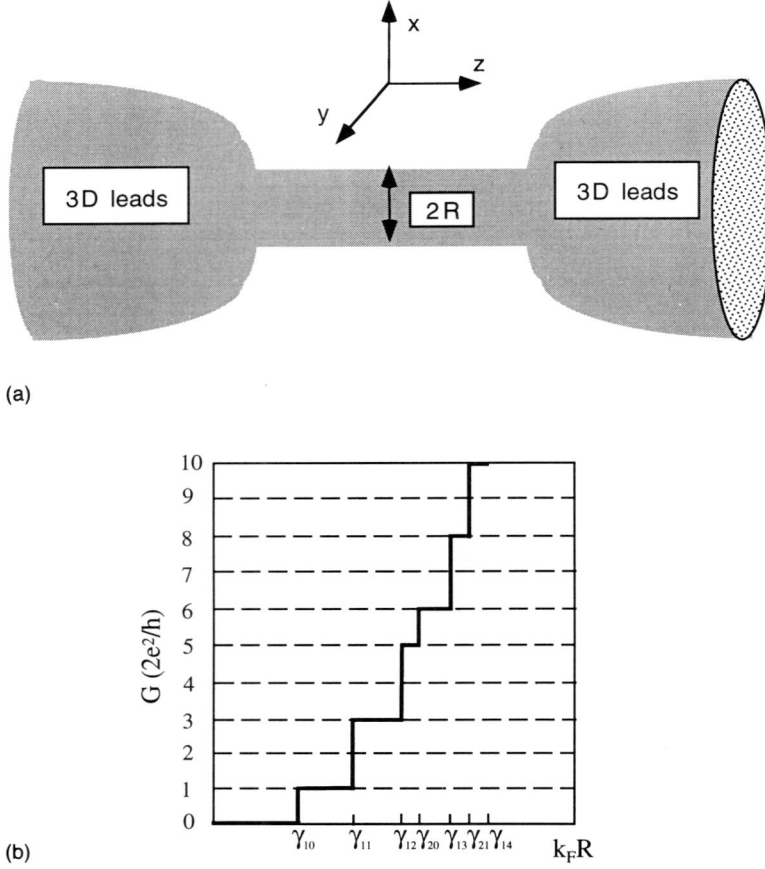

(a)

(b)

Figure 5.3 (a) Schematics of a long cylindrical constriction between two three-dimensional reservoirs. (b) The conductance of the above contact plotted vs the dimensionless parameter $k_F R$. Roots of the Bessel function γ_{mn} are marked on the *x*-axis in increasing order. (After [19].)

In equilibrium, all states of the one-dimensional wire are filled up to E_F, and no net current flows since there are as many left-going as right-going electrons. While a voltage V applied over the one-dimensional wire creates a chemical potential difference between the left and right reservoirs, we can calculate the total net current by integrating the net current dI_{mn} carried by the transverse mode (m,n) between $E_{mn}(k)$ and $E_{mn}(k) + dE_{mn}(k)$, which should be the product of electron charge, electron velocity

$$\frac{1}{\hbar}\frac{dE}{dk}$$

and the number of electrons within the energy interval

$$\left(\rho_{mn}(E)dE_{mn}(k) = \frac{2}{2\pi}dk\right),$$

where the numerator 2 describes spin degeneracy:

$$dI_{mn} = e \frac{dE_{mn}(k)}{dk} \frac{2dk}{2\pi} \{f(k, E_F + eV)\} - \{f(-k, E_F)\}$$

$$= \frac{2e}{h} dE_{mn}(k)\{f(k, E_F + eV) - f(-k, E_F)\} \tag{5.3}$$

where $f(E)$ is the Fermi–Dirac distribution function

$$\frac{1}{e^{(E-E_F)/kT} - 1},$$

and $f(k, E_F + eV) - f(-k, E_F)$ describes the population difference between the left-going and right-going states induced by the applied voltage.

At zero temperature

$$f(E) = \begin{cases} 1 & \text{when } 0 < E < E_F \\ 0 & \text{when } E > E_F. \end{cases} \tag{5.4}$$

and thus

$$I_{mn} = \frac{2e}{h} \int_{E_F}^{E_F + eV} dE_{mn}(k)$$

$$= \frac{2e^2}{h} V; \tag{5.5}$$

$$G_{mn} = \frac{2e^2}{h}.$$

The above result has fundamental significance since the conductance of one mode consists of only fundamental constants with the absence of dependence on specific transverse modes, material used, or geometric parameters. We define $G_0 = 2e^2/h$, which is called the fundamental conductance unit. This quantum effect is of similar importance as the quantized Hall resistance in the quantum Hall effect and the quantized voltage increments in the ac Josephson effect in the presence of radiation; the latter effects are accurate to a very high degree of precision and are therefore useful as standards. However, the total number of conducting modes still depends on how many confined transverse modes are below the Fermi energy, and the conductance of the wire is given by the fundamental conductance unit multiplied by the number of conducting modes. At finite temperature and in a real one-dimensional wire, there is reflection and scattering around the constriction and tunnelling through non-conducting modes. A generalized description is given by Landauer's formula [8]:

$$G = \frac{2e^2}{h} \sum_{m,n} T_{mn}, \tag{5.6}$$

where the transmission probability T_{mn} for mode (m,n) accounts for reflection, scattering, and tunnelling.

In Fig. 5.3(b) the conductance is plotted vs $k_F R$ for $m = 0, 1, 2, 3, 4$ and $n = 1, 2$. For every non-zero m, modes (m,n) and $(-m, n)$ have the same energy and thus the two modes open or close simultaneously. For $m = 0$, there is no degeneracy. As a result, the conductance increases by $2e^2/h$ when a $m = 0$ mode becomes conducting and increases by $2(2e^2/h)$ when a $m \neq 0$ channel opens. In other words, because of the cylindrical symmetry, the conductance takes 0, 1, 3, 5, 6, 8, ... times the fundamental conductance unit. In a quantum point contact realized in a 2DEG system, there is no such symmetry and thus conductances are 0, 1, 2, 3, 4, 5, ... times the fundamental conductance unit both theoretically and experimentally.

More realistic geometric shapes of the contact have been considered by many people [19,20,21,22], among which a hyperbolic geometry is of great interest, as shown in Fig. 5.4. Exact quantum mechanical calculations have been derived and the basic result is plotted in Fig. 5.5, where conductance is plotted vs constriction size and the opening angle ϑ_0. When ϑ_0 is 0, we get a cylindrical conducting wire and the transitions between conductance plateaus take sharp steps. When ϑ_0 is $\pi/2$, the constriction becomes a conducting hole between two reservoirs and the sharp transitions between conductance plateaus smear out.

When we apply the basic theory to the quantum point contact realized in metallic systems, there is a lot of room for doubt. Since the Fermi wavelength in metal is of the order of atomic radius, discrete atomic rearrangement in the point contact can also increase or decrease the conductance in the order of G_0. A number of experiments [4,11–18] show that the transition from contact to tunnelling, where the electrodes lose contact, takes place at $2e^2/h$. Measurements in the contact regime do not always show clear horizontal conductance plateaus at integer multiples of $2e^2/h$ when the constriction size is varied. Often conductance steps are observed that are not of the 'right' value and link non-integer conductance plateaus. Recent work has indicated that the conductance quantization, if at all present, becomes apparent only after averaging over many contact configurations. This has been shown clearly in measurements on sodium at 4.2 K [18].

A related experimental problem is the stability required to maintain an atomic-scale contact constant in size. Often conductance traces have to be recorded in milliseconds

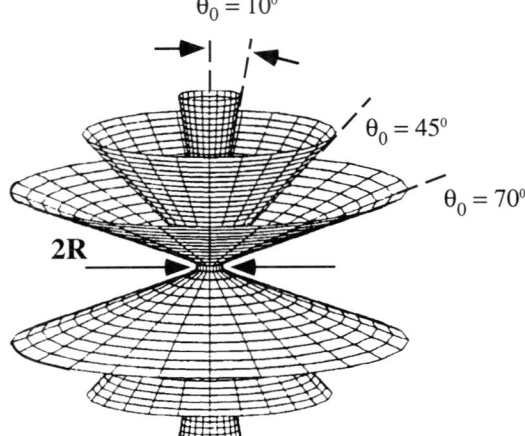

Figure 5.4 Contact with hyperbolic geometry. R is the radius of the narrowest section and ϑ_0 is the asymptotic opening angle, which can vary from 0 (a cylindrical wire) to 90° (a circular hole). (After [22].)

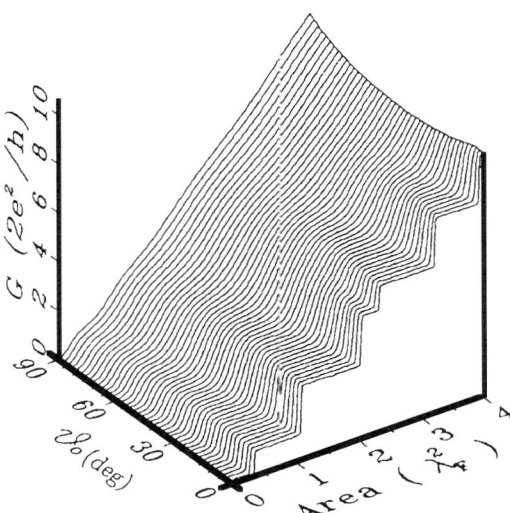

Figure 5.5 Theoretical result of the conductance of the contact in Fig. 5.4 plotted as a function of the area of the narrowest section and the asymptotic opening angle ϑ_0. As the opening angle increases, sharp transitions between higher conductance plateaus smear out gradually. (After [22].)

or less since the contact changes its size in this time scale. Some experiments show gradual transitions between conductance levels, possibly resulting from this small recording time, while other experiments show abrupt steps. This has been the basis for some controversy in the past since an abrupt conductance step favours a different physical mechanism from a gradual transition [17].

We present measurements on conductance quantization with MCB for the first time at room temperature. Our results show more prominent and more horizontal conductance pleateaus separated by sharp transition regions than the past experiments at low temperature. Compared with the basic theory presented above, there are subtle deviations in the quantum numbers of present conductance plateaus. We discuss specific mechanisms for the deviation.

2.2   *Results*

Contacts of gold and copper (99.99%) are adjusted at room temperature. In order to reduce surface contamination the electrode material is fractured in a vacuum system (10^{-7} Torr), which uses an oil-free absorption/ion pump combination. Two terminal conductance measurements are performed by biasing the junction at 26 mV and recording the current while contact size changes. No qualitative differences in the recordings are observed by lowering the bias voltage. After fracture of the metal wire, the electrodes are carefully brought close together in the tunnelling regime until suddenly a large contact with a typical resistance of 100 Ω is formed. The large contact cross-section is then reduced in size by increasing the piezo voltage until the conductance is about 10 times $2e^2/h$. The piezo voltage is now fixed and the contact is allowed to relax by itself. The conductance decreases spontaneously in time, presumably due to surface diffusion of atoms away from the contact. This effect is shown for two contacts of gold and copper in Fig. 5.6 in the last stages of the contact while the conductance is measured with a sampling rate of 100 Hz. Remarkably it appears that the conductance attains specific values near n times $2e^2/h$, with n an integer. Transitions between the levels occur abruptly. At some point the neck becomes so thin that the two electrodes can be bridged by a single atom after which a jump to the tunnelling regime occurs and the conductance becomes zero neglecting vacuum tunnelling. This jumps usually occurs near $n = 1$, and sometimes near $n = 2$ or $n = 3$. In general, we observe that copper contacts relax into the tunnelling regime within 4 s, and gold contacts within 9 s. Some plateaus show structure and intrinsic noise which considerably exceeds our measurement accuracy. The noise amplitude is observed to increase for higher n.

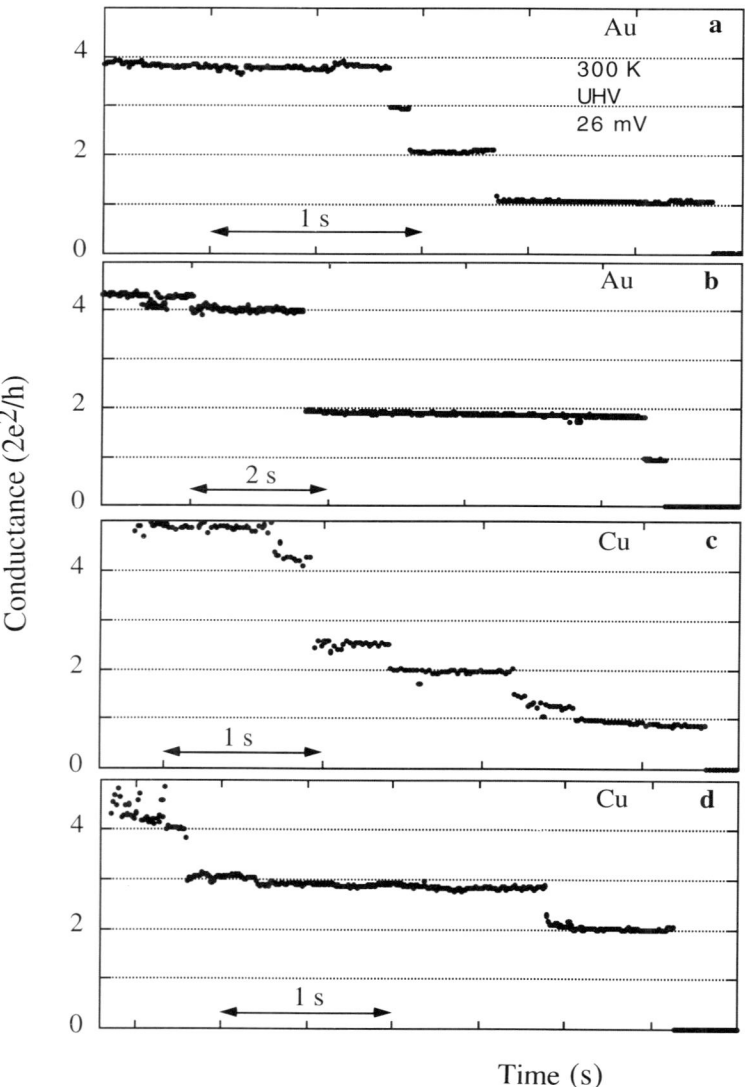

Figure 5.6 Conductance traces of (a, b) gold and (c, d) copper, showing clear plateaus separated by abrupt discrete steps. After adjustment of a contact with a conductance of about 10 times $2e^2/h$, the contact spontaneously reduced its size. During the displayed time interval the junction evolved from a few atom point contacts to a tunnelling junction after breaking. Both plateaus near integer values and plateaus far off are shown to be present.

The integer conductance values of the various horizontal plateaus such as those in Fig. 5.6 suggest the presence of conductance quantization in Au and Cu at room temperature. Caution is needed in claiming conductance quantization on the basis of individual traces since some conductance plateaus can be found which are not equal to, and are sometimes even half of the integer values (see e.g. panel (c) in Fig. 5.6). The abrupt steps between the plateaus are caused by sudden atomic rearrangements in the constriction. In each individual scan the precise atomic geometry of the constriction is different and the scans do not reproduce in detail. In order to give clear evidence for conductance quantization in these constrictions, we must show that the integer conductance values prevail over non-integer ones. To this end, a summation is performed over a large number of conductance traces, thus sampling a wide range of individual contact geometries. This is not possible for the conventional 2DEG experimental situation since the fabrication fixes the contact structure. The only selection criterion we applied for the traces was that the contact should relax into the tunnel regime within 9 s for gold. Such a criterion was applied because after several hours the result became anomalous, presumably due to contamination by adsorbates. The observed linear I–V curves of the selected contacts give us confidence that the contact is metallic. A total of 72 gold measurements (four samples of gold) have been used to construct a suitable set. A histogram of the observed conductance values is shown in Fig. 5.7. The histograms of individual samples are very similar to the ones shown here. Pronounced peaks around integer values of $2e^2/h$ are evident. Gold shows a narrow $n = 1$ first peak clearly separated from $n = 2$. For $n > 2$ a background starts to rise, but a number of peaks for $n > 2$ can still be resolved clearly. Comparing it with the calculation shown in Fig. 5.5, we assume the background for $n > 2$ means the contact realized is not perfectly cylindrical, but the open angle is still small. This histogram shows that even in the absence of exact quantization in individual traces, the effect can be recovered by averaging over many different contact configurations.

The width of the peaks show that many plateaus occur at values just below or above

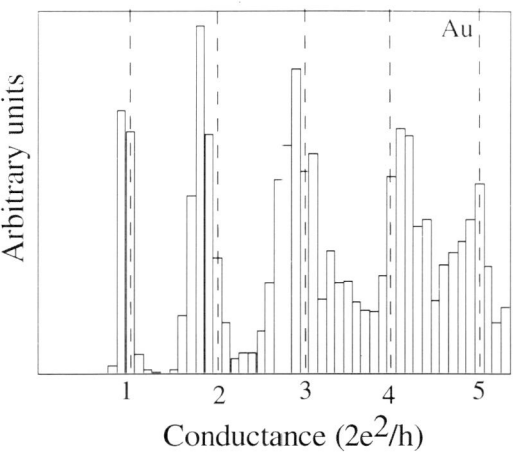

Figure 5.7 The histograms of gold and copper reflect data of many conductance traces like the ones in Fig. 5.6. Both histograms are obtained from several mechanically controllable break junctions (MCB). The fundamental unit is divided into 10 sections; the total number of data points of all traces in a specific section is represented by a bar.

n times $2e^2/h$, suggesting that the deviation from integers occurs randomly. However, a limited number of traces can be mapped exactly onto integers for multiple plateaus by the subtraction of a constant resistance value. This is shown in Fig. 5.8 where the plateaus *n* = 2 and *n* = 4 fall exactly on the integer values after 315 Ω is subtracted from all data points. A similar correction is commonly used in experiments on 2DEG semiconductor devices.

In contrast to the present room temperature result, in recent MCB experiments on Cu at 4.2 K similar averaging revealed peaks only near *n* = 1 and *n* = 3 and additional smaller peaks at non-integer conductance values. While the absence of *n* = 2 plateau at low temperature is thought to be the fingerprint of a cylindrical contact, it contradicts our belief that at room temperature thermal energy can help atoms to find the most energy-favourable positions and thus the contact should tend to form cylindrical and low defect neck. Krans [20] suggests that the origin could be that contamination causes defects and thus breaks the cylindrical symmetry at room temperature. In fact the plateaus at room temperature are much more horizontal and closer to the multiples of $2e^2/h$ than the plateaus observed at 4.2 K, which suggests that the contact does have less defects at 300 K than at 4.2 K. In addition, most traces were taken within 1 h after the initial fracture and they do not show a gradual change in quality.

We suggest another mechanism which we would like to call 'parallel contacts'. Although a histogram is a good tool to average over many individual contact geometries, it averages over specific physics for each individual trace. Figure 5.9 shows several typical plots for gold samples; (a) and (b) are the same traces as shown in Fig. 5.6(a) and (b). We notice that the *n* = 3 plateau is substantially shorter than the rest of the plateaus in the same trace, which means the prominent plateaus are *n* = 0, 1, 2, 4. This is exactly the case for (b). Trace (c) shows plateaus at *n* = 0, 1, 2, 4, 6, 8, 10. Around half of all traces show clearly *n* = 0, 1 and 2 plateaus, and then it jumps by two units to *n* = 4, 6. Trace (d) displays different behaviour: we can see *n* = 0, 1, 3, consistent with low temperature experiments. We argue that for traces (a), (b) and (c), when the large contact forms and then reduces due to the thermal-assisted diffusion, at a certain stage

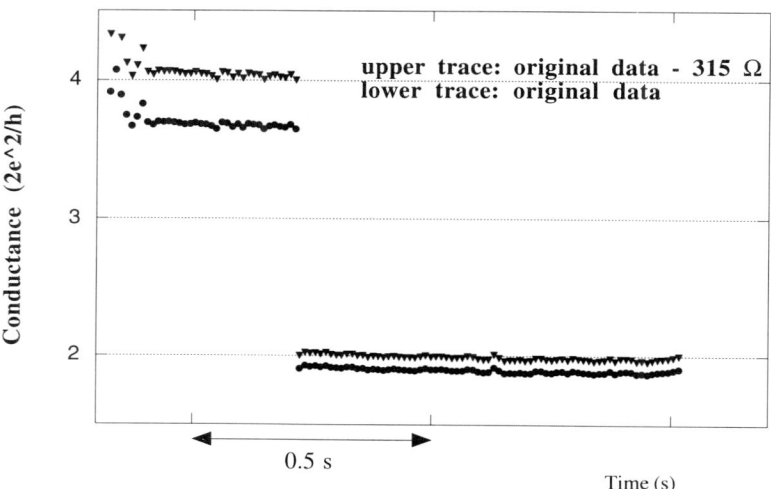

Figure 5.8 For a number of traces multiple plateaus can be mapped onto integer values by subtraction of a constant resistance from the data.

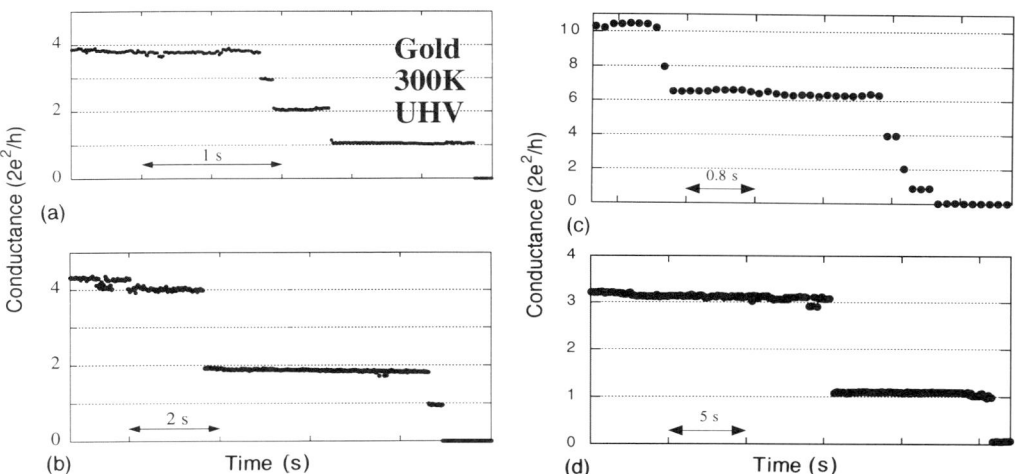

Figure 5.9 Conductance traces for gold. (a) and (b) are the same traces shown in Figs 5.6(a) and (b). (a) Prominent plateaus only at $n = 0, 1, 2, 4$ and a short plateau at $n = 3$. (b) $n = 0, 1, 2, 4$ plateaus with the absence of $n = 3$. (c) Plateaus $n = 0, 1, 2, 4, 6, 8, 10$. (a), (b) and (c) suggest there are parallel contacts when the traces are taken. (d) Plateaus near $n = 0, 1, 3$, consistent with the single cylindrical contact model.

there are two parallel smaller contacts. Each contact is cylindrical and thus conductance reduces by two units at each step for large n. When the conductance is at plateau $n = 2$, there is only one conducting mode in each contact and thus the conductance starts to decrease by one unit. For trace (d), we believe there is only one contact, which can be well explained with the cylindrical contact model. Similar behaviour is observed for copper.

So far, only the $n < 5$ conductance region has been discussed. The behaviour for larger conductance values is shown in Fig. 5.10 for a trace of gold and copper. As in Fig. 5.6, an increase in noise is observed for larger n. At these larger values the noise amplitude can easily exceed $2e^2/h$, which prohibits the assignment of specific plateau values. Often switching behaviour occurs between two neighbouring conductance levels. Another type of noise is observed to be more random (see $n = 7$ plateau in the copper trace). Noise due to the thermal vibrations of the atoms will have a time period 10^{-13} s and will not be resolved experimentally. Plausible sources of noise are fluctuations in the geometry whereby one or a few atoms switch between energetically equifavorable configurations. This mechanism will depend strongly on temperature, consistent with the absence of noise at low temperatures.

We define conductance quantization as a conductance staircase with plateaus at integer multiples of $2e^2/h$, as opposed to simply conductance jumps between arbitrary values. In the case of a perfectly homogeneous constriction of adiabatically varying cross-section, each electron propagating along the constriction stays in a particular transverse mode. In this idealized system the transmission coefficients for all incident channels are either zero or one, and a quantized conductance n times $2e^2/h$, where n is the number of propagating modes in the narrowest part of the constriction, then follows from the Landauer's formula [8]. Small imperfections in the geometry of the contact will cause channel mixing, whereby the electrons in the constriction undergo scattering between different propagating channels travelling in the same direction. In that case the

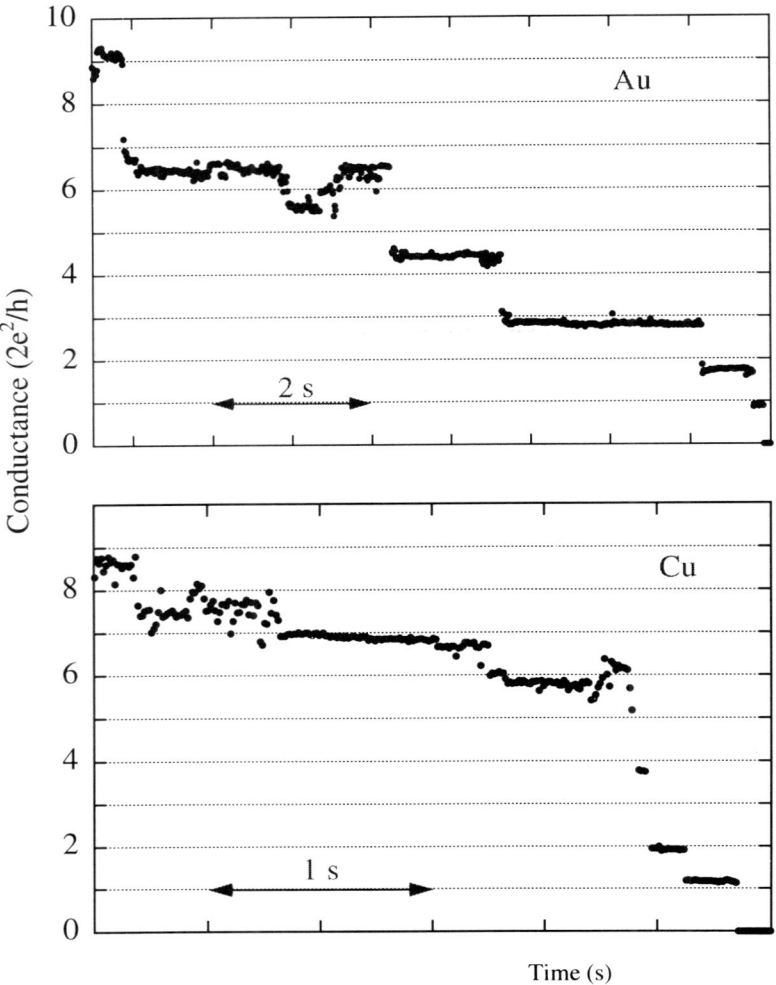

Figure 5.10 These two conductance traces illustrate that the noise level at a plateau increases for larger n.

total conductance will still be n times $2e^2/h$, even though the transmission coefficient for individual incident channels may be fractional [23]. In the presence of larger imperfections, back scattering (transitions between states travelling in opposite directions) may develop. Back scattering generally causes departures from conductance quantization. In the jellium picture, back scattering may result from non-adiabaticity and irregularities in the confining profile of the constriction. In reality, back scattering may also result from defects in the internal atomic structure of the contact. Theoretically, internal disorder has been studied only recently within a tight-binding model [24–27]. None of our traces show conductance quantization in the strict sense, defined above. However certain trace segments come very close to showing the effect, such as the last three plateaus in (a) and (b) in Fig. 5.6. At low temperatures, no comparable horizontal plateaus at integer conductance values larger than $n = 1$ have been observed in individual traces [4,16–18]. Furthermore, the present room temperature histograms show peaks at all integer conductance values for $n < 6$, in contrast to corresponding low temperature results on the same metals. At any temperature, in its mechanical evolution the contact seeks energetically favourable geometries. At high temperatures the higher kinetic energy of atoms enables the contact to explore a wider range of structures in its search for the most favourable geometry. Recent theoretical work [24] shows that this has two important consequences. First, the contact tends to be more uniform in cross-section and shape. Second, internal structural defects created during the evolution of the contact heal more rapidly, improving the crystallinity and internal order in the contact. Both effects improve the conditions for conductance quantization and explain the experimental observation that the quantization occurs more rapidly at elevated temperature.

As a conclusion to this section, our room temperature results show clearer signs of conductance quantization in individual traces than seen in previous measurements. After averaging over many contact configurations the effect at room temperature appears to be different from that at low temperature, especially for $n > 1$, where more peaks are observed at integer conductance values. We also find persistent deviations from the effect in individual traces. These deviations carry important information about the structure of small metallic contacts.

3 Break junctions as tools to investigate conduction through molecules

Charge transport in, and the measurement of the conductance of, single organic molecules is an intriguing, experimentally challenging, and long sought goal. These measurements have been performed on benzene-1,4-dithiolate connected between proximal metallic gold contacts. The metal–molecule–metal configuration presents the molecular embodiment of a system analogous to a quantum dot [28–36], with the potential barriers replaced by the contact barrier of the gold–thiolate endgroups. The results show an apparent Coulomb gap measurable at room temperature. Contrary to previous measurements on molecular systems done by STM [37–39], this approach presents statically stable contacts to the molecule and thus does not involve an STM vacuum tunnelling gap or a gold cluster, making the molecule or surface contacts the dominant resistance of the system and allowing for direct measurement of the molecule conductance. Compared with previous experiments with evaporated metal top contact–molecules–metallic bottom contact configurations in which tens of thousands of parallel molecules are active [40,41], our approach has the advantage that the number of active molecules can be as few as one.

Experiments are conducted at room temperature with the use of the MCB. The main operating principle is shown in Fig. 5.11(a). A glass tubing is assembled onto the substrate and then filled with 1 mM solution of benzene-1,4-dithiol in tetrahydrofuran (THF) under an argon atmosphere [42]. The gold filament is broken inside the solution and the interelectrode distance is adjusted to ~8 Å with the help of previously calibrated piezo voltage change : tunnelling distance change ratio. The benzene-1,4-dithiol molecules are self-assembled onto the two facing gold electrodes. The THF solvent is removed prior to the measurements described below. Due to thermal gradients as the THF is removed, the static dimensional stability of the MCB is disturbed, requiring the tips to be withdrawn, then returned to measure the molecule(s) adsorbed on the surfaces, as schematically shown in Fig. 5.11(b).

Figure 5.12(a) shows the highly repeatable room temperature $I(V)$ and $G(V)$ charac-

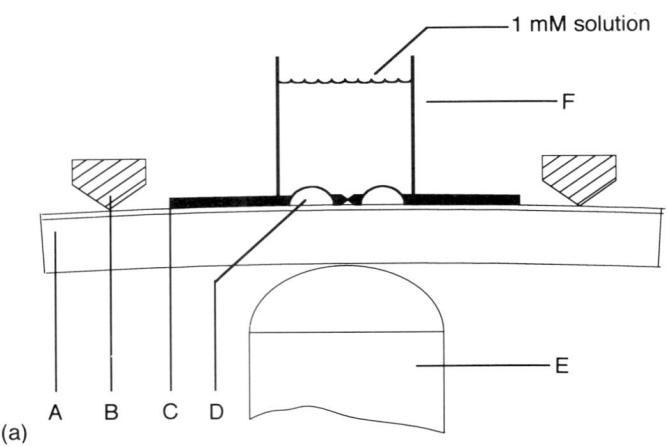

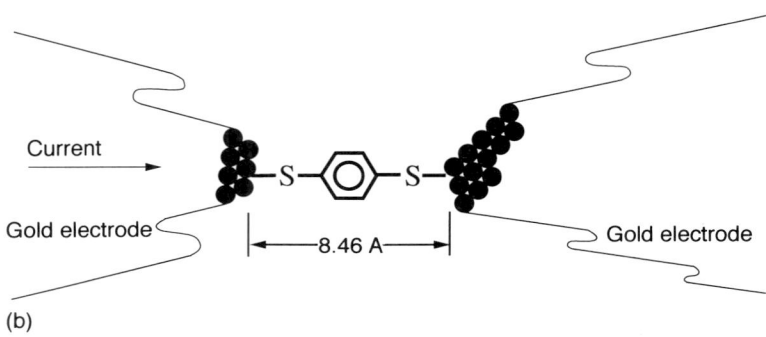

Figure 5.11 (a) A schematic drawing of the set-up for molecule depositions with the bending beam (A), the counter supports (B), the notched gold wire (C), the glue contacts (D), the piezoelement (E), and the glass tube (F) containing the solution. (b) A schematic drawing of a benzene-1,4-dithiolate self-assembled between the two facing gold electrodes formed after breaking the junction.

teristics observed at the onset of conductance as the tips are brought together. The tip spacing is ~8 Å as determined by previous calibration; however, calibration shift due to the solvent evolution cannot be eliminated. This is compared with the approximate molecule length of 8.46 Å. This was calculated using a molecular mechanics (MM2) force field by measuring to the centre of the two gold radii minus the covalent radii of both gold atoms. An apparent Coulomb blockage gap of ~0.7 V is observed in nearly all cases. The first derivative shows two steps in both bias directions with the lower step ~22.2 MΩ (0.045 μS) and the higher step ~13.3 MΩ (0.075 μS), indicative of a Coulomb staircase. It is noted that the high Fermi energy of the gold contacts (~2 eV) compared with the low energies of semiconductor quantum dot systems (< 100 meV) precludes the observation of negative differential resistance here [43], as is often seen in semiconductor systems [28–34]. We also note that a control experiment with unevolved THF solvent alone (i.e. without the benzene-1,4-dithiolate) exhibited a resistance of 1–2 GΩ (linear up to 10 V) independent of electrode spacing; and when evolved, regular vacuum tunnelling with much higher resistance.

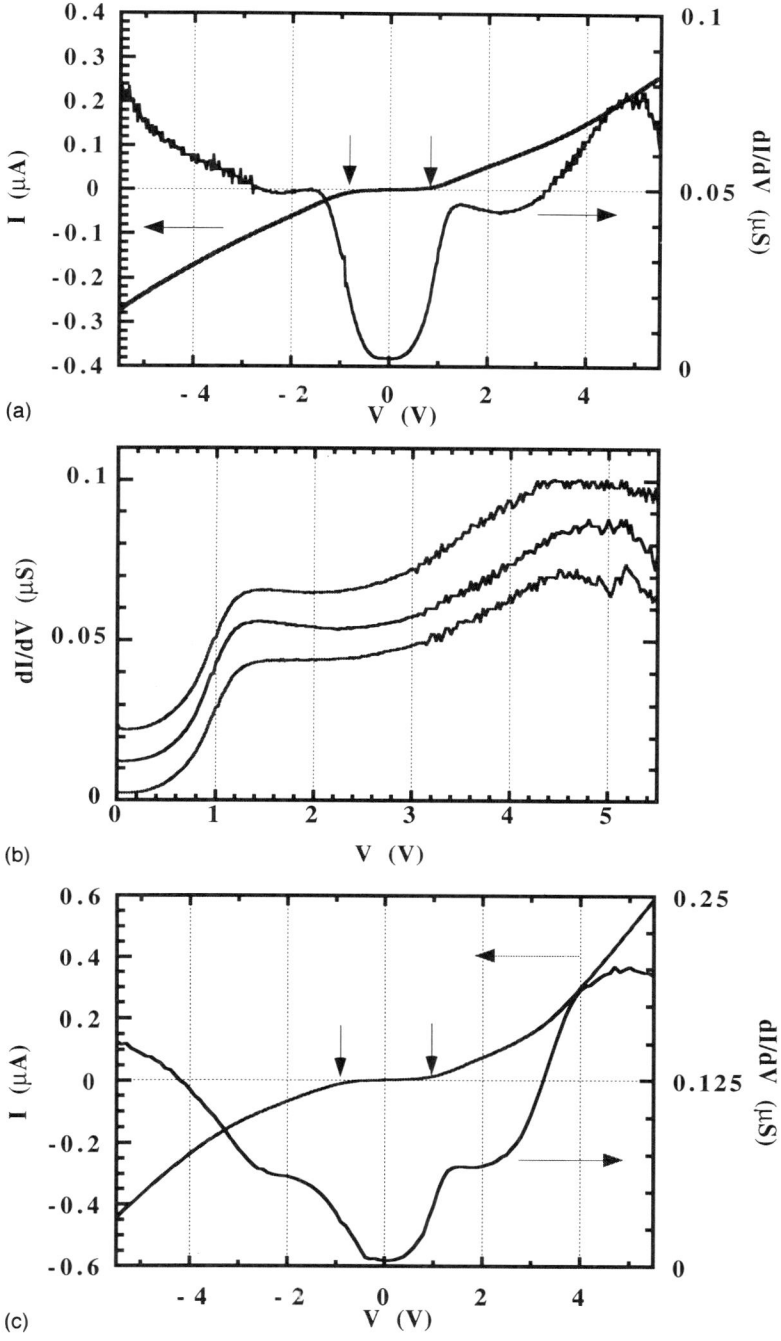

Figure 5.12 (a) Typical $I(V)$ characteristics which illustrate a Coulomb blockade gap of 0.7 V, and the first derivative $G(V)$ which shows a step-like structure. (b) Three independent $G(V)$ measurements, offset for clarity, illustrating reproducibility of the conductance values. The middle curve is the same data as in (a). (c) An $I(V)$ and $G(V)$ measurement illustrating conductance values approximately twice that of the observed minimum conductance values.

Figure 5.12(b) shows the repeatability of independent measurements (tips/oligomer brought into contact on each measurement) on this system, illustrating the reproducibility of the results even with arbitrarily different though atomically sharp contacts. Offsets of 0.01 μS for the middle curve and 0.02 μS for the top curve are used for clarity. The first Coulomb step for these three measurements gives values of 22.2, 22.2, and 22.7 MΩ (top to bottom); the next step gives values of 12.5, 13.3, and 14.3 MΩ, a slightly less constant value (the middle curve is the same data as in Fig. 5.12a). This is compared with a value of 9 MΩ [38] and 18 ± 12 MΩ [39] derived from previous results on an ensemble of similar molecules contacted to a gold nanocrystal. A value greater than ~22 MΩ was not observed in our measurements; however, values less than this maximum were occasionally observed. Figure 5.12(c) shows $I(V)$ and $G(V)$ measurements of one observation that gives a value of ~14 MΩ for the first step, and 7.14 MΩ (negative bias) and 5 MΩ (positive bias) for the second step, which are approximately half the value of the maximum resistance values. These values, and the fact that the apparent gap is the same (as expected for identical yet independent Coulomb blockade paths), suggests a configuration of two non-interacting self-assembled molecules in parallel. This further substantiates that the threshold resistance of a single molecule is ~22 MΩ.

The apparent Coulomb gap may be an overestimation of the real gap, as the alignment of the Au contact Fermi level to the HOMO–LUMO gap of the molecule is not known. Naively using the apparent Coulomb gap, an experimental capacitance of 1.1×10^{-19} F is obtained. Although the charge transport through the molecule is in principle a many-body effect, as a first step we can estimate the capacitance of the aryl group with a crude model: a 4.5 Å metallic sphere bound 2.0 Å from proximal metallic planes with intervening vacuum barriers gives a capacitance of 0.4×10^{-19} F, very close to the experimentally derived 1.1×10^{-19} F. A lower limit of the gap can be estimated from the requirement that $E_{gap}(\text{Coulomb}) \gg kT$; assuming $10\,kT$, this would infer a E_{Fermi}–LUMO gap of 0.45 eV.

In summary, we present measurements on a stable contacted molecular system which exhibits a well-defined resistance and apparent room temperature Coulomb blockade. The definitive demonstration of Coulomb blockade requires a third gate electrode, not possible in the present configuration. The number of active molecules could be as few as one. A better theoretical understanding of the threshold resistance of this system, the apparent Coulomb gap derived from the capacitance of a single molecule configuration, and the determination of the contact Fermi level–LUMO gap alignment, is needed to compare to the experimental values of ~22 MΩ and ~0.7 V, respectively.

Despite the above-mentioned success, the geometrical parameters of the conventional break junctions vary from junction to junction because they are made by hand, and further improvement of the stability should improve the reliability of this system and enable wider applications, which can be achieved by shrinking the junction size. As an improved version, silicon micromachined break junctions have been demonstrated and are described in the next section.

4   Microfabrication of a mechanically controllable break junction in silicon

Micromachining in silicon is an ongoing effort to provide ever smaller devices for the active part of a sensor. Currently, it is straightforward to produce suspended beams, small springs, and vibrating or rotating structures on a chip. A tunnel transducer (e.g. an STM) [44] is compatible with further miniaturization and possesses an astonishing sensitivity to displacements. When a vacuum tunnel gap between two metallic electrodes is increased by 1 Å, the tunnel resistance increases approximately by an order of magnitude. This has been realized by a number of groups who have used tunnel sensors in devices. The extreme sensitivity of these sensors on position displacements however implies that the practical range of operation is limited to distances smaller than 5 Å since at larger distances the resistance becomes almost infinite and unmeasurable.

In conventional STM embodiments, one electrode is usually mounted on a flexible lever, which can be moved by an electrical signal. The tunnel gap is kept constant with the use of a feedback system, necessary since temperature fluctuations, (acoustic) vibrations or other disturbances will otherwise change the vacuum gap over distances much larger than the practical range. An accelerometer, magnetometer, and an infrared sensor [45,46] and integrated STM [47,48] have been successfully developed with these kind of tunnel sensors in feedback operation. Despite these successes we have used a different approach and constructed an inherently stable tunnel sensor. In order to improve the static stability of conventional MCBs further, we fabricated 100 times smaller junctions in silicon in a way that the electrode separation during operation remains in the practical range of about 5 Å. Due to the extreme stability of this device it can be operated without feedback; however it may also be used in a feedback loop. It is inherently stable, adjustable, and compatible with silicon technology. Detailed measurements are shown, in both the contact and tunnel regimes.

The principle of operation and a schematic perspective and cross-sectional view of the device are shown in Fig. 5.13. The starting material is a ⟨100⟩ oriented 250-μm-thick silicon wafer with an oxide layer of 400 nm. Standard electron beam lithography is used to define a pattern in a polymethyl methacrylate (PMMA) bilayer used for the evaporation of an adhesion layer (10 Å Ti) and 800 Å of gold onto the oxide. The gold film has a shape as indicated in Fig. 5.13(a). Next a photolithographically defined thick layer of aluminium is evaporated everywhere on the oxide except over a distance u, centred around the smallest gold feature. The next step uses the gold and aluminium films as a mask to etch through the SiO_2 into the Si with a CF_4/O_2 plasma (Fig. 5.13b). The aluminium is then removed using a standard wet etch. The last step is a wet etch of the exposed Si area using a pyrocatechol–ethylene–diamine mixture [49]. Since the two cantilevers are aligned with the ⟨110⟩ direction in the substrate, a triangular pit is etched into the silicon, bounded by the SiO_2 edges and the ⟨111⟩ surfaces. Rapid undercutting at the convex corners [49] by this etchant assures that the two cantilevers are free standing after the etching process. The final device consists of two small cantilever beams (2.5 μm long, 4 μm wide) connected with a 100-nm-wide wire over a length L_{eff} (Fig. 5.13c).

The device is mounted against two counter supports, approximately 20 mm apart, in a break junction configuration. A force is exerted on the backside via the piezo element which is moved towards the device using a course adjustment screw (Fig. 5.13d). The silicon beam is strained, resulting in an elongation of the top layer. The elongation of u is concentrated on L_{eff} resulting in the fracture of the gold wire while the Si substrate stays intact (even though gold is more ductile than silicon). The piezo element has a

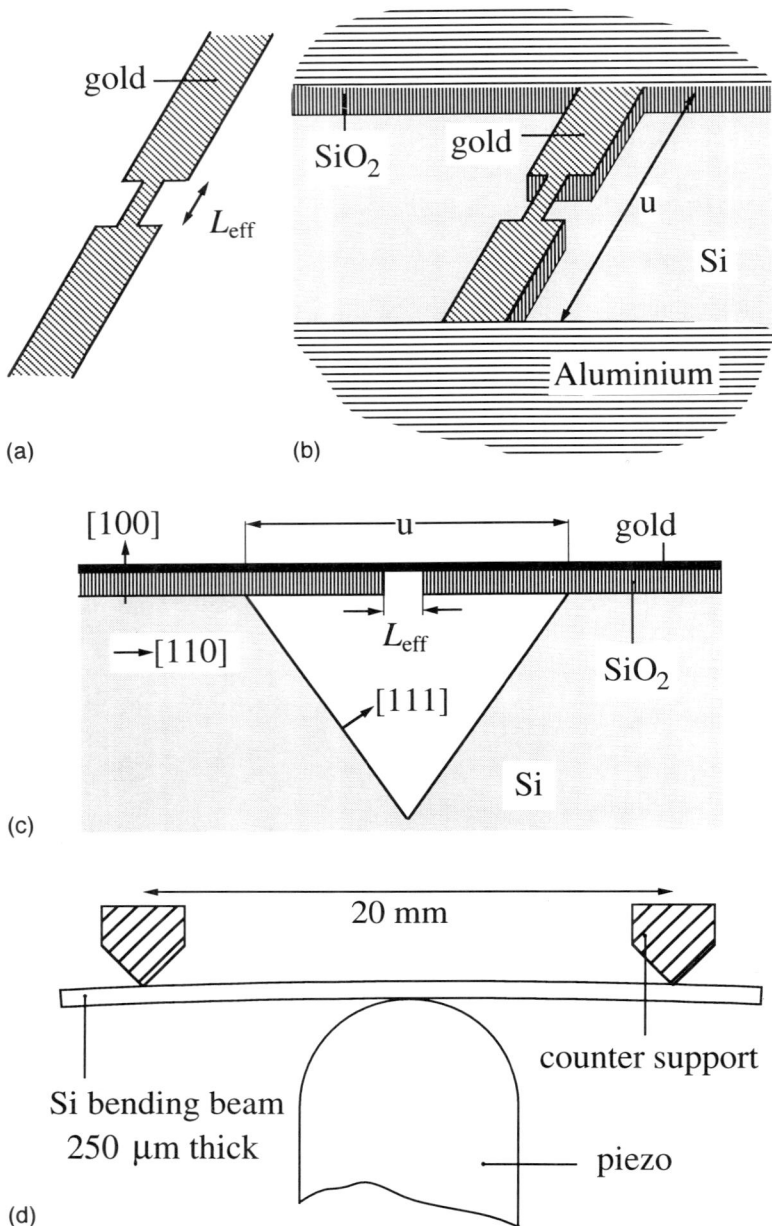

Figure 5.13 (a) The gold wire defined by e-beam lithography. The smallest width of the wire is 100 nm, L_{eff} is about 200 nm. (b) Both the aluminium and gold film are used as an etching mask to etch through the oxide into Si. (c) A cross-section along the gold wire after the pit is etched into the silicon. Silicon etching is stopped at the concave corners and the intersection between the $\langle 111 \rangle$ crystallographic surface and the SiO_2 edges. (d) The mounting configuration of the silicon bending beam in a break junction set-up.

maximum elongation of 5 µm and is used for fine adjustment of either atomic size contacts or vacuum barrier tunnel junctions between the fractured gold electrodes. Figure 5.14 shows an SEM photograph of a device before the bridging wire is broken. A slight undercut of the SiO_2 below the gold cantilevers can be seen, and the 100-nm-wide bridging wire is totally undercut. The etched pit into the Si (Fig. 5.14a) is bounded by a

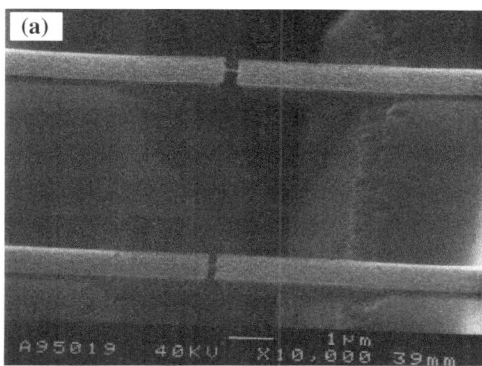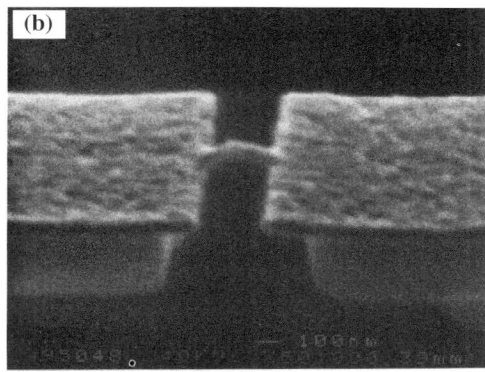

Figure 5.14 (a) Scanning electron microscope (SEM) images of two devices suspended above a triangular pit in the Si substrate before the connecting wire is broken in the break junction set-up. Each device shows two SiO$_2$ cantilevers which are covered by gold and bridged by a small gold wire. (b) A magnified view of the connecting wire prior to breaking.

relatively rough SiO$_2$ edge, caused by the photolithography step. Some of the undercut below the SiO$_2$ layer results from this roughness and enlarges u to about 10 μm.

Experiments are performed at room temperature in the vacuum system (10^{-7} Torr) described above. Figure 5.15 illustrates the long-term stability and the exponential dependence of the tunnel current I_t on the vacuum barrier gap distance of this device. The junction is biased at 100 mV while a triangular voltage wave is applied to the piezo element (lower curve in Fig. 5.15). The variation in the piezo length induces a variation in the gap distance resulting in a change of the tunnel resistance (top curve in Fig. 5.3). The exponential dependence of I_t on the gap distance s is given by $I_t \propto \exp - \alpha\sqrt{\phi}s$ with $\alpha = 1.025$ Å^{-1} eV$^{-1/2}$ and ϕ the work function of the gold electrodes. As the electrodes are displaced over about 2 Å the tunnel current changes over almost two

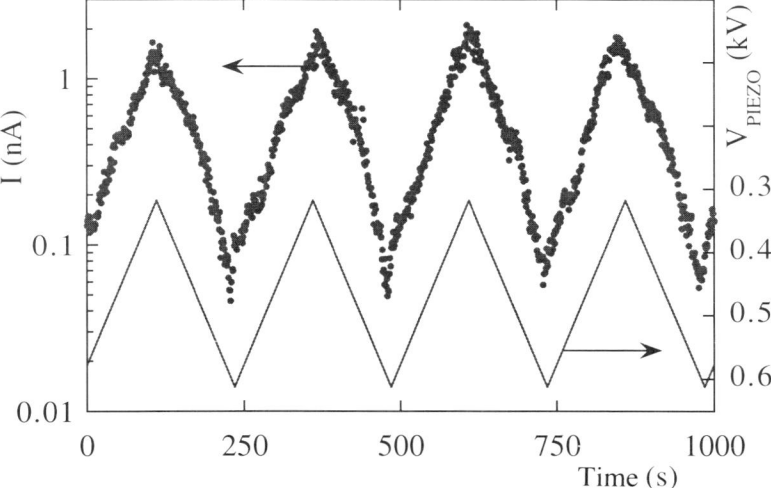

Figure 5.15 The piezo voltage is changed in a triangular wave (the lower curve). The almost linear behaviour of the tunnel current on a logarithmic scale reflects the exponential dependence on electrode separation. Note the large time scale, indicating the long-term stability of the junction.

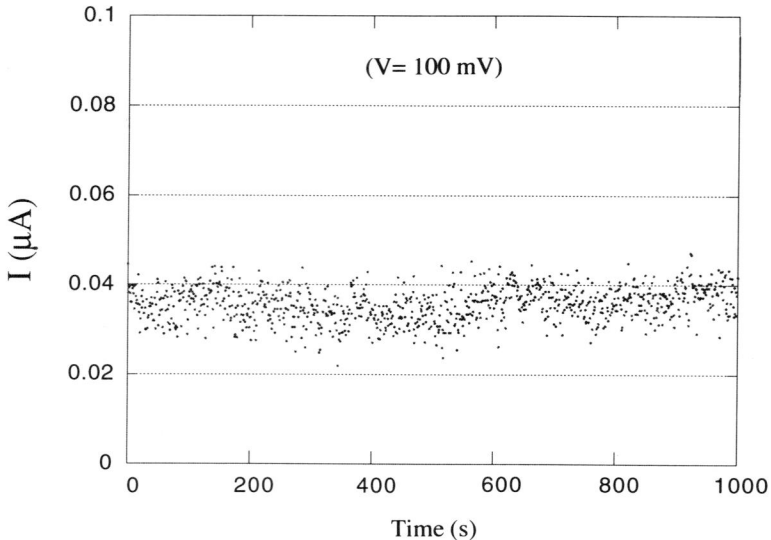

Figure 5.16 Tunnel current plotted vs time with fixed piezo voltage. The bias voltage is 100 mV. The resistance is about 3.3×10^6 Ω. The current noise amplitude implies about 3 pm fluctuations in the tunnel gap distance.

orders of magnitude. The reason for this exceptional stability is the smallness of u which determines the reduction factor (the ratio between the piezo elongation and the induced electrode separation). From the known piezo elongation and assuming an exponential dependence of the tunnel current with $\phi = 4$ eV, we can calculate that our micromachined break junctions have a reduction factor of 10 000, which is about two orders higher than that of conventional MCBs. In the tunnel regime the current noise amplitude, which depends on the tunnel resistance, is determined at a 100 mV bias for tunnel resistances between 100 kΩ and 10 MΩ with a 1 kHz bandwidth as shown in Fig. 5.16. In this resistance range the experimental value for the current noise amplitude is measured for 1000 s and implies about 3 pm fluctuations in the tunnel gap distance. Although we do not know the exact origin of these fluctuations, a detailed noise analysis should include the thermal agitation of the cantilever. When the electrodes are brought close enough together, a contact is formed. Conductance quantization is observed, shown in Fig. 5.17, similar to that with conventional MCBs. However, during the breaking process, if we fix the piezo voltage, conductance will stay at a plateau without the presence of spontaneous relaxation. The traces in Fig. 5.17 were taken while adjusting the piezo voltage to reduce the contact size. Molecule depositions and measurements are in progress with these new junctions.

5 Conclusion

In conclusion, we have presented conductance quantization effects with both conventional and silicon micromachined break junctions. Since the Fermi wavelength of a metallic system is of the order of atomic radius, the last conductance plateau before tunnelling corresponds to a single atom bridged between two metallic leads. Molecule depositions of benzene-1,4-dithiol have been demonstrated to have a profound influ-

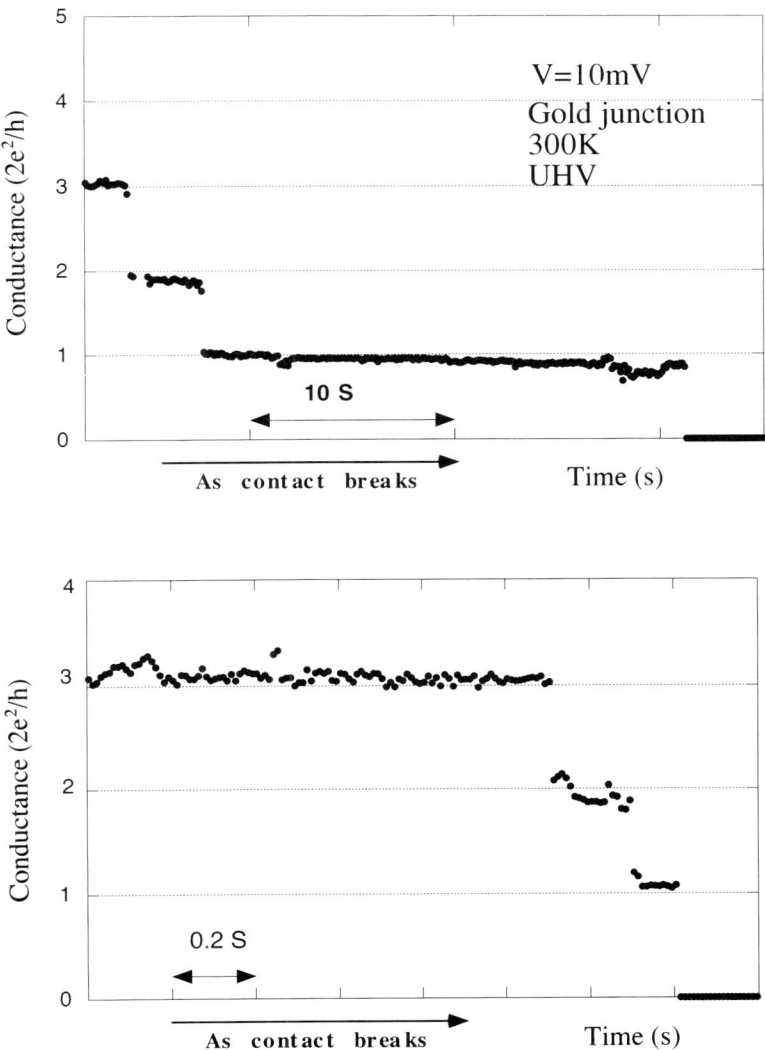

Figure 5.17 Two conductance traces recorded with an atomic-scale contact, with the cross-section reduced as a function of time. Conductance plateaus are found to be near multiples of the fundamental conductance unit.

ence on the *I–V* characteristics of atomic-sized structures: $I(V)$ measurements at room temperature demonstrate a highly reproducible apparent Coulomb blockade gap ~0.7 V. The $G(V)$ curve shows two steps in both bias directions with the first step ~0.045 μS (22 MΩ) and the second step ~0.075 μS (13.3 MΩ). The number of conducting molecules can be as few as one. In the improved version, the silicon micromachined break junctions have been demonstrated to be ultra-stable and highly adjustable. The tunnelling distance can be adjusted to increase or decrease by 1 Å with 3 pm stability without a feedback system.

Acknowledgement. We acknowledge DARPA for financial support. We are grateful to L. Jones II and D.R. Lombardi for their help in molecule deposition. We thank J.M. Krans for his assistance in the study of conductance quantization with conventional MCBs at room temperature. We acknowledge M.R. Deshpande and J.W. Sleight for their help in e-beam lithography and B.J. Vleeming for his assistance in building the high vacuum system.

6 References

1 Muller CJ. Ph D thesis, University of Leiden, 1991.

2 Muller CJ, van Ruitenbeek JM, de Jongh LJ. *Physica C* 1992; **191**: 485.

3 Moreland J, Ekin JW. *J Appl Phys* 1985; **58**: 3888.

4 Muller CJ, van Ruitenbeek JM, de Jongh LJ. *Phys Rev Lett* 1992; **69**: 140.

5 Vleeming BJ, Muller CJ, Koops MC, de Bruyn Ouboter R. *Phys Rev B* 1994; **50**: 16741.

6 van der Post N, Peters ET, Yanson IK, van Ruitenbeek JM. *Phys Rev Lett* 1994; **73**: 2611.

7 Yanson IK, Fisum VV, Hesper R *et al. Phys Rev Lett* 1995; **74**: 302.

8 Landauer R. *IBM J Res Dev* 1957; **1**: 223.

9 van Wees BJ, van Houten H, Beenakker CWJ *et al. Phys Rev Lett* 1988; **60**: 848.

10 Wharam DA, Thornton TJ, Newbury R *et al. J Phys C* 1988; **21**: L209.

11 Agrait N, Rodrigo JG, Vieira S. *Phys Rev B* 1993; **47**: 12345.

12 Pascual JI, Mendez J, Gomez-Herrero J, Baro AM, Garcia N, Vu Thien Binh. *Phys Rev Lett* 1993; **71**: 1852.

13 Pascual JI, Mendez J, Gomez-Herrero J *et al. Science* 1995; **267**: 1793.

14 Oleson L, Lœgsgaard E, Stensgaard I *et al. Phys Rev Lett* 1994; **72**: 2251.

15 Oleson L, Lœgsgaard E, Stensgaard I *et al. Phys Rev Lett* 1995; **74**: 2147.

16 Krans JM, Muller CJ, Yanson IK, Govaert Th CM, Hesper R, van Ruitenbeek JM. *Phys Rev B* 1993; **48**: 14721.

17 Krans JM, Muller CJ, van der Post N *et al. Phys Rev Lett* 1995; **74**: 2146.

18 Krans JM, van Ruitenbeek JM, Fisun VV, Yanson IK, de Jongh LJ. *Nature* 1995; **375**: 767.

19 Bogachek EN, Zagoskin AN, Kulik IO. *Sov J Low Temp Phys* 1990; **16**: 796.

20 Krans JM. PhD thesis, University of Leiden, 1994.

21 Zagoskin AM, Kulik IO. *Sov J Low Temp Phys* 1990; **16**: 533.

22 Torres JA, Pascual JI, Saenz JJ. *Phys Rev B* 1994; **49**: 16581.

23 Butcher PN, McInnes JA. *J Phys Cond Matt* 1995; **7**: 745.

24 Bratkovsky AM, Sutton AP, Todorov TN. *Phys Rev B* 1995; **52**: 5036.

25 Todorov TN, Sutton AP. *Phys Rev Lett* 1993; **70**: 2138.

26 Mujica V, Kemp M, Ratner M. *Abstr Am Chem Soc* 1993; **206**: 134.

27 Kemp M, Mujica V, Ratner M. *Abstr Am Chem Soc* 1993; **206**: 135.

28 Reed MA, Randall JN, Aggarwal RJ, Matyi RJ, Moore TM, Wetsel AE. *Phys Rev Lett* 1988; **60**: 535.

29 Meirav U, Kastner M, Wind S. *Phys Rev Lett* 1990; **65**: 771.

30 Dellow MW, Beton PH, Langerak CJGM *et al. Phys Rev Lett* 1992; **68**: 1754.

31 Kouwenhoven LP, Johnson AT, van der Vaart NC, Harmas CJPM, Foxon CT. *Phys Rev Lett* 1991; **67**: 1626.

32 Tewordt M, Martín-Moreno L, Law VJ *et al. Phys Rev B* 1992; **46**: 3948.

33 Su B, Goldman V, Cunningham J. *Phys Rev B* 1992; **46**: 7644.

34 Klein DL, McEuen PL, Katari JEB, Roth R, Alivisatos AP. *Appl Phys Lett* 1996; **68**: 2574.

35 Black CT, Ralph DC, Tinkham M. *Phys Rev Lett* 1996; **76**: 688.

36 Grabert H, Martinis JM, Devoret MH eds. *Single Charge Tunneling.* New York: Plenum, 1991.

37 Bumm LA, Arnold JJ, Cygan MT *et al. Science* 1996; **271**: 1705.

38 Dorogi M, Gomez J, Osifchin R, Andres RP, Reifenberger R. *Phys Rev B* 1995; **52**: 9071.

39 Andres RP, Bein T, Dorogi M *et al. Science* 1996; **272**: 1323.

40 Fischer CM, Burghard M, Roth S, von Klitzing K. *Appl Phys Lett* 1995; **66**: 3331.

41 Grabert H. *Z Phys B* 1991; **85**: 319.

42 Tour JM, Jones II L, Pearson DL *et al. J Am Chem Soc* 1995; **117**: 9529.

43 Mujica V, Kemp M, Roitberg A, Ratner M. *J Chem Phys* 1996; **104**: 7296.
44 Bocko MF, Stephenson KA, Koch RH. *Phys Rev Lett* 1988; **61**: 726.
45 Kenny TW, Waltman SB, Reynolds JK, Kaiser WJ. *Appl Phys Lett* 1991; **58**: 100.
46 Rockstad HK, Kenny TW, Reynolds JK, Kaiser WJ, Gabrialson Th B. *Sens Actuators A* 1994; **43**: 107.
47 Lutwuche MI, Wada Y. *Appl Phys Lett* 1995; **66**: 2807.
48 Xu Y, Mcdonald NC, Miller SA. *Appl Phys Lett* 1995; **67**: 2305.
49 Peterson KE. *IEEE Trans Electron Devices* 1978; **ED-25**: 124.

6

Electron Hopping Transport in Electrochemically Active, Molecular Mixed Valent Materials

R.H. TERRILL* and R.W. MURRAY†

*Department of Chemistry, University of Illinois, Urbana, IL 61820, USA

†Kenan Laboratories of Chemistry, University of North Carolina, Chapel Hill, NC 27599-3290, USA

1   Introduction

Molecular electronics covers a huge arena of chemical phenomena that transduce a molecular property into an electrical or optical one of potential usefulness. It is not really a new subject. Perhaps the oldest example of an artificial molecular electronics device involving electrochemistry is a battery; the oldest batteries were constructed mainly from compounds of inorganic elements while batteries made of substantially organic molecular materials have become part of the contemporary battery field. In both old and new electrochemical versions of molecular electronics, electron transport through some bulk solid or semi-solid phase is part of the overall energy transduction phenomenon.

Molecular materials in which facile electron transport occurs fall into two general categories: (i) materials with band electronic structures that exhibit electronic delocalization over considerable distances, so-called conducting polymers; and (ii) materials composed of distinct and spatially well defined and constrained electron donor–acceptor (D–A) or redox sites between which transport occurs by electron hopping. There can be intermediate situations, such as the common one in which a delocalized material's electron transport is dominated by hopping across defects in the delocalized structure.

This article will confine itself to redox materials, category (ii). Redox materials exhibit electronic conductivities less dramatically large than those of delocalized, conducting polymers. Polymer films used in electrophotography [1] contain localized electronic state donors or acceptors and comprise a large body of technologically important category (ii) materials. Our focus in this article will be on those category (ii) materials where studies of electrochemical reactivity have been done in parallel with those on solid state electron transport, and on the measurement tools used for their investigation.

A premier reason for investigating solid state D–A electron transport is the window that it provides on electron self-exchange reaction dynamics in the solid state. Equation 6.1 illustrates how electron hopping in a mixed valent D–A solid is equivalent to an electron self-exchange reaction.

$$D_{position\ X} + A_{position\ Y} \rightarrow A_{position\ X} + D_{position\ Y} \tag{6.1}$$

where k_{ex} is the bimolecular electron self-exchange rate constant ($M^{-1}\ s^{-1}$) for the D–A reaction pair. A huge body of knowledge and understanding exists about electron self-exchanges between donors and acceptors in solutions, but in the solid and semi-solid state, both data and the extent of understanding about electron self-exchanges are, in comparison, primitive. Consider, for example, electron self-exchange between ferrocene and ferrocenium (Fc/Fc^+) in a fluid solution. Much is known about this reaction [2–4] with regards to outer and inner sphere barriers, solvent dynamics control, etc., but for the same reaction in a solid, relatively little is known about such microscopic molecular factors or what changes have occurred to make the Fc/Fc^+ reaction slower (as it is; see Table 6.1) in the solid-state version.

Our laboratory over the past several years [5–17] has worked to develop and verify experimental methodologies for measuring k_{ex} in solid-state molecular materials (such as polyvinylferrocene). The emphasis has been on redox sites that have also been investigated as monomers in fluid solution, and whose properties in polymeric forms are amenable to electrochemical investigation. Both (i) the establishment of analytical methods and (ii) the choice of redox materials for which a comparison between monomer, polymer and solid state forms is possible are preludes to exploring microscopic factors that control solid-state electron self-exchange dynamics. This article will focus primarily on experimental methodologies, illustrated with some specific results, and will conclude with a description of a molecular electronics device recently fashioned from electron-hopping materials.

Electrically conductive, localized electronic state materials (references are to examples) may be classified as:

1 polymeric versions of redox sites [18–20];
2 polymers which contain covalently attached pendant redox sites [16];
3 mixed valent, inorganic crystals and polycrystalline pressed pellets [21];
4 ion exchange polymers with redox counterions [15];
5 polymers with dissolved redox monomers [68];
6 polymer electrolytes with dissolved redox monomers or polymers [22];
7 melts of redox monomers (redox hybrid poly-ethers) [23–26];

Figure 6.1 gives examples that will be used later in illustrations of experiments. Section 2 deals with materials in which the covalent bonding, crystal packing, etc., prevents physical diffusion of redox sites from occurring over macroscopic distances, and with the methodologies used to study them. Section 2 applies to the first three of the above classes of materials, and to the others in circumstances of negligible levels of physical diffusivity. For convenience, we will call these collectively redox polymers. In the latter four classes of materials, physical diffusion of the redox sites is in principle possible, since they are not covalently attached to a lattice. Whether such physical diffusion is actually significant depends on the material and the experiment. For example, physical diffusion of the redox counterion in the ion exchange polymer Nafion is important when the Nafion is water-swollen [27], but not so when the Nafion is rigorously dry [15]. Physical diffusion occurs in melts of redox hybrid polyethers but presumably not below their glassing temperatures. The experimental methods used to measure electron self-exchange reaction dynamics when physical diffusion is important are taken up in Section 3.

2   Redox polymers and experimental methodologies

In redox polymers, the donor and acceptor sites are covalently bonded to a lattice such that the molecular sites have no macroscopic (meaning many multiples of polymer or

Table 6.1 Summary of solid- and semi-solid-state electron self-exchange rate constants.

Compound	Environment	Technique/geometry*	Redox couple	k_{ex} (M^{-1} s^{-1})†	E_A (kcal)	Ref.
Viologen pyrrole polymer‡	Wet, CH_3CN/Bu_4NClO_4	dcdx/sw	$2^+/1^+$	8.4×10^3	12	[9]
Viologen pyrrole polymer‡	Wet, CH_3CN/Bu_4NClO_4	dcdx/sw	$1^+/0$	1.6×10^5	9	[9]
Viologen pyrrole polymer‡	Vapour, Ch_3CN	dcdx/sw	$1^+/0$	2.7×10^5	—	[9]
Viologen pyrrole polymer‡	Dry, N_2	dedx/sw	$1^+/0$	1.1×10^5	—	[9]
Me viologen monomer	Dilute solution	—	$2^+/1^+$	8.4×10^8	—	[71]
Viologen sily/polymer‖	Wet, H_2O/electrol.	dcdx/of/mediat.	$2^+/1^+$	2.3×10^5	—	[72]
Viologen sily/polymer‖	Wet, H_2O/electrol.	dcdx/of/mediat.	$1^+/0$	2.3×10^6	—	[72]
Viologen pyrrole polymer¶	Wet, CH_3CN/electrol.	dcdx/ida	$2^+/1^+$	2.9×10^5	10	[73]
Viologen EO polymer§	Wet, CH_3CN, THF	dcdx/ida	$2^+/1^+$	$11 \pm 7 \times 10^6$	—	[77]
Viologen EO polymer§	Wet, CH_3CN, THF	dcdx/ida	$1^+/0$	$9 \pm 6 \times 10^7$	—	[77]
Viologen EO polymer§	Dry, N_2	dedx/ida	$2^+/1^+$	$8 \pm 4 \times 10^6$	—	[77]
Ferrocene monomer	Wet, Ch_3CN/electrol.	—	$0/1^+$	9×10^6	—	[3]
Polyvinylferrocene	Wet, CH_3CN/electrol.	dcdx/ida(A)**	$0/1^+$	$1.6 \pm 1 \times 10^6$	—	[65]
Polyvinylferrocene	Dry, N_2	dcdx/ida(A)**††	$0/1^+$	2.8×10^5	—	[65]
Polyvinylferrocene	Dry, N_2	dcdx/ida(A)**	$0/1^+$	$3.7 \pm 1 \times 10^5$	—	[65]
Polyvinylferrocene	Dry, N_2	dcdx/ida(B)**	$0/1^+$	$5.8 \pm 2 \times 10^4$	—	[65]
Polyvinylferrocene	Dry, N_2	dedx/ida(B)**‡‡	$0/1^+$	$1.7 \pm 0.4 \times 10^5$	—	[65]
$[Os(vbpy)_3]$ monomer	Wet, Ch_3CN/electrol.	—	$2^+/3^+$	2.2×10^7	—	[75]
$Poly[Os(vbpy)_3](BF_4)_{2.5}$	Wet, CH_2Cl_2	dcdx/sw	$2^+/3^+$	$1.0 \pm 0.1 \times 10^6$	—	[14]
$Poly[Os(vbpy)_3](BF_4)_{2.5}$	Vapour, CH_2Cl_2/electrol.	dcdx/sw	$2^+/3^+$	$6.2 \pm 1.2 \times 10^5$	—	[14]
$Poly[Os(vbpy)_3](BF_4)_{2.5}$	Dry, N_2	dcdx/sw	$2^+/3^+$	$3.0 \pm 0.8 \times 10^5$	—	[14]
$Poly[Os(vbpy)_3](BF_4)_{2.5}$	Dry, N_2	dedx/sw§§	$2^+/3^+$	$4 \pm 2 \times 10^5$	—	[14]
$Poly[Os(vbpy)_3](BF_4)_{2.5}$	Dry, N_2	dedx/sw‖‖	$2^+/3^+$	3.9×10^5	9.1	[1]
$Poly[Os(bpy)_2(vpy)_2](BF_4)_{2.5}$	Dry, N_2	dedx/sw/ac	$2^+/3^+$	4.2×10^5	8.3	[10]
$Poly[Os(bpy)_2(vpy)_2](ClO_4)_{2.5}$	Wet, CH_3CN/electrol.	dcdx/sw	$2^+/3^+$	$2 \pm 1 \times 10^6$	—	[7]
$Poly[Os(bpy)_2(vpy)_2](ClO_4)_{2.5}$	Dry, N_2	dcdx/sw	$2^+/3^+$	$1 \pm 0.4 \times 10^6$	—	[7]
$Poly[Os(bpy)_2(vpy)_2](ClO_4)_{2.5}$	Dry, N_2	dedx/sw¶¶	$2^+/3^+$	$8.2 \pm 4 \times 10^5$	8.0 ± 0.7	[7]
$Poly[Os(bpy)_2(vpy)_2](ClO_4)_{2.5}$	Vapour, CH_3CN	dcdx/sw	$2^+/3^+$	$3.5 \pm 1.5 \times 10^6$‡‡	—	[5]
$Poly[Os(bpy)_2(vpy)_2](ClO_4)_{1.5}$	Vapour, CH_3CN	dcdx/sw	$2^+/1^+$	$5.4 \pm 2.7 \times 10^7$‡‡	—	[5]
$Poly[Os(bpy)_2(vpy)_2](ClO_4)_{0.5}$	Vapour, CH_3CN	dcdx/sw	$1^+/0$	$5.8 \pm 2.9 \times 10^8$‡‡	—	[5]
$Poly[Os(bpy)_2(vpy)_2](ClO_4)_{2.5}$	Dry, N_2	dcdx/sw	$2^+/3^+$	$1.2 \pm 0.6 \times 10^6$‡‡	—	[5]
$Poly[Os(bpy)_2(vpy)_2](ClO_4)_{1.5}$	Dry, N_2	dcdx/sw	$2^+/1$	$4.7 \pm 2.3 \times 10^8$‡‡	—	[5]
$Poly[Os(bpy)_2(vpy)_2](ClO_4)_{2.5}$	Wet, CH_3CN/electrol.	dcdx/sw	$2^+/3^+$	$4.7 \pm 1.9 \times 10^6$‡‡	—	[5]
$Poly[Os(bpy)_2(vpy)_2](ClO_4)_{1.5}$	Wet, CH_3CN/electrol.	dcdx/sw	$2^+/1^+$	$2.3 \pm 0.7 \times 10^7$‡‡	—	[5]
$Poly[Os(bpy)_2(vpy)_2](ClO_4)_{0.5}$	Wet, CH_3CN/electrol.	dcdx/sw	$1^+/0$	$3.7 \pm 0.6 \times 10^8$‡‡	—	[5]
$Poly[Os(bpy)_2(vpy)_2](Tos)_{2.5}$	Wet, CH_3CN/electrol.	dcdx/sw	$2^+/3^+$	$6 \pm 2 \times 10^5$‡‡	—	[5]
$Poly[Os(bpy)_2(vpy)_2](Tos)_{1.5}$	Wet, CH_3CN/electrol.	dcdx/sw	$2^+/1^+$	$2.5 \pm 1.5 \times 10^6$‡‡	—	[5]
$Poly[Fe(o\text{-}NH_2)TPP]$§§	Wet, CH_3CN/Py/electrol.	dcdx/sw	$Fe(2^+/3^+)$	2×10^6‖‖	—	[6]
$Poly[Fe(o\text{-}NH_2)TPP]$§§	Vapour, Py/CH_3CN vapour	dcdx/sw	$Fe(2^+/3^+)$	7×10^5‖‖	—	[6]
Nafion-$[Os(bpy)_3]$ in H_2O	Wet, H_2O/electrol.	dcdx/of/mediat.	$2^+/3^+$	4.2×10^4	—	[27]
Nafion-$[Os(bpy)_3]$, dry	Dry, N_2	dedx/ida^i	$2^+/3^+$	$1.1 \pm 1 \times 10^4$	11.2	[15]
Tri(p-methylphenyl)amine	CH_2Cl_2	dedx/ida	$0/1^+$	1.6×10^7	—	[76]
Polytetraphenylbenzidine¶¶,[a]	Wet, CH_3CN/electrol.	dcdx/ida	$0/1^+$	4.5×10^6	7.1	[19]
Polytetraphenylbenzidine¶¶,[b]	Wet, CH_3CN/electrol.	dcdx/ida	$0/1^+$	1.1×10^9	3.9	[19]
Polytetraphenylbenzidine¶¶,[a]	Dry, N_2	TOF/sw	$0/1^+$	$\approx$ above	10–12	[19]
Polytetraphenylbenzidine¶¶,[a]	Vapour, CH_3CN	dcdx/ida	$0/1^+$	7×10^6	12.7	[20]
Polytetraphenylbenzidine¶¶,[a]	Dry, N_2	TOF/sw	$0/1^+$	3.1×10^7	12.2	[20]

Continued on p. 218

Table 6.1 (*Continued*).

Compound	Environment	Technique/geometry*	Redox couple	k_{ex} (M^{-1} s^{-1})†	E_A (kcal)	Ref.
$KMnO_4$–K_2MnO_4[c]	PTFE contacts, N_2	dedx/sw/ac	$Mn^{VI/VII}$	$\sim 1 \times 10^5$ s^{-1}	12 ± 1	[21]
$KFe^{2+}Fe(II)(CN)_6^{4-}$[d]	Dry, N_2	dedx/sw/ac	$Fe^{II/III}$	$\sim 4 \times 10^5$	17 ± 2	[58]
$KCo^{II}Fe^{III}(CN)_6 \cdot 14H_2O$	Dry, N_2	dedx/sw/ac	$Fe^{II/III}$	$\sim 1 \times 10^5$	11 ± 1	[59]
$KV^{III}Fe^{II}(CN)_6 \cdot 14H_2O$	Dry, N_2	dedx/sw/ac	$Fe^{II/III}$	$\sim 1 \times 10^6$	16 ± 2	[59]
$MeBlueFe^{II}(CN)_6 \cdot 14H_2O$	Dry, N_2	dedx/sw/ac	$Fe^{II/III}$	$\sim 5 \times 10^3$	11 ± 1	[59]
Prussian Blue	Dry, N_2	dcdx/ida	$Fe^{II/III}$	2×10^6	—	[8]
$Fe(bpyCO_2MPG350_2)_3ClO_4)_2$[c]	Pure melt/electrol.	dcdx/μ-disk	$2^+/3^+$	4×10^4	15	[25]
$Fe(bpyCO_2MPG350_2)_3ClO_4)_2$[c]	Pure melt/no electrol.	dcdx/μ-disk	$2^+/3^+$	4×10^5	8.4	[25]

* SW, sandwich; ida, interdigitated array electrode; of, open face, i.e. electrode|polymer|solution; mediat., electron transport-limited oxidation of a mediator in the solution.

† Where Eqn [6.9] applied, and de/dx used, rate constant is ρk_{ex}.

‡ N,N'-bis(3-pyrrol-1-yl-propyl)-4,4'-bipyridinium tetrafluoroborate, porous Au sandwich.

§ Copolymer of tetraethyleneglycol-di-p-tosylate and 4,4'-bipyridine.

‖ N,N'-bis-[3-(trimethoxysilyl)propyl]-4,4'-bipyridinium dibromide.

¶ 1-Methyl-1'-(6-(pyrrol-1-yl)hexyl)-4,4'-bipyridinium hexafluorophosphate.

** IDAs with two different gap/finger sizes [IDA(A) and IDA(B)] gave slightly different results.

†† $\rho = 10$; $^i\rho = 5.4$; $^j\rho = 2.8$; $^k\rho = 3.0$; $^l\rho = 7.3$.

‡‡ k_{ex} calculated from D_e results using $C_T = 1.3$ M and $\delta = 1.08$ nm.

§§ Electrochemically polymerized Fe-tetra(o-amino)phenylporphyrin.

‖‖ Ref. [6] result corrected for factor of 6 in Eqn 6.4.

¶¶ Structure shown in Fig. 6.1.

[a] Solvent cast films, fresh.

[b] Solvent cast films, electropolymerized.

[c] From dielectric relaxation frequency in ε' vs ε'' plot. Also, conc. of Mn sites not clear since Mn^{VII} and Mn^{VI} are crystallographically distinct, so rate is in s^{-1}.

[d] Conductivity predicted by intervalence (visible) spectra via Hush theory.

[e] Corresponds to $n = 7$ in the Co redox hybrid polyether of Fig. 6.1.

crystalline lattice units) physical diffusivity. Short-range, microscopic site mobility (meaning over fractional or a few lattice dimensions) can be expected to be present when the material is amorphous and above its glass transition. The spatial dimensions, frequency, and energies of microscopic site motions are in fact probable crucial factors in donor–acceptor electron transfer reaction dynamics, but these features are hard to evaluate. The solvation state provides some useful qualitative guides [5,9]. Microscopic site mobility is expected to be least in a dried redox polymer, and to increase, successively, when the polymer is contacted by the vapour or liquid of a plasticizing solvent or an especially strongly solvating material.

Redox polymers are typically studied as thin films, and thus must possess properties permitting their processing into thin films. Some redox polymer materials are prepared as bulk polymers and cast as films for electron transport study. Much thinner films can be prepared directly on electrodes by electrochemical polymerization [28]. The former polymers require relatively large samples, but with larger samples one can also apply other chemical and physical analyses to probe the polymer's detailed structure. Very thin films of electropolymerized polymers are harder to analyse structurally, making

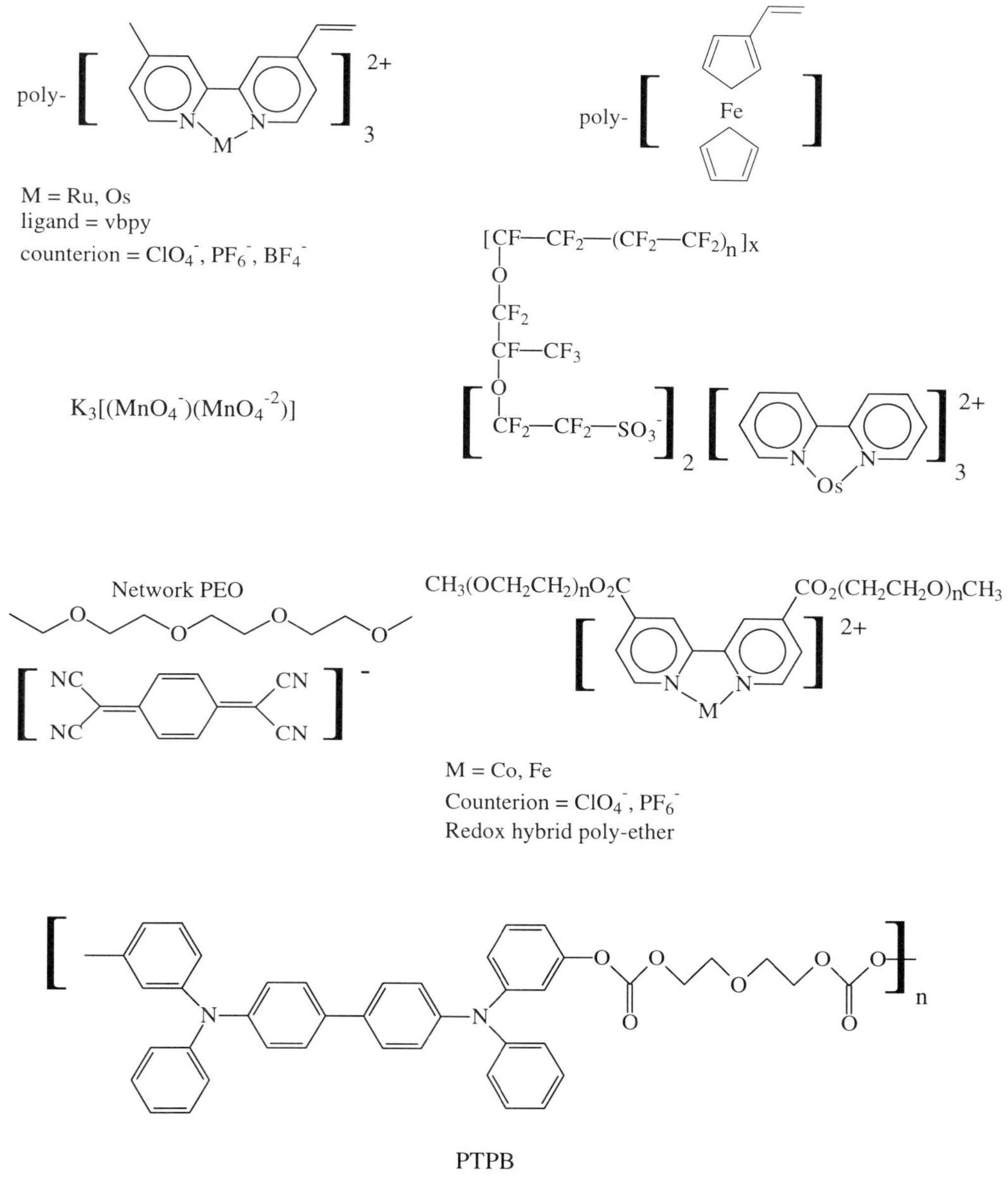

Figure 6.1 Examples of redox materials used in solid-state electron transport investigations.

design of clean and straightforward electropolymerization reactions crucial to knowing the redox site structures. Very thin polymer films offer the advantages of short electrolysis times and convenience of access to large electric field strengths.

Equation 6.1 is a one-electron, bimolecular reaction, and thus the nominally expected relation between current flow (electron transport rate, or electronic conductivity) and the electron self-exchange rate, irrespective of direction, is

$$\text{electron flux} = i/FA = C_D C_A k_{ex} \tag{6.2}$$

where C_D and C_A are, respectively, donor and acceptor concentrations, F is the

Faraday, and A is the area of the contacting electrodes. There are three important aspects of this relation.

1 How does the electron flux depend on the product of concentrations $C_A C_D$?

2 How does the flux depend on the absolute concentrations C_A and C_D (roughly equivalent to the 'doping level')?

3 For any given set of concentrations, what are the factors (including the microscopic ones) that control the value of k_{ex}?

Considering first the concentration dependencies (1 and 2), in a fluid solution it is typical that at any concentration of donor and acceptor, the reactants can freely explore possible reactions by diffusing from reaction partner to partner on time scales faster than the overall electron transfer reaction velocity. In that case, Eqn 6.2 applies and k_{ex} behaves as a *concentration-independent* reaction rate constant. In a solid or semi-solid such as a redox polymer, however, such freedom of partnering is restricted by the limited physical mobilities of donor and acceptor sites, and the bimolecularity of Eqn 6.2 may be obscured. This (in principle) particularly occurs when the sites have zero physical mobility (either macroscopic or microscopic). Then, the rules [29–31] of *static percolation* come into play, and at total site $(C_A + C_D)$ concentrations lower than the 'percolation threshold', the electron flux of Eqn 6.2 should fall to zero. For mixtures of donor, acceptor, and diluent (inactive or blank) sites, for example, and excepting long-range electron transfers, the percolation threshold is about 40 mol % of (donor + acceptor).

In the other extreme, a sufficiently large microscopic site mobility allows sites to explore more donor and acceptor electron transfer partners than just their most immediate neighbours, and donor–acceptor reactions can occur even when the sites have been strongly diluted. Equation 6.2 then applies much as in a fluid solution. This is called the mean field condition. Intermediate behaviour is expected for intermediate redox site mobilities, and is called bounded diffusion [29] or dynamic percolation [32,33]. There are known [27] experimental examples of mean field electron transport behaviour for redox polymers, but to date none that exactly [25] meet the expectations of static percolation.

However mobile are the sites, Eqn 6.2 also predicts that the electron transport rate should vary with the product $C_A C_D$, which maximizes at a 1 : 1 proportion of donor and acceptor. That is, the maximum electronic conductivity in a redox polymer is obtained by electrolysis at the polymer's electrochemical formal potential $E^{0\prime}$, which gives 1 : 1 mixed valent material, i.e. the concentrations $C_A = C_D$. This prediction has been verified [15,16] in several materials and seems satisfactorily established for ratios of C_A and C_D concentrations lying between 10 : 1 and 1 : 10. Experiments at much higher or lower concentration ratios, or 'doping levels', are difficult, since it is difficult to establish the smaller of the two concentrations with certainty. Data are sparse in this important area.

It is worth observing that a mixture of donor and acceptor sites in a redox material is different from a doped semiconductor. It is correct to speak of doping a redox material with electron (donor) or hole (acceptor) states, but under the typical conditions of $0.01 < [C_D]/[C_A] < 100$, there are no such things as majority and minority carriers in a mixed valent redox material. Hole mobility and electron mobility are synonymous in the context of Eqn 6.1. Also, redox materials do not have band gaps in the sense of the energy differences between localized 'valence band' and delocalized 'conduction band' energy levels. Finally, donor and acceptor concentrations in redox materials are very high relative to dopant concentrations typical of semiconductors. Because redox conductivity is generally low, redox polymers are almost always studied at high site concentration.

Turning to the third factor above, for any given combination of site concentrations and site mobilities there are a host of microscopic factors, besides concentration, that control the actual value of k_{ex}. These form the basis of fundamental understanding of solid-state electron transfer chemistry, which at present is poor in comparison with that in fluid solutions. In solid and semi-solid redox materials, *chemical* components of thermal barrier energies (i.e. energetics of dipolar reorganizations and of nuclear coordinate displacements that comprise, in the parlance of Marcus theory [34–37], the reorganizational barriers), electronic coupling or adiabacity characteristics, collision rates, dynamics of dipolar relaxation, association processes, etc., will variously exert control over electron hopping rates, just as they are known to do in dilute fluid solutions. It is presently difficult to single out roles of individual microscopic factors in solid-state results, and this will require exploration of a broad range of redox materials and of materials with systematically variable structures. The comparisons in Table 6.1 show, however, that these microscopic factors, in collective behaviour, generally lead to slower electron transfer reaction rates in solids and semi-solids compared with the fluid solution state.

Kinetic dispersity is relatively unfamiliar in fluid solutions, but is an important aspect of electron hopping transport in the solid and semi-solid state [38]. (It is known in fluid solutions in the phenomenon of dynamic fluorescence quenching [39–43].) In kinetically disperse redox polymers, a range of electron transfer rates will be operative within any given ensemble of donor and acceptor sites. This dispersity can conceivably arise from several factors. First, if the time constants for averaging of the redox sites' environments are longer than those for electron hopping, then all redox sites in an amorphous matrix will not have identical environments [44], which will lead to a range of reorganizational barrier energies among other things. For the same reason, there may be a distribution of site energies (i.e. in electrochemical terms in their $E^{0'}$ values, which effectively is then a kinetic dispersity of thermodynamic origin). Finally, a spread of the distances separating reaction partners can, from the exponential relation [37] between rate and distance, create a range of hopping rates.

The problem of a dispersion of electron hopping rates is well known [38] in the literature of charge transport through films of electrophotographic materials, and appears in the smearing out of the leading edge in current transients observed in injected electron time-of-fight (TOF) experiments. Numerous theoretical formalisms for dispersity have been devised, a recent and rather promising one of which is due to Bassler and co-workers [45]. In their 'disorder theory', in a manifold of localized electronic states (redox sites), a Gaussian distribution is assumed to exist in the electron hopping site energies (equivalent to a distribution in $E^{0'}$) and in the distances separating them. The energies of successive electron hops and of adjacent sites are implied to be uncorrelated (in spite of the high site concentrations involved). This theory represents many features of TOF experiments, but has not been applied so far to the steady-state electron transport measurements discussed in this chapter.

Kinetic dispersity also appears in electron transfer reactions through self-assembled and somewhat disordered molecular monolayers on electrodes [46,47]. The electrode reactant in this case experiences a partially rigidified environment and thus has kinship to electron transfers in amorphous materials. The dispersity in the self-assembled monolayer reactions, like that in redox polymer solids and semi-solids, is most apparent under conditions of large reaction free energy. The monolayer kinetic dispersity has been analysed [48] based on a Gaussian distribution of site energies (i.e. values of $E^{0\prime}$, a typical $\sigma(E^{0\prime})$ in an experimental example was 0.03 eV), of activation barrier energies, and of site–site distances. The assumptions in that analysis were analogous to those in the Bassler formalism [45], but the situations contain important differences in that dispersive monolayer rate constants are averaged in parallel whereas electron transport in a redox polymer involves some kind of serial averaging of the electron hopping rate.

Exploration and further understanding of these general ideas requires of course suitable experimental approaches, to which we turn next. The reaction free energy that drives electron hopping in a mixed valent material by Eqn 6.1 in a net direction can be supplied either by a spatial gradient of concentrations of donor and acceptor sites, or by a gradient of electrical potential. Over the past several years, we have focused on the experimental conditions for applying these two modes to a variety of mixed valent redox polymers, and for reconciling electron transport data obtained by the two modes for a given mixed valent substance [7,9,10,14,16].

The concentration–distance and potential–distance diagrams in Fig. 6.2 illustrate generic aspects of concentration and electrical gradient-driven electron transport, in a mixed valent film (both donor and acceptor present) sandwiched as a thin film between two electrodes. The concentration–distance diagram of Fig. 6.2(a) is drawn for the 1 : 1 mixed valent case with spatially uniform donor and acceptor concentrations. The mixed valency is introduced by whatever means convenient or available. For example, the film can be placed in an electrolyte solution and electrolysed at the average formal potential $E^{0\prime}$ of the redox sites; then the film, now 1 : 1 mixed valent in donor and acceptor sites, is removed from that solution and exposed to the desired gas or solvent environment. When a potential bias ΔE is applied to the two electrodes, current flows by electron hopping as in Eqn 6.1, but the nature of the current–potential curve observed varies according to whether the ionic components of the mixed valent D–A film are (on the experimental timescale) physically mobile (left hand diagrams in Fig. 6.2) or not (right hand diagrams in Fig. 6.2). These two conditions lead respectively to concentration gradient and electrical gradient-based measurements.

2.1 *Ionically conductive films*

If the mixed valent redox polymer sample is ionically conductive on the experimental time scale, application of a potential bias ΔE will prompt electrolytic reduction at the negative electrode/redox polymer interface, and oxidation at the positively biased interface. A current transient ensues, decaying to a steady-state value when linear (more of less) gradients of donor and acceptor site concentrations are established between the electrodes as shown in Fig. 6.2(b). The currents exhibit limiting values at potentials sufficiently large that dC/dx in the concentration profile is maximized (as in Fig. 6.2b). The limiting current is described by

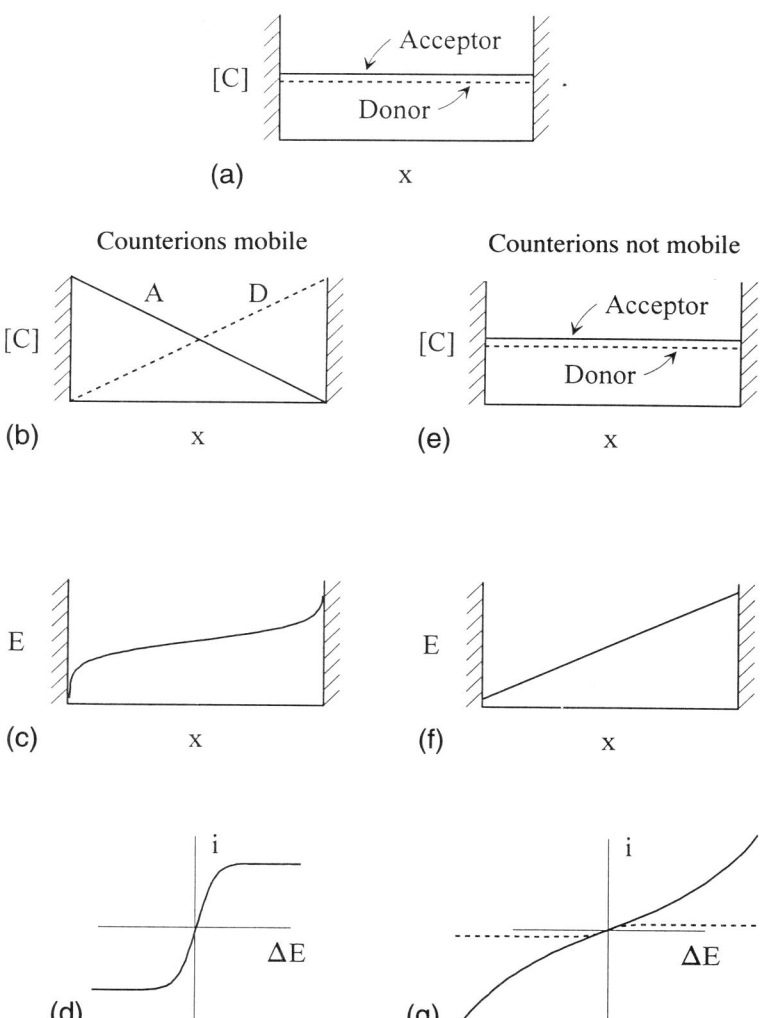

Figure 6.2 Schematic representations of concentration, potential, and current–potential responses to the application of a potential bias ΔE to a mixed valent redox polymer sandwiched between two (ion-blocking) electrodes (a) for the case in which the polymer is (b–d) ionically conductive and (e–g) not ionically conductive.

$$i_{\lim} = \omega^0 FAD_e \frac{C_T}{d}, \tag{6.3}$$

where C_T is the redox site concentration, d the interelectrode gap distance, A their area, ω^0 is a small migration correction [49] which is neglected and D_e is the 'electron diffusion coefficient' which by a cubic lattice model [29,50,51] is related to the self-exchange rate constant k_{ex} between the donor and acceptor sites,

$$D_e = \frac{k_{ex}\delta^2 C_T}{6}, \tag{6.4}$$

where intersite distance δ can be estimated using a cubic packing model in which $C_T = C_D + C_A$ and $\delta = [C_T N]^{-1/3}$ and N is Avogadro's number. If the sign of the applied potential bias is reversed, after a current transient, a steady-state current is again obtained, of opposite sign (Fig. 6.2d), and in which the donor–acceptor concentration gradients drawn in Fig. 6.2(b) are reversed. Equations 6.3 and 6.4 still apply.

For a sufficiently slow potential sweep a sigmoidal current–potential response is obtained (Fig. 6.2d). At intermediate potentials, concentration gradients of smaller and less-than-limiting slope are formed. These gradients are determined by the relationship between the applied ΔE and the interfacial donor and acceptor concentrations at the negatively and positively biased interfaces as expressed in the Nernst equation

$$\Delta E = \frac{RT}{nF} \ln \left[\frac{C_D}{C_A} \right]_- - \frac{RT}{nF} \ln \left[\frac{C_D}{C_A} \right]_+ . \tag{6.5}$$

The shape of the current–potential curve can be derived accordingly. At sufficiently small applied potential bias, the Nernst relation can be linearized [7,15] and the (ohmic) slope of the current–potential curve expressed as a conductivity

$$\sigma = \frac{id}{A\Delta E} = \frac{F^2 D_E C_D C_A}{C_T RT} = \frac{F^2 \delta^2 C_D C_A k_{ex}}{6RT} , \tag{6.6}$$

which provides an additional experimental tie to the bimolecular rate constant k_{ex}. Values of k_{ex} derived for the same redox polymer by analysing both limiting currents and small potential bias conductivities with Eqns 6.3 and 6.6 have been shown to agree [14].

Attendant to the redox site concentration gradient-forming electrolyses described above, counterions present within the mixed valent redox polymer film become spatially redistributed in order to provide charge compensation to the donor and/or acceptor sites. While the redox concentration gradients are formed by electron redistribution, the counterions must physically diffuse through the polymer to meet the electroneutrality requirement. As a consequence of this diffusional requirement, achieving the above steady-state concentration profiles and currents involves a substantial time constant that depends on the mixed valent film thickness and the counterion mobility in the given redox polymer lattice. The time constant can be regarded as a faradaic pseudo-capacitance. In the context of molecular electronics, the slow time constant of electrolytically based current–potential responses is, of course, a liability.

While the steady-state linear redox site concentration–distance profiles of Fig. 6.2(b) are linear, the distribution of electrochemical potentials within the mixed valent film is decidedly non-linear. (It is of course, strictly speaking, actually the electrochemical potential gradient, caused by the redox site concentration gradient, that drives the electron hopping reactions, not the concentration gradient itself.) Large electrochemical potential gradients should at first thought enhance electron transport rates, but Fig. 6.2(d) shows that currents do not increase at large ΔE potential biases. This aspect of electron transport in ionically conductive redox polymers can be understood again by recourse to the Nernst relation, written [52] for the distance-dependent pairs of donor and acceptor concentrations:

$$E_X = \frac{RT}{F} \ln \left[\frac{C_{D,X}}{C_{A,X}} \right] . \tag{6.7}$$

Figure 6.2(c) shows such a profile of electrochemical potentials for a potential bias ΔE producing the limiting current. Two aspects of this figure are of importance. First, changing ΔE alters the interfacial concentrations according to Eqn 6.5, and enhances the gradient of electrochemical potential immediately next to the interface, but as ΔE surpasses RT/F it has a quickly vanishing effect on the gradient in the interior of the film. This, qualitatively, is the reason that larger ΔE values do not yield larger currents. Secondly, the gradient of electrochemical potential in the interior of the film is relatively shallow, i.e. the electron hopping reactions there are promoted by very small free energy gradients. Intersite free energy differences at the film's centre are considerably less than kT_{298}. A typical intersite value would be 10^{-4} eV, or $<10^3$ V cm^{-1}. Even smaller electrochemical potential gradients are at work in the ohmic region of the Fig. 6.2(d) current–potential curve, by 10–10^3-fold depending on ΔE. These small gradients are relevant to the apparent lack of kinetic dispersity effects in electron hopping rate results from concentration gradient-driven experiments.

Equations 6.3 and 6.4 also apply to limiting currents obtained from electrolytic potential bias of redox polymer films in electrolyte solutions [53]. Electron transport through solvent-swollen polymer films on electrodes and sandwiched between them has been extensively investigated in electrolyte media. The ideas of using electrolytically generated concentration gradients of redox sites in polymers that are contacted only by bathing gases [54] were in fact stimulated by the earlier electrochemical work in electrolyte solutions, which provides an important data base for comparison of k_{ex} results.

Figure 6.3(a) is an example [14] of a concentration gradient-based current–potential response, for 60–70-nm-thick films of the electrochemically polymerized metal complex [Os(vbpy)$_3$]$^{2+}$ (Fig. 6.1), sandwiched between electrodes comprised of the tip of a Pt wire and an overlying evaporated Au film. The 1 : 1 poly-[Os$^{II/III}$(vbpy)$_3$](ClO$_4$)$_{2.5}$ form of the polymer film was prepared by electrolysis in an electrolyte solution (0.1 M Bu$_4$NBF$_4$/CH$_3$CN), then rinsed and dried under N$_2$. For the measurement shown, the potential was swept at a very slow rate (5 mV s^{-1}) in order to not out-run the electroneutrality-preserving diffusion of BF$_4^-$ counterions across the film. (Even then, the slight current hysteresis on the limiting current regions reflects small deviations from true steady-state conditions.) Rate constants averaging $2.7(\pm 0.8) \times 10^5$ M^{-1} s^{-1} were obtained from responses like that shown. In accord with Eqn 6.6, a virtually identical $k_{ex} = 2.8 \times 10^5$ M^{-1} s^{-1} was obtained [14] from the small ΔE, ohmic region of the current–potential curve.

Figure 6.4(a) shows another example [6] of a concentration gradient-based experiment, using a film of the redox polymer polytetraphenylbenzidine (PTPB; Fig. 6.1) coated on an Au interdigitated array (IDA) consisting of 50 finger sets 5 μm wide with 5 μm interfinger gaps. (The film thickness in this geometry is set by the IDA geometry.) The film had been half-oxidized so as to contain equal quantities of neutral and radical cation triarylamine sites, and then removed to a CH$_3$CN vapour bath. The limiting current of Fig. 6.4(a) gave an electron diffusion coefficient of 5×10^{-8} cm^2 s^{-1}. With Eqn 6.4 and $\delta = 11.7$ Å this converts to a self-exchange rate constant of 7×10^6 M^{-1} s^{-1}.

Table 6.1 summarizes literature data for these and other electron transfer dynamics results. The 'technique/geometry' column refers to concentration gradient experiments using limiting currents and Eqn 6.3 as 'dcdx'. 'Geometry' refers to the experimental

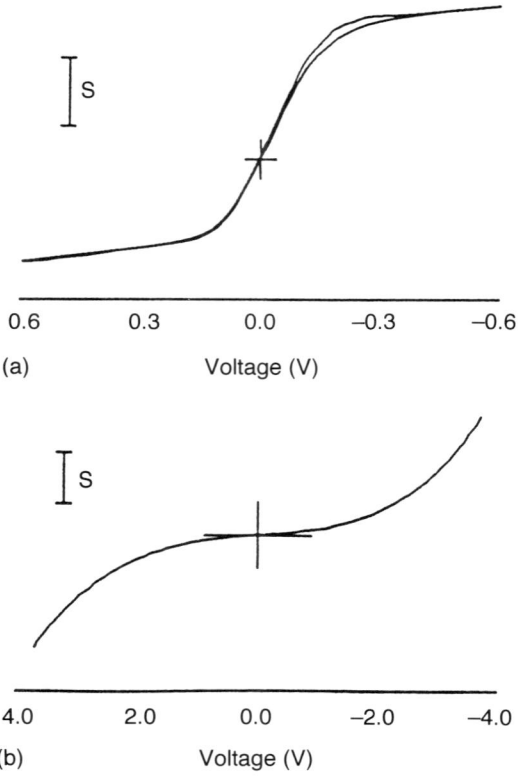

Figure 6.3 Current–potential responses of a thin mixed valent electropolymerized film of the redox polymer poly[Os(vbpy)$_3$](BF$_4$)$_{2.5}$ under N$_2$. (a) A potential sweep at 5 mV s^{-1}, $S = 1.3 \times 10^{-2}$ A cm^{-2}; (b) a potential sweep at 64 000 V s^{-1}, $S = 1.6$ A cm^{-2}. (a) reflects development of concentration gradients in the film, and limiting currents according to Eqn 6.3; (b) reflects electrical gradient-driven electron transport; the curve shown is fitted using Eqn 6.9. (Adapted from Figs 4 and 5 of [14], copyright American Chemical Society.)

manner in which an electrode–film–electrode structure is made, which is specified in the footnote. The 'sandwich geometry' was used in Fig. 6.3 and provides measurements on exceptionally thin redox polymer films, but has been difficult to apply to redox polymer films beyond the electrochemically polymerized metal polypyridine complexes. Table 6.1 contains results (see column labelled 'environment') where the mixed valent films are exposed to a variety of bathing media, including examples of using electrolyte solution contacts (solvent–electrolyte), organic vapours, and dried under N$_2$ or other suitable inert gas. Table 6.1 also contains rate constants for the redox monomer constituents for some of the polymers, and we see that uniformly the values of polymer-phase k_{ex} are smaller.

2.2   *Ionically non-conductive films*

2.2.1   MIXED VALENT FILMS

If the mixed valent redox polymer film of Fig. 6.2(a) is not ionically conductive, application of a potential bias, for reasons of electroneutrality requirements [55], does

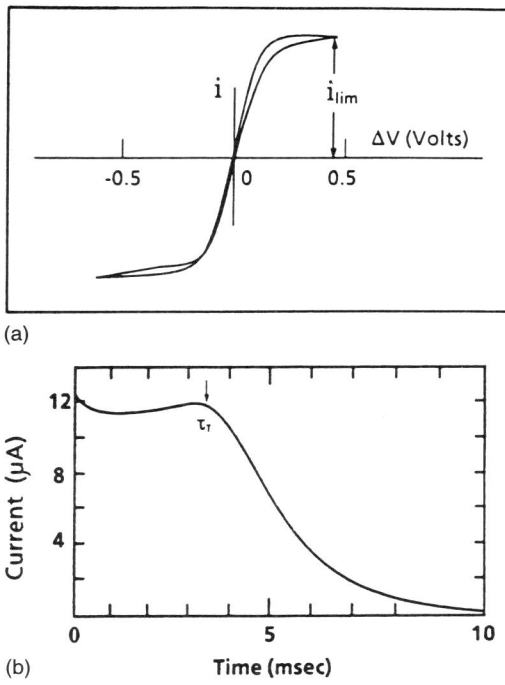

(a)

(b)

Figure 6.4 Responses of films of polytetraphenylbenzidine (Fig. 6.1): (a) as a 1 : 1 mixed valent neutral/radical cation film on an IDA, subjected to slow potential sweeps, in a plasticizing bath of CH_3CN vapour; (b) current transient observed in a time-of-flight determination of drift mobility in a dry, 11-μm-thick non-mixed valent film, under an impressed field of 3×10^4 V cm^{-1}. The transit time is shown as τ_T. (Reproduced with permission from [20] (a), copyright American Physical Society, and [19] (b), copyright American Chemical Society.)

not provoke change in the spatial distribution of donor and acceptor sites i.e. there is no electrolysis and the uniformly distributed concentrations remain so; Fig. 6.2(e). The applied potential thus distributes itself more or less linearly (assuming no significant interfacial resistance effects) across the film (Fig. 6.2(f)), and the current–potential response takes the shape shown in Fig. 6.2(g). At low potential bias ΔE, the conductivity is ohmic-like (linear I–E curve) and can be stated in terms of the electron self-exchange rate constant,

$$\sigma = \frac{id}{A\Delta E} = \frac{F^2\delta^2 C_D C_A (\rho k'_{ex})}{6RT}, \tag{6.8}$$

which is the same as Eqn 6.6 (excepting for ρ, explained below). In the present context, the product $\rho k'_{ex}$ of Eqn 6.8 is equated to k_{ex} of Eqn 6.6. Instead of using linear potential sweeps as in Fig. 6.2, the electron conductivity can alternatively be assessed by small amplitude ac impedance; experimentally identical values of σ are obtained [15] by the two approaches.

At larger potential bias for an ionically non-conductive film, corresponding roughly to gradients of 10^4 V cm^{-1} and up, the microscopic potential drop between adjacent donor and acceptor sites becomes sufficiently large that the classical exponential increase in rate with applied (reaction) free energy become appreciable. The current–potential response then rises in a super-ohmic fashion as in Fig. 6.2(g). Although the potential gradients required to produce the super-ohmic behaviour are large, the estimated microscopic gradient between adjacent redox sites, $\Delta E\delta/d$, is still small . For example, for $\delta = 1.6$ nm, and $\Delta E/d = 10^5$ V cm^{-1}, the intersite driving force is only 16 mV, about 40% less than kT_{298}.

Expressions for solid-state electron transport incorporating the exponential relation were given some years ago [1]. We have used the exponential relation in the notation of contemporary Marcus electron transfer theory [34–37], which has in general terms enjoyed great success in application to electron transfers in dilute fluid solutions, as a vehicle to assess microscopic aspects of solid-state electron transfers. Marcus theory gives reasonably good (not exact) fits to experimental current–potential data [10,14–16] when cast in the form

$$i = \frac{FA\delta C_D C_A k'_{ex}}{6}\left(e^{\frac{\rho F\delta\Delta E}{2dRT}} - e^{\frac{-\rho F\delta\Delta E}{2dRT}}\right). \tag{6.9}$$

Equation 6.8 follows from linearization of Eqn 6.9, which is the same as Marcus theory except for the parameter ρ. Marcus theory was not written to account for the dispersive nature (*vide supra*) of electron hopping rates in amorphous solids and semi-solids, which shows up in experimental current–potential curves rising more rapidly with increasing ΔE than anticipated by Eqn 6.9. We have accounted for this effect empirically with the fitting factor ρ added to the exponential of Eqn 6.9. Ideally unity, ρ in practice ranges from 2 to ca 10.

In principle, identical rate constants should be obtained from application of Eqns 6.3 and 6.9 to the same redox polymer (i.e. investigated under conditions of significant and insignificant ionic conductivity, using concentration gradient and the electrical gradient-driven electron transport, respectively. The electron self-exchange rate should not depend on the source of free energy driving it.) In early comparisons [10], rate constants k_{ex} obtained by applying Eqn 6.3 to concentration gradient-based limiting currents seemed larger than those of k'_{ex} obtained using Eqn 6.9 to analyse current–potential responses shaped like Fig. 6.2(g). After exploration of several redox polymers, it became apparent [14,16] that the *product* of k'_{ex} and the phenomenological ρ factor in Eqn 6.9 is the appropriate parameter to compare to the k_{ex} results of Eqn 6.3.

Figure 6.2(g) also shows a current–potential response (dotted line) like that in Fig. 6.2(d) but now on the same current scale. This emphasizes that electrical gradient-driven electron transport rates are much larger than those obtained with concentration gradients. The underlying reason for the difference was explained in connection with Fig. 6.2(c).

Opportunities to perform both concentration gradient and electrical gradient experiments on the same mixed valent redox polymer film involve satisfying a delicate balance of factors, since for the former experiment the film must be ionically conductive, and for the latter it must not be. For the metal complex polymer poly[Os$^{II/III}$(vbpy)$_3$](ClO$_4$)$_{2.5}$ and related electropolymerized metal polypyridine films, for example, this was accomplished by manipulating experimental time scales. Thus, a concentration gradient experiment could be conducted [5,7] for a dry poly[Os$^{II/III}$(vbpy)$_3$](ClO$_4$)$_{2.5}$ film with very slow potential sweeps (5 mV s^{-1}; Fig. 6.3(a)), allowing time for counterion redistribution. For the same dry film, using potential sweeps sufficiently fast as to make the current–potential result sweep rate independent ($>5 \times 10^4$ V s^{-1}) means that the counterion migration is insignificant on that time scale, relative to electron migration. Figure 6.3(b) shows this result, which has a current–potential response like that anticipated from Eqn 6.9. Comparison of the current sensitivities for Figs 6.3(a) and 6.3(b) shows that the latter currents are >100-fold larger than the limiting current of Fig. 6.3(a). Treatment of a number of curves like that in Fig. 6.3(b) with Eqn 6.9 gave average rate constants $k'_{ex} = 1.4 \times 10^5$ M^{-1} s^{-1} and $\rho = 2.8$ or $\rho k'_{ex} = 4 \times 10^5$ M^{-1} s^{-1} which is quite close to the $k_{ex} = 2.7 \times 10^5$ M^{-1} s^{-1} obtained (Fig. 6.3a) for the same redox polymer sample with concentration gradients.

When a mixed valent poly[Os$^{II/III}$(vbpy)$_3$](ClO$_4$)$_{2.5}$ film is exposed [5] to a solvent vapour (such as CH$_2$Cl$_2$), both its electronic hopping and ionic conductivity rise, and curves like Fig. 6.3(a) can be obtained with larger potential sweep rates. However, it also becomes a practical impossibility to attain potential sweep rates that can out-run the counterion polarization, and electrical gradient experiments are no longer possible. The opposite circumstance occurs [16] for mixed valent films of polyvinylferrocene (Fig. 6.1), which exhibit very low ionic conductivity. Electrical gradient experiments are easy in this instance, but concentration gradient measurements require intermediate exposure of the film to a plasticizing organic vapour bath to achieve a steady current at each (constant) applied ΔE, followed by drying the film under that ΔE bias. Difficulties like these have constrained direct comparisons of concentration and electrical gradient-driven electron transport to a limited number of mixed valent materials. Those for the metal polypyridines, polyvinylferrocene, viologen polymers, and polytetraphenylbenzidine, data given in Table 6.1, are the only available examples.

The proposal that the necessary inclusion of the fitting factor ρ in Eqn 6.9 reflects kinetic dispersity in electron hopping rates within the redox polymer made it important to compare data to solid-state electron hopping models more explicitly designed to account for kinetic dispersity. A model examined [14,16] for this purpose was that of Scher and Montroll [56] and Pfister [57] (SMP), which represents a distribution of electron hopping rates in a film of thickness L using a distribution in hopping distances.

$$i = \frac{FA\delta C_D C_A}{6} \left[k_{ex,SMP} L^{1-\frac{1}{\varepsilon}} \right] \left[e^{\frac{F\delta\Delta E}{2dRT}} - e^{\frac{-F\delta\Delta E}{2dRT}} \right]^{\frac{1}{\varepsilon}}, \tag{6.10}$$

where ε is a parameter for the degree of kinetic dispersion. The units of $k_{ex,SMP}$ are M^{-1} s^{-1} cm$^{(1/\varepsilon-1)}$, a complication discussed in more detail by Pfister [57]. In a disordered system $\varepsilon < 1$; in a crystalline one $\varepsilon = 1$ and Eqn 6.10 reduces to the classical Marcus expression. Examination of Eqn 6.10 shows [16] that the dispersive rate constant is the parameter $k_{ex} L^{1-1/\varepsilon}$.

Current–potential curves for dry, mixed valent films of poly[Os$^{II/III}$(vbpy)$_3$](ClO$_4$)$_{2.5}$ like that in Fig. 6.3(b) were analysed [14] by both Eqns 6.9 and 6.10. Fitting of Eqn 6.9 gave $\rho k'_{ex} = 4 \times 10^5$ M^{-1} s^{-1} for the Os$^{II/III}$ electron self-exchange reaction whereas fitting to Eqn 6.10 gave a very similar value of $k_{ex} L^{1-1/\varepsilon} = 4.5 \times 10^5$ M^{-1} s^{-1} with $\varepsilon = 0.50$. Similar agreement [16] was found for electrical gradient data obtained with films of mixed valent polyvinylferrocene on interdigitated array electrodes, where the dispersity parameter ε of ca 0.95 indicated modest dispersity but ρ was in the range 5–10. Functionally, the empirical ρ parameter in Eqn 6.9 seems to satisfactorily account for kinetic dispersity for the purpose of study of the average electron transfer rate constant. It is not intended for use as a parameter to investigate the dispersity itself;

the SMP model and the more contemporary one by Bassler *et al.* [45] are better suited for that purpose.

Finally, capacitances associated with the response of mixed valent films to applied electrical gradients have been little studied as such but are qualitatively much smaller than the large faradaic pseudo-capacitances associated with forming concentration gradients. Time constants for achieving steady current responses to electrical gradients are also many orders of magnitude smaller.

A special note should be paid to investigations of mixed valent films that are inorganic in character (Table 6.1, near bottom). Mixed valent inorganic crystals fall into four broad categories ranging from insulators to metallic conductors, depending, most likely, on the crystallographic distinguishability of the sites of differing valence. Important to the present discussion are those materials which exhibit intermediate or semiconductor-scale conductivity, as well as intervalence transfer bands in the visible, so called type II and IIIA materials. Of such materials, those based on 'Prussian Blue' chemistry [8,58,59] involve redox transformations that are relatively well defined but films that are also structurally disordered and even partially amorphous. The latter feature allows some ionic conductivity that enhances experiments on these materials. Unique, however, seems to be a permanganate mixed valent material whose electron transport properties have been studied [21] within a single crystal (Fig. 6.1). Figure 6.5 shows the ac impedance response (as a Cole–Cole plot) of this material, from which an electron hopping rate of $10^5 \, s^{-1}$ was calculated. The detailed picture of the crystal structure within which this electron transfer occurred would seem to harbour many opportunities to probe the fundamentals of solid-state electron transfer energetics. One would wish to vary the site–site distances, and the surrounding structure and its dipolar characteristics, and hopefully this can be done. However, it has been classically difficult to produce systematic variations of crystalline mixed valent materials.

2.2.2   NON-MIXED VALENT FILMS

Ionically non-conductive redox polymer films are also investigated as non-mixed valent, typically ion-free materials, using well-known [1] electron and hole injection TOF methods. The transit time or TOF of photoinjected charge has been a standard parameter used for electron transport studies in materials of electrophotographic interest. In this experiment, the layer of redox material, termed a charge transport layer and consisting of a redox monomer or polymer, often as a solid solution in a binder polymer, is sandwiched between a metal contact and an electrode structure with which

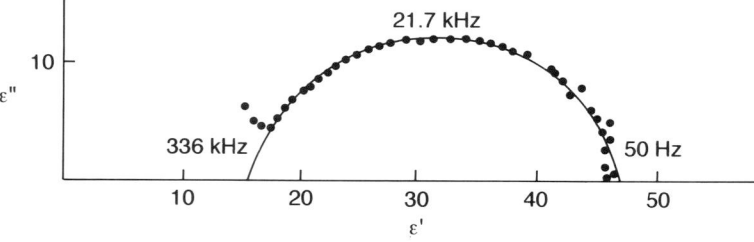

Figure 6.5 Cole–Cole plot from ac impedance measurements on mixed valent single crystal $K_3(MnO_4)_2$. (Adapted from [21], copyright Royal Society of Chemistry.)

electrons or holes (depending on the applied bias) can be electro- or photo-generated and injected into the transport layer. When, for example, the electrode contact is a positively biased (ΔE) transparent metal film, a light pulse generates a layer of 'holes', or electron acceptor, next to the film contact, which by Eqn 6.1 is transported across the film and recorded as a current front such as that shown in Fig. 6.4(b) for the tetraphenylbenzidine polymer PTPB (Fig. 6.1). The 'drift mobility' $\mu_{(E)}$ of the electrons crossing the film is evaluated from the transit time τ_T by the relation

$$\mu_{(E)} = \frac{d}{\tau_T E},$$
(6.11)

where d is film thickness and E is the electrical field gradient ($\Delta E/d$). Values of $\mu_{(E)}$ (cm^2 V-s^{-1} or cm^{-1} s per V cm^{-1}) depend on ΔE and temperature as well as on the specific redox material, its concentration, and the binder material. Connections [19,20] to the preceding relations can be made from the relation of the drift mobility at zero field, $\mu_{(E=0)}$, to the electron diffusion coefficient D_e as made with Eqn 6.4 and the Einstein equation,

$$\mu_{E=0} = D_e \frac{e}{k_B T}.$$
(6.12)

$\mu_{(E=0)}$ is thus an alternate representation of the electron self-exchange rate constant k_{ex} and should in principle exhibit an associated dependency on the donor or acceptor concentration as given in Eqn 6.4 and be subject to effects of microscopic mobility.

TOF measurements are difficult [19,20] for small electrical field gradients ($<10^3$ V cm^{-1}), but drift mobilities at zero electrical gradients $\mu_{(E=0)}$ are desired in studies of the transport dynamics. Experimental values of $\mu_{(E)}$ taken in experiments have to be extrapolated to zero field and the field dependency of electron drift mobility has consequently been an extensively investigated subject. Although a microscopic basis is so far lacking, plots of $\ln[\mu_{(E)}]$ against the square root of the electrical gradient ($E^{1/2}$) are linear for a wide range of charge transport materials [19,20,38]. For example [38], solid solutions of the electron donor triphenylamine in a polycarbonate host give TOF-based drift mobilities that at large E are linear in $\ln[\mu_{(E)}]$ vs $E^{1/2}$ whereas over a comparable field strength range $\ln[\mu_{(E)}]$ vs E plots exhibit decreasing slopes with increasing E. The $E^{1/2}$ field dependency is not however universal; there are numerous exceptions especially at smaller field strengths. Clearly the field dependence of drift mobility is at some level, material specific. A lucid discussion of this topic, in the context of disorder effects, has recently appeared [38].

We close this section by briefly enumerating some differences between TOF investigations and those which are based on mixed valent films.

1 TOF charge transport layers are not only univalent (not mixed valent) but also typically have no significant ionic content. The charge-compensating counterions of mixed valent films are part of the donor–acceptor environment, and being strong dipoles, potentially can influence reorganizational barriers for electron transfers in both the inner and outer sphere sense. In the specific case of polytetraphenylbenzidine, Facci et al. [20], by agreement between TOF drift mobility results with a concentration gradient experiment on mixed valent PTPB (making the connection through Eqn 6.12),

showed that the counterion effect was not dominating. On the other hand, in redox melts k_{ex} can be very dependent on the ionic content of the melt (T. Masui and R.W. Murray, unpublished results, 1995). Generalities with respect to ion dipole effects are clearly not warranted.

2 Concentration gradient measurements of k_{ex} involve, relative to TOF measurements, very small electrochemical potential gradients. As a result, influences of kinetic dispersity may be largely (completely?) avoided, which means the window of observations on microscopic chemical influences on k_{ex} is relatively unclouded by the problems of the representation of dispersity. From the standpoint of understanding the underlying chemistry of solid-state electron transfers, this is an important consideration.

3 A TOF measurement is of the time required for a film-transiting electron whereas measurements of mixed valent films involve a steady-state electron flux. TOF transit times are often in the ms range, whereas transit times in concentration gradient can be measured in seconds. Such differences in time scale could conceivably serve to probe relaxation phenomena in the redox medium.

4 The univalent films used in TOF experiments are ideally devoid of other electron donors or acceptors (and of mixed valency), and the amount of charge injected in this experimental motif is very small relative to the total number of acceptor or donor sites in the film, which means that tiny amounts of impurity states ('traps') can have a substantial effect on electron transit times. Trapping effects are a common concern in TOF experiments as a result. Mixed valent films, on the other hand, have both large electron flux and large populations of electron donor and acceptor states, and impurity trap effects are substantially ameliorated, simplifying examination of concentration dependencies as expressed in Eqn 6.2.

3   Redox melts and experimental methodologies

These materials, examples of which are found in Fig. 6.1, include solutions of redox molecules in polymer electrolytes [22,23], as well as undiluted molecular melts of redox species [24,25]. Those under investigation in our laboratory all involve polyethers as components. The bulk of our investigations into these materials have employed ultra-microelectrode voltammetry, discussed further below. The donor and acceptor species in these materials exhibit an experimentally (electrochemically) discernible level of physical (macroscopic) diffusivity, but this physical mobility is comparable to or much lower than the mobility that the electron achieves by hopping between more slowly moving donor and acceptor species. Under these conditions electron transport can be described by the mean-field case of electron self-exchange described above, and the so-called Dahmns–Ruff relation [60–62] should apply to the apparent net diffusivity (D_{app}) of the redox molecule

$$D_{app} = D_{phys} + \frac{k_{ex}\delta^2 C_T}{6}, \tag{6.13}$$

where D_{phys} is the physical diffusion coefficient of the D or A species, C_T is the total donor or acceptor concentration and δ is the distance between them at the instant of electron transfer, and k_{ex} is the electron self-exchange rate constant. The right hand term of Eqn 6.13 is identical to that of the lattice model for electron diffusion, Eqn 6.4.

Like steady-state concentration gradient-*based* measurements on mixed valent redox polymers, evaluation of k_{ex} in redox melts involves a diffusion coefficient measurement. Relative to the case of redox polymers, site concentrations in redox melt materials are conveniently adjustable. Adjusting site concentration provides an experimental handle on the average electron hop distance, δ, and the transport parameter $k_{ex}\delta^2$, and a powerful method for comparing experiment and theory. The voltammetric methodologies applied to redox melts to date have dictated that the bulk of the redox melt be initially not mixed valent; and electrolysis is used to create a thin film of mixed valent material around a working electrode. The physical diffusion and electron hopping transport described by Eqn 6.13 takes place within this 'diffusion layer'. Another significant difference is that macroscopic physical diffusivity, not of concern in redox polymers, requires explicit evaluation in redox melts using Eqn 6.13. This latter factor places an important requirement on the chemical design of melt materials suitable for detailed solid-state electron transfer dynamics investigations.

Within the redox materials that have been examined [22–26] in the melt and polymer electrolyte solution state (Fig. 6.1), a large range of values of D_{phys} and δ have been encountered. The physical diffusivity of sites can furthermore be manipulated by the addition of electrolyte (such as $LiClO_4$ to a polyether, which transiently cross-links by ether–Li^+ coordination and decreases D_{phys}) or of small molecules (increases D_{phys} by 'diffusion plasticization'). These materials offer some experimental advantages in examining the relationships between physical site mobility, site concentration, and electron transfer rate constant.

Experimental approaches for measuring k_{ex} and related parameters in redox melts have been developed under the methodological label of solid-state voltammetry [63]. Measurements of the diffusivities D_{app} and D_{phys} rely on the use of microelectrodes, i.e. voltammetric electrodes of very small dimensions, with both potential sweeps and steps. The production of mixed valent layers around electrodes by electrochemical reactions depends on the ionic conductivity of the solid or semi-solid medium surrounding the electrode. The ionic conductivity of polyethers containing dissolved $LiClO_4$ is appreciable, but in hybrid redox polyethers (Fig. 6.1) is often abysmal. Microelectrodes alleviate this problem in several ways, the most obvious of which is that small currents flow at a miniaturized interfacial area, making the ohmic product problem iR_{unc} small (R_{unc} is uncompensated resistance). In this way electrochemical voltammetry has been accomplished [65] in hybrid redox polyethers in which the physical diffusivity of the redox molecule (the Co bipyridine complex in Fig. 6.1) is less than 10^{-15} cm^2 s^{-1}.

Redox materials of interest for solid-state electron transfer rate studies using solid-state voltammetry and Eqn 6.13 must exhibit very slow physical diffusivities (D_{phys}) compared with the rate of electron diffusion (D_e). Solutions of redox species in polyethers often exhibit this property. In order to concurrently measure D_{app} and D_{phys} in Eqn 6.13, a chemical with three acessible oxidation states is used, i.e. one possessing two donor–acceptor pairs, the k_{ex} for one of which is much smaller than the other. D_{app} for the slower donor–acceptor couple is used as an estimate of the D_{phys} for the faster one. Polyether electrolyte solutions of the $[Co(bpy)_3]^{2+/3+}$ and $[Co(bpy)_3]^{2+/1+}$ redox couples, and those of tetracyanoquinodimethane ($TCNQ^{1-/2-}$ and $TCNQ^{1-/0}$, Fig. 6.1) are examples [22,23] of this chemical design.

An interesting but complicating feature of such experiments, however, is that electron transfers of the faster redox pair (i.e. such as in the electrogenerated mixed valent layer of $TCNQ^{1-/0}$ around the electrode [22,23]) may not occur at contact/collision. Longer distance electron transfers will occur if their time constant is comparable to that of the reactants diffusing together to contact. Since the value of k_{ex} is then distance dependent, both the rate constant k_{ex} and electron transfer distance δ become variables that depend on D_{phys}. (At smaller D_{phys}, δ is larger and k_{ex} correspondingly smaller, and the value of D_e is as a result never very different from that of D_{phys} even though D_{phys} may vary by many orders of magnitude.)

A more straightforward solid-state voltammetric measurement of k_{ex} becomes possible when the redox material is an *undiluted* melt. Recent work has shown [8,31,32] that attachment of short polyether chains to metal bipyridine complexes (including $[Co(bpy)_3]^{2+}$; Fig. 6.1) converts them into room temperature melts in which the concentration of redox centres ranges up to 0.6 M (depending on the diluting length of the attached polyether tail) [24–26,65–67]. We call these new materials hybrid redox polyethers. Their redox site concentrations are much larger than could be achieved by dissolving underivatized redox solutes in polyether solvents. While the hybrid redox polyethers are generally miscible with polyether polymer electrolyte solvents, it is more interesting to use them in their native, neat state, i.e. as redox molecules to which the solvent is attached. The attached polyether chains can effect electrolyte dissolution when that is necessary or desirable. Then, in the undiluted hybrid redox polyether molecules, ambiguity about the distance of electron transfer δ is substantially removed, being fixed by the molecular dimensions of the melt species.

Room temperature melts of the metal complex hybrid polyether $[Co\{bpy(CO_2 (CH_2CH_2O)_3CH_3)_2\}](PF_6)_2$ (Fig. 6.1) provide an example [24] of an undiluted electron transport medium. Figure 6.6 shows microelectrode-based solid-state voltammetry in which potential is swept positively, oxidizing the complex ($Co^{II/III}$) and negatively, reducing it ($Co^{II/I}$). D_{app} for the former voltammetric wave was ca 10^4 smaller than the latter, from which by using the D_{app} for the $Co^{II/III}$ reaction (which has very slow electron transfer rates) as a measure of D_{phys} in Eqn 6.13, k_{ex} for the $Co^{II/I}$ couple was evaluated as 2×10^6 M^{-1} s^{-1}. (We have since found that the sample used for this measurement contained some excess, plasticizing ligand, and the value of k_{ex} in the more highly purified melt is smaller.)

Potential steps have special usefulness in solid-state voltammetry [63], in being less affected by uncompensated resistance effects than are current–potential responses to potential sweeps. In a chronoamperometry experiment, the potential is stepped onto the 'plateau' of the redox reaction's voltammetric wave. The current–time response follows the Cottrell equation [69]

$$i = \frac{FAD_{app}^{1/2}C_T}{\pi^{1/2}t^{1/2}}. \tag{6.14}$$

The diffusion coefficient, D_{app} is obtained from a plot of current vs $t^{1/2}$, within a time interval short enough that diffusion near the electrode can be approximated as planar. Other straightforwardly applied relations are used [63] when the near-electrode depletion zone becomes partially radial in character. For the typical 10-μm-diameter microdisk electrode, radial diffusion effects can become noticeable on a minute time

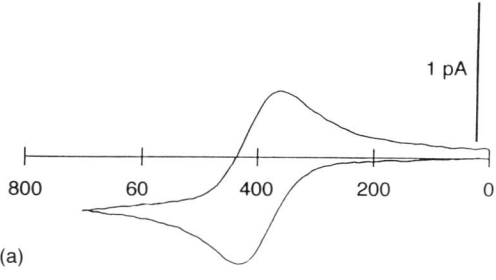

(a)

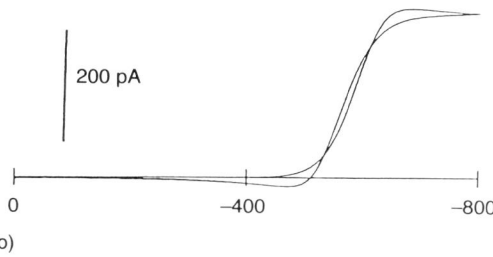

(b)

Figure 6.6 Solid-state cyclic voltammetry at 5-μm radius Pt microdisks and 50 μ V s^{-1} potential sweep rate, in the metal complex hybrid polyether melt [Co{bpy(CO$_2$(CH$_2$CH$_2$O)$_3$CH$_3$)$_2$}$_3$](ClO$_4$)$_2$, dry and under N$_2$. Potentials relative to Ag wire pseudo-reference electrode. (a) Positive potential scan through the Co$^{II/III}$ oxidation wave, $D_{app} = 3 \times 10^{-13}$ cm^2 s^{-1}; (b) negative potential scan through the Co$^{II/I}$ reduction wave, $D_{app} = 6 \times 10^{-9}$ cm^2 s^{-1}. (Adapted from Fig. 1 of [24], copyright American Chemical Society.)

scale when D_{app} exceeds ca 10^{-9} cm^2 s^{-1}. The chronoamperometric experiment is generally the preferred tool for measuring D_{app} values in solid-state voltammetry.

Hybrid redox polyethers are relatively recently discovered [66], and the generality of their usefulness in exploring solid-state electron transfer dynamics is yet to be established. The necessary measurement of D_{phys} promises to give, however, a new window on the importance of microscopic mobility in solid-state electron transfers. For example, using the metal complex hybrid polyethers [Co{bpy(CO$_2$(CH$_2$CH$_2$O)$_n$CH$_3$)$_2$}$_3$](PF$_6$)$_2$ and [Fe{bpy(CO$_2$(CH$_2$CH$_2$O)$_n$CH$_3$)$_2$}$_3$](PF$_6$)$_2$ (Fig. 6.1), we have established that longer polyether 'tails' increase D_{phys} but decrease k_{ex} of the Fe$^{II/III}$ reaction, whereas addition of LiClO$_4$ electrolyte (see Table 6.1) decreases both D_{phys} and k_{ex}. (J. Long and R.W. Murray, unpublished results, 1994) The latter effect may involve restriction of microscopic mobility of reacting Fe$^{II/III}$ donor–acceptor pairs whereas the former may involve a distance of electron transfer effect.

Mixtures of the metal complex hybrid polyethers [Co{bpy(CO$_2$(CH$_2$CH$_2$O)$_n$CH$_3$)$_2$}$_3$](PF$_6$)$_2$ and [Fe{bpy(CO$_2$(CH$_2$CH$_2$O)$_n$CH$_3$)$_2$}$_3$](PF$_6$)$_2$ in which $n = 7$ (Fig. 6.1) also provide an opportunity to examine concentration dependencies as were discussed above for Eqn 6.2. Microelectrode voltammetry of the Fe$^{II/III}$ and Co$^{II/III}$ couples in their respective pure melts (containing LiClO$_4$ electrolyte) shows [25] that D_{app} in the Fe complex melt (5×10^{-11} cm^2 s^{-1}) is 10^3-fold faster than the extremely slow physical diffusion (5×10^{-14} cm^2 s^{-1}) in the Co complex melt. Using the Co$^{II/III}$ D_{app} to approximate D_{phys} of the Fe complex (the two metal complexes are isostructural) in melt

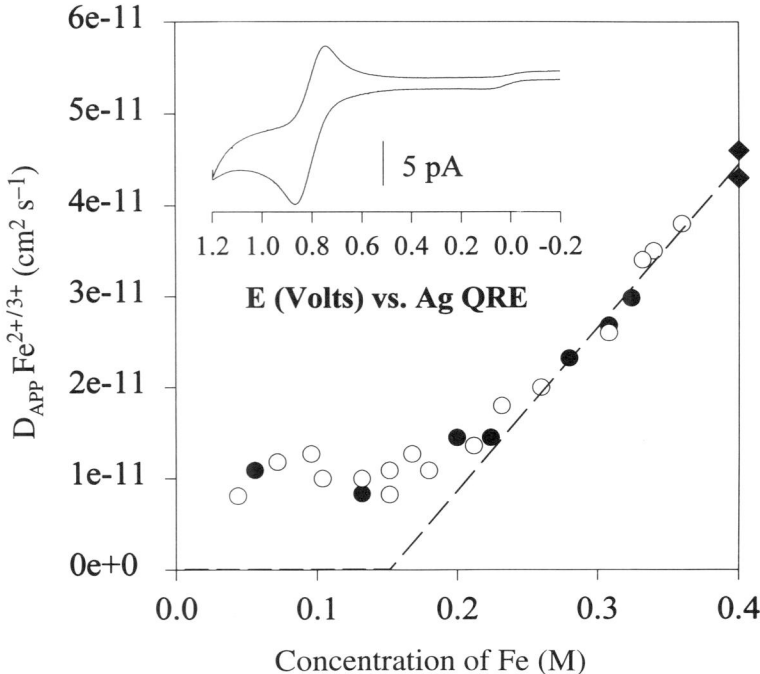

Figure 6.7 D_{app} for electron hopping between the $Fe^{III/II}$ states of the metal complex hybrid polyether melt $[Fe\{bpy(CO_2(CH_2CH_2O)_7CH_3)_2\}_3](ClO_4)_2$ in melt mixtures of it and the diluting, isostructural metal complex hybrid polyether $[Co\{bpy(CO_2(CH_2CH_2O)_7CH_3)_2\}_3](ClO_4)_2$, as determined by chronoamperometry at 5-μm radius Pt microdisk electrodes. D_{app} for the $M^{II/III}$ reaction in undiluted $[Co\{bpy(CO_2(CH_2CH_2O)_3CH_3)_2\}_3](ClO_4)_2$ was 5×10^{-14} cm^2 s^{-1} whereas that in undiluted $[Fe\{bpy(CO_2(CH_2CH_2O)_3CH_3)_2\}_3](ClO_4)_2$ was 5×10^{-11} cm^2 s^{-1}. (Adapted from Figs 1 and 5 of [25], copyright American Chemical Society.)

mixtures of the two complexes, enables study of the electron hopping rate in the Fe complex as it is diluted with the Co complex. Figure 6.7 shows results [25] for D_{app} for the Fe complex as it is diluted with the Co complex melt. The differences in physical and electron diffusivity in this material make it analogous to a redox polymer, and indeed at high concentrations of Fe complex the decrease in electron hopping rate with dilution follows the expectation of static percolation [29]. The high concentration data extrapolate to a threshold of about 0.4 mole fraction Fe complex. At low concentrations, however, the electron hopping rates do not decrease to negligible values (e.g. near the very small D_{phys}), as would be expected in static percolation, but exhibit a gently sloping plateau. The reason for the latter behaviour is not entirely clear, but one important implication [44] is that the macroscopic physical diffusivity D_{phys} may seriously underestimate the microscopic mobility enjoyed by the Fe complex.

4   A diode structure based on electrochemically generated mixed valent films containing frozen redox gradients

We finish this chapter with an example [70] of a molecular electronic device the conception of which was a direct result of striving for an understanding of electron

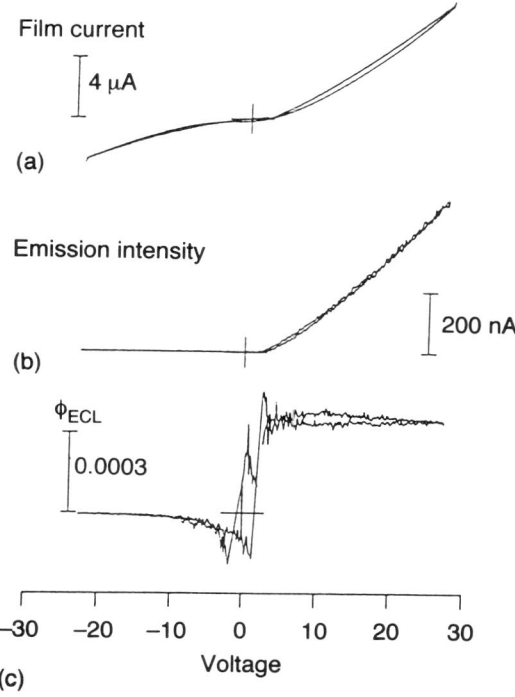

Figure 6.8 (a) Current, (b) PMT output measuring light emission, and (c) photon/electron emission efficiency for a ±25 V potential sweep around an interfinger bias of 2.6 V for a dry (0°C) film of poly[Ru(vbpy)₃](PF₆)₂ on an IDA and containing electrolytically prepared serial concentration gradients of $Ru^{II/III}$ and $Ru^{II/I}$ states. (Adapted from Fig. 3 of [70], copyright American Chemical Society.)

transport in mixed valent redox polymers. Figure 6.8(a) shows a current–potential response of a film of the redox polymer poly[Ru(vbpy)₃](ClO₄)₂ that was first electrochemically polymerized onto an interdigitated array electrode assembly then biased at a ΔE of 2.6 V in an electrolyte solution, rinsed, dried, and cooled. The effect of the pretreatment was to introduce, by disproportionation, two adjoining sets of concentration gradients of redox sites, those of $Ru^{III/II}$ originating at the positively biased electrode, and those of $Ru^{II/I}$ originating at the other. At the intersection of these gradients, Ru^{III} and Ru^{I} sites undergo energetic electron transfers, creating excited states that luminesce (Fig. 6.8b), and a region that under reverse bias (Figure 6.8a, negative potentials) becomes depleted in those sites and is thus rectifying. These responses are stable only so long as the concentration gradients generated by the initial 2.6 V polarization are preserved; thus the importance of cooling the film to reduce counterion mobility and to freeze the concentration gradients in place. This device, while prepared by electrochemistry, does not involve mass transport leading to electrolysis in its operation, and so is qualitatively different from electrogenerated chemiluminescence.

Acknowledgements. The research from this laboratory that is described in this chapter was supported by grants from the National Science Foundation, the Department of Energy, and the Office of Naval Research.

5 References

1 Mort J, Pai DM. *Photoconductivity and Related Phenomena.* Amsterdam: Elsevier, 1976.
2 Weaver MJ, McManis GE. *Acc Chem Res* 1990; **23**: 294.
3 Weaver MJ. *Chem Rev* 1992; **92**: 463.
4 Fawcett WR, Opallo M. *Angew Chem Int Ed Engl* 1994; **33**: 2131.
5 Jernigan JC, Murray RW. *J Am Chem Soc* 1987; **109**: 1738.
6 White BA, Murray RW. *J Am Chem Soc* 1987; **109**: 2576.
7 Jernigan JC, Surridge NA, Zvanut ME, Silver N, Murray RW. *J Phys Chem* 1989; **93**: 4260.
8 Feldman BJ, Murray RW. *Anal Chem* 1986; **58**: 2843.
9 Dalton EF, Murray RW. *J Phys Chem* 1991; **95**: 6383.
10 Surridge NA, Zvanut ME, Keene RF, Sosnoff CS, Silver MF, Murray RW. *J Phys Chem* 1992; **96**: 962.
11 Feldman BJ, Murray RW. *Inorg Chem* 1987; **26**: 1702.
12 Feldman BJ, Feldberg SW, Murray RW. *J Phys Chem* 1987; **91**: 6558.
13 Chidsey CED, Feldman BJ, Lundgren C, Murray RW. *Anal Chem* 1986; **58**: 601.
14 Sosnoff CS, Sullivan MG, Murray RW. *J Phys Chem* 1994; **98**: 13643.
15 Terrill RH, Sheehan PE, Long VC, Washburn S, Murray RW. *J Phys Chem* 1994; **98**: 5127.
16 Sullivan MB, Murray RW. *J Phys Chem* 1994; **98**: 4343.
17 Feldman BJ, Murray RW. *Inorg Chem* 1987; **26**: 1702.
18 Potts KT, Usifer DA, Guadalupe A, Abruña HD. *J Am Chem Soc* 1987; **109**: 3961.
19 Facci JS, Abkowitz M, Linburg W, Knier F, Yanus J, Renfer D. *J Phys Chem* 1991; **95**: 7908.
20 Abkowitz MA, Facci JS, Limburg WW, Yanus JF. *Phys Rev B* 1992; **46**: 6705.
21 Hursthouse MB, Quillin KC, Rosseinsky DR. *J Chem Soc Faraday Trans* 1992; **88**: 3071.
22 Watanabe M, Wooster TT, Murray RW. *J Phys Chem* 1991; **95**: 45.
23 Wooster TT, Watanabe M, Murray RW. *J Phys Chem* 1992; **96**: 5886.
24 Velazquez CS, Hutchinson JE, Murray RW. *J Am Chem Soc* 1993; **115**: 7896.
25 Long JW, Murray RW. *J Phys Chem* 1997; **101** (in press).
26 Velazquez CS, Murray RW. *J Electroanal Chem* 1995; **396**: 349.
27 Sharp M, Lindholm B, Lind E-L. *J Electroanal Chem* 1989; **274**: 35.
28 Murray RW. In: Murray RW, ed. *Molecular Design of Electrode Surfaces.* New York: Wiley, 1992.
29 Blauch DN, Savéant J-M. *J Am Chem Soc* 1992; **114**: 3323.
30 Blauch DN, Savéant J-M. *J Phys Chem* 1993; **97**: 6444.
31 Stauffer D. *Introduction to Percolation Theory.* London: Taylor & Francis, 1985.
32 Druger SD, Nitzan A, Ratner MA. *J Chem Phys* 1983; **79**: 3133.
33 Druger SD, Ratner MA, Nitzan A. *Mol Cryst Liq Cryst* 1990; **190**: 171.
34 Marcus RA. *Annu Rev Phys Chem* 1964; **15**: 155.
35 Marcus RA. *J Chem Phys* 1965; **43**: 679.
36 Sumi H, Marcus RA. *J Chem Phys* 1986; **84**: 4894.
37 Marcus RA, Sutin N. *Biochim Biophys Acta* 1985; **811**: 265.
38 Borsenberger PM, Magin EH, van der Auweraer M, de Schryver FC. *Phys Stat Sol* 1993; **140**: 9.
39 Burshtein AI, Kapinus EI, Kucherova I, Yu Morozov VA. *J Lumin* 1989; **43**: 291.
40 Song L, Dorfman RC, Swallen SF, Fayer MD. *J Phys Chem* 1991; **95**: 3454.
41 Dorfman RC, Liu Y, Fayer MD. *J Phys Chem* 1990; **94**: 8007.
42 Eads DD, Dismer BG, Fleming GR. *J Chem Phys* 1990; **93**: 1136.
43 Szabo A. *J Chem Phys* 1989; **93**: 6929.
44 Cicerone MT, Ediger MD. *J Chem Phys* 1995; **103**: 5684.
45 Bassler H. *Mol Cryst Liq Cryst* 1994; **252**: 11.
46 Richardson JN, Rowe GK, Carter MT *et al. Electrochim Acta* 1995; **40**: 1331.
47 Richardson JN, Peck SR, Curtin LS *et al. J Phys Chem* 1995; **99**: 766.
48 Rowe GK, Carter MT, Richardson JN, Murray RW. *Langmuir* 1995; **11**: 1797.

49 Savéant J-M. *J Electroanal Chem* 1988; **242**: 1.

50 Laviron E. *J Electroanal Chem* 1980; **112**: 1.

51 Andrieux CP, Savéant J-M. *J Electroanal Chem* 1980; **111**: 377.

52 Terrill RH, Hatazawa T, Murray RW. *J Phys Chem* 1995; **99**: 16676.

53 Majda M. In: Murray RW, ed. *Molecular Design of Electrodes Surfaces*. New York: Wiley, 1992.

54 Jernigan JC, Chidsey CED, Murray RW. *J Am Chem Soc* 1985; **107**: 2824.

55 Jernigan JC, Murray RW. *J Phys Chem* 1987; **91**: 2030.

56 Scher N, Montroll EW. *Phys Rev B* 1975; **12**: 2455.

57 Pfister G. *Phys Rev B* 1977; **16**: 3676.

58 Rosseinski DR, Tonge JS, Berthelot J, Cassidy JF. *J Chem Soc Faraday Trans* 1987; **83**: 231.

59 Rosseinski DR, Tonge JS. *J Chem Soc Faraday Trans* 1987; **83**: 245.

60 Ruff I, Botar L. *J Chem Phys* 1985; **83**: 1292.

61 Ruff I, Botar L. *Chem Phys Lett* 1986; **126**: 348.

62 Ruff I, Botar L. *Chem Phys Lett* 1988; **149**: 99.

63 Wooster TT, Longmire ML, Zhang H, Watanabe M, Murray RW. *Anal Chem* 1992; **64**: 1132.

64 Wightman RM, Wipf DO. In: Bard AJ, ed. *Electroanalytical Chemistry: A Series of Advances*, Vol. 15. New York: Marcel Dekker, 1989: 267.

65 Poupart MW, Velazquez CS, Hassett K *et al. J Am Chem Soc* 1994; **116**: 1165.

66 Haas O, Velazquez CS, Porat Z, Murray RW. *J Phys Chem* 1995; **99**: 15279.

67 Pinkerton M, LeMest I, Zhang H, Watanabe M, Murray RW. *J Am Chem Soc* 1990; **112**: 3730.

68 Young RH, Sinicropi JA, Fitzgerald JF. *J Phys Chem* 1995; **99**: 9497.

69 Bard AJ, Faulkner LR. *Electrochemical Methods*. New York: Wiley, 1980: 143.

70 Maness KM, Terrill RH, Meyer TJ, Murray RW, Wightman RM. *J Am Chem Soc* 1995; **118**: 10609.

71 Bock CR, Connor JA, Guitierrez AR *et al. Chem Phys Lett* 1979; **61**: 522.

72 Lewis TJ, White HS, Wrighton MS. *J Am Chem Soc* 1984; **106**: 6947.

73 Shu CF, Wrighton MS. *J Phys Chem* 1988; **92**: 5221.

74 Mikkelsen KV, Ratner MA. *Chem Rev* 1987; **37**: 113.

75 Chan M-S, Wahl AC. *J Phys Chem* 1982; **86**: 126.

76 Koshechko VG, Titov VE, Pokhodenko VD. *Theoret Exp Chem* 1983; **19**: 161.

77 Terrill RH, Murray RW. *J Phys Chem* 1997; **101** (in press).

7 Electrical 'Wiring' of Glucose Oxidase in Electron Conducting Hydrogels

R. RAJAGOPALAN* and A. HELLER†

*Applied Materials Inc., 3100 Bowers Avenue, Santa Clara, CA 95054, USA

†Department of Chemical Engineering, The University of Texas at Austin, Austin, TX 78712-1062, USA

1 Introduction

The transport of electrons and ions through redox polymers has been the subject of intensive study during the past two decades [1]. A subgroup of these materials, consisting of redox hydrogels, formed by cross-linking water-soluble redox polymers, is of particular interest in the context of biosensors, among which glucose sensors and hydrogen peroxide sensors are of particular importance, the first because of their relevance to management of diabetes, the second because of the utility of peroxidase labels in biorecognition reactions of DNA and immunoreactants, and the enzymatic transduction of glucose, lactate, choline and other biochemical fluxes to hydrogen peroxide in clinical chemistry [2–8].

When redox enzymes are integrated in the cross-linked redox hydrogel, electrons are transported via the redox polymer network between the enzymes' reactive centres and electrodes (Fig. 7.1). An example of a cross-linkable enzyme-connecting redox polymer is poly(4-vinylpyridine) partially complexed with osmium bis (2,2'-bipyridine) chloride and partially quaternized with 2-bromoethylamine (designated POsEA) [3] (Fig. 7.2). The dynamic electron-relaying properties of such a polymer are of importance in defining its current-carrying capacity from enzyme redox centres to electrodes. Realization of high current density electrodes, made with $1–20\ \mu$m-thick, enzyme-loaded hydrogel films, requires that both electron diffusion as well as substrate and product permeation be rapid. Water-soluble substrates and products permeate rapidly through the hydrogels, where their solubilities and diffusion coefficients approach those in water.

The electron diffusion coefficients in diepoxide-cross-linked polycationic POsEA hydrogels depend on the pH, on the nature of the counterion and on the ionic strength of the contacting aqueous solution. Specifically, electron diffusion increases when the polymer network is charged by protonation of its free pyridines and decreases either if Cl^-, a hydrophilic counterion, is replaced by ClO_4^-, a less hydrophilic counterion, or when the ionic strength is raised.

Transport of electrons, by self-exchange between redox centres of the polymer network, involves both charge propagation along the polymer's backbone (through σ and other chemical bonds) and collisions between segments of the folded polymer. The colliding segments may be spatially in each other's proximity, even if separated by a long sequence of bonds. When propagating along bonds in chains electrons hop between neighbouring redox sites, traversing occasionally cross-linker segments. Such hopping, while possible, results in a much longer path and in a more resistive route than the combination of hopping between neighbouring redox sites of chains and electron transferring collisions between redox polymer segments. Electron transport involving both σ bonds and hopping through space together accounts better for observed electron

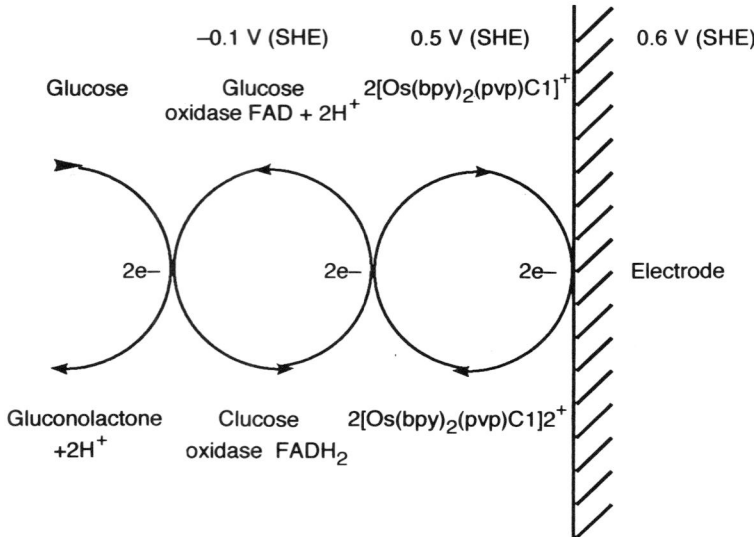

Figure 7.1 Electro-oxidation of glucose on a redox hydrogel modified electrode. The redox hydrogel consists of a cross-linked network of $[Os(bpy)_2Cl]^{+/2+}$ complexed poly(4-vinylpyridine) derivative (PVP) and glucose oxidase (bpy = 2,2-bipyridine). When the $FAD/FADH_2$ redox centres of glucose oxidase are electrically 'wired' by the redox hydrogel to the electrode, glucose is electro-oxidized to gluconolactone at potentials positive of the redox potential of the redox polymer. The current is proportional to the turnover rate of the 'wired' enzyme. When the turnover rate is glucose flux limited, the current is proportional to the flux and thus to the glucose concentration. Reprinted with permission from *Acc. Chem. Res.*, Vol. 23, No. 5, 1990, American Chemical Society.

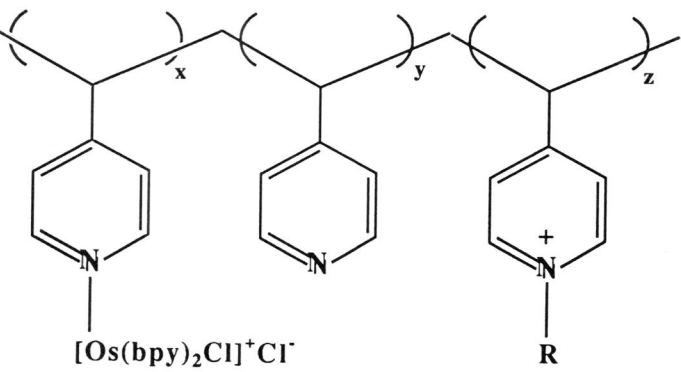

POs	: x = 1, y = 5, z = 0
POs-Me1	: x = 1, y = 4, z = 1, R = CH_3
POs-Me2	: x = 1, y = 3, z = 2, R = CH_3
POs-Me3	: x = 1, y = 2, z = 3, R = CH_3
POs-Me4	: x = 1, y = 1, z = 4, R = CH_3
POs-Ac1	: x = 1, y = 4, z = 1, R = CH_2CONH_2
POs-Ac3	: x = 1, y = 2, z = 3, R = CH_2CONH_2
POsEA	: x = 1, y = 4, z = 1, R = $CH_2CH_2NH_2$

Figure 7.2 Structure and nomenclature of the redox polymers.

transfer rates in proteins than either electron transfer exclusively along tortuous σ bond sequences or exclusively along the shortest route where electrons propagate along the shortest path by jumping across spaces between protein segments [9–11]. Because the exponential decay of the rate of electron transport along bonds of a polymer's backbone with distance is less steep than the decay of the rate of electron transport between chains, an optimum exists where the resistance along the longer route along the backbone and the shorter collision routes are equal.

Segmental motion of the polycationic polymer backbone involves and is slowed by the associated displacement of anions [12–16]. The optimal balance between electron routing along redox polymer chains and routing between chains depends on the flexibility of the redox polymer backbone [17], the nature of the redox centres [18–22], and their density [23].

2 Theoretical models

Dahms–Ruff theory [24–26], which requires a 'mean-field' approximation to hold, predicts a linear decrease in D_{app} with decreasing concentration of redox centres. The expression for electron diffusion coefficient in this case is given by:

$$D_{app} = D_{phys} + \frac{1}{6} k_{ex}\delta^2 C_{RT},$$

(6.1)

where D_{phys} is the diffusion coefficient for physical movement of the redox species, k_{ex} is the bimolecular rate constant for electron transport by self exchange, δ is the centre-to-centre distance between the electron-transferring redox species at the time of electron transfer and C_{RT} is the concentration of the redox species. The model of Saveant and Laviron [27,28] for rigorously fixed redox centres in a matrix and extended electron transfer predicts an exponential decay of D_{app} with redox centre concentration, according to Marcus theory as given by Eqn 6.2:

$$D_{app} = \frac{1}{6} k_{ex}\delta^2 C_{RT}.$$

(6.2)

In between the above two extremes is the case of a polymer having redox centres that are tethered, i.e. are free to move within a small radius about an equilibrium position. Blauch and Saveant [29] developed a 'bounded' diffusion model to predict D_{app} for this case. The expression for D_{app} for the limiting case where the displacement of the redox centres is rapid and extensive is given by:

$$D_{app} = \frac{1}{6} k_{ex}(\delta^2 + 3\lambda^2)C_{RT},$$

(6.3)

where λ is the displacement of the redox centre about equilibrium position. As will be shown here, in the case of cross-linked redox hydrogel it is λ, which is a measure of the local segmental motion, that determines the rate of electron transport. The electron diffusion coefficients increase when the gel swells, i.e. is hydrated, because upon hydration the fluidity of chain segments is increased and better mobility of chain segments increases λ. As will be seen, hydration and swelling of cross-linked redox polymers depends on the charge density on the polymer, the extent of cross-linking and the counterion. The charge density depends, in polymers that are polybases or poly-acids, on the pH. The apparent electron diffusion coefficients depend on the concentration of redox centres because the charge density and segmental fluidity are both functions of the concentration.

3   Dependence of D_{app} on pH

The dependence of D_{app} on pH for POsEA cross-linked with 5.0 wt% poly(ethylene glycol)diglycidyl ether (PEGDGE) is shown in Fig. 7.3 [30]. D_{app} increases from 4.5×10^{-9} to 1.6×10^{-8} cm^2 s^{-1} when the pH is lowered from 7.0 to the pK_a of pyridine [2]. The protonated and positively charged gel visibly expands. Although the expansion decreases the concentration of the Os sites, segmental motion of the polymer is facilitated. Electrochemical studies have revealed that with the pyridine nitrogens of POsEA protonated and the polymer swollen at pH 2.0, the distances between the Os sites was large and the sites did not interact. At pH > 4 the pyridine nitrogens are not protonated, the polymer is less swollen and the distance between the Os sites is small enough for the sites to interact.

4   Dependence of D_{app} on cross-linking

The D_{app} values for the POsEA polymer at various cross-linker concentrations are summarized in Table 7.1 [30]. In the network made with 5.0 wt% cross-linker that could swell when protonated, D_{app} was pH dependent. D_{app} was, however, almost independent of pH when the polymer film was made with 25 wt% PEGDGE and could not swell because about 32% of the pyridines were cross-linked. At such extreme cross-linking the electron transfer-affecting segmental motion of the polymer backbone was restricted.

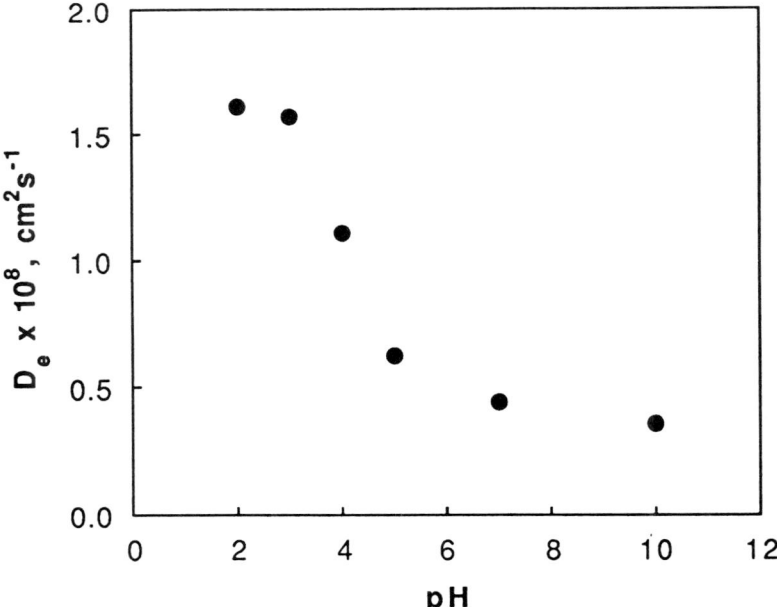

Figure 7.3 Effect of ionic strength on the electron diffusion coefficient (a) and the half-wave potential (b) of POsEA. The conditions are the same as in Figure 7.1. Reprinted with permission from *J. Phys. Chem.*, Vol. 97, No. 42, 1993, American Chemical Society.

Table 7.1 Effect of the extent of cross-linking on the electron diffusion coefficient (D_e).

Cross-linker (wt%)	D_e (cm^2 s^{-1})	
	pH 3.0	pH 7.0
5	1.6×10^{-8}	4.5×10^{-9}
25	4.2×10^{-9}	3.2×10^{-9}

5 Effect of the ionic strength and the anionic species on D_{app}

D_{app} decreases upon increasing the ionic strength, as shown in Fig. 7.4 [30]. The decrease of D_{app} with increasing ionic strength is attributed to association between $[Os(bpy)_2vpyCl]^{+/2+}$ and chloride counterions, leading to excessive cross-linking. Such cross-linking, like its covalent counterpart, prevents swelling, reduces the fluidity of the chain segments and thereby the rate of the electron transferring collisions. For Nafion films with only electrostatically bound $Os(bpy)_3^{2+/3+}$ ion, Anson and Savéant [31–33] suggested a model involving, at high ionic strengths, charge propagation by dissociation of ion pairs, electron transfer, and association of pairs. This model is applicable for low dielectric constant domains in the interior of Nafion. Association in POsEA is suggested by the fact that the polymer is soluble in 0.1 M NaCl yet precipitates from 1.0 M NaCl [30]. It would explain the observed decrease by 20 mV /decade^{-1} in the apparent half-wave potential of $Os^{2+/3+}$ upon increasing the NaCl concentration (Fig. 7.4b) [2,34,35].

When the hydrated chloride anion is replaced by the hydrophobic perchlorate anion that is not hydrated, the electron transport become sluggish [15,17,36]. Upon replacement of Cl$^-$ by ClO$_4^-$ the half-wave potential of POsEA in 0.1 M perchlorate was shifted cathodically by 40 mV to 0.245 V (SCE). This shift resulted from a change in the equilibrium constant for pairing of $[Os(bpy)_2vpyCl]^{+/2+}$ cation and the anion or from a related change in the solubility of the oxidized redox polymer [35,37,38]. In the ClO$_4^-$ system ion paring is enhanced. POsEA, that is soluble in 0.1 M NaCl, precipitates from 0.1 M NaClO$_4$ [30].

6 pH dependence of D_{app} in quaternized poly(vinylpyridine)-based redox polymers

It was seen that an increase in the charge density on the polymer backbone obtained upon lowering the pH led to an increase in D_{app} in cross-linked redox hydrogels. Hydrogels with high D_{app} near pH 7.0 are, however, required for sensors based on electron-conducting hydrogels that operate *in vivo*. For use in such biosensors, redox polymers based on poly(vinylpyridine) can be quaternized with methyl or with acetamide groups to form the polymers listed in Fig. 7.2 [39]. The variation of D_{app} with pH as a function of the fraction of the quaternized polymers is shown in Fig. 7.5 [39]. While D_{app} depends strongly on pH in the case of the non-quaternized POs, and remains pH dependent for polymers with a low degree of quaternization, it increases with increasing quaternization and becomes pH independent when a half to two-thirds of the pyridine rings are quaternized. Thus, increasing the charge on the polymer network, by either quaternization or protonation, enhances swelling and with it the rate of collisional charge transfer between chain segments. However, only quaternization leads to high D_{app} across the entire 2.0–7.0 pH range. Some D_{app} values are summa-

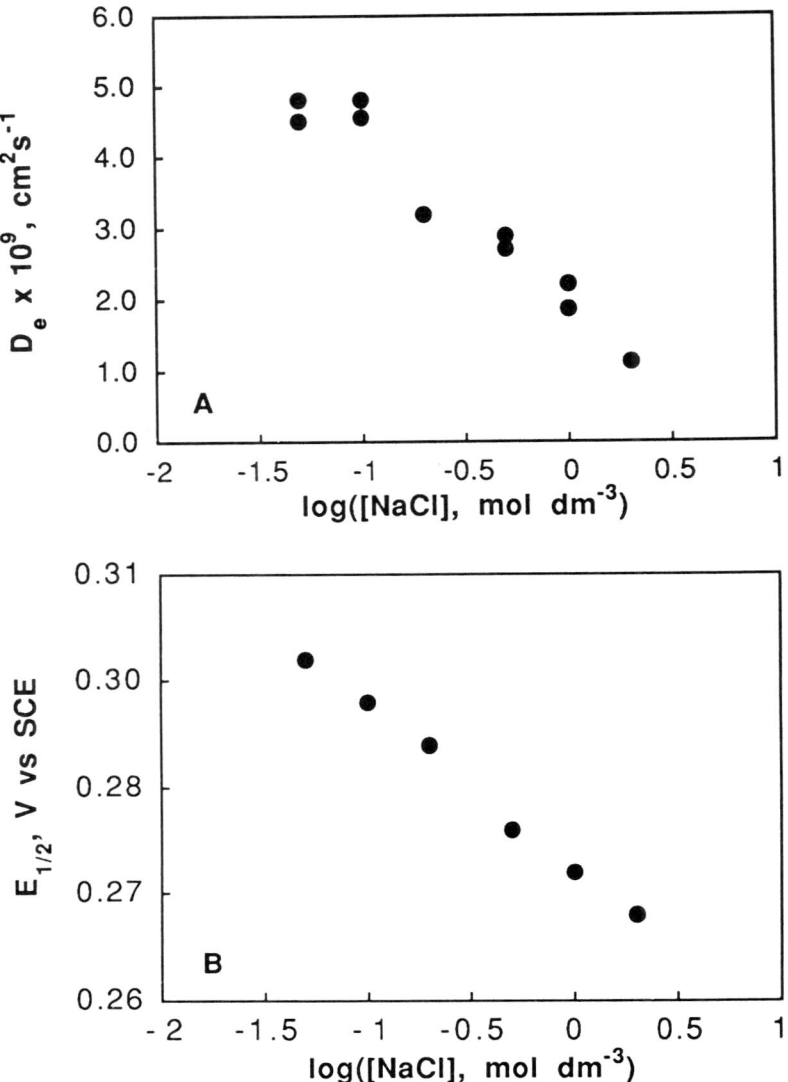

Figure 7.4 pH dependence of the electron diffusion coefficient of POsEA cross-linked with 5.0 wt% poly (ethylene glycol) diglycidyl ether (PEGDGE). 20 mM phosphate buffer; 0.1 M NaCl; 2.0 mV s^{-1}. Reprinted with permission from *J. Phys. Chem.*, Vol. 97, No. 2, 1993, American Chemical Society.

rized in Table 7.2 [39]. The D_{app} value for POs–Me$_3$ is almost identical to that of POs–Me$_4$ near 3.9×10^{-8} cm^2 s^{-1}, the maximum for the series.

The increase in D_{app} upon quaternization is not unique to methyl groups. As seen in Fig. 7.6, quaternization by the larger acetamide groups also increases D_{app} [39].

Quaternization of the polymers decreases the concentration of redox sites in the film because it leads to swelling. Very substantial swelling is seen by environmental SEM [39]. Although the concentration of redox sites decreases, the segmental motion of the polymer increases upon hydration, and with it λ increases. The increase in λ, the effective range of segmental motion of the polymer chains, more than compensates for the decrease in the concentration of redox sites and D_{app} increases.

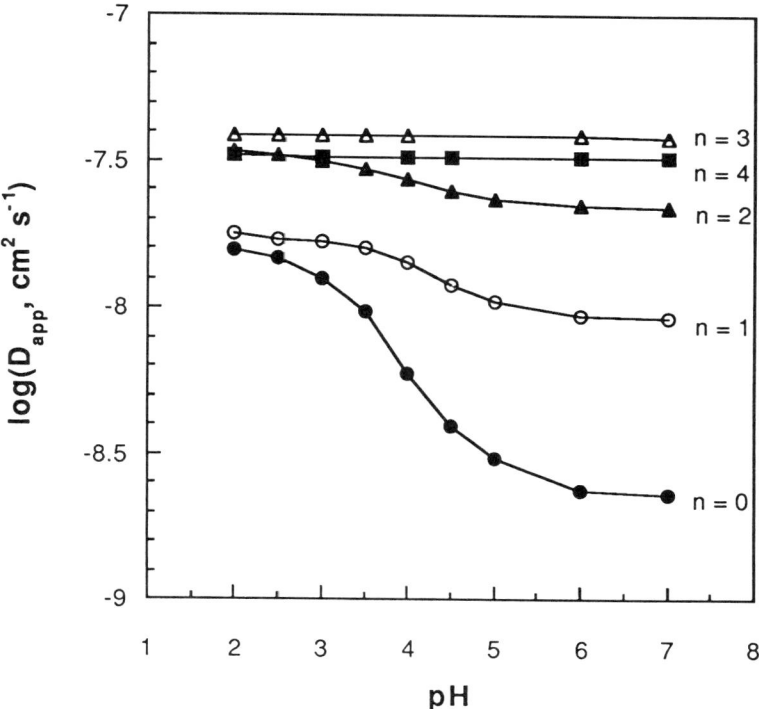

Figure 7.5 Variation of the apparent electron diffusion coefficient, D_{app}, with pH for POs (non-quaternized, solid circles), POs–Me$_1$ (one ring in six quaternized, open circles), POs–Me$_2$ (one-third of the rings quaternized, solid triangles), POs–Me$_3$ (half the rings quaternized, open triangles), and POs–Me$_4$ (two-thirds of the rings quaternized, solid squares). The values were derived from IDA voltammograms using 5 wt% cross-linked PEGDGE polymers in 20 mM phosphate, 0.1 M NaCl, at 5.0 mV s^{-1} scan rate. Reprinted with permission from *J. Phys. Chem.*, Vol. 99, No. 14, 1995, American Chemical Society.

7 Dependence of D_{app} on cross-linking for quaternized redox polymers

The dependence of D_{app} on the weight fraction of cross-linker in the film for non-quaternized POs is shown in Fig. 7.7 [39]. In the case of the little or not quaternized polymers that are not swollen at pH 7.0, D_{app} does not vary significantly with the extent of cross-linking. Evidently, in the case of the redox polymer that is neither protonated nor quaternized, the rate of charge propagation is low and is not influenced by the extent of cross-linking. However, at pH 2.0, where the polymer is protonated, D_{app}

Table 7.2 Electron diffusion coefficients of 5 wt% cross-linked redox polymer films in which a varying fraction of the pyridine rings was quaternized with methyl groups, at pH 2.0 and 7.0.

Redox polymer	Fraction of quaternized pyridines	D_{app} (cm^2 s^{-1}) pH 2.0	pH 7.0
POs	0/6	1.6×10^{-8}	2.3×10^{-9}
POs–Me$_1$	1/6	1.6×10^{-8}	9.2×10^{-9}
POs–Me$_2$	1/3	3.4×10^{-8}	2.2×10^{-8}
POs–Me$_3$	1/2	3.9×10^{-8}	3.8×10^{-8}
POs–Me$_4$	2/3	3.3×10^{-8}	3.2×10^{-8}

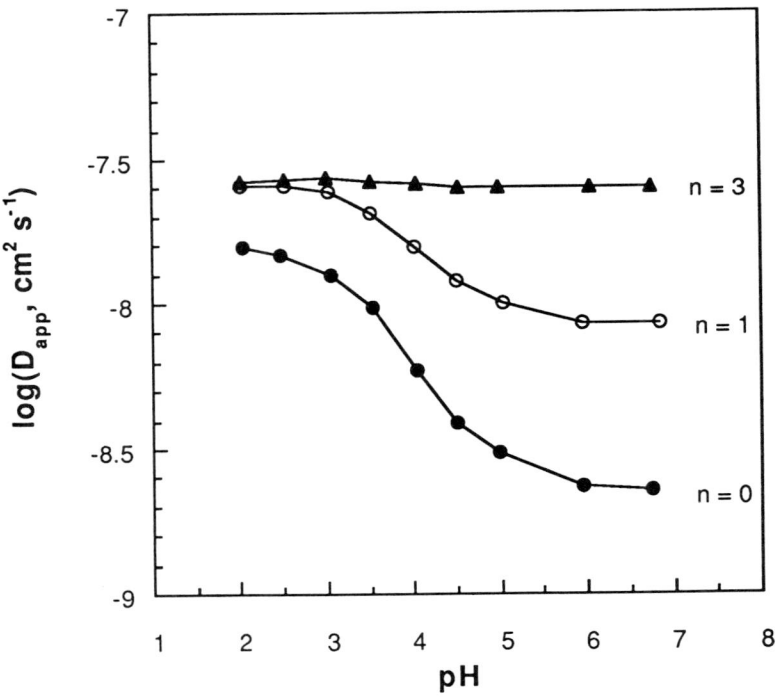

Figure 7.6 Variation of apparent electron diffusion coefficient, D_{app}, with pH for POs (non-quaternized, solid circles), POs–Ac$_1$ (one ring in six quaternized with acetamide groups, open circles), and POs–Ac$_3$ (half the rings quaternized with acetamide groups, solid triangles); conditions as in Fig. 7.4. Reprinted with permission from *J. Phys. Chem.*, Vol. 99, No. 14, 1995, American Chemical Society.

decreases with the extent of cross-linking. At very high cross-linking, D_{app} remains low and almost independent of pH. Increasing the extent of cross-linking lowers the fluidity of the polymer, i.e. makes the polymer more rigid and λ, in Eqn 7.3, becomes smaller as the extent of cross-linking is increased. Although cross-linking by epoxidation quaternizes the pyridine rings [40], the favourable effect of such quaternization is more than negated by the reduced fluidity of the polymer chains.

8 Dependence of D_{app} on loading with redox species

Plots of the dependence of D_{app} on pH for POs, POs$_3$ and POs–Me$_1$ polymers are shown in Fig. 7.8 [39]. D_{app} increases with the loading of redox sites, consistent with the predictions of Eqns 7.1 and 7.2. However, an increase in concentration of the redox sites also leads to an increase in the charge on the polymer backbone. Though the loading of POs$_3$ was twice that of POs, the total charge on reduced POs$_3$ was also twice that on POs.

To distinguish between the effects of charge and redox centre concentration on D_{app}, the D_{app} values for POs, POs$_3$ and POs–Me$_1$ are compared. POs$_3$ and POs–Me$_1$ have the same electrostatic charge, but the loading of redox sites in POs$_3$ is about twice that on POs–Me$_1$. As seen in Fig. 7.8 [39], when the polymers are not substantially protonated (pH $\geq$ 4.0), the D_{app} values of POs$_3$ and POs–Me$_1$ are nearly identical. At pH 2.0, the D_{app} values for all the three polymers converge. Thus, at the relevant redox

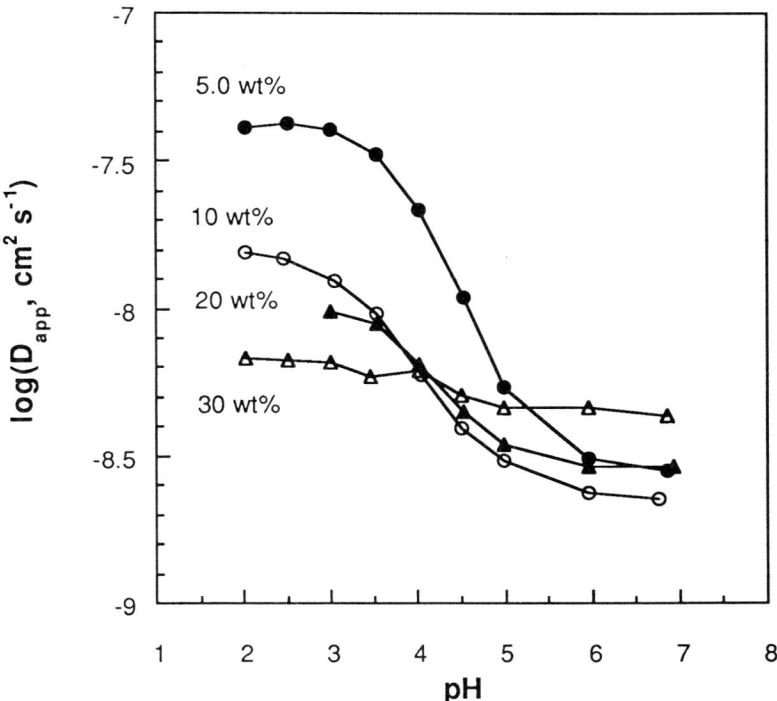

Figure 7.7 Dependence of the apparent electron diffusion coefficient, D_{app}, on the extent of cross-linking of non-quaternized POs. Conditions as in Fig. 7.4, except for the wt% of cross-linker in the films, which is shown. Reprinted with permission from *J. Phys. Chem.*, Vol. 99, No. 14, 1995, American Chemical Society.

centre loadings the electrostatic charge on the polymer backbone, rather than loading of redox sites, determines D_{app}.

9 Dependence of D_{app} on enzyme loading

Figure 7.9 shows the dependence of D_{app} on enzyme loading for redox polymers differing in their extent of quaternization by methyl groups and for POsEA [41]. D_{app} is highest for POs–Me$_3$ and lowest for POsEA at any enzyme loading. Data for 0 wt% and 60 wt% glucose oxidase (GOX) loading are summarized in Table 7.3 [41]. A similar dependence of D_{app} with enzyme loading has been observed by Surridge and coworkers for POs. [42]. Electron transport in the enzyme-loaded redox hydrogel is best described by the model of tethered redox centres that are mobile but only within a finite and small volume element (Eqn 7.3) [29]. Simulations for certain cases (e.g. $D_{phys}/D_e = 0.01$) reveal a dependence of D_{app} on C_{RT} that is close to that in Fig. 7.9.

The decrease in D_{app} upon increasing the weight fraction of enzyme may be caused by electrostatic bonding of the polycationic redox polymer by the polyanionic GOX, leading to a more rigid network, or by the decrease in the concentration of the osmium centres upon dilution of the network with GOX. Electrostatic complex formation has been shown to be necessary for electrical communication between glucose oxidase and redox hydrogels based on poly(vinylpyridine) [43]. Because the extent of swelling of pure redox polymer films, without enzymes, is directly proportional to the density of

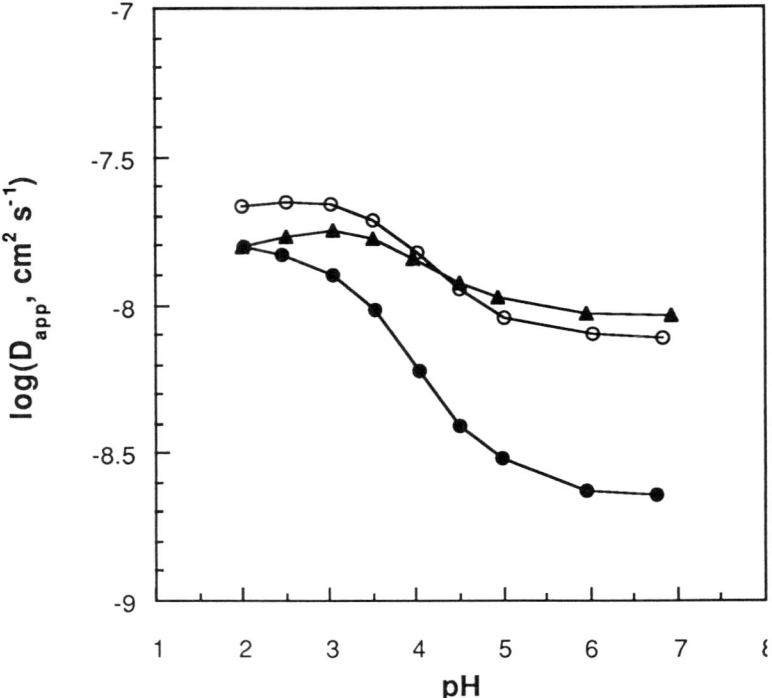

Figure 7.8 Dependence of the apparent electron diffusion coefficient, D_{app}, on the density of redox sites and on the total charge on the cross-linked polymer. POs (non-quaternized, one sixth of the rings $[Os(bpy)_2Cl]^{+/2+}$ complexed, solid circles), POs_3 (one-third of the rings $[Os(bpy)_2Cl]^{+/2+}$ complexed, open circles), and $POs-Me_1$ (one-sixth of the rings $[Os(bpy)_2Cl]^{+/2+}$ complexed, solid triangles). POs_3 and $POs-Me_1$ have the same electrostatic charge when the osmium centres are reduced. 5 wt% cross-linking, conditions as in Fig. 7.4. Reprinted with permission from *J. Phys. Chem.*, Vol. 100, No. 9, 1996, American Chemical Society.

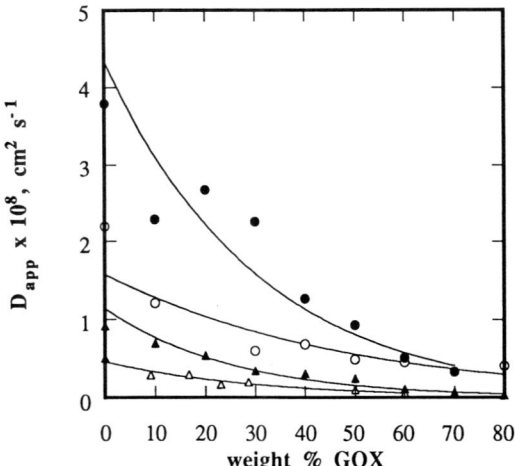

Figure 7.9 Dependence of the electron diffusion coefficient on enzyme loading for $POs-Me_3$ (solid circles), $POs-Me_2$ (open circles), $POs-Me_1$ (solid triangles), and POsEA (open triangles). The solid lines represent an exponential curve fit through the data points. All IDA electrodes were made with 2.5 µg polymer and 10 wt% Poly (ethylene glycol) diglycidyl ether (PEGDGE). Reprinted with permission from *J. Phys. Chem.*, Vol. 100, No. 9, 1996, American Chemical Society.

Table 7.3 Values of the apparent electron diffusion coefficients for redox hydrogels made of polymers varying in their extent of quaternization without GOX and at 60 wt% GOX. Reprinted with permission from *J. Phys. Chem.*, Vol. 100, No. 9, 1996, American Chemical Society.

Polymer	D_{app} (cm^2 s^{-1})	
	0 wt% GOX	60 wt% GOX
POsEA	5.1×10^{-9}	5.72×10^{-10}
POs–Me$_1$	9.2×10^{-9}	1.06×10^{-9}
POs–Me$_2$	2.19×10^{-8}	4.55×10^{-9}
POs–Me$_3$	3.79×10^{-8}	5.13×10^{-9}

positive charges on the polymer backbone, the enzyme-containing redox polymer films are unlikely to swell to the same degree as the pure polymer. As the polymer complexes with the enzyme, the charge on the polymer backbone is neutralized. The network becomes more rigid and the segmental motion of the polymer chains, and hence the values of D_{app}, are diminished. Thus, both electrostatic cross-linking and dilution of redox centres contribute to the decrease in D_{app} with the weight fraction of enzyme.

10   Dependence of j_{max} on enzyme loading

Figure 7.10 [41] shows the dependence of j_{max} on the weight fraction of enzyme in the film. For the various polymers j_{max} increases with weight fraction at low enzyme loading, reaches a maximum at 30–40 wt% GOX and then decreases [3,5,7,8]. j_{max} varies in the order POs–Me$_2$ > POs–Me$_1$ > POsEA. At 40 wt% GOX, the values of j_{max} are 944, 722 and 548 µA cm^{-2}, respectively, for the three polymers when thin films are applied.

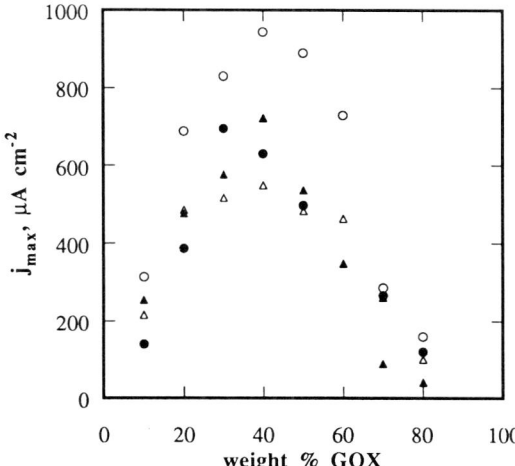

Figure 7.10 Dependence of j_{max} on GOX loading for POs–Me$_3$ (solid circles), POs–Me$_2$ (open circles), POs–Me$_1$ (solid triangles), and POsEA (open triangles). Electrodes were made with 2.5 µg polymer and 10 wt% poly (ethylene glycol) diglycidyl ether (PEGDGE). Argon; 0.5 V (SCE); 1000 rpm. Reprinted with permission from *J. Phys. Chem.*, Vol. 100, No. 9, 1996, American Chemical Society.

11   Conclusions, applications, and outlook

Electron conduction, with the electron diffusion coefficient at 25°C exceeding 10^{-8} cm^2 s^{-1}, is realized in redox hydrogels. When enzymes are integrated by covalent bonding into the redox polymer network of the hydrogels, their redox centres can be electrically 'wired' to electrodes coated by the gels. Because the conductive hydrogels are permeable to water-soluble biochemicals, high (10^{-3} A cm^{-2}) current densities are reached in 'wired' enzyme-modified electrodes when the rate of enzyme turnover is rapid (>100 s^{-1}). In this case, practical biosensors can be miniaturized to ~100 μm diameter, and research devices to 1–10 μm diameter. Miniature sensors based on wired glucose oxidase now measure the glucose concentration in as little as ~0.5 μl of blood serum and plasma. They enable the reduction of the volume of the fluid withdrawn when a diabetic patient measures his or her glucose concentration. More importantly, subcutaneously implanted, continuously operating miniature (0.29 μm diameter) glucose sensors, the presence of which is not felt by the diabetic patient, are in hand. These sensors, based on hydrogels in which both the electron-transporting polymer and enzyme are immobilized through covalent binding, have no leachable components. Furthermore, unlike earlier subcutaneously implanted glucose sensors, the 'wired' glucose oxidase sensors do not require molecular oxygen for their operation. They are at this time the only ones that can be calibrated, while implanted, by withdrawal of a single sample of fluid and its independent assay [44,45]. Such calibration is made possible by the selectivity of the sensors and the linear variation of their current output with glucose concentration through the entire physiological range. The novel microsensors enable physiological studies of the transient dynamic differences between subcutaneous and venous blood glucose concentrations. The physiological studies have already led to the finding that the transport of glucose from the blood to the subcutaneous tissue is a facilitated process, i.e. is controlled by membrane channels that close at high glucose concentration, wherefore the delay between the rise or maxima of blood and subcutaneous glucose levels is glucose dose dependent, increasing at high glucose doses [46].

The first 'wired' enzyme sensors of chemicals and biochemicals other than glucose are now commercially available. Hydrogen peroxide in biological fluids is detected by a sensor developed by E. Heller & Co. of Alameda with BAS of West Lafayette, Indiana. The sensor utilizes a 'wired' peroxidase on which hydrogen peroxide is electroreduced to water at 0 V (SCE). Unlike the earlier used platinum sensor for hydrogen peroxide, on which H_2O_2 was electro-oxidized to O_2 at 0.7 V (SCE), the wired enzyme sensor is not fouled in the biological environment and its signal to noise ratio is superior to that of platinum by a factor of 10. A wired thermostable peroxidase electrode has been operated at temperatures as high as 75°C [47]. Wired lactate oxidase microsensors have been developed by E. Heller & Co. for continuous reporting of the state of tissue oxygenation to surgeons during surgery. The 'wiring' of a DNA-labelling peroxidase has recently enabled amperometric sensing of DNA hybridization [48]. Overall, the 'wiring' of enzymes forms the basis for unprecedentedly small *in vitro* and *in vivo* biosensors. As better electron-conducting hydrogels evolve, their minaturization will continue and will match the shrinking dimensions of devices in integrated circuits.

Acknowledgements. This work was supported by the Office of Naval Research, National Science Foundation, and the Robert A. Welch Foundation.

References

1 Murray RW. In: Bard AJ, ed. *Electroanalytical Chemistry.* New York: Marcel Dekker, 1984.
2 Gregg BA, Heller A. *Anal Chem* 1990; **95**: 5970.
3 Gregg BA, Heller A. *Anal Chem* 1990; **95**: 5976.
4 Heller A. *J Phys Chem* 1992; **96**: 3579.
5 Katakis I, Heller A. *Anal Chem* 1992; **64**: 1008.
6 Ohara TJ, Rajagopalan R, Heller A. *Anal Chem* 1993; **24**: 3512.
7 Ohara TJ, Rajagopalan R, Heller A. *Anal Chem* 1994; **94**: 2451.
8 Rajagopalan R, Ohara TJ, Heller A. In: Usmani AM, Akmal N, eds. *Diagnostic Biosensor Polymers* ACS Symposium Series 556. Washington, DC: American Chemical Society, 1994.
9 Beratan DN, Onuchic JN, Hopfield JJ. *J Chem Phys* 1987; **86**: 4488.
10 Betts JN, Beratan DN, Onuchic JN. *J Am Chem Soc* 1992; **114**: 4043.
11 Onuchic JN, Beratan DN. *J Chem Phys* 1990; **92**: 722.
12 Lyons MEG, Fay HG, Vos JG, Kelly AJ. *J Electroanal Chem* 1988; **250**: 207.
13 Forster RJ, Kelly AJ, Vos JG, Lyons MEG. *J Electroanal Chem* 1989; **270**: 365.
14 Forster RJ, Vos JG. *Macromolecules* 1990; **23**: 4372.
15 Forster RJ, Vos JG, Lyons MEJ. *J Chem Soc Faraday Trans* 1991; **87**: 3761.
16 Forster RJ, Vos JG. *J Electrochem Soc* 1992; **139**: 1503.
17 Oh SM, Faulkner LR. *J Electroanal Chem* 1989; **269**: 77.
18 Surridge NA, Zvanut ME, Keene FR *et al. J Phys Chem* 1992; **96**: 962.
19 Chidsey CED, Feldman BJ, Lundgren C, Murray RW. *Anal Chem* 1986; **58**: 601.
20 Dalton EF, Surridge NA, Jernigan JC *et al. Chem Phys* 1990; **141**: 143.
21 Feldman BJ, Murray RW. *Anal Chem* 1986; **58**: 2844.
22 Feldman BJ, Feldberg SW, Murray RW. *J Phys Chem* 1987; **91**: 6558.
23 Facci JS, Schmehl RH, Murray RW. *J Am Chem Soc* 1982; **104**: 4959.
24 Dahms H. *J Phys Chem* 1968; **72**: 377.
25 Ruff I, Friedrich VJ. *J Phys Chem* 1971; **75**: 3297.
26 Ruff I, Friedrich K, Csalliag K. *J Phys Chem* 1971; **75**: 3303.
27 Andreauix CP, Saveant JM. *J Electroanal Chem* 1980; **111**: 377.
28 Laviron E. *J Electroanal Chem* 1980; **112**: 1.
29 Blauch DN, Saveant JM. *J Phys Chem* 1993; **97**: 6444.
30 Aoki A, Heller A. *J Phys Chem* 1993; **97**: 11014.
31 Anson FC, Blauch DN, Saveant JM, Shu CF. *J Am Chem Soc* 1991; **103**: 1922.
32 Saveant JM. *J Phys Chem* 1988; **92**: 4526.
33 Saveant JM. *J Phys Chem* 1988; **92**: 1011.
34 Inzelt G. *Electrochim Acta* 1989; **34**: 83.
35 Inzelt G, Szabo L. *Electrochim Acta* 1986; **31**: 1381.
36 Dalton EF, Murray RW. *J Phys Chem* 1991; **95**: 6383.
37 Oosawa F. *Polyelectrolytes.* New York: Marcel Dekker, 1971.
38 Marcus Y. *Ion Solvation.* New York: Wiley, 1985.
39 Aoki A, Rajagopalan R, Heller A. *J Phys Chem* 1995; **99**: 5102.
40 Potter WG. *Epoxide Resins.* London: The Plastics Institute, 1958.
41 Rajagopalan R, Aoki A, Heller A. *J Phys Chem* 1996; **100**: 3719.
42 Surridge NA, Diebold ER, Chang J, Neudeck GW. In: Usmani AM, Akmal N, eds. *Diagnostic Biosensor Polymers.* ACS Symposium Series 556. Washington, DC: American Chemical Society, 1994.
43 Katakis I, Ye L, Heller A. *J Am Chem Soc* 1994; **116**: 3617.
44 Csöregi E, Quinn CP, Schmidtke DW *et al. Anal Chem* 1994; **66**: 3131.

45 Csöregi E, Schmidtke DW, Heller A. *Anal Chem* 1995; **67**: 1240.
46 Quinn CP, Pathak CP, Heller A, Hubbell JA. *Biomaterials.* 1995; **16**: 389.
47 Vreeke MS, Yong KT, Heller A. *Anal Chem* 1995; **67**: 4247.
48 De Lumley-Woodyear T, Campbell CN, Heller A. *J Am Chem Soc* 1996; **118**: 5504.

8 Resonant Tunnelling and Molecular Rectification in Langmuir–Blodgett Films

S. ROTH, M. BURGHARD and C.M. FISCHER

Max-Planck-Institut für Festkörperforschung, Heisenbergstr. 1, D-70569 Stuttgart, Germany

1 Introduction

The trend in solid-state electronics is towards smaller and smaller devices, from microelectronics to nanoelectronics and further, and logically the process must end with molecular electronics [1–6]. If guided by the concepts of present-day semiconductor technology, one would attempt to design and synthesize molecules with transistor functions, test these molecules, then arrange and interconnect them, finally synthesizing a molecular computer. It is more likely, however, that the concepts will be generalized and that molecular transistors will not necessarily be the basic units of molecular electronics, as is already evident in the discussion of cellular automata [7], neural networks [8], holographic memories [9] and others.

The present contribution deals with an initial step towards molecular electronics: rectifying molecules, rectifying molecular layers, and resonant tunnelling through molecular orbitals in such layers.

Evidently, single molecules cannot be clamped by alligator clips or soldered to copper leads. There are attempts to bridge a narrow gap in a microlithographic metal strip by chain-like molecules [10] or to approach a molecule with the tip of a scanning tunnelling microscope (STM) [11,12] and thus to measure current voltage characteristics and rectifying properties of individual molecules. Our approach is to incorporate the molecules into well-ordered thin films (Langmuir–Blodgett (LB) films) and to sandwich these films between evaporated metal electrodes. As a result, a large number of molecules will be contacted simultaneously. The route to measurements on single molecules would then be by dilution, i.e. by mixing active molecules with inert molecules so that in the future experiments would be carried out on only a very few molecules, or perhaps on only a single molecule [13].

2 The concept of molecular rectification

The standard semiconductor rectifier is a p–n junction. In Fig. 8.1 we show the well-known band structure of a semiconductor together with relevant molecular orbitals of an organic molecule with conjugated double bonds (such as benzene, phthalocyanine, polyacetylene, etc.). In a semiconductor the bonding orbitals form the valence band and the anti-bonding orbitals form the conduction band. In a molecule with conjugated double bonds the respective molecular orbitals are denoted π and π^*. A p–n junction is formed by doping. The energy bands are shifted, upwards on the p and downwards on the n side, due to the electrostatic charges of the dopant ions, which repel or attract the electrons. In an analogous way the energetic position of the molecular π and π^* orbitals also depends on the chemical environment, which modifies

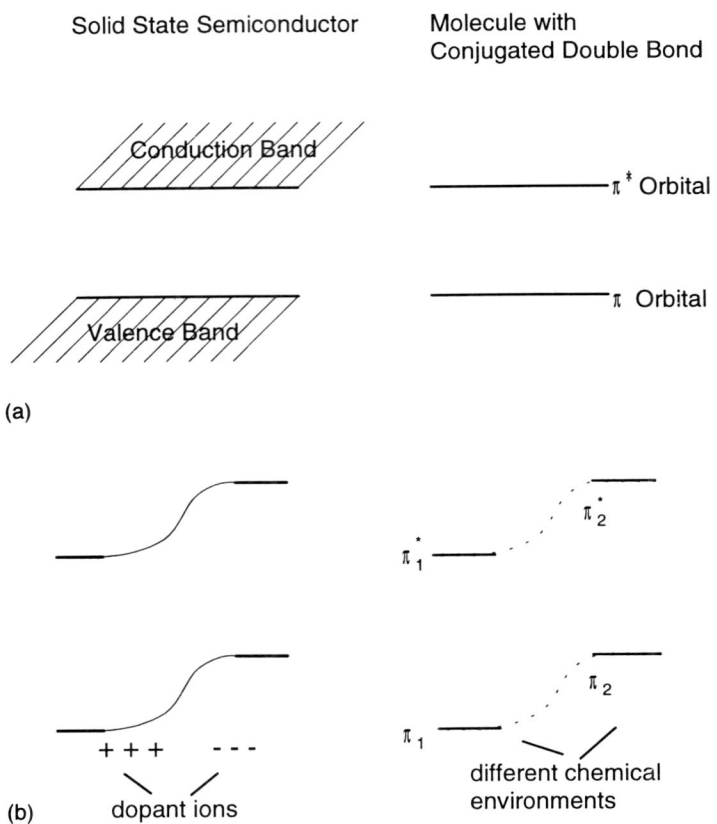

Figure 8.1 Analogy between band structure in a semiconductor p–n junction and the molecular orbitals of σ-bridged conjugated systems: (a) undoped semiconductor and π, π* orbitals; (b) p–n junction and σ-bridged conjugated systems.

electron affinity and ionization energy. As a consequence, the molecular orbitals of conjugated systems in different chemical environments resemble the band structure in a p–n junction.

In 1974 Aviram and Ratner [14] proposed a 'rectifying molecule' composed of tetrathiafulvalene (TTF) and tetracyanoquinodimethane (TCNQ) entity connected by a saturated bridge. The bridge is denoted by σ, as opposed to the unsaturated π electron systems of TTF and TCNQ. The species TTF is a well-known organic electron donor and TCNQ a well-known acceptor. Single crystals with alternating TTF and TCNQ stacks can be grown, and in such a 'charge transfer salt' electrons are transferred from the donor to the acceptor, leaving the respective conjugated systems partially charged. In the crystal, molecular orbitals within a stack interact to form one-dimensional bands and due to partial band filling TTF–TCNQ is a one-dimensional organic metal. In the 1970s TTF–TCNQ was very popular as a possible candidate for high temperature superconductivity, and an article on 'giant conductivity' [15] in this material has become the most often quoted paper of the journal *Solid State Communications*. An introductory book on the physics of one-dimensional metals is given in [16].

The chemical structure of the Aviram–Ratner molecule is shown in Fig. 8.2. Figure 8.3 shows the electronic-level scheme, for zero bias, and forward and reverse bias. The molecule is sandwiched between two metal contacts with work function φ.

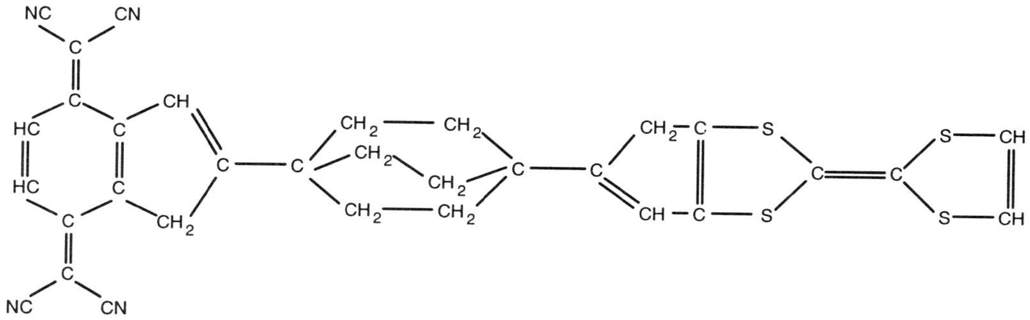

Figure 8.2 Chemical structure of the σ-bridged donor–acceptor molecule proposed by Aviram and Ratner as a molecular rectifier [14].

There are thin external tunnelling barriers A and C between the contacts and the π orbitals, and there is a thicker barrier B corresponding to the σ bridge. At zero bias the electrons have to tunnel through the barriers A, B, and C as well as through the layer occupied by the TTF and TCNQ entities. But at a small forward bias (Fig. 8.3(b)) the level π_1^* is in resonance with the Fermi level of the metal contact on the left hand side and at the same bias or a bias value close nearby, π_2 is in resonance with the Fermi level on the right hand side. Electrons and holes are injected into the TTF and TCNQ systems and recombine via tunnelling through barrier B. In reverse bias a much higher voltage has to be applied to bring π_1 in resonance, when a first step in the current–voltage characteristic will occur. A second step is expected for resonance with π_2^* at the right hand side. The analysis shows that the Aviram–Ratner molecule behaves very similar to a p–n junction in a conventional semiconductor.

3   Scanning tunnelling microscope experiments

Soon after the invention of the STM [17] this device was used to contact single molecules and verify molecular rectification [11]. The experimental set-up is shown in Fig. 8.4.

A gold bottom electrode is evaporated onto a silicon wafer. Then a monolayer of Aviram–Ratner molecules is deposited by a self-assembly technique and finally the tip of an STM is lowered towards the rectifying molecules. It is hoped that the tunnelling current will select the closest molecule and that the current–voltage characteristic of the set-up is determined by the molecular properties of the organic layer. The current–voltage characteristic obtained by Aviram et al. [11] is shown in Fig. 8.5. Indeed, typical rectifying behaviour is observed. The merit of the experiment, however, is less in the proof of molecular rectification but rather in the demonstration that such a proof would be within experimental range. The asymmetry observed by Aviram et al. is not necessarily dominated by the molecule properties in the organic layers. If the metals used as top and bottom electrodes have different work functions, this will lead to an asymmetric I/V characteristic by itself. Even if nominally the same metal is used for the bottom electrode and for the STM tip, the work function might be quite different, as the work function of a metal strongly depends on the crystallographic orientation of the surface exposed and on the quality of the surface [18]. In fact, a more careful

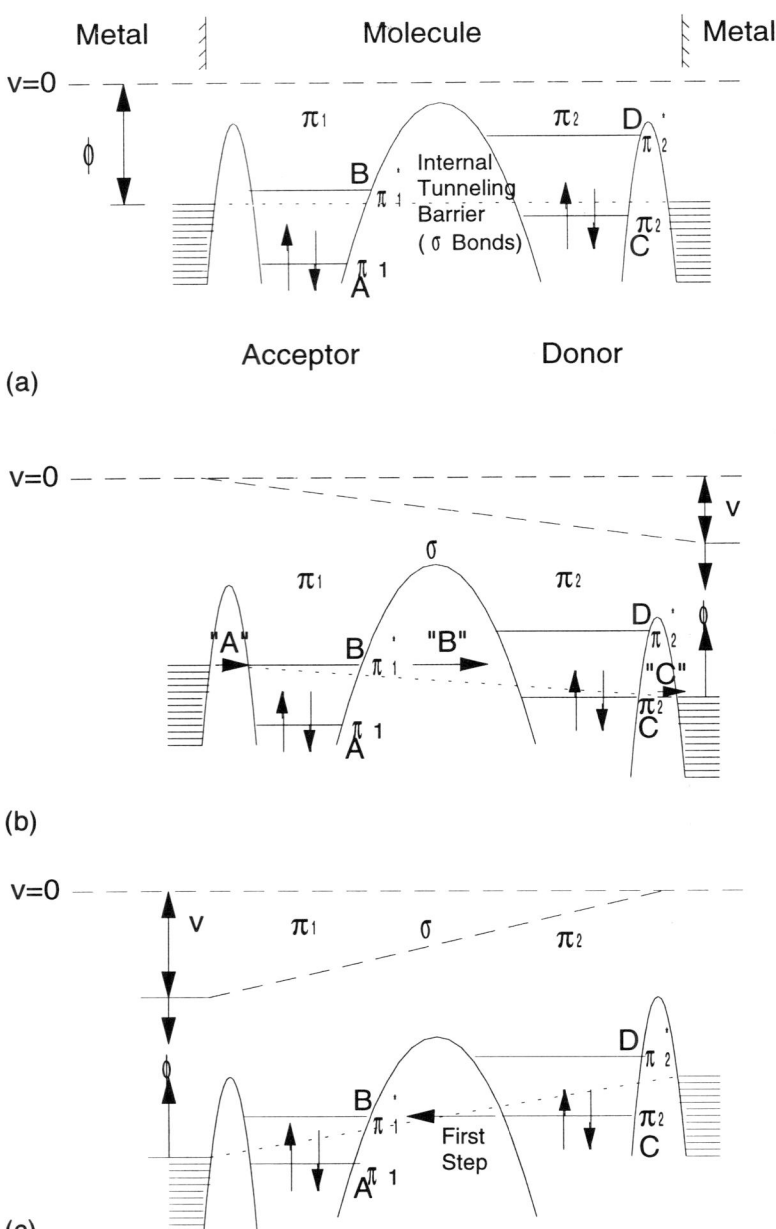

Figure 8.3 Level scheme of an Aviram–Ratner molecule sandwiched between two metal electrodes: (a) at zero bias; (b) forward bias; (c) reverse bias (after Aviram and Ratner [14]).

preparation of the rectifying molecular layer has led to an asymmetry reduction in the I/V characteristic instead of the expected increase [19].

Very recently Stabel *et al.* [12] have investigated epitactic layers of alkylated hexabenzocoronen on highly oriented pyrolytic graphite. Again, using the STM tip they have measured current–voltage characteristics. Two types of characteristics were observed, as shown in Fig. 8.6: symmetric curves when the tip was positioned on the molecule's alkyl segment, and rectifying curves for tip positions over the molecules aromatic core. The symmetry of curve (a) is taken as evidence that in this case the

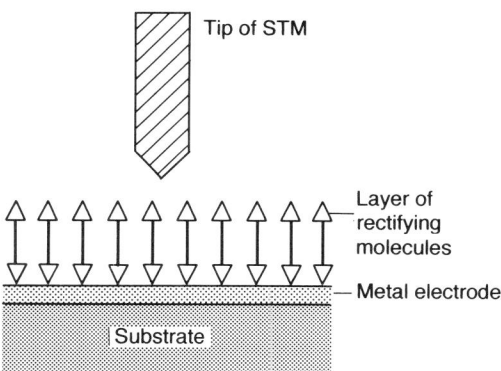

Figure 8.4 Experimental set-up to measure molecular rectification using a scanning tunnel microscope (STM) (after Aviram et al. [11]).

electrode configuration alone does not lead to rectification and that the observed asymmetry is really caused by the molecule's conjugated system. As in our LB film, the rectifying behaviour is not caused by the original Aviram–Ratner mechanism, but is attributed to the geometrical asymmetry of the conjugated system with respect to the electrodes.

4 Langmuir–Blodgett films

The LB technique is a well-established method to produce ordered molecular layers [20,21]. The essential features are illustrated in Fig. 8.7(a): amphiphilic molecules (i.e. molecules with a hydrophilic and a hydrophobic part) are dissolved in an organic solvent. The solution is poured over ultrapure water and the molecules form a monomolecular layer on the water surface (Langmuir film). By the motion of a swimming barrier this layer is compressed until it is completely dense. The force acting on the barrier is measured and plotted vs the film area. This yields the surface

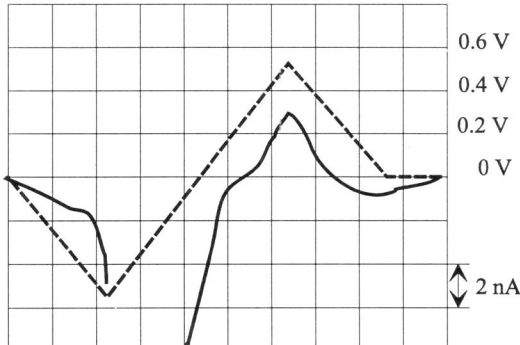

Figure 8.5 Current–voltage characteristics of rectifying molecules contacted by the tip of a scanning tunnel microscope (STM), as measured by Aviram et al. [11]. Shown are the oscilloscope traces of the applied voltage (dashed line), which varies between ± 0.5 V, and of the current response (solid line). In reverse bias the current rises to 3 nA at 0.5 V, whereas it leaves the screen ($I \gg 8$ nA) for forward bias.

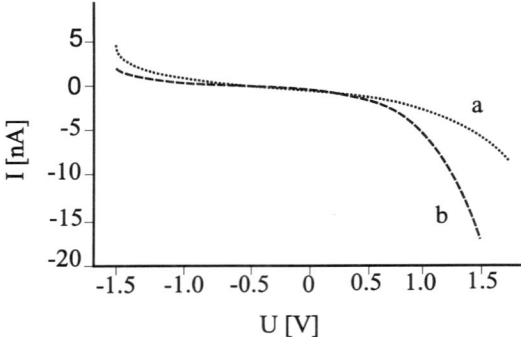

Figure 8.6 Current–voltage characteristics of alkylated hexabenzocoronen adsorbed on highly oriented pyrolytic graphite. Curve (a) reversed over the alkyl part of the molecules; curve (b) reversed over the aromatic part. (After Stabel *et al.* [12].)

pressure/area diagrams, of which a typical example is shown in Fig. 8.8. The first kink when coming from the large area side corresponds to the formation of a liquid-like monolayer, the second kink to the formation of a solid-analogue phase, and at the third kink the molecules begin to stack one on top of the next. From the initial first kink

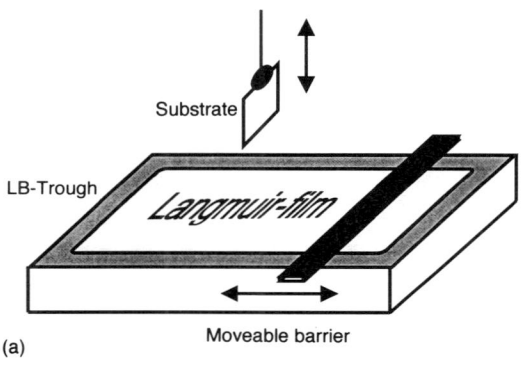

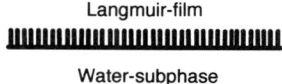

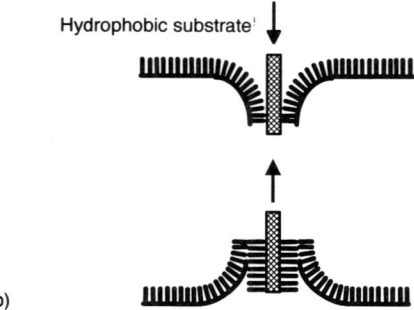

Figure 8.7 LB technique: (a) compression of monolayer at the air–water interface and (b) subsequent transfer onto solid support upon dipping through the monofilm.

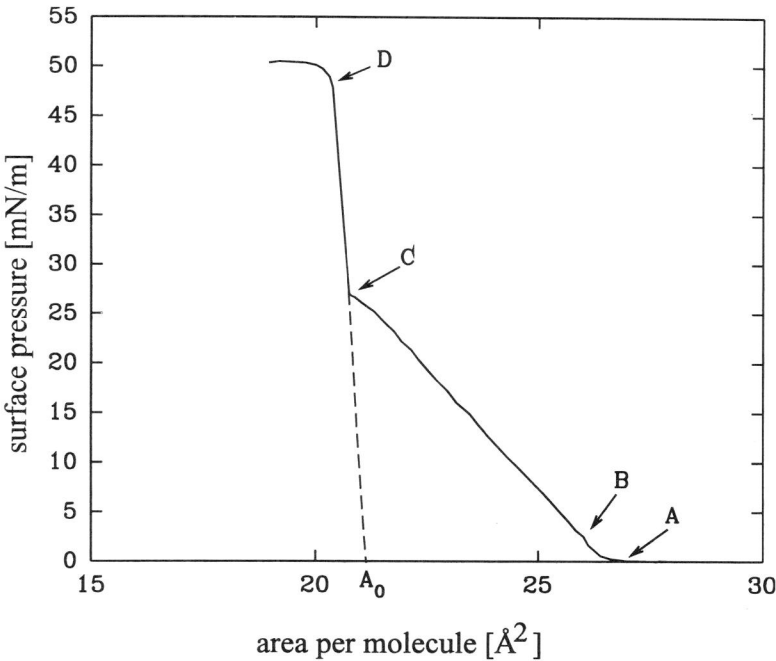

Figure 8.8 Surface pressure/area isotherm of stearic acid on acidic water subphase. Up to A, a gas-like phase exists, between B and C, a liquid-expanded film and between C and D, a two-dimensional solid is formed (collapse occurs at D).

position the area occupied by a molecule can be calculated, and this information usually gives initial evidence of how molecules are oriented in the film.

If a solid substrate (glass, silicon, etc.) is dipped into the Langmuir film, the film can be transferred to the substrate (LB film) and from the floating film area decrease the transfer rate can be checked. Figure 8.7(b) shows the case where by dipping and withdrawing of the substrate a bilayer is formed, and multiple dips lead to multilayers. In this case a centrosymmetric Y-type LB film is created. If transfer occurs only upon downstroke, a non-centrosymmetric head-to-tail arrangement of the monolayers is formed (X-type film); for the case of exclusive transfer upon upstroke, a Z-type film is obtained.

Electrical transport through LB layers has been studied since the early 1980s [22]. In 1985 Fujihira *et al.* [23] measured light-induced currents through LB monolayers from molecules functionalized with donor and acceptor moieties. Since about 1990 the Thin Film and Interface Group at Exeter [24–27] has reported on rectifying properties of LB films from the zwitterionic molecule $C_{16}H_{33}$–γQ3CNQ. The chemical structure of this molecule is shown in Fig. 8.9. In contrast to the Aviram–Ratner molecular σ-bridge (Fig. 8.2), this molecule has a π-bridge between the conjugated systems and due to strong coupling via the π electrons there is charge transfer already in the ground state and hence the molecule has a permanent dipole moment. The LB film of five monolayers is sandwiched between a silver bottom electrode on a glass substrate and magnesium pads as top electrodes. Via additional silver layers and gallium–indium eutectic droplets the magnesium pads are connected to gold wires and a sophisticated

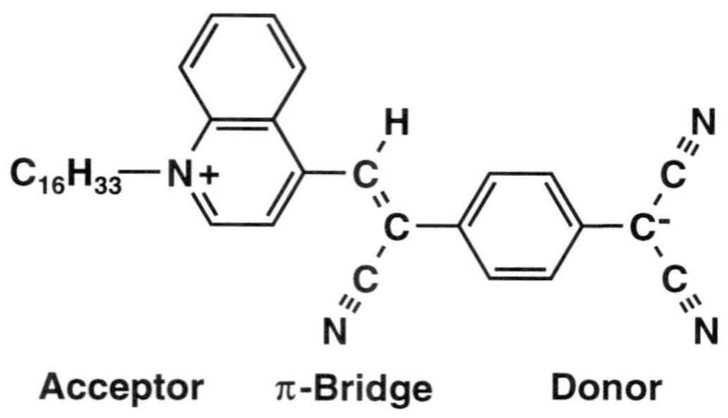

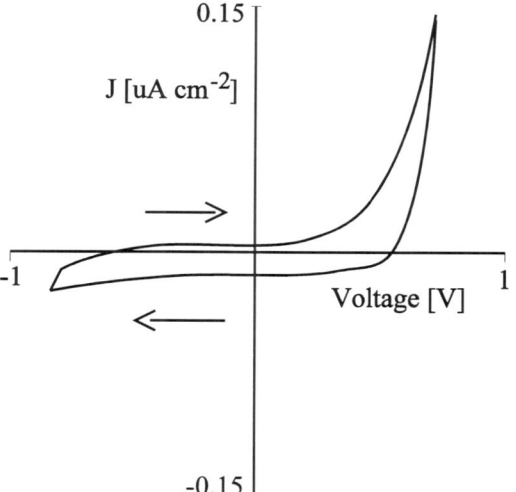

Acceptor   **π-Bridge**   **Donor**

Figure 8.9 Chemical structure of $C_{16}H_{33}-\gamma Q3CNQ$.

contact geometry allows for four-probe measurements of the current/voltage characteristics.

A current/voltage characteristic from [28] is reproduced in Fig. 8.10. The rectifying
behaviour is clearly seen. To establish that the zwitterionic molecules are responsible
for rectification and not the electrodes, which are made of metals with different work
functions, and to exclude the influence of Schottky barriers at the interfaces to the
electrodes, the experiment was repeated with a new set-up, in which the electrodes were
separated from the active zwitterionic layer by inert layers of ω-tricosenic acid.
Rectification was still observed as only the current density was lower due to the
additional resistance of the inert layers. When tricosenic acid alone was used, without
zwitterionic molecules, the current–voltage characteristics were symmetrical. This is
strong evidence for molecular rectification. Further evidence is obtained from the fact
that rectification can be suppressed by 'bleaching' the zwitterions, i.e. by chemically
modifying the film in such a way that there is no charge transfer from donor to acceptor
in $C_{16}H_{33}-\gamma Q3CNQ$. So evidently rectification is associated with the zwitterionic
molecules molecular nature. The detailed role of these molecules, however, is not yet
clear. Reviews on the 'Quest for unimolecular rectifiers' have been written by Metzger
and Panetta [29–31].

Figure 8.10 Current density–voltage curve from an Ag/seven monolayer $C_{16}H_{33}-\gamma Q3CNQ$/Mg
junction. The voltage sweep rate is set at 44 mV s^{-1}.

5   The Stuttgart approach

For a semiconductor physicist, evaporated gold electrodes would be the most obvious choice for contacting layered devices. But gold has a tendency to form small filaments which penetrate into the LB films [32]. This will happen noticeably at film imperfections, such as pinholes and grain boundaries. Therefore the idea arose to miniaturize the experimental set-up and to evaporate such small contact pads that the measurement is carried out between the pinholes. (The philosophy is related to the biophysical gigaseal technique by Neher and Sakmann [33], where a small piece of biological membrane is contacted so that on the statistical average there is only one conduction channel per contact.)

Miniaturization is the key idea of the Stuttgart approach for contacting electroactive LB films, but the experimental success is established by the combination of the following five requirements [34–36].

1 Optimization of molecular structure and of the parameters for the LB film transfer and extensive characterization of the films by physico-chemical methods (pressure–area isotherms for the films on water, IR and UV/visible optical investigations of the films on the substrates, X-ray diffraction, UPS, and surface plasmon attenuation).

2 Technique for 'soft' evaporation of gold top contacts so that the delicate films are not damaged.

3 Miniaturization of the device area, so that only pinhole-free sections are contacted.

4 Multiplexation, i.e. implementation of many devices on a chip, so that even at low yield rate there are at least some devices without short circuits (at the beginning of these investigations the yield was less than 10%).

5 Measurements at liquid helium temperature, so that leakage currents are frozen out and current/voltage characteristics show well-defined structures.

Figure 8.11 shows schematically the sample mounting. There are 50 gold stripes (only three are shown in the figure) which are evaporated onto a properly cleaned BK7 glass substrate. They act as bottom electrodes. The stripes are typically 20 μm wide and 18 nm thick. The interspacing is 30 μm. The substrate is then dipped into the LB trough to deposit the LB film. The bottom gold stripes must be as thin as possible, otherwise the LB film would break at the edges. Film rupture is one of the biggest problems in this experiment. Finally another narrow gold strip (100 μm) is evaporated as top electrode. Where top and bottom electrodes cross, an area of about 2×10^{-5} cm^2 (2000 μm^2) is defined as the effective microdiode. The LB film pinhole density must be so small that there is a fair chance to find microdiodes without short circuit (which will occur when gold particles fill the holes). Depending on the molecules used for the LB

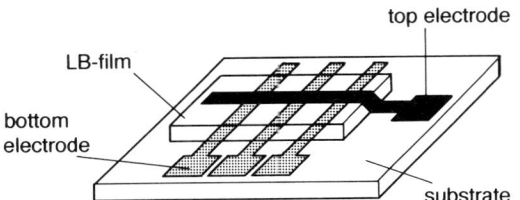

Figure 8.11 Typical mounting of microstructured tunnel junction. The active junction area is given by the intersection of bottom and top contact.

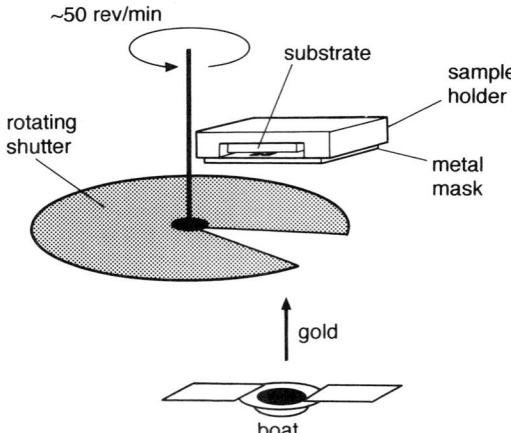

Figure 8.12 Experimental set-up inside the evaporation chamber. A rotating shutter between source and target reduces the effective evaporation rate. Between the 'gold pulses' the organic film can thermally relax.

films and on the electrode widths, yields of pinhole-free devices vary between 10 and 50%.

Another important aspect is the 'soft' evaporation technique for the top electrode [37]. The delicate film must be protected against thermal degradation during the evaporation process. Very slow evaporation will reduce the thermal load, but it does not lead to continuous metal films, therefore the evaporation is 'by pulses'. A shutter rotates between source and target, as indicated in Fig. 8.12, and between pulses the film is allowed to thermally relax.

Figure 8.13 schematically illustrates the structure of a microdiode. We see gold as top and bottom electrodes and the organic molecules in between. The π electron systems in the double bonds are indicated as large discs and the aliphatic chains as zig-zag lines. These chains are the hydrophobic parts of the molecules and facilitate film formation. The figure shows a so-called homostructure of identical monolayers. In a heterostructure different layers are present, e.g. some with donor molecules and others with acceptor molecules. These terms are taken from semiconductor technology, where for

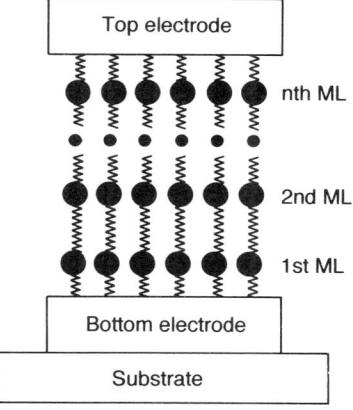

Figure 8.13 Homostructure of identical monolayers. The π systems of the molecules are indicated as discs and the aliphatic chains as zig-zag lines.

example GaAs–AlAs heterostructues are grown by molecular beam epitaxy. In the context of this work, in addition to homostructures and heterostructures, there is a third type of important structure, i.e. D-σ-A and D-π-A structures, where donor and acceptor moieties are part of the same molecule and connected by a σ bridge (Fig. 8.2, Aviram–Ratner molecule) or a π bridge (Fig. 8.9). Particularly with respect to the 'unimolecular rectifier' these bridge structures are relevant. To complete these considerations we also have to mention the buffer layers, which correspond to the insulating 'I' layers in silicon technology, and which in the LB technique occasionally are used to separate the active organic layers from the electrodes.

6 Langmuir–Blodgett film characterization

For the experimental results described in this chapter, three types of molecules are of importance: substituted phthalocyanines, perylenes and a D-σ-A molecule. The first two materials exhibit good chemical and thermal stability which makes them useful candidates to withstand the application of a top gold electrode.

The palladium phthalocyanine derivative PcPd structure is shown in Fig. 8.14. There are eight pentyloxy substituents at the central π system peripheral positions. Of note is that this molecule lacks the typical long-chain amphiphilic structure and is characterized by a predominantly hydrophobic nature. Initially the molecules must be dispersed uniformly upon spreading onto the water subphase. A number of conclusions can be drawn from surface pressure/area isotherms such as the one for PcPd depicted in Fig. 8.15. The isotherm exhibits a relatively steep rise which demonstrates a close packing of the molecules. Further, the extrapolated area per molecule is obtained to $90 \, \text{Å}^2$, in good agreement with a true monolayer of molecules. The molecules are standing edge-on tilted by 20–30° against the surface normal [38]. This behaviour is in contrast to that of many other phthalocyanine materials where the effective molecular

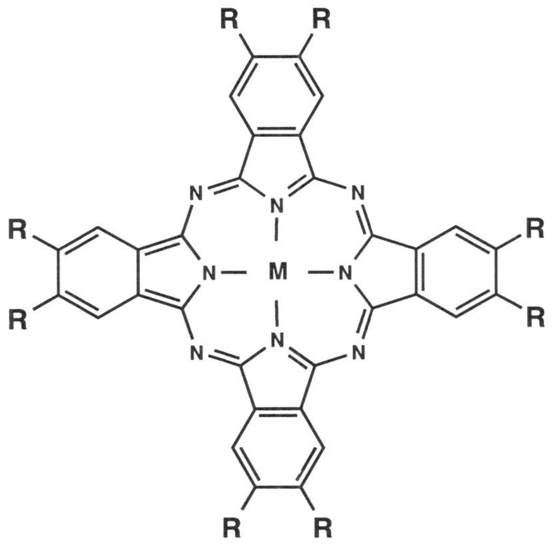

PcPd: M=Pd, R=OC₅H₁₁

Figure 8.14 Chemical structure of octa-substituted palladium phthalocyanine PcPd.

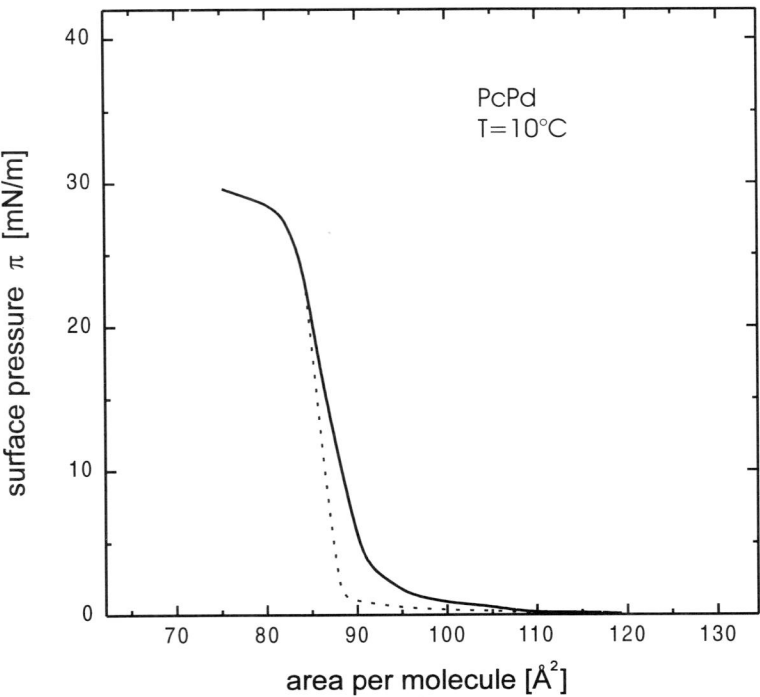

Figure 8.15 Surface pressure/area isotherm for PcPd on pure water subphase at $T = 10°C$.

area is considerably smaller than the expected minimal value, indicative of (partial) multilayer structure formation. Relaxation of the monolayer after compression to 22.5 mN m^{-1} follows the dotted curve and indicates some irreversible aggregation occurring upon first compression. However, this effect is small when compared with the monolayer behaviour of similar phthalocyanine compounds. The PcPd monofilm can be transferred with high efficiency as a Y-type LB film onto many different substrates. Comparison of the absorption spectrum of the molecules in solution and in the LB film (Fig. 8.16) provides information on the ring systems arrangement. The chloroform solution band shows a vibronic structure and is characterized by a strong Q_{0-0} peak at 663 nm; this peak is broadened and blue-shifted to about 600 nm in the film spectrum. This indicates the formation of cofacial stacks and, according to classical exciton theory, from the blue shift it follows that the ring planes are tilted less than about 35° against the surface normal. However, the shoulder in the region of 650–670 nm in the film spectrum provides evidence that there are also some molecules in a monomer-like environment. Consequently, there emerges the picture of well-ordered stacked regions embedded in a more amorphous matrix consisting of only weakly coupled molecules. STM investigations support such a film structure. If we speak of ordered LB films, we also have to show ordering in the growth direction. This can be done by small angle X-ray scattering experiments. The X-ray reflection curve for a 20 monolayer LB film of PcPd is given in Fig. 8.17. Only one Bragg reflection is obtained despite the presence of palladium atoms as strong scattering centres, demonstrating that there is some ordering although less than observed in the case of, e.g. fatty acid salt LB films. On the other hand, the well-developed Kiessig fringes demonstrate a high film smoothness and homogeneous covering of the substrate. The monolayer thickness is calculated to

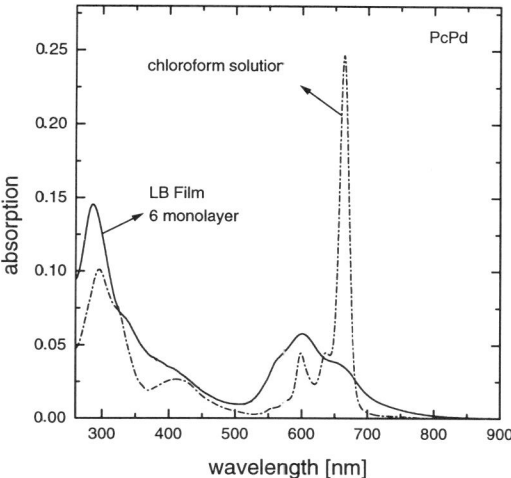

Figure 8.16 Optical absorption spectra of PcPd in chloroform solution (broken line) and a six monolayer Langmuir–Blodgett (LB) film (full line).

21.3 Å, a value which is consistent with the molecules standing edge-on and preferentially upright. The π-systems are thus separated by aliphatic ('insulating') layers of about 6 Å acting as tunnelling barriers.

The chemical structure of the two perylene derivatives is depicted in Fig. 8.18. Again the same principle is met: a rigid π system is provided with side chains in the periphery and already very subtle changes in these substituents have a pronounced influence on the monolayer behaviour, the transferability as well as the defect density in the LB films. Consequently, the molecular structures and LB transfer conditions have to be very carefully optimized. In addition, the four sulphur atoms of PTCDI–SPent (Pent = pentyl) were introduced with the aim of providing binding sites for impinging gold atoms of the top electrode. In a certain sense, this means an inversion of the self-assembly technique where the affinity of sulphur to a gold substrate is explored for the build-up of an organic monolayer [39].

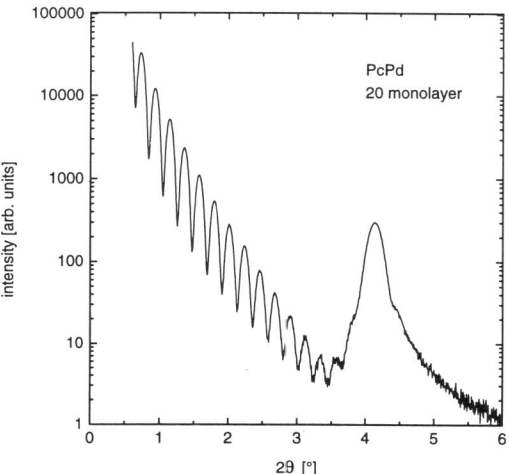

Figure 8.17 Small angle X-ray scattering curve for a 20 monolayer LB film of PcPd on Si(111).

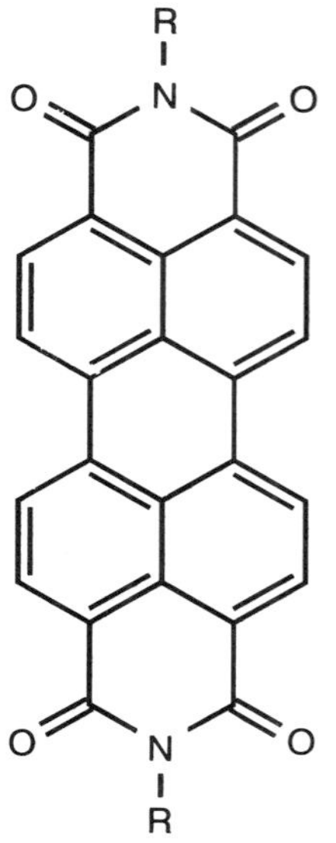

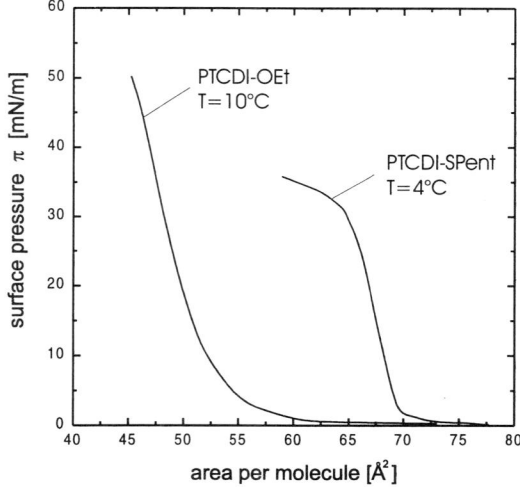

PTCDI-OEt: R=CH₂-CH(OEt)₂

PTCDI-SPent: R=CH₂-CH(SPent)₂

Figure 8.18 Chemical structure of the two different PTCDI derivatives.

The surface pressure/area isotherms (Fig. 8.19) yield a molecular area of 54 Å^2 for PTCDI–OEt (Et = ethyl) and 68 Å^2 for PTCDI–SPent. From these values it follows that we are dealing with true monolayers, i.e. no formation of three-dimensional crystallites takes place upon spreading. Further, the molecules cannot stand upright on their short edge but must be oriented with their long axis tilted against the surface normal. As PTCDI–OEt bears four short ethoxy substituents, the monofilm is relatively rigid and can only be transferred by a horizontal technique. In contrast, PTCDI–SPent can be very smoothly transferred as a Y-type film by vertical dipping. The LB film's optical

Figure 8.19 Surface pressure/area isotherms of PTCDI–OEt ($T = 10°$C) and PTCDI-SPent on pure water subphase.

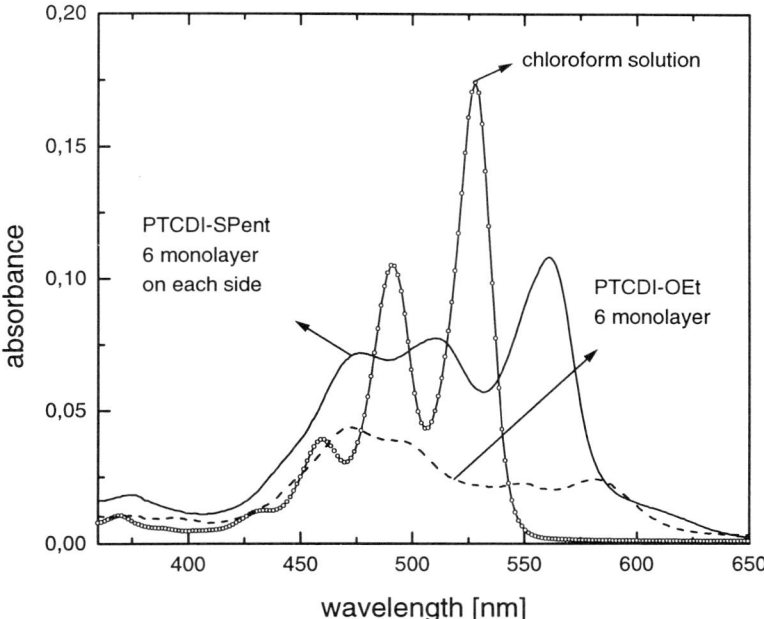

Figure 8.20 Optical absorption spectra of PTCDI derivatives in chloroform solution (hollow circles), six monolayer Langmuir–Blodgett (LB) film of PTCDI–OEt (broken line) and PTCDI–SPent (full line), respectively.

absorption spectra are characterized by red- and blue-shifted bands compared with the chloroform solution band at 528 nm (Fig. 8.20). For the PTCDI-SPent film, the absorption is about twice that of the PTCDI–OEt film because in the latter case only one side is covered by the horizontal dipping. The large shifts in excitation energy indicate that the peripheral substituents still allow a strong coupling of the π-systems, a factor which is of importance for the monofilms stability on the water subphase. It can further be concluded that the molecules pack in a herringbone-like arrangement with two translationally non-equivalent molecules per unit cell. The small angle X-ray diffraction curves of the LB films of PTCDI–OEt (Fig. 8.21) and PTCDI–SPent

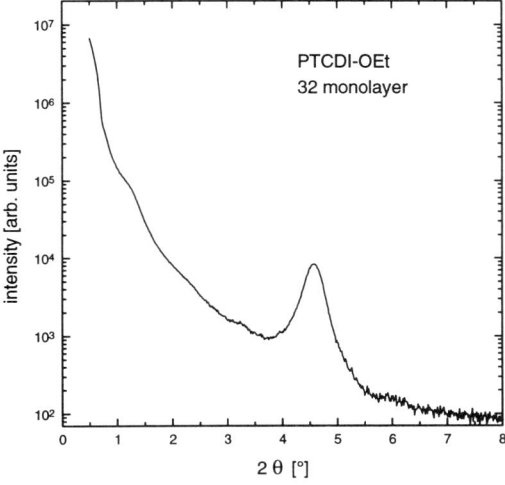

Figure 8.21 X-ray reflectivity curve for 32 monolayer LB film of PTCDI–OEt on Si(111).

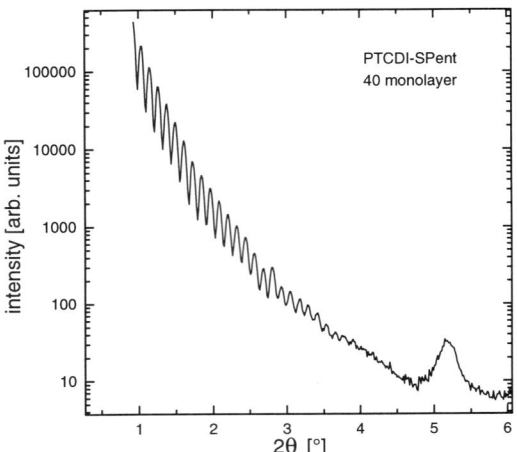

Figure 8.22 X-ray reflectivity curve for 40 monolayer LB film of PTCDI-SPent on Si(111).

(Fig. 8.22) illustrate order in the vertical direction by exhibiting one Bragg peak. The occurrence of well-defined Kiessig fringes for the PTCDI–SPent film (but not in the case of PTCDI–OEt) indicates that at least for the present molecular system the vertical transfer technique is more preferable. In each case, the repeat distances obtained from the Bragg reflections confirm that the molecules are arranged in slipped stacks with the molecular long axes considerably tilted against the substrate normal.

The D-σ-A molecule T_3–V^{2+} (Fig. 8.23) consists of a terthiophene T_3 electron donor unit that is connected through a $-(CH_2)_{11}-$ bridge to the viologen V^{2+} electron acceptor part. This molecule forms a close-packed monolayer at the air–water interface with a molecular area of 20 Å^2, which points to the molecules standing upright in an elongated conformation. The transferability of the monolayer, however, is only poor. Thus, T_3–V^{2+} has to be mixed with other assisting molecules, in this case in a 3 : 1 ratio with methyl stearate (a classical long chain compound). A monolayer of this mixture can be more smoothly transferred onto surfaces such as the LB film of PTCDI–SPent (which will be used as charge transport barrier), although there still remain problems in building up well-ordered films consisting of a greater number of monolayers (transfer proceeds in a non-centrosymmetric manner as Z-type). X-ray scattering results on such multilayer films display only a weak and broad Bragg reflection leading to a monolayer thickness of 34 Å. Figure 8.24 shows the optical absorption spectrum of a T_3–V^{2+} LB film. The band at 260 nm is attributed to the viologen unit, while the one at 352 nm belongs to the π–π* transition of the terthiophene part. The latter band is blue-shifted by 10 nm with respect to the solution (CHCl$_3$: DMSO 3 : 1) band; this indicates the formation of H-aggregated stacks of the terthiophene units.

7 Tunnelling experiments

7.1 Tunnel junctions incorporating one type of molecule

The first experiment deals with tunnel junctions which consist of only one type of molecule. For sample 1 we transferred 10 monolayers of PcPd. Samples 2 and 3 consist of 20 monolayers of PTCDI–OEt and 10 monolayers of PTCDI–SPent, respectively

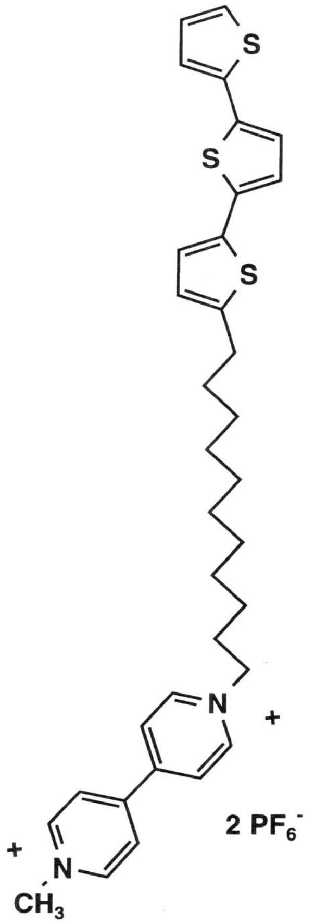

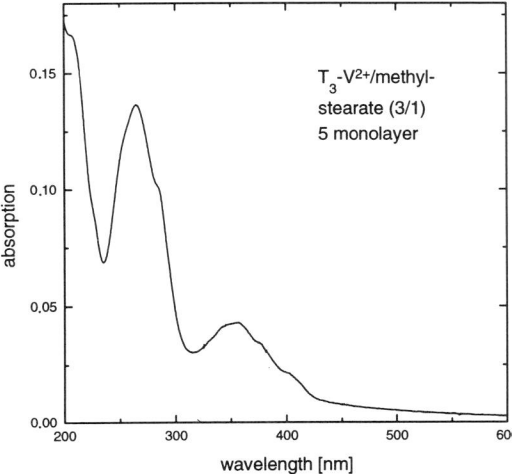

T$_3$-V^{2+}

Figure 8.23 Structure of the D-σ-A molecule, T$_3$–V^{2+}.

Figure 8.24 Optical absorption spectrum of five monolayer Langmuir–Blodgett (LB) film of T$_3$–V^{2+} : methyl stearate (3 : 1).

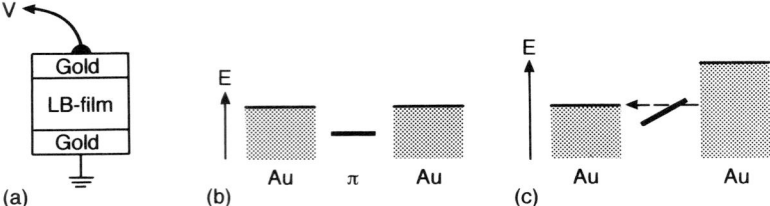

(a)   (b)   (c)

Figure 8.25 Schematic view of a homostructure as located between the electrodes. (a) Sandwich structure; (b) energetic diagram of the Fermi levels and the π orbitals. When applying a voltage to the electrodes, the π levels are shifted and get in resonance with the Fermi edge of the positively charged electrode at a certain voltage (c).

(Fig. 8.25a). The I/V characteristics measured at liquid helium temperature are shown in Fig. 8.26. All curves exhibit abrupt current increases at symmetric voltage values. In the case of 10 monolayers of PcPd, these thresholds are observed at ± 0.27 V. The 20 monolayer PTCDI–OEt junction shows current onsets at ± 0.55 V; in the curve of the 10 monolayer PTCDI–SPent junction these occur at ± 0.68 V.

To explain the abrupt increases in current, we consider the respective molecular orbitals. A schematic energy diagram is shown in Fig. 8.27, as obtained by comparison with literature values of similar compounds. The bonding π levels are a few tenths of an eV below the Fermi level of gold, as indicated in Fig. 8.25(b). The anti-bonding π* levels are above, with typical π–π* gaps of some 2 eV. When applying a voltage to the outer electrodes, the electric field will shift the energy levels. From a certain bias voltage on the π levels become resonant with the positively charged electrode. At this voltage electrons can resonantly tunnel from the π levels to the electrode. The hole created in the π levels recombines with an electron from the negatively charged electrode. Since the molecular orbital is the highest occupied, one can also speak of resonant hole injection from the electrodes into the π states. Due to this resonant tunnelling process, the current through the junction increases. This situation is depicted in Fig. 8.25(c). If

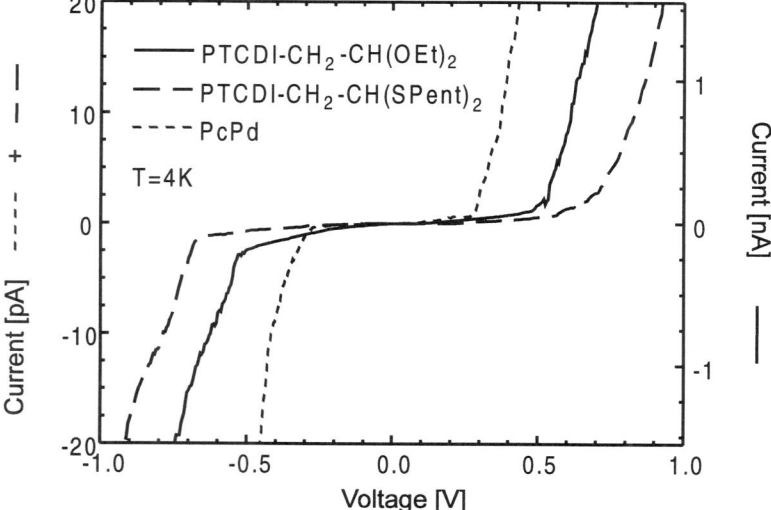

Figure 8.26 I/V characteristics of three different homostructures incorporating 20 monolayer PTCDI–OEt, 10 monolayer PTCDI–SPent and 10 monolayer PcPd, respectively. All curves were measured at 4 K.

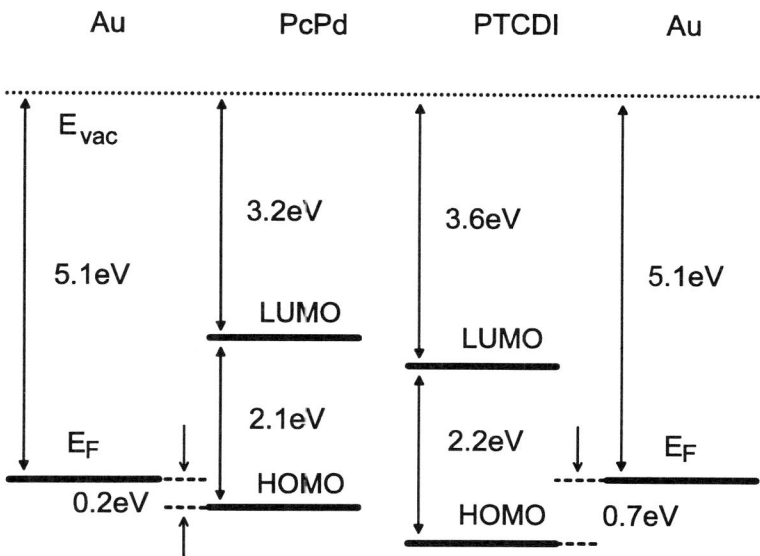

Figure 8.27 Schematic energy diagram of the π and π* orbitals of the PcPd and PTCDI derivatives as obtained from literature.

we compare the observed voltage thresholds with the energy differences between the gold Fermi edge and the π levels of the respective molecule, we observe a good correlation. These results suggest the use of different molecular systems as 'functional units' in more complex organic heterostructures.

7.2 *Bimolecular rectifier*

The simplest heterostructure would be a structure consisting of two different molecular layers. For our experiment we use a heterostructure of six monolayers of PcPd and six monolayers of PTCDI–OEt with the PcPd directly transferred to the substrate and the PTCDI–OEt layer on top of it (Fig. 8.28). Similar to the original idea of Aviram and Ratner, such an asymmetric structure acts as a rectifier, if we interpret PTCDI–OEt as the acceptor part, PcPd as donor part, and the intermolecular aliphatic chains as σ bridge [40].

The I/V characteristic measured on this double layer structure at liquid helium temperature is shown in Fig. 8.29. The junction clearly exhibits rectifying behaviour. To explain the details of the I/V characteristic, both the energetic position of the π molecular levels and the molecule's geometric location have to be considered. The closer a molecule is placed to an electrode, the more the Fermi level must be shifted by the field. At the voltages of -0.3 V and $+0.2$ V the π level of PcPd will get into resonance and the current will increase. Since at these voltages the π states of PTCDI–OEt are still below the Fermi level, they do not yet contribute to resonant tunnelling and therefore the PTCDI–OEt layers act as barriers. Consequently the observed current increase is only modest (Fig. 8.28b). At -0.4 V and $+0.9$ V much stronger current increases occur, because now the π states of PTCDI–OEt are additionally available for hole tunnelling (Fig. 8.28c).

The rectifying behaviour is a direct consequence of asymmetric hole tunnelling

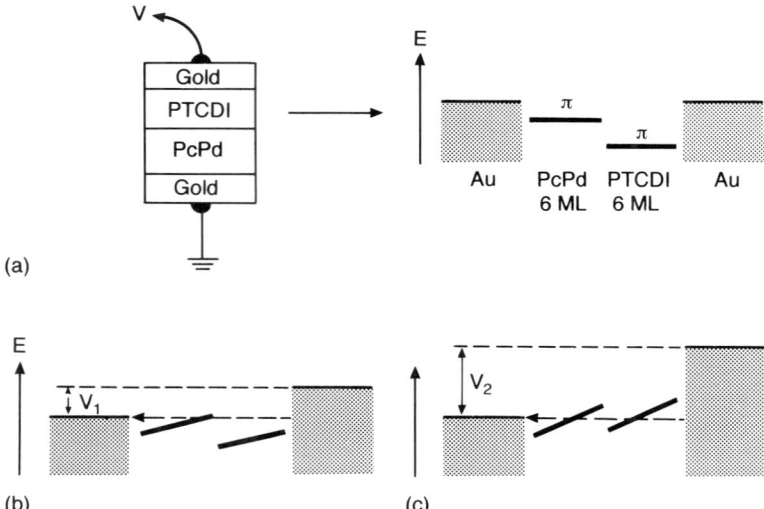

(a)

(b)   (c)

Figure 8.28 (a) Schematic view of the bimolecular rectifier and energetic position of the corresponding π orbitals. (b) At low voltages only the PcPd molecules contribute to resonant tunnelling. (c) At higher voltages the PTCDI molecules get in resonance as well and the current strongly increases.

processes through the highest occupied molecular orbitals of the two different molecular species. This interpretation is manifested by the fact that the tunnel junctions which incorporate only one type of molecule have shown symmetric I/V characteristics. Due to this model, however, the asymmetric characteristic is based on a transport mechanism through the π states only without involving the π* states, in contrast to the original model of Aviram and Ratner.

In addition to the asymmetric thresholds for the strong current increases, the I/V characteristics of the molecular rectifier show in the positive voltage region four equidistant steps in the current with a voltage spacing of 140 mV. Such behaviour is strongly reminiscent of Coulomb charging and single-electron effects of quantum dots in inorganic semiconductors [41]. In these systems usually a semi-classical approach is used to describe single-electron tunnelling through small-capacitance islands. Then the

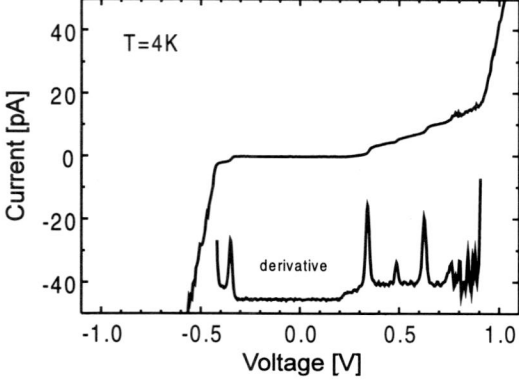

Figure 8.29 I/V characteristic of the bimolecular rectifier, measured at 4 K. Strong increases in current occur at -0.4 V and $+0.9$ V. For low positive voltages four equidistant steps are observed.

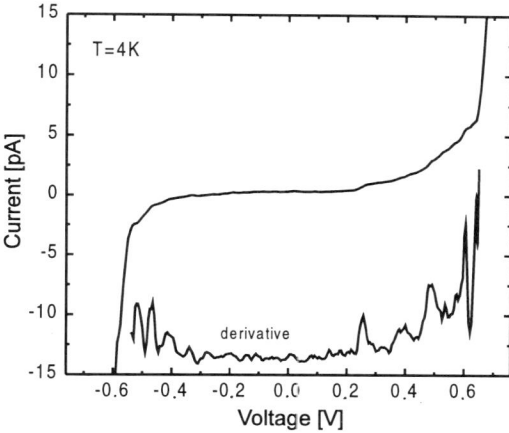

Figure 8.30 I/V characteristic of a bimolecular rectifier where PTCDI–OEt is replaced by PTCDI–SPent, measured at 4 K.

voltage spacing ΔV between equidistant steps is given by $\Delta V = e/C$, where C is the small island capacitance [42]. Following this relation, the observed voltage spacing would lead to a capacitance of the order of 10^{-18} F. In comparison to typical capacitances observed in inorganic quantum dot structures, this value is about three orders of magnitude lower. In the present case the corresponding island must be much smaller than an artificial quantum dot and therefore is assumed to be of molecular dimensions. The nature of quantum dots is not yet clear. It is tempting to assume that such a small capacitance is displayed by a single molecule or a molecular stack, but then well-defined tunnel barriers are needed to further fulfil the requirements of the simple model of Coulomb charging effects. Certainly, the actual origin of the observed equidistant steps is much more complex and cannot be explained at the present stage.

In Fig. 8.30 the I/V characteristic of a different bimolecular tunnel junction is shown, but now PTCDI–OEt is replaced by PTCDI–PSent. At 4.2 K the curve exhibits similar rectifying behaviour as observed on the junction in Fig. 8.29. Even the equidistant steps in the range of positive voltages can be reproduced, as underlined by the first derivative of the curve.

7.3 *Organic double barrier structures*

The previous experiment on the molecular rectifier has shown that the 'functional' molecular films presented in Section 7.1 are suitable to build multilayer structures in the sense that they lead to tunnel characteristics which can be interpreted in terms of specific resonant tunnelling processes through molecular states. To obtain a more precise understanding of the resonant tunnelling phenomena, an organic double barrier structure with an A–B–A layer sequence was prepared, guided by the following idea. The tunnel junctions incorporating only one type of molecules show symmetric current increases at certain voltage thresholds. Below these thresholds the junctions are highly insulating, i.e. the respective molecular layer acts as a good tunnelling barrier [43]. If we now use molecule A with a relatively low π state as barrier molecule and incorporate a layer of a different molecule B with an energetically higher π level between two barrier layers of molecule A, we obtain a double barrier structure as schematically shown in

Fig. 8.31(a). When applying a voltage to the electrodes, the molecular orbitals' relative energetic position will be shifted until the π states of molecules B open resonant tunnelling channels. At this voltage the molecules π states A are still lower than the Fermi edge and act as tunnel barriers. The resonant tunnelling processes through the molecular layer B contribute to the total current through the junction and should be observable in the I/V characteristic. If the distance of molecule B from the electrodes is known, the energetic position of the π bonding levels of molecule B with respect to the Fermi level of the gold electrodes can be determined from the kink in the current–voltage characteristic (Fig. 8.31b).

Double barrier structures were prepared by using PTCDI–SPent as barrier molecule (A) and PcPd as incorporated molecule (B). We use four monolayers for each barrier with two monolayers of PcPd in the middle. The I/V characteristic of this device, measured at 4.2 K, is shown in Fig. 8.32. At ±0.65 V two major increases in current are observed, which are interpreted as opening of the π states of PTCDI–SPent barriers for resonant tunnelling. This assignment is supported by a direct comparison to the I/V characteristic of an identically fabricated junction with 10 monolayers of PTCDI–SPent, which is also shown in Fig. 8.32. The current–voltage characteristic of 10 monolayers of PTCDI–SPent shows abrupt increases at identical voltage values. In contrast to the pure PTCDI–SPent junction, the I/V characteristic of the double barrier structure shows two additional symmetric steps at ±0.32 V. These steps are attributed to the resonant tunnelling processes through the π states of the PcPd layer. The additional steps occur at symmetric values $\pm V_s$. Therefore we conclude that the drop of the applied potential is identical across each of the two barriers. Assuming that the potential drop occurs only at the insulating PTCDI–SPent barriers, the potential at the PcPd layer is at half of the voltage between these electrodes. From $V_s = 0.32$ V we learn that the π level of PcPd is 0.16 eV below the Fermi level of the gold electrode.

If we consider a work function of the gold electrodes of 5.1 eV [44], we derive an energy of the π level relative to the vacuum level of 5.26 eV, which is in the range of ionization potentials typically obtained for substituted phthalocyanines from redox potential measurements and photoelectron spectroscopy [45]. The fact that the additional increases in current occur at symmetric voltages proves that the two PTCDI–SPent barriers are of the same thickness, i.e. that the fabrication process of junctions and in particular the evaporation of the top contact do not significantly influence the organic layers arrangement. The presented double barrier structure turns out to be a useful tool for precise determination of molecular orbital energy levels by resonant tunnelling spectroscopy.

7.4 *Unimolecular rectifier*

The experiment on the double barrier structure has shown that PTCDI–SPent forms LB layers with well-defined thicknesses and can be used as a transport barrier. In the following experiment we use a similar double barrier structure to investigate resonant tunnelling through a donor/acceptor-substituted molecule, namely the T_3–V^{2+}. Again, four monolayers of PTCDI–SPent are used for each barrier; one single monolayer of T_3–V^{2+} is incorporated in between. The arrangement of the molecular layers and the molecular orbitals' energetic scheme inside the junction is shown in Fig. 8.33. The

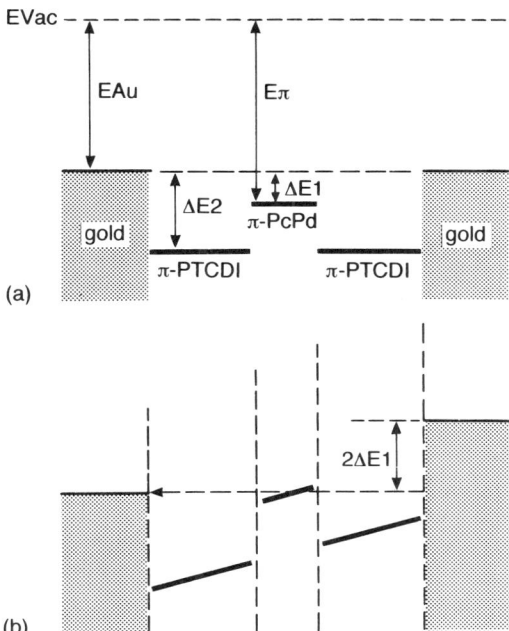

(a)

(b)

Figure 8.31 (a) Organic double barrier structure built up from two PTCDI–SPent barrier layers and a PcPd layer incorporated between them. (b) When applying a voltage to the electrodes the first increases in current are expected when the π orbitals at the PcPd get in resonance.

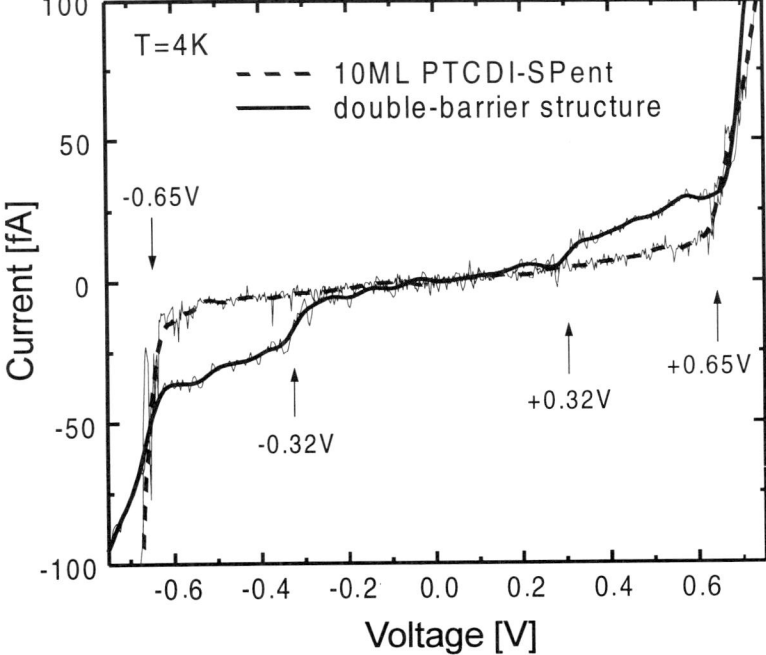

Figure 8.32 I/V characteristic of the double barrier structure (solid line), measured at 4 K. At ±0.32 V symmetric onsets in current occur which are not present in a pure 10 monolayer PTCDI–SPent homostructure (dashed line).

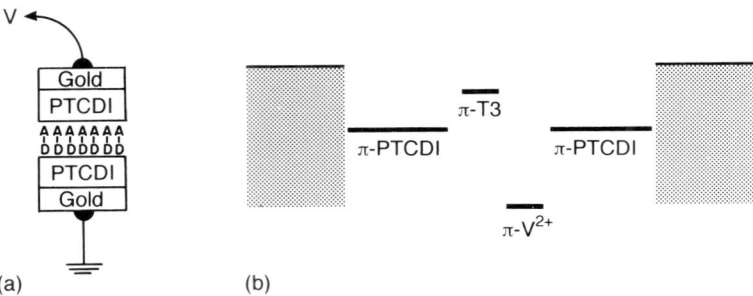

(a)   (b)

Figure 8.33 (a) Schematic view of the unimolecular rectifier. The first increases in current are expected due to resonant tunnelling processes through the π orbitals of the T_3-donors. The V^{2+} acceptor π states are energetically lower than the PTCDI–SPent π state barriers and therefore do not contribute to the transport (b).

π level of the V^{2+} part is much lower than the π level of the PTCDI–SPent barrier molecules and therefore does not contribute to resonant tunnelling. The π* states of all corresponding molecules can be neglected due to the same consideration, i.e. the only molecular orbital relevant to resonant tunnelling is the π level of the donor part T_3.

Figure 8.34 shows the I/V characteristic of the junction, measured at 4.2 K. As in Section 7.3, two major current onsets are observed at ±0.65 V which are attributed to the resonant opening of the PTCDI–SPent π state barriers. In addition, two asymmetric major increases in current occur at the voltages –0.53 V and +0.28 V. These steps are interpreted in terms of resonant tunnelling through the donor π states. (Because of the low position of the acceptor π level, the acceptor moiety acts only as a spacer.) Since the donor part of the D/A molecule is asymmetrically located inside the double barrier structure, the potential drop is different for positive and negative voltages. Within the 'window of observation' which is given by threshold voltages of the barrier layers, a clear asymmetric I/V characteristic is observed. So we clearly observe molecular rectification, but the mechanism is not completely identical to that proposed by Aviram and Ratner. In contrast to the Aviram–Ratner case and also in contrast to p–n junctions

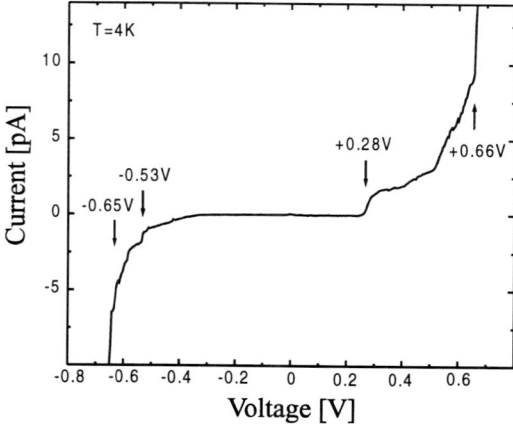

Figure 8.34 I/V characteristic of the unimolecular rectifier. At ±0.65 V the well-known increases in current occur which are due to resonant tunnelling through the π orbitals of the PTCDI–SPent barriers. Within this 'window of observation' two asymmetric kinks in the current are observed at – 0.53 V and + 0.28 V.

in conventional semiconductors, only bonding π levels (analogous to the valence band) are involved. The anti-bonding π^* levels would only participate in transport at higher applied voltages.

8   Conclusion and outlook

The five features of the Stuttgart approach — optimized film preparation and characterization, gentle electrode deposition, device miniaturization, multiplexation, and measurements at liquid helium temperature — have enabled us to measure current–voltage characteristics of monolayers of functionalized molecules in Langmuir–Blodgett films and to demonstrate resonant tunnelling through molecular orbitals, molecular rectification, and Coulomb charging of molecule-sized objects. Thus we have established a novel technique for physico-chemical measurements. As their inorganic counterpart, organic molecular rectifiers, tunnel diodes and quantum dots can be explored as possible constituents of molecular electronic devices and the knowledge obtained from future investigations along these lines will help to assess the technological relevance of organic molecular materials as a replacement or as a complement for silicon in nanoelectronics.

Acknowledgements. We would like to thank K. von Klitzing, M. Hanack and W. Göpel, as well as S. Blumentritt, S. Curran, C. Müller-Schwanneke, and G. Philipp for many stimulating discussions, and our colleagues in Stuttgart and Tübingen and notably in the Sonderforschungsbereich 'Molekularelektronik' of the Deutsche Forschungsgemeinschaft and the European Communities Espirit Network NEOME (New Electroactive Organic Materials for Electronics) for their fruitful cooperation.

Note in proof. The above discussion assumes perfect molecules, no chemical reactions at the organic–inorganic interface and no space charges which might build up from electron exchange between the electrodes and defect states. From surface potential measurements with Kelvin probes we now know that this assumption is not valid. Space charges do build up and this will lead to shifts of the molecular levels. As a consequence, some of the features in the current–voltage characteristics will have to be interpreted in a more complicated way.

9   References

1 Carter FL. *Molecular Electronics Devices*. New York: Marcel Dekker, 1982.

2 Carter FL. *Molecular Electronics Devices*, Vol. II. New York: Marcel Dekker, 1987.

3 Carter FL, Siatkowski RE, Wohltjen H. *Molecular Electronic Devices*. Amsterdam: North Holland, 1988.

4 Aviram A. *Molecular Electronics — Science and Technology*. New York: United Engineering Trustees, 1989.

5 Hong FT. *Molecular Electronics — Biosensors and Biocomputers*. New York: Plenum Press, 1989.

6 Ashwell GJ. *Molecular Electronics*. New York: Wiley, 1992.

7 Roth S, Mahler G, Shen YQ, Coter F. *Synth Metals*. 1989; **28**: C815.

8 Müller B, Reinhardt J, Strickland MT. *Neural Networks*. Heidelberg: Springer Verlag, 1995.

9 Bräuchle C, Hampp N, Oesterhelt D. *Adv Mater* 1991; **3**: 420.

10 Itoua SJ-L. PhD thesis, Montpellier 1995.

11 Aviram A, Joachim C, Pomerantz M. *Chem Phys Lett* 1988; **146**: 490.

12 Stabel A, Herwig P, Müllen K, Rabe J. *Angew Chem* 1995; **107**: 1768.

13 Kuhn H. *Thin Solid Films* 1989; **178**: 1.

14 Aviram A, Ratner MA. *Chem Phys Lett* 1974; **29**: 277.

15 Cohen MJ, Coleman LB, Garito AF, Heeger AJ. *Phys Rev B* 1974; **10**: 1298.

16 Roth S. *One-Dimensional Metals — Physics and Materials Science*. Weinheim: VCH Verlags-gesellschaft, 1995.

17 Binnig G, Rohrer H. *Rev Mod Phys* 1987: **59**: 615.

18 Lüth H. *Surfaces and Interfaces of Solids*. Berlin: Springer Verlag, 1993.

19 Metzger RM. In: Blank M, ed. *Electricity and Magnetism in Biology and Medicine*. San Francisco: San Francisco Press, 1993: 175.

20 Tredgold RH. *Order in Thin Organic Films*. Cambridge: Cambridge University Press, 1994.

21 Roberts G. *Langmuir–Blodgett Films*. New York: Plenum Press, 1990.

22 Polymeropoulos EE, Möbius D, Kuhn H. *Thin Solid Films* 1980; **68**: 173.

23 Fujihira M, Nishiyama K, Yamada H. *Thin Solid Films* 1985; **132**: 77.

24 Ashwell GJ, Sambles JR, Martin AS, Parker WG, Szablewski M. *Chem Soc Chem Commun* 1990; **19**: 1374.

25 Ashwell GJ. *Thin Solid Films* 1990; **186**: 155.

26 Ashwell GJ, Dawnay EJC, Kuczynski AP *et al.* *J Chem Soc Faraday Trans* 1990; **86**: 1117.

27 Ashwell GJ, Dawnay EJC, Kuczynski AP. *Chem Soc Chem Commun* 1990; **19**: 1355.

28 Martin AS, Sambles JR, Ashwell GJ. *Phys Rev Lett* 1993; **70**: 218.

29 Metzger RM, Panetta CA. *New J Chem* 1991; **15**: 209.

30 Metzger RM. In: Aviram A, ed. *Molecular Electronics — Science and Technology*, Vol. 262. American Institute of Physics Conference Proceeding, 1992: 85.

31 Metzger RM. In: Birge B, ed. *Biomolecular Electronics*, Vol. 240. American Chemical Society Advances in Chemistry Series, 1994: 81.

32 Herdt GC, Czanderma AW. *J Vac Sci Technol A* 1994; **12**: 2410.

33 Sakmann B, Neher E. *Single-Channel Recording*. New York: Plenum, 1983.

34 Fischer CM, Burghard M, Roth S, von Klitzing K. *Europhys Lett* 1994; **28**: 129.

35 Burghard M, Fischer CM, Schmelzer M, Roth S, Haisch P, Hanack M. *Synth Metals* 1994; **67**: 193.

36 Fischer CM, Burghard M, Roth S. *Mater Sci For* 1995; **191**: 149.

37 Fischer CM, Burghard M, Roth S, von Klitzing K. *Appl Phys Lett* 1995; **66**: 3331.

38 Burghard M, Schmelzer M, Roth S, Haisch P, Hanack M. *Langmuir* 1994; **10**: 4265.

39 Ulman A. *An Introduction to Ultrathin Organic Films*. San Diego: Academic Press, 1991.

40 Fischer CM, Burghard M, Roth S. *Synth Metals* 1995; **71**: 1975.

41 Su B, Goldman VJ, Cunningham JE. *Phys Rev B* 1992; **46**: 7644.

42 Grabert H, Devoret MH. *Single Charge Tunneling*, NATO ASI Series B 294. New York: Plenum Press, 1992.

43 Burghard M, Fischer CM, Schmelzer M, Roth S, Hanak M, Göpel W. *Chem Mater* 1995; **7** (November).

44 Michaelson HB. *J Appl Phys* 1977; **48**: 4792.

45 Schlettwein D, Armstrong NR. *J Phys Chem* 1994; **98**: 11771.

9 Optical Properties of Semiconductor Nanocrystals (Quantum Dots)

D.J. NORRIS,* M.G. BAWENDI† and L.E. BRUS‡

*Department of Chemistry and Biochemistry, University of California, San Diego, 9500 Gilman Drive, La Jolla, CA 92093-0340, USA,

†Department of Chemistry, Massachusetts Institute of Technology, 77 Massachusetts Avenue, Cambridge, MA 02139, USA,

‡Department of Chemistry, Columbia University, 3000 Broadway, New York, NY 10027, USA

1 Introduction

One possible route to 'molecular electronics' is to incorporate molecular-like behaviour in standard semiconductor materials. This approach has led to the investigation of nanometre-scale semiconductor structures over the past several decades. Interest in new opto-electronic materials has driven the development of thin semiconductor sheets (quantum wells), narrow rods (quantum wires), and nanometre-scale crystallites (quantum dots). Absorption of a photon by these materials promotes an electron from the valence band into the conduction band, creating an electron–hole pair. When the size of these semiconductor structures is comparable to or smaller than the natural length scale of the electron–hole pair, it is confined in one (wells), two (wires), or all three dimensions (dots). The spatial confinement leads to interesting molecular-like optical behaviour as the motion of the carriers is quantized by the boundary of the material. Since these materials are structurally identical to the bulk crystal, these properties arise solely due to their finite size. This phenomenon, known as the 'quantum size effect', is both fundamentally and practically important. An investigation of these quantum structures addresses a size-regime intermediate between molecular and solid-state behaviour. In addition it may lead to unique properties for potential applications.

Due to the complete confinement of the carriers, the quantum size effect is the most dramatic in quantum dots. When a semiconductor nanocrystal is embedded in an insulating material, as illustrated in Fig. 9.1, the photoexcited carriers reside in a potential well in all three dimensions and the conduction and valence bands are quantized due to the finite size of the dot [1–3]. In contrast to the bulk absorption spectrum, which is a continuum above the band gap of the semiconductor (E_g^s) [4], quantum dot spectra exhibit a series of discrete electronic transitions. Because of this behaviour these materials are sometimes referred to as 'artificial atoms'. Since the energies of the transitions are affected by the amount of confinement, the optical spectra of quantum dots are also strongly dependent on the size of the crystallite.

In this chapter we address the optical properties of semiconductor quantum dots and discuss recent developments in understanding their basic photophysics. Two prototypical quantum dot systems are described. First, in Section 2 the absorption and emission behaviour of cadmium selenide (CdSe) dots are discussed. Due to significant advances in fabrication methods [5], CdSe is now one of the most well understood quantum dot materials. Extremely high quality CdSe dots can be produced and much new information has been obtained for this system. We begin Section 2 by describing the basic

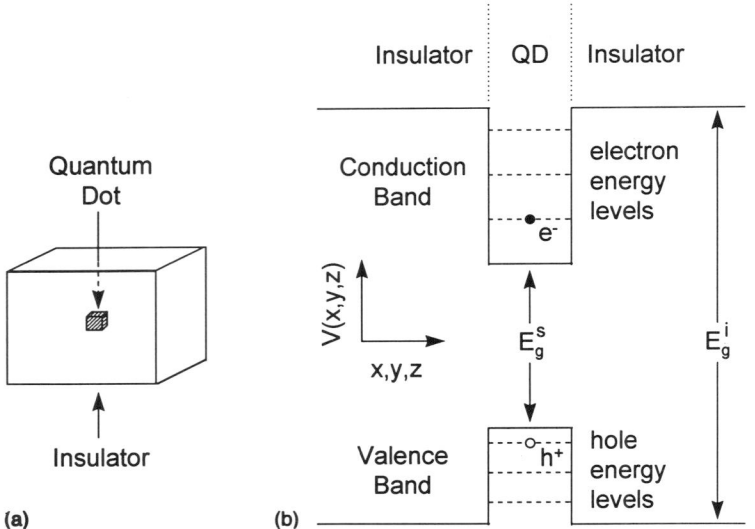

Figure 9.1 (a) Quantum dot (QD) structure. A nanometre-scale semiconductor crystallite is embedded inside an insulating material. (b) Potential well formed in any one dimension (x, y, or z) in the conduction and valence bands. The energy levels of the excited carriers (electrons and holes) become quantized due to the finite size of the semiconductor.

concepts necessary to understand the optical properties of CdSe quantum dots. We then address two long-standing questions in strongly confined quantum dot systems:

1 the size dependence of the optical spectrum;

2 the origin of the long-lived luminescence.

Second, in Section 3 we discuss the optical properties of silicon (Si) quantum dots. These dots are particularly interesting not only because of silicon's technological significance, but also because Si, in contrast to CdSe, is an indirect gap semiconductor. In such materials, optical transitions at the band gap are dipole forbidden and occur only via less efficient vibrationally assisted processes [4]. However, since this selection rule arises due to the long-range translational symmetry of the bulk material, it is modified in finite size dots where translational invariance is no longer valid. Therefore, band gap absorption and luminescence become weakly dipole-allowed in Si quantum dots. The possibility that Si quantum dots might be a useful optical material has attracted widespread interest, especially in connection with the quantum and luminescence properties of porous Si thin films made by electrochemical etching. In Section 3 our present imperfect understanding of these systems is discussed.

2 CdSe quantum dots

2.1 *Basic concepts*

2.1.1 'PARTICLE-IN-A-SPHERE'

For direct gap semiconductor quantum dots such as CdSe much of the basic photophysics can be explained by the simple 'particle-in-a-sphere' model [1–3]. This model considers an arbitrary particle of mass m_0 inside a spherical potential well of radius a,

$$V(r) = \begin{cases} 0 & r < a \\ \infty & r > a. \end{cases} \tag{9.1}$$

Following Flügge [6], the Schrödinger equation is solved yielding wavefunctions

$$\Phi_{n,l,m}(r, \theta, \phi) = C \frac{j_l(k_{n,l}r)Y_l^m(\theta, \phi)}{r}, \tag{9.2}$$

where C is a normalization constant, $Y_l^m(\theta, \phi)$ is a spherical harmonic, $j_l(k_{n,l}r)$ is the lth order spherical Bessel function, and

$$k_{n,l} = \frac{\alpha_{n,l}}{a} \tag{9.3}$$

with $\alpha_{n,l}$ the nth zero of j_l. The energy of the particle is given by

$$E_{n,l} = \frac{\hbar^2 k_{n,l}^2}{2m_0} = \frac{\hbar^2 \alpha_{n,l}^2}{2m_0 a^2}. \tag{9.4}$$

Due to the symmetry of the problem the eigenfunctions (Eqn 9.2) are simple atomic-like orbitals which can be labelled by the quantum numbers $n(1, 2, 3. . .)$, $l(s, p, d. . .)$, and m. The energies (Eqn 9.4) are identical to the kinetic energy of the free particle, except that the wavevector, $k_{n,l}$, is quantized by the spherical boundary condition. Note also that the energy is proportional to $1/a^2$ and therefore is strongly dependent on the size of the sphere.

At first glance, this model may not seem useful for the quantum dot problem. The particle above is confined to an empty sphere, while the quantum dot is filled with semiconductor atoms! However, by a series of approximations the quantum dot problem can be reduced to the 'particle-in-a-sphere' form (Eqn 9.1). The photoexcited carriers (electrons and holes) may then be treated as particles inside a sphere of constant potential.

First, the bulk conduction and valence bands are approximated by simple isotropic bands within the effective mass approximation [7]. According to Bloch's theorem, the bulk wavefunctions can be written as

$$\Psi_{nk}(\mathbf{r}) = u_{nk}(\mathbf{r})\exp(i\mathbf{k} \cdot \mathbf{r}) \tag{9.5}$$

where u_{nk} is a function with the periodicity of the crystal lattice and the wavefunctions are labelled by the band index n and wavevector $\mathbf{k}$. The energy of these wavefunctions are typically described in a 'band diagram', a plot of E vs $\mathbf{k}$. Although band diagrams are in general quite complex and difficult to calculate, in the effective mass approximation the bands are assumed to have simple parabolic forms near extrema in the band diagram. For example, since CdSe is a direct gap semiconductor, both the valence band maximum and conduction band minimum occur at $k = 0$ (see Fig. 9.2). In the effective mass approximation, the energy of the conduction ($n = c$) and valence ($n = v$) bands are approximated as

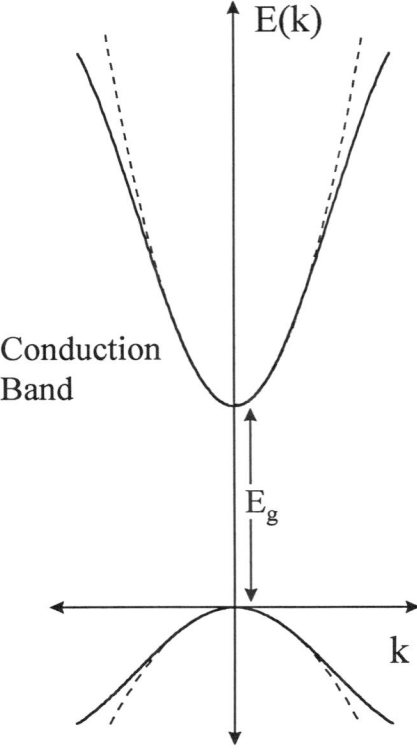

Conduction
Band

E_g

k

Valence Band

Figure 9.2 Simple two-band model for bulk direct gap semiconductors. The real band structure (solid lines) is approximated by parabolic bands (dashed lines) at $k = 0$ in the effective mass approximation. The curvature of the bands reflects the 'mass' of the electrons in the conduction band (CB) and the holes in the valence band (VB).

$$E_k^c = \frac{\hbar^2 k^2}{2m_{\text{eff}}^c} + E_g$$

$$E_k^v = -\frac{\hbar^2 k^2}{2m_{\text{eff}}^v}, \tag{9.6}$$

where E_g is the semiconductor bandgap and the energies are relative to the top of the valence band. In this approximation the carriers behave as free particles with an 'effective mass', $m_{\text{eff}}^{c,v}$. Graphically, the effective mass accounts for the curvature of the conduction and valence bands at $k = 0$. Physically, the effective mass attempts to incorporate the complicated periodic potential felt by the carrier in the lattice. This approximation allows us to completely ignore the semiconductor atoms in the lattice and treat the electron and hole as 'free particles'.

If the effective mass approximation is combined with a spherical boundary condition then each of the carriers in the quantum dot problem is in the 'particle-in-a-sphere' form (Eqn 9.1). However, this step requires that we treat the quantum dot as a 'bulk' sample. In other words, we assume that the single particle (electron or hole) wavefunction can be written in terms of Bloch functions (Eqn 9.5). This approximation, sometimes called the· *envelope function approximation* [8,9], is valid when the dot diameter is much larger than the lattice constant of the material. To satisfy the boundary condition the single particle (sp) wavefunction is then written as a linear combination of Bloch functions

$$\Psi_{sp}(\mathbf{r}) = \sum_k C_{nk} u_{nk}(\mathbf{r}) \exp(i\mathbf{k} \cdot \mathbf{r}) \tag{9.7}$$

with expansion coefficients, C_{nk}. If we assume that the functions u_{nk} have a weak k dependence then Eqn 9.7 can be rewritten as

$$\Psi_{sp}(\mathbf{r}) = u_{n0}(\mathbf{r}) \sum_k C_{nk} \exp(i\mathbf{k} \cdot \mathbf{r}) = u_{n0}(\mathbf{r}) f_{sp}(\mathbf{r}), \tag{9.8}$$

where $f_{sp}(\mathbf{r})$ is the single particle 'envelope function'. In the tight-binding approximation [or linear combination of atomic orbitals (LCAO) approximation] the periodic function u_{n0} is written as a sum of atomic wavefunctions, φ_n,

$$u_{n0}(\mathbf{r}) \approx \sum_i C_{ni} \varphi_n(\mathbf{r} - \mathbf{r}_i), \tag{9.9}$$

where the sum is over lattice sites and n represents the conduction band or valence band for the electron or hole, respectively. The functions u_{n0} are known from the bulk material and the quantum dot problem is reduced to determining the envelope functions for the single particle wavefunctions, f_{sp}.

As depicted in Fig. 9.3, during quantum dot optical transitions two particles are created, an electron and a hole. If we assume spherically shaped dots with an infinitely high potential barrier at the dot boundary, the envelope functions of these particles are given by the 'particle-in-a-sphere' solutions, Eqn 9.2. The energy of each carrier is then described by Eqn 9.4 with the free particle mass m_0 replaced by $m_{eff}^{c,v}$.

However, in this treatment we completely ignore the Coulombic attraction between the electron and the positively charged hole. In the bulk material this interaction creates bound hydrogenic-like states, or excitons [4]. Since this effect is also present in quantum dots [10], a third approximation, the *strong confinement approximation* [1–3], is used to justify the neglect of this term. According to Eqn 9.4, the confinement energy of each carrier scales as $1/a^2$. The Coulomb interaction scales as $1/a$. In sufficiently small dots the quadratic confinement terms dominates. This condition is satisfied when the size of the quantum dot is much smaller than the size of the bulk exciton. In this strong confinement size regime the electron and hole are treated independently and each is described as a 'particle-in-a-sphere' [1–3]. The Coulomb term may then be added as a first-order energy correction, E_C. Therefore, using Eqns 9.4 and 9.8 the electron–hole pair (ehp) states are written as:

$$\Psi_{ehp}(\mathbf{r}_e, \mathbf{r}_h) = \Psi_e(\mathbf{r}_e)\Psi_h(\mathbf{r}_h) = u_c f_e(\mathbf{r}_e) u_v f_h(\mathbf{r}_h)$$

$$= C \left[u_c \frac{j_{L_c}(k_{n_c,L_c} r_e) Y_{L_c}^{m_c}}{r_e} \right] \left[u_v \frac{j_{L_h}(k_{n_h,L_h} r_h) Y_{L_h}^{m_h}}{r_h} \right] \tag{9.10}$$

with energies

$$E_{ehp}(n_h L_h n_e L_e) = E_g + \frac{\hbar^2}{2a^2} \left\{ \frac{\alpha_{n_h,L_h}^2}{m_{eff}^v} + \frac{\alpha_{n_c,L_c}^2}{m_{eff}^c} \right\} - E_c. \tag{9.11}$$

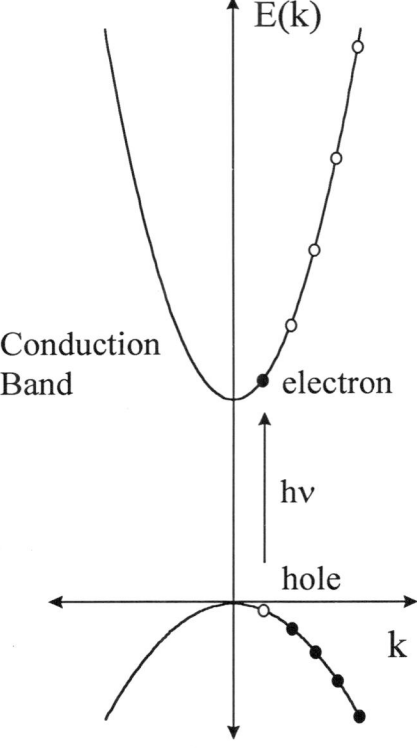

Figure 9.3 Optical transitions in finite size semiconductor quantum dots are discrete due to the quantization of the bulk band diagram. An electron is promoted into the conduction band, creating a hole in the valence band.

The states are labelled by the quantum numbers $n_h L_h n_e L_e$. For example, the lowest pair state is written as $1S_h 1S_e$. For pair states with the electron in the $1S_e$ level, the first-order Coulomb correction, E_c, is $1.8e^2/\varepsilon a$, where ε is the dielectric constant of the semiconductor [3]. Equations 9.10 and 9.11 are usually referred to as the 'particle-in-a-sphere' solutions to the quantum dot spectrum.

2.1.2   OPTICAL TRANSITION PROBABILITIES

For a particular quantum dot pair state the transition probability is given by the dipole matrix element

$$P = |\langle \Psi_{ehp} | \mathbf{e} \cdot \hat{p} | 0 \rangle|^2, \tag{9.12}$$

where $|0\rangle$ is the vacuum state, $\mathbf{e}$ is the polarization vector of the light, and $\hat{p}$ is the momentum operator. In the strong confinement approximation where the carriers are treated independently, Eqn 9.12 is commonly rewritten in terms of the single particle states

$$P = |\langle \Psi_e | \mathbf{e} \cdot \hat{p} | \Psi_h \rangle|^2. \tag{9.13}$$

Since the envelope functions are slowly varying in terms of **r**, the operator $\hat{p}$ acts only on the unit cell portion (u_{nk}) of the wavefunction. Equation 9.13 is simplified to

$$P = |\langle u_c|\mathbf{e}\cdot\hat{p}|u_v\rangle|^2|\langle f_e|f_h\rangle|^2. \tag{9.14}$$

In the 'particle-in-a-sphere' model this yields

$$P = |\langle u_c|\mathbf{e}\cdot\hat{p}|u_v\rangle|^2\delta_{n_e,n_h}\delta_{L_e,L_h} \tag{9.15}$$

due to the orthonormality of the 'particle-in-a-sphere' envelope functions. Therefore, simple selection rules ($\Delta n = 0$ and $\Delta L = 0$) are obtained in the 'particle-in-a-sphere' model.

2.1.3 THE REAL BAND STRUCTURE

In the above model the bulk conduction and valence bands are approximated by simple parabolic bands (Fig. 9.2). However, in general the band structure of II–VI semiconductors, such as CdSe, is more complicated. While the conduction band is well approximated by the simple treatment, the valence band is not. The valence band arises from Se 4p atomic orbitals and is 6-fold degenerate at $k = 0$, including spin. (In contrast, the conduction band arises from Cd 5s orbitals and is only 2-fold degenerate at $k = 0$.) When this degeneracy is lifted, valence band substructure occurs. For quantum dot pair states, this structure strongly modifies the results of the 'particle-in-a-sphere' model [11].

For convenience CdSe is often approximated as having the diamond-like band structure illustrated in Fig. 9.4. Due to strong spin–orbit coupling ($\Delta = 0.42$ eV in CdSe [12]) the valence band degeneracy at $k = 0$ is split into $p_{3/2}$ and $p_{1/2}$ sub-bands, where the subscript refers to $J = l + s$ ($l = 1$, $s = 1/2$). Away from $k = 0$ the $p_{3/2}$ band is further split into $J_m = \pm 3/2$ and $J_m = \pm 1/2$ sub-bands. The three sub-bands are referred to as either the heavy hole, light hole, and split-off hole sub-bands, or the A, B, and C sub-bands, as shown in Fig. 9.4.

For CdSe the structure in Fig. 9.4 is an approximation for two reasons.

1 Figure 9.4 ignores the crystal field splitting which occurs in CdSe crystals with a wurtzite (or hexagonal) lattice. This lattice, with its unique 'c' axis, has a crystal field which lifts the degeneracy of the A and B bands at $k = 0$ as shown in Fig. 9.5. This 'A–B splitting' is small (25 meV [10]) in bulk CdSe and is often neglected in quantum dot calculations. However, we discuss below how this term can cause additional splittings in the quantum dot transitions.

2 While the diamond structure has inversion symmetry, the hexagonal CdSe lattice does not. In detailed calculations this lack of inversion symmetry leads to linear terms in k which further split the A and B sub-bands in Fig. 9.5 away from $k = 0$ [13]. Since these linear terms are extremely small, they are generally neglected and are ignored below.

2.1.4 THE $k \cdot p$ METHOD (PRONOUNCED K-DOT-P)

Due to the complexity of the real band structure accurate quantum dot calculations require one to go beyond the 'particle-in-a-sphere' model and incorporate a better

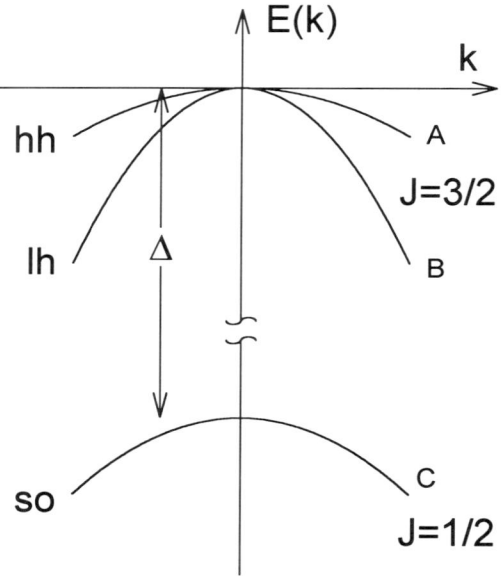

Figure 9.4 Simplified illustration of valence band structure at $k = 0$ for diamond-like semiconductors. Due to spin–orbit coupling (Δ) the valence band is split into two bands ($J = 3/2$ and $J = 1/2$) at $k = 0$. Away from $k = 0$, the $J = 3/2$ band is further split into the $J_m = \pm 3/2$ heavy hole (hh or A) and the $J_m = \pm 1/2$ light hole (lh or B) sub-bands. The $J = 1/2$ band is referred to as the split-off (so or C) band.

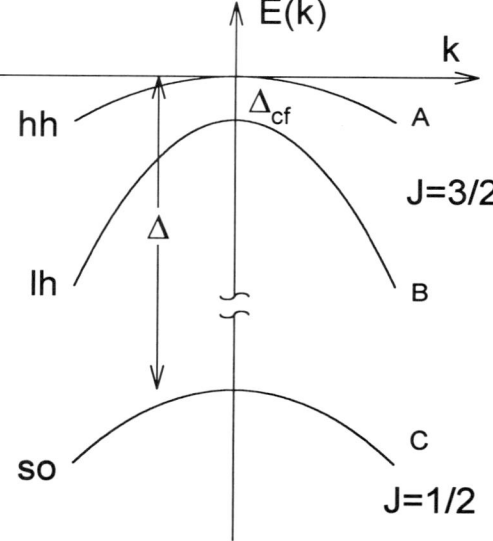

Figure 9.5 Simplified illustration of the bulk valence band for wurtzite (hexagonal) CdSe near $k = 0$. Due to the crystal field of the hexagonal lattice the A and B bands are split by Δ_{cf} (25 meV) at $k = 0$.

description of the bulk bands. Although bulk band structure can be determined with high accuracy using a variety of computational methods, these methods rarely provide analytical expressions for the description of the bands. For this purpose a much simpler, but often sufficient approach, the $k \cdot p$ method, is used [14]. In this method the bulk bands are expanded analytically around a particular point in k-space, typically $k = 0$.

Around this point the band energies and wavefunctions are then expressed in terms of the periodic functions u_{nk} and their energies E_{nk}.

To derive general expressions for u_{nk} and E_{nk} within this approach one considers the Bloch functions in Eqn 9.5. These functions are solutions of the Schrödinger equation for the single particle Hamiltonian

$$H_0 = \frac{p^2}{2m_0} + V(x),$$ (9.16)

where $V(x)$ is the periodic potential of the crystal lattice. Using Eqns 9.5 and 9.16 it is simple to show that the periodic functions, u_{nk}, satisfy the equation

$$\left[H_0 + \frac{1}{m_0}(k \cdot p) \right] u_{nk} = \lambda_{nk} u_{nk},$$ (9.17)

where

$$\lambda_{nk} = E_{nk} - \frac{k^2}{2m_0}.$$ (9.18)

Since u_{n0} and E_{n0} are assumed known, Eqn 9.17 can be treated in perturbation theory around $k = 0$ with

$$H' = \frac{(k \cdot p)}{m_0}.$$ (9.19)

Then using non-degenerate perturbation theory to second order* one obtains the energies

$$E_{nk} = E_{n0} + \frac{k^2}{2m_0} + \frac{1}{m_0^2} \sum_{m \neq n} \frac{|\mathbf{k} \cdot \mathbf{p}_{nm}|^2}{E_{n0} - E_{m0}}$$ (9.20)

and functions

$$u_{nk} = u_{n0} + \frac{1}{m_0} \sum_{m \neq n} u_{m0} \frac{\mathbf{k} \cdot \mathbf{p}_{mn}}{E_{n0} - E_{m0}}$$ (9.21)

with

$$\mathbf{p}_{nm} = \langle u_{n0} | \mathbf{p} | u_{m0} \rangle.$$ (9.22)

The summations in Eqns 9.20 and 9.21 are over all bands $m \neq n$. As one might expect the dispersion of band n is due to coupling with nearby bands.

Therefore, with the $k \cdot p$ approach analytical expressions can be obtained which describe the bulk bands to second order in k. While our discussion here outlines the general method, the approach must be slightly modified for CdSe. First, for the CdSe valence band, degenerate perturbation theory must be used. In this case the valence band must be diagonalized before coupling with other bands is considered. Second, we

---

*Inversion symmetry is assumed. In CdSe the lack of inversion symmetry introduces terms linear in k. These terms are small and generally neglected.

have neglected spin–orbit coupling terms. However, these terms are easily added as can be seen in Kittel [14].

2.1.5 THE KANE MODEL

In some bulk systems it is necessary to go beyond second order in the $k \cdot p$ approach to properly describe the bands. However, in higher orders this approach is often cumbersome and Kane [15,16] developed an alternate procedure for bulk semiconductors which is also widely used in confined systems. In the Kane model a small subset of bands are treated exactly by explicitly diagonalizing Eqn 9.17 (or the equivalent expression with the spin–orbit interaction included). This subset usually contains the bands of interest, e.g. the valence band and conduction band. Then the influence of outlying bands is included within the second-order $k \cdot p$ approach. Due to the exact treatment of the bands of interest, the dispersion of each band is no longer strictly quadratic as in Eqn 9.20. Therefore, the Kane model better describes band 'nonparabolicities'.

2.1.6 THE LUTTINGER HAMILTONIAN

For bulk diamond-like semiconductors the 6-fold degenerate valence band is described by the Luttinger Hamiltonian [17,18]. This expression, a 6 by 6 matrix, is derived within the context of degenerate $k \cdot p$ perturbation theory [19]. The Hamiltonian is commonly simplified further using the spherical approximation [20–22]. In this case only terms of spherical symmetry in the Luttinger Hamiltonian are considered. 'Warping' terms of cubic symmetry are neglected and, if desired, treated as a perturbation. For quantum dot theories which include the valence band degeneracies, the Luttinger Hamiltonian is the initial starting point to obtain the hole eigenstates and their energies. We note that since CdSe is wurtzite, as discussed above, use of the Luttinger Hamiltonian for CdSe quantum dots is an approximation. It does not include the crystal field splitting that is present in wurtzite CdSe.

2.2 *Experimental*

2.2.1 SAMPLES

Although we focus here on quantum dot spectroscopy, we cannot overemphasize the importance of sample quality in obtaining useful optical information. Fortunately, tremendous progress has been made in recent years in the fabrication of II–VI quantum dots. In particular, extremely high quality CdSe dots can now be prepared according to the method of Murray *et al.* [5], or variations thereof [23]. This procedure uses a wet chemical (organometallic) synthesis to fabricate the crystallites. The size distribution is then further narrowed by a size-selective precipitation step. Highly crystalline, nearly monodisperse (<4% r.m.s.) dots are obtained with well-passivated surfaces. Without further modification these dots exhibit strong band edge luminescence with a quantum yield greater than 0.1 and measured as high as 0.9 at 10 K. At room temperature the quantum yield is typically 10% [5]. The intensity of 'deep trap' emission,* which dominates the luminescence behaviour of dots prepared by many other methods, is very weak in these samples.

Due to their versatile surface chemistry, the dots can be dispersed in a variety of host materials, such as solvents, polymers, and other semiconductors. In fact, very recently Guyot-Sionnest [24] has reported that CdSe quantum dots encapsulated in a higher band gap semiconductor such as ZnS exhibit room temperature emission quantum yields greater than 0.5. This development, along with the recent demonstrations of both disordered CdSe quantum dot glasses [25] and three-dimensionally ordered CdSe quantum dot arrays [26] are providing exciting new classes of quantum dot materials for future study.

With the synthetic method of Murray *et al.* [5] high quality samples ranging from ~15 Å to ~120 Å in mean diameter are easily produced. Therefore, with such samples one of the original and basic experimental questions about quantum dots — how their electronic spectra evolve with size in the strong confinement regime — may be addressed. Early work on this question [27–34] was constrained by difficulties in preparing high quality, monodisperse samples. Inhomogeneities such as distributions in size and shape which conceal the higher transitions prevented a more complete investigation. More recent studies [35–38] which do examine quantum dots of sufficient quality to resolve many of the higher states are restricted to one [35–37] or a few [38] sizes. The newer samples allow the size dependence question to be more satisfactorily addressed. In addition, their spectra exhibit fine structure in the lowest exciton feature which has been crucial to explaining the luminescence behaviour of these materials [39–41].

2.2.2 SPECTROSCOPIC METHODS

Despite the high quality of these samples, residual sample inhomogeneities still remain which broaden absorption and emission features and conceal transitions. To reduce these effects several spectroscopic techniques may be used. For absorption information the most common technique has been transient differential absorption spectroscopy (TDA), also called pump-probe or hole-burning spectroscopy [35,39,42–53]. This technique measures the absorption change induced by a spectrally narrow pump beam. TDA effectively increases the resolution of the spectrum by optically exciting and bleaching a narrow subset of the quantum dots. The bleach spectrum reveals absorption information with inhomogeneous broadening greatly reduced. However, pump-induced absorption features which overlap with the bleach features of interest complicate the analysis. More recently, many groups have begun to utilize a simpler optical technique which avoids this problem, photoluminescence excitation (PLE) [35,54–57]. Like TDA, PLE selects a narrow subset of the sample distribution and absorption informa-

*'Deep trap' emission is luminescence which is strongly red-shifted from the HOMO–LUMO transition (or 'band edge' in solid-state terminology) of the quantum dot. While the true origin of this luminescence is unknown, it is generally assumed that its presence indicates defect and/or impurity states which are deep in the gap.

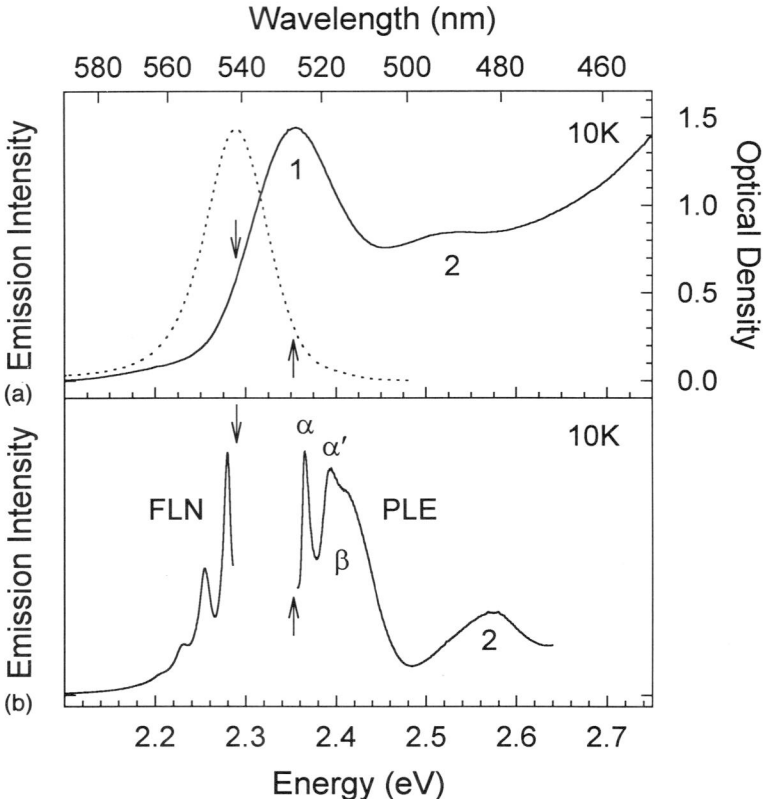

Figure 9.6 (a) Absorption (solid line) and full luminescence (dotted line) spectra for ~19 Å effective radius CdSe quantum dots. (b) Fluorescence line narrowing (FLN) and photoluminescence excitation (PLE) spectra for the same sample. An LO–phonon progression is observed in FLN. Both narrow (α, α′) and broad (β) absorption features are resolved in PLE. The downward (upward) arrows denote the excitation (emission) position used for FLN (PLE). (Adapted from [41].)

tion with increased resolution is obtained. When PLE is combined with fluorescence line narrowing (FLN) spectroscopy, both absorption and emission information representative of a 'single dot' is revealed.

Both PLE and FLN techniques are demonstrated in Fig. 9.6 along with absorption and emission results for a 19 Å effective radius CdSe quantum dot sample.* On this scale only the lowest two exciton features are observed in the absorption spectrum (solid line in Fig. 9.6a). The emission spectrum (dotted line in Fig. 9.6a) is obtained by exciting the sample well above its first transition so that emission occurs from the entire sample distribution. This inhomogeneously broadened emission feature is referred to as 'full luminescence'. If instead a subset of the sample distribution is excited, a significantly narrowed and structured emission spectrum is revealed. For example, when the sample in Fig. 9.6 is excited on the low energy side of its first absorption feature (downward arrows in Fig. 9.6) a vibration (longitudinal optical (LO) phonon) progres-

*Sizes reported are estimated from extensive size-dependent transmission electron microscopy (TEM) and small angle X-ray measurements and are based on the energy of the first absorption peak. We define the effective radius of our prolate dots as $a = [(b^2c)^{1/3}]/2$ where b and c are the short and long axes, respectively.

sion is clearly resolved. This FLN spectrum can be used to extract a model 'single dot' emission lineshape.

PLE can similarly be used to extract 'single dot' absorption information by monitoring a narrow spectral band (upward arrows) of the full luminescence while scanning the excitation energy. As seen in Fig. 9.6(b), both absorption features present in Fig. 9.6(a) are obtained with higher resolution. In addition structure is observed with the lowest absorption feature. As we show below these features (α and β) represent fine structure present in the lowest exciton and have important implications for quantum dot emission. However, before discussing the exciton fine structure, in the next section we use PLE results on a large series of CdSe samples to describe the size dependence of the exciton structure.

2.3 *Exciton structure vs size*

2.3.1 EXPERIMENTAL RESULTS

While the absorption and PLE spectra in Fig. 9.6 show only the two lowest exciton features, current samples reveal much more exciton structure. For example, in Fig. 9.7 PLE results for a 28 Å radius CdSe sample are shown along with its absorption and full luminescence spectra. These data cover a larger spectral range than Fig. 9.6 and show more of the exciton spectrum. To determine how the exciton structure evolves with quantum dot size, PLE data can be obtained for a large sample series. Seven such spectra are shown in Fig. 9.8. The quantum dots are arranged (top to bottom) in order of increasing radius from ~15 to ~43 Å. Quantum confinement clearly shifts the transitions blue (>0.5 eV) with decreasing size. The high quality of these dots also allows the resolution of as many as eight absorption features in a single spectrum.

By extracting peak positions from PLE data such as Fig. 9.8 the quantum dot spectrum as a function of size is obtained. Figure 9.9 plots the final size-dependent spectrum determined in [57]. Although dot radius (or diameter) is not used as the *x*-axis, Fig. 9.9 still represents a 'size-dependent' plot. The *x*-axis label, the energy of the first excited state, is a strongly size-dependent parameter. It is also much easier to measure accurately than quantum dot size. For the *y*-axis the energy relative to the first excited state is used so that the plot focuses on the higher excited states. Therefore, Fig. 9.9 summarizes the size dependence of the first 10 transitions for CdSe quantum dots in the strong confinement regime.

2.3.2 COMPARISON WITH THEORY

In order to understand this result one could begin with the simple 'particle-in-a-sphere' model outlined above (see Section 2.1.1). The complicated valence band structure, shown in Fig. 9.4, could then be included by considering each sub-band (A, B, and C) as a simple parabolic band. In such a zero-order picture each bulk sub-band would lead to a ladder of 'particle-in-a-sphere' states for the hole, as shown in Fig. 9.10. Quantum dot transitions would occur between these hole states and the electron levels arising from the bulk conduction band. However, we find that this simplistic approach fails to describe the experimental absorption structure. In particular two avoided crossings are

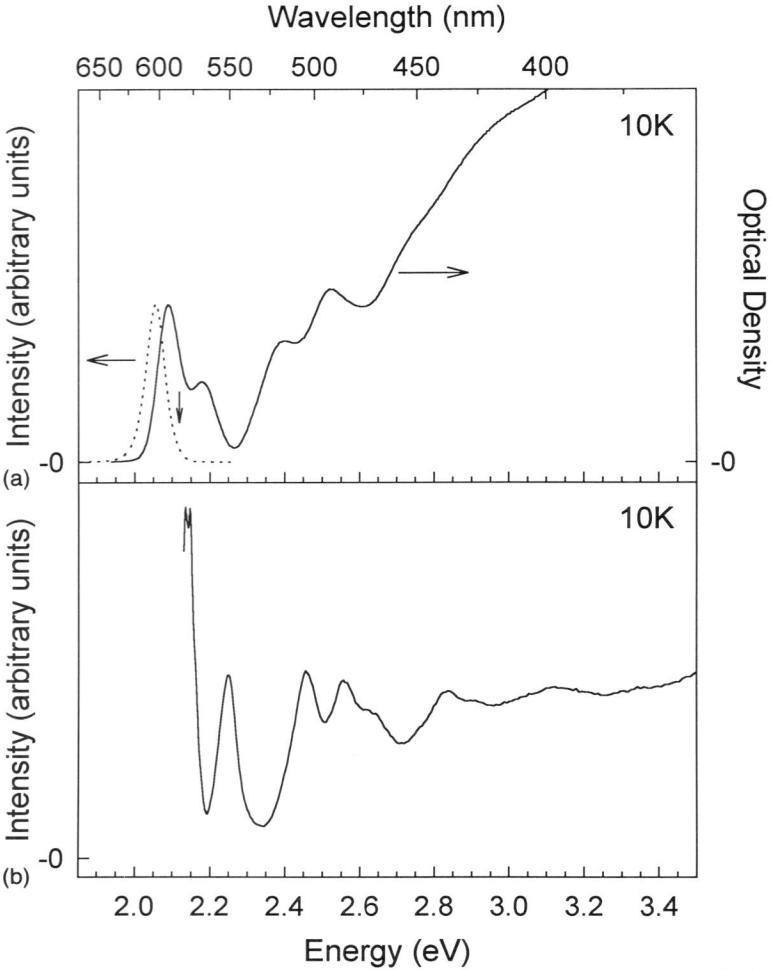

Figure 9.7 (a) Absorption (solid line) and full luminescence (dashed line) spectra for ~28 Å radius CdSe quantum dots. In luminescence the sample was excited at 2.655 eV (467.0 nm). The downward arrow marks the emission position used in photoluminescence excitation (PLE). (b) PLE scan for the same sample. (Adapted from [57].)

present in Fig. 9.8 (between features (e) and (g) at ~2.0 eV and between features (e) and (c) above 2.2 eV) and these are not predicted by this 'particle-in-a-sphere' model.

The problem lies in the assumption that each valence sub-band produces its own independent ladder of hole states. In reality the hole states are mixed due to the underlying quantum mechanics. To help understand this effect, we summarize all of the relevant quantum numbers in Fig. 9.11. The total angular momentum of either the electron or hole (F_e or F_h) has two contributions (i) a 'unit cell' contribution (J) due to the underlying basis which forms the bulk bands; and (ii) an envelope function contribution (L) due to the 'particle-in-a-sphere' orbital. Above we assume that the quantum numbers describing each valence sub-band (J_h) and each envelope function (L_h) are conserved. However, when the Luttinger Hamiltonian, which describes the bulk valence band, is combined with a spherical potential, mixing between the bulk valence bands occurs. This effect, which was first shown for bulk impurity centres [20–22], also mixes quantum dot hole states [11,38,58–61]. Only parity and the total

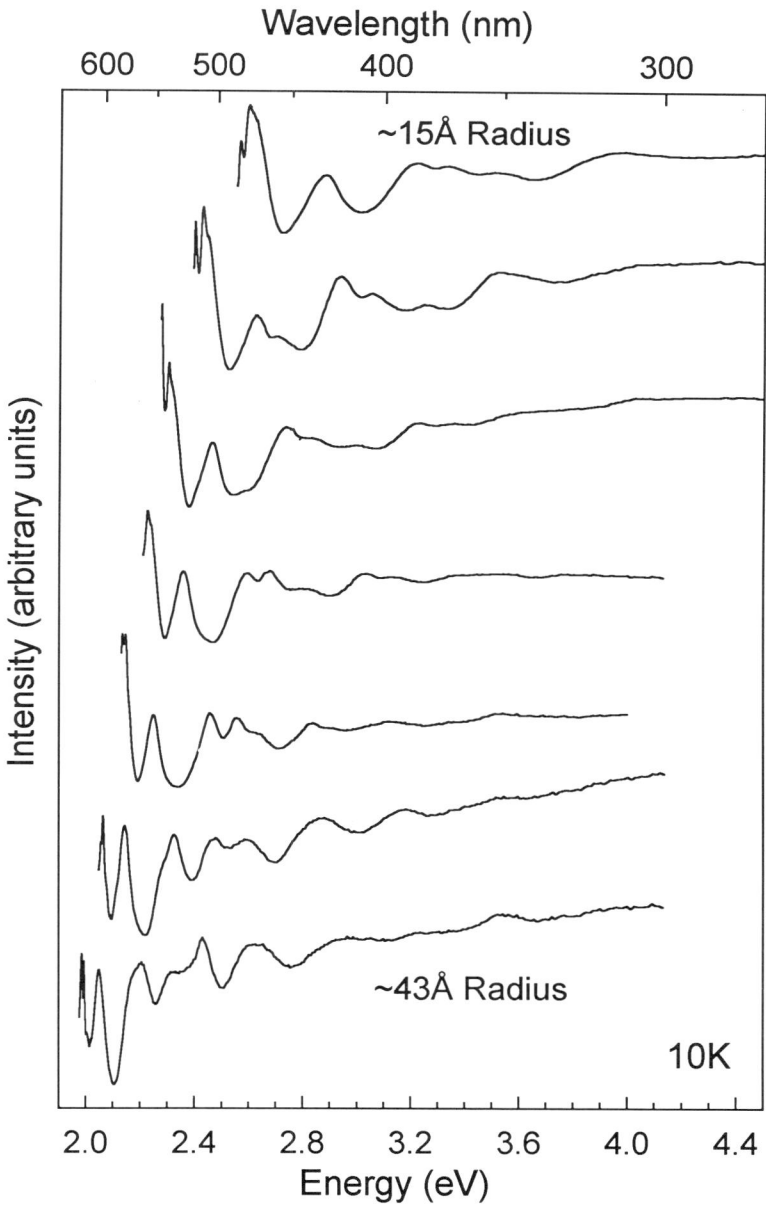

Figure 9.8 Normalized photoluminescence excitation (PLE) scans for seven different size quantum dot samples. Size increases from top to bottom and ranges from ~15 to ~43 Å in radius. (Adapted from [57].)

hole angular momentum (F_h) are good quantum numbers. Neither L_h nor J_h are conserved. Therefore, each quantum dot hole state is a mixture of the three valence sub-bands (valence band mixing) as well as 'particle-in-a-sphere' envelope functions with angular momentum L_h and $L_h + 2$ ('S–D mixing'). The three independent ladders of hole states shown in Fig. 9.10 are coupled. The electron levels which originate in the simple conduction band are not affected by the valence band complexities and are well described by the 'particle-in-a-sphere' ladder.

When these effects are included theory correctly describes the observed size-

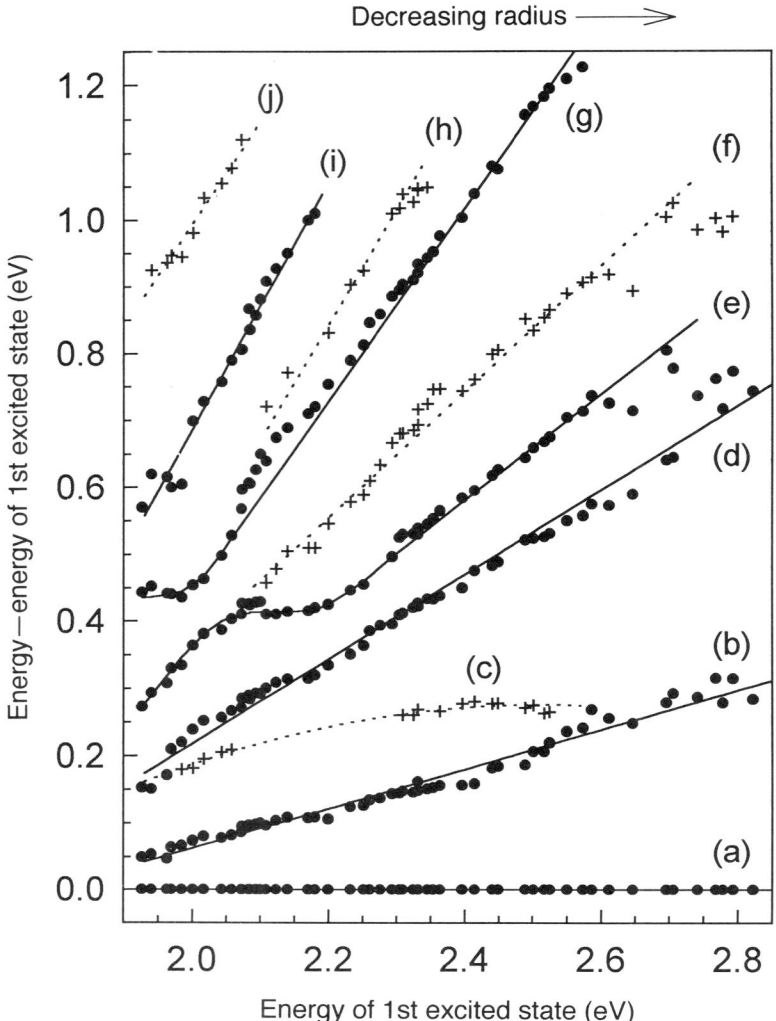

Figure 9.9 Transition energies (relative to the first excited state) vs the energy of the first excited state. Peak positions are extracted from photoluminescence excitation (PLE) data such as that shown in Fig. 9.8. Strong (weak) transitions are denoted by circles (crosses). The solid (dashed) lines are visual guides for the strong (weak) transitions to clarify their size evolution. (Adapted from [57].)

dependent absorption structure. Using the approach of Efros [38,58] in which the energies of the hole states are determined by solving the Luttinger Hamiltonian and the electron levels are calculated within the Kane model, strong agreement with the data is obtained, as shown in Figs 9.12 and 9.13. Figure 9.12 compares theory with the lowest three transitions which exhibit simple size-dependent behaviour (i.e. no avoided crossings). Figure 9.13 shows the avoided crossing regions. The transitions can be assigned and labelled by modified 'particle-in-a-sphere' symbols which account for the valence band mixing discussed above [57].

Although theory clearly predicts the observed avoided crossings, Fig. 9.13 also demonstrates that theory underestimates the repulsion in both avoided crossing regions, causing theoretical deviation in the predictions of the $1S_{1/2}1S_e$ and $2S_{1/2}1S_e$

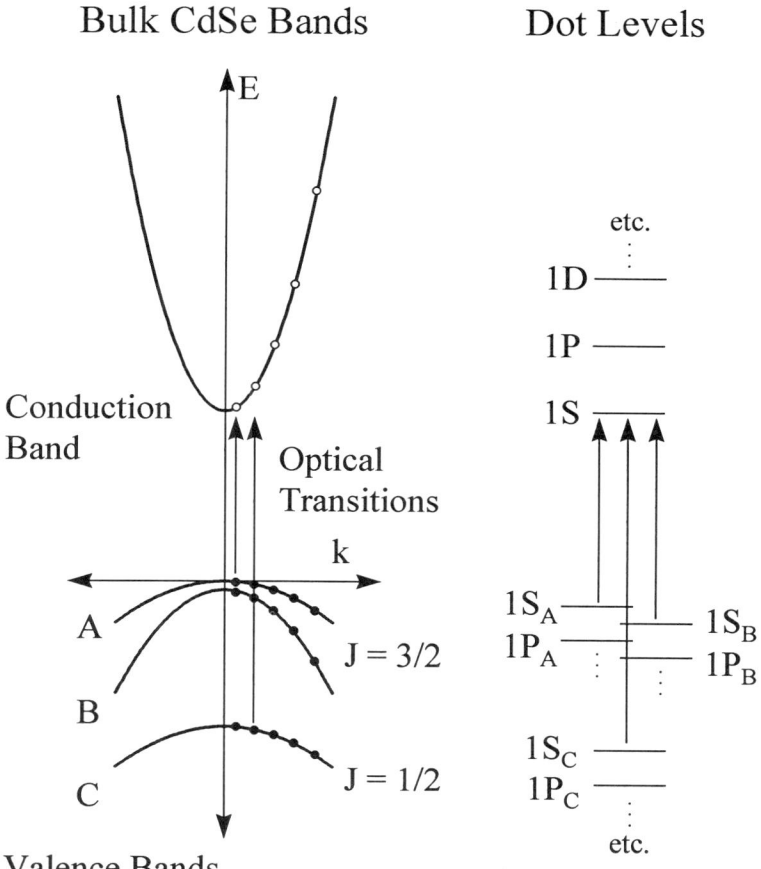

Figure 9.10 A simplistic approach to describing the quantum dot spectrum. If we consider each valence sub-band independently, each contributes a ladder of 'particle-in-a-sphere' states for the hole. The transitions then occur between these hole states and the electron levels arising from the conduction band. However, this zero-order model fails to predict the observed structure due to mixing of the different hole ladders, as discussed in the text.

transitions. This discrepancy is most likely due to the Coulomb mixing of the electron–hole pair states, which is ignored by the model (via the strong confinement approximation). If included this term would further couple the $nS_{1/2}1S_e$ transitions such that these states 'push off' each other more strongly. In addition the Coulomb term would cause the $1S_{1/2}1S_e$ and $2S_{1/2}1S_e$ states to avoid one another through their individual repulsion from the strongly allowed $1P_{3/2}1P_e$. A current area of theoretical research is to include a more rigorous treatment of the Coulomb mixing and correct these discrepancies.

Despite these discrepancies, however, theory is clearly on the right track. Therefore, since the transitions are now assigned, we can use the model to understand the physics behind the avoided crossings. As discussed above, in the zero-order picture of Fig. 9.10 each valence sub-band contributes a ladder of hole states. Due to spin–orbit splitting (see Fig. 9.5) the C band ladder is offset 0.42 eV below the A and B band ladders. This leads to possible resonances between hole levels from the A and B bands with C band levels. Since the levels are spreading out with decreasing dot size, resonance conditions are satisfied only in certain size dots. Figure 9.14 demonstrates the two resonances responsible for the observed avoided crossings. For simplicity we treat the A and B

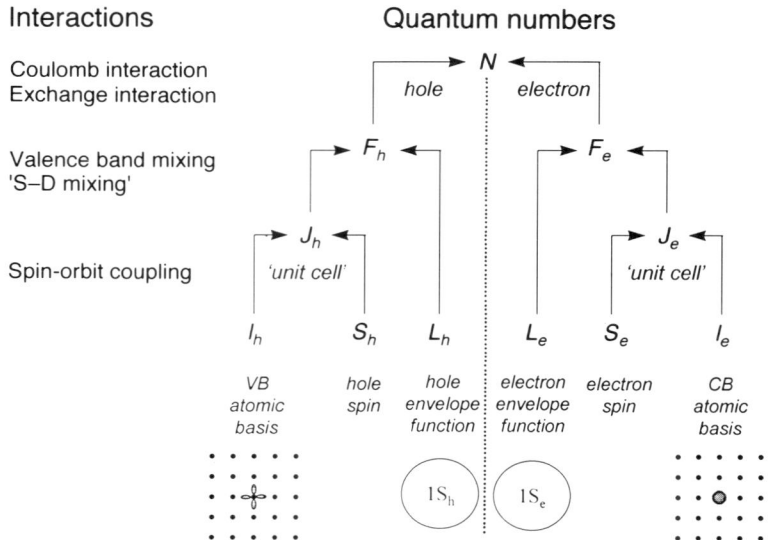

Figure 9.11 Summary of quantum dot quantum numbers and important interactions. The total electron–hole pair angular momentum (N) has contributions due to both the electron (F_e) and hole (F_h). Each carrier's angular momentum (F) may then be further broken down into a 'unit cell' component (J) due to the atomic basis (l) and spin (s) of the particle and an envelope function component (L) due to the 'particle-in-a-sphere' orbital.

bands together. For example, in Fig. 9.14(a) the $2D$ level from the A and B bands is resonant with the 1S level from the C band. The size dependence of these levels is plotted in Fig. 9.14(c). Due to both valence band mixing and S–D mixing, these resonant conditions lead to the observed avoided crossings. Although this description is based on the simple 'particle-in-a-sphere' model of Fig. 9.10, the explanation is consistent with a more detailed analysis [57].

2.4 *Exciton fine structure*

2.4.1 PERTURBATIONS TO THE SPHERICAL MODEL

In the previous section, we have shown that the exciton structure of CdSe quantum dots is now fairly well understood. In general the size-dependent absorption features are described by quantum dot effective mass models which incorporate the complexities of the CdSe valence band. For convenience these models assume spherical dots and work within the spherical band approximation [20–22], since more sophisticated treatments have not been required to explain experimental results. These models predict that the lowest energy electron–hole pair state ($1S_{3/2}1S_e$ — which we refer to as the 'band edge exciton') is 8-fold degenerate. However, recent theoretical work which extends the 'spherical model' to include the effects of the hexagonal lattice [62], the non-spherical shape [63], and the electron–hole exchange interaction [64–66] has predicted that exciton fine structure should be present. In this case the initially 8-fold degenerate band edge exciton is split into five sublevels [40].

This exciton fine structure is shown in the energy level diagram of Fig. 9.15. To

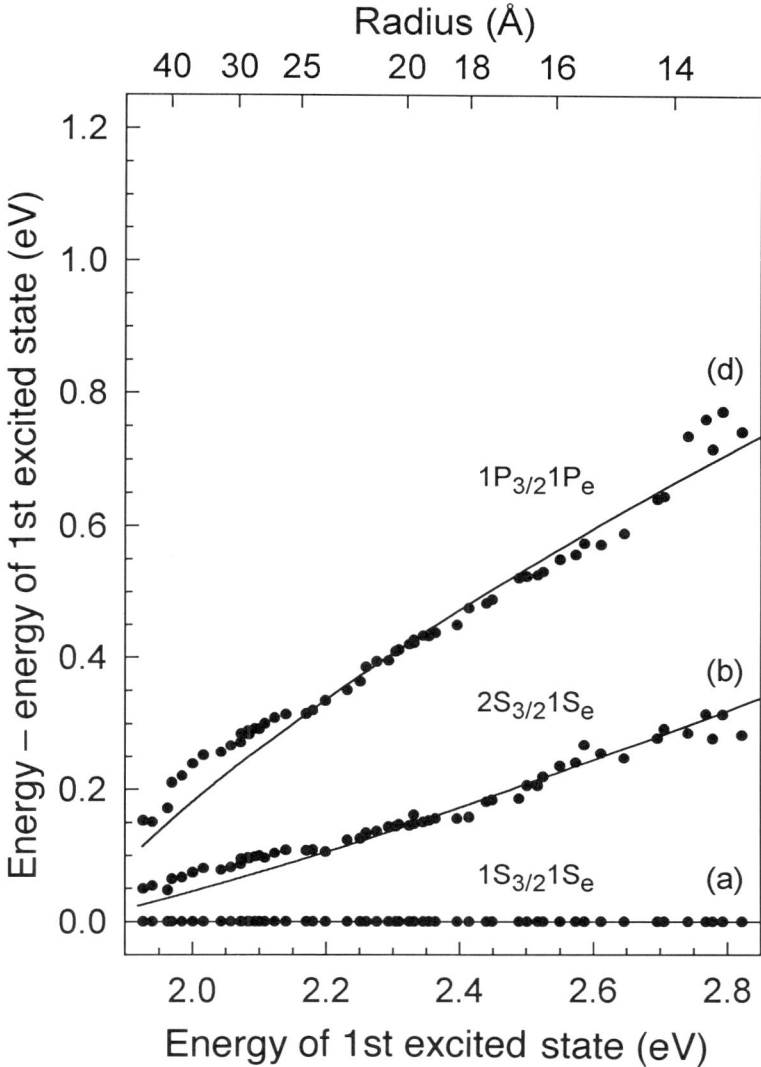

Figure 9.12 Theoretically predicted pair states (solid lines) assigned to features (a), (b), and (d) in Fig. 9.9. The experimental data is shown for comparison (circles). (Adapted from [57].)

describe the structure we consider two limits. On the left of Fig. 9.15 the effect of the anisotropy of the crystal lattice and/or the non-spherical shape of the crystallite dominates. This corresponds to the bulk limit where the exchange interaction between the electron and hole is negligible (0.15 meV [67]). The band edge exciton is split into two 4-fold degenerate states, analogous to the bulk 'A–B splitting' (see Fig. 9.5). The splitting occurs due to the reduction from spherical to uniaxial symmetry. However, since the exchange interaction is proportional to the overlap between the electron and hole, in small dots this term is strongly enhanced due to the confinement of the carriers [64]. Therefore, the right of Fig. 9.15 represents the 'small dot' limit where the exchange interaction dominates. In this case the important quantum number is the total angular momentum, N (see Fig. 9.11). Since $F_h = 3/2$, and $F_e = 1/2$, the band edge exciton is split into a 5-fold degenerate $N = 2$ state and a 3-fold degenerate $N = 1$ state. In the middle of Fig. 9.15 the correlation diagram between these two limits is shown.

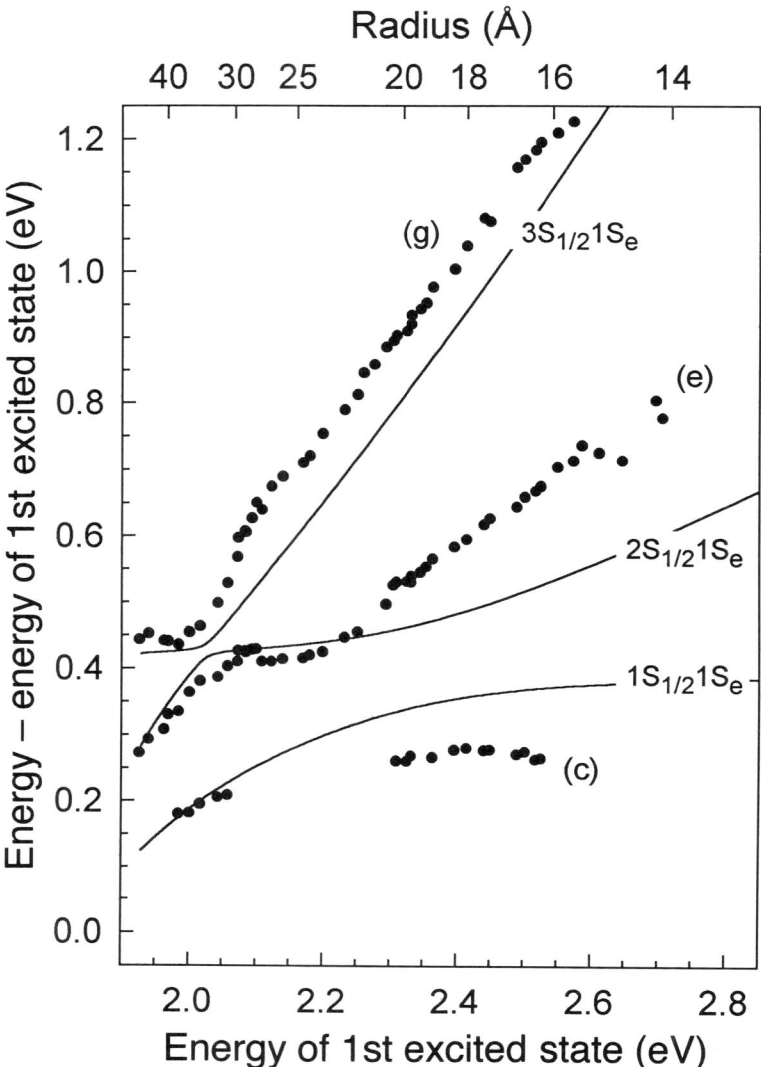

Figure 9.13 Theoretically predicted pair states (solid lines) assigned to features (c), (e), and (g) in Fig. 9.9. The experimental data are shown for comparison (circles). (Adapted from [57].)

When both effects are included, the good quantum number is the projection of N along the unique crystal axis, N_m. The five sublevels are then labelled by $|N_m|$: one sublevel with $|N_m| = 2$, two with $|N_m| = 1$, and two with $|N_m| = 0$. Levels with $|N_m| > 0$ are 2-fold degenerate.

To quantify these effects, the anisotropy and exchange terms can be added to the spherical model within the framework of perturbation theory [40]. Figure 9.16 shows the calculated size dependence of the exciton fine structure. The five sublevels are labelled by $|N_m|$ with superscripts to distinguish upper (U) and lower (L) sublevels with the same $|N_m|$. Their energy, relative to the 1^L sublevel, is plotted vs effective radius. The enhancement of the exchange interaction with decreasing dot size is clearly present.

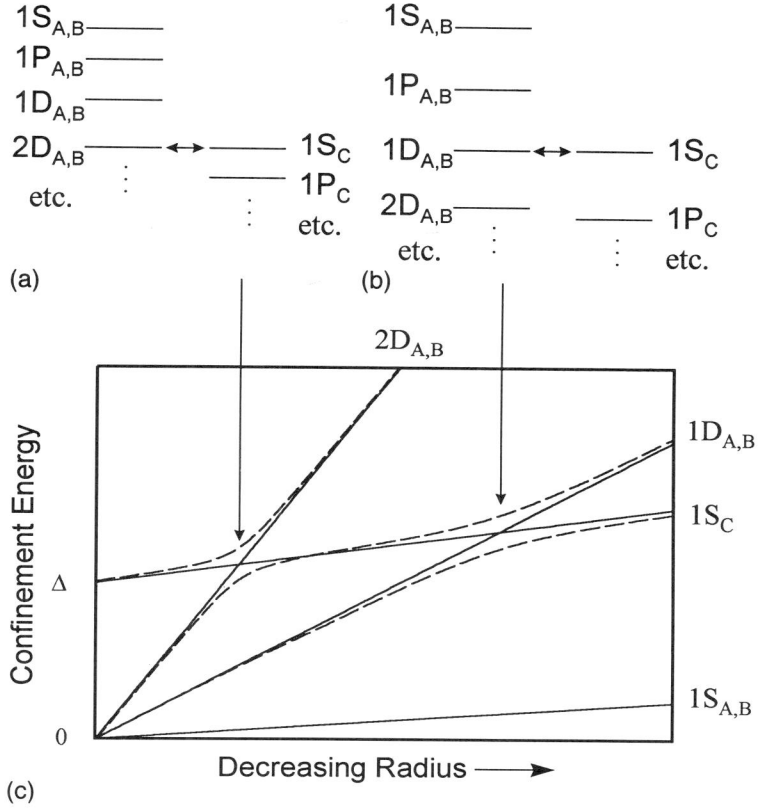

Figure 9.14 Origin of the observed avoided crossings. For a particular quantum dot size a resonance occurs between a hole level from the A and B bands (combined for simplicity) and a hole level from the C band. Energy level diagrams for the hole states are shown in (a) and (b) for the two resonances responsible for the observed avoided crossings. (c) The energy of the hole states vs decreasing radius. The solid (dashed) lines represent the levels without (with) the valence band and S–D mixing.

2.4.2 IMPLICATIONS OF EXCITON FINE STRUCTURE: THE 'DARK EXCITON'

While exciton recombination in bulk II–VI semiconductors occurs with a ~1 ns lifetime [68], CdSe quantum dots exhibit a ~1 μs radiative lifetime at 10 K. This effect is perhaps not surprising in early samples which were of poor quality and emitted weakly via 'deep trap' fluorescence. However, more recent samples which are of much higher quality and emit strongly at the band edge also have long radiative lifetimes [40,69,70]. To explain this behaviour, quantum dot emission has been rationalized by many researchers (including the authors) as a 'surface effect' [35,69–73]. The anomalous emission lifetime has been explained by localization of the photoexcited electron [71–73] and/or hole [35,69,70] at the dot/matrix interface. Once the carriers are localized in surface 'traps', the decrease in carrier overlap increases the recombination time. The influence of the surface on quantum dot emission was considered reasonable since these materials have such large surface-to-volume ratios (e.g. in a ~30 Å diameter dot ca one-third of the atoms are on the surface). This 'surface model' can explain the

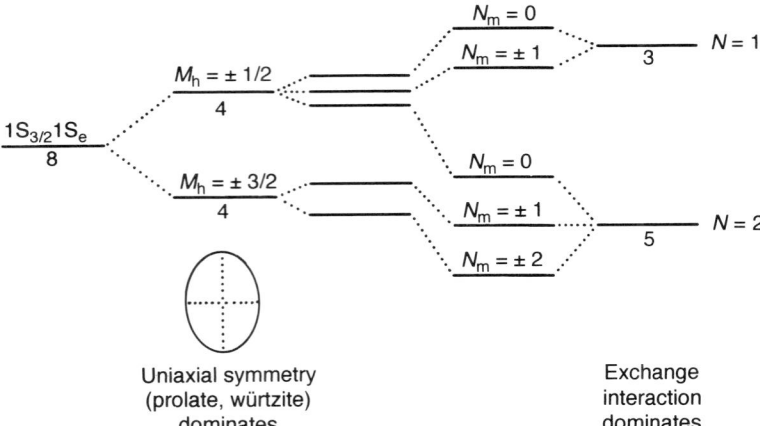

Figure 9.15 Energy level diagram describing the exciton fine structure. In the spherical model the band edge exciton ($1S_{3/2}1S_e$) is 8-fold degenerate. This degeneracy is split by the non-spherical shape of the dots, their hexagonal (wurtzite) lattice, and the exchange interaction. On the left of the diagram, the effects of the non-spherical shape and hexagonal lattice dominate and the band edge exciton is split into two 4-fold degenerate states. This corresponds to the bulk limit. On the right, the exchange interaction dominates and the band edge exciton is split into a 5-fold degenerate $N = 2$ state and a 3-fold degenerate $N = 1$ state. This corresponds to the 'small dot' limit. The middle represents a correlation diagram between these two extremes.

long radiative lifetimes, luminescence polarization results [70], and even the unexpectedly high longitudinal optical (LO) phonon coupling observed in emission [69].

However, the presence of band edge exciton fine structure provides an alternative explanation for the anomalous emission behaviour [40,66,74]. Emission from the lowest band edge state, $|N_m| = 2$, is optically forbidden in the electric dipole approxi-

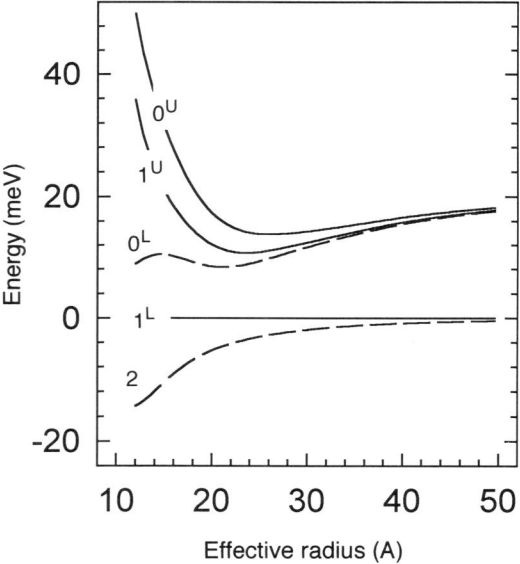

Figure 9.16 Calculated band edge exciton ($1S_{3/2}1S_e$) structure vs effective radius. The sublevels are labelled by $|N_m|$ with superscripts to distinguish upper (U) and lower (L) sublevels with the same $|N_m|$. Positions are relative to 1^L. Optically active (passive) levels are shown as solid (dashed) lines. (Adapted from [41].)

mation. Relaxation of the electron–hole pair into this state, referred to as the 'dark exciton', can explain the long radiative lifetimes observed in CdSe quantum dots, previously attributed to surface traps. Since two units of angular momentum are required to return to the ground state from the $|N_m| = 2$ sublevel, this transition is one-photon forbidden. However, less efficient, phonon-assisted transitions can occur, explaining the stronger LO–phonon coupling of the emitting state and the long radiative lifetimes (at 10 K). In addition polarization effects observed in luminescence [70] can be rationalized by relaxation from the 1^L sublevel to the dark exciton, as shown by Chamarro *et al.* [66].

2.4.3 SPECTROSCOPIC EVIDENCE FOR EXCITON FINE STRUCTURE

We mentioned in Section 2.2.2 that the PLE spectra of high quality samples exhibit additional absorption structure within the lowest exciton feature. For example, in Fig. 9.6(b) a narrow feature (α), its LO phonon replica (α'), and a broader feature (β) are observed. To test whether this structure is consistent with the predicted fine structure, a size-dependent study is shown in Fig. 9.17. For eight different size CdSe samples the band edge FLN/PLE structure is measured. In each FLN/PLE pair the FLN excitation and PLE emission energies are the same and the data are plotted relative to this energy. The actual excitation/emission positions are indicated with arrows in the full luminescence spectra, shown in Fig. 9.18. From the experimental data an underlying lineshape can be extracted for each sample [41]. The single dot lineshape obtained for the sample from Fig. 9.6 is shown in Fig. 9.19. To be consistent with the exciton fine structure the emitting state must be assigned to the dark exciton. The narrow absorption feature (α) is then assigned to the 1^L sublevel since the 0^L sublevel is optically passive. The broader feature (β) is assigned to a combination of the 1^U and the 0^U sublevels.

The assignment of β to a combination of 1^U and 0^U is further supported by data from the larger samples in which a third absorption feature is observed. In Fig. 9.20 the band edge region of FLN/PLE data for a ~44 Å effective radius sample is shown. While in Fig. 9.6(b) three band edge states are resolved — a narrow emitting state, a narrow absorbing state (α), and a broad absorbing state (β) — in Fig. 9.20 four band edge states are present — a narrow emitting state and three narrow absorbing states (α, β_1 and β_2) [41]. In this case β_1 and β_2 can be assigned to the individual 1^U and 0^U sublevels.

To quantitatively test these assignments we compare the experimental results with the predictions of theory in Fig. 9.21. Figure 9.21(a) shows the size dependence of the calculated band edge structure. Figure 9.21(b) shows the position of the absorbing (filled circles and squares) and emitting (open circles) features from Fig. 9.17 and TDA results [39], relative to the narrow absorption feature α (1^L). For larger samples both the positions of β_1 and β_2 (pluses) and their weighted average (squares) are shown. Comparison with theory indicates that the model accurately reproduces many aspects of the data. Both the splitting between $|N_m| = 2$ and 1^L (the Stokes shift) and the splitting between 1^L and the upper states (1^U and 0^U) are described reasonably well. This result is particularly significant since, although the predicted structure strongly depends on the theoretical input parameters [40], only literature values were used in the theoretical calculation.

Further quantitative evidence for our assignments is obtained from the oscillator

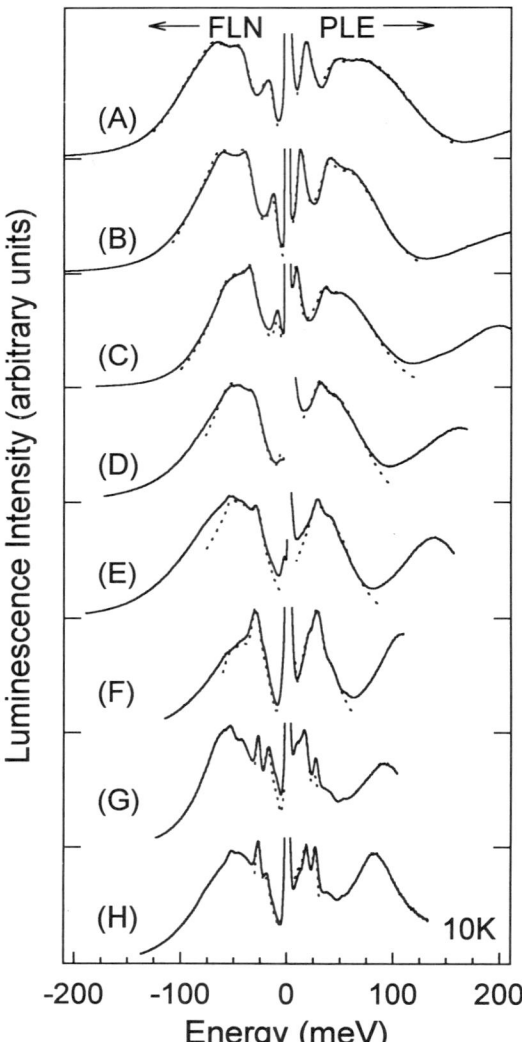

Figure 9.17 The size dependence of band edge fluorescence line narrowing/photoluminescence excitation (FLN/PLE) spectra. In each pair of FLN/PLE results (solid lines) the FLN excitation and the PLE emission energies are the same and indicated by arrows in the full luminescence spectra in Fig. 9.18. The PLE (FLN) data are plotted relative to the emission (excitation) energy. Dotted lines show the best fit used to extract band edge parameters shown in Fig. 9.21. (Adapted from [41].)

strengths of the optically allowed sublevels. In Fig. 9.21(c) the predicted oscillator strength of the optically active sublevels is shown. The strength of the upper states (1^U and 0^U) is combined since these states are not individually resolved in many of our samples. The experimental values are plotted in Fig. 9.21(d). Reasonable agreement between experiment and theory is observed, again with no fitting parameters.

To understand the size dependence of the oscillator strengths we consider two opposing limits. In large dots the states converge to A- and B-like excitons (as in Fig. 9.21a), each possessing half of the total band edge oscillator strength. Therefore, we expect 1^L and the combined upper states (1^U and 0^U) to each approach 0.5 in large sizes. In small dots the exchange interaction dominates and the crystal field and non-spherical shape effects become negligible. In this limit (right hand side of Fig. 9.15)

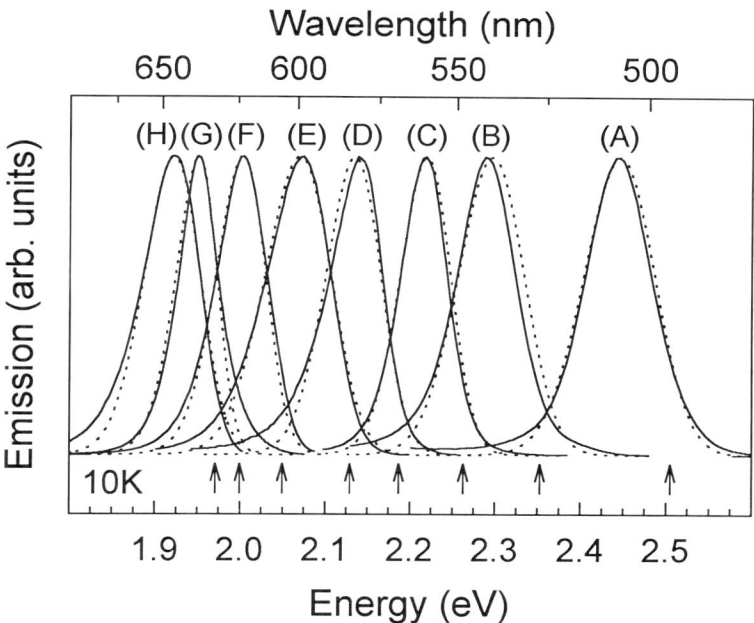

Figure 9.18 Full luminescence spectra for the size series shown in Fig. 9.17 (solid lines). Arrows indicate the fluorescence line narrowing (FLN) excitation positions and photoluminescence excitation (PLE) emission positions (eV). Dotted lines show the best fit used to extract band edge parameters shown in Fig. 9.21. (Adapted from [41].)

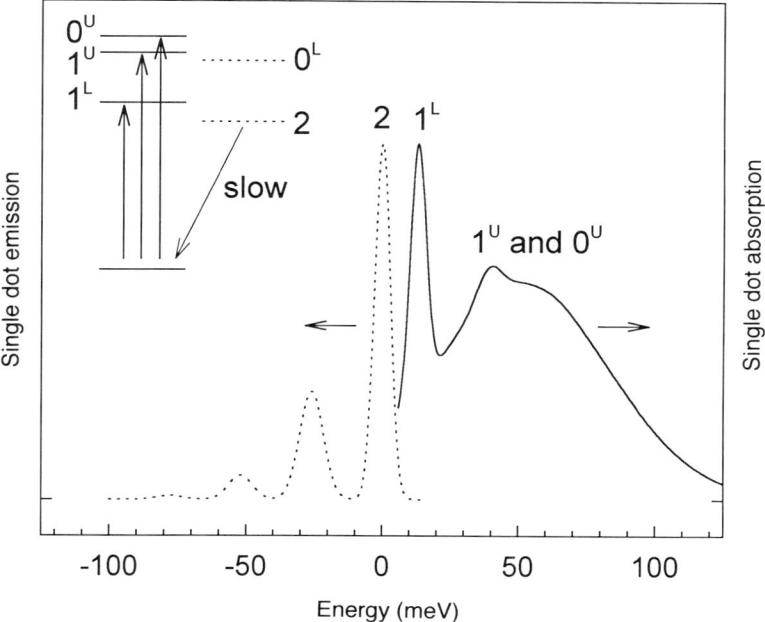

Figure 9.19 Single dot absorption (solid line) and emission (dotted line) structure extracted for the sample shown in Fig. 9.6 including longitudinal optical (LO)–phonon coupling. An energy level diagram illustrates the band edge exciton structure. The sublevels are labelled as in Fig. 9.16. Optically active (passive) levels are shown as solid (dotted) lines. (Adapted from [41].)

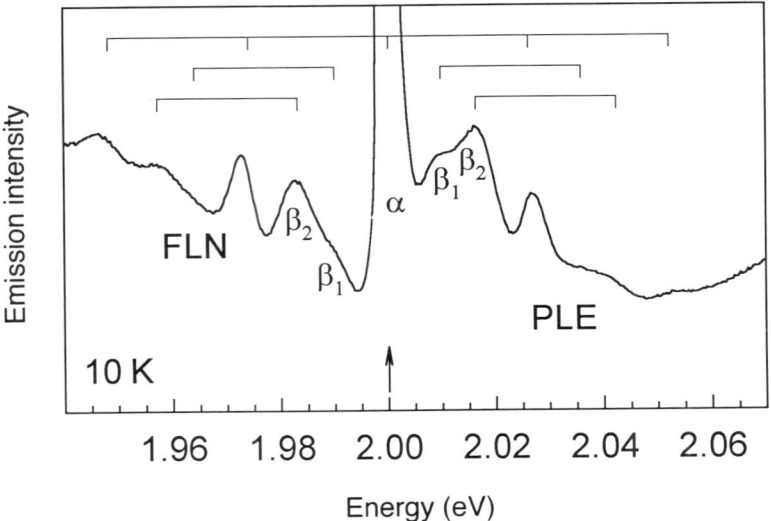

Figure 9.20 Normalized fluorescence line narrowing (FLN) and photoluminescence excitation (PLE) data for a ~44 Å effective radius sample. The FLN excitation and PLE emission energies are the same and are designated by the arrow. Although emission arises from a single emitting state and its LO–phonon replicas, three overlapping LO–phonon progressions are observed in FLN due to the three band edge absorption features (α, β_1 and β_2). Horizontal brackets connect the FLN and PLE and features with their LO–phonon replicas. (Adapted from [41].)

the sublevels converge to the optically forbidden 5-fold degenerate $N = 2$ state, and the optically allowed 3-fold degenerate $N = 1$ state. Since 1^L is correlated to the $N = 2$ state in the large exchange limit, we expect it to be only weakly allowed in small dots. 1^U and 0^U converge to the $N = 1$ state and therefore carry nearly all of the oscillator strength.

2.4.4   EVIDENCE FOR THE 'DARK EXCITON'

Another piece of evidence supporting the existence of the fine structure and its role in the long radiative lifetimes is found in measurements of the emission dynamics. In small dots at cryogenic temperatures the separation between the dark exciton and the 1^L sublevel is large relative to kT. Therefore, as thermalization processes are highly efficient, the photoexcited dot relaxes into the lowest sublevel (the dark exciton). The long (µs) emission is consistent with recombination from this weakly emitting state. Since a strong magnetic field couples the dark exciton to the optically allowed sublevels [40], the emission lifetime should decrease in the presence of a magnetic field. In Fig. 9.22(a) the magnetic field dependence of the emission decays for a ~12 Å radius sample are shown. With increasing magnetic field, the luminescence lifetime clearly decreases. Since the quantum yield remains essentially constant, this result is consistent with the presence of the dark exciton [40].

Another interesting effect of the magnetic field is its influence on the vibrational spectrum. In Fig. 9.22(b) the magnetic field dependence of the FLN spectrum is shown. A dramatic increase in the relative strength of the zero phonon line is observed with increasing field. While this effect is difficult to rationalize within the surface model, it is readily explained by the presence of the dark exciton [40]. In the absence of other relaxation mechanisms the dark exciton would have an infinite lifetime since the

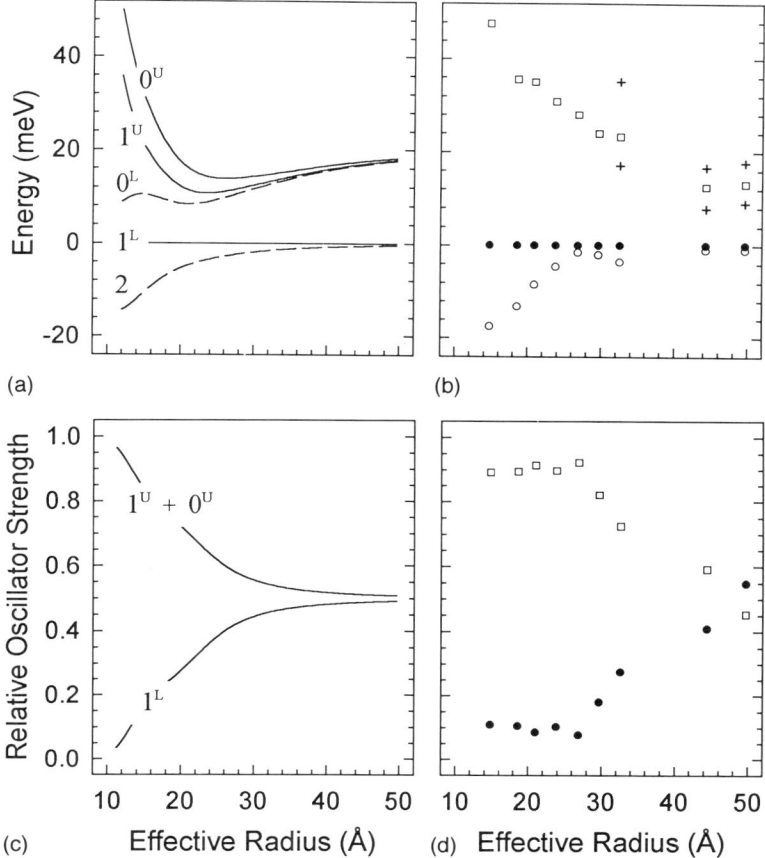

Figure 9.21 (a) Calculated band edge exciton ($1S_{3/2}1S_e$) structure vs effective radius as in Fig. 9.16. (b) Position of the absorbing (filled circles and squares) and emitting (open circles) features from Fig. 9.7 and transient differential absorption spectroscopy (TDA) results from [39]. In samples F–H both the positions of β_1 and β_2 (pulses) and their weighted average (squares) are shown. (c) Calculated relative oscillator strength of the optically allowed band edge sublevels vs effective radius. The combined strength of 1^U and 0^U is shown. (d) Observed relative oscillator strength of the band edge sublevels: 1^L (filled circles) and the combined strength of 1^U and 0^U (squares). (Adapted from [41].)

photon cannot carry an angular momentum of 2 within the electric dipole approximation. However, other less efficient mechanisms are present. For example, the dark exciton can recombine via a LO-phonon-assisted, momentum-conserving transition [74,75]. In this case the higher phonon replicas are enhanced relative to the zero phonon line at zero field. As the field strength is increased the dark exciton becomes partially allowed due to mixing with the optically allowed sublevels. Consequently, the strength of the zero phonon line increases.

2.5 *Summary of CdSe results*

The agreement between theory and experiment shown above indicates that the basic physics behind the absorption and emission behaviour of CdSe quantum dots is understood. The size dependence of the exciton structure and the origin of the

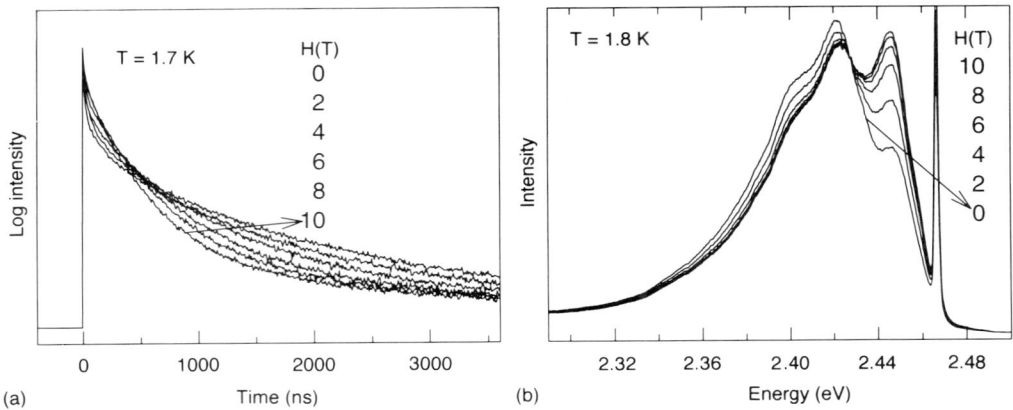

Figure 9.22 Magnetic field dependence of (a) emission decays recorded at the peak of the luminescence (2.346 eV) with an excitation energy of 2.736 eV, and (b) fluorescence line narrowing (FLN) spectra excited at the band edge (2.467 eV) for 12 Å radius dots and normalized to their one phonon line. A small amount of the excitation laser is included to mark the pump position. Experiments were carried out in the Faraday geometry (magnetic field parallel to the light propagation vector). (Adapted from [40].)

long-lived luminescence can now be quantitatively explained by theoretical models. As a prototypical system these explanations may have implications for other quantum confined semiconductor materials. Since the presence of band edge fine structure explains many optical properties previously attributed to surface trapping of the carriers, the role of the surface in the photophysics of these materials is less clear. Whether any discrepancies observed above between experiment and theory can be attributed to the influence of the surface remains an open question.

3 Si quantum dots

3.1 *Radiative rates*

As a result of the extensive work on high quality samples from the Murray *et al.* [5] CdSe synthesis, the spectroscopy of CdSe is well understood. The assignment of the exciton fine structure is proof that the emitting state results from three-dimensional confinement of bulk wurtzite CdSe. The excited state wavefunction is inside the nanocrystal, with a node on the surface in zero order, both before and after structural relaxation between photon absorption and luminescence. In this one case, our level of understanding now approaches that of other classes of large molecules, e.g. aromatic hydrocarbons.

Understanding of Si nanocrystals is far less advanced, and significant questions remain. Investigation of Si is important since this indirect gap material is of overwhelming significance in the electronics and telecommunications industry. Figure 9.23 shows a simplified band structure of Si, which has a diamond lattice. The valence bands are quite similar to those of CdSe (Fig. 9.4), with a maximum at the Brillouin zone centre ($k = 0$). However, the conduction band is different with a minimum at the * point (see Fig. 9.23) near the zone boundary. In Eqn 9.5 the dependence of u_{nk} on k across the Brillouin zone is strong. In the valence bands of both Si and CdSe, u_{nk} is p-like at the zone centre as shown in Fig. 9.11, and develops sp hybridization as k increases. In

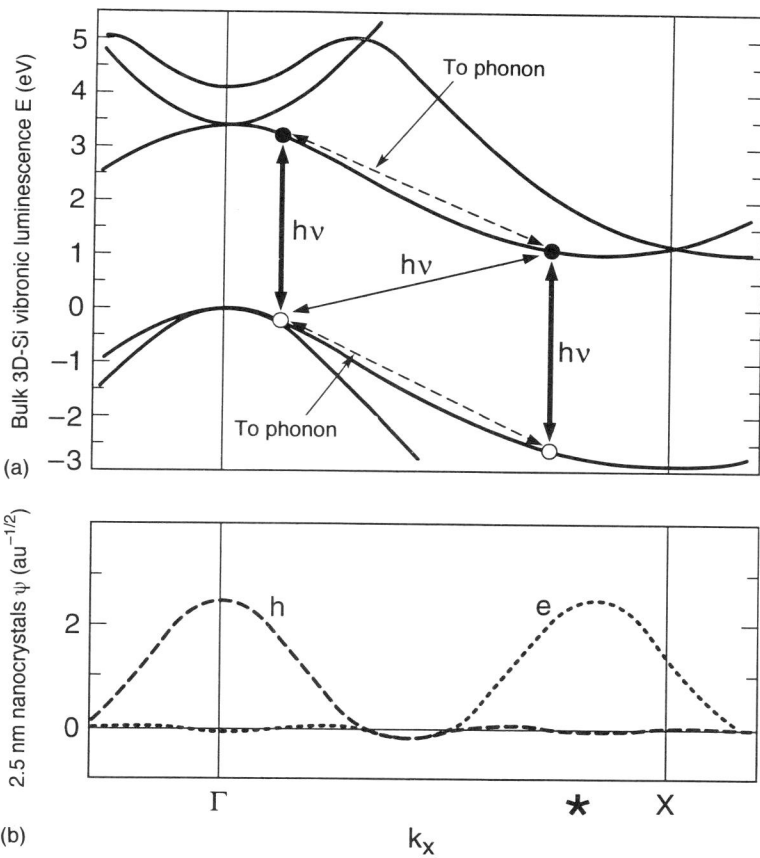

Figure 9.23 (a) Si band structure showing transverse optic (TO) vibronic luminescence mechanism. Broad arrows are allowed transitions, and the narrow arrow is the weak vibronically induced transition; (b) wavevector distributions for confined electron and hole in 2.5 nm nanocrystal, adapted from [76].

CdSe, the conduction band is s-like and nearly isotropic at the zone centre, and thus the direct gap transition is p- to s-like within the unit cell. Optical excitation is a type of charge transfer transition from a p orbital on Se to an empty s orbital on Cd. In covalent Si, the conduction minimum at * is an sp hybrid along a 001 direction. It is 6-fold spatially degenerate in view of the possible directions (including sign) of k. If one considers additionally the spin and valence band degeneracies previously described for CdSe, then the lowest indirect gap exciton is 48-fold degenerate. A complex exciton fine structure, dependent upon both shape and size, should be present.

Non-vertical electronic transitions of bulk Bloch states, such as that across the Si indirect gap in Fig. 9.23, are forbidden under dipole selection rules because of the phase difference between electron and hole wavefunctions from one unit cell to the next. This selection rule is a consequence of k being an exact quantum number in an infinite periodic lattice. In a finite nanocrystal, k is not exact and the transition becomes weakly dipole allowed. Figure 9.23(b) shows the Fourier wavevector superpositions necessary to localize an electron or hole inside a 2.5 nm diameter Si nanocrystal [76]. The tails of the two distributions overlap, and this is the source of the purely electronic transition dipole which should develop in small nanocrystals.

Just as in molecules, and in the forbidden $N = 2$ component of CdSe nanocrystals

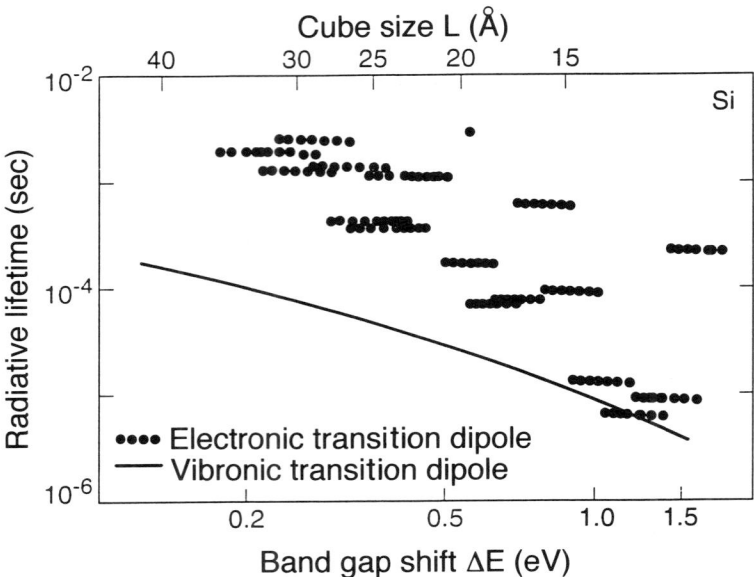

Figure 9.24 Purely electronic radiative lifetimes (dots) and vibronic radiative lifetimes (line) vs size on upper axis. For a given size, several lifetimes are possible depending upon shape, as shown in [76].

discussed above, a forbidden electronic transition can be weakly vibronically allowed through participation of a symmetry-lowering vibration. In bulk crystalline Si the transition is vibronically induced by a transverse optic (TO) vibration that compensates for the k mismatch, as shown in Fig. 9.23. In bulk semiconductors, electron–phonon coupling is generally weak because of the delocalized nature of the wavefunctions, and especially weak in non-polar Si. Simple theories of quantum confinement predict the coupling increases as d^{-3} in nanocrystals, as the wavefunctions become more compact [77]. Thus vibronically induced emission, which occurs in parallel to electronic emission, is also predicted to increase in nanocrystals, as shown in Fig. 9.24. Emission rates are a function of shape as well as size, because of the directional degeneracy of the confined electron. In this calculation, the transition is predicted to remain an indirect gap — that is, the vibronic rate dominates the purely electronic rate — down to the smallest sizes near 1.5 nm [76].

For comparison one might consider the hydrogenic exciton of bulk Si, bound by the weak, screened Coulomb attraction between electron and hole at liquid helium temperatures. In this species, about 3 nm in diameter, the transition is experimentally indirect with a singlet, vibronically induced luminescence rate Γ_r of 2×10^4 s^{-1} [78,79]. This is a slow rate — somewhat longer than the vibronically induced fluorescence rates of the lowest excited states of formaldehyde and benzene, and the $N = 2$ component of CdSe. An allowed transition, such as in the $N = 1$ CdSe component, would have a rate near 10^9 s^{-1}.

3.2   Porous silicon

Much of the interest in Si nanocrystals luminescence comes from the recent discovery of quantum confinement properties and high yield photoluminescence in porous Si thin films [80–83]. As schematically shown in Fig. 9.25, anodic electrochemical etching of wafer Si in alcoholic HF solution creates a yellowish surface film of micron thickness. Etching occurs as holes flow to the wafer surface and oxidize surface Si atoms in contact with HF. This film is composed of an open, irregular network of Si wires with undulating diameter, and/or Si nanocrystals partially fused to their neighbours. The etching process naturally creates electrically passivating H atom termination on newly uncovered surface Si atoms. Eighty per cent porous films have larger optical band gaps than the 1.1 eV band gap of bulk Si, and show efficient red photoluminescence near 2.0 eV. The typical particle size is in the 1–10 nm range, and these optical effects are attributed to quantum confinement. The high luminescence yield naturally suggests the idea that porous Si light-emitting diodes could be grown on Si microcircuits, and has attracted widespread industrial interest.

The electrochemical properties of these films are especially interesting. Lehmann and Goselle originally proposed that the etching process is self-terminating in the quantum confinement size regime [84]. Electrochemical etching stops when the hole confinement energy in a nanocrystal becomes larger than the bulk Si hole energy at a particular applied voltage. Etching alternately can occur for holes generated by direct light absorption in the film, and here the final size can be controlled by the wavelength of the optical excitation. Porous Si liquid junction cells show strong electroluminescence with appropriate electron injecting redox carriers in the electrolyte [85–89]. The colour is voltage tunable, apparently as the Fermi level moves across the nanocrystal electron

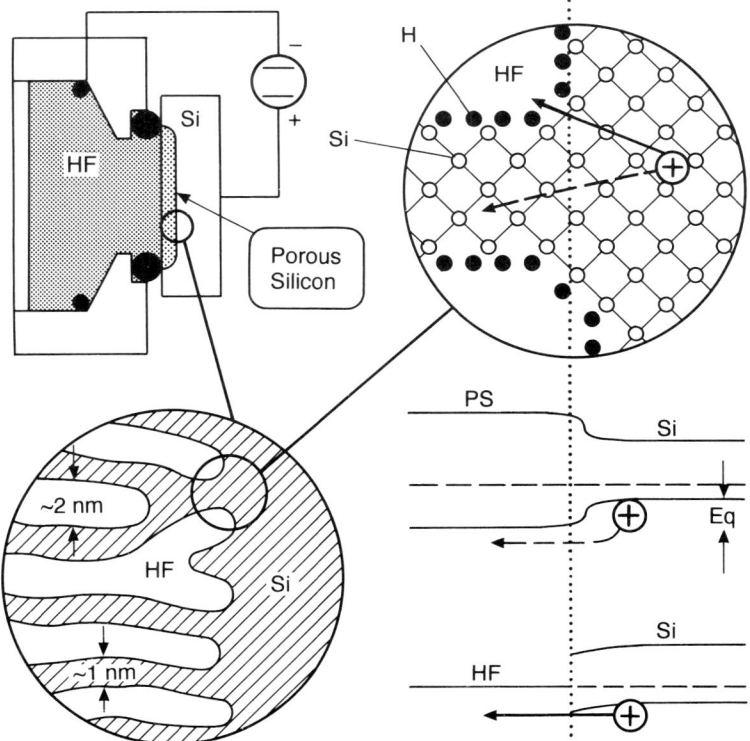

Figure 9.25 Structural and electronic schematic diagram of porous Si films on Si crystal, adapted from [93].

affinity size distribution in the porous-Si film. For all these reasons, as well as the possibility of new electronic devices made from nanometer scale Si, there is considerable interest in understanding the properties of individual Si nanocrystals.

3.3   SILICON NANOCRYSTAL GENERATION AND CHARACTERIZATION

At present the preparation of Si nanoparticles is primitive with respect to the Murray *et al.* CdSe organometallic synthesis. The situation is somewhat similar to the state of experimentation a decade ago in the CdSe work. Si nanocrystals with a passivating oxide surface layer about 0.8 nm thick have been made in a two-stage aerosol apparatus operating near 1000°C [90]. Gaseous disilane is thermally decomposed in about 1.5 atm of flowing He gas. Diamond lattice Si particles nucleate, grow, anneal, and become faceted at such temperatures in the gas phase. They are then briefly (30 ms) oxidized, cooled while flowing, and collected as a colloid in ethylene glycol. There is no size control except through the initial concentration of disilane, and a wide size distribution is made. High pressure liquid chromatography and size-selective precipitation methods are used to partially narrow the distribution for optical experiments [91]. For example, Fig. 9.26 shows broad luminescence spectra and size distributions of two fractions created by size-selective precipitation, from a Si nanocrystal colloid that initially showed photoluminescence in the 600–900 nm range. In bulk Si crystals, band gap luminescence occurs weakly at about 1120 nm. One fraction contains smaller single

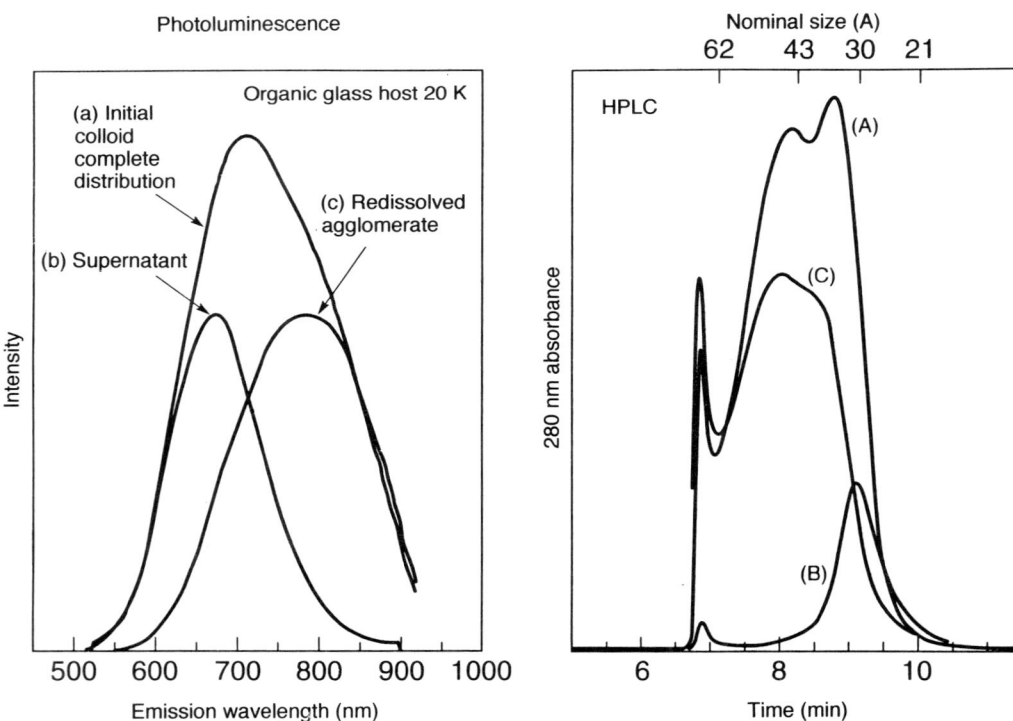

Figure 9.26 Left hand side: spectrally corrected Si nanocrystal luminescence spectra in organic glass (350 nm excitation wavelength, 20 K). Right hand side: corresponding HPLC chromatograms with approximate logarithmic size calibration. Relative intensities are arbitrary, as adapted from [91].

nanocrystals and emits at 650 nm in the red. The absolute quantum yield of this sample is 5.8% at room temperature, increasing to near 50% below 50 K. The other fraction contains larger nanocrystals, and aggregates, and emits near 800 nm.

Physical characterization is difficult. In larger nanocrystals direct TEM imaging is possible, but in smaller particles the contrast is poor. Si K shell X-ray near-edge absorption data in smaller nanocrystals that emit near 2.0 eV confirm the Si core oxide shell structure, and show a 1.2–1.5 nm Si core diameter [92]. The total diameter including oxide is about 2.5 nm. The identification and characterization of Si particles below 2.5 nm in diameter remains a very difficult problem. Such particles are especially difficult to detect in the presence of larger Si particles.

Ethylene glycol nanocrystal colloids are optically clear, and relatively crack-free organic glasses with embedded Si nanocrystals can be made by freezing between sapphire plates. In CdSe optical characterization, hole burning (TDA) and size-selective PLE and FLN methods probe optical properties within the size distribution. The band edge exciton oscillator strength is so small in Si that a direct hole burning experiment would be very difficult. However, photoexcitation experiments are feasible due to the high luminescence quantum yield. Figure 9.27 shows that the PLE excitation spectrum of 600 nm luminescence at ~15 K is a featureless, monotonically increasing continuum, characteristic of an indirect gap material with band gap the same as the monitored 600 nm luminescence wavelength [93,94]. This spectrum actually shows less structure than the equivalent, calculated spectrum for bulk diamond lattice Si, in that

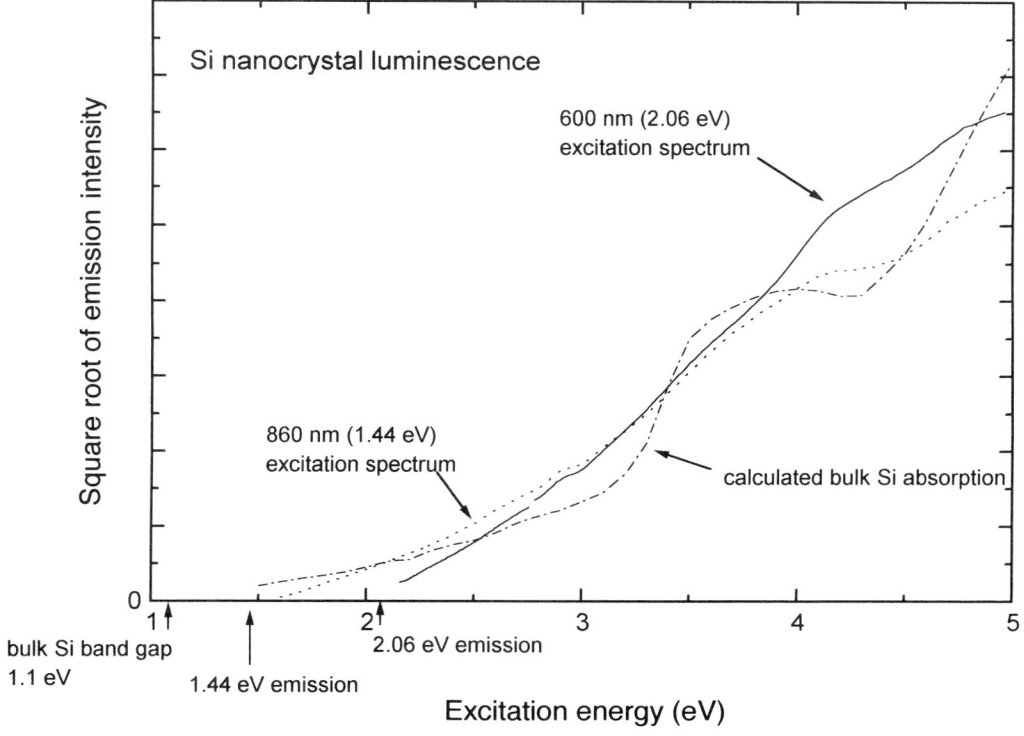

Figure 9.27 Square root of 600 nm and 860 nm luminescence intensity from Si nanocrystals at ~15 K as a function of excitation radiation energy. For comparison, the figure also shows the square root of the optical absorption cross-section of a small Si sphere calculated from the electric dipole term of Mie theory using bulk Si dielectric constant data, as adapted from [93].

the broad ultraviolet feature due to the 3.4 eV direct gap is almost washed out in nanocrystals. This result should be contrasted with the equivalent Figs 9.6(b) and 9.7(b), which shows discrete structure in CdSe. In contrast to Si, CdSe nanocrystals show more discrete structure than bulk CdSe.

The ultraviolet-excited, full photoluminescence spectrum is a broad featureless band, peaking near 700 nm in the particular sample in Fig. 9.28. As in CdSe, possible structure is obscured by sample size broadening. CW laser excitation at 710 nm excites luminescence from just the larger nanocrystals in the distribution, and this spectrum is narrower and exhibits partial TO phonon structure. However, this photoluminescence structure is different from the strong emission observed just below the laser excitation line for CdSe nanocrystals in Fig. 9.6. In Si the luminescence is weak near the laser excitation energy, and increases for Stokes shifts of 1 and 2 TO phonons.

This difference is a signature of the difference between a direct and indirect gap transition. In Si the luminescence data can be fit, as a function of excitation wavelength, by a near-Gaussian band gap size distribution centred at 650 nm, and a single

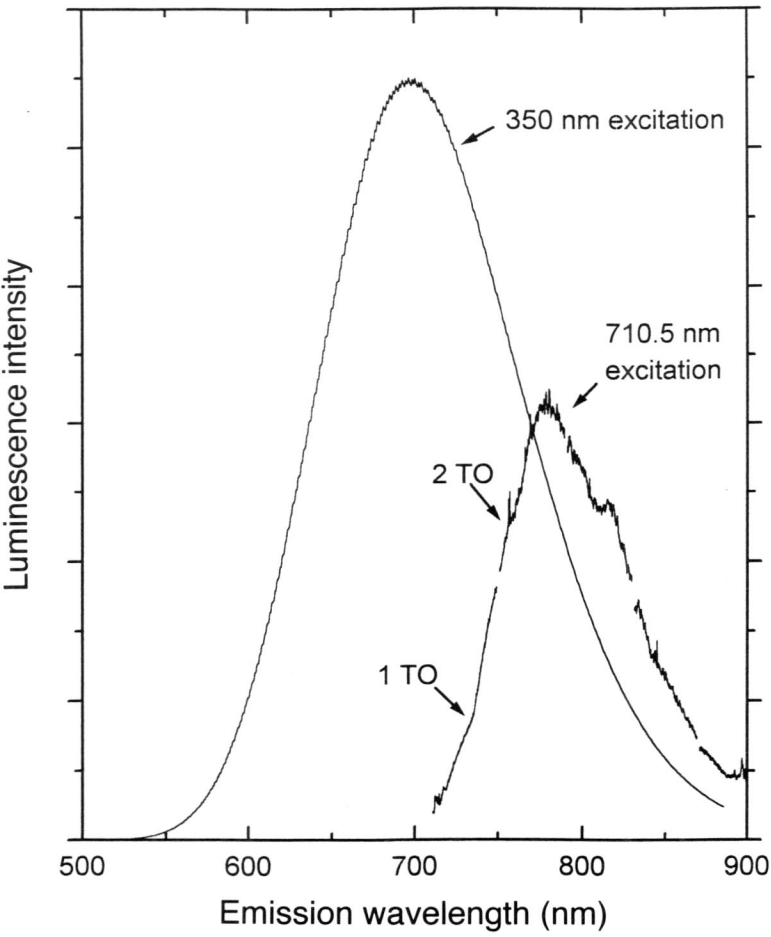

Figure 9.28 Comparison of the 350 nm excited, low resolution emission spectrum of Si nanocrystals at ~15 K, and the higher resolution (0.2 nm) emission spectrum excited at 710.5 nm. Weak thresholds are observed at one and two times the transverse optic (TO) phonon frequency in the 710.5 spectrum, as adapted from [93].

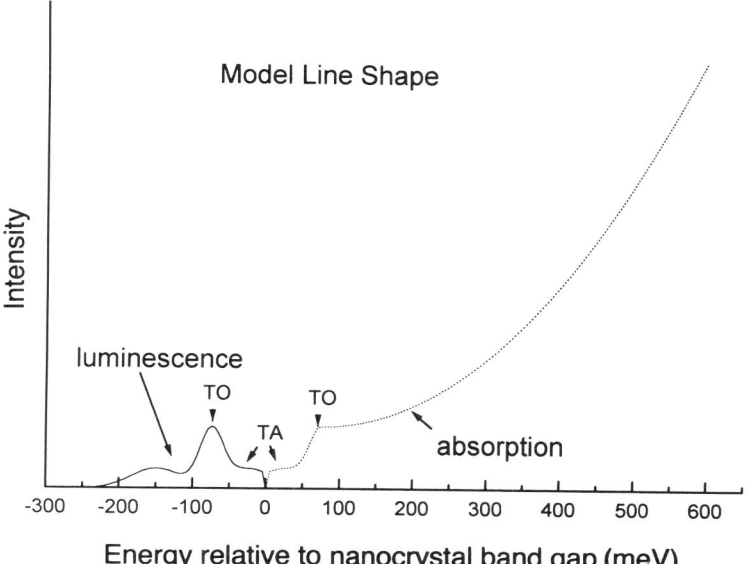

Figure 9.29 Si single nanocrystal model luminescence and absorption spectra, derived from fitting spectroscopic data in [93]. The zero of energy corresponds to the exciton energy of the single nanocrystal.

nanocrystal spectrum as shown in Fig. 9.29. The strongest features are TO vibronic features present in both absorption and emission. There is a weaker vibronic feature in the TA phonon position, and essentially no intensity on the band gap zero phonon line. In absorption there is also rising continuous absorption that corresponds to absorption into the higher electronic states of an indirect gap material. The spectra and assignments are very similar to those of p-Si [75,95,96], and confirm the indirect gap nature of Si nanocrystals emitting near 650 nm. This spectrum should be compared with the CdSe single dot spectrum in Fig. 9.19.

The lifetimes are very long; the 630 nm lifetime τ increases from $\sim 5 \times 10^{-5}$ s at 293 K to $\sim 2.5 \times 10^{-3}$ s at 20 K [91]. In CdSe the lifetime is on the order of 10^{-8} s at 293 K. In a molecule, or in a nanocrystal with just one electron–hole pair, the decay rate (inverse lifetime) is $\tau^{-1} = \Gamma_r + \Gamma_{nr}$. The first term is the radiative rate, and the second term is the competing non-radiative rate. The luminescence quantum yield is defined by $QY = \Gamma_r/(\Gamma_r + \Gamma_{nr})$. From measurement of both QY and τ, both Γ_r and Γ_{nr} are determined as a function of temperature. At 20 K, Γ_{nr} is negligible with respect to Γ_r, which itself is quite slow, $\sim 10^3$ s^{-1}. Γ_r increases by an order of magnitude as temperature increases to about 150 K, and then plateaus. Γ_{nr} increases by several orders of magnitude as temperature increases, so at room temperature the quantum yield has decreased to $\sim 5\%$. In both p-Si and nanocrystal Si, the non-radiative process is not understood.

3.4 *Photophysics analysis, and comparison with bulk crystalline Si*

The indirect gap nature of the spectra, the absence of a significant Stokes shift beyond the TO thresholds, the very long lifetimes, and the similarity of hydride and oxide terminated nanocrystal/porous Si data, all support assignment of the red luminescence to the volume-confined band gap exciton, just as in CdSe. There are some differences between predicted and observed properties. Present theory predicts a diameter larger by a factor of 2 than the 1.2–1.5 nm size deduced for nanocrystals emitting near 2.0 eV [97–99]. Also, the very long radiative rates described above appear to be even longer than predicted for volume-confined indirect excitons. Perhaps the wavefunction is polarized by a rough surface or non-spherical shape in some way we do not presently understand. Distributions of shape and surface roughness will obscure the data in laser excitation spectra. In fact, close analysis of the present Si nanocrystal spectra shows partially resolved sharp phonon transitions in the TO phonon regions. Improved syntheses and methods are needed in order to obtain high quality fine structure data as described above for CdSe.

At liquid helium temperatures, the Coulombic exciton in bulk Si has similarities to the quantum confined electron–hole pair in a nanocrystal. Both are spatially confined systems, yet both are indirect with radiative lifetimes at least 10^4 longer than that of an allowed transition. In both cases the competing multiphonon radiationless transition Γ_{nr} is negligible, and the quantum yield of luminescence, once the bound pair is formed, is near unity. Multiphonon internal conversion, of the type often seen in molecules, is slow as a consequence of the weak electron–phonon coupling found in a covalent, strongly bound, defect-free lattice.

At room temperature, however, bulk crystalline Si is fundamentally different from nanocrystal Si. In bulk Si, electron–hole pairs dissociate, and the ensemble of individual carriers interact. The carriers are mobile over macroscopic distances, and recombination can be efficiently catalysed by defects and impurities present in very low concentrations. In the perfect lattice, all recombination processes are slow, and carriers at very low densities live as long as 40 ms at 293 K [100]. At normal carrier densities, however, the recombination process is the three-body electronic non-radiative Auger process

$$e + h + (e \text{ or } h) \rightarrow (e \text{ or } h) \qquad (9.23)$$

with 1.1 eV of kinetic energy. The third carries away the band gap recombination energy, and then rethermalizes. This process shortens the effective lifetime by orders of magnitude.

The luminescence quantum yield increases in nanocrystal Si and porous Si at 293 K, not because Γ_r increases, but because Γ_{nr} decreases, with respect to bulk crystalline Si. In a material made of well-passivated Si nanocrystals, an electron–hole pair in one nanocrystal is electrically isolated from pairs in other nanocrystals, and the Auger process is significantly decreased. In addition, a rare impurity or lattice defect can only quench luminescence in one nanocrystal. In nanocrystal Si, an electron and hole are superimposed for long periods of time, not by their Coulomb attraction but by confinement. Because the pair remains spatially superimposed, and competing non-radiative processes are depressed, the QY is far larger in nanocrystal Si than in bulk Si. In almost any electronic material, photoluminescence increases as mobility decreases.

This work allows us to recognize nanocrystal size regimes in Fig. 9.30. Spectroscopic properties can be broken into molecular, quantum dot (e.g. nanocrystal), polariton, and bulk regions. These labels represent the evolution of molecular to unit cell structures, discrete electronic states to continuous bands, and weak dipole scattering into strong

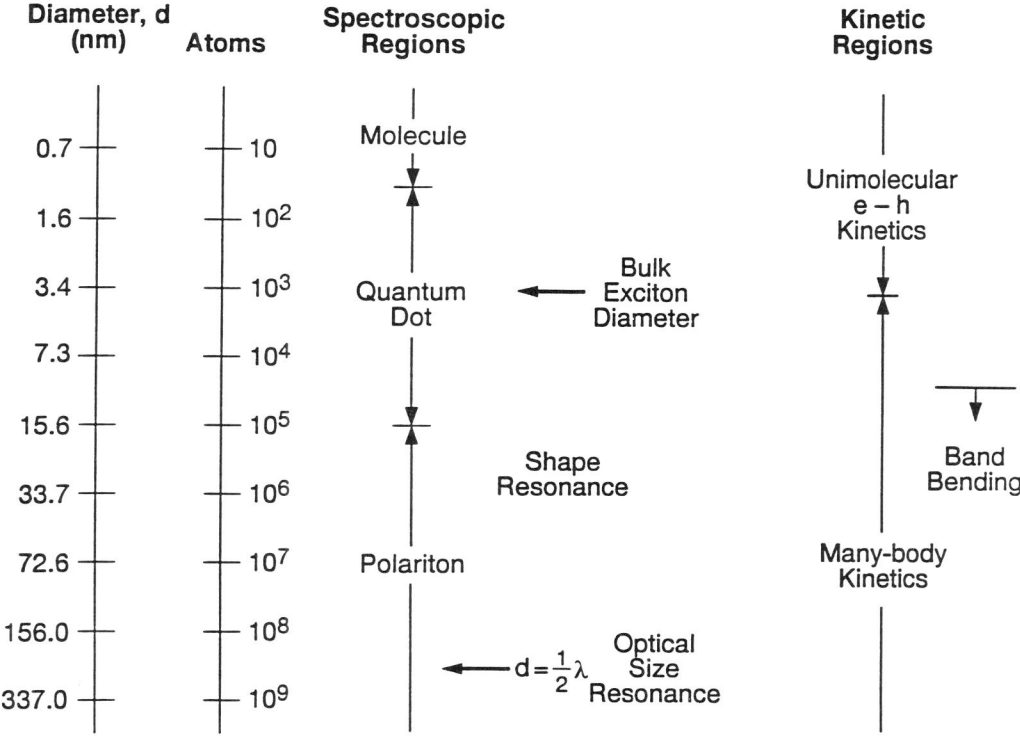

Figure 9.30 Schematic representation of size regimes used for describing spectroscopic and excited state kinetic properties of semiconductor nanocrystals, as adapted from [93].

polariton electromagnetic scattering. Bulk optical properties appear only when size is much larger than an optical wavelength.

In the nanocrystal or quantum dot regime, CdSe nanocrystals show discrete excited state optical absorption, while in Si the optical absorption is continuous. Two effects contribute to this difference: first, the spacing of quantized conduction band states is much larger in CdSe than in Si, because of the very small CdSe electron effective mass $\sim 0.11\ m_e$. Second, in CdSe only a few transitions, out of the many possible, appear strongly because of electric dipole selection rules. In Si, all possible transitions appear to be present at similar intensity due to vibronic interaction. In Si the discrete yet dense spectrum appears continuous, as individual transitions overlap with respect to their line widths, at least in the present Si nanocrystal samples. Comparison of rock salt (indirect) and wurtzite (direct) CdSe spectra leads quite unambiguously to a similar conclusion [101].

This work on Si, and previous work on AgBr nanocrystals [102–106], show that size regimes exist in kinetic properties as well. At typical excitation intensities, large crystallites contain several dissociated pairs and exhibit many interacting carrier recombination kinetics. Small nanocrystals can exhibit size exclusion of bulk lattice defects and impurities, and also do not show surface band bending if dopants are excluded due to small size. Unexcited small nanocrystals are intrinsic dielectric particles. At typical excitation intensities, unimolecular decay of a single, quantum confined electron–hole pair is observed.

3.5   *Electron affinities and transport in nanocrystal materials*

If a nanocrystal contains an extra electron and thus has a net charge, an electric field exists in the nanocrystal and interacts with the local environment. In p-Si of high porosity this external field dominates both single nanocrystal energetics [2,3,107–112] and transport phenomena [113], and provides a way of 'tuning' the transport behaviour via the polarity of the media in the pores. The electron affinity of a Si nanocrystal decreases by ΔA from the approximately 4.5 eV affinity of an electron on the conduction band edge in bulk Si:

$$\Delta A = KE(d) + \langle 1S|P(r)|1S \rangle. \tag{9.24}$$

$KE(d)$ is the kinetic energy, which can be treated by the effective mass approximation in the simplest approximation. The second term is the loss of dielectric polarization energy, averaged over a 1S wave function in the nanocrystal; it tends to be larger than $KE(d)$ if the local environment has a low dielectric constant. The second term can be approximated as [110]

$$\langle 1S|P(r)|1S \rangle \cong \left(\frac{e^2}{d} \right) \left(\frac{1}{\varepsilon_{out}} - \frac{1}{\varepsilon_{Si}} \right) + \delta\Sigma. \tag{9.25}$$

Here ε_{out} refers to the medium outside the nanocrystal, and

$$\delta\Sigma = \frac{0.94e^2}{\varepsilon_{Si}d} \left(\frac{\varepsilon_{out}\varepsilon_{Si}}{\varepsilon_{out} + \varepsilon_{Si}} \right). \tag{9.26}$$

Transport in p-Si· can be modelled as the hopping of electrons among touching nanocrystals of variable diameter. If an electron jumps from a 2 nm Si nanocrystal to a 4 nm nanocrystal in vacuum, the exothermicity is

$$\Delta G = \Delta A(4 \text{ nm}) - \Delta A(2 \text{ nm}) = -0.5 \text{ eV}, \tag{9.27}$$

which must be dissipated in some coupled 'vibrational' degree of freedom.

In Si an electron is weakly coupled to Si acoustic modes by the deformation potential. The internal acoustic reorganization energy λ is small — for 2 nm Si, λ is only ~12 meV. The external λ for water polarization (if present) outside the nanocrystal is much larger, ~400 meV [113]. In the Marcus theory of solvent influence on kinetics, the electron hopping rate at room temperature is

$$k\,(\text{s}^{-1}) \propto \exp(-G_{act}/kT), \tag{9.28}$$

where the activation energy is given by [114–116]

$$G_{act} = (\lambda + \Delta G)^2/4\lambda. \tag{9.29}$$

For fast, activationless transfer, the negative exothermicity ΔG must equal the total λ, which is positive. Thus for fast electron transfer from a 2 nm to a 4 nm nanocrystal, λ must be ~0.5 eV.

Figure 9.31 is a plot of log(k) vs the diameter of the electron accepting nanocrystal, for a fixed 2 nm donor size. For a 2 nm acceptor in vacuum, the transfer is resonant and

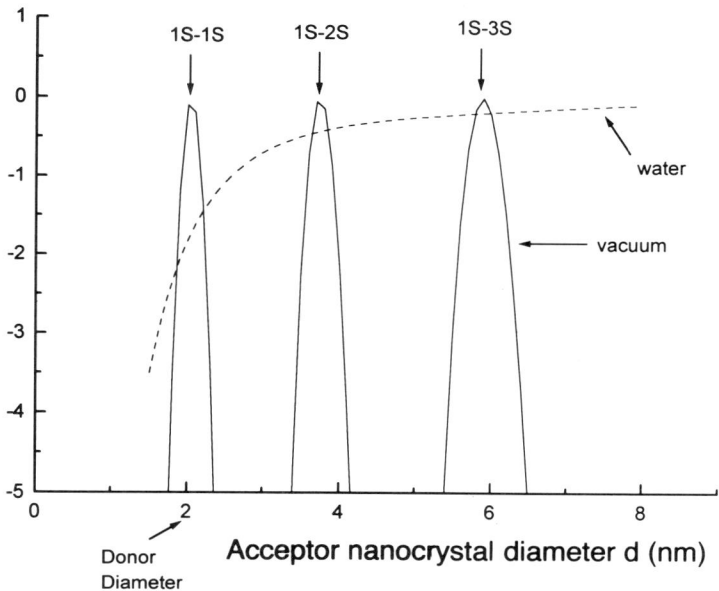

Figure 9.31 Log of the unimolecular electron hopping rate in water and in vacuum, for a 2 nm Si donor nanocrystal, as a function of acceptor nanocrystal size, as adapted from [113].

the rate is fast. It is also fast for specific larger sizes where resonant transfer to excited $2S$ and $3S$ states is possible. For other sizes transfer is very slow due to a huge activation energy as λ is small. In water where λ is large, the resonant nature of transfer is lost. In water, transfer is fast for all cases except resonant transfer between nanocrystals of equal size. In this case, the polarization of the water around the initial nanocrystal tends to 'self-trap', or stabilize, the carrier. Interstitial water has a huge enabling effect on electron hopping, if the material is highly porous and the size distribution is rather wide. Water also has a major effect on electron injection into a nanocrystal already containing a hole, creating an exciton which then may decay radiatively. In this process two charged nanocrystals convert into two neutral nanocrystals, and a large polarization energy is dissipated during injection.

General speaking, dry p-Si behaves electrically like a resonant tunnelling device. However, wet p-Si shows fast, non-resonant electron transfer similar to that commonly seen in biological and molecular processes in polar environments. This Marcus solvation effect on kinetics demonstrates yet another molecule-like property of nanocrystals.

4 Concluding remarks

Most nanocrystal research in the past has focused on understanding the chemistry and physics of individual nanocrystals. We now can outline, in broad form, the evolution from molecular properties at small size to solid-state properties at large size. In the coming decade, synthetic methods should improve to yield high quality nanocrystals of all major semiconductors, in gram amounts with controlled and variable surface chemistry, as is presently the situation with CdSe. Nanocrystals are new classes of large molecules in the context of chemistry. In the context of molecular electronics, nanocrystals are building blocks. Our challenge is to now invent new nanometre devices with unique properties, that go beyond the 'canonical' devices — diodes, transistors, and wires — that industry already knows how to miniaturize, organize, and assemble very efficiently. New functionality can alternately come from new material properties. New materials made with nanocrystals are now appearing. Inexpensive, large area solar cells have been made from porous TiO_2 with a liquid junction [117]. The combination of nanocrystals with conductive organic polymers promises to significantly increase design possibilities in photovoltaic and electroluminescent diodes [118–120]. Finally, the crystallization of nanocrystals into opal-like structures [26] promises to create a new class of semiconducting materials with properties tunable via interstitial doping, surface chemistry, and nanocrystal size. This area of materials science should have a rich future.

Acknowledgements. The work presented in this review is the result of a number of collaborations. D.J.N. and M.G.B. especially thank our colleagues Al. L. Efros, C.B. Murray, and M. Nirmal, who have contributed significantly to the results described. L.E.B. gratefully acknowledges a continuous, stimulating collaboration with many excellent colleagues at AT&T Bell Laboratories over the past decade. Much of the nanocrystal research at Murray Hill was performed by former post-doctoral fellows: J.J. Macklin, K.A. Littau, M.G. Bawendi, and A.P. Alivisatos. D.J.N. benefited from fellowships from NSF and Arthur D. Little. M.G.B. thanks the Lucille and David Packard Foundation and the Alfred P. Sloan Foundation for fellowships. This work was funded in part by the NSF-MRSEC program (DMR-94-00034), by the NSF (DMR-91-57491), and by AT&T Bell Laboratories.

5   References

1   Efros AlL, Efros AL. *Sov Phys Semicond* 1982; **16**: 772.
2   Brus LE. *J Chem Phys* 1983; **79**: 5566.
3   Brus LE. *J Chem Phys* 1984; **80**: 4403.
4   Pankove JI. *Optical Processes in Semiconductors.* New York: Dover, 1971: Ch. 3.
5   Murray CB, Norris DJ, Bawendi MG. *J Am Chem Soc* 1993; **115**: 8706.
6   Flügge S. *Practical Quantum Mechanics*, Vol. 1. Berlin: Springer, 1971: 155.
7   Kittel C. *Introduction to Solid State Physics.* New York: Wiley, 1986: Ch. 8.
8   Bastard G. *Wave Mechanics Applied to Semiconductor Heterostructures.* New York: Wiley, 1988.
9   Altarelli M. In: Stella A, ed. *Semiconductor Superlattices and Interfaces.* Amsterdam: North Holland, 1993: 217.
10   Bányai L, Koch SW. *Semiconductor Quantum Dots.* Singapore: World Scientific, 1993.
11   Xia JB. *Phys Rev B* 1989; **40**: 8500.
12   Hellwege KH, ed. *Landolt-Bornstein Numerical Data and Functional Relationships in Science and Technology, New Series, Group III*, Vol. 17b. Berlin: Springer-Verlag, 1982.
13   Aven M, Prener JS. *Physics and Chemistry of II–VI Compounds.* Amsterdam: North Holland, 1967: 41.
14   Kittel C. *Quantum Theory of Solids.* New York: Wiley, 1987: Chs 9, 14.
15   Kane EO. *J Chem Phys Solids* 1957; **1**: 249.
16   Kane EO. In: Zawadzki W, ed. *Narrow Band Semiconductors. Physics and Applications, Lecture Notes in Physics*, Vol. 133. Berlin: Springer Verlag, 1980.
17   Luttinger JM, Kohn W. *Phys Rev* 1955; **97**: 869.
18   Luttinger JM, *Phys Rev* 1956; **102**: 1030.

19 Bir GL, Pikus GE. *Symmetry and Strain-Induced Effects in Semiconductors.* New York: Wiley, 1974.

20 Lipari NO, Baldereschi A. *Phys Rev Lett* 1970; **42**: 1660.

21 Baldereschi A, Lipari NO. *Phys Rev B* 1973; **8**: 2697.

22 Ge'lmont BL, D'yakonov MI. *Sov Phys Semicond* 1972; **5**: 1905.

23 Bowen Katari JE, Colvin VL, Alivisatos AP. *J Phys Chem* 1994; **98**: 4109.

24 Hines MA, Guyot-Sionnest P. *J Phys Chem* 1996; **100**: 468.

25 Kagan CR, Murray CB, Nirmal M, Bawendi MG. *Phys Rev Lett,* 1996; **76**: 1517.

26 Murray CB, Kagan CR, Bawendi MG. *Science* 1995; **270**: 1335.

27 Ekimov AI, Efros AlL, Onushchenko AA. *Solid State Commun* 1985; **56**: 921.

28 Ekimov AI, Onushchenko AA. *JETP Lett* 1984; **40**: 1136.

29 Rossetti R, Hull R, Gibson JM, Brus LE. *J Chem Phys* 1985; **82**: 552.

30 Ekimov AI, Onushchenko AA, Efros AlL. *JETP Lett* 1986; **43**: 376.

31 Chestnoy N, Hull R, Brus LE. *J Chem Phys* 1986; **85**: 2237.

32 Alivisatos AP, Harris AL, Levinos NJ, Steigerwald ML, Brus LE. *J Chem Phys* 1988; **89**: 4001.

33 Ekimov AI, Efros AlL, Ivanov MG, Onushchenko AA, Shumilov SK. *Solid State Commun* 1989; **69**: 565.

34 Wang Y, Herron N. *Phys Rev B* 1990; **42**: 7253.

35 Bawendi MG, Wilson WL, Rothberg L *et al. Phys Rev Lett* 1990, **65**: 1623.

36 Peyghambarian N, Fluegel B, Hulin D *et al. IEEE J. Quantum Electron* 1989; **25**: 2516.

37 Esch V, Fluegel B, Khitrova G *et al. Phys Rev B* 1990; **42**: 7450.

38 Ekimov AI, Hache F, Schanne-Klein MC *et al. J Opt Soc Am B* 1993; **10**: 100.

39 Norris DJ, Bawendi MG. *J Chem Phys* 1995; **103**: 5260.

40 Nirmal M, Norris DJ, Kuno M, Bawendi MG, Efros AlL, Rosen M. *Phys Rev Lett* 1995; **75**: 3728.

41 Norris DJ, Efros AlL, Rosen M, Bawendi MG. *Phys Rev B,* 1996; **53**: 16347.

42 Hilinski EF, Lucas PA, Wang Y. *J Chem Phys* 1988; **89**: 3435.

43 Roussignol P, Ricard D, Flytzanis C, Neuroth N. *Phys Rev Lett* 1989; **62**: 312.

44 Peyghambarian N, Fluegel B, Hulin D *et al. IEEE J Quantum Electronics* 1989; **QE-25**: 2516.

45 Park SH, Morgan RA, Hu YZ, Lindberg M, Koch SW, Peyghambarian N. *J Opt Soc Am B* 1990; **7**: 2097.

46 Esch V, Fluegel B, Khitrova G *et al. Phys Rev B* 1990; **42**: 7450.

47 Kang KI, Kepner AD, Gaponenko SV, Koch SW, Hu YZ, Peyghambarian N. *Phys Rev B* 1993; **48**: 15449.

48 Kang K, Kepner AD, Hu YZ *et al. Appl Phys Lett* 1994; **64**: 1487.

49 Alivisatos AP, Harris AL, Levinos NJ, Steigerwald ML, Brus LE. *J Chem Phys* 1988; **89**: 4001.

50 Woggon U, Gaponenko S, Langbein W, Uhrig A, Klingshirn C. *Phys Rev B* 1993; **47**: 3684.

51 Gaponenko SV, Woggon U, Saleh M *et al. J Opt Soc Am B* 1993; **10**: 1947.

52 Norris DJ, Nirmal M, Murray CB, Sacra A, Bawendi MG. *Z Phys D* 1993; **26**: 355.

53 Norris DJ, Sacra A, Murray CB, Bawendi MG. *Phys Rev Lett* 1994; **72**: 2612.

54 Hoheisel W, Colvin VL, Johnson CS, Alivisatos AP. *J Chem Phys* 1994; **101**: 8455.

55 de Oliveira CRM, de Paula AM, Plentz Filho FO, Medeiros Neto JA *et al. Appl Phys Lett* 1995; **66**: 439.

56 Rodrigues PAM, Tamulaitis G, Yu PY, Risbud SH. *Solid State Commun* 1995; **94**: 583.

57 Norris DJ, Bawendi MG. *Phys Rev B,* 1996; **53**: 16338.

58 Grigoryan GB, Kazaryan EM, Efros AlL, Yazeva TV. *Sov Phys Solid State* 1990; **32**: 1031.

59 Vahala KJ, Sercel PC. *Phys Rev Lett* 1990; **65**: 239.

60 Sercel PC, Vahala KJ. *Phys Rev B* 1990; **42**: 3690.

61 Koch SW, Hu YZ, Fluegel B, Peyghambarian N. *J Crystal Growth* 1992; **117**: 592.

62 Efros AlL. *Phys Rev B* 1992; **46**: 7448.

63 Efros AlL, Rodina AV. *Phys Rev B* 1993; **47**: 10005.

64 Takagahara T. *Phys Rev B* 1993; **47**: 4569.

65 Nomura S, Segawa Y, Kobayashi T. *Phys Rev B* 1994; **49**: 13571.

66 Chamarro M, Gourdon C, Lavallard P, Ekimov AI. *Jpn J Appl Phys* 1995; **34**: (Suppl 34-1): 12.

67 Kochereshko VP, Mikhailov GV, Ural'tsev IN. *Sov Phys Solid State* 1983; **25**: 439.

68 Henry CH, Nassau K. *Phys Rev B* 1970; **1**: 1628.

69 Nirmal M, Murray CB, Bawendi MG. *Phys Rev B* 1994; **50**: 2293.

70 Bawendi MG, Carroll PJ, Wilson WL, Brus LE. *J Chem Phys* 1992; **96**: 946.

71 O'Neil M, Marohn J, McLendon G. *J Phys Chem* 1990; **94**: 4356.

72 Eychmüller A, Hässelbarth A, Katsikas L, Weller H. *Ber Bunsenges Phys Chem* 1991; **95**: 79.

73 Hässelbarth A, Eychmüller A, Weller H. *Chem Phys Lett* 1993; **203**: 271.

74 Calcott PDJ, Nash KJ, Canham LT, Kane MJ, Brumhead D. *J Lumin* 1993; **57**: 257.

75 Calcott PDJ, Nash KJ, Canham LT, Kane MJ, Brumhead D. *J Phys Condens Matter* 1993; **5**: L91.

76 Hybertsen MS. *Phys Rev Lett* 1994; **72**: 1514.

77 Schmitt-Rink S, Miller DAB, Chemla DS. *Phys Rev B* 1987; **35**: 8113.

78 Haynes JR, Lax M, Flood WF. *Proc Int Conf Semicond Phys Prague* 1961; **423.**

79 Cuthbert JD. *Phys Rev B* 1970; **1**: 1552.

80 Canham LT. *Appl Phys Lett* 1990; **57**: 1046.

81 Koyama H, Araki M, Yamamoto Y, Koshida N. *Jpn J Appl Phys* 1991; **30**: 3606.

82 Lehmann V, Gösele U. *Appl Phys Lett* 1991; **58**: 856.

83 Petrova-Koch V, Muschik T, Kux A, Meyer BK, Koch F, Lehmann V. *J Appl Phys* 1992; **61**: 943.

84 Lehmann V, Gösele U. *Adv Mater* 1992; **4**: 114.

85 Halimaoui A, Oules C, Bomchil G et al. *Appl Phys Lett* 1991; **59**: 304.

86 Bressers PMMC, Knapen JWJ, Meulenkamp EA, Kelly JJ. *Appl Phys Lett* 1992; **61**: 108.

87 Bsiesy A, Muller F, Ligeon M et al. *Phys Rev Lett* 1993; **71**: 637.

88 Ligeon M, Muller F, Herino R et al. *J Appl Phys* 1993; **74**: 1265.

89 Kooij ES, Despo RW, Kelly JJ. *Appl Phys Lett* 1995; **66**: 2552.

90 Littau KA, Szajowski PF, Muller AJ, Kortan RF, Brus LE. *J Phys Chem* 1993; **97**: 1224.

91 Wilson WL, Szajowski PF, Brus LE. *Science* 1993; **262**: 1242.

92 Schuppler S, Friedman SL, Marcus MA et al. *Phys Rev Lett* 1994; **72**: 2648.

93 Brus L, Szajowski P, Wilson W, Harris T, Schuppler S, Citrin P. *J Am Chem Soc* 1995; **117**: 2915.

94 Schuppler S, Friedman SL, Marcus MA et al. *Phys Rev B* 1995; **52**: 4910.

95 Suemoto T, Tanaka K, Nakajima A, Itakura T. *Phys Rev Lett* 1993; **70**: 3659.

96 Suemoto T, Tanaka K, Nakajima A. *J Phys Soc Jpn* 1994; **63** (Suppl B): 1900.

97 Takagahara T, Takeda K. *Phys Rev B* 1992; **46**: 15578.

98 Proot JP, Delerue C, Allan G. *Appl Phys Lett* 1992; **61**: 1948.

99 Delley B, Steigmeier EF. *Phys Rev B* 1993; **47**: 1397.

100 Yablonovitch E, Allara DL, Chang CC, Gmitter T, Bright TB. *Phys Rev Lett* 1986; **57**: 249.

101 Tolbert SH, Herhold AB, Johnson CS, Alivisatos AP. *Phys Rev Lett* 1994; **73**: 3266.

102 Johansson KP, McLendon GP, Marchetti AP. *Chem Phys Lett* 199; **179**: 32.

103 Johansson KP, Marchetti AP, McLendon GP. *J Phys Chem* 1992; **96**: 2873.

104 Marchetti AP, Johansson KP, McLendon GP. *Phys Rev B* 1993; **47**: 4268.

105 Chen W, McLendon G, Marchetti A, Rehm JM, Freedhoff M, Myers C. *J Am Chem Soc* 1994; **116**: 1585.

106 Kanzaki H, Tadakuma Y. *Solid State Commun* 1991; **80**: 33.

107 Babic D, Tsu R, Greene RF. *Phys Rev B* 1992; **45**: 14150.

108 Tsu R, Babic D. *Appl Phys Lett* 1993; **64**: 1806.

109 Delerue C, Allan G, Lannoo M. *Phys Rev B* 1993; **48**: 11024.

110 Lannoo M, Delerue C, Allan G. *Phys Rev Lett* 1995; **74**: 3415.

111 Chazalviel J, Ozanam F, Dubin V. *J Phys I France* 1994; **4**: 1325.
112 Martin E, Delerue C, Allan G, Lannoo M. *Phys Rev B* 1994; **50**: 18258.
113 Brus LE. *Phys Rev B* 1996; **53**: 4649.
114 Marcus R. *J Chem Phys* 1956; **24**: 966.
115 Marcus R. *Faraday Disc Chem Soc* 1960; **29**: 21.
116 Miller RJD *et al. Surface Electron Transfer Processes.* New York: VCH Publishers, 1995: Chs 1, 4.
117 O'Regan B, Gratzel M. *Nature* 1991; **353**: 737.
118 Colvin V, Schlamp M, Alivisatos AP. *Nature* 1994; **370**: 6488.
119 Dabbousi B, Bawendi M, Onitsuka O, Rubner M. *Appl Phys Lett* 1995; **66**: 1316.
120 Wang Y, Herron N. *Chem Phys Lett* 1992; **200**: 71.

10 Molecular and Supramolecular Nanostructures and Nanomachines

S.J. LANGFORD, F.M. RAYMO and J.F. STODDART

School of Chemistry, University of Birmingham, Edgbaston, Birmingham B15 2TT, UK

1 Learning from nature

Self-assembly [1–4], self-organization [5–7], and self-replication [8–11] are synthetic paradigms widely employed by nature to produce kinetically and thermodynamically stable molecular structures and supramolecular arrays on both the cellular and subcellular levels. These structures and superstructures are assembled from relatively simple modular subunits which are encoded with the information necessary to promote their own assembly into stable three-dimensional architectures. Precise recognition features within each of the subunits drive the assembly process by means of cooperative non-covalent bonding interactions, thus allowing alterations or corrections to be performed without the potential problems of irreversible bond formation (e.g. breaking a covalent carbon–carbon bond) whilst being synthetically efficient under extremely mild conditions. Functioning molecular and supramolecular machines able to perform specific functions are constructed in living organisms by means of such genealogically directed synthetic processes.

Chemists are now beginning to realize (i) the potential of the intelligent synthetic processes witnessed in biology and (ii) the role played by non-covalent bonding interactions in the information storage and chemical reactivity of biological systems. Numerous examples of artificial self-assembling chemical systems ranging from the almost forgotten and earliest examples of crystallization, through metal chelation and template-directed synthesis, to model systems for self-replication have been developed and investigated. Figure 10.1 lists some of these chemical systems which have been self-assembled through either covalent or non-covalent bonding interactions against some examples of biological self-assembly. One of the major aims of this area of research is to generate wholly synthetic and complex nanometer-scaled molecular and

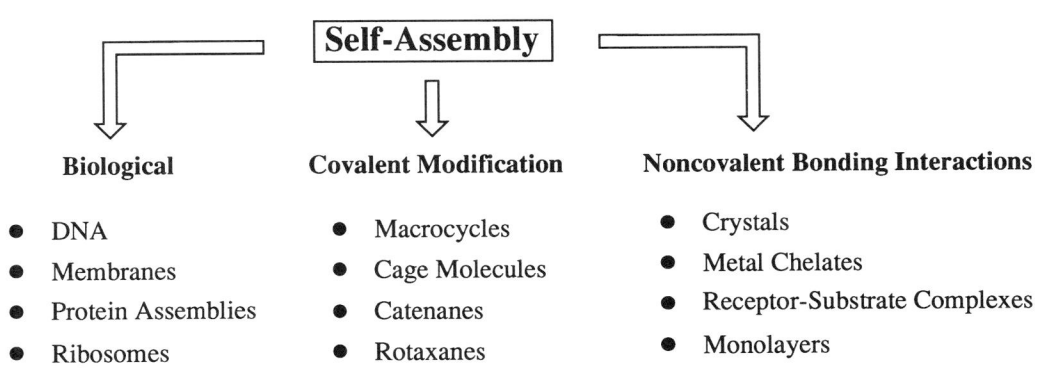

Figure 10.1 Examples of naturally occurring and artificial chemical systems self-assembled through covalent and non-covalent bonding interactions.

supramolecular architectures able to process and store information, thus mimicking the many naturally occurring systems.

2 Nanochemistry

The extension of the concept of a 'device' to the molecular level is of interest not only as a curiosity-driven line of research, but also for the growth of nanoscience and the development of nanotechnology [12–15]. Progress in the field of constructing molecular devices may have important implications for biomimetic engineering, molecular-based logic, computer technologies, sensory techniques, and the development of artificial transmembrane channels and molecular carrier systems. Two approaches — namely 'top-down' and 'bottom-up' — have been devised in order to generate functioning systems on the nanometre-scale (1 nm = 10 Å). The so-called 'top-down' approach has been employed for many years as an artificial means of conveniently miniaturizing devices and electronic components. This methodology involves dismantling elemental or molecular arrays down to the nanometre scale and it is rapidly approaching its limits of resolution. Further miniaturization under this approach may lead also to problems associated (i) with electronic tunnelling and heat dissipation and, ultimately, (ii) with the difficulty and cost of fabrication. Hence, there is a need for a chemical science that will enable the construction of mechanical, electrical, or photochemical devices from simple atomic and molecular components — i.e. a 'bottom-up' approach.

Supramolecular chemists are looking to adopt the so-called 'bottom-up' approach to generate functioning structures and superstructures from simple molecular subunits by exploiting the synthetic paradigms of self-assembly and self-organization. Thus, bridging the gap (Fig. 10.2) between the traditional chemistry scale (Ångstrom scale) and the

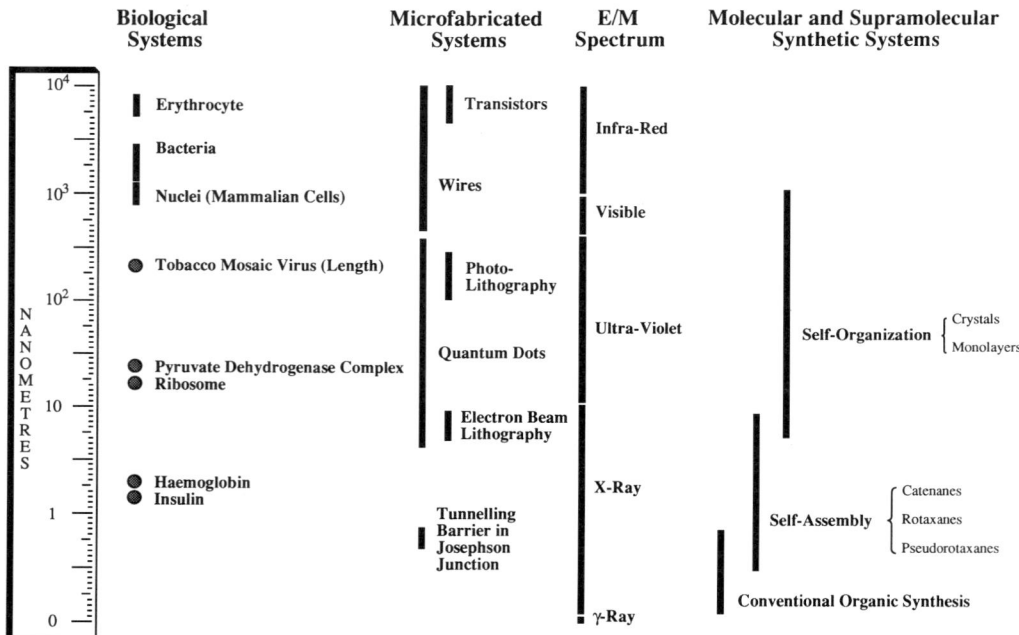

Figure 10.2 Comparison of the sizes associated with some biological, microfabricated, and artificial molecular and supramolecular systems.

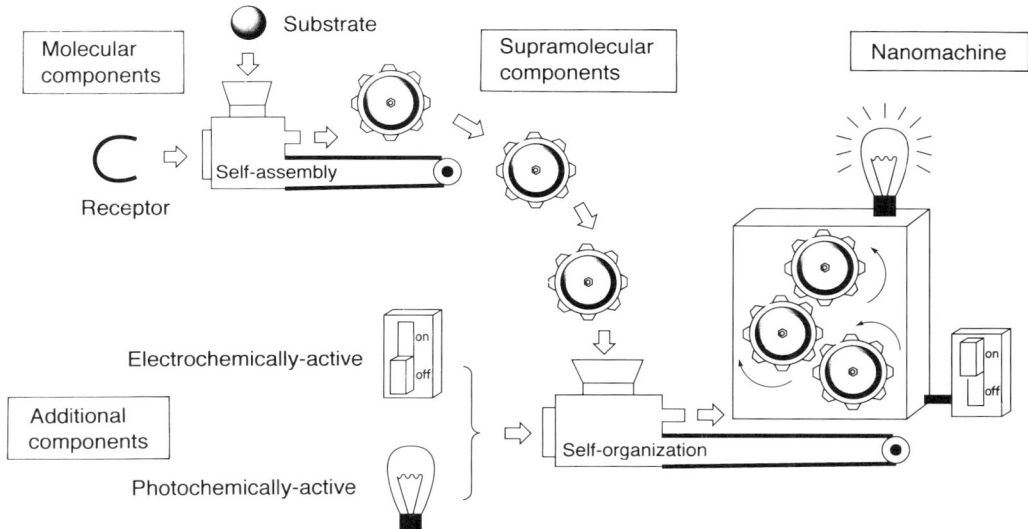

Figure 10.3 Construction of nanomachines from molecular and supramolecular components as a result of self-assembly and self-organization processes.

limits of the electronics industry scale (nanometre-scale) will be achieved by using favourable non-covalent bonding interactions.

While there are still many problems associated with the technology needed to make this type of approach plausible for commercialization, the most fundamental of challenges, i.e. synthesis and fabrication, are now being addressed. The advantages associated with the bottom-up method relate to the device mechanism being located within a well-defined molecular assembly which (i) is cheap to synthesize, (ii) can be built up if necessary into a larger supramolecular array on the nanoscale, and (iii) is perhaps tunable through functionalization by chemical manipulation. Furthermore, the reversibility of non-covalent interactions allows error checking and recovery. By contrast, molecular systems such as those found in today's nanofabrication industry are constructed by inherently non-reversible processes, and so the effects of a defect cannot be altered once they have occurred.

The basis of a general supramolecular approach to synthesizing nanometre-scaled devices and machines can be summarized as illustrated schematically in Fig. 10.3. Under the guide of molecular recognition, a preorganized receptor combines with a suitable substrate to form a supramolecular assembly whose properties are quite different from the sum of the two components. Additional components, that are perhaps electrochemically and/or photochemically active, can then be added (if the orginal component does not already possess such units) through the processes of self-assembly or self-organization. It is envisaged that the overall process brings together all the components in a definite way to yield nanomachines that may hold their device-like characteristics through either their static or dynamic behaviour.

3 Self-assembling supramolecular complexes

In the search for a suitable receptor for the bipyridinium-based herbicide paraquat (PQT), we synthesized [16] the hydroquinone-based macrocyclic polyether, bis-*p*-

phenylene-34-crown-10 (BPP34C10). Upon mixing BPP34C10 with an equimolar amount of PQT·2PF$_6$ in acetone at room temperature (Fig. 10.4), a deep orange colour develops on account of charge transfer interactions between the π-electron-rich hydroquinone units lining the cavity of the macrocyclic polyether and the π-electron-deficient bipyridinium unit, indicating that the supramolecular complex self-assembles spontaneously [17]. Characterization of the resulting complex was achieved in solution by nuclear magnetic resonance (NMR) spectroscopy, fast-atom bombardment (FAB) mass spectrometry, and absorption spectroscopy in the visible region, as well as in the solid state by X-ray crystallography. The X-ray crystallographic analysis revealed that the PQT^{2+} unit is inserted centrosymmetrically inside the cavity of BPP34C10. This pseudorotaxane-like geometry is achieved as a result of a range of non-covalent bonding interactions including (i) [C—H···O] hydrogen bonding interactions between the protons in the α-positions with respect to the nitrogen atoms in the paraquat dication and the polyether oxygen atoms and (ii) π–π stacking interactions between the hydroquinone units of BPP34C10 and the bipyridinium unit of PQT^{2+}.

The generality of this inclusion phenomenon observed [18–20] for a wide range of macrocyclic polyether hosts differing in either the length of the polyether chains and/or the constitution of the π-electron-rich aromatic units and bipyridinium-based guests led us to ask whether the same types of interactions would still hold on reversing the roles of the recognition sites possessed by host and guest. Thus, a tetracationic cyclophane, cyclobis(paraquat-*p*-phenylene)tetrakis(hexafluorophosphate) (**3**·4PF$_6$), incorporating two π-electron-deficient bipyridinium units separated by *p*-xylylene spacers holding them apart (ca 7 Å, i.e. about the same interplanar separation observed between the two hydroquinone rings of BPP34C10), was synthesized [16] and its ability to bind π-electron-rich guests demonstrated [16]. The development of an efficient route to the synthesis of **3**·4PF$_6$, and, hence, our entry into catenanes and rotaxanes, unfolded in the following step-wise manner. Initially, the tetracationic cyclophane was formed in a rather modest yield (12%) by reacting **1**·2PF$_6$ and 1,4-bis(bromomethyl)benzene (**2**) under high dilution conditions in acetonitrile (Fig. 10.5). However, complexation between **3**·4PF$_6$ and π-electron-rich guests, such as **4**, is easily achieved in solution (K_a = 2000 M^{-1} in acetonitrile) by means of: (i) π–π stacking interactions between the π-donor of the guest and the π-acceptors of the host; (ii) [C—H···O] hydrogen bonding between the polyether oxygen atoms and the α-protons on the bipyridinium units; and (iii) edge-to-face T-type interactions involving the hydroquinone ring protons and the

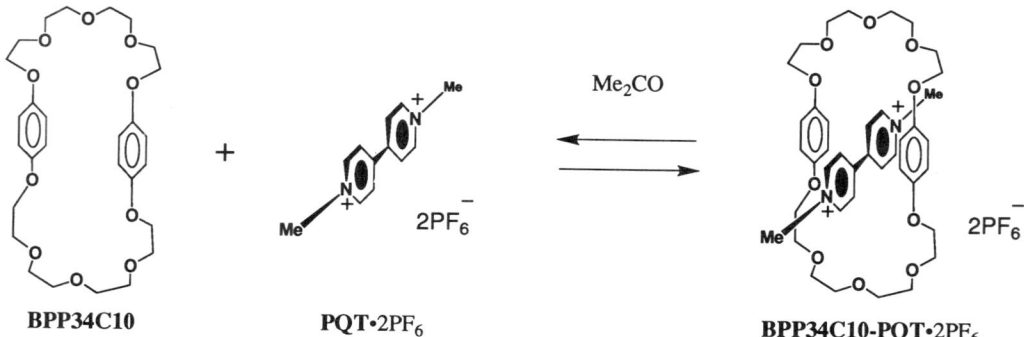

Figure 10.4 Spontaneous self-assembly in solution of a pseudorotaxane.

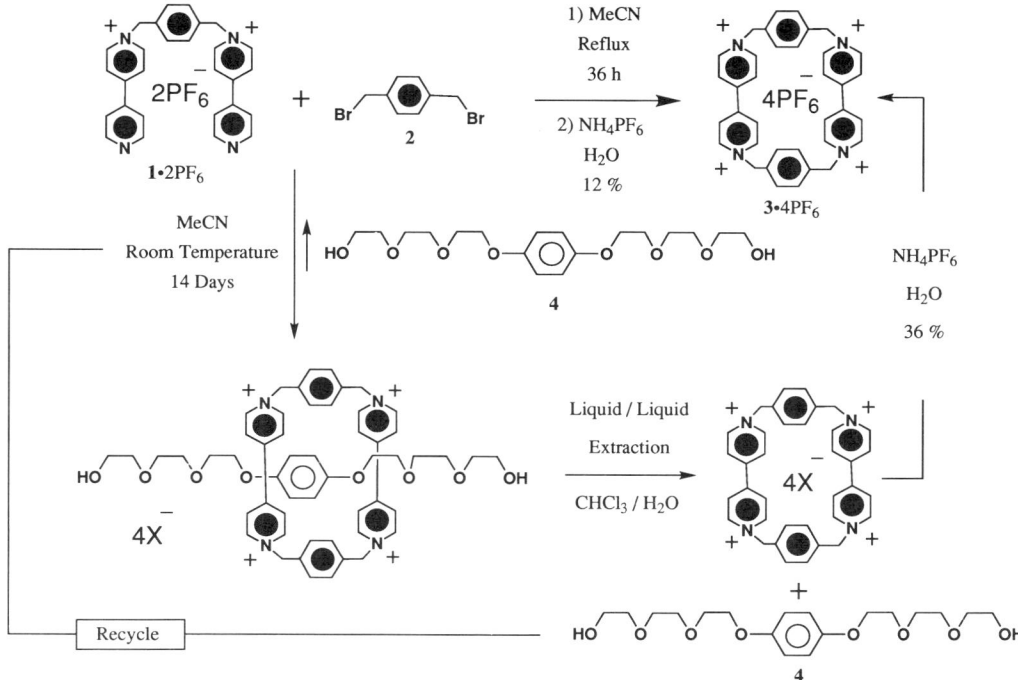

Figure 10.5 Template-directed synthesis of the tetracationic cyclophane $3 \cdot 4PF_6$.

p-xylylene spacers. Thus, the acyclic polyether **4** can be employed as a template in the synthesis of the tetracationic cyclophane $3 \cdot 4PF_6$. Reaction of $1 \cdot 2PF_6$ and **2** in acetonitrile in the presence of **4** affords a 3-fold increase in the yield (36%) of the resulting cyclophane $3 \cdot 4PF_6$. The template preorganizes the components of the π-electron-deficient tetracationic cyclophane decreasing the energy of activation associated with the macrocyclization, for which a pseudorotaxane-like supramolecular complex is self-assembled. Liquid–liquid continuous extraction of the template — which can be recycled — into chloroform releases the *free* cyclophane $3 \cdot 4PF_6$, after counterion exchange.

These results demonstrate that, by encoding the individual components with the right stereoelectronic information — i.e. incorporating appropriate complementary recognition sites — a range of supramolecular complexes can be self-assembled. These approaches became the basis for the efficient template-directed syntheses of catenanes and rotaxanes.

4   Self-assembling catenanes and rotaxanes

The self-assembly of the supramolecular pseudorotaxane-like complexes shown in Figs 10.4 and 10.5 led to the development [21–23] of efficient template-directed syntheses of catenanes and rotaxanes. Reaction of $1 \cdot 2PF_6$ with **2** — i.e. the components of the tetracationic cyclophane $3 \cdot 4PF_6$ — in the presence of an excess of the macrocyclic polyether BPP34C10 in acetonitrile during 48 h afforded a deep red coloured precipitate (Fig. 10.6). Purification by column chromatography, followed by counterion exchange (NH_4PF_6/H_2O) of the resulting red solid afforded [16] the [2]catenane $5 \cdot 6PF_6$

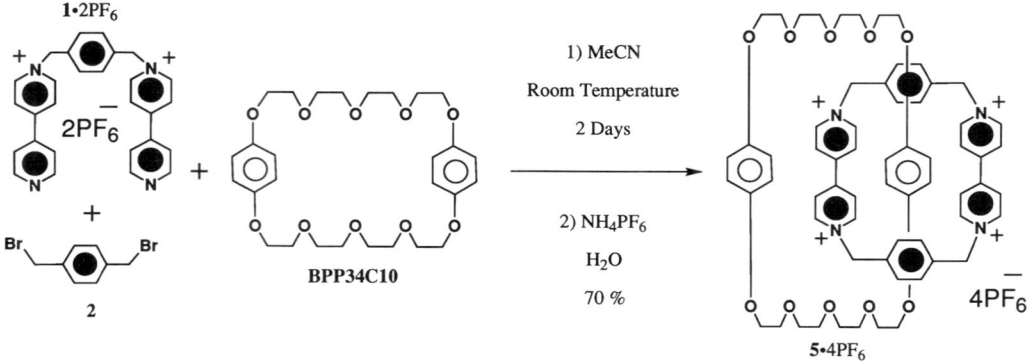

Figure 10.6 Self-assembly of the [2]catenane $5 \cdot 4PF_6$.

in a yield of 70%. The macrocyclic polyether BPP34C10 present in the reaction mixture acts as a template promoting the macrocyclization of the tetracationic cyclophane and yielding a molecular compound in which the two macrocyclic components are inter-locked, as demonstrated unequivocally [16] by X-ray crystallographic analysis.

A synthetic protocol very similar to that responsible for the template-directed synthesis of the tetracationic cyclophane $3 \cdot 4PF_6$ was employed to self-assemble rotax-anes. The template was turned into a permanent part of the molecular assembly by termination of both its ends with bulky stoppers. The preformed tetracationic cyclo-phane $3 \cdot 4PF_6$ and the linear hydroquinone-based polyether 4 self-assemble spontane-ously in acetonitrile solution, affording a pseudorotaxane-like complex in which the linear guest is inserted through the cavity of the macrocycle. Subsequent addition (Fig. 10.7) of triisopropylsilyl chloride and lutidine (a sterically hindered base which allows reaction to occur without adverse effects on the bipyridinium unit) affords [16]

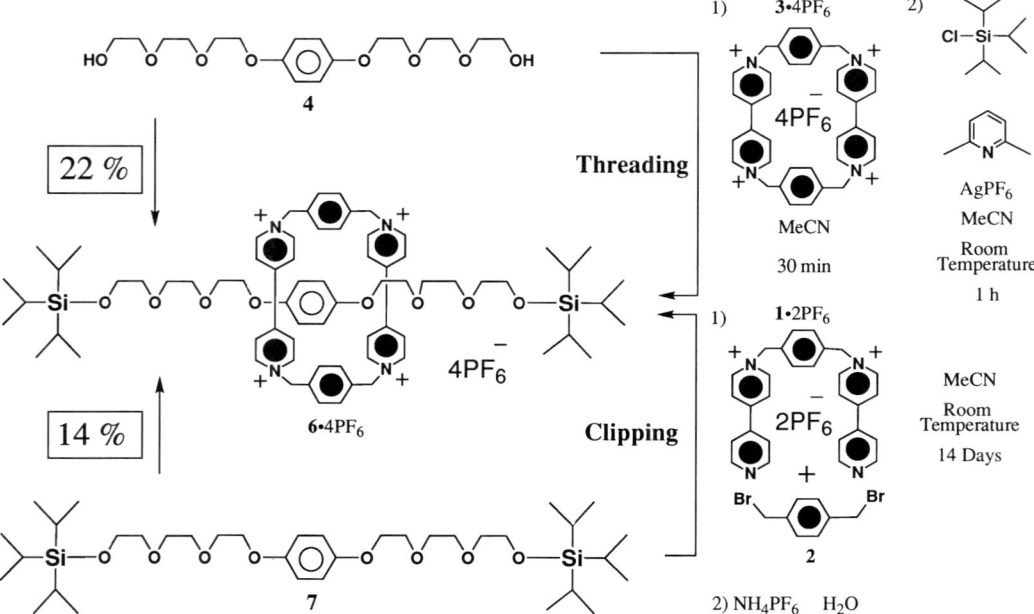

Figure 10.7 Self-assembly of the [2]rotaxane $6 \cdot 4PF_6$.

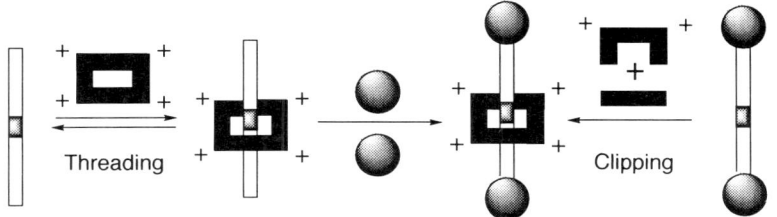

Figure 10.8 Threading and clipping approaches to self-assembling rotaxanes.

the [2]rotaxane **6**·4PF$_6$ in 22% yield by this method. Alternatively, the preparation of the [2]rotaxane **6**·4PF$_6$ was achieved [16] by the so-called clipping procedure in which the macrocyclization of the tetracationic cyclophane is performed in the presence of the dumbbell-shaped template **7**. However, in the case of the clipping approach the yield is only 14%, presumably as a result of unfavourable steric interactions between the stoppers and the forming cyclophane.

Figure 10.8 summarizes schematically the threading and clipping approaches to the self-assembly of rotaxanes such as **6**·4PF$_6$ incorporating a π-electron-rich dumbbell-shaped component and a π-electron-deficient tetracationic cyclophane.

5   From self-assembly to self-organization

In order to employ self-assembling molecular compounds, such as catenanes and rotaxanes, and self-assembling supramolecular complexes, such as pseudorotaxanes, as chemical sensors and/or information storage devices, their organization into defined nanoscaled supramolecular arrays is required. In solution, the individual molecular assemblies or supramolecular complexes are randomly oriented. On the contrary, precise spatial arrangements of the molecules or supermolecules are achieved in the solid-state or in self-assembled monolayers, such that the molecules or supermolecules can be addressed individually and/or in groups. Furthermore, robustness — i.e. simpler manipulation — and cooperativity, which may lead to amplification of a molecular signal, will also be ensured.

The non-covalent bonding interactions driving the self-assembly of catenanes, rotaxanes and pseudorotaxanes in solution live on in the final assemblies governing their dynamic properties (*vide infra*) and dominating their self-organization in the solid state. The X-ray crystal structure (Fig. 10.9) of the [2]catenane **5**·4PF$_6$ reveals [16] continuously stacked [2]catenanes aligned along one of the crystallographic axes. The π-donors — namely, hydroquinone rings — and π-acceptors — namely, the bipyridinium units — alternate intramolecularly, as well as intermolecularly, affording an infinite one-dimensional π–π stacking motif. The interplanar separation between the complementary aromatic units is ca 3.5 Å. This regular stacking means that approximately 10 aligned [2]catenane molecules afford a monodirectional array 10 nm long!

The crystal packing motif suggested that, by appropriate design of the individual molecules, defined superstructures (e.g. stacks and two-dimensional sheets) may be self-organized with high degrees of control in the solid state, once again as a result of non-covalent bonding interactions.

The tetracationic cyclophane **8**·4PF$_6$ (Fig. 10.10) possesses complementary

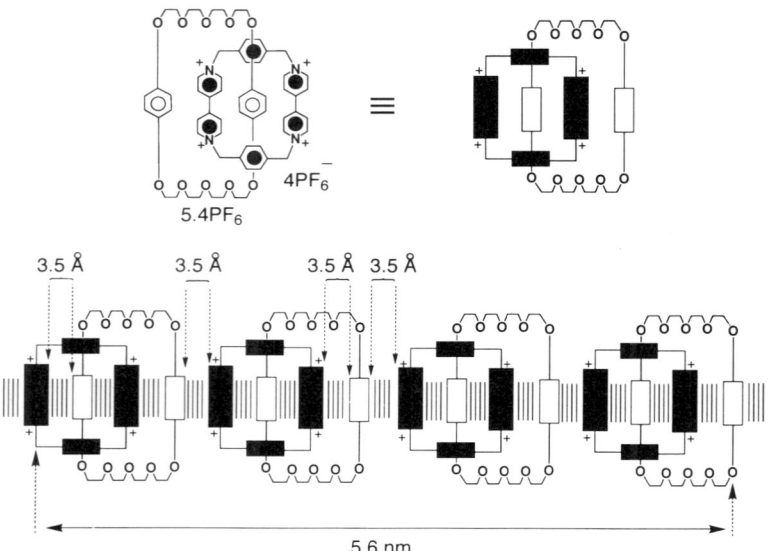

Figure 10.9 Self-organization in the solid state of the [2]catenane **5**·4PF$_6$.

π-electron-rich and π-electron-deficient sides and has the dimensions of a near perfect square. Crystallization of **8**·4PF$_6$ from an acetonitrile solution, on vapour diffusion of *i*-propylether into it, afforded [24] the highly ordered two-dimensional array shown in Fig. 10.10. The self-organization of **8**·4PF$_6$ into such a superstructure is a consequence of the complementarity between the π-electron-rich hydroquinone and π-electron-deficient bipyridinium sides of the cyclophane which stack intermolecularly with a distance of 4 Å. As a result, the length of the side of the square-like matrix which is represented schematically in Fig. 10.10 is 3.8 nm! These mosaic-like layers propagate infinitely in two dimensions and stack one on top of the other with the cavities of the cyclophanes perfectly in register, affording, in the third crystallographic direction,

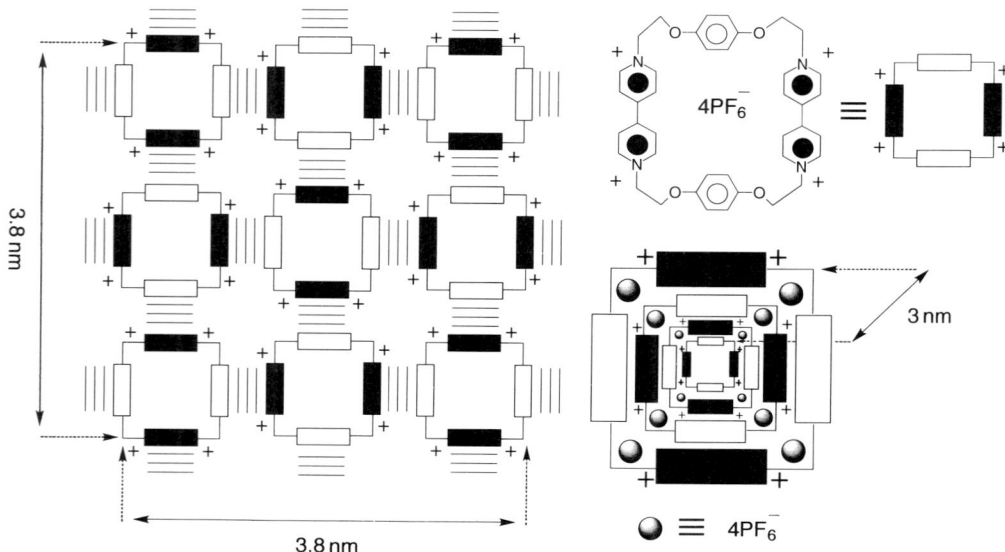

Figure 10.10 Self-organization in the solid state of a three-dimensional nanoscaled superstructure.

infinite channels running perpendicular to the mean planes of the layers which are separated by a distance of 9.8 Å — i.e. approximately 1 nm! The resulting three-dimensional superstructure behaves as an organic molecular sponge and is able to include within its channels small organometallic guests, such as ferrocene, without altering its structure.

Self-organization of these polycationic compounds can be achieved also at the air–water interface by employing monoanionic phospholipids as counterions. The tetracationic cyclophane 3^{4+} was 'attached' [25] (Fig. 10.11) to the air–water interface by employing the monoanionic derivative **9**, possessing a negatively charged head linked to a long lipophilic tail. The complexity of the self-organized monolayer is increased by introducing the acyclic derivative **10** consisting of a π-electron-rich head group containing a polyether chain intercepted by a 1,5-dioxynaphthalene unit linked to a lipophilic tail. The resulting supramolecular assembly consists of a monolayer of pseudorotaxane-like units, anchored electrostatically to the interface by the lipophilic tails of the counterions and the guest inserted into the cavity of the tetracationic cyclophane. The π–A isotherms of the three component mixture — namely, the cyclophane, the phospholipids, and the guest — were shown to be very similar to the π–A isotherms of the two component mixture — namely, the cyclophane and the phospholipids. Furthermore, loss of fluorescence of the 1,5-dioxynaphthalene unit and a charge transfer band arising from the π–π interactions between the bipyridinium units and the 1,5-dioxynaphthalene ring were observed. These results support the evidence of the self-assembly of pseudorotaxane-like complexes and their self-organization into a monolayer at the air–water interface. Similarly, [2]catenanes incorporating long alkyl chains, attached to the p-xylylene spacers of their tetracationic cyclophane components, have been self-organized [26] at the air–water interface.

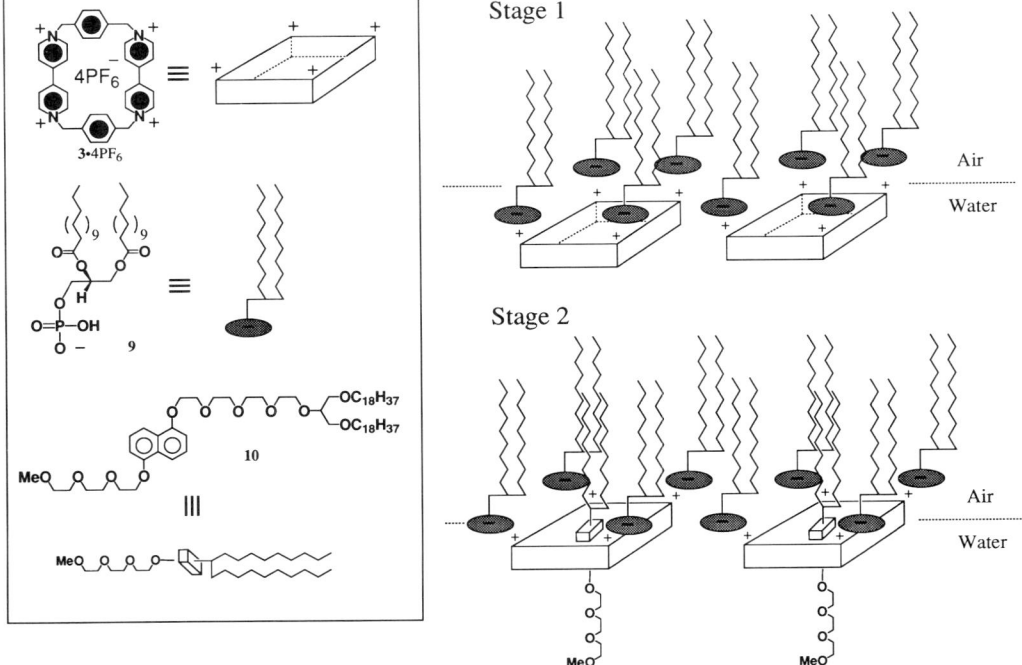

Figure 10.11 Self-organizing monolayers at the air–water interface.

6 Molecular machines

The efficient self-assembly of pseudorotaxanes, rotaxanes, and catenanes can be achieved with high degree of control in solution by exploiting non-covalent bonding interactions. Furthermore, their self-organization into defined three-dimensional super-structures on the nanometre-scale can be achieved also in either the solid state or at an air–water interface. Thus, a methodology, mimicking the synthetic processes witnessed in nature to generate precisely artificial nanostructures, is already in hand. However, in order to construct chemical systems possessing device-like characteristics, shape is not the only requisite that has to be engineered and realized in a controlled manner. More importantly, potential molecular machines have to possess functions.

Pseudorotaxanes, rotaxanes, and catenanes can display (Fig. 10.12) changes of the relative positions of their component parts under certain sets of conditions. Pseudoro-taxanes are supramolecular complexes which can dissociate reversibly into *free* host and *free* guest. Rotaxanes incorporating two recognition sites within the dumbbell-shaped component can behave as molecular shuttles, i.e. the macrocycle component can shuttle from one recognition site to the other. This kind of translational isomerism is also associated with catenanes: circumrotation of one macrocyclic component through the cavity of the other leads from one translational isomer to the other. As a result, external stimuli — chemical, electrochemical, and/or photochemical — can be employed in order to control reversibly the interconversion from state 0 to state 1 and vice versa. Furthermore, if these two states can be easily distinguished — e.g. by their spectroscopic behaviour — a binary switch at a molecular level can be generated. In summary, the structural features of pseudorotaxanes, rotaxanes, and catenanes provide the possibility of writing information into the chemical system — i.e. applying an external stimulus — and subsequently of reading it out — i.e. analysing the response of the molecular or supramolecule.

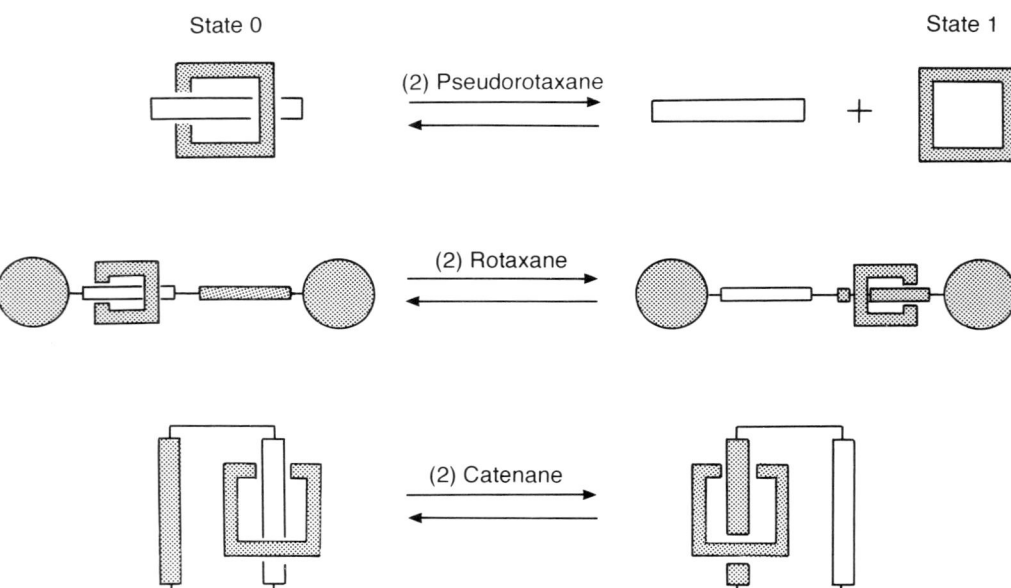

Figure 10.12 Pseudorotaxanes, rotaxanes, and catenanes as binary switches.

7 Pseudorotaxanes

In order to investigate the possibility of generating nanomachines in the shape of supramolecular complexes, a range of pseudorotaxanes have been self-assembled [27] and their properties investigated. The acyclic polyether derivative **11**, incorporating a 1,5-dioxynaphthalene unit, is bound [28] with pseudorotaxane-like geometry by the tetracationic macrocycle **3·4Cl** in a variety of solvents, e.g. acetonitrile, acetone, and water. As a result of a charge transfer interaction between the bipyridinium units and the 1,5-dioxynaphthalene ring within the pseudorotaxane, a purple colour appears upon mixing the two components in solution. Furthermore, quenching of the fluorescence associated with the 1,5-dioxynaphthalene ring is observed upon complexation as a result of the interaction between the complementary aromatic units within the complex. Thus, the threading of the tetracationic cyclophane on to the polyether thread-like guest can be monitored by following the increasing intensity of the charge transfer band, while decomplexation can be monitored by following the increasing intensity of the fluorescence band. Dismembering of the supramolecular species can be achieved by reducing the tetracationic cyclophane, thus eradicating its π-accepting ability. An equimolar solution of the tetracationic cyclophane **3·4Cl** and the π-electron-rich guest **11** in deoxygenated water was prepared [28] (Fig. 10.13) and a photosensitizer — namely, 9-anthracene carboxylic acid — as well as a sacrificial electron donor — namely, triethanolamine — were added. The appearance of a purple colour indicates the spontaneous self-assembly of the pseudorotaxane-like supramolecular complex **3–11·4Cl**. Irradiation of the solution causes photoinduced electron transfer from the photosensitizer to the tetracationic component of the pseudorotaxane. The reduction of the cyclophane results in dismembering of the supramolecular complex, as proven by the appearance of the fluorescence of the free guest **11**. However, addition of oxygen

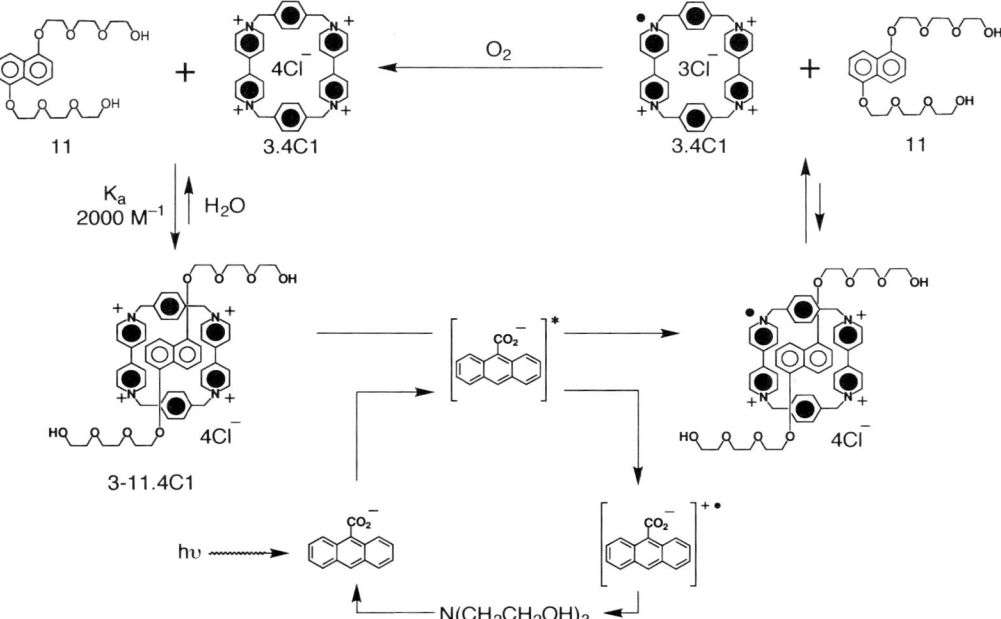

Figure 10.13 A photochemically controllable pseudorotaxane.

restores the original oxidation state of the cyclophane, allowing recombination of host and guest to form the supramolecular complex. Also, the oxidized 9-anthracene carboxylic acid is reduced by the sacrificial electron donor and can be recycled. In summary, by means of photochemical and chemical stimuli, reversible interconversion between pseudorotaxane — state 0 — and free components — state 1 — is achieved.

The complexation of a π-electron-deficient guest by a π-electron-rich macrocyclic host has been employed similarly to generate a photochemically active supramolecular complex. The bis(hexafluorophosphate) salt **12**·2PF$_6$ is strongly bound [29] by the macrocyclic polyether, bis-1,5-dioxynaphthalene-38-crown-10 (1/5DN38C10), with a pseudorotaxane-like geometry. Although the free host and free guest are luminescent molecules, their association into a complex results in the quenching of luminescence and in the appearence of a charge transfer band. On mixing the guest **13**·2PF$_6$ and the host 1/5DN38C10 in acetonitrile (Fig. 10.14), the corresponding pseudorotaxane is self-assembled [29] spontaneously, as indicated by noticeable changes in both the emission and absorption spectra — namely, the disappearance of the luminescence of the two components and appearance of a broad charge transfer band, respectively. The chemically driven dismembering of the pseudorotaxane can be achieved by exploiting the ability of amines to form charge transfer complexes with the dicationic compound **12**$^{2+}$. Thus, addition of hexylamine to the acetonitrile solution of the pseudorotaxane causes an instantaneous colour change from red to green. This transformation is accompanied by: (i) the appearance of the charge transfer absorption band of the 1 : 2 adduct between **12**$^{2+}$ and hexylamine; (ii) the recovery of the emission band of 1/5DN38C10; (iii) the appearance of an emission band associated to the 1 : 2 complex **12**$^{2+}$–hexylamine; and (iv) the decrease in intensity of the charge transfer band associated to the pseudorotaxane. These changes are consistent with the formation of a 1 : 2 adduct between **12**$^{2+}$ and hexylamine which promotes the dismembering of the pseudorotaxane. The process can be reversed quantitatively by protonation of hexylamine by adding trifluoroacetic acid (TFA) to the solution, as shown by: (i) the disappearance of the absorption and emission bands of the 1 : 2 **12**$^{2+}$–hexylamine

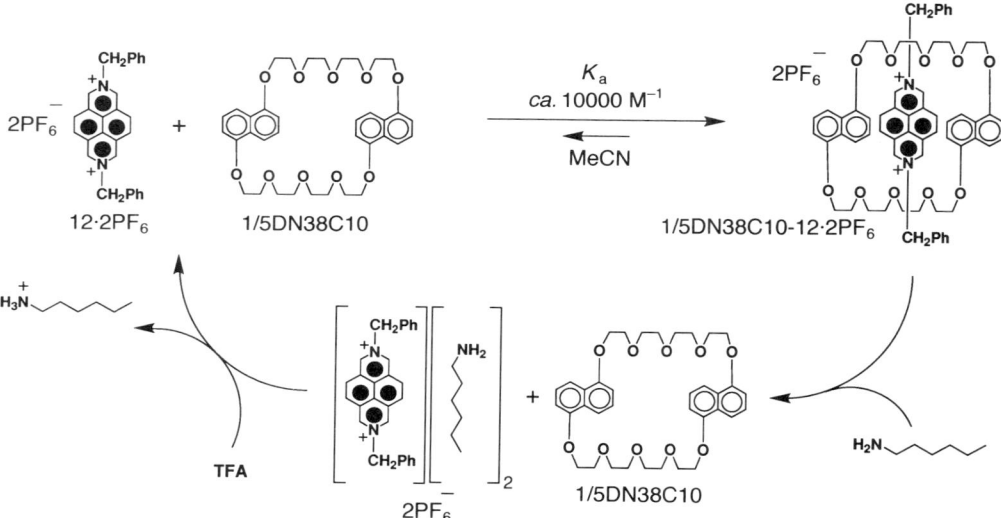

Figure 10.14 A chemically controllable pseudorotaxane.

adduct, (ii) the decrease of the emission intensity of 1/5DN38C10; and (iii) the reappearance of the charge transfer absorption band of the pseudorotaxane. These results demonstrate that the self-assembly and dismembering of a [2]pseudorotaxane can be controlled by external stimuli. These systems represent prototypes, at the supramolecular level, for a simple molecular machine in which the changes in the relative positions of the components can be followed by readily monitorable differences in the absorption and luminescence spectra.

8 Rotaxanes

In order to generate a controllable molecular shuttle, such as the one schematically represented in Fig. 10.12, a [2]rotaxane incorporating two recognition sites within its dumb-bell-shaped component encircled by one macrocyclic component has to be self-assembled. Furthermore, chemically, electrochemically, and/or photochemically active recognition sites have to be introduced to ensure that the macrocyclic component can be modulated by external stimuli.

A series of pseudorotaxanes incorporating the π-electron-deficient bipyridinium-based cyclophane $3 \cdot 4PF_6$ and either benzidine- or biphenol-based π-electron-rich guests were employed [30] as model compounds. Association constants of 800 and 100 M^{-1} were calculated in acetonitrile for the complexation of the benzidine-based **14** and the biphenol-based **13** guests by the tetracationic cyclophane, respectively. Thus, a marked preference for the self-assembly of the pseudorotaxane $[3–14] \cdot 4PF_6$ incorporating the benzidine unit is observed under competitive conditions. Furthermore, benzidine easily undergoes protonation and reversible oxidation — i.e. the benzidine unit is both chemically and electrochemically active. Competitive binding experiments (Fig. 10.15) between the tetracationic cyclophane $3 \cdot 4PF_6$ and the two guests were undertaken in acetonitrile to illustrate the possible reversible control of the self-assembly process. Absorption spectroscopy in the visible region revealed a charge transfer band centred on 690 nm corresponding to the pseudorotaxane $[3–14] \cdot 4PF_6$. A dramatic decrease of the intensity of the band at 690 nm and the corresponding appearance of the charge transfer band centred on 480 nm associated with $[3–13] \cdot 4PF_6$ was observed upon addition of TFA to the solution. These results suggest that protonation of the benzidine unit occurs, dismembering the pseudorotaxane $[3–14] \cdot 4PF_6$ and pushing the equilibrium in favour of $[3–13] \cdot 4PF_6$. Two factors are mainly responsible for such a transformation: (i) strong electrostatic repulsion between the diprotonated form of the guest **15** and the tetracationic cyclophane $3 \cdot 4PF_6$; and (ii) the reduced electron density on the aromatic unit of **14** which decreases its π-donor ability destabilizing the complex. Addition of a base — namely, pyridine — to the solution restores the initial equilibrium by deprotonating the benzidine units. These results demonstrate clearly the reversible switching action of the trimolecular system achieved by controlling the protonation state of the benzidine-based guest **14**.

The next step forward was to incorporate the benzidine and biphenol units into a molecular system — namely, a [2]rotaxane comprised of the tetracationic cyclophane, as the macrocyclic component, and the benzidine and biphenol units, as the two recognition sites incorporated within the dumbbell-shaped component. Indeed, the [2]rotaxane $16 \cdot 4PF_6$ was self-assembled [31] by means of a threading approach (*vide*

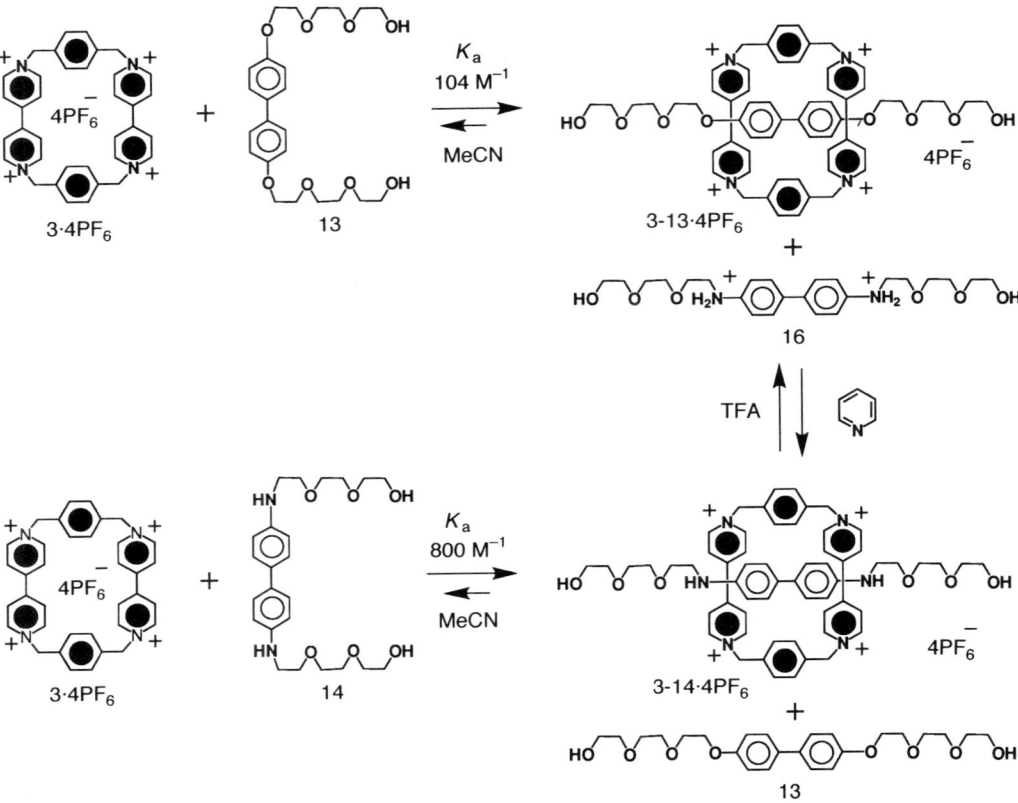

Figure 10.15 Reversible control of the self-assembly of the pseudorotaxanes **3–13**·4PF$_6$ and **3–14**·4PF$_6$.

supra). The tetracationic cyclophane resides preferentially on the benzidine recognition site, as a result of strong non-covalent interactions. However, protonation of the benzidine unit with TFA forces the cyclophane to move on to the biphenol recognition site. Addition of pyridine deprotonates the benzidine unit, allowing the macrocycle to shuttle back on to it. Alternatively, the shuttling process can be controlled reversibly by electrochemical stimuli. Upon electrochemical oxidation of the benzidine station, the tetracationic cyclophane moves away from the now positively charged benzidine unit: electrochemical reduction of the benzidine restores the original equilibrium state. In summary, the [2]rotaxane **16**·4PF$_6$ is a molecular shuttle which can be controlled by means of chemical or electrochemical stimuli (see Fig. 10.16), allowing controlled interconversion between state 0 and state 1.

9   Catenanes

Molecular compounds possessing device-like characteristics can be generated also in the shape of catenanes. Figure 10.12 shows a schematic representation of a [2]catenane incorporating two interlocked macrocyclic components which can function as a binary switch. One of the macrocycles possesses two appropriate recognition sites such that, if one of them is located inside, the other has to be alongside the cavity of the second macrocycle. Interconversion of the two recognition sites can be achieved by the

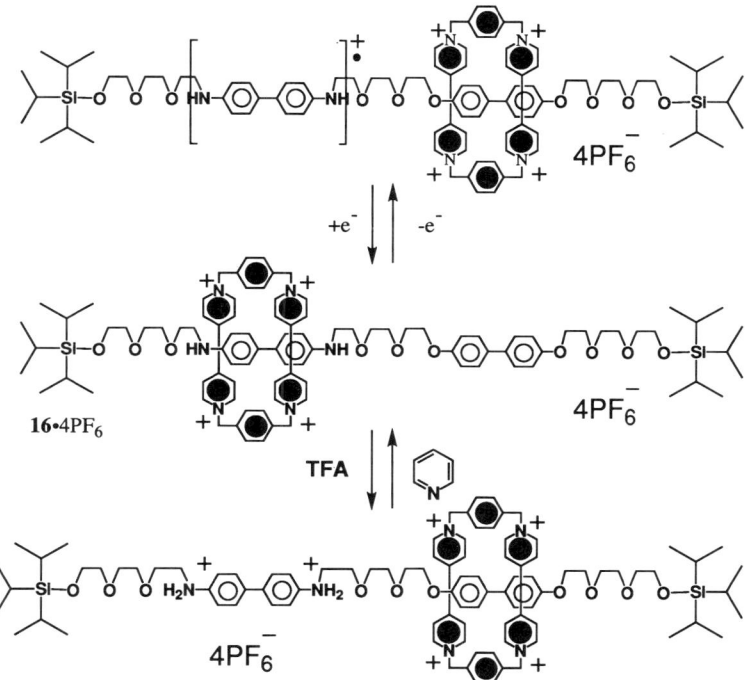

Figure 10.16 A chemically and electrochemically controllable [2]rotaxane.

circumrotation of one macrocycle through the cavity of the other. By employing chemically, electrochemically, and/or photochemically active units as the two recognition sites, reversible control of the switching process can be achieved.

In order to generate a controllable [2]catenane, a series of catenanes incorporating structural modifications with respect to the original one $\mathbf{5} \cdot 4PF_6$ were self-assembled [32,33]. In particular, the possibility was envisaged of introducing asymmetry into the π-electron-rich and/or the π-electron-deficient macrocyclic components by simply replacing one of their aromatic recognition sites — namely, a hydroquinone ring or a bipyridinium unit. Three respresentative examples (Fig. 10.17) are the [2]catenanes $\mathbf{16} \cdot 4PF_6$, $\mathbf{17} \cdot 4PF_6$, and $\mathbf{18} \cdot 4PF_6$. Asymmetry has been introduced into $\mathbf{16} \cdot 4PF_6$ by replacing one of the hydroquinone rings of the π-electron-rich macrocyclic polyether by a 1,5-dioxynaphthalene unit. On the contrary, asymmetry has been introduced in $\mathbf{17} \cdot 4PF_6$ by replacing one of the bipyridinium units of the π-electron-deficient cyclophane by a bis(4-pyridinium)*trans*-ethylene unit. The [2]catenane $\mathbf{18} \cdot 4PF_6$ incorporates simultaneously both modifications, i.e. the π-electron-rich and π-electron-deficient macrocyclic components can each incorporate two different aromatic recognition sites.

The degree of control of the dynamic processes occuring within such mechanically interlocked molecular structures is illustrated by the behaviour of the [2]catenane $\mathbf{17} \cdot 4PF_6$. Variable temperature ^{1}H-NMR spectroscopy demonstrated [33] that the translational isomers **I** and **II**, shown in Fig. 10.18, exist in acetone solution at − 60°C. However, the ratio is 92 : 8 in favour of the one bearing the bipyridinium recognition site inside the cavity of the hydroquinone-based macrocycle. The affinity of the π-electron-rich macrocyclic polyether for the bipyridinium recognition site decreases significantly upon its reduction, a change which can be performed electrochemically,

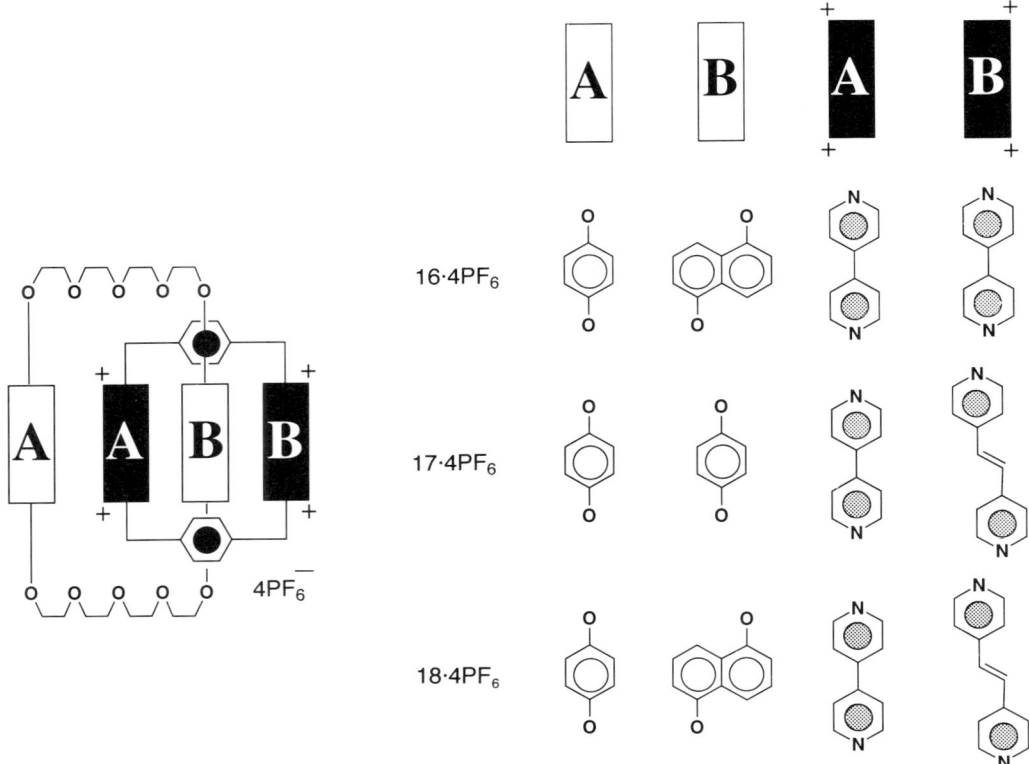

Figure 10.17 [2]Catenanes incorporating two different recognition sites within the π-electron-rich and/or the π-electron-deficient macrocyclic components.

leaving unaltered the bis(4-pyridyl)*trans*-ethylene unit. As a result, circumrotation of the π-electron-deficient cyclophane through the cavity of the macrocyclic polyether occurs in order to place the bis(4-pyridyl)*trans*-ethylene recognition site inside its cavity. Thus, electrochemical reduction (process A, Fig. 10.19) of the bipyridinium recognition site is followed by the circumrotation (process B, Fig. 10.19) of the π-electron-deficient cyclophane, thus replacing the reduced bipyridinium unit with the bis(4-pyridyl)*trans*-ethylene recognition site. By oxidizing (process C, Fig. 10.19) the bipyridinium unit back to the original oxidation state, the initial equilibrium between the two translational isomers **I** and **II** is restored. Thus, the [2]catenane **17**·4PF$_6$, incorporating two different recognition sites — namely, a bipyridinium and a bis(4-pyridinium)*trans*-ethylene unit — within its π-electron-deficient macrocyclic component, is a molecular switch which can be controlled reversibly by means of electrochemical stimuli.

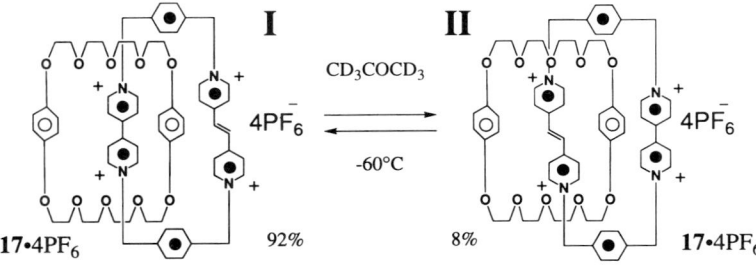

Figure 10.18 Translational isomerism associated with the [2]catenane **17**·4PF$_6$.

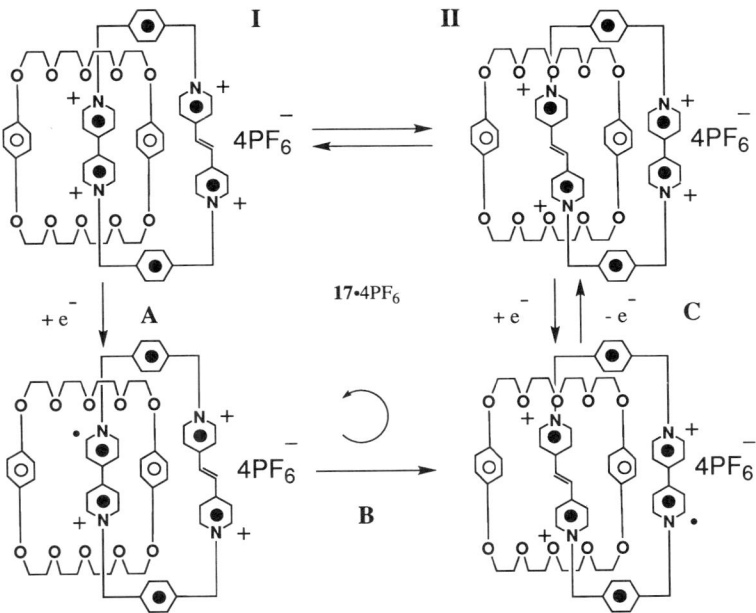

Figure 10.19 An electrochemically controllable [2]catenane.

10   Conclusions

Chemists are learning from nature the principles of self-assembly and self-organization to construct efficiently and with a high degree of control nanoscaled molecular structures and supramolecular arrays. These synthetic paradigms can now be employed to generate, from simple molecular components, functioning artificial chemical systems possessing device-like characteristics — namely, molecular machinery [34–37]. Indeed, a number of supramolecular and molecular species have been self-assembled by careful design of their modular components and by exploiting a range of non-covalent bonding interactions. The discrete assemblies generated in solution can be self-organized into superstructures possessing engineered three-dimensional geometries of nanometre size. Thus, synthetic methodologies to generate artificial chemical systems possessing defined forms have already been conceived and demonstrated. The next step is the introduction of specific functions into such systems so that nanomachines which are able to store and process information can be generated.

The structural features possessed by pseudorotaxanes, rotaxanes, and catenanes suggest the possibility of generating molecular and supramolecular devices possessing switching capabilities at a molecular level. Indeed, a number of chemically, electrochemically, and/or photochemically controllable systems have been self-assembled and self-organized, and their properties investigated. In summary, efficient synthetic methodologies to construct engineered systems at the nanoscopic level are already available and systems able to act as devices at a molecular level have been generated at an elementary level. We believe that, by taking advantage of the results already achieved, researchers will be able to construct, in the near future, a new generation of engineered and functioning artificial chemical systems — namely, nanomachines.

Acknowledgements. This work was supported in the UK by the Engineering and Physical Sciences Research Council and by the European Community Human Capital and Mobility Programme.

11 References

1 Lindsey JS. *New J Chem* 1991; **15**: 153.

2 Whitesides GM, Mathias JP, Seto CT. *Science* 1991; **254**: 1312.

3 Lawrence DS, Jiang T, Levett R. *Chem Rev* 1995; **95**: 2229.

4 Philp D, Stoddart JF. *Angew Chem Int Ed Engl* 1996; **35**: 1154.

5 Ringsdrof H, Schlarb B, Venzmer J. *Angew Chem Int Ed Engl* 1988; **27**: 113.

6 Ahlers M, Muller W, Reichert A, Ringsdorf H, Venzmer J. *Angew Chem Int Ed Engl* 1990; **29**: 1269.

7 Lehn JM. *Angew Chem Int Ed Engl* 1990; **29**: 1304.

8 Rebek Jr J. *Acc Chem Res* 1990; **23**: 399.

9 Famulok M, Nowick JS, Rebek Jr J. *Acta Chem Scand* 1992; **46**: 315.

10 Orgel LE. *Nature* 1992; **358**: 203.

11 von Kiedrowski G. *Bioorg Chem Front* 1993; **3**: 113.

12 Ozin GA. *Adv Mater* 1992; **4**: 612.

13 Merkle RC. *Nanotechnology* 1993; **2**: 134.

14 Tomalia DA, Durst HD. *Top Curr Chem* 1993; **165**: 193.

15 Drexler KE. *Annu Rev Biophys Biomol Struct* 1994; **23**: 377.

16 Anelli PL, Ashton PR, Ballardini R *et al. J Am Chem Soc* 1992; **114**: 193.

17 Allwood BL, Spencer N, Shariari-Zavareh H, Stoddart JF, Williams DJ. *J Chem Soc Chem Commun* 1987; 1064.

18 Ashton PR, Slawin AMZ, Spencer N, Stoddart JF, Williams DJ. *J Chem Soc Chem Commun* 1987; 1066.

19 Stoddart JF. *Pure Appl Chem* 1988; **60**: 467.

20 Ashton PR, Philp D, Reddington MV, Slawin AMZ, Spencer N, Stoddart JF. *J Chem Soc Chem Commun* 1991; 1680.

21 Philp D, Stoddart JF. *Synlett* 1991; 445.

22 Amabilino DB, Stoddart JF. *Pure Appl Chem* 1993; **65**: 2351.

23 Pasini D, Raymo FM, Stoddart JF. *Gazz Chim Ital* 1995; **125**: 431.

24 Ashton PR, Claessens CG, Hayes W *et al. Angew Chem Int Ed Engl* 1995; **34**: 1862.

25 Ahuja RC, Caruso PL, Mobius D *et al. Langmuir* 1993; **9**: 1534.

26 Preece JA, Stoddart JF. *Nanobiology* 1994; **3**: 149.

27 Amabilino DB, Anelli PL, Ashton PR *et al. J Am Chem Soc* 1995; **117**: 11142.

28 Ballardini R, Balzani V, Credi A *et al. Angew Chem Int Ed Engl* 1996; **35**: 978.

29 Ballardini R, Balzani V, Gandolfi MT *et al. Angew Chem Int Ed Engl* 1993; **32**: 1301.

30 Córdova E, Bissell RA, Spencer N, Ashton PR, Stoddart JF, Kaifer AE. *J Org Chem* 1993; **58**: 6550.

31 Bissell RA, Córdova E, Kaifer AE, Stoddart JF. *Nature* 1994; **369**: 133.

32 Ashton PR, Blower M, Philp D *et al. New J Chem* 1993; **17**: 689.

33 Ashton PR, Ballardini R, Balzani V *et al. J Am Chem Soc* 1995; **117**: 11171.

34 Stoddart JF. *Chem Br* 1991; **27**: 714.

35 Urry DW. *Angew Chem Int Ed Engl* 1993; **32**: 819.

36 Raymo FM, Stoddart JF. In: Kahn O, ed. *Magnetism: A Supramolecular Function.* Dordrecht: Kluwer, 1996; **33**.

37 Raymo FM, Stoddart JF. In: Catlow CRA, ed. *New Trends in Materials Chemistry.* Dordrecht: Kluwer, 1997: in press.

11 Micro- and Nanofabrication Techniques Based on Self-assembled Monolayers

P.F. NEALEY*, A.J. BLACK†, J.L. WILBUR† and G.M. WHITESIDES†

*Department of Chemical Engineering, University of Wisconsin–Madison, 1415 Johnson Dr., Madison, WI 53706, USA

†Department of Chemistry, Harvard University, Cambridge, MA 02138, USA

1 Introduction

New fabrication techniques are required to continue the trend in the microelectronics industry to work at smaller scale. Optical lithography [1], the most widespread technique for the fabrication of microelectronic devices, is approaching the lower limit (~100 nm) in the size of features that it can produce. Smaller features (50 nm) can be produced with electron beam writing or atom-by-atom manipulation. These processes, however, are currently linear and will require significant development for high-volume commercial applications. Opportunities exist to develop fabrication techniques that are based on different principles to reach the scale of 50 nm. Desirable attributes of any new fabrication technique include low cost (capital and operational), low environmental impact, and the ability to make complex structures of appropriate scale reproducibly with low levels of defects.

Molecular- or atomic-scale devices based on the electronic properties of single molecules represent a plausible lower limit in size. Efforts to fabricate these devices have defined the emerging field of molecular electronics. Even if useful devices of scale 1–30 nm can be synthesized, however, a host of problems present themselves for the application of the molecular-scale devices. Structures between 100 nm and 1 nm will still be needed, for example, to make internal and external interconnects, or for use in new architectures such as cellular automata to access the molecular-scale devices.

Molecular self-assembly is a potential strategy for fabrication of structures with dimensions that range from nanometres to millimetres [2,3]. In molecular self-assembly, subunits of molecular dimensions, whose structures can be controlled with atomic resolution, spontaneously form molecularly ordered aggregates [4,5] with certain dimensions of 1–100 nm. Self-assembly leads to equilibrium structures that are at (or close to) thermodynamic minimum [6,7], and result from multiple, weak, reversible interactions such as hydrogen bonds, ionic bonds, and van der Waals forces between subunits. The information that determines the final structure of the aggregate is coded in the structure and properties of the subunits. As a strategy for fabrication, self-assembly is thus automatically defect rejecting [5], and self-registering on a scale that is too small for current techniques for microfabrication and too large for conventional organic synthesis. The principles of self-assembly are observed in nature [4–6,8,9]; processes such as protein folding and aggregation [10] and formation of DNA double helices [11] serve as biological models for the potential of self-assembly in microfabrication. Table 11.1 shows some examples of self-assembling systems.

Self-assembled monolayers (SAMs) represent the most widely studied and best developed class of non-biological self-assembling systems. SAMs are highly ordered

Table 11.1 Self-assembling systems.

System	Smallest dimension	Refs
Self-assembled monolayers	1–2 nm	[12–18]
Langmuir–Blodgett films	1–2 nm	[14,15]
Liquid crystals	1 nm	[19]
Hydrogen-bonded aggregates	1.5 nm	[20]
Templated crystals	0.1 nm	[21,22]
Block copolymers	1–2 nm	[23]
Systems in nature	1 nm	[4–6,8–11]

molecular assemblies of long-chain alkanes that chemisorb on the surfaces of appropriate solid materials. The structure of SAMs, effectively two-dimensional crystals with controllable chemical functionality, makes them ideal model systems for the investigation of wetting, adhesion, lubrication, corrosion, protein adsorption, and cell attachment [12–18].

This chapter describes the principles, achievements, and potential of micro- and nanofabrication techniques that are based on self-assembled monolayers; the systems that are discussed have at least one dimension of molecular scale. We have demonstrated applications for SAMs in microelectronics that include passivation of surfaces, use of SAMs for ultrathin resists and masks, directed deposition of materials on surfaces patterned with SAMs, and the use of chemical and biochemical functionality on the surface of SAMs for sensors.

2 Self-assembled monolayers

2.1 *Alkanethiols on gold*

Self-assembled monolayers of alkanethiolates on gold (RSH/Au) are the best studied system of self-assembled monolayers [24,25]. The RSH/Au system is attractive because of: (i) ease of fabrication; (ii) degree of perfection; (iii) chemical stability under ambient laboratory conditions; (iv) availability of materials; and (v) flexibility in chemical — and thus surface — functionality. Two important factors may limit the use of current hydrocarbon-based SAMs of alkanethiolates on gold in microelectronic devices: (i) they are thermally unstable above 100°C [26] and (ii) the high rate of diffusion of gold into Si; this property makes gold structures on silicon largely incompatible with current techniques of microfabrication [27]. SAMs of alkanethiolates on gold, however, can serve as models or prototypes for new techniques or applications that will be generally applicable to new classes of SAMs designed specifically for microprocessing and microfabrication.

2.1.1 PREPARATION AND GENERAL STRUCTURE

Self-assembled monolayers of alkanethiolates on gold form by spontaneous adsorption of alkanethiols $[X(CH_2)_nSH]$ [12–18,25,28–33] or dialkyldisulfides $[X(CH_2)_nS–S(CH_2)_mY]$ [33,34] onto gold, either from solution or the vapour phase (Eqn 11.1).

$$X\text{-}R\text{-}SH + Au(0)_n \rightarrow X\text{-}R\text{-}S^- Au(I) \cdot Au(0)_n + 1/2 H_2 \tag{11.1}$$

$$1/2 (X\text{-}R\text{-}S)_2 + Au(0)_n \rightarrow X\text{-}R\text{-}S^- Au(I) \cdot Au(0)_n.$$

Dialkylsulfides $[X(CH_2)_n S(CH_2)_m Y]$ also form SAMs [35], but are significantly less reactive than alkanethiols, and form SAMs of poorer quality. Assembly of thiolates may also be induced by electrochemical oxidation of alkanethiols in solution [36] using gold as an electrode. Thiols are generally preferred over disulphides and sulphides for the preparation of SAMs because they are more soluble, react with the surface of the gold $\sim 10^3$ faster [33,37], and yield monolayers of better quality. In addition, they are easier to synthesize (many are available commercially).

Typically, thin (20–50 nm) films of gold are used as substrates for SAMs; SAMs also form on colloidal gold [38–40]. The gold films are usually deposited by electron beam or thermal evaporation onto a solid, flat support such as a polished silicon wafer (with a native oxide), a sheet of mica, or a glass slide. Because gold does not wet these surfaces, an adhesion promoter such as titanium, chromium, or an appropriate organic material [41,42] is used.

2.1.2 STRUCTURE AT THE ATOMIC SCALE

The mechanistic details [32] of the reaction of thiols or disulphides with the gold surface are not completely understood. The fate of the hydrogen atom in the case of alkanethiol adsorption, for example, and the exact structure of the resulting monolayer are not known [43]. On Au(111) the sulphur atoms are thought to form an overlayer commensurate with the Au atoms with structure $(\sqrt{3} \times \sqrt{3})R30°$. Recent X-ray diffraction studies, however, suggested that the species on the surface is a disulphide [44] (rather than a thiolate [15,16]). The alkyl chains extend perpendicularly from the plane of the surface in a nearly all-*trans* configuration. To maximize the van der Waals interactions between adjacent methylene groups (~ 1.5–2 kcal mol^{-1} per CH$_2$) [15,16], the chains tilt at an angle of 30° with respect to the surface normal (Figure 11.1(a)). These van der Waals forces (~ 20–30 kcal mol^{-1} for C$_{16}$SH) and the strength of the sulphur–gold bond (~ 44 kcal mol^{-1}) drive formation of the monolayer. The resulting structure on the atomic scale is ordered: for alkyl chains with $n > 6$, the monolayer is quasi-crystalline in two dimensions. Table 11.2 summarizes the characterization of SAMs of alkanethiols on gold.

2.1.3 DEFECTS

Figure 11.1(b) shows high resolution scanning tunnelling microscopy (STM) images (from Delamarche *et al.* [63,64]) of dodecanethiol adsorbed on Au(111). The sample in the large image was annealed at 100°C for 10 h in air. A step in the gold surface runs from the upper right corner to the centre of the image where it disappears with a screw dislocation. The occassional black lines that parallel it are depressions in the monolayer where neighbouring thiols have either tilted over or migrated to cover missing lines of thiols. Although molecular resolution is apparent, the c(4*2) rectangular superlattice structure is not clearly discernible because of the large scan size. The inset shows a similar sample that was not annealed. The gold terrace has five depressions: these are

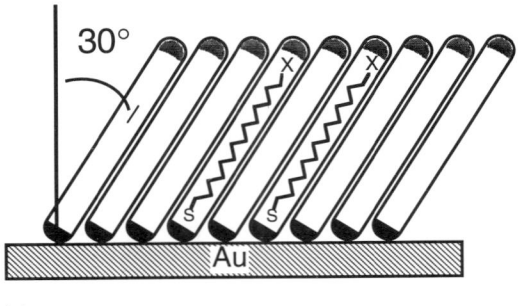

(a)

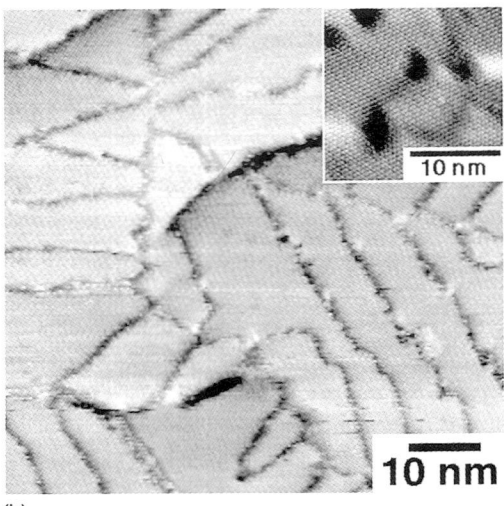

10 nm

10 nm

(b)

Figure 11.1 (a) Schematic illustration of the molecular-level structure of a self-assembled monolayer of *n*-alkanethiolates on gold. Modification of the head group, X, allows control of the monolayer's surface properties, and the thickness of the monolayer depends on the length of the alkyl chain. (b) Scanning tunnelling micrograph of dodecanethiol adsorbed on Au(111) that shows a gold step, a screw dislocation in the centre, and depressed lines in the monolayer due to the accommodation of missing lines of thiols by vicinal molecules. The inset shows five depressions (gold pits 2.4 Å deep) in the centre that are linked by domain boundaries. The monolayer packing corresponds to a phase of the c(4*2) rectangular superlattice. These images were provided by Delamarche *et al.* [63]. See text for references.

Table 11.2 Properties of self-assembled monolayers of thiols on gold.

Monolayer property	Technique	Refs
Structure	Surface Raman scattering	[45]
	Transmission (high energy) electron diffraction	[46]
	Low energy helium diffraction	[47–51]
	X-ray diffraction	[44,47,48,52,53]
	Infrared spectroscopy	[29,54–59]
	Scanning tunnelling microscopy	[49,60–71]
Composition	X-ray photoelectron spectroscopy	[25,33,55,56,72]
Wettability	Contact angle	[13,31,33,73–77]
Thickness of the adsorbed layer(s)	Ellipsometry	[29,32]
Degree of perfection and electrical properties	Electrochemical methods	[29,43,60,75,78–84]

pits in the gold, 2.4 Å deep, linked by domain boundaries (slightly depressed regions, where the packing is distorted). It is important to note, however, that the holes are still covered with SAMs.

Defect densities are a primary determinant of the suitability of SAMs for use in micro- and nanofabrication. Estimates for the minimum number of defects for SAMs on Au over relatively large areas (several cm^2) range from two to several thousand pinholes per cm^2; the latter value is more realistic [16,85; X.M. Zhao, J.L. Wilbur and G.M. Whitesides, unpublished results]. A stringent test using a wet-chemical etchant to amplify the defects in SAMs gives 90 defects mm^{-2} as a minimum value for the defect density for a SAM of hexadecanethiolate (HDT) on 20 nm thick gold.

With chemical control over the length of the methylene chains, the height of the monolayer perpendicular to the plane of the surface can be controlled with 0.1 nm scale precision. Organic and inorganic synthesis also facilitates the incorporation of different functional groups into, and at the termini of, the alkyl chains: hydrocarbons, fluorocarbons, acids, esters, amides, alcohols, nitriles, and ethers are a few of the many possible functional groups. Mixed monolayers — either with regard to alkyl chain length or chemical functionality — afford further control over surface modification and are prepared by simultaneous coadsorption of two different thiols.

Figure 11.2 shows how interfacial properties such as wetting and adhesion are controlled by SAMs of varying functionality. Water preferentially condenses on regions of a surface patterned with SAMs terminated by hydroxy (OH) groups (patterning techniques are discussed below); the water drops do not form (prior to coalescence) on the regions of the surface covered with SAMs terminated by methyl (CH$_3$) groups.

SAMs are capable of passivating certains surfaces, and of protecting these surfaces from chemical attack. SAMs of alkanethiols on Cu [86] and GaAs [87,88] have been shown to inhibit oxidation by retarding oxygen transport to the surface. Fe and Ni have

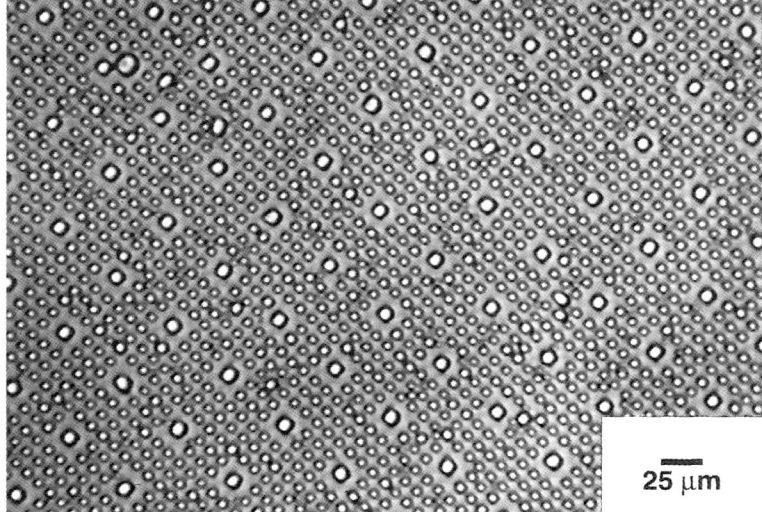

25 μm

Figure 11.2 Water droplets formed by condensation on regions of a surface (dark) that are covered with a self-assembled monolayer (SAM) that has a hydrophilic head group (e.g. -OH). Water does not condense on other regions of the surface (light) that are covered with an SAM that has a hydrophobic head group (e.g. -CH$_3$).

Table 11.3 Substrates and ligands that form self-assembled monolayers.

Surface	Ligand	Refs
SiO_2, glass	$RSiCl_3$, $RSiOR_3$	[95–100]
Si	$[RCOO]_2$ (neat)	[101]
Si	$RCH = CH_2$, $[RCOO]_2$	[102]
Ag	RSH	[47,48,54,103–105]
GaAs	RSH	[87,106,107]
Cu	RSH	[55,86,108,109]
InP	RSH	[110]
Au	R_3P	[111]
Au	RSO_2H	[112]
Pt	RNC	[113]
Pt	RSH, ArSH	[45,113–116]
Metal oxides	RCOOH	[52,73,117–127]
Metal oxides	RCONHOH	[128]
ZiO_2	RPO_4H_2	[129,130]
In_2O_3/SnO_2 (ITO)	RPO_2H_2	[131]

also been protected from corrosion by other types of thin films [89]. SAMs are also excellent insulators both in, and perpendicular to, the plane of the monolayer: a well-formed C_{18} SAM reduces the current passing across a gold electrode by roughly 10^6 (relative to an underivatized surface), with the residual current attributed to defects. SAMs have also been designed for studies of electrochemistry: ferrocene-terminated monolayers have been used in fundamental studies of the dependence of the rate of electron transport through SAMs on the thickness of the SAM [90]. Many other thiol ligands have been made in order to enhance the electrical properties of SAMs. These include thiol-terminated polythiophene and polyphenylene chains [91,92] and SAMs of polypyrrole-terminated alkanethiols [93,94].

2.2   *Other systems*

Table 11.3 summarizes the wide variety of ligands and substrates that may be used in the formation of self-assembled monolayers.

3   Patterns in two dimensions: microcontact printing (μCP)

3.1   *μCP of alkanethiolates on gold*

Some applications of SAMs in micro- and nanofabrication require that they can be patterned in the plane of the substrate. Techniques to pattern SAMs include microcontact printing (μCP) [85,132,133], micromachining [37,134,135], photolithography/liftoff [136], photochemical patterning [137,138], photo-oxidation [139–142], focused ion beam writing [143], electron-beam writing [87,144–154], STM writing [62,145, 149,155,156], and microwriting with a pen [157,158]. Of these methods, microcontact printing (μCP) is the most versatile, and will be discussed in detail for SAMs of alkanethiolates on gold.

Figure 11.3 shows a schematic of the μCP process. An elastomeric stamp with

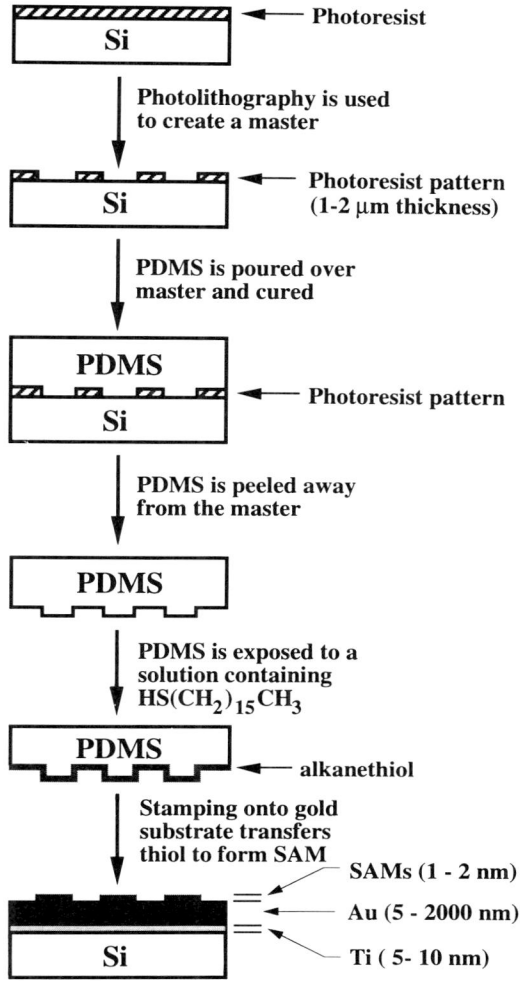

Figure 11.3 Schematic of procedure for microcontact printing.

three-dimensional relief is formed by casting polydimethyl siloxane elastomer (PDMS, Dow-Corning Corporation, Sylgard 184) on a master. Masters are typically prepared by photolithography. Other techniques for the fabrication of masters include electron beam lithography and micromachining; commercial objects with existing relief structure (diffraction gratings, for example) can also be used. Microcontact printing involves several steps.

1 The stamp is inked by applying a dilute ($\sim$1 mM) ethanolic solution of alkanethiol to the surface of the stamp with a cotton swab.

2 The stamp is blown dry for 10–20 s with a stream of dry nitrogen.

3 The stamp is placed on a gold-coated substrate (light pressure is sometimes applied to assure conformal contact between the stamp and the gold).

4 SAMs are formed in the regions of contact between the stamp and the surface.

5 The stamp is peeled from the surface.

The underivatized regions of gold can be selectively etched, after μCP, to produce structures of gold. These gold structures can be used subsequently as resists for etching the underlying substrates. Alternatively, SAMs of the same or a different alkanethiolate can be formed on the underivatized regions by a second μCP step, or by washing the surface with a dilute solution of an alkanethiol.

Several properties of the elastomeric PDMS stamp are critical for the high resolution and the degree of perfection of the patterned SAMs formed by μCP [2,159].

1 In most cases, SAMs are only formed on the surface where there is conformal contact of the stamp with the gold (for appropriate alkanethiols and contact times, see below). PDMS is deformable enough such that contact is achieved even on surfaces with significant roughness [160].

2 The elastomer is sufficiently rigid that the relief structure of the stamp retains its shape.

3 PDMS has a low interfacial free energy (22.1 dynes cm^{-1}) [161] so it does not adhere irreversibly to the surface of the gold.

4 PDMS swells in ethanol; this characteristic allows the stamp to absorb the alkanethiol ink.

The details of the process by which alkanethiol transfers from the stamp to the substrate are not completely understood. Because the stamps are elastomeric in nature, μCP offers some immediate advantages over traditional patterning techniques such as photolithography: patterns can be introduced on curved surfaces [162].

A number of alkanethiols can be used for μCP on gold surfaces. Hexadecanethiol (HDT) is particularly well suited for μCP because it is autophobic and has low volatility. Non-autophobic alkanethiols [163–166] such as $HS(CH_3)_{15}COOH$ can spread reactively after application to the gold in air and do not yield high resolution patterns smaller than 1–2 μm. To reduce reactive spreading, non-autophobic alkanethiols have been stamped under water [167]. Alkanethiols shorter than HDT can be too volatile for use in μCP: SAMs can form from alkanethiol vapour in regions not in contact with the stamp.

Figure 11.4(a) shows a scanning electron microscopy (SEM) image of a master that was prepared by photolithography. Figure 11.4(b) shows an SEM image of patterned SAMs formed with μCP using an elastomeric stamp that was cast from the master shown in Fig. 11.4(a). These images represent the level of complexity, perfection, and scale of features that can be produced routinely by μCP. Figure 11.4(c) shows a lateral force micrograph [160] of a test pattern of SAMs terminated by two different alkanethiols. SAMs terminated by CH_3 (dark stars in the image) were formed by μCP with HDT; SAMs terminated by COOH were deposited (in regions of the gold not derivatized by μCP) by washing the substrate with a dilute, ethanolic solution of $HS(CH_3)_{15}COOH$. The edge resolution between regions terminated by CH_3 and COOH was 50 nm [160]. The smallest pattern generated to date with μCP is 200 nm lines separated by 200 nm spaces [168].

Microcontact printing is an attractive technique because it is simple, inexpensive, and flexible. The cost of PDMS elastomer is negligible, many stamps can be cast from one master, and each stamp can be used hundreds of times. The process is inherently parallel in that large areas (many cm^2) can be patterned at once. Routine access to a clean room is not required, although occasional use of a clean room is convenient to make the photolithographic masters. Areas of future research in the development of μCP include the generation of smaller scale patterns (< 100 nm), minimization of the number of defects in the patterns, and the registration and alignment of patterns over large areas or in multiple printing steps.

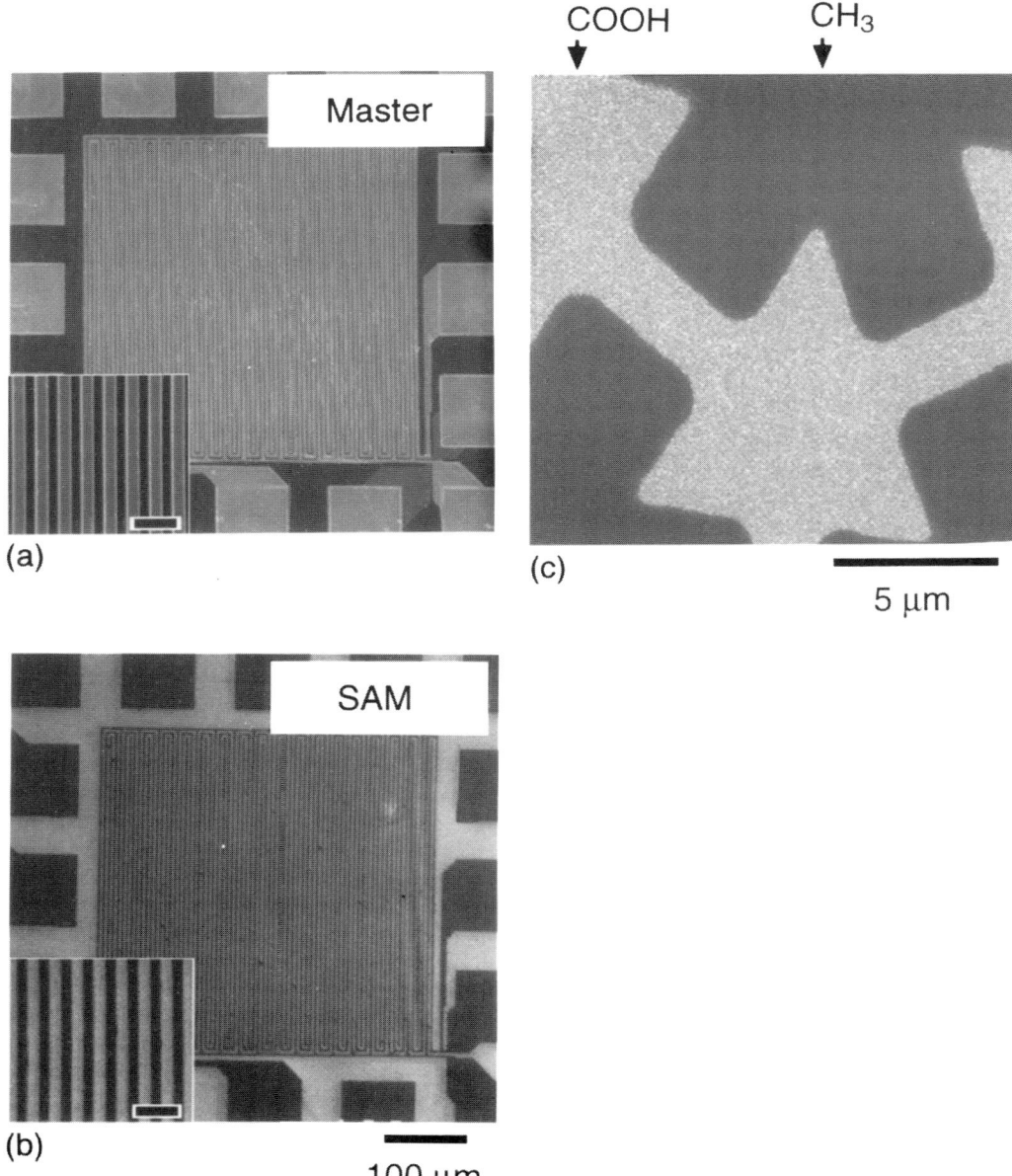

Figure 11.4 (a) Scanning electron micrograph (SEM) of a master with three-dimensional relief that was used to cast stamps for microcontact printing (μCP). The contrast in the image results from height differences between regions, and also from different materials on the surface in the relief structure. (b) SEM image of a patterned self-assembled monolayer (SAM) formed by microcontact printing. The dark regions of the surface are covered with SAMs, and the light regions are underivatized gold. The scale bars in the inserts correspond to 10 μm. (c) Lateral force micrograph of a patterned SAM. Relatively high frictional forces between the probe tip and the surface are detected in regions (light) covered with a carboxy-terminated SAM, and relatively low frictional forces are detected over regions (dark) covered with a methyl-terminated SAM.

Feature sizes of approximately 100 nm in patterned SAMs have been fabricated by adapting the μCP process. In one procedure, mechanical compression of the stamp produced features ca five times smaller than those originally cast from the master [169]. In another process, stamps with small feature sizes were cast from masters prepared by anisotropic etching of silicon [168]. Alternatively, controlled reactive spreading of 4 non-autophobic alkanethiols also produced features of dimension 100 nm [170].

3.2   *μCP of other materials*

Microcontact printing (μCP) has been used in systems other than alkanethiols on gold. Patterned SAMs have been formed with alkanethiols on copper [171], alkanethiols on silver [172], and with alkyltrichlorosilanes on Si/SiO$_2$ and glass [173]. The latter system is particularly useful because: (i) coinage metals (which are often incompatible materials for silicon microelectronic devices) are not necessary; (ii) SAMs formed from alkyltrichlorosilanes have higher thermal stability than SAMs of alkanethiolates on gold; (iii) patterned SAMs on glass are desirable for optical applications. SAMs formed from alkyltrichlorosilanes take longer to form and are not as highly ordered as SAMs of alkanethiolates on gold; they are also not as effective as chemical resists [173]. The edge resolution in μCP of alkyltrichlorosilanes on Si/SiO$_2$ or glass is currently ~200 nm [174] compared with the edge resolution of 20–50 nm [175] observed in the alkanethiolate/gold system.

Microcontact printing of liquids containing suspended palladium colloids [175] has also been demonstrated. Features with dimensions on the micron- and submicron-scale were prepared with edge resolution of approximately 100 nm. Colloids can be deposited on a variety of substrates including glass, Si/SiO$_2$, and polymers. Following deposition, the palladium can act as a catalyst for electroless deposition of metals. These experiments indicate that μCP may be a general technique that can be used in systems other than SAMs.

4   Fabrication of three-dimensional structures

4.1   *SAMs as ultrathin resists*

Problems in optical lithography concerning depth of focus, optical transparency, shadowing and undercutting for the fabrication of features of sub-50-nm dimensions result from the inability to produce sufficiently thin films of photoresist (<20 nm). The use of SAMs as resists potentially resolves these problems because the thickness of the monolayer is ~1–2 nm. SAMs are capable of protecting the underlying substrate from corrosion by wet-chemical etchants [176]. Relative capabilities are determined by the length of the alkyl chain and the nature of the terminal functional group [134].

Figure 11.5 shows a series of metal structures that were fabricated using patterned SAMs of HDT formed by μCP as resists. The underivatized regions were etched by a solution of ferricyanide, or with a basic, oxygenated cyanide etch [3,172,176]. The derivatized sections were barely affected by the etch in the time required to remove the underivatized gold (Figs 11.5a–c) or silver (Fig. 11.5d). Silver etches more quickly and

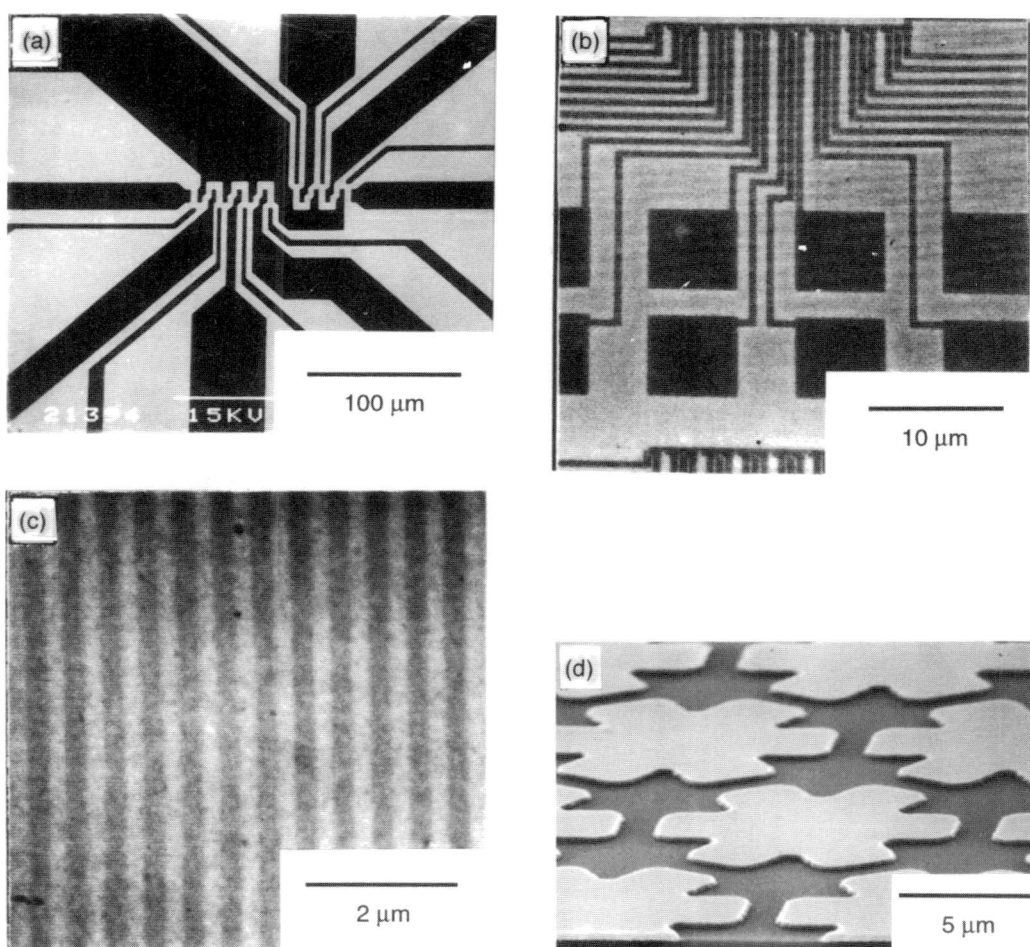

Figure 11.5 Metal structures [(a, b, c) gold, (d) silver] fabricated using patterned self-assembled monolayers [SAMs; formed by microcontact printing (μCP)] as resists. The underivatized metal is etched with a solution of ferric cyanide (a, d) or with a basic oxygenated cyanide solution (b, c). The 200 nm lines of gold separated by 200 nm wides spaces of Si/SiO$_2$ are the smallest structures fabricated to date by μCP.

with better resolution than gold [172]. This technique is capable of producing 200 nm features (Fig. 11.5c) [168] with edge resolution of ~20 nm.

Microcontact printing of patterned SAMs and selective etching was used to fabricate gold features on non-planar surfaces [162]; the capability to work on non-planar substrates is beyond the capabilities of routine photolithography. Figure 11.6 shows clearly resolved features of gold with dimensions of a few microns on curved surfaces with diameter 500 μm (Fig. 11.6a–b) and diameter 50 μm (Fig. 11.6c).

The gold (or silver) features can act as secondary resists for further substrate etching [85,168,174,177,178]. Complex silicon structures were fabricated with features as small as 100 nm. A typical fabrication process involves the following steps; (i) titanium (1.5–5 nm) and then gold (15–100 nm) are evaporated onto a Si [99] wafer; [ii] a patterned SAM is formed by μCP; (iii) underivatized regions of gold are etched with a basic ferricyanide solution to create a gold structure based on the pattern of the SAM;

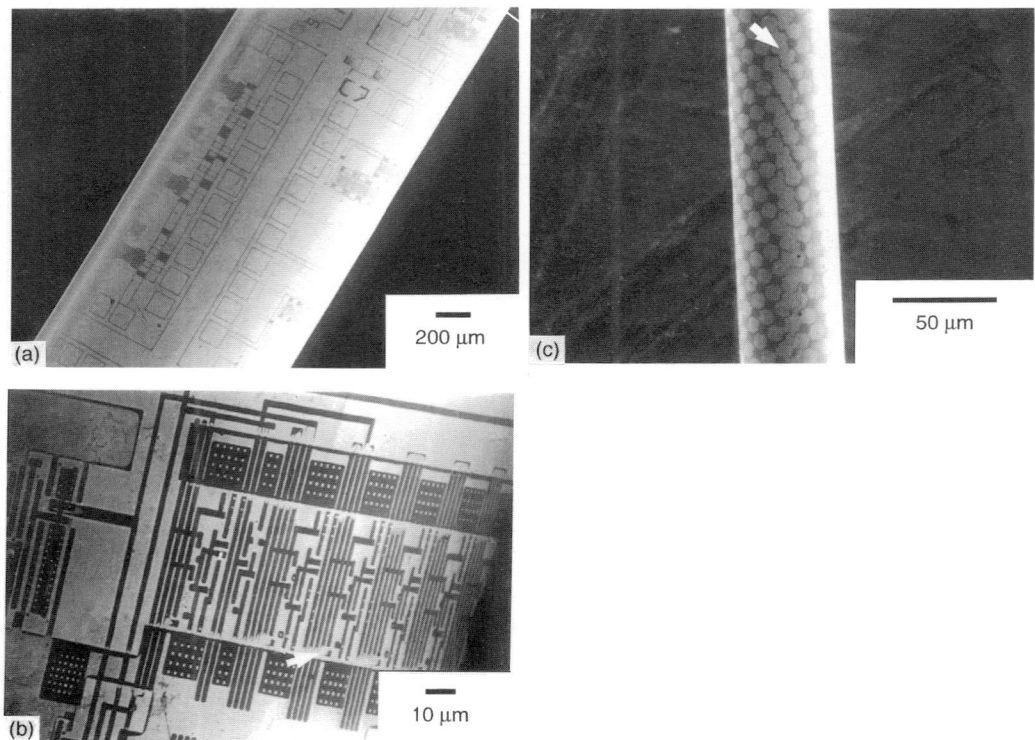

Figure 11.6 (a, b) Scanning electron micrograph (SEM) image of gold microstructures formed by microcontact printing (μCP) with HDT on a gold-coated capillary ($r = 500$ mm), followed by etching. A number of defects are apparent in (b) (white arrow); these defects originated in the master used to cast the stamp. (c) SEM image of patterned gold structures formed on a 50 mm diameter gold-coated glass fibre. There was a stripe (white arrow) where the capillary was printed twice when it rolled more than 360°. Light regions in these images correspond to gold; dark regions are Ti/TiO_2/glass where etching removed the gold. [Reprinted with permission from Jackman RJ, Wilbur JL, Whitesides GM. *Science* 1995; **269**: 664. Copyright (1995) American Association for the Advancement of Science].

(iv) the titanium layer and native SiO_2 layer are removed by dissolution in 1% HF; (v) the silicon wafer is etched anisotropically (15% solution by volume of 4 M KOH in *i*-PrOH, at 60°C for time periods proportional to the desired etch depth) using the gold pattern as a resist; and (vi) the remaining gold and titanium are removed by exposure to aqua regia (1 : 1 HNO_3 : HCl). Figure 11.7(a) shows a gold grid fabricated by μCP and subsequent etching of regions not covered with SAMs. Figure 11.7(b) shows the resulting pyramid-shaped 'pits' in the substrate after anisotropic etching of silicon. Figures 11.7(c)–(d) are SEM images of more complex silicon structures that were formed with this same process.

Microlithography using neutral metastable argon atoms [179] has been investigated as a technique for patterning and fabricating surface features with high resolution (<100 nm); beams of the neutral atoms have de Broglie wavelengths of order 0.01 nm, and can in principle be focused to a spot that is limited by the size of the atom. Conventional resists cannot be patterned with these beams because damage from the metastable argon atoms is restricted to a surface layer less than 0.5 nm thick. This depth, however, is sufficient to damage SAMs of HDT on gold such that an aqueous

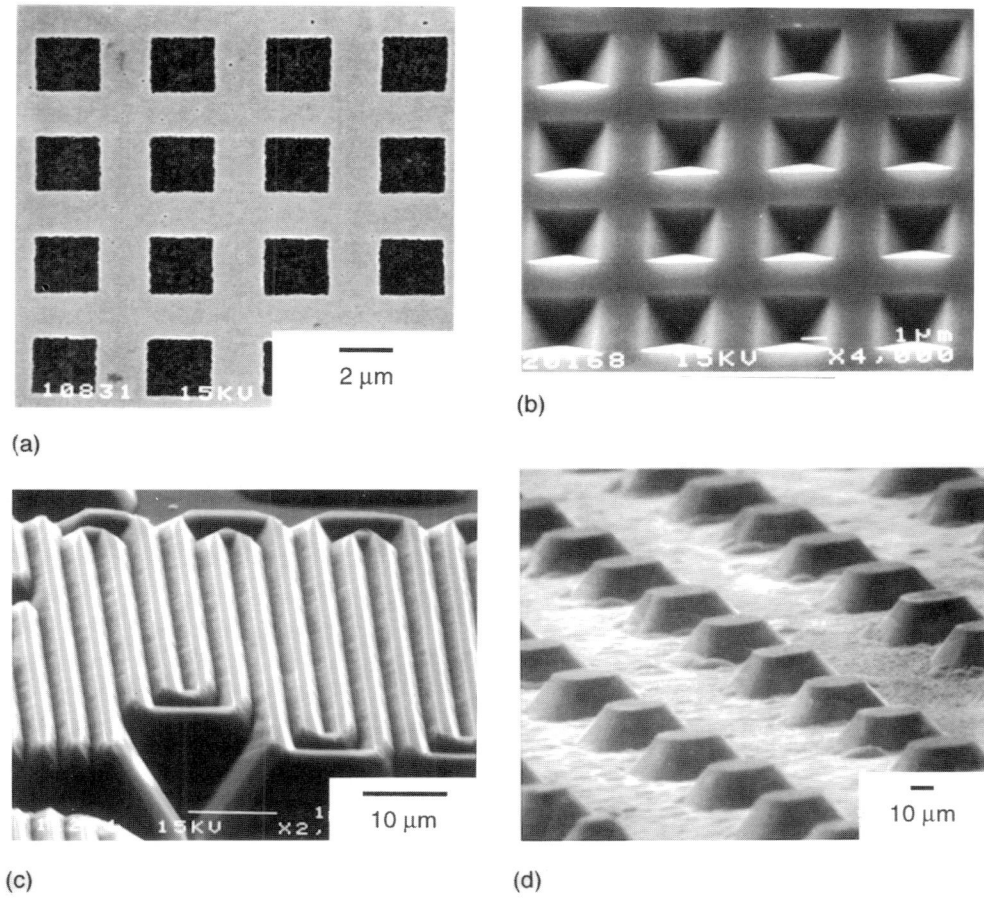

Figure 11.7 (a) Scanning electron micrograph (SEM) of a grid of gold fabricated by microcontact printing (μCP) and selective etching of gold. The dark regions are the underlying silicon. (b) Selective anisotropic etching of the silicon in (a) results in 'pits' that are inverted pyramids. (c–d) SEMs of etched silicon structures using gold patterns (formed by μCP and etching) as resists.

ferricyanide solution will etch the metal under the exposed regions. Figure 11.8 shows a schematic of the process, an SEM image of a copper TEM grid used as a mask (in contact with the surface of the substrate), and an SEM image of the gold structure that was fabricated using the TEM grid as a mask.

4.2   *Templated adsorption*

SAMs are essentially two-dimensional structures. The techniques for fabrication described above focused on the use of SAMs as resists; the pattern in the SAMs was transferred to the underlying substrate by selective etching to produce a structure with three-dimensional features. Materials can also be assembled on the surface of patterned SAMs to build three-dimensional structures.

Patterned SAMs that consist of regions of bare gold and regions covered by SAM have been used as templates for assembling polymeric structures [180] and metal structures [150,151] with electrochemistry, and metal structures with chemical vapour

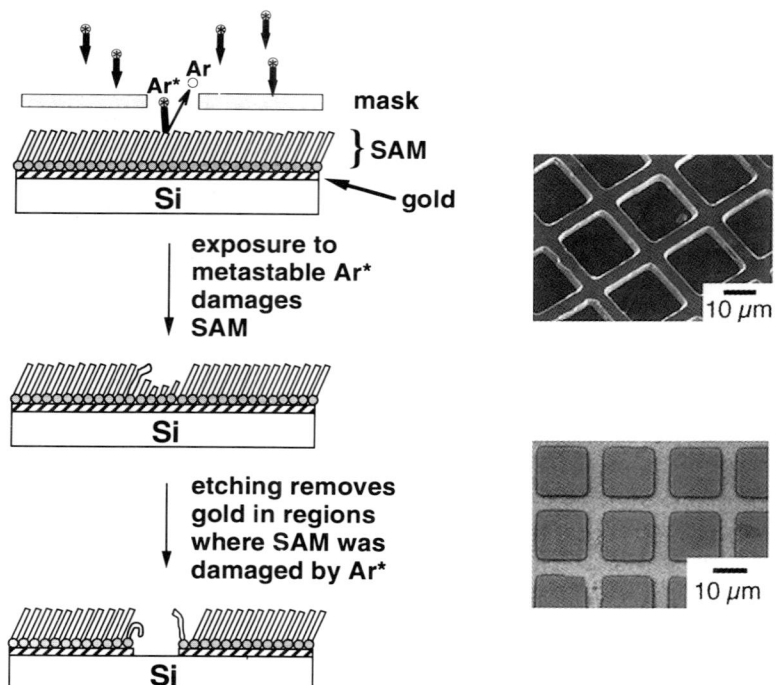

Figure 11.8 The scheme represents exposure of self-assembled monolayers of dodecanethiol (∼1.5 nm thick) on gold (20 nm with 1.5 nm of titanium) to a beam of metastable Ar, followed by wet-chemical etching. The SEM pictures are of the copper TEM grid (∼10 nm thick) used as a mask (in contact with the surface of the substrate), and the resulting fabricated gold structure.

deposition [156,181–183] and electroless deposition [85]. The insulating properties of SAMs inhibit deposition in these processes in selected regions. Figure 11.9(a) shows a schematic of a procedure to pattern a silicon structure with SAMs of $CH_3(CH_2)_{17}SiCl_3$ by μCP with a flat stamp on a surface with relief. Copper is deposited on regions not covered with SAMs by chemical vapour deposition (Fig. 11.9b) [183].

Other assembly processes take advantage of our ability to pattern SAMs with different organic functional groups. Different regions of the substrate can be tailored to have different surface free energies and different wettabilities. For example, a surface patterned with SAMs terminated by a methyl group and SAMs terminated by a carboxy group has regions that are extremely hydrophobic (CH_3) and hydrophilic (COOH), respectively. When such a surface is exposed to water vapour at low temperatures or at high humidity, water droplets preferentially condense on the hydrophilic regions (see Fig. 11.2). These droplets or 'condensation figures' have been used to image patterned SAMs [158], as sensors for monitoring humidity and temperature [85], and as optical diffraction gratings [184].

Figure 11.10 shows a schematic of a process to fabricate solid three-dimensional structures based on the different wetting properties of patterned SAMs. Hydrophilic regions are formed on a gold surface by microcontact printing with $HS(CH_2)_{15}COOH$. All other regions of the gold surface are made hydrophobic by immersing the sample into a dilute solution of HDT in ethanol. The sample is then slowly pulled from a prepolymer liquid (for example, UV-curable polyurethanes, Norland Optical Adhesives 60, 61 or 81) through an interface into air. The prepolymer liquid preferentially wets

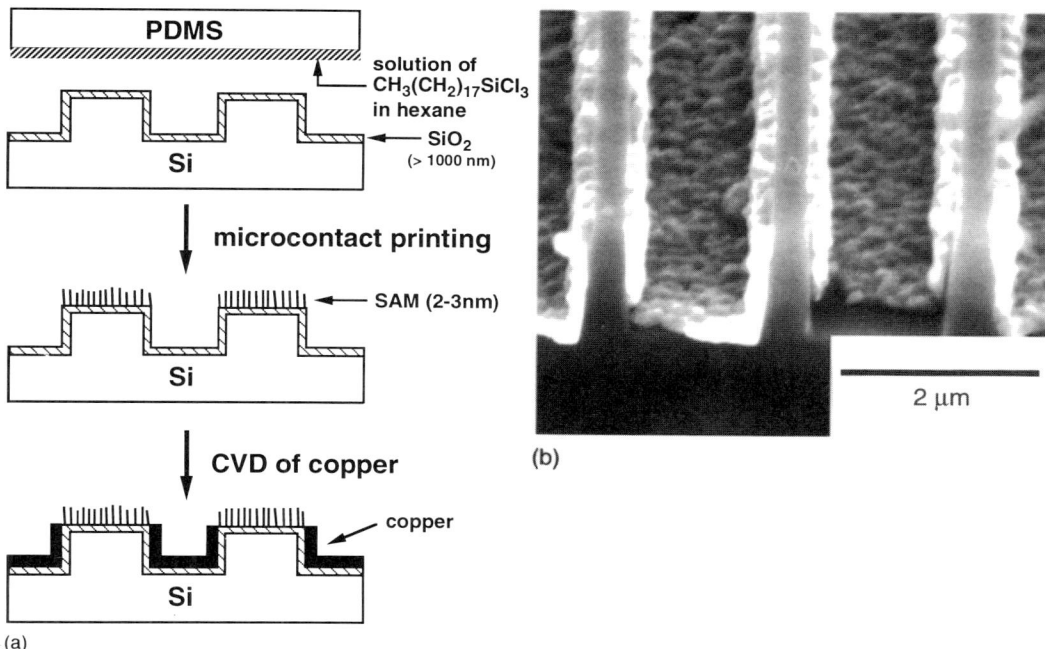

Figure 11.9 (a) Schematic outline of the procedure used for patterning alkylsiloxanes on a non-planar substrate. Nucleation and deposition of copper only occurs on those regions underivatized by the self-assembled monolayer in the following process of selective CVD. (b) Scanning electron micrograph of a microstructure of copper fabricated by CVD on a substrate whose surface was derivatized by a monolayer of octadecyltrichlorosilane. The specimen has been fractured in cross-section to reveal the copper morphology and coverage — here ~200 nm. Copper deposits only in recessed regions not covered by alkylsiloxane (the tops of the ridges).

the hydrophilic regions of the surface, assumes a shape to minimize its free energy, and does not adhere to the hydrophobic regions. The structure is solidified by curing the polymer in place by exposure to UV light. The smallest structures we have fabricated in this way have lateral dimensions of ~3 μm, and rise above the plane of the sample by a few microns. Optical waveguides with lengths as long as 2–3 cm have been fabricated using a similar process [185]. Figure 11.11(a) and (b) shows these waveguides in cross-section, and Fig. 11.11(c) shows the multimode output from a waveguide.

Hydrocarbon liquids can also be assembled on hydrophobic regions of patterned SAMs by pulling the substrate from the liquid through an interface into water [159,186] (instead of air). Again, the process of assembly is controlled by the minimization of free energies. In the case of assembling the polyurethane prepolymers on patterned SAMs, the final structures rise higher above the surface when fabricated on hydrophobic regions in water than when fabricated on hydrophilic regions in air because the surface tension of water is greater than the surface tension of air. Figure 11.11(d) shows an atomic force micrograph of an array of polymeric lenses [159] that were assembled on hydrophobic regions of a patterned SAM in water. The pattern in this case was circles (~10 μm diameter) of SAMs of HDT in a background of SAMs of $HS(CH_2)_{15}OH$.

As discussed earlier, palladium colloids can be patterned by μCP on a variety of surfaces [175]. Figure 11.12 shows a schematic of the procedure for patterning the colloids, and an SEM image depicting the scale and perfection of the copper structures

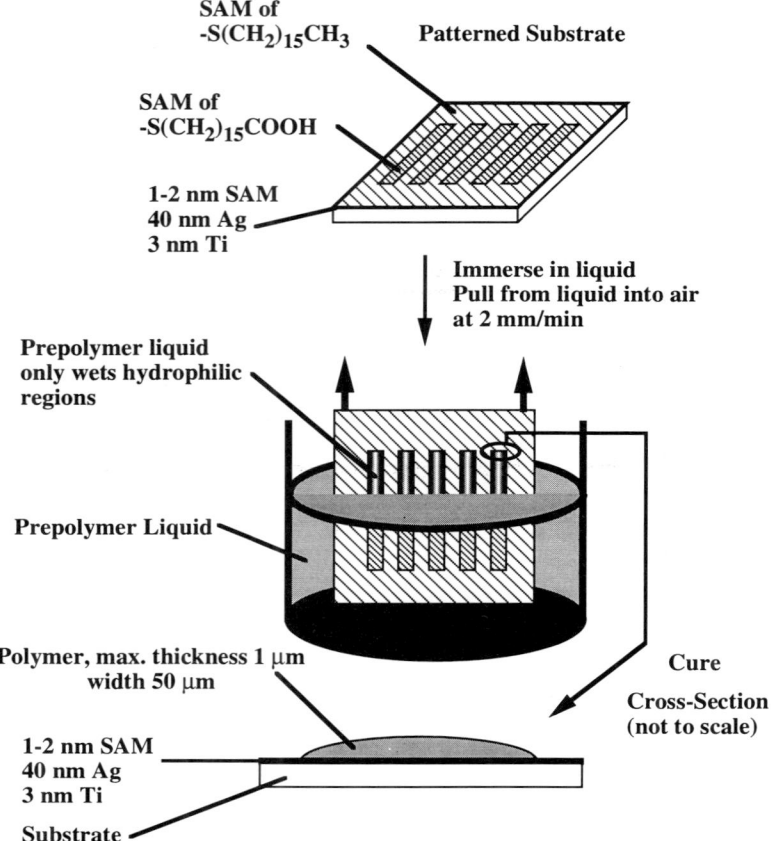

Figure 11.10 Schematic of the process of assembling liquids on patterned self-assembled monolayers (SAMs). The liquid preferentially wets the hydrophylic regions (carboxy-terminated SAM) of the surface, and de-wets from the hydrophobic regions (methyl-terminated SAM) of the surface. Liquid prepolymers are cured after assembly to fabricate solid three-dimensional structures.

that can be fabricated by using the colloids as templates for electroless deposition of copper.

4.3   *Micromoulding in capillaries (MIMIC)*

Patterning techniques are well established for semiconductors and metals, but are relatively underdeveloped for organic polymers, with the notable exception of the specialized polymers used in photolithography. Methods to pattern polymers based on the preferential wetting of patterned SAMs by liquid polymer precursors were discussed above. Micromoulding in capillaries (MIMIC) [187] is another technique to pattern polymers that uses many of the materials described above, but achieves patterning by completely different principles. Figure 11.13 shows a schematic of the MIMIC process. An elastomeric stamp is placed in conformal contact with a substrate to form a mould that consists of a network of channels. A low viscosity polymer precursor is placed in contact with openings to the network, and the channels fill by capillary action. After curing the polymer precursor, the elastomeric stamp is removed. The patterned

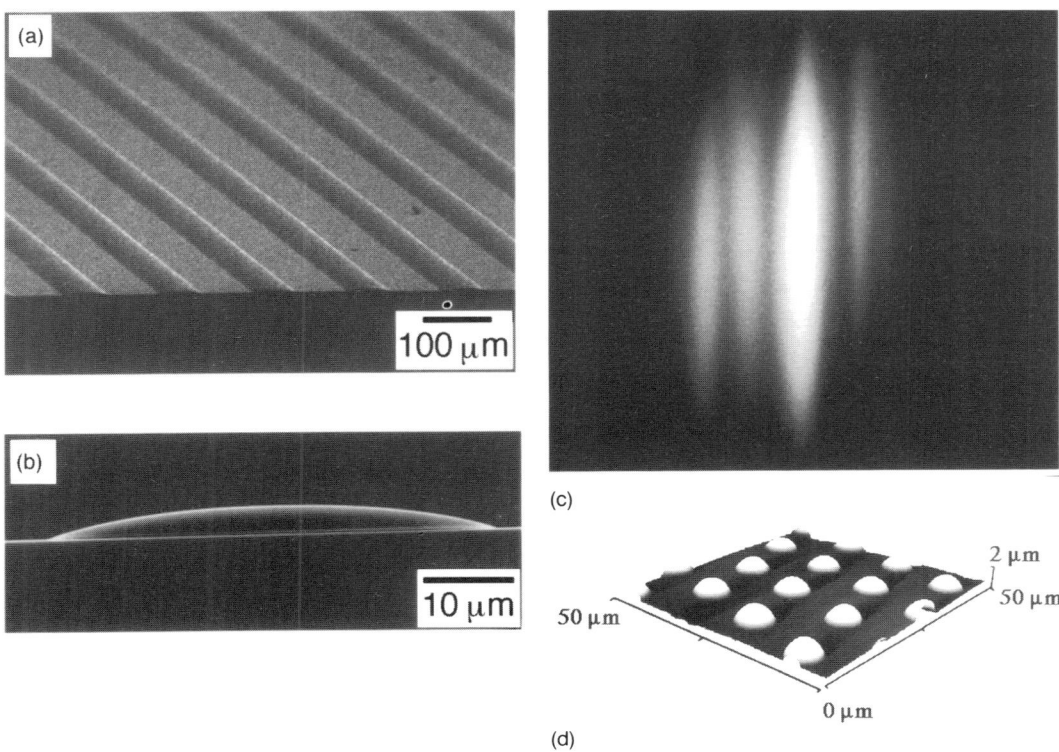

Figure 11.11 (a, b) Scanning electron micrograph (SEM) images of polymeric waveguides that were fabricated by assembling prepolymer liquids (Norland Optical Adhesive 60, polyurethane) on patterned self-assembled monolayers (SAMs) (as in Fig. 11.9). (c) Projection of the laser output from a waveguide 1 cm in length. (d) AFM image (constant force mode) of an array of polyurethane lenses that were assembled on a patterned SAM.

polymer does not adhere to the stamp, and remains on the support. Under certain conditions, the patterned polymer was subsequently detached from the substrate to produce a free standing film. Figure 11.14 shows examples of polymeric structures that were fabricated using MIMIC.

As a method to produce polymeric microstructures, MIMIC [187] holds several advantages over conventional photolithography. Photolithography requires the use of specialized polymers and photosensitizers; MIMIC is applicable to most polymers with low viscosity. Photolithography requires three steps: (i) spin coating to form a film; (ii) patterning (by exposure to light); and (iii) developing exposed (or unexposed) regions to fabricate microstructures. MIMIC forms and patterns polymeric microstructures in a single step. In photolithography, the polymeric microstructure formed in each exposure must be the same thickness; with MIMIC, patterning structures with multiple thicknesses is possible. Like the masks employed in conventional photolithography, the stamps used as templates for MIMIC can be reused.

4.4 *Micromachining of electrodes*

Gold electrodes were fabricated using a combination of micromachining and SAMs [134]. In one procedure, SAMs of HDT were formed over the entire surface of a

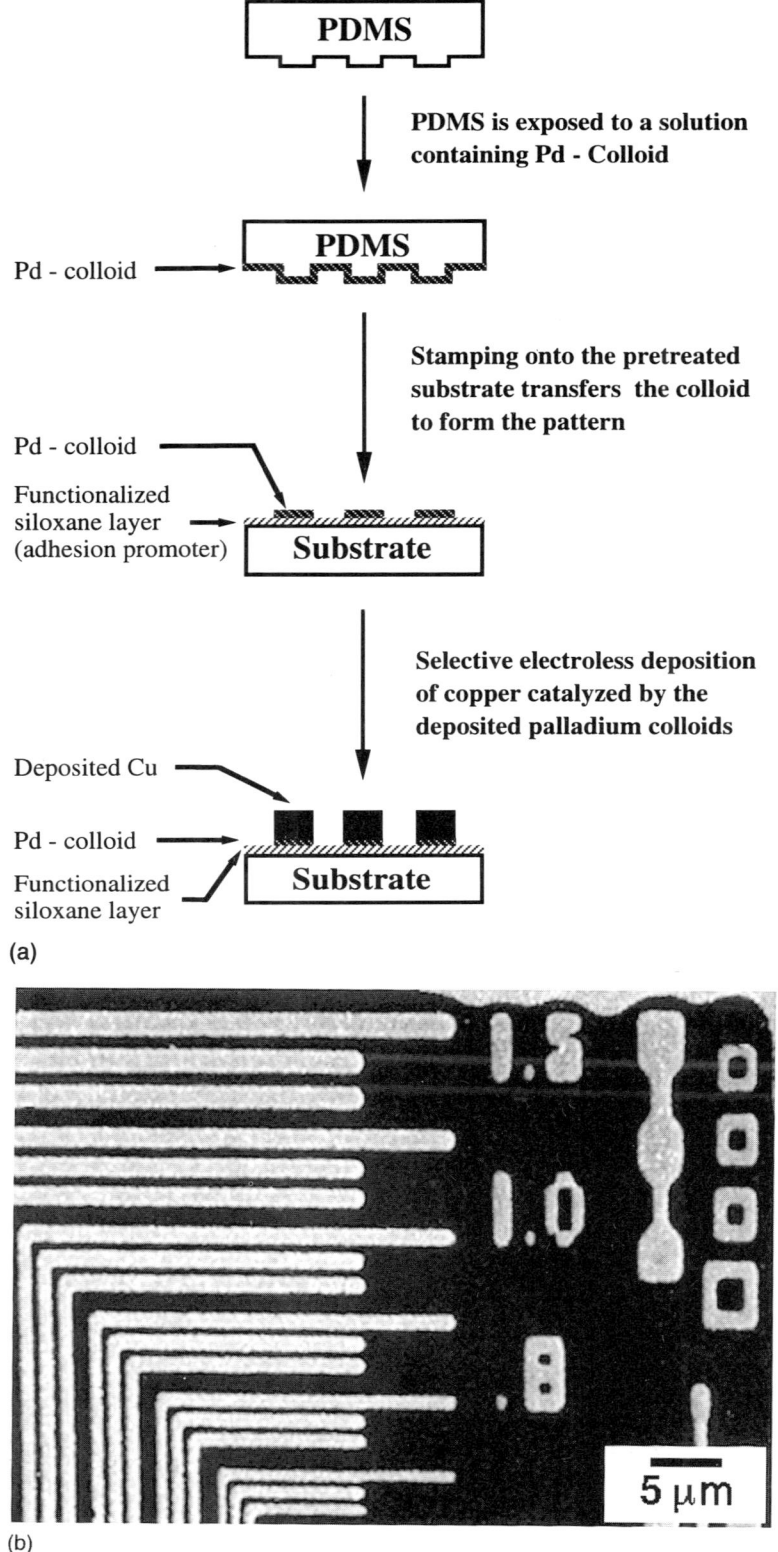

(a)

(b)

Figure 11.12 (a) Schematic illustration of the process to pattern surfaces with Pd colloids with microcontact printing. The Pd colloids are used to catalyse the electroless deposition of Cu on specific regions of the substrate. (b) Scanning electron micrograph (SEM) image of Cu structures that were fabricated with the process described above.

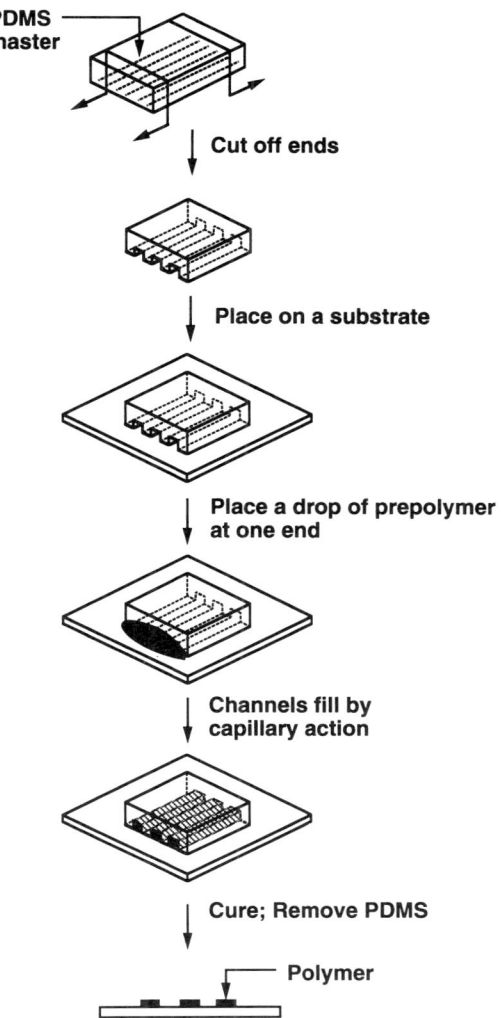

Figure 11.13 Schematic diagram of capillary micromoulding of a pattern of parallel, rectangular channels. It is not necessary for both ends of the channels to be open: even if one end is closed, the channels fill with the prepolymer. The trapped air seems to escape by diffusing through the polydimethyl siloxane (PDMS) master.

gold-coated substrate, and bare gold was exposed in patterns of lines on the substrate by scratching the surface using the tip of a surgical scalpel blade. Large area (>0.1 mm^2) arrays of band type electrodes (1 μm wide) were fabricated in this way; the exposed gold forms the electrochemically active surface of the electrode. In a second procedure, SAMs of HS(CH$_2$)$_2$OH were deposited on the gold surface, lines were micromachined, and the exposed gold was derivatized with HDT. The regions of gold covered with HS(CH$_2$)$_2$OH were subsequently etched with an aqueous solution of CN$^-$ saturated with O$_2$, and the gold wires that remained were micrometre wide, centimetre long, and 100 nm thick. Paired electrodes with electrochemically active areas as small as 100 nm × 1 μm were fabricated from these structures by machining a 1 μm gap into the supported gold wire (still covered with an electrically insulating SAM of HDT). The exposed gold on either side of the gap is electrochemically active.

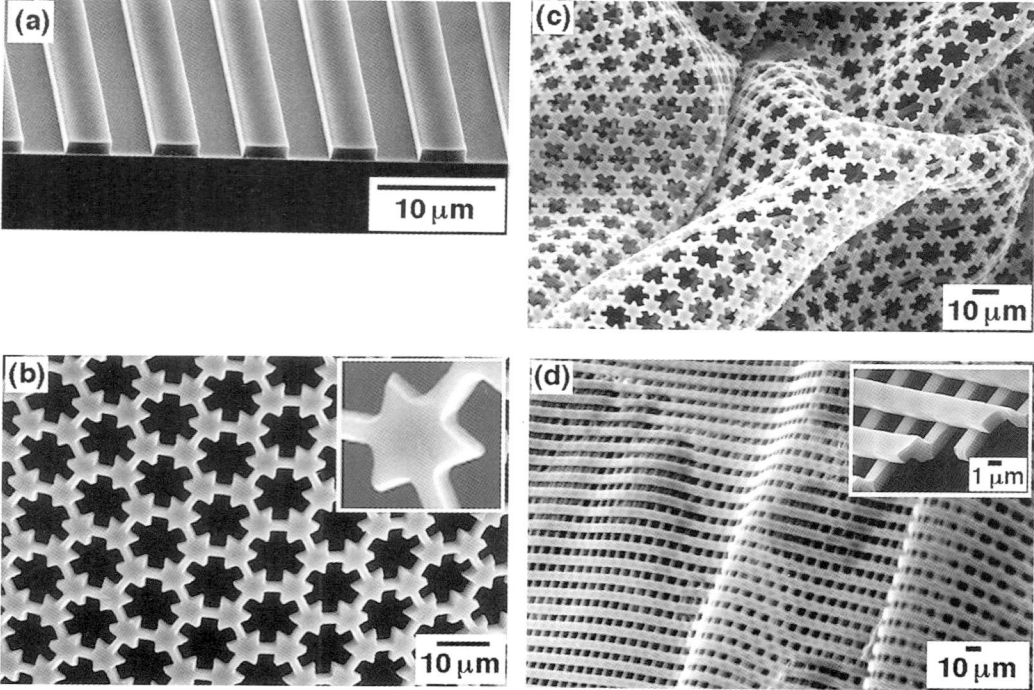

Figure 11.14 Patterns fabricated by micromoulding in capillaries (MIMIC). Patterns were sputtered with gold before imaging by scanning electron microscopy. The polymer used was photocured poly(methylmethacrylate). (a) An oblique image of a fractured sample showing rectangular slabs of polymer on a gold film supported on Si/SiO$_2$. (b) Image (captured at 30°) of more complex patterns on a silicon wafer. (c) A free-standing film: the same film as in (b) was released from the surface by dissolving the SiO$_2$ layer in NH$_4$F/HF. The folding in this region of the sample was accidental, but illustrates the flexibility of the film. (d) A free-standing polymeric structure fabricated in channels formed by conformal contact of two polydimethyl siloxane (PDMS) masters like those in (a), using an orientation in which the grooves of the two masters were perpendicular. The two layers of the channels formed one interconnected polymeric structure (inset). [Reprinted with permission from Kim E, Xia Y, Whitesides GM. *Nature* 1995; **376**: 581. Copyright (1995) Macmillan Magazines Limited.]

5   Conclusions and future directions

Molecular self-assembly is a useful strategy for micro- and nanofabrication, and potentially for other applications in molecular electronics. Equilibrium structures are formed that are at, or close to, thermodynamic minimum [6,7]. These structures tend to be self-healing and defect rejecting [5], and the size of the structures can be in the range of 1–100 nm, a range difficult to address with current fabrication techniques. In particular, systems that by definition have at least one dimension of molecular scale — self-assembled monolayers — have demonstrated applications in microelectronics and micro- and nanofabrication that include passivation of surfaces, use of SAMs for ultrathin resists and masks, directed deposition of materials on surfaces patterned with SAMs, and the use of surface functionality for sensors.

Several issues remain to be resolved, however, before SAMs find real applications in microelectronics. Alkanethiolates on gold are the most developed system of SAMs; the potential applications and techniques for fabrication that we described are based on this system. Unfortunately, gold is incompatible with silicon processing [27], and SAMs of alkanethiolates on gold are not sufficiently thermally stable for many applications. New systems of SAMs need to be developed that are more compatible with current processes and materials for the fabrication of microelectronic devices. At present, the density of defects in metal structures formed by chemical etching using patterned SAMs as resists is too high for applications in microelectronics, although it is acceptable for many optical systems. The formation and distribution of these defects must be understood to reduce their numbers. To take full advantage of SAMs as ultrathin resists or to use SAMs to fabricate quantum devices (lateral dimensions less than 50 nm [188]), techniques to pattern SAMs at sub-50-nm scales (in the plane of the monolayer) must be developed. SAMs are patterned routinely by microcontact printing [85,132,176] with features of micron dimensions, and these structures form the basis of fabrication techniques that are immediately suitable for applications in optics [159,185] and biotechnology [189–191], if not in microelectronics.

Acknowledgements. We thank Dr Emmanuel Delamarche and Dr Hans Biebuyck for providing the images in Fig. 11.1. The research was supported in part by the Office of Naval Research, ARPA and NSF (PHY 9312572). This work made use of MRSEC Shared Facilities supported by the National Science Foundation (DMR-9400396). J.L.W. gratefully acknowledges a postdoctoral fellowship from the NIH.

6   References

1 Moreau WM. *Semiconductor Lithography: Principles and Materials.* New York: Plenum Press, 1988.
2 Wilbur JL, Whitesides GM. In: Timp G, ed. *Nanotechnology.* Washington, DC: AIP, in press.
3 Xia Y, Zhao XM, Whitesides GM. *Microelectronic Engineering,* in press.
4 Whitesides GM, Mathias JP, Seto CT. *Science* 1991; **254**: 1312.
5 Lindsey JS. *New J Chem* 1991; **15**: 153.
6 McGrath KP, Kaplan DL. *Mater Res Soc Symp Proc* 1994; **330**: 61.
7 Kim E, Whitesides GM. *J Am Chem Soc,* submitted.
8 Varner JE, ed. New York: Alan R. Liss, 1988.
9 Kossovsky N, Millett D, Sponsler E, Hnatyszyn HJ. *Bio/Technology* 1993; **11**: 1534.
10 Creighton TE. *Proteins: Structure and Molecular Properties.* New York: Freeman, 1983.
11 Sanger W. *Principles of Nucleic Acid Structure.* New York: Springer-Verlag, 1986.
12 Bain CD, Whitesides GM. *Angew Chem Int Ed Engl* 1989; **28**: 506.
13 Whitesides GM, Laibinis PE. *Langmuir* 1990; **6**: 87.
14 Ulman A. *J Mater Educ* 1989; **11**: 205.
15 Ulman A. *An Introduction to Ultrathin Organic Films.* San Diego, CA: Academic Press, 1991.
16 Dubois LH, Nuzzo RG. *Annu Rev Phys Chem* 1992; **43**: 437.
17 Bard AJ, Abruna HD, Chidsey CED *et al. J Phys Chem* 1993; **97**: 7147.
18 Whitesides GM, Gorman CB. In: Hubbard AT, ed. *Handbook of Surface Imaging and Visualization.* Boca Raton: CRC Press, in press.
19 de Gennes P-G. *The Physics of Liquid Crystals,* 2nd edn. *International Series of Monographs on Physics,* Vol. 83. New York: Oxford University Press, 1993.
20 Simanek EE, Mathias JP, Seto CT *et al. Acc Chem Res* 1995; **28**: 37.
21 Archibald DD, Mann S. *Nature* 1993; **364**: 430.

22 Heywood B, Mann S. *Chem Mater* 1994; **6**: 311.

23 Noshay A, McGrath JE. *Block Copolymers: Overview and Critical Survey.* New York: Academic Press, 1977.

24 Bain CD, Evall J, Whitesides GM. *J Am Chem Soc* 1989; **111**: 7155.

25 Bain CD, Whitesides GM. *J Am Chem Soc* 1989; **111**: 7164.

26 Ulman A. *Adv Mater* 1991; **3**: 298.

27 Li J, Seidel TE, Mayer JW. *Mater Res Soc Bulletin* 1994; **XIX**: 15.

28 Li T, Weaver MJ. *J Am Chem Soc* 1984; **106**: 6107.

29 Porter MD, Bright TB, Allara DL, Chidsey CED. *J Am Chem Soc* 1987; **109**: 3559.

30 Bain CD, Whitesides GM. *Science* 1988; **240**: 62.

31 Bain CD, Whitesides GM. *J Am Chem Soc* 1988; **110**: 3665.

32 Bain CD, Troughton EB, Tai Y-T, Evall J, Whitesides GM, Nuzzo R. *J Am Chem Soc* 1989; **111**: 321.

33 Bain CD, Biebuyck HA, Whitesides GM. *Langmuir* 1989; **5**: 723.

34 Nuzzo RG, Allara DL. *J Am Chem Soc* 1983; **105**: 4481.

35 Troughton EB, Bain CD, Whitesides GM, Nuzzo RG, Allara DL, Porter MD. *Langmuir* 1988; **4**: 365.

36 Weisshaar DE, Lamp BD, Porter MD. *J Am Chem Soc* 1992; **114**: 5860.

37 Abbott NL, Folkers JP, Whitesides GM. *Science* 1992; **257**: 1380.

38 Leff DV, Ohara PC, Heath JR, Gelbart WM. *J Phys Chem* 1995; **99**: 7036.

39 Brust M, Walker M, Bethell D, Schiffrin DJ, Whyman R. *J Chem Soc Chem Commun* 1994; 801.

40 Sondag-Huethorst JAM, Schonenberger C, Fokkink LGJ. *J Phys Chem* 1994; **98**: 6826.

41 Goss CA, Charych DH, Majda M. *Anal Chem* 1991; **63**: 85.

42 Wasserman SR, Biebuyck H, Whitesides GM. *J Mater Res* 1989; **4**: 886.

43 Krysinski P, Chamberlin RV, II, Majda M. *Langmuir* 1994; **10**: 4286.

44 Fenter P, Eberthardt A, Eisenberger P. *Science* 1994; **266**: 1216.

45 Pemberton JE, Bryant MA, Joa SL, Garvey SD. *Proc Spie Int Soc Opt Eng* 1992.

46 Strong L, Whitesides GM. *Langmuir* 1988; **4**: 546.

47 Fenter P, Eisenberger P, Li J *et al. Langmuir* 1991; **7**: 2013.

48 Laibinis PE, Lewis NS. *Chemtracts: Inorg Chem* 1992; **4**: 49.

49 Camillone NI, Esienberger P, Leung TYB *et al. J Chem Phys* 1994; **101**: 11031.

50 Camillone NI, Chidsey CED, Liu GY, Scoles G. *J Chem Phys* 1993; **98**: 3503.

51 Chidsey CED, Liu GY, Scoles G, Wang J. *Langmuir* 1990; **6**: 1804.

52 Samant MG, Brown CA, Gordon JGI. *Langmuir* 1993; **9**: 1082.

53 Fenter P, Li J, Eisenberger P, Ramanarayanan TA, Liang KS. *Mater Res Soc Symp Proc* 1992.

54 Walczak MM, Chung C, Stole SM, Widrig CA, Porter MD. *J Am Chem Soc* 1991; **113**: 2370.

55 Laibinis PE, Whitesides GM, Allara DL, Tao YT, Parikh AN, Nuzzo RG. *J Am Chem Soc* 1991; **113**: 7152.

56 Nuzzo RG, Dubois LH, Allara DL. *J Am Chem Soc* 1990; **112**: 558.

57 Evans SD, Urankar E, Ulman A, Ferris N. *J Am Chem Soc* 1991; **113**: 4121.

58 Arndt T, Schupp H, Schrepp W. *Thin Solid Films* 1989.

59 Sun L, Kepley LJ, Crooks RM. *Langmuir* 1992; **8**: 2101.

60 Sun L, Crooks RM. *Langmuir* 1993; **9**: 1951.

61 Creager SE, Hockett LA, Rowe GK. *Langmuir* 1992; **8**: 854.

62 Kim YT, Bard AJ. *Langmuir* 1992; **8**: 1096.

63 Delamarche E, Michel B, Gerber CH *et al. Langmuir* 1994; **10**: 2869.

64 Delamarche E, Michel B, Kang H, Gerber CH. *Langmuir* 1994; **10**: 4103.

65 Schoenenberger C, Sondag HJAM, Jorritsma J, Fokkink LGJ. *Langmuir* 1994; **10**: 611.

66 Stranick SJ, Kamna MM, Krom KR, Parikh AN, Allara DL, Weiss PS. *J Vac Sci Technol, B* 1994; **12**: 20004.

67 Stranick SJ, Parikh AN, Tao YT, Allara DL, Weiss PS. *J Phys Chem* 1994; **98**: 7636.

68 Bucher JP, Santesson L, Kern K. *Langmuir* 1994; **10**: 979.

69 Gregory BW, Dluhy RA, Bottomley LA. *J Phys Chem* 1994; **98**: 1010.

70 Sondag HJAM, Schonenberger C, Fokkink LGJ. *J Phys Chem* 1994; **98**: 6826.

71 Wolf H, Ringsdorf H, Delamarche E *et al*. *J Phys Chem* 1995; **99**: 7102.

72 Folkers JP, Laibinis PE, Whitesides GM. *Langmuir* 1992; **8**: 1330.

73 Allara DL, Atre SV, Elliger CA, Snyder RG. *J Am Chem Soc* 1991; **113**: 1852.

74 Laibinis PE, Hickman JJ, Wrighton MS, Whitesides GM. *J Am Chem Soc* 1990; **112**: 570.

75 Chidsey CED, Loiacono DN. *Langmuir* 1990; **6**: 682.

76 Tidswell IM, Rabedeau TA, Pershan PS, Folkers JP, Baker MV, Whitesides GM. *Phys Rev B: Condens Matter* 1991; **44**: 10869.

77 Hautman J, Klein ML. *Mater Res Soc Symp Proc* 1992.

78 Kim JH, Cotton TM, Uphaus RA. *J Phys Chem* 1988; **92**: 5575.

79 Chidsey CED, Loiacono DN. *Langmuir* 1990; **6**: 709.

80 Uosaki K, Sato Y, Kita H. *Langmuir* 1991; **7**: 1510.

81 De LHC, Donohue JJ, Buttry DA. *Langmuir* 1991; **7**: 2196.

82 Hickman JJ, Ofer D, Zou C, Wrighton MS, Laibinis PE, Whitesides GM. *J Am Chem Soc* 1991; **113**: 1128.

83 Chidsey CED, Bertozzi CR, Putvinski TM, Mujsce AM, Thorp HH. *Chemtracts: Inorg Chem* 1991; **3**: 27.

84 Sabatani E, Cohen BJ, Bruening M, Rubinstein I. *Langmuir* 1993; **9**: 2974.

85 Kumar A, Biebuyck HA, Whitesides GM. *Langmuir* 1994; **10**: 1498.

86 Laibinis PE, Whitesides GM. *J Am Chem Soc* 1992; **114**: 9022.

87 Tiberio RC, Craighead HG, Lercel M, Lau T, Sheen CW, Allara DL. *Appl Phys Lett* 1993; **62**: 476.

88 Sheen CW, Shi JX, Maartensson J, Parikh AN, Allara DL. *J Am Chem Soc* 1992; **114**: 1514.

89 Stratmann M, Wolpers M, Losch R, Volmer M. *Bull Electrochem Soc* 1992; **8**: 52.

90 Chidsey CDD. *Science* 1991; **251**: 919.

91 Tour JM. *Adv Mat* 1994; **6**: 190.

92 Pearson DL, Schumm JS, Jones LI, Tour JM. *Polym Prep* 1994; **35**: 202.

93 Willicut RJ, McCarley RL. *Langmuir* 1995; **11**: 296.

94 Sayre CN, Collard DM. *Langmuir* 1995; **11**: 302.

95 Wasserman SR, Tao YT, Whitesides GM. *Langmuir* 1989; **5**: 1074.

96 Wasserman SR, Whitesides GM, Tidswell IM, Ocko BM, Pershan PS, Axe JD. *J Am Chem Soc* 1989; **111**: 5852.

97 Sagiv J. *J Am Chem Soc* 1980; **102**: 92.

98 Netzer L, Sagiv J. *J Am Chem Soc* 1983; **105**: 674.

99 Maoz R, Sagiv J. *J Colloid Interface Sci* 1984; **100**: 465.

100 Hoffmann H, Mayer U, Krischanitz A. *Langmuir* 1995; **11**: 1304.

101 Linford MR, Chidsey CED. *J Am Chem Soc* 1993; **115**: 12631.

102 Linford MR, Fenter P, Eisenberger PM, Chidsey CED. *J Am Chem Soc* 1995; **117**: 3145.

103 Laibinis PE, Fox MA, Folkers JP, Whitesides GM. *Langmuir* 1991; **7**: 3167.

104 Chang SC, Chao I, Tao YT. *J Am Chem Soc* 1994; **116**: 6792.

105 Li W, Virtanen JA, Penner RM. *J Phys Chem* 1994; **98**: 11751.

106 Nakagawa OS, Ashok S, Sheen CW, Maertensson J, Allara DL. *Jpn J Appl Phys* 1991; **30**: 3759.

107 Bain CD. *Adv Mater* 1992; **4**: 591.

108 Laibinis PE, Whitesides GM. *J Am Chem Soc* 1992; **114**: 1990.

109 Smith EL. *Report* 1992; 18.

110 Gu Y, Lin Z, Butera RA, Smentkowski VS, Waldeck DH. *Langmuir* 1995; **11**: 1849.

111 Uvdal K, Person I, Liedberg B. *Langmuir* 1995; **11**: 3145.

112 Chadwick JE, Myles DL, Garrell RL. *J Am Chem Soc* 1993; **115**: 10364.

113 Lee TR, Laibinis PE, Folkers JP, Whitesides GM. *Pure Appl Chem* 1991; **63**: 821.

114 Black AJ, Wooster TT, Geiger WE, Paddon-Row MN. *J Am Chem Soc* 1993; **115**: 7924.

115 Shimazu K, Sato Y, Yagi I, Uosaki K. *Bull Chem Soc Jpn* 1994; **67**: 863.

116 Hines MA, Todd JA, Guyot SP. *Langmuir* 1995; **11**: 493.

117 Bigelow WC, Pickett DL, Zisman WA. *J Colloid Interface Sci* 1946; **1**: 513.

118 Timmons CO, Zisman WA. *J Phys Chem* 1965; **69**: 984.

119 Golden WG, Snyder CD, Smith B. *J Phys Chem* 1982; **86**: 4675.

120 Allara DL, Nuzzo RG. *Langmuir* 1985; **1**: 45.

121 Schlotter NE, Porter MD, Bright TB, Allara DL. *Chem Phys Lett* 1986; **132**: 93.

122 Laibinis PE, Hickman JJ, Wrighton MS, Whitesides GM. *Science* 1989; **245**: 845.

123 Chen SH, Frank CW. *Langmuir* 1989; **5**: 978.

124 Chau L-K, Porter MD. *Chem Phys Lett* 1990; **167**: 198.

125 Tao YT, Lee MT, Chang SC. *J Am Chem Soc* 1993; **115**: 9547.

126 Smith E, Porter MD. *J Phys Chem* 1993; **97**: 8032.

127 Ahn SJ, Mirzakhojaev DA, Son DH, Kim K. *Bull Korean Chem Soc* 1994; **15**: 369.

128 Folkers JP, Gorman CB, Laibinis PE, Buchholz S, Whitesides GM. *Langmuir* 1995; **11**: 813.

129 Cao G, Hong H-G, Mallouk TE. *Acc Chem Res* 1992; **25**: 420.

130 Katz HE. *Chem Mater* 1994; **6**: 2227.

131 Gardner TJ, Frisbie CD, Wrighton MS. *J Am Chem Soc* 1995; **117**: 6927.

132 Kumar A, Whitesides GM. *Appl Phys Lett* 1993; **63**: 2002.

133 Wilbur JL, Kumar A, Biebuyck HA, Kim E, Whitesides GM. *Nanotechnology* 1995; in press.

134 Abbott NL, Rolison DR, Whitesides GM. *Langmuir* 1994; **10**: 2672.

135 Lopéz GP, Biebuyck HA, Harter R, Kumar A, Whitesides GM. *J Am Chem Soc* 1993; **115**: 10774.

136 Kleinfeld D, Kahler KH, Hockberger PE. *J Neurosci* 1988; **8**: 4098.

137 Rozsnyai LF, Wrighton MS. *J Am Chem Soc* 1994; **116**: 5993.

138 Wollman EW, Kang D, Frisbie CD, Lorkovic IM, Wrighton MS. *J Am Chem Soc* 1994; **116**: 4395.

139 Calvert JM, Georger JH, Peckerar MC, Pehrsson PE, Schnur JM, Schoen PE. *Thin Solid Films* 1992; **211**: 359.

140 Tarlov MJ, Burgess DR, Gillen G. *J Am Chem Soc* 1993; **115**: 5305.

141 Dressick WJ, Calvert JM. *Jpn J Appl Phys* 1993; **1**: 5829.

142 Huang J, Dahlgren DA, Hemminger JC. *Langmuir* 1994; **10**: 626.

143 Gillen G, Wight S, Bennett J, Tarlov MJ. *Appl Phys Lett* 1994; **65**: 534.

144 Sondag-Huethorst JAM, van Helleputte HRJ, Fokkink LGJ. *Appl Phys Lett* 1994; **64**: 285.

145 Lercel MJ, Redinbo GF, Craighead HG, Sheen CW, Allara DL. *Appl Phys Lett* 1994; **65**: 974.

146 Marrian CRK, Perkins FK, Brandow SL, Koloski TS, Dobisz EA, Calvert JM. *Appl Phys Lett* 1994; **64**: 390.

147 Rieke PC, Tarasevich BJ, Wood LL *et al. Langmuir* 1994; **10**: 619.

148 Mino N, Ozaki S, Ogawa K, Hatada M. *Thin Solid Films* 1994; **243**: 374.

149 Perkins FK, Dobisz EA, Brandow SL *et al. J Vac Sci Technol B* 1994; **12**: 3725.

150 Lercel MJ, Redinbo GF, Pardo FD *et al. J Vac Sci Technol B* 1994; **12**: 3663.

151 Sondag-Huethorst JAM, van Helleputte HRJ, Fokkink LGJ. *Appl Phys Lett* 1994; **64**: 285.

152 Lercel MJ, Redinbo GF, Rooks M *et al. Microelectr Eng* 1995; **27**: 43.

153 Rieke PC, Baer DR, Fryxell GE, Engelhard MH, Porter MS. *J Vac Sci Technol A* 1993; **11**: 2292.

154 Baer DR, Engelhard MH, Schulte DW, Guenther DE, Wang L-Q, Rieke PC. *J Vac Sci Technol A* 1994; **12**: 2478.

155 Ross CB, Sun Li, Crooks RM. *Langmuir* 1993; **9**: 632.

156 Schoer JK, Ross CB, Crooks RM, Corbitt TS, Hampden SMJ. *Langmuir* 1994; **10**: 615.

157 Lopéz GP, Biebuyck HA, Whitesides GM. *Langmuir* 1993; **9**: 1513.

158 Lopéz GP, Biebuyck HA, Frisbie CD, Whitesides GM. *Science* 1993; **260**: 647.

159 Biebuyck HA, Whitesides GM. *Langmuir* 1994; **10**: 2790.

160 Wilbur JL, Biebuyck HA, MacDonald JC, Whitesides GM. *Langmuir* 1995; **11**: 825.

161 Chaudhury MK, Whitesides GM. *Langmuir* 1991; **7**: 1013.

162 Jackman RJ, Wilbur JL, Whitesides GM. *Science* 1995; **269**: 664.

163 Bain CD, Whitesides GM. *Langmuir* 1989; **5**: 1370.

164 Hare EF, Zisman WA. *J Phys Chem* 1955; **59**: 335.

165 de Gennes P-G. *Rev Mod Phys* 1985; **57**: 827.

166 Holmes-Farley SR, Reamey RH, McCarthy TJ, Deutch J, Whitesides GM. *Langmuir* 1985; **1**: 725.

167 Biebuyck HA, Whitesides GM. *Langmuir* 1994; **10**: 4581.

168 Wilbur JL, Kim E, Xia Y, Whitesides GM. *Adv Mater*, in press.

169 Xia Y, Whitesides GM. *Adv Mater* 1995; **7**: 471.

170 Xia Y, Whitesides GM. *J Am Chem Soc* 1995; **117**: 3274.

171 Xia Y, Kim E, Mrksich M. *Chem Mater*, submitted.

172 Xia Y, Kim E, Whitesides GM. *J Electrochem Soc*, submitted.

173 Xia Y, Mrksich M, Kim E, Whitesides GM. *Langmuir*, in press.

174 Wilbur JL, Kumar A, Kim E, Whitesides GM. *Adv Mater*, 1994; **7–8**: 600.

175 Hidber PC, Helbig W, Kim E, Whitesides GM. *Langmuir*, submitted.

176 Kumar A, Biebuyck HA, Abbott NL, Whitesides GM. *J Am Chem Soc* 1992; **114**: 9188.

177 Kim E, Kumar A, Whitesides GM. *J Electrochem Soc* 1995; **142**: 628.

178 Whidden TK, Ferry DK, Kozicki MN *et al. Nanotechnology* 1994; in press.

179 Berggren KK, Bard A, Wilbur JL *et al. Science* 1995; **269**: 1255.

180 Gorman CB, Biebuyck HA, Whitesides GM. *Chem Mater* 1995; **7**: 526.

181 Potochnik SJ, Hsu DSY, Calvert JM, Pehrsson PE. *Mater Res Soc Symp Proc: Advanced Metallization for Devices and Circuits: Science, Technology and Manufacturability* 1994; **337**: 429.

182 Potochnik SJ, Pehrsson PE, Hsu DSY, Calvert JM. *Langmuir* 1995; **11**: 1841.

183 Jeon NL, Nuzzo RG, Xia Y, Mrksich M, Whitesides GM. *Langmuir,* in press.

184 Kumar A, Whitesides GM. *Science* 1994; **263**: 60.

185 Kim E, Whitesides GM, Lee LK, Smith SP, Prentiss M. *Adv Mater* 1995; in press.

186 Gorman CB, Biebuyck HA, Whitesides GM. *Chem Mater* 1995; **7**: 252.

187 Kim E, Xia Y, Whitesides GM. *Nature* 1995; **376**: 581.

188 Reiss H. *J Chem Phys* 1951; **19**: 482.

189 Prime KL, Whitesides GM. *Science* 1991; **252**: 1164.

190 Mrksich M, Whitesides GM. *TIBTECH* 1995; **13**: 228.

191 Singhvi R, Kumar A, Lopez GP *et al. Science* 1994; **264**: 696.

12 The Design of Starburst Dendrimer Electron Transfer Systems

J.N. ONUCHIC,* S.M. RISSER,† S.S. SKOURTIS‡ and D.N. BERATAN‡

*Department of Physics, University of California, San Diego, La Jolla, CA 92093, USA,

†Department of Physics, Texas A 8 M-Commerce, TX 75429, USA,

‡Department of Chemistry, University of Pittsburgh, Pittsburgh, PA 15260, USA

1 Introduction

Starburst dendrimers are macromolecules built of repeat units that are each linked to three or more nearest neighbours, rather than two as in conventional organic polymers (Fig. 12.1). The chemistry of the repeat unit, the number of connections to neighbours, and the number of generations synthesized around a central core allow remarkable control of the size (from tens to hundreds of Ångstroms), shape, flexibility, surface chemistry, and solubility. The mechanical, electronic and optical properties of dendrimer-based materials are expected to be diverse and adjustable [1–7]. Applications in areas such as drug delivery, photocatalysis, and structural materials are being considered [1–7]. These materials also provide an opportunity to enhance our understanding of macromolecular electronic structure and reactivity. Starburst dendrimers allow for systematic investigations of functions such as molecular recognition, information storage, and electron transfer because they provide such a high degree of control of molecular structure and surface.

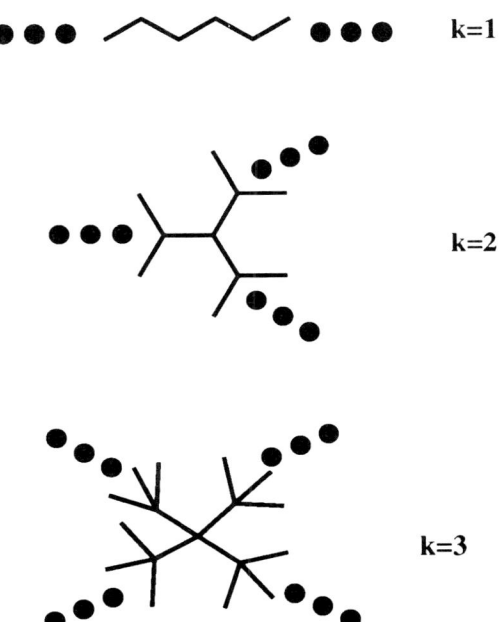

Figure 12.1 The connectivity of a linear polymer ($k = 1$) and dendrimers (with $k = 2$ and $k = 3$). The vertices may be occupied by numerous chemical groups (substituted benzene rings, amides, etc.). These chemical groups may be rigid or flexible.

We will describe the unusual electronic structure and electronic coupling mediation properties expected of starburst dendrimers (compared with their linear polymer counterparts) [8]. These properties are largely a consequence of the special molecular connectivity in starburst dendrimers. Macromolecules, from conducting polymers to the photosynthetic reaction centre proteins, perform their electron transport functions in distinct ways determined by their own unique electronic structure. In addition to identifying the novel electron mediation properties of starburst dendrimers, we show that the introduction of disorder to the molecules influences the electronic properties of starburst dendrimers in a rather different manner from that in the analogous linear molecules. The electronic properties of these new materials, and the ability to tune the properties upon the introduction of disorder, may allow new applications in synthetic charge separation systems, optoelectronics, and ultra-small electronics.

2   Bridge-mediated electron transfer

The rate of electron transfer between distant sites bound to a macromolecule is a sensitive probe of the electronic interactions within that molecule. Electron transfer (ET) rates (k_{ET}) between distant weakly interacting sites are proportional to the square of the interaction between the sites:

$$k_{ET} = \frac{2\pi}{\hbar} |T_{DA}|^2 (FC). \tag{12.1}$$

FC is a nuclear factor associated with the activation barrier to the reaction [9,10] and T_{DA} is the coupling between the donor and acceptor sites mediated by the bridge. The electronic coupling between donor and acceptor in an orthogonal basis is

$$T_{DA} = \sum_{ij} V_{Di} G_{ij} V_{jA}, \tag{12.2}$$

where the coupling elements between the donor (acceptor) and bridge sites $i(j)$ are $V_{Di(Aj)}$. G is the Green's function associated with the bridge Hamiltonian H:

$$\hat{G} = (E - \hat{H})^{-1}, \tag{12.3}$$

and the basis set is assumed orthonormal. When one bridge orbital couples to the donor and a separate bridge orbital couples to the acceptor, the electronic interaction between donor and acceptor mediated by the bridge is directly proportional to the Green's function element connecting the two bridge sites. There is great interest in developing effective Hamiltonians, H, that are appropriate for the electron transfer problem [11–20].

3   Through-bond and through-space coupling interactions control macromolecule electron transfer

The bridge structure dependence of electron transfer reactions is determined by the energetics of the donor and acceptor and the strength of their communication via the chemical bridge. In macromolecules, the bridge is a dense composite of atoms participating in bonded and non-bonded interactions. The qualitative nature of the bridge-mediated electronic coupling is set by the rate of decrease of the non-bonded (or 'through-space') interactions compared with the rate of decay of bond-mediated interactions.

In electron transfer proteins such as cytochrome c, chains of covalent bonds dominate the electronic coupling. Occasional 'through-space' couplings are essential as well. Coupling decay across through-space contacts is larger compared to decay across covalent bonds of similar length. However, these non-bonded contacts can provide essential short-cuts and significantly enhance coupling. Our 'pathway' strategy for protein analysis identifies the key bonded and through-space interactions between protein donor and acceptor sites that give rise to the donor–acceptor coupling. There is much qualitative similarity between the coupling mechanism in proteins and starburst dendrimers (particularly those connected with amide linkages). The connectivity of dendrimers, however, is much simpler than in proteins. Because of the ring structure in some amino acid side chains and the extensive protein hydrogen-bonded structure, a large number of loops enter the connectivity of proteins. While loops may exist in dendrimer repeat units, extensive loops beyond a single repeat unit do not commonly occur (we will see for this reason that the possibility of constructing hydrogen-bonded dendrimers is of considerable interest). Proteins fold into 'unique' three-dimensional structures, determined by primary sequence, while most dendrimers exist in a rather broad distribution of accessible geometries. The goal of this work is to map out the structural dependence of dendrimer electronic coupling interactions, particularly with respect to their very special branched structure. Our preliminary calculations show just how important this connectivity will be with respect to electron transport processes.

4 Electronic coupling in macromolecules

Before describing how tunnelling matrix elements are calculated for dendrimers, we summarize the pathway method that we developed for similar calculations in proteins. Until the middle 1980s, calculations of matrix elements in proteins were performed without taking into account the details of the protein medium [10]. The protein was treated as an isotropic linear barrier and therefore the tunnelling matrix element was assumed to decay exponentially with the separation distance between donor and acceptor.

Since this approximation is an oversimplification, in 1987 we decided to develop the simplest possible model that would take into account the protein electronic structure [14]. Pathways was developed based on the fact that the electronic wave function for the redox sites decays much more softly through covalent bonds than through empty space. In addition, the decay through the lone pair of a hydrogen bond is similar to that across a covalent bond. For this reason, as the electron tunnels between the donor and acceptor, its route is dominated by covalent and hydrogen bonds. Through-space jumps will be used only when the covalent alternative is extremely long. In the limit that a single pathway tube dominates, the matrix element and the tunnelling matrix element can be written as the following product

$$T_{DA} \propto \prod_{i_C} \varepsilon_C \prod_{i_{HB}} \varepsilon_{HB} \prod_{i_{TS}} \varepsilon_{TS}, \tag{12.4}$$

where ε_C is the wave function decay through a covalent bond, ε_{HB} through a hydrogen bond and ε_{TS} through space [11–15,21–22].

The fact that this tunnelling matrix element can be written as a product of individual decays permits extensive analysis of different proteins and an easy understanding of the rate mechanism. More detailed quantum calculations [15,21,22] have shown that as long as the matrix element is dominated by a single pathway tube, i.e. a pathway of orbitals plus additional ones appended to the core, this form for T_{DA} is valid. Interference arising from scattering in the core pathway and the appended orbitals exists, but these effects can be renormalized and the final result is a product of effective decays through the orbitals of the core pathway. This picture breaks down, however, if several tubes are needed to calculate the matrix element properly. Interference among these tubes is then important, and a more sophisticated analysis is needed to describe the tunnelling mechanism [22].

The single pathway picture, however, has successfully explained electron transfer in numerous experimental systems [23,24]. One of the best examples is the chemically modified proteins of Gray and co-workers. In cytochrome *c*, electron transfer rates were measured from the Fe to different Ru complexes bound to histidines at the protein surface [25]. The experimental rates show that the simple model with an exponential decay with donor–acceptor separation is inadequate. In one of the cases, where a long through-space jump exists in the main pathway, the tunnelling matrix element is much smaller than would be otherwise expected for that donor–acceptor separation. The pathways analysis predicted this effect even before the experiments were performed. This single pathway tube description fully explains the full collection of rates in these modified cytochromes. More recent experiments and calculations in modified azurin provide examples where multiple tube effects appear [22,26]. Even though the tunnelling matrix element is not a product of decay factors here, the pathway description for individual tubes still is the appropriate way to describe the protein mediation.

As described above, dendrimers are similar to proteins, but have a much simpler three-dimensional structure. For this reason, the formalism we have developed for proteins provides a natural framework to analyse this new problem. Also, due to the simplicity of dendrimers, provides a natural system in which to develop molecular electronics applications.

5 Electronic coupling in idealized dendrimers

In a bridge with translational symmetry (dendrimers *do not* possess translational symmetry as we discuss below), the wavefunction and the bridge-mediated coupling (T_{DA}) change by an energy-dependent factor ε per repeat unit [11–15,21,22,27–32]:

$$T_{DA} \propto \varepsilon^N, \tag{12.5}$$

where $|\varepsilon| \leqslant 1$ (localized donor and acceptor states) and N is the number of repeat units between donor and acceptor. This is a simple consequence of Bloch's theorem. The relation between ε and E, the electronic energy [11], is:

$$\varepsilon + 1/\varepsilon = f(E). \tag{12.6}$$

Bloch's theorem does not apply to starburst dendrimers directly because of the lack of translational symmetry. Yet, in analogy with the coefficients that would appear in the corresponding 1D material [27], one can assign wavefunction coefficients. The generalized relation between decay and electronic energy in starburst dendrimer structures is then [8]:

$$k\varepsilon + 1/\varepsilon = f(E) \tag{12.7}$$

for $k+1$ chains branching from each node. For one orbital (ϕ) on each site with interaction β between neighbouring sites in a tight-binding representation ($\beta = \langle \phi_i | H | \phi_{i \pm 1} \rangle$), $f(E) = E/\beta$.

The decay factor $\varepsilon(E)$ for the dendrimer connects formally to the familiar 1D Bloch relation (Eqn 12.7, $k = 1$). Dividing the equation by $\sqrt{k}$ produces

$$\varepsilon' + 1/\varepsilon' = E/\beta' \tag{12.8}$$

with $\varepsilon' = \sqrt{k}\varepsilon$ and $\beta' = \sqrt{k}\beta$. The scaled decay factor and coupling in this equation for dendrimers produces an equation exactly analogous to that in the periodic 1D case ($k = 1$). Within the bands, in the long chain limit, $|\varepsilon'| = 1$. The dendrimer decay factor is rescaled by the purely real term $\sqrt{k}$, so that coupling inside of the dendrimer band (in the long chain limit) drops by the factor $\sqrt{k}$ per repeat unit. For the finite dendrimers, inside of the band, the coupling drops by $\sqrt{k}$ per unit but also oscillates with distance by an approximately sinusoidal factor. Because the dendrimer band width in the simple case described here is $4|\beta'|$, the width expands as k increases. As a result, for *fixed* energies outside of the band, electronic coupling decay is softer in larger k systems.

The relationship between $|\varepsilon|$ and E appears in Fig. 12.2 for the one orbital per unit 1D and dendrimer structures of infinite extent. In a 1D polymer ($k = 1$), the band states have $\varepsilon = \exp(-i\theta)$ and $|\varepsilon| = 1$. In the gaps between the bands $|\varepsilon| < 1$. Thus, states in the band (with energy spectrum $E = 2\beta \cos \theta$ for the one orbital per site nearest neighbour

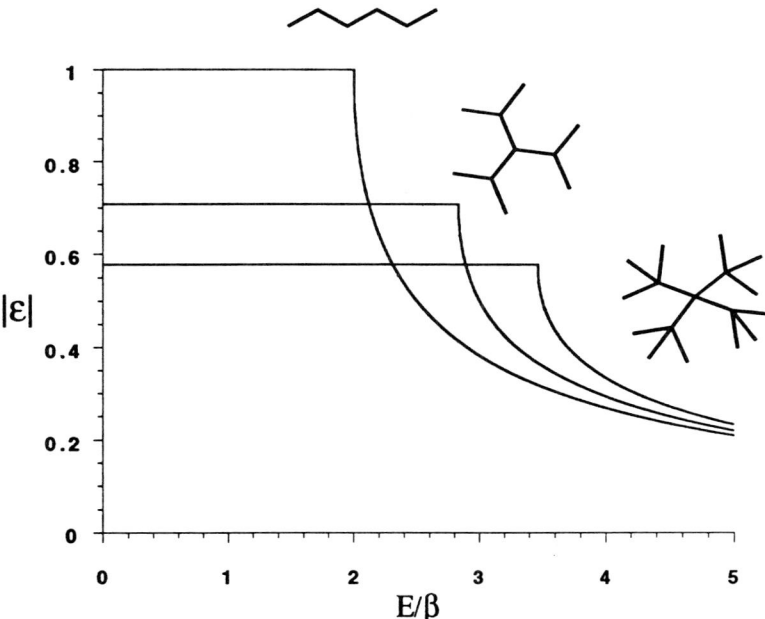

Figure 12.2 The energy-dependent electronic coupling decay factor, ε, in a dendrimer ($k = 2, 3$) and a 1D material ($k = 1$). For a fixed energy inside the band, the magnitude of ε is larger for smaller k values, while outside the band the decay is softer for larger k.

tight-binding model) are delocalized and those outside of the band decay by an amount defined by the roots of the quadratic equation relating E to ε (Eqn 12.7, $k = 1$). For idealized infinite dendrimers, in the band $\varepsilon = 1/\sqrt{k}\exp(-i\theta)$ and $|\varepsilon| = 1/\sqrt{k}$. This decay factor leads to localized states, even within the band of the dendrimer. In the band gaps, $|\varepsilon| < 1/\sqrt{k}$. For idealized Bethe lattices, these results have been known for some time [27–32] but the relevance to starburst dendrimers was apparently not appreciated. The density of states and band structure is complicated for multiorbital repeat units, but the value of $|\varepsilon|$ in the bands is retained for each material of a given connectivity, k. This value of $|\varepsilon|$ in the bands is an upper limit of that found for energies in the gaps.

Electron transfer rates between localized states scale with the square of the electronic coupling mediated by the bridge ($|T_{DA}|^2$, Eqn 12.5). As such, the resulting electron transfer rate for dendrimers is predicted to decrease by a factor of at least $1/k$ for each additional unit between the donor and acceptor. This leads to the simple prediction that if analogous (extended configuration) 1D and starburst materials are synthesized, the coupling and electron transfer rate across a fixed number of units in the 1D material would be expected to exceed that in the starburst material for energies in the band of the 1D material, while outside of that energy range the starburst materials are expected to have the larger coupling and faster electron transfer rates. This is understood as arising from the simple rescaling of the dendrimer (in band) decay per bond factor by $1/\sqrt{k}$ compared with the 1D material.

The prior results are strictly valid for coupling interactions mediated by bridges of infinite extent, while actual starburst dendrimers are finite. The number of generations that can actually be grown out from a core depends on the rigidity of the repeat unit and its size. Yet, the arguments that we have made for dendrimers of infinite extent apply surprisingly well to finite systems as well. Within the band of the finite system the decay is precisely $1/\sqrt{k}$ per unit, while in the finite systems the decay is $1/\sqrt{k}$ modulated by a sinusoidal function of distance. The oscillation frequency is energy dependent, and this behaviour parallels the sinusoidal variation of wavefunction magnitude in finite 1D systems.

6   Electronic coupling in disordered dendrimers

So far, we have assumed identical coupling interactions between adjacent units of the dendrimer. Real systems will surely have some distribution of couplings. We will describe two features of dendrimer-mediated interactions that arise when disorder in the intersubunit interactions is introduced. The first is for donor and acceptor states in the 'gap' and the second is for donor and acceptor states in the 'band'.

As a simple example, we set the coupling between adjacent units equal to one of two values, β_s (strong) or β_w (weak), and examine the electronic coupling from the centre of the dendrimer to the surface for an energy outside the dendrimer energy band. For a dendrimer of five generations, we vary the composition of the couplings from 100% β_w to 100% β_s, and determine the maximum value of the coupling to the dendrimer surface (considering all sites on the surface). In these examples, E is far enough from the band that the coupling to a surface site is approximately $\Pi_{i=1}^{5}(\beta_i/E)$. Figure 12.3 shows the probability of finding a maximally coupled pathway (i.e. a pathway consisting

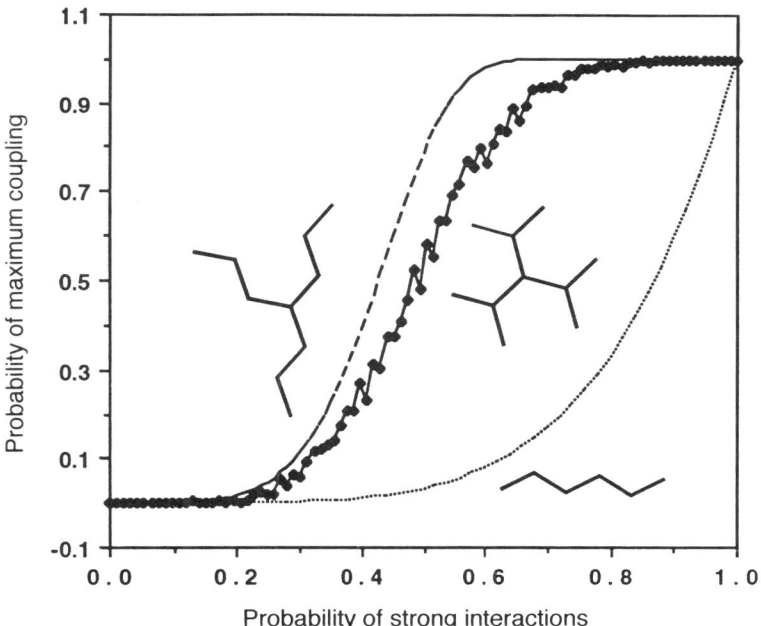

Figure 12.3 A 'pathway phase transition' plot showing the probability of finding a path with maximum coupling vs the probability that a given interaction bond is strong (shown for a linear polymer, a $k = 2$ dendrimer, and a star polymer with independent chains radiating from a core) [8].

entirely of the large β_s interactions) to the surface for a given fraction of β_s composition. Figure 12.3 also shows the corresponding plot for a simple 1D chain and for a theoretical star polymer (many independent branches joined at a single core) with the number of branches set equal to the number of surface sites for the five-generation dendrimer (48 surface sites) [8]. We call this growth of strong paths to the dendrimer surface a 'pathway phase transition'.

What would the simplest experimental signature of an electron transfer pathway phase transition be? Consider an ensemble of dendrimer molecules, each with an electron donor species at the core (that can be activated by photoexcitation) and one or many acceptors bound to the surface. The observed decay kinetics of the prepared electron transfer systems will change from single exponential to multi-exponential as disorder increases (influenced, perhaps, by temperature, solvent, etc.). Bimolecular quenching probes of pathway phase transitions can be visualized too. A single pathway to the surface might be introduced via structural changes with a coupling several orders of magnitude larger than that from weaker paths. This single path could facilitate electron transfer to diffusing acceptors in solution. This quenching mechanism would not be accessible in the absence of the strong paths. The multiple branches of the dendrimer and the larger number of surface sites give a large probability of finding some strongly coupled surface sites, even though a substantial fraction of the surface sites are weakly coupled.

Consider the role of disorder when the donor/acceptor energy (E) is inside the dendrimer band. In a periodic 1D polymer, such a donor (acceptor) state is delocalized and the bridge-mediated donor–acceptor interaction between any two sites in the chain does not decay with distance. In contrast, for a finite or infinite dendrimer, the magnitude of the decay per bridge unit is $1/\sqrt{k}$. Consider the case of two couplings β_s and β_w, where the strong coupling interactions can only occur along a single specified pathway. In this case, the dendrimer can behave locally like a 1D material. This sort of configuration of the dendrimer eliminates the democratic $1/\sqrt{k}$ amplitude splitting along the backbone and results in a coupling that changes only by the sinusoidal factor ε. This latter type of coupling behaviour is consistent with that of a 1D system. In this regard, disorder can play a qualitatively different role in a starburst material than in an idealized 1D chain. In the starburst, disorder can make individual pathways mediate coupling like simple 1D chains. While disorder in a 1D chain generally decreases electronic coupling between distant sites, disorder in a dendrimer can create 1D paths (with either wide or narrow effective band widths) that can substantially increase the coupling between two points [27].*

7 A model for the coupling in disordered dendrimers

In the previous section we considered a very simple model of disorder for the inter-site couplings (binary–alloy type). The effect of disorder was to create a 'pathway phase transition'. We emphasize that alternative models of disorder should also be considered to analyze further such effects. A very useful model is Gaussian disorder, particularly for analytical calculations. It is different from the binary–alloy disorder (although the qualitative arguments are the same), and it can be used to make formal analogies between electron transfer and the statistical mechanics of spin glasses. Thus, many of the analytical tools developed for spin glass statistical physics can be borrowed in this research [33,34].

Consider a statistical ensemble of dendrimers, where the site energies and the inter-site couplings satisfy probability distributions $P(\{E_i\})$ and $P(\{\beta_{ij}\})$. If the time scale of fluctuations in $\{E_i\}$ and $\{\beta_{ij}\}$ is very slow with respect to the time scale of electron transfer (this will be true for a small dendrimer embedded in solvent), the electron transfer rate will be given by $\langle k_{ET}(\{E_i\}, \{\beta_{ij}\})\rangle$, where the angular brackets denote an average over the distributions of energies and couplings ($P(\{E_i\})$, $P(\{\beta_{ij}\})$). Another situation which requires such averaging is an ensemble of dendrimers embedded in a frozen inhomogeneous host matrix. For these situations we are interested in computing $\langle|G_{ij}(E)|^2\rangle$ or the average of some function of $G_{ij}(E)$. $G_{ij}(E) = \langle i|(E - \hat{H})^{-1}|j\rangle$, where $|i\rangle$, $|j\rangle$ are one-electron states localized on sites i and j, respectively.

Let us assume that the fluctuations at different sites are statistically independent and Gaussian. Namely, $P(\{E_i\}) = \Pi_i P_i(E)$ or/and $P(\{\beta_{ij}\}) = \Pi_{ij} P_{ij}(\beta)$, where $P_i(E) \propto \exp\{-(E - E_0)^2/2\sigma_E^2\}$ and $P_{ij}(\beta) \propto \exp\{-(\beta - \beta_0)^2/2\sigma_\beta^2\}$. This assumption enables one to treat the problem (to some extent) analytically and it also describes any physical process that induces uncorrelated energy fluctuations. For site-energy fluctuations, the Gaussian model is a fair approximate description of solvent effects on the dendrimer.

* The probability that a star polymer (a polymer with independent linear chains linked only to one core unit) of N chains consisting of m repeat units each has at least one branch with all strong interactions is $P = 1 - (1 - x^m)^N$ where the fraction of strong interaction is x. For the case of a 1D polymer, this probability is $P = x^m$.

For nearest neighbour coupling fluctuations, the physical interpretation of the Gaussian model is not so straightforward. One could think of the latter fluctuations as arising from rotations of subunits. If there is no congestion between them, the rotation of each subunit may approximately obey a Gaussian probability distribution, due to collisions with solvent molecules. However, if there is congestion, these rotations are not fully stochastic. In spite of this difficulty, there is an important advantage in modelling nearest neighbour couplings as independent Gaussian random variables. The disordered dendrimer ET problem can be mapped into the 'random energy model' (REM) of spin glasses [28–35] (because of the starburst dendrimer topology). In the REM, energy fluctuations are measured relative to k_BT while in the dendrimer they are measured relative to the tunnelling energy. Therefore, extremely useful models of spin glass statistical physics and critical phenomena can be employed for this ET problem [36].

It should be emphasized that for systems as small as starburst dendrimers it is easy to numerically invert $E - \hat{H}$ and thus to exactly compute the Green's function matrix element between any sites of the dendrimer. This enables one to introduce disorder in an exact manner so as not to be limited by analytical methods, or by oversimplifying approximations such as total statistical independence and Gaussian probability distributions. However, the feasibility of an exact computational analysis provides further motivation for constructing simplified spin glass models of ET in starburst dendrimers. Such models can readily be tested against detailed computations, and in relation to the 'pathway phase transition'.

8 Conclusions

Within the bands of finite idealized starburst dendrimers, the electronic coupling between distant sites decays on the average as $1/\sqrt{k}$, controlled simply by the number of branches per node. The decay per unit in the gaps is faster than $1/\sqrt{k}$, and the precise value is determined by the electronic energy and the dendrimer structure. These predictions are general for both infinite and finite dendrimers, and can be tested in bimolecular and unimolecular electron transfer experiments.

The ability to tune electronic structure and degree of localization with geometrical disorder (a function of temperature, solvent, dendrimer size, repeat unit chemistry, etc.) suggests a very rich electron transfer chemistry for these materials.

In both energy regimes (inside and outside of the dendrimer energy bands), the addition of disorder in coupling between dendrimer units produces dramatic changes in the electronic coupling and the resulting electron transfer rates between chromophores at the centre and the surface of the dendrimer. The connectivity of the dendrimer may give rise to strongly coupled sites on the surface, even in the presence of significant disorder in the inter-site coupling. Just as the folded structure of a protein leads to hot and cold spots for electron transfer on the protein surface [37,38], disorder is expected to generate hot and cold spots for electron transfer on a starburst dendrimer surface. Manipulation of the coupling between sites along a specific pathway also can have a dramatic effect on the coupling along that pathway. In some cases, quasi-1D behaviour along a specific path can be induced (for example, when contacts in that path are strong and contacts linking the path to other bonds is weak). This sensitivity to the coupling exists at energies both inside and outside of the dendrimer bands.

These predictions can be tested in a number of ways, including the construction of dendrimers with varied repeat unit, connectivity, and attachment point of donor/acceptor chromophores. Comparison with the charge transport properties of linear oligomers and polymers can be made as well. Transitions between coupling pathway regimes might be accessible in starburst dendrimers that have sufficient configurational freedom. With multiple acceptor chromophores, one can imagine probing transient charge transfer states to look for this transition. Many organic materials — organic semiconductors, conductors, and light emitting diodes — are based on unsaturated linear polymers. The electronic properties of analogous materials assembled in a dendrimer geometry could be varied in a controlled manner by altering molecular structure, connectivity, and disorder.

The ability to control the electronic communication between sites in macromolecules should assist in the design of new organic and organometallic electronic materials. Recent progress using designed molecules to develop molecular materials for artificial photosynthesis, optoelectronics, and ultra-small electronics shows the central role that bridge-mediated electronic coupling plays [38–42]. The 'pathway control' of electronic interactions described here might be exploited to manipulate charge separation and hold times, conductivities, electronic hyperpolarizabilities, or other useful properties.

Acknowledgements. The work in San Diego is supported by the National Science Foundation (MCB-9316186). In Pittsburgh, this work is supported by the National Science Foundation (CHE-9257093) and the Department of Energy (DE-FG36-94G010051). Research in Commerce is supported by the Research Corporation.

9   References

1   Tomalia DA, Naylor A, Goddard III WA. *Angew Chem Int Ed Engl* 1990; **29**: 138.

2   Xu ZF, Moore JS. *Angew Chem* 1993; **32**: 246.

3   Gopidas KR, Leheny AR, Caminati G, Turro NJ, Tomalia DA. *J Am Chem Soc* 1991; **113**: 7335.

4   Kim YH, Webster OW. *J Am Chem Soc* 1990; **112**: 4592.

5   Hawker CJ, Lee R, Fréchet JMJ. *J Am Chem Soc* 1991; **113**, 4583.

6   Turro NJ, Barton JK, Tomalia DA. *Acc Chem Res* 1991; **24**: 332.

7   Morenobondi MC, Orellana G, Turro NJ, Tomalia DA. *Macromolecules* 1990; **23**: 910.

8   Risser SM, Beratan DN, Onuchic JN. *J Phys Chem* 1993; **97**: 4523.

9   Marcus RA, Sutin N. *Biochim Biophys Acta* 1985; **811**: 265.

10 Hopfield JJ. *Proc Natl Acad Sci USA* 1974; **71**: 3640.

11 Beratan DN, Hopfield JJ. *J Am Chem Soc* 1984; **106**: 1584.

12 Onuchic JN, Beratan DN. *J Am Chem Soc* 1987; **109**: 6771.

13 Onuchic JN, de Andrade PCP, Beratan DN. *J Chem Phys* 1991; **95**: 1131.

14 Beratan DN, Onuchic JN, Hopfield JJ. *J Chem Phys* 1987; **86**: 4488.

15 Regan JJ, Risser SM, Beratan DN, Onuchic JN. *J Phys Chem* 1993; **97**: 13083.

16 Shephard MJ, Paddon-Row MN, Jordan KD. *Chem Phys* 1993; **176**: 289.

17 Curtiss LA, Naleway CA, Miller JR. *Chem Phys* 1993; **176**: 387.

18 Gruschus JM, Kuki A. *J Phys Chem* 1993; **97**: 5581.

19 Siddarth P, Marcus RA. *J Phys Chem* 1993; **97**: 2400.

20 Stuchebrukhov AA. *J Chem Phys* 1996; **104**: 8424.

21 Lippard SJ, Berg JM. *Principles of Bioinorganic Chemistry.* Mill Valley, CA: University Science Books, 1994.

22 Regan JJ, DiBilio AJ, Langen R et al. Chem Bio 1995; 2: 489.
23 Bertini I, Gray HB, Lippard SJ, Valentine J. Bioinorganic Chemistry. Mill Valley, CA: University Science Books, 1994.
24 Skourtis SS, Regan JJ, Onuchic JN. J Phys Chem 1994; 98; 3379.
25 Wuttke DS, Bjerrum MJ, Winkler JR, Gray HB. Sceince 1992; 256: 1007.
26 Langen R, Chang IJ, Germanas JP, Richards JH, Winkler JR, Gray HB. Science 1995; 268: 1733.
27 Ziman JM. Models of Disorder. New York: Cambridge University Press, 1979.
28 Economou EN. Green's Functions in Quantum Physics, 2nd edn. New York: Springer-Verlag, 1990.
29 Bethe H. Proc Roy Soc A 1935; 216: 45.
30 Bethe H. Proc Roy Soc A 1935; 216: 150, 552.
31 Thorpe MF, Weaire D. Phys Rev B 1971; 4: 3518.
32 Thorpe MF. In: Thorpe MF, ed. Excitations in Disordered Systems. New York: Plenum, 1982: 85–107.
33 Mezard M, Parisi G, Virasoro MA. Spin Glass Theory and Beyond. World Scientific Lecture Notes in Physics, Vol. 9. Singapore: World Scientific, 1987.
34 Chowdhury D. Spin Glasses and Other Frustrated Systems. Princeton Series in Physics. Princeton, NJ: Princeton University Press, 1986.
35 Derrida B. Phys Rev B 1981; 24: 2613.
36 Pande VS, Onuchic JN. Phys Rev Let 1997; 78: 146.
37 Beratan DN, Betts JN, Onuchic JN. Science 1991; 252: 1285.
38 Onuchic JN, Beratan DN, Winkler JR, Gray HB. Annu Rev Biophys Biomol Struct 1992; 21: 349.
39 Burroughes JH, Bradley DDC, Brown AR et al. Nature, 1990; 347: 539.
40 O'Neil MP, Niemczyk MP, Svec WA, Gosztola D, Gaines GL, Wasielewski MR. Science 1992; 257: 63.
41 Gust D, Moore TA. Science 1989; 244: 35.
42 Hopfield JJ, Onuchic JN, Beratan DN. Science 1988; 241: 817.

13 Conjugating Polymer Superlattice and Porphyrin Arrays Connected with Molecular Wires

T. SHIMIDZU

Kansai Research Institute, Kyoto Research Park, 17 Chudoji-minumi-machi, Shimogyo-ku, Kyoto 600, Japan

1 Introduction

Both quantum functional materials and molecular devices are of interest as ultimate functional materials. The former show a novel property which is specific to the structure and the latter give the functional material of smallest characteristic size. In this paper, as examples, conjugating polymer superlattices and porphyrin arrays connected with molecular wires are described.

2 Conjugating polymer superlattice

Since almost all conjugated polymers are organic semiconductors, structural control, such as compositional control of a copolymer thin film, corresponds to manipulation of the band structure of the thin film. With structural control on a 10–100 Å scale, functions due to carriers confined in the well structure, i.e. quantum-size effect, should be observable. In the past few decades, ever since the proposal by Esaki and Tsu in 1970 [1], inorganic semiconductor superlattices and multiple quantum wells have been extremely active research subjects in semiconductor physics and material science. Many new physical phenomena, such as negative differential resistance, have been discovered, and many novel device concepts, such as high electron mobility transistors, have been developed. These advances were based on ultrathin layer structures of semiconductor materials constructed by epitaxial techniques under ultrahigh vacuum, molecular beam epitaxy and metal organic chemical vapour deposition. The most important feature, which has produced this great activity, is energy quantization of the electronic structure by mesoscopic size modulation [2]. This not only changes the energy scheme but also alters the density of states and restricts electron motion within the layer plane, leading to a lower dimensional electron system: the electronic structure is changed fundamentally. This feature gives rise to new transport, optical and magnetic properties. Furthermore, because one can control and largely prescribe all these features by adjusting parameters of the heterostructure, periodicities of the layers and band discontinuities by composition of the mixed crystal, new materials with desired physical properties can be specifically designed.

In this chapter, realization of such 'wavefunction engineering' in conjugated polymer thin films, by novel fabrication methods for compositional modulation in conducting polymer films, is described.

Any function which is specific to the overall structure of the macromolecule, such as superlattice behaviour or size quantization due to ultrafine particles, is called an ultimate function of a material, and can be created through proper fabrication methods.

It is realized by nanometre-scale control of the material. The fabrication of such ultimate functional materials is known as ψ engineering, or quantum technology, and has developed as a new field of advanced materials.

Ultrathin conjugating polymer multilayers have structures suitable to exhibit non-classical functions, and can be connected to many devices including quantum materials, non-linear optical devices, and superconducting electronic devices. Superlattices are, in general, fabricated with inorganic and organic molecules in high vacuum systems. Contrastingly, polymerization followed by the *in situ* deposition of the polymer makes such superlattice structure fabrication possible.

Electropolymerization is one of the most interesting methods to control the copolymer composition in a molecular or chain sequence. Accordingly, if the electropolymerized material is electroconductive and insoluble, a heterolayered structure and/or a sloped structure with conducting polymers can be constructed on the electrode. The potential-programmed electropolymerization method (PPEP) is utilized for modulating the composition of conducting polymer composite thin films in the depth direction [3,4]. By this method, nanometre-scale compositional control of composite thin films of conducting polymers can be obtained, permitting the fabrication of alternate layered and graded structures.

Monomers such as pyrrole, thiophene, and their derivatives, can be electropolymerized so that the corresponding conducting polymer thin films are obtained on the surface of a working electrode. The growth rate of the film thickness is proportional to current i. Figure 13.1 shows the i–E curves for electropolymerization of pyrrole and 3-methylthiophene in a $LiClO_4/CH_3CN$ solution. Pyrrole and 3-methylthiophene can be electropolymerized by applying a potential greater than ca 0.8 V (vs RHE) and ca 1.4 V, respectively. Accordingly, the corresponding copolymerization occurred at potentials higher than ca 1.4 V. The rate of the monomer uptake in the copolymer is similar to the rate estimated from the i–E curves obtained during homopolymerization.

In general, the PPEP method consists of the electropolymerization of a mixture of two monomers under potentiostatic control in accordance with the appropriate potential sweep function. The function is programmed in advance from the current fraction curves for each monomer, which leads to a definite copolymer composition and control of layer thickness. The resulting conducting polymer layer has a compositionally modulated depth structure corresponding to the applied potential sweep function, e.g. a

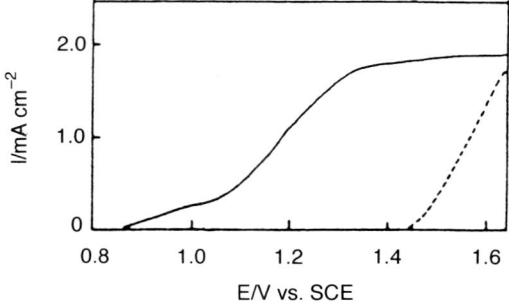

Figure 13.1 The i–E curves for electropolymerization of pyrrole (25 mM; ——) and 3-methylthiophene (50 mM; -----) in CH_3CN containing 0.1 M $LiClO_4$. Potential sweep rate, 8.3 mV s^{-1}.

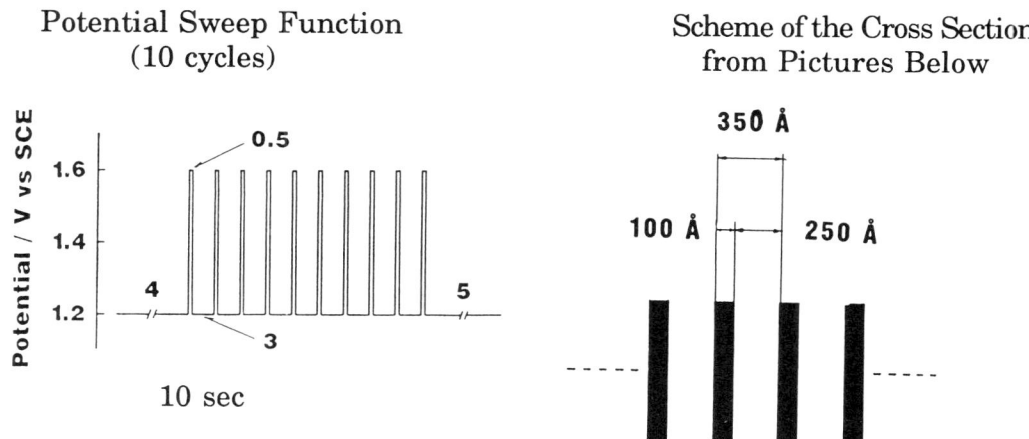

Potential Sweep Function
(10 cycles)

Scheme of the Cross Section
from Pictures Below

SEM Pictures of Cross Section

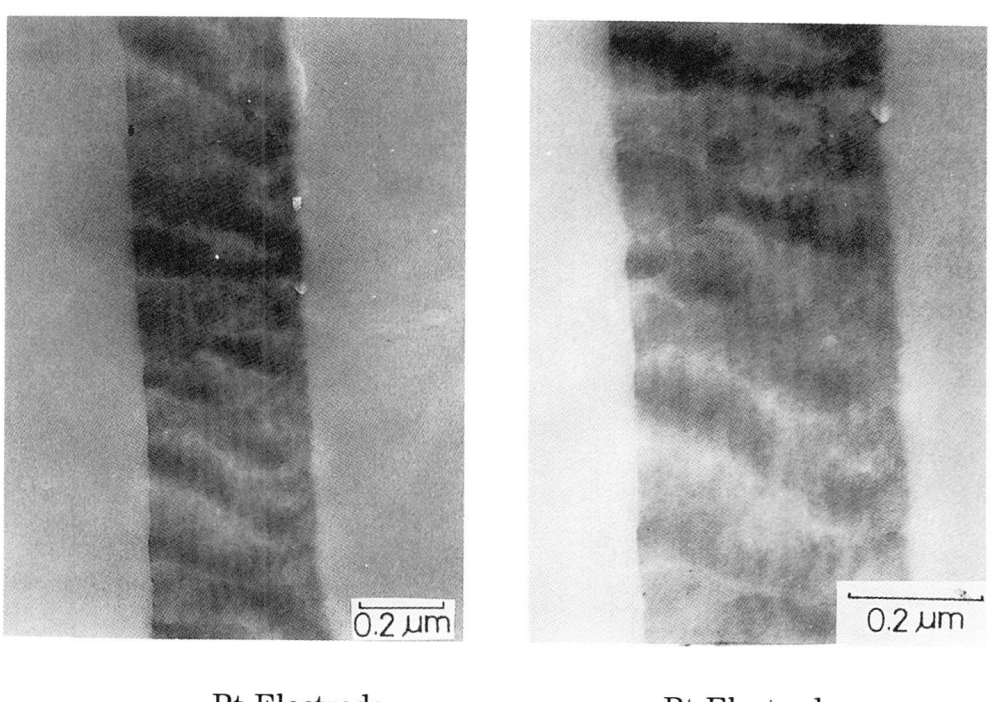

Pt Electrode

Pt Electrode

Figure 13.2 Ultrathin conducting polymer heterolayers by the potential sweep programmed electropolymerization of pyrrole and 3-methylthiophene: potential sweep programmes and TEM pictures of their cross-sections.

layered structure for a rectangular (square wave) function and a graded structure for a varying function, such as a sawtooth wave.

By applying the present PPEP method to the copolymerization of pyrrole and 3-methylthiophene, various kinds of the conducting polymer heteromultilayers were fabricated. Figure 13.2 shows the resulting SEM cross-section and EPMA line analysis on sulphur (reflecting thiophene content). The layer on the electrode side, from which the

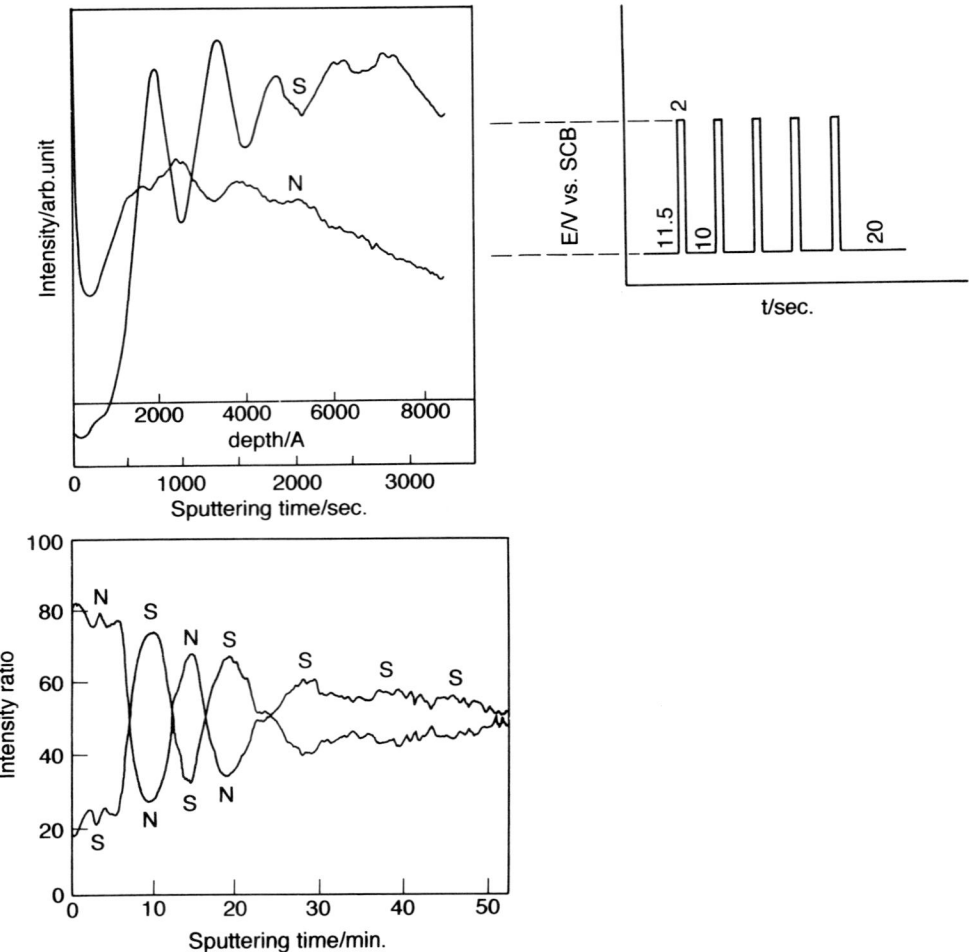

Figure 13.3 Heterolayers of polypyrrole and poly(pyrrole-co-3-methylthiophene). Potential sweep function profile analysed by AES and depth profile analysed by SIMS. N from pyrrole and S from 3-methylthiophene components.

material grew, showed a clear and flat interface. The depth profile of the resulting conducting polymer multilayers was also evaluated by SIMS, AES, TEM, and EPMA (Fig. 13.3) [5]. Alternate layered structures were fabricated by a rectangular potential wave, stair-like structures by a step sweep function, and triangular sloped structures by a triangular sweep function; triangularly sloped structures resulted from sawtooth potential waves.

Figure 13.4 shows the band structures of several homopolymers and pyrrole–bithiophene copolymers, which serve as representative polymer systems [6,7]. A combination of these homopolymers and/or copolymers leads to the fabrication of various superlattice structures. The electrochemical preparation of both homopolymer multiheterolayers and copolymer multiheterolayers results in a superlattice. But, in the former case, change of polymerization solution is necessary. On the contrary, electrochemical copolymerization results in the fabrication of copolymer multiheterolayers by simply changing the applied electrode potential. The present electrocopolymerization method produces compositionally modulated copolymer heterolayers and is considered

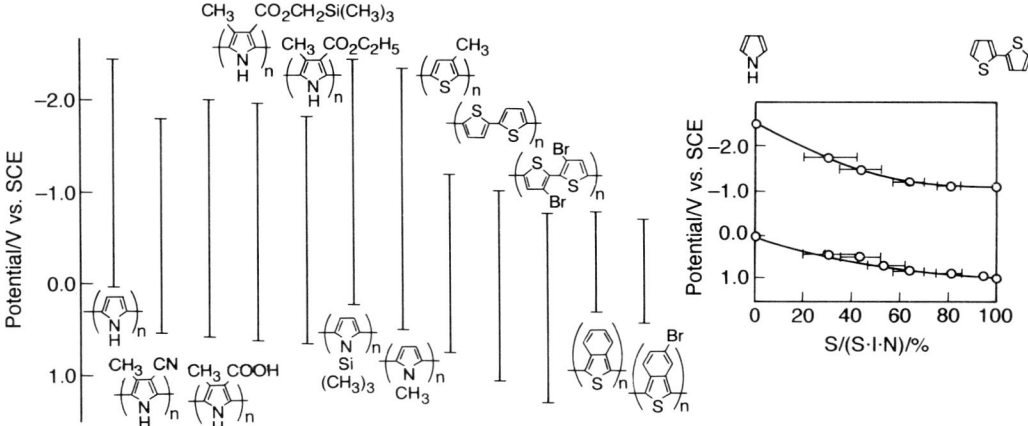

Figure 13.4 Band structures (E_c and E_v) of conjugating homopolymers (left) and copolymer (right).

one of the most fascinating methods to fabricate organic superlattices.

The copolymer multiheterolayer fabrication was carried out on a rotating HOPG disk electrode (working electrode, 1000 r.p.m) which leads to a flat and sharp interface having a 10 Å resolution. The electrocopolymerization of the mixture of pyrrole (2.5×10^{-4} M) and bithiophene (2.5×10^{-2} M) by a rectangular potential sweep having limits of 1.0 and 1.4 V gave superlattice multilayers whose dedoped layers were expected to be a type II superlattice [1]. The barrier layer was composed of 33% bithiophene and 67% pyrrole while the well layer was composed of 87% bithiophene and 13% pyrrole. In this superlattice, ΔE_c is 0.58 V and ΔE_v is 0.41 V, as shown in Fig. 13.5. The layers with 87% thiophene content serve as an electron well while those containing 33% thiophene act as a barrier.

Photoluminescence spectra of the dedoped copolymer (pyrrole–bithiophene) films whose thiophene content was higher than 50% consisted of three peaks around 2.0, 1.8, and 1.7 eV, corresponding to phonon side bands at 10 K. These peaks correspond to the radiative relaxation of self-trapped excitons. The peak at the highest energy reflects the copolymer's band gap. Actually, the peak positions observed in the spectra of copolymer films shifted to higher energy as thiophene content in the film decreased, and peak positions showed good agreement with E_g as estimated in Fig. 13.5. The copolymer containing a thiophene fraction less than 50% did not, however, show photoluminescence.

The photoluminescence of the multilayers having 60 Å of 87% thiophene layer and 100 Å of 33% thiophene layer (10 layers) shifted to higher energy compared with that of the bulk (87% thiophene content) copolymer film. Quantitative analysis is presented in Fig. 13.6 [6,7].

The photoluminescence of the above-mentioned multilayers shifted to higher energy as the thickness of the well layer (L_w) became smaller than 120 Å, even when the barrier thickness (L_b) remained constant (100 Å) and when the ratio $L_w : L_b$ was constant (0.6). On the other hand, the bulk thin layer did not show a significant energy shift. Such a shift to higher energies is considered to be the result of the confinement of excited electrons in the quantum well layer. We have also found a good fit of experimental results to the Kronig–Penney model, which derives the energy-wave vector relationship

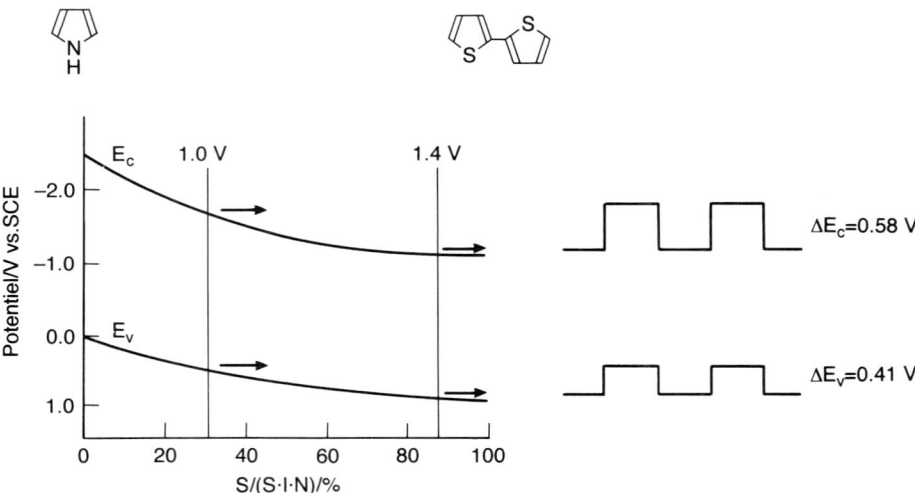

Figure 13.5 Fabrication of type II conducting polymer heterolayer superlattice by the electro-copolymerization of pyrrole (2.5×10^{-10} M) and bithiophene (2.5×10^{-10} M) CH$_3$CN solution. Sweep potentials, copolymer compositions and ΔE_c, ΔE_v of the resulting heterolayers. The lines correspond to band gaps.

in rectangular-type potential profile by assuming that $m^* = 0.6m_e$ where m_e is the electron mass (depicted as the solid line in Fig. 13.6).

The Kronig–Penney model is given as:

$$\cos k(L_w + L_b) = \cos \frac{\sqrt{2m^*E}}{\hbar} L_w \cosh \frac{\sqrt{2m^*(V_0 - E)}}{\hbar} L_b$$

$$+ \left(\frac{V_0}{2E} - 1 \right)\left(\frac{V_0}{E} - 1 \right)^{-1/2} \sin \frac{\sqrt{2m^*E}}{\hbar} L_w \sinh \frac{\sqrt{2m^*(V_0 - E)}}{\hbar} L_b, \quad (13.1)$$

where V_0 and m^* are barrier height and effective mass, respectively.

It is noteworthy to mention that other multilayers have also shown a similar phenomenon.

Consequently these spectral behaviours suggest that the conjugating polymer hetero-layer fabricated by the present PPEP method shows a quantum size effect. This result also leads to the expectation that many other novel functional materials and devices whose properties reflect their designed structure may be fabricated by this method.

3 Porphyrin arrays connected with a molecular wire

The nanofabrication of molecular semiconductors for the construction of molecular photoelectronic devices is an important area of research [8–12]. The incorporation of multiple redox centres into conducting molecular systems is a useful approach for trial construction. For example, the incorporation of a photosensitizer and a suitable electron donor and/or acceptor into a polymeric chain has been proposed as a molecular system based on photoinduced electron transfer [12]. However, the production of such polymers containing a number of large aromatic moieties or metal complexes is difficult because of the intractable solubility and flexibility, which limit the

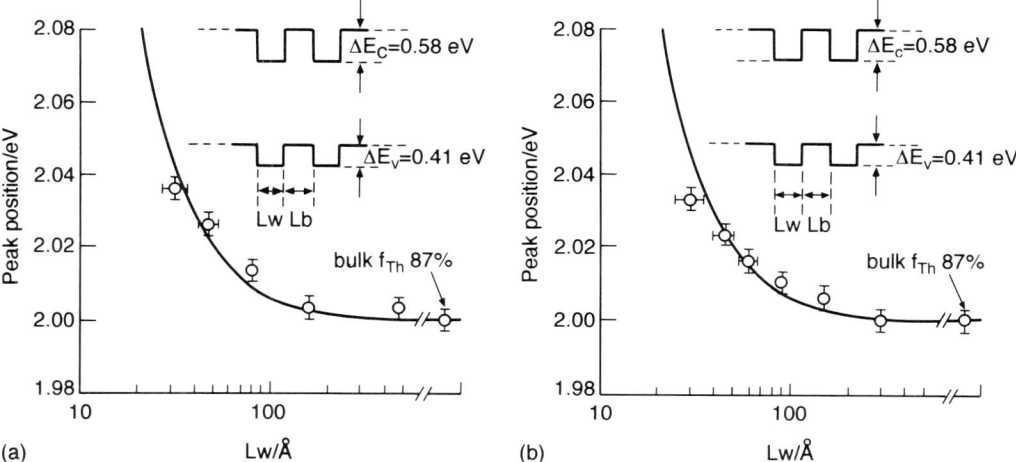

Figure 13.6 Structure of type II heterolayer superlattice and emission peak shift as a function of layer thickness. Solid line is estimated from Kronig–Penney model. (a) $L_w/L_b = 0.6$; (b) $L_b = 100$ Å const.

possibilities for controlled fabrication on terminal electrodes. In order to overcome these difficulties, electrochemical polymerization is useful, since the polymer is deposited directly on the terminal electrode. With this in mind, we have synthesized a series of one- or two-dimensional porphyrin arrays connected with conjugating and insulating wires which can be polymerized by normal electrochemical oxidation for the former and by esterification for the latter. On polymerization, one- or two-dimensional polymers, with porphyrin moieties separated by ordered oligothienyl molecular wires can be obtained. In addition, lateral polymerization can be combined with axial polymerization to afford three-dimensional molecular systems.

Such construction of intramolecular systems whose photoactive molecules are linked with a conducting molecular wire is an important subject in the realization of molecular electronic or photonic devices. For such objectives, systematization of donor–photosensitizer–acceptor triad molecules into large molecular systems is one of the feasible approaches. This is because the exquisite incorporation of the photoactive moiety and a suitable electron donor and/or acceptor into a conducting polymeric chain is useful for various molecular systems based on photoinduced electron transfer [12]. Recently, a symmetrical donor–acceptor–donor triad molecule was polymerized by normal electrochemical oxidation, which led to one-dimensional (1D) donor–acceptor polymers with porphyrin moieties separated by an ordered oligothienyl molecular wire, a 1D porphyrin array [13].

The oligothiophenes, which are easily dimerized by electrochemical oxidation, were used not only as molecular wire but also as the coupling elements [13]. Phosphorus(V) porphyrins [P(V)TPP], which have strong oxidizing powers and are stable to electrochemical oxidation, were used as the photoactive moiety [14,15]. Since P(V)TPP can form two stable axial bonds on the central phosphorus atom, P(V)TPP triads having two oligothiophene moieties in the axial direction can be synthesized easily [16]. Three different P(V)TPP derivatives (see Table 13.1) containing two thienylalkoxy or oligothienylalkoxy groups at the axial positions of the central phosphorus atom were

Table 13.1 Absorption and emission properties of bis(thienylalkoxy)- and bis(oligothienylalkoxy) phosphorus(V) porphyrin triad molecules in CH_3CN.

Porphyrin*	Absorption, λ_{max} (nm)				Fluorescence			
	Axial group	Soret	$Q(1,0)$	$Q(0,0)$	λ_{max} (nm)		τ_s† (ns)	ϕ_f
a	232	430	559	599	613	668	4.1	2.7×10^{-2}
b	310	428	558	600	615	668	< 0.5	5.1×10^{-4}
c	357	431	561	604	622	673	0.8	3.9×10^{-3}

* See Fig. 13.7.
† The fluorescence lifetime of diethoxyphosphorus(V) tetraphenylporphyrin triad molecules.

synthesized by the reaction of the dichlorophosphorus(V) tetraphenylporphyrin and the corresponding thienyl or oligothienyl alcohols [13]. The resulting triad molecules gave normal P(V)TPP absorption spectra which contained characterized thienyl or oligothienyl absorption peaks.

All the derivatives have similar fluorescence originating from the P(V)TPP moiety, but their lifetimes and the relative quantum yields of fluorescence depend on the axial substituents (Table 13.1). In particular, the fluorescence was strongly quenched in the latter two cases (**b** and **c**) compared with diethoxyphosphorus(V) tetraphenylporphyrin (τ_s = 4.4 ns) in the absence of any thienyl moieties. Taking into account the energy levels of the P(V)TPP and the oligothienyl moieties, the fluorescence quenching can be attributed to the photoinduced electron transfer from the oligothienyl moieties to the P(V)TPP [13]. If fluorescence is quenched, the oxidation potential of the oligothienyl moieties is sufficiently low compared with the reduction potential of the singlet excited state of the P(V)TPP. These results suggest that the reductive electron transfer occurs in **b** and **c**, as depicted in Fig. 13.7. An important point is that the P(V)TPP is able to act as a good photoinduced hole generator in the donor–acceptor molecules with oligothienyl moieties and is therefore expected to play a similar role in donor–acceptor polymers.

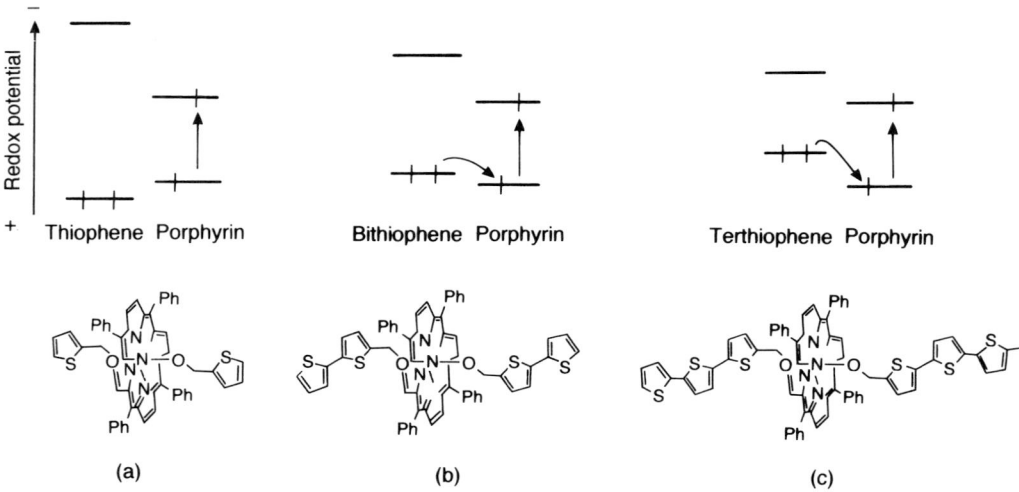

Figure 13.7 Schematic representation of relationships between photoinduced electron transfer and corresponding energy levels of phosphorus(V) porphyrin and oligothienyl axial groups.

Both P(V)TPP derivatives **b** and **c** were polymerized by electrochemical oxidation, whereas **a** was scarcely polymerized. Consequently, poly-**b** and poly-**c**, 1D porphyrin arrays, were electrochemically deposited on the indium tin oxide (ITO) electrode at potentials >1.2 V and 0.9 V (vs SCE), respectively. The peak current observed around − 0.4 V, which was assigned to the redox reaction of the P(V)TPP moieties, increased and thus signalled the deposition of the porphyrin polymer onto the electrode.

The photoconductivity of the present polymer was measured using a sandwich cell composed of ITO/polymer/gold as depicted in Fig. 13.8. $i–E$ curves of both poly-**b** and poly-**c** show that contact between the polymer and the electrode is ohmic. In these polymers, it is confirmed that a Schottky junction was not formed at the interface with either the ITO or gold. The dc conductivity of the polymers in the dark was 1.2×10^{-9} S cm^{-1} and 5.1×10^{-8} S cm^{-1} for poly-**b** and poly-**c**, respectively. Interestingly, the conductivity of both poly-**b** and poly-**c** was strongly enhanced upon photoirradiation, but this was strongly dependent on light intensity. A polymer having little orientation and prepared in a micropore film showed significantly greater photoconductivity. This implies that photoinduced carrier formation occurs efficiently in these donor–acceptor polymers [17,18].

Two-dimensional (2D) phorphyrin arrays were also prepared by electropolymerization of phosphorus(V) porphyrin derivatives containing four oligothienyl groups at the meso-position of porphyrin ring. These 2D porphyrin arrays showed similar functions as the 1D porhyrin arrays.

On the other hand, porphyrin arrays connected with an insulating molecular wire [19,20] showed that both photoexcited singlet and triplet states are localized in a short molecular wire but not localized in a long one. This suggests photoinformation storing capability at a certain porphyrin ring which in turn suggests a photoinformation storing molecular system. Figure 13.9 shows a schematic picture of photoinformation storing using localized and delocalized excitations.

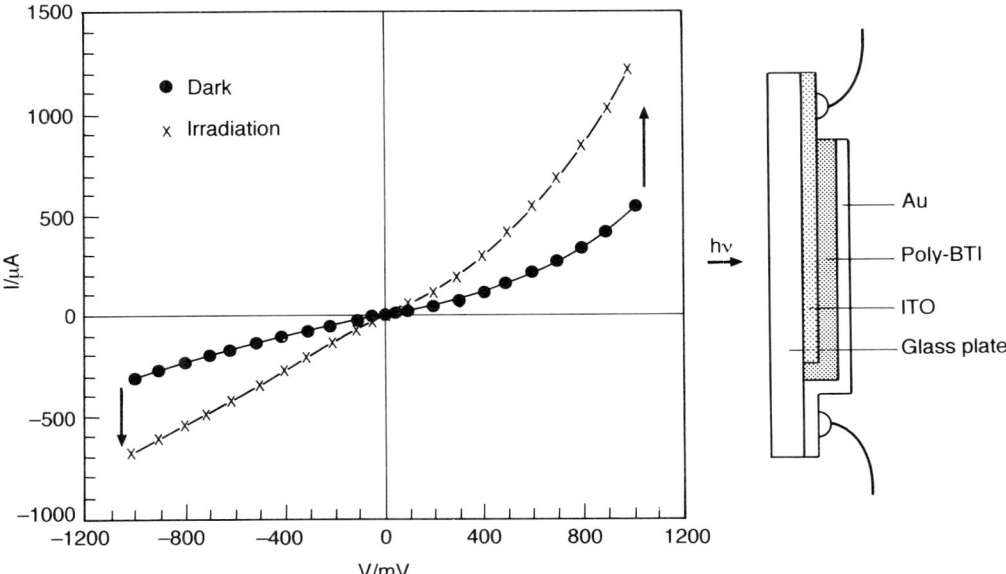

Figure 13.8 The $I–V$ curves of poly-**b** (poly-BTI) in the dark and under photoirradiation.

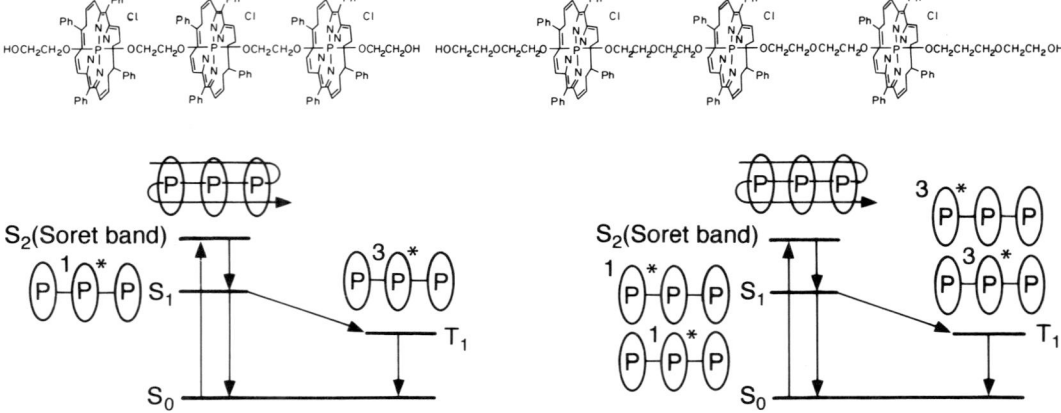

Figure 13.9 Singlet and triplet photoexcited states of 1D porphyrin arrays connected with insulating molecular wires of short and long chains.

The present photoactive porphyrin polymers containing oligothienyl molecular wires are useful not only for their fine homopolymeric fabrication but also for the hybridization with other conducting and/or insulating polymers, which can be used to realize 3D porphyrin arrays, a prototype molecular device and an artificial photoneuron and photo- and electroactive brain.

4 Conclusion

The materials and their synthetic methods discussed in this chapter are considered as candidates for molecular engineering material. Further development and improvement of molecular manipulation can be anticipated.

Acknowledgement. This work was partially supported by a Grant-in-Aid for Scientific Research on New Program from the Ministry of Education, Science and Culture of Japan.

5 References

1 Esaki L, Tsu R. *IBM J Res Dev* 1970; **14**: 61.
2 Chang LL, Esaki L. *Phys Today* 1992; **45**: 36.
3 Shimidzu T. *React Polym* 1989; **11**: 177.
4 Iyoda T, Toyoda H, Fujitsuka M *et al. J Phys Chem* 1991; **95**: 5215.
5 Iyoda T, Toyoda H, Fujitsuka M *et al. Thin Solid Films* 1991; **205**: 258.
6 Fujitsuka M, Nakahara R, Iyoda T, Shimidzu T, Tsuchiya H. *J Appl Phys* 1993; **74**: 1283.
7 Shimidzu T, Iyoda T, Toyoda H, Fujitsuka M, Nakahara R. *Synth Metals* 1993; **55**: 1335.
8 Aviram R, Ratner MA. *Chem Phys Lett* 1974; **29**: 281.
9 Wrighton MS. *Science* 1986; **231**: 32.
10 Chidsey CED, Murray RW. *Science* 1986; **231**: 25.
11 Simon J, Tournilhac F, Andre J-J. *New J Chem* 1987; **11**: 383.
12 Hopefield JJ, Onuchic JN, Beratan DN. *Science* 1988; **241**: 817.
13 Segawa H, Nakayama N, Shimidzu T. *J Chem Soc Chem Commun* 1992; 784.
14 Sayer P, Gouterman M, Connell CR. *J Am Chem Soc* 1977; **99**: 1082.
15 Carrano CJ, Tsutsui M. *J Coord Chem* 1977; **7**: 79.

16 Segawa H, Kunimoto K, Nakamoto A, Shimidzu I. *J Chem Soc Perkin Trans* 1992; **1**: 939.

17 Segawa H, Nakahara R, Iyoda T, Shimidzu T. *J Appl Phys* 1993; **74**: 1283.

18 Shimidzu T, Segawa H, Wu F, Nakayama N. *J Photochem Photobiol A Chem* 1995; **92**: 121.

19 Segawa H, Kunimoto K, Taniguchi M, Shimidzu T. *J Am Chem Soc* 1994; **116**: 11193.

20 Susumu K, Kunimoto K, Segawa H, Shimidzu T. *J Phys Chem* 1995; **99**: 29.

14 Subwavelength Molecular Exciton Probes

W. TAN* and R. KOPELMAN†

*Department of Chemistry and The Brain Institute, University of Florida, Gainesville, FL 32611, USA

†Department of Chemistry, University of Michigan, Ann Arbor, MI 48109-1055, USA

1 Introduction

Scanning probe microscopy (SPM) has become one of the most important techniques in scientific research. The field has evolved rapidly over the past decade. There are many different types of SPM; among them an optical microscope, near-field scanning optical microscopy (NSOM), based on near-field optics (NFO), has begun to gain recognition in several fields, including microscopy, spectroscopy, photofabrication, single molecule detection and imaging, and biochemical sensing [1–3]. During the early development stage of NFO, a new microscopy, molecular exciton microscopy (MEM), was proposed [4]. It is based on subwavelength molecular exciton probes (MEPs).

Since the inception of subwavelength molecular exciton probes, progress has been made from their preparation to application. In this chapter, we will summarize our efforts in these areas. Different molecular designs have been utilized for the preparation of MEPs. Many different methodologies for the preparation of MEPs have been developed. The most notable among these may be that of photonanofabrication, based on NFO and chemical synthesis, which has been successful for controllable preparation of subwavelength MEPs. Tens of nanometre to single molecular cluster exciton light sources have been prepared with a wide variety of fluorescence materials. These MEPs have also seen their preliminary applications in solving physical and analytical chemistry problems. Exciton light probes have demonstrated their advantages in microscopy, spectroscopy, and in probe-to-sample interaction studies. Probe-to-sample energy transfer and probe-to-sample interfacial heavy atom quenching experiments have been carried out successfully with MEPs. MEPs have higher sensitivity in fluorescence detection compared with direct photon excitation, and the detection limit has been improved by up to 50 times in the analysis of rhodamine B molecules. In addition, the development of optical supertips will be discussed. Molecular exciton probes hold the promise for optical microscopy and spectroscopy with single molecular spatial resolution, nanosecond temporal resolution and single molecular biochemical sensitivity. This single molecule probe-based microscope can be considered as one of the 'Holy Grails' in modern physical and biomedical sciences.

2 Near-field optics

Most optical microscopes involve conventional optics (now called 'far-field optics') where the standard rules of interference and diffraction lead to the Abbè diffraction limit [5] on the resolution of optical imaging. This limit is approximately $\lambda/2$, where λ is the wavelength of the light. Even electron and X-ray microscopy do not overcome this limit, except that their λ is significantly shorter. Their better absolute resolution comes, however, at the price of low penetration depth and/or high ionizing radiation, which causes severe

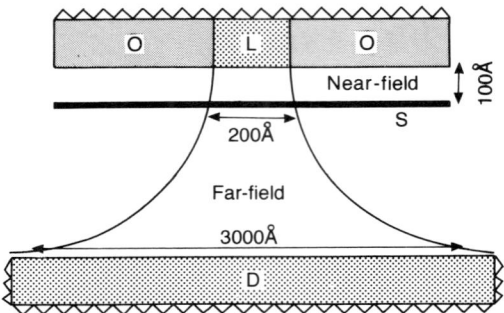

Figure 14.1 Schematic drawing of near-field optics. L, subwavelength light source; O, opaque material; S, sample; D, far-field optical detector.

damage to many samples. In near-field optics, the diffraction limit is overcome, using subwavelength light sources and samples positioned very close to them (i.e. in the 'near-field'). The realization of near-field optics is through optical nanoprobes. The principle is shown schematically in Fig. 14.1 The near-field apparatus consists of a near-field light source, sample in the near-field and a far-field detector. To form a subwavelength optical probe, light is directed to an opaque screen containing a small aperture. The radiation emanating through the aperture and into the region beyond the screen is first highly collimated, with dimension equal to the aperture size, which is independent of the wavelength of the light employed. The region of collimated light is known as the 'near-field' region. The highly collimated emissive photons only occur in the near-field regime. To generate a high resolution image, a sample has to be placed within the near-field region of the illuminated aperture. The aperture then acts as a subwavelength-sized light probe which can be used as a scanning tip to generate an image. That is why this optical microscopy is called near-field scanning optical microscopy (NSOM) [6,7].

Unlike the harsher probe–sample interactions of scanning tunnelling microscopy (STM) or AFM, imaging in NSOM is via the interaction of light with the surface by either a simple refraction/reflection contrast, or by quantum optics effects such as the absorption and fluorescence mechanisms. The advantages of NSOM are its non-invasive nature, its ability to look at non-conducting and soft surfaces, and the addition of a spectral dimension, the latter of which does not exist in either STM (room temperature) or AFM. This potential for extracting spectroscopic information from a nanometre-sized area makes it particularly attractive for biomedical research and materials science. The resolution of an NSOM image is limited by the size of the light probe. Thus the light source is the 'heart' of the NFO technique. It has to be small, intense, durable, and spatially controlled. A tiny but intense light source has been a problem. Originally, nanofabricated orifices (holes) have been used as light sources. However, the phonon throughput is very limited. With the advent of active subwavelength light sources [8], one can get a very high light throughput using NFO nanoprobes [9].

2.1   NFO nanoprobes

There are two major probes used in NFO: metal-coated glass micropipettes [10] and nanofabricated optical fibre tips [9,11,12]. Both probes are easily fabricated to sizes of approximately 50 nm, and the smallest nanofabricated optical fibre tip reported to date is about 20 nm [13]. The fabrication of miniaturized optical probes has enabled the development and application of NFO in a wide variety of fields. NFO nanoprobes can be classified into three different kinds: passive optical probes, such as coated micropipettes or small holes on a screen [10], semi-active light sources, such as optical fibre tips [13], and active light sources, such as nanometer crystal light sources [8,14,15]. The micrographs of these three different kinds of light sources are shown in Fig. 14.2. Both optical fibre tips and micropipettes have been used in NFO applications. The problem in using these probes is the conflict between smallness and light intensity. Compared with a hollow micropipette tip, an optical fibre tip is orders of magnitude brighter, easily coupled to an optical source, and at least as mechanically sturdy as a micropipette. At the same time, the optical properties of an active or semi-active material also lead to some advantages for optical fibre tips. Using a classical optics description, we note that the higher refractive index (n) of the fibre material reduces the photon wavelength inside it to $\lambda = \lambda_0/n$, compared with the wavelength in vacuum or air (λ_0). This reduces significantly the diffraction of the light at the orifice. In principle, as λ approaches the optical absorption of the dielectric, n increases, and eventually becomes a complex quantity [54]. Thus the optical fibre tip exhibits a crossover with wavelength

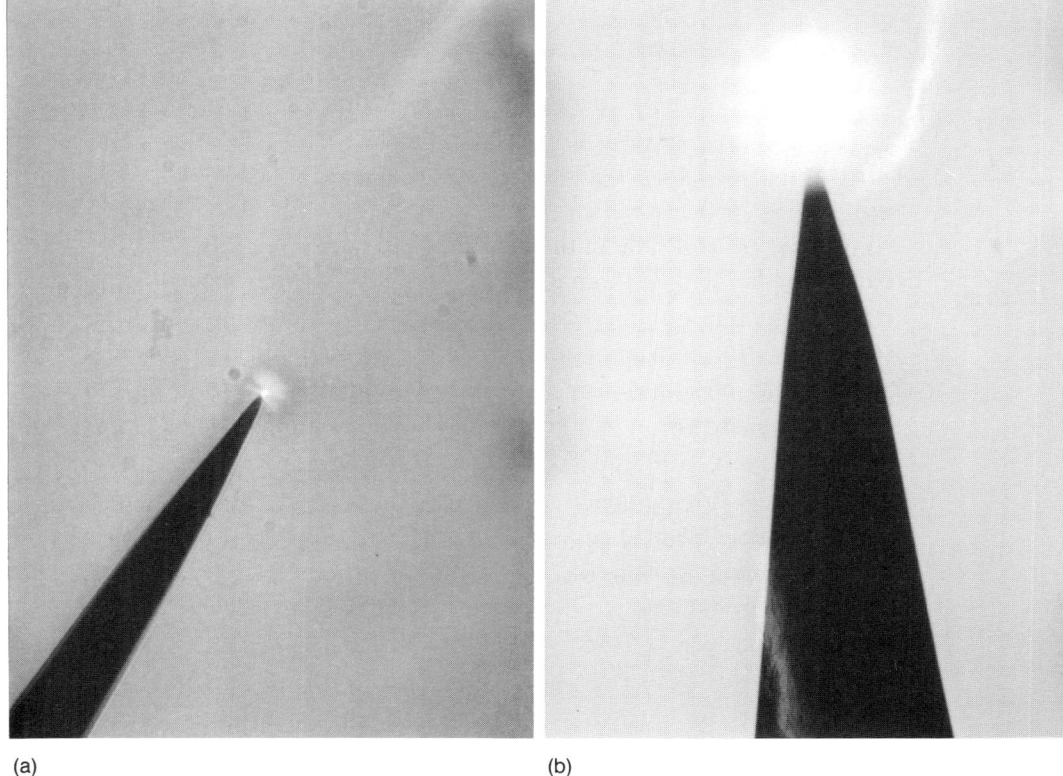

(a)   (b)

Figure 14.2 Micrographs of two kinds of light sources. (a) Conventional optical micrograph (about $600\times$) of an optical fibre tip light source: aluminium-coated optical fibre tip excited with 442 nm He/Cd laser line. Note that the tip is unresolved and overexposed. (b) Optical micrograph of micropipette light source: perylene crystal tip in aluminium-coated micropipette. Magnification about $600\times$. The 50 nm supertip is unresolved but overexposed.

from a passive to an active photon tip. This may be one of the reasons for the optical fibre tip's high light throughput.

NSOM has gained increasing attention in scientific research and development. Most of the development of NFO technique has been centred on its application in microscopy. Different subwavelength light and exciton probes have been prepared by micropipettes and nanofabricated optical fibre tips. Different modes of operation and signal detection systems have been designed. A variety of contrast mechanisms (e.g. fluorescence and shear force) have been developed, yielding images with a high degree of fidelity, as well as additional information content. The best resolution has been claimed to be about 12 nm (with 514 nm light) [13]. Presumably this was achieved with a 20 nm diameter aperture. A signal of 50 nanowatts has been claimed for an 80 nm aperture [9]. Also, NSOM has been successfully applied to single molecule localization, detection, and related studies [16]. By using different microscopic and nanoscopic samples, spectroscopic studies have been carried out on heterogeneous samples and quantum well structures [17]. A new near-field nanotechonology, photonanofabrication, has been developed to prepare nanometre light and exciton probes [18,19]. The large variety of nanoprobes should lead to new applications in NFO. We expect that NFO will be further developed and NFO-based microscopy, spectroscopy, and biochemical sensing techniques will become conventional tools in scientific research and development.

2.2 *Near-field quantum optics effects*

A special point of both theoretical and practical interest is the increased photoexcitation cross-section of optically active molecules in the near-field range. Several of our experiments indicate a higher per-photon molecular absorption and photopolymerization [14,19] in the near-field regime, compared with far-field configurations. In other words, a single molecule in the near field attenuates (e.g. absorbs) light more efficiently than in the far field, with an equal photon flux. There is, however, the question of how to define the photon flux over a region much smaller than the wavelength. This factor is important for both the photofabrication process and the utilization of such probes. For example, NFO has been applied in two different ways in biochemical sensing: (i) for the nanofabrication of subwavelength optical fibre biochemical sensors, and (ii) for the applications of these sensors. In the first case, by applying NFO, we successfully demonstrated a new concept of near-field fabrication, in which the synthesis maps the electromagnetic far-field or near-field profile of the light source. The second utilization of NFO occurs during the operation of these sensors. In NFO, the molecular absorption probability is much higher than that in far-field optics where about one billion photons are needed to excite one isolated molecule [1,2]. This near-field effect is important for both the photonanofabrication process and the utilization of such optical probes in biochemical analysis. Similar differences between near-field and far-field quantum optics occur for various non-linear optics effects [20].

Near-field excitation [1,2,21] is a concept introduced for the excitation of subwavelength optical and excitonic probes. Effects unique to the near field are shown to be involved in the excitation by a subwavelength light source [1,2]. There are three major reasons to introduce this new concept. First, geometrically, NFO encompasses a very small space, thus only those molecules inside the near-field region will be excited efficiently. The molecules to the left and right of the near-field light source will not be excited at all. This is in contrast to an evanescent wave spreading over a large region. Second, in the near-field region the light intensity flux is much higher than that in the far-field of the same light source. Thus the excitation will be much stronger. Third, we believe that in the near-field region the molecular absorption probability is much higher than in the far-field region (see above). Moreover, in the special case, when a molecule is within the Förster radius [22] of an active near-resonance light source, the molecular excitation probability is near unity. Accordingly, the probability in the near-field region should be between unity and that in far-field optical excitation.

Theoretically, the problem of near-field optical absorption has not yet been solved. However, we give here some qualitative arguments. From a quantum mechanics point of view, the electromagnetic wavefunction of light in the near-field region is constrained. This may imply that the overlap between the molecular wavefunction and this constrained wavefunction should be significantly better. Thus a higher excitation probability is expected. A classical analogy is the effect of the refractive index on the Einstein absorption coefficient. The absorption coefficient increases with λ^{-3}, where λ is the real wavelength. Thus we believe that any spatial squeezing of a photon due to high refractive index or due to the near-field conditions are comparable. This leads to a much enhanced cross-section.

Also, the mechanism of light–matter interaction may be different in the far- and near-field regimes [11,12], leading to different spectral selection rules and in particular to an enhanced cross-section of light absorption (and thus fluorescence). We expect a failure of the transition multiple expansion approach, including the transition dipole, when the photon is not much larger than the molecule. These phenomena are an extra bonus for near-field detection. In addition, good optical fibre tips do *not* affect light polarization and not even cross-correlation [1].

2.3   *The problem of light propagation through an ultrasmall aperture*

In NSOM, the capability of fabricating ultrasmall optical probes is limited and often the extremely small tips have very weak intensity for imaging. All the passive light sources are typically apertures letting light through, and when the size of the aperture gets to be significantly below that of a wavelength, most of the light will be diffracted or reflected back, rather than transmitted [23,24]. This resolution limitation derives from the less than ideal characteristics of the aperture. There are no propagating electromagnetic modes in the subwavelength cylindrical metallic waveguide such as a metal-coated micropipette, which has a thin layer of metal deposited on its outside tapered surface to protect light from escaping and to miniaturize the size of the light passage aperture. For instance, for a hole in a metal plate, for apertures of 500 Å or less, the intensity goes down superexponentially with aperture. This occurs because the metal surrounding the hole has to be at least about 500 Å thick [10] in order to be opaque enough to define a hole. The metal with the largest opacity in the visible region is aluminium for which the light penetration length is $d = 500$ Å [10] when the wavelength is 5000 Å. With a given minimal thickness, the aperture radius (r) dependence of the emerging light intensity, I, is described by [23]:

$$I \sim r^6 \exp(-L/r),\tag{14.1}$$

where L is the length (thickness) of the aperture. If we start with typical values of $L = 1500$ Å and $r = 400$ Å, then reducing r by one order of magnitude reduces I by about 20 orders of magnitude. The above assumes that the wavelength λ is much larger than the aperture radius:

$$\lambda \gg r.\tag{14.2}$$

L has to be significantly larger than the penetration length of the radiation into the metal, therefore the smallest practical value for L is 500 Å (using aluminium). This prevents us from making small optical probes. Furthermore, there is a heating problem at the tip end. Both optical fibre probes and micropipette probes are excellent light conductors only with the metal coating on the outside surface of the probes. The smaller the aperture size, the higher the light intensity needed at the tip to get the same throughput of light. This results in a large heating effect which causes damage to the metal coating on the tip side surface, and with a very small size therefore limits us in making smaller tips. We thus have a limitation in making probes deliver enough photons.

2.4   *Limitations*

In NFO, there are still many technical problems to be solved — from understanding the contrast mechanism to the control of photo-bleaching (a standard problem in fluorescence microscopy). The NFO scanning probe has been a major problem for the development of NFO for applications in physical and biomedical sciences. The necessary high scanning speeds are presently limited by the probe intensity. In NSOM, the size of the light source determines the resolution of the image, provided that it can be scanned in the near-field region. However, the requirements of smallness and intensity are in direct conflict. The smaller the aperture of an optical probe, the weaker the light throughput from the probe. We believe that the best achievable spatial resolution of NFO is about 100 Å with standard probes. While very recently, resolutions of 8–10 Å have been claimed [NFO-4 Conference, Feb., 1997], these involve probes that are not likely to work with soft molecular samples. The goal of nanometre-resolved microscopy, spectroscopy, and biochemical sensing is to push biochemical analysis much closer to the non-invasive detection and manipulation of single molecules, radicals or ions, as well as the determination of a molecule's precise coordinates, a characterization of its structural conformation, its internal dynamics and energetics, all as a function of time and environmental perturbations. By NFO alone, this goal is not going to be realized (see Table 14.1).

It has long been a challenge to mimic nature's photosystems with sophisticated molecularly engineered nanostructures [1]. An interesting application has been the subwavelength light source, based on exciton transporting molecular nanocrystals [8]. The molecular excitonics approach [8,15] can 'focus' the light within one molecule, i.e. 5–10 Å. We note that quantum optics effects take over at such small sizes, i.e. there is a highly localized excitation, confined inside a molecule, which may result in energy

Table 14.1 Comparison of optical microscopies.

Optical technique	Interaction	Resolution limit	Implementation
Far-field optics	Conventional diffraction limited	2500–5000 Å	Lens
Near-field optics	Evanescent wave intensity limited	100–200 Å	Fibre optic tip Micropipette tip
Molecular exciton microscopy	• Excitation transport	• 3.5–10 Å	• Molecular donor supertip
	• Spin–orbit coupling	• 2.5 Å	• Molecular sensor supertip

transfer, electron transfer, and other non-radiative processes, as well as light emission. This brings us to the topic of molecular excitations as probes.

3 Molecular excitations

Molecular excitations have been well studied and applied in a number of areas [22]. We are using molecular excitons for the creation of a new class of scanning probes for superimaging microscopy. Based on NSOM and molecular excitons, a novel microscopy, MEM [4], has been proposed. MEM seems very promising in a wide variety of scientific research, but the scanning tip fabrication is not that straightforward. There have been several methods suggested for overcoming the above-mentioned NSOM handicap, i.e. the severe losses of light in subwavelength cavities.

1 Transforming the photons into physically smaller energy quanta, such as excitons, using luminescent materials.

2 Overcoming Eqn 14.1 by overcoming Eqn 14.2. In practice this means an effective reduction of λ. It is well known that inside dielectric materials λ is reduced from the vacuum (or air) value by a factor equal to the refractive index n. One can thus choose appropriate transparent materials with high n. The resultant reduction in λ effectively pushes down the diffraction limit — a well-known trick, i.e. the use of lenses with oil rather than air. In general, n has a value of about 1.5–2, at best. However, using materials at the edge of their absorption wavelengths (or, appropriately, shifting the wavelength to this absorption edge) may effectively increase n by an order of magnitude.

3 Using the wave-guide (or 'coax') principle. The use of appropriately designed optical interfaces creates an 'optical fibre' in the subwavelength regime of evanescent waves [25].

In this chapter, we mainly discuss the first approach: converting photons into excitons and using exciton probes for applications.

3.1 *Molecular excitation and subwavelength probes*

Exciton probes are scanning probes using excitons as the energy source, similar to electrons in electron microscopy and photons in optical microscopy. There are technical difficulties in making smaller optical probes for NSOM. However, a passive light source can be converted into an active light and exciton source by luminescent materials with the help of the propagation of molecular excitons within crystals. Using this property of luminescent materials, one can develop a subwavelength exciton source by growing a suitable crystal within the subwavelength confines of a micropipette. With this approach, energy can be guided directly to the aperture at the tip instead of being allowed to propagate freely in the form of an electromagnetic wave. Such a material can be excited through an electrical or radiative process to produce an abundance of excitons that allows light to be effectively propagated through the bottleneck created by the subwavelength dimensions of the tip near the aperture. The excitons can be generated directly at the tip or within the bulk of the material and allowed to diffuse to the tip via an excitonic transfer [22]. In either case, these excitons will either undergo a radiative decay producing a tiny source of light at the very tip of the micropipette, or have other non-radiative processes to transfer its energy [22]. The excitonic throughput is basically independent of the wavelength and is a linear function of the cross-section of the aperture [15], instead of a superexponential one.

There are problems when exciton probes are used as light sources. One such concern is the durability of the active material. For instance, the anthracene crystals used in the first exciton light source [8] deteriorated significantly within minutes (resulting in long-time amplification factors [21] of about 3, compared with several orders of magnitude, initially). Photochemistry, thermochemistry, etc. are the obvious 'culprits'. For instance, it is well known that in the presence of oxygen and ultraviolet light, anthracene is oxidized to anthraquinone [22]. Obviously, one can either avoid the oxygen, reduce the temperature, or replace the material with a more stable one. The problem is thus transformed to one of molecular engineering.

3.2 *Molecular exciton microscopy*

The concept of active light sources enables a totally new mode of NSOM, based not on the blocking or absorption of photons but rather on quenching directly the energy quanta that otherwise would have produced photons [1]. MEM is schematically shown in Fig. 14.3. Even though MEM has not been fully implemented and developed [21], it holds the promise of single molecule spatial resolution with single molecular biochemical sensitivity. For instance, a thin, localized gold film (or cluster) can quench an excitation (or exciton) that would have been the precursor of photons. Furthermore, a single atom or molecule on the sample could quench (i.e. by energy transfer) the excitations located at the tip of the light source. For simplicity, we assume that the active part of the light source is a single atom, molecule or crystalline site, serving as the 'tip of the tip'. Such quenching energy transfer from the excitation source's active part (donor) to the sample's active part (acceptor) is the best hope, currently, for single atom or molecule resolution and sensitivity. This technique basically is a quantum optics microscopy.

MEM is conceptually quite similar to STM [26]. The excitons 'tunnel' from the tip to the sample. However, there is no driving voltage or field. Rather, it is the energy transfer matrix element which controls the transfer efficiency. Some of these matrix elements allow for the highest sensitivity to distance, higher than that of STM and

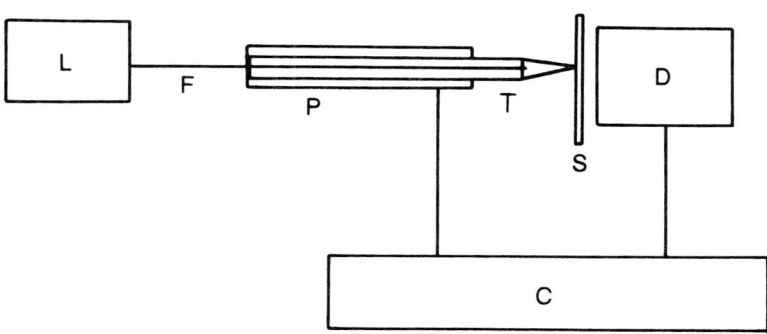

C: Computer controlled
  force feedback system
D: Light detector
L: Light source
S: Sample

F: Optical fibre
P: Piezo-electric tube
T: Scanning exciton tip

Figure 14.3 Schematic drawing of molecular exciton microscopy (MEM).

comparable to that of AFM. In addition, the most striking result of this direct energy transfer is its ultrahigh sensitivity to isolated or single molecular chromophores. The quantum–optics energy transfer is highly efficient within the range of the 'Förster radius'. Thus, one single excitation could be 'absorbed' by the sample acceptor. In contrast, based on the Beer–Lambert law, about a billion photons are needed to excite a single acceptor in the absence of other acceptors. Here, the extinction coefficient is assumed to be high, with an oscillator strength between 0.1 and 1. Furthermore, as the distance range is limited to about 10 nm for the direct energy transfer, MEM is as much a near-field technique as STM or AFM, i.e. very sensitive in the single digit nm range and much less sensitive beyond 10 nm. However, in combination with conventional NSOM, the range can be extended to about 200 nm. Thus MEM is a technique which is able to 'zoom in' from macroscopic to nanoscopic distances. Obviously such 'zooming in' enhances the speed of operation. It also allows for a much more universal range of samples, from metal spheres and clusters to soft, *in vitro* biological units. In addition, MEM can use fluorophores to enhance contrast, sensitivity, and resolution with the help of NSOM. It can also be used in conjuction with lateral force feedback, in the same way as NSOM. We would like to emphasize that MEM is still in its infancy. More development will be needed before practical application becomes feasible. Some of the exciton probe results do lend credentials to our MEM concept. The potential strengths of the MEM are: (i) high resolution; (ii) ability to image only surface molecules; (iii) low cost of construction, maintenance and operation; (iv) ability to quickly locate interesting areas of a specimen at low resolution; (v) ability to use a wide variety of substrates; and (vi) a simple contrast mechanism (i.e. the 'external heavy atom effect') to discriminate among atoms of different atomic numbers.

4   Subwavelength molecular exciton probe

Scanning tips used in various SPMs are different in their physical structures and in their imaging principles. The single most important element in NSOM and MEM is still the probe itself. An ideal light source for MEM has to be small, intense, durable, and spatially controllable. These requirements add extra difficulties for their development. There are two major techniques in the preparation of subwavelength molecular exciton probes, as shown in Fig. 14.2. The first one uses optical fibre probes and bonds luminescent materials covalently to the top surfaces of probes by photonanofabrication [18,19], while the second one uses micropipettes or nanofabricated optical fibre tips to hold luminsecent materials at their tips [8,14]. Nanometre molecular exciton probes are prepared with organic and inorganic crystals or molecularly doped polymers grown inside the very tip of a micropipette from a solution. We have used a great variety of crystals and doped polymers. The luminescent materials involved are anthracene, perylene, 9,10-diphenylanthracene (DPA), 9,10-dimethylanthracene (DMA), pyrene, fluorescein and its derivatives, rhodamine series dyes, various aminoanthracenes, tetracene, DCM [21], BASF dyes [21], rubrene, dendrimers [27,28], uranyl compunds, CsCl, zinc sulphide (ZnS) and cadmium sulphide (CdS) as well as dye-doped polymers. These active probes can be excited by different laser lines from UV to visible. The techniques used in the crystal growth inside the micropipette are crystal growth from solution, from melt, from vapour, from chemical reactions [21] and from other sources. The physical sizes of the subwavelength molecular exciton probes are as small as $0.02\,\mu m$, and mostly in the range of 0.05–$0.5\,\mu m$. Tiny polymer aggregates for microscopic studies have also been prepared.

4.1 *Molecular design*

The design of molecular exciton light sources basically has to answer the following pertinent questions. How to prepare an exciton tip with extremely small or even molecular dimension? How to prepare an ideal exciton tip in which there is a high concentration of excitons at the tip surface? How to prepare an optically stable scanning tip? How to reproducibly prepare exciton tips? All these questions may have to be answered by molecular engineering. There is an analogy in STM tip fabrication. STM has been used to take images of semiconductor and metal surfaces which show the positions of individual atoms in real space [26]. In order to get atomic resolution, the probe itself must be as sharp as a single atom. At first glance, it might appear that tip fabrication must be done with atomic precision. Actually the STM tip is not achieved by atomic precision fabrication but by 'natural selection' [29]. This 'natural selection' can only partially be applied in the exciton probe production. An exciton tip has to be a high intensity site and has to be optically stable, thus it is different from an STM tip. An exciton tip needs excitons produced or maintained on the tip surface. The energy quanta are transferred from the luminescent materials first being excited by laser light. There are two major fabrication concerns. The first has to do with the physical nature of the scanning tip, i.e. the smallness and the location of the active centre. The second is related to the electronic and optical nature of the probe. It deals with how to prepare an optically stable scanning active centre with high throughput of excitons. Thus the designing principle for exciton probes is much more complicated and the preparation is much more demanding.

4.2 *Nanofabrication of molecular exciton probes*

The exciton probe is an active light emitter which employs molecular excitons. The subwavelength tip of a metal-coated micropipette is filled with a molecular crystal, such as anthracene, perylene, tetracene, or a doped polymer. Figure 14.4 illustrates such an exciton probe. The incoming photons propagate through the supermicron portion of the micropipette and get absorbed by the crystal. The absorbed photons generate excitons that propagate (diffuse) towards the submicrometre tip. We note here that the diffusion length of organic crystals is somewhat controversial. For example, the values given for the diffusion length of anthracene differ by a factor of 200, from about 500 Å [22] (perpendicular to the ab plane) to room-temperature data of about 10 µm [30]. Due to the overlap of the absorption and the emission spectra at room temperature, the emission–reabsorption process is fairly efficient. The exciton transport is largely controlled by the geometry of the crystal, i.e. the cross-section of the tip's cavity. Thus, the efficiency decreases geometrically, rather than exponentially, with smaller aperture size.

The key for the preparation of usable exciton sources is the preservation of high intensity of light for a small enough scanning tip. We have adopted a wide spectrum of designing strategies for their fabrication. We have used organic compounds, especially aromatic ring molecules, inorganic crystals, multiple component crystals, and luminescent molecule-doped polymers. These probes produce different energy levels of excitons and can be used in multiple systems for excitation and imaging. Based on energy quanta transmitting mechanisms, we have adopted same energy level transferring and high to low energy transferring mechanisms. To be more specific, the first approach uses a single luminescent material, while the second uses two or more components and their excited states have different energy levels. The energy quanta are transferred from higher energy sites to lower ones with a high efficiency. The lower energy sites are the real scanning tips. This requires that these lower energy sites to be on the tip end surface.

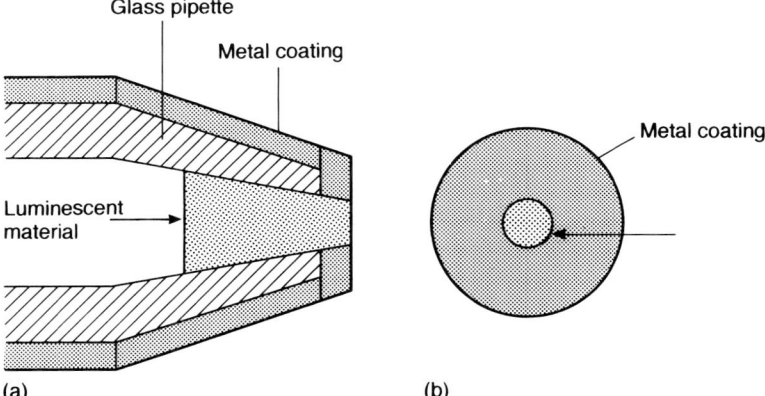

(a)   (b)

Figure 14.4 (a) Schematic drawing of active light and exciton source in micropipette. (b) Top surface cross-section view.

4.2.1   MULTISTEP MICROPIPETTE PULLING

The first step in making molecular exciton probes is the fabrication of micropipettes and optical fibre tips, from which exciton probes are prepared. Micropipettes have been widely used in biology and in electrochemical sensors [31,32]. They were adapted for NSOM as subwavelength optical probes first in the early 1980s and have played an important role in the initial development of NSOM [6]. There are commercially available micropipette pullers, such as P-87 and P-2000 from Sutter Instrument Co., the latter with infra-red laser heating replacing the electric heater filaments. The requirements for micropipettes used in our preparation are quite different from most of their conventional usage. Not only small but also very short shank micropipettes are crucial for nanometre luminescent material growth. Overall, there have been severe problems in the nanofabrication of optical probes, mainly in reproducibility, light throughput and further miniaturization. Since the geometry of a tapered probe significantly affects photon throughput [33], a systematic study of throughput as a function of geometry has been carried out. We have developed a multiple step pulling methodology to fabricate small diameter and short shank micropipettes [34] for exciton source preparation. However, there is a direct conflict between the smallness of the tip diameter and the short shank requirement.

In general, the pulling process consists of a few steps: heating, pulling, cooling, and breaking [32]. In multistep pulling, the first few pulling cycles are designed to produce a rapid taper over a desired distance without breakage. In P-87, the parameters for each pulling cycle, i.e. heat, pull, velocity, and time, are programmed into the puller. Up to 16 sets of parameters can be entered, and the process evolves step by step until the final breakage occurs. If the micropipette has not parted by the last step, the programme cycles back to the first set of parameters and the procedure is repeated. In our selection of the parameters of a multistep programme, usually no pulling force is added (i.e. pull = 0) to the first few steps until the last breakage step. This assures the execution of the whole programme. The heat parameter usually goes with this pattern: high on the first step, then decreasing continuously to the last step. In the last step, heat is chosen in such a way that it will assure the desired size. For some programmes, the heat value for the last step is higher than previous steps. Generally, it will be much lower than the value that would be used in a one-step programme for the similarly desired size. This is the main reason why a multistep programme produces a much shorter shank than that from a one-step programme for the same tip size. In a multistep programme the last step is still the most crucial one. As in a one-step pulling, after inserting the capillary tube and programming the puller, the remainder of the process is controlled by the puller microprocessor. We constructed different multistep (with four, five, or six steps) programmes for pulling both borosilicate and aluminosilicate glass tubes.

By varying the parameters in each step as well as the glass type and tube dimensions, a wide variety of micropipette shapes and sizes were achieved. According to SEM micrographs, the smallest tips pulled for the exciton probe experiments were about 200 Å inner diameter with a usable shank. Usually the ratio of tip shank length over that of tip diameter is improved by two orders of magnitude. For one-step pulling, the ratio for a 0.05 mm tip diameter is around 20 000, while that for the same diameter tip pulled by multiple pulling cycles is about 200. Micropipettes have to be metallized on their outsides before being used as subwavelength exciton probes. We have also studied other factors in pulling feasible micropipettes [21]. Micropipette pulling is a combination of various factors, and the relationship between the tip diameter and the taper length is complicated. However, our multistep pulling programme has helped to achieve the specially desired micropipettes for exciton light source preparation.

4.2.2   FIBRE TIP NANOFABRICATION

In fibre tip nanofabrication, the fibre end structure tapers uniformly from the original fibre to a subwavelength tip with a flat end surface perpendicular to the fibre axis. The taper on one end of an optical fibre is created with heating and stretching or by chemical etching, then truncating the extreme tip and coating the tapered region with metal to seal it optically. Basically, the pulling apparatus consists of a P-87 micropipette puller and a 25 W CO_2 infrared laser. The CO_2 infrared laser beam replaces the electric filament in the puller to heat the optical fibre. By using appropriate pulling programme and laser power, optical fibres can be tapered to subwavelength diameters, from 0.02 to about 0.5 μm. The optical fibre probe (see Fig. 14.2) delivers light very efficiently to the aperture since all the radiation remains bound to the core until a few microns from the tip end. Compared with micropipettes, optical fibre tips have a much higher light throughput with the same size. However, micropipettes are transparent to short-wave radiation (deep UV or soft X-rays) while fibres are not.

4.2.3   CRYSTAL GROWTH IN A MICROPIPETTE FROM SOLUTION

Crystal growth in a micropipette from solution has been the major methodology used in exciton source preparation. There is a special effect in the micropipettes we have pulled. These tubes have an Omega Dot inside [32], in which a glass fibre is fused along the inner bore of the capillary tubing to facilitate the filling of the micropipette tips with liquid solutions. Even for tips as small as 100 Å there is little difficulty in filling the tip with organic solution as well as other relatively viscous solutions. In filling with organic solution for crystal growth, the micropipette is held vertically, pointing upwards, while a tiny drop of solution is drawn into the tip. The strong capillary action and the Omega Dot effect immediately pull the liquid up into the tip. The solvent is then allowed to evaporate out from the tip, with the organic crystal precipitating at the end of the tip. By varying the concentration of the solution and other factors, such as temperature or solvent, the size of the deposited crystal can be controlled. The crystal can be illuminated either by directing light through a micropipette [14], or alternatively by having an external beam shining on the crystal at the micropipette tip [35]. With the second method large amounts of energy can be brought directly on the spot where the illumination is desired, with the upper limit being imposed only by the onset of photochemical bleaching in the crystal or the sample.

We have grown crystals from solutions of a large variety of organic and inorganic compounds. The most successful ones were perylene, DPA and BASF dyes. For perylene and BASF dyes, the success rate for crystals deposited at the tip could be as high as 70%. In the following we will summarize the results we have obtained in examining the effects of a few factors on crystal growth from solution inside micropipettes.

First, we have tested the concentration factor. For the concentrations used, the optimum one for exciton source preparation is a moderate concentration. In the perylene/benzene solution case, a concentration around 0.005 M is the best for crystal growth. There are two major reasons: first, this concentration provides crystal growth at the tip end, without excessive nucleation on the wall and taper area; second, there is a high percentage of cases of a single crystal formed at the tip end. For the exciton light source, the size of the crystal is as important as its location. Only those crystals deposited at the tip end are considered usable ones. Thus the nucleation position is critical. For crystal growth in micropipettes, the evaporation of solvents determines the nucleation for new crystals. If there is no temperature change and no decrease in the amount of solvent, no nucleation should occur for an unsaturated solution. However, the solvent inside a micropipette usually has a high evaporation rate, probably due to its high surface area and surface tension. As the solvent evaporates, the solution becomes more and more concentrated and nucleation begins. The key to control the position of the new nucleus is to control the position and the number of nucleations. Both are directly related to the concentrations of the filled solutions.

Five different temperatures, 0, 25, 38, 60, and 200°C, have been used for certain exciton probe fabrications. These micropipettes were viewed under a fluorescence microscope. At 200°C, the evaporation of benzene is so rapid that too many nucleations occur simultaneously. Thus a large number of tiny crystals are formed not only at the tip, but also along the micropipette shank area. Sixty, 38 and 25°C are all useful for crystal growth under the present conditions, but in order to have a clean micropipette (here we mean that there is only one crystal at the tip), 60°C is better. For most preparation done at 25°C, two crystals are formed: one is relatively large (micron sized) and positioned in the middle of the taper area, and the other is tiny (nanometre sized) and positioned at the tip end. The lower the temperature, the longer time it takes for nucleation. At 0°C, it appears that only one nucleation happens, but the position of this nucleation is often not at the tip end.

As is well known, the solvent plays a crucial role in crystal growth from solution in any kind of container. The evaporation rate of a solvent determines the nucleation of new crystals inside a micropipette. Different solvents have various saturated vapour pressures, and exercise different rates of evaporation. In selecting a good solvent for crystal growth in micropipettes, the following considerations should apply.

1 The solute should have a reasonable solubility at the crystal growth temperature. Usually higher solubility is preferred since it gives a wider choice of solute concentrations.
2 The solvent should not be highly volatile at the temperature of crystal growth so that rapid evaporation can be avoided.
3 The solvent should carry minimal amounts of solute with it during evaporation.
4 The solvent should be pure enought to avoid impurities which may affect the optical properties, i.e. forming energy traps.

4.2.4 CRYSTAL GROWTH FROM MELTS OR VAPOUR

Melting techniques have been used for organic compounds whose melting points are lower than 300°C. Two types of micropipettes are used: one is the usual micropipette and the other has a narrow capillary tube before the tapered tip region. This narrow capillary tube is a section of the micropipette which was heated and stretched to about one-quarter of its original inner diameter. This was done by using the aforementioned multistep pulling procedure [21]. For crystal growth from melts, the chemical powders are pushed down to the narrow passage and the open end is sealed. The micropipette is placed in a heater with the tip right at the heater edge. The tiny capillary figuration seems to provide some advantages.

1 A tiny diameter passage controls the amount of crystal grown at the tip. With regular tips, there are always larger crystals than needed or many crystals form along the wall.
2 The powder sublimes and the vapour is forced to fly down to the tip and recrystallize there. A narrow passage plays a role of 'injecting' the vapour directly into the tip end.
3 Since the original powders are far away from the tip, the recrystallized crystal at the tip should be purer.

We have also used the vapour deposition method. The crystals are pushed down into the middle of a micropipette and heated to produce vapour. Then the vapour is forced to go to the tip end where the temperature is lower than the melting point. Good crystal tips have been prepared for DPA and BASF dyes this way.

4.2.5   DIFFUSION-CONTROLLED A + B REACTION TO PRODUCE AN EXCITON
PROBE AT THE MICROPIPETTE TIP

We have tried both conventional techniques and a variety of new techniques. One of the new ones is a diffusion-controlled A + B reaction at a micropipette tip. Chemical kinetics in low dimensions also usually involves active sites or surfaces which lower significantly the activation energy barriers. At the same time, there is essentially no convective stirring in low dimensions and the diffusion process is less efficient due to the compactness of the Brownian motion [4]. This phenomenon was first demonstrated experimentally with a reactant segregation in diffusion-controlled reactions in non-convective media [36]. The two reactant species (A and B) remain well separated throughout and a depletion zone between the two grows in time.

We have borrowed this reaction kinetics principle for practical applications in crystal growth in micropipettes. The basic idea is illustrated in Fig. 14.5. The micropipette is filled with a solution, reactant A. Then the micropipette is positioned contacting the top surface of another reactant, B. Suppose A and B are each in a liquid phase and react, giving product, C, which will not dissolve in the solvents for A or for B. Therefore precipitation is expected at the tip. At the contact, there will be a depletion of reactants A and B, which is important to get only one nucleation. If the concentrations of A and B are properly proportioned, A will not diffuse into B, nor B into A. Usually, we use higher concentrations of reactant A to prevent crystal growth inside the micropipettes. Thus the A + B reaction at the interface of the tip is an ideal method to control the location of crystal growth in micropipettes. We have searched for proper reactions for crystal growth in micropipettes which meet the following requirements: (i) fast enough to ensure the diffusion-limited condition; (ii) a precipitate is formed; (iii) no side reactions (for a pure nanocrystal); and (iv) the product has excellent optical and excitonic properties. Unfortunately, not many reactions meet such requirements and, furthermore, if a reaction is performed in a volatile organic solvent, the evaporation of the solvent ruins the experiments since the diffusion process is very slow.

The principle of reactions between ions, was first demonstrated with inorganic nano-crystals of AgCl. They were grown by having a silver nitrate solution in a micropipette

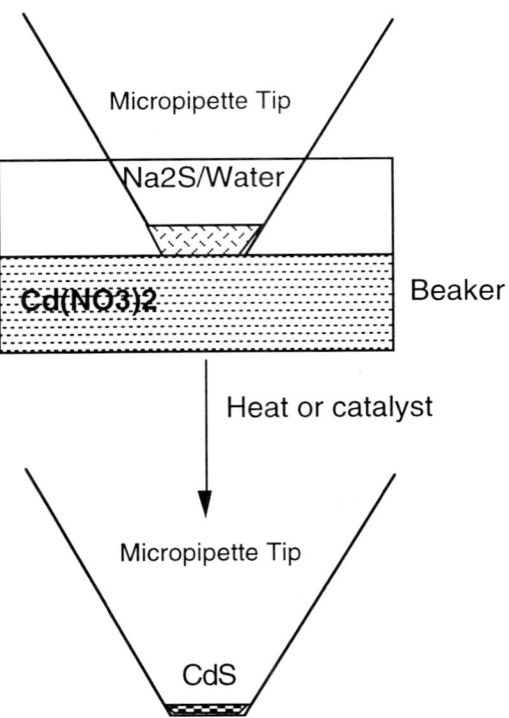

Figure 14.5 Schematic drawing of diffusion controlled A + B reaction in a micropipette.

and a sodium chloride solution in a beaker. This is similar to the experiments on complex formation in a capillary [36], but here no gel is needed to prevent convection in the nanocapillary. After growing AgCl, we synthesized a series of inorganic compounds whose optical and excitonic properties are more interesting, i.e. CdS, ZnS, CsCl, and UO_2HPO_4.

The A + B reactions are controlled by the concentrations of reactant A and B, the contact between the micropipette and solution B, and the temperature of B in the beaker. There are two kinds of contacts between the micropipette and the beaker solution: touching the surface and submerging the micropipette. Different concentration combinations have also been used to achieve the best results. In addition, we have used liquid–gas phases for A + B reactions. We have successfully prepared CdS crystals at the tip end of micropipettes. First a $CdCl_2$ solution is filled into a micropipette, then the micropipette is capped on the open side and suspended inside a container where an H_2S atmosphere is generated by adding HNO_3 to a concentrated Na_2S solution. Some crystals are found at the micropipette tip. The higher the pressure of H_2S gas, the better results obtained for the exciton probe preparation. Overall, using A + B reaction for crystal growth does help to prepare inorganic exciton sources. We believe that this technique has the potential to be a more controllable method for crystal growth in micropipettes. It could also be an excellent nanoscopic sampling method [21].

4.2.6 DOPED POLYMER IN MICROPIPETTES

The basic idea in this approach is to use doped polymer as the nanometre luminescent material deposited at the micropipette tip. The advantages are the miniaturization of the effective probe size and the enhancement of photostability of the exciton source. The effective probe size is smaller than the physical size of the tip which is to be filled with a doped polymer. The polymer can also be used as energy donor and the doped dye as acceptor. Since the polymer is the bulk of the light source (usually the doped polymer has a dye : polymer ratio of 1% (wt/wt)), the incident light will be well absorbed by the polymer to produce enough energy quanta which then are transferred to the acceptor dye molecules. The preparation procedure of doped polymer micropipettes is similar to that with organic crystals. Dye-doped polymer solution is sucked into a micropipette to grow nanometre-sized luminescent material at the tip. A few organic dyes have been used: perylene, DPA, tetracene, and anthracene. The polymer solution is well homogenized and not too viscous to be sucked into a micropipette. The dye concentrations are low, ranging from 0.1% to 3% (wt/wt). We have prepared many micropipettes with different combinations, i.e. different concentrations of DPA/PMMA, tetracene/PMMA, and perylene/PMMA. Similar combinations have been used for polystyrene. Tips made from doped polymers have a very intense light emission from the tip end.

4.2.7 FIBRE TIP EXCITON PROBE

Nanofabricated optical fibre tips have also been used for direct crystal growth or doped polymer growth on their tip end surfaces. This, however, is not the same as in photonanofabrication (see below). The major advantage of this approach is the much higher incident light throughput to the tip than that from a micropipette. Also, the fibre tip can be made smaller than a micropipette. In principle, an optical fibre tip can be treated chemically, so as to produce specific affinity to absorb specific chemicals on its surface. We have attached various molecular crystals to the nanofabricated fibre optic tips. A fibre tip is made to touch the designated organic or doped polymer solution, and picks up some luminescent materials from the solution on its tip end surface. Compared with photonanofabrication, this is a random experiment: there is little control of where and how much luminescent material would be picked up, but we have always been able to prepare designated exciton sources. We have controlled the tip positioning when touching the solution, the dye concentrations and the vicosity of the solution. We have grown perylene, tetracene, DPA, and fluorescein crystals from both organic solutions and doped polymer solutions. One perylene fibre tip has been characterized and its fluorescence image is shown in Fig. 14.6. We have also activated the fibre tip surface before dipping. There are some advantages when the fibre tip is silanized for crystal growth from organic solutions [21].

4.2.8 PHOTONANOFABRICATION

NFO enables a revolution in nanofabrication techniques. Photonanofabrication [18,19] is a novel nanofabrication technique based on NFO. It can produce nanometre-sized optical and exciton probes with or without specific biochemical sensitivity. Probes with a specific chemical sensitivity are automatically fibre-optic chemical sensors (FOCS). Using the NFO principle, photonanofabrication controls the size of the luminescent material grown at the top of a light transmitter, such as a micropipette or optical fibre tip, by photochemical reactions. These reactions are initiated and driven by an

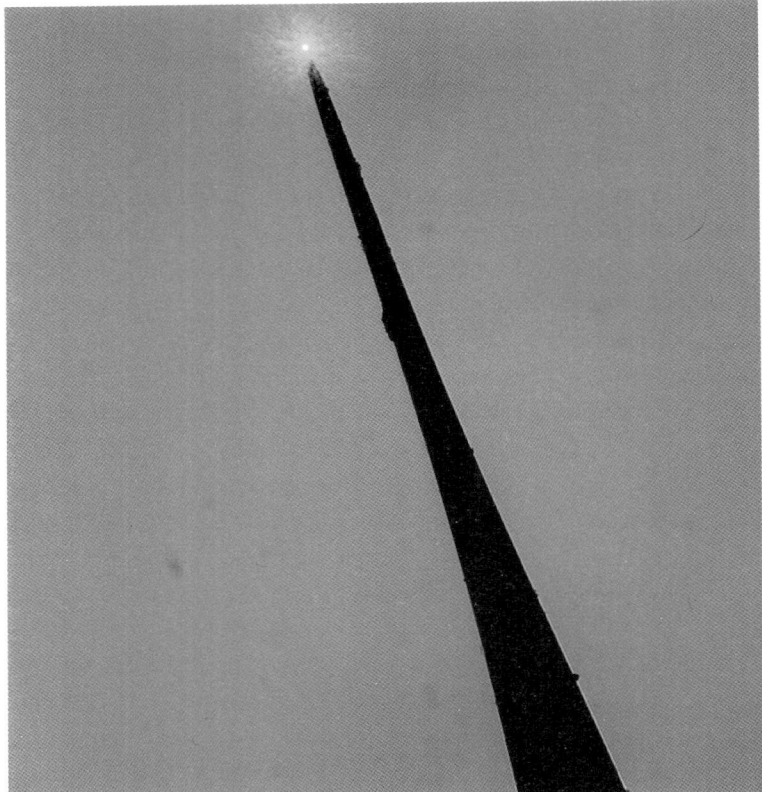

Figure 14.6 Optical fibre tip perylene exciton probe.

appropriate wavelength of light. The luminescent material is synthesized only in the presence of light and is 'bonded' only to the area where light is emitted. The key to photonanofabrication is a near-field photochemical reaction, in which the electromagnetic waves of the light source are mapped by the photochemical process. The size of the luminescent probe is defined by the light-emitting aperture and is independent of the wavelength of light used to promote the chemical synthesis. The photochemical reaction only occurs in the near-field region [19], where the photon flux and the absorption cross-section are the highest.

To illustrate the principle of photonanofabrication, we here describe the near-field photopolymerization process by which submicrometre optical fibre pH sensors have been prepared. After silanization of the metal-coated fibre optic tip, the photopolymerization is controlled by the light emanating from the near-field light source. The size of the light source and the near-field evanescent photon profile control the size and shape of the immobilized photoactive polymer. The pH sensors are prepared by incorporating a fluoresceinamine derivative, acryloylfluorescein (FLAC), into an acrylamide–methylenebis(acrylamide) copolymer that is attached covalently to a silanized fibre tip surface by photopolymerization [19]. The polymerization process is shown schematically in Fig. 14.7. The size of the polymer grown on the aperture of the optical fibre tip is equal to or smaller than that of the aperture. The ultimate goal of our photonanofabrication technique is to produce optical, exciton, and sensor probes with molecular size for NSOM, NSOS, MEM, and FOCS by controllable molecular engineering.

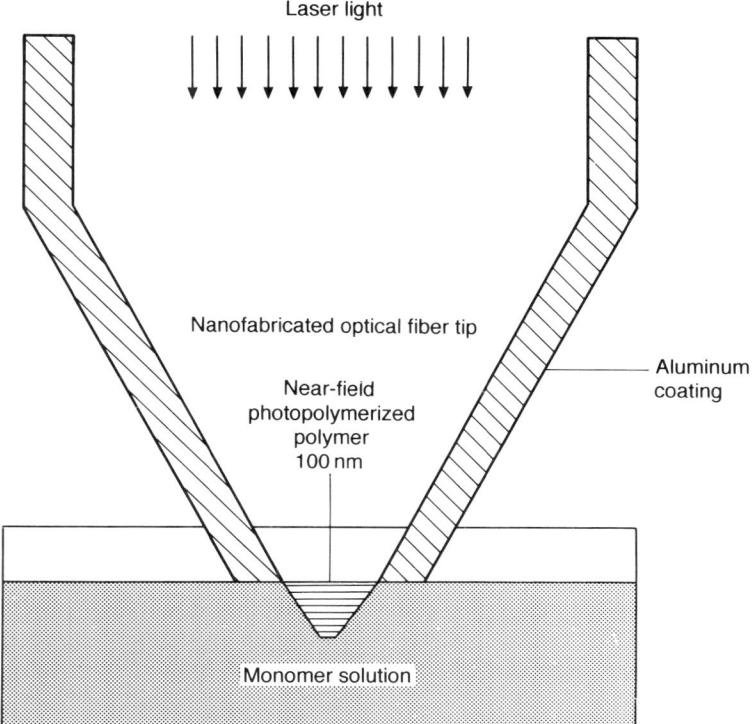

Figure 14.7 Schematic drawing of near-field photopolymerization.

4.2.9 MULTI-DYE PHOTOPOLYMERIZATION

There is a great potential for measuring two or more parameters with one miniaturized probe in biochemical analysis [37]. Multidye-doped polymer and multistep photopolymerization have been tested with the fluorescent dye-doped polymer approach [38]. By using a multi-dye solution for photochemical synthesis, we have prepared multifunctional optic probes with micrometre to submicrometre size. These probes emit multiwavelength photons and thus have a multiple sensitivity potential, provided that either internal calibration or the scheme to build supertip sensors [1], based on energy transfer, is used. There are two different ways to prepare a multidye sensor. The first is to use two or more dye molecules with polymerizable functional groups upon photochemical promotion. Thus a cross-linked copolymer will be synthesized and all the dyes are covalently bonded to the surface of the light probe. The second way is to use a dye-doped polymer, where the dyes are either bonded covalently to the probe or 'trapped' inside the polymer [38]. In preliminary experiments, we used double-dye systems, such as rhodamine B (RhB) with FLAC, or fluo-3 (a calcium-sensitive dye from Molecular Probes, Inc.) with FLAC, to prepare sensors and optical nanoprobes.

In Fig. 14.8, we show the spectra of a few RhB/FLAC polymer probes prepared with different RhB concentrations in the monomer solutions. In the five emission spectra for RhB/FLAC polymer fibre tips, it can be seen that even with a very low concentration of RhB (10^{-7} M), there is still significant RhB emission in the spectrum. Compared with the pure FLAC polymer spectrum (on the left) there are always some red shifts. The higher the concentration of RhB, the larger the red shift, and the RhB emission

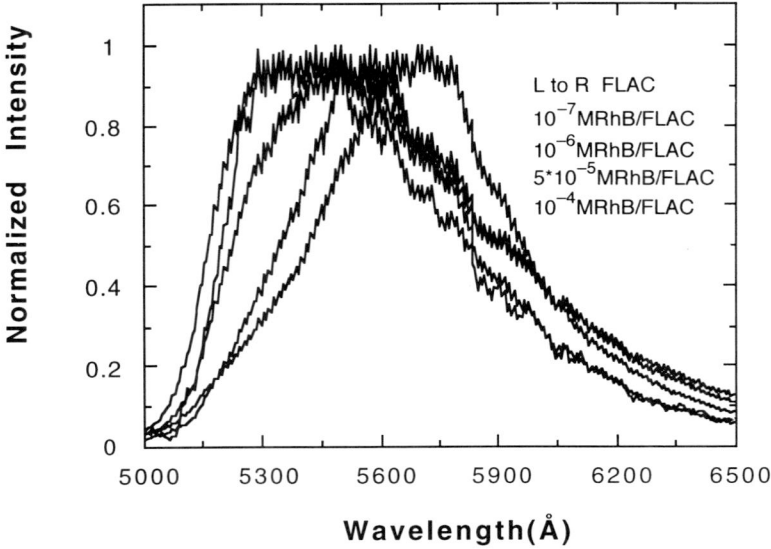

Figure 14.8 Fluorescence spectra of rhodamine B-doped acryloylfluorescein (FLAC) multidye optic probe at room temperature.

becomes more and more dominant in the spectra. The disappearance of the FLAC spectral peak clearly indicates energy transfer from FLAC to RhB since the laser excitation of 488 nm is the absorption maximum. This demonstrates that we can incorporate two dyes into the copolymer, and thus may be able to create multisensitivity miniaturized optical sensor probes.

4.2.10   MULTISTEP PHOTOPOLYMERIZATION

By using multistep photochemical synthesis [15], we have further miniaturized optical probes to sizes much smaller than the sizes of the original light conductors. We developed a multistep near-field photopolymerization to fabricate smaller probes. In the above-described RhB inside FLAC polymer system, what is important is the distribution of RhB inside the polymer on top of the fibre tip. It is reasonable to assume that RhB is homogeneously distributed inside the polymer, thus there is essentially no effective size reduction for the probe if RhB fluorescence is targeted. However, performing a multistep polymerization will assure us a size reduction in preparing the doped polymer probe. We note that a cone-shaped polymer on the fibre tip, as shown in Fig. 14.7, is usually obtained in the near-field photonanofabrication, which makes miniaturization possible [14].

The basic principle of the multistep nanofabrication process is illustrated in Fig. 14.9. We have used the combination of FLAC and RhB as an example for multistep photopolymerization. In the beginning of the multistep polymerization, only FLAC monomer solution is used for near-field photo polymerization. By controlling the reaction time, one should be able to know how much polymer is grown on a fibre tip. When the polymer is grown to a certain thickness, the RhB solution is added to the polymerization solution. We keep the disturbance to the polymerization process to a

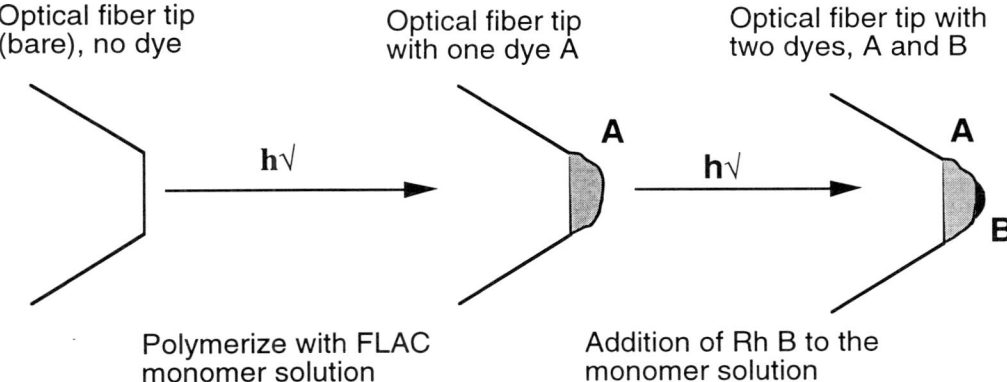

Figure 14.9 Schematic drawing of multistep photonanofabrication with rhodamine (Rh) B and FLAC.

minimum, and then the polymerization process on the fibre tip is continued as usual. Thus we obtain a polymer tip with RhB only at the cone-shaped polymer tip [6,31]. In multistep photochemical synthesis, the location of the active centre is controlled so that it is only at the very tip of the probe. Therefore a probe made in this way should have the following advantages: (i) RhB is permanently 'trapped' in the polymer; (ii) size reduction is realized by time controlling; (iii) RhB is only on the top surface of the probe; (iv) by increasing the concentration of FLAC in the monomer solution, we are able to make an efficient donor–acceptor energy transfer system, where FLAC is the energy donor site and RhB is the acceptor.

4.2.11   MULTIPLE COMPONENT LIGHT SOURCES

The optical properties of binary systems, such as pyrene, perylene, anthracene, and rhodamine dyes, have been widely studied [39,40]. However, most of the studies of perylene-doped anthracene systems emphasized the dynamics of the excitation processes and the stimulated emission of the mixed crystals. Here our study emphasizes the practical properties, such as optical stability, luminescence intensity, and emission characteristics. We are exploring them for exciton source preparation.

Binary systems of anthracene and perylene, and similar pair systems, have been produced, by several methods, for studying optical stability, luminescence intensity, and emission properties. These solid solutions and crystallite aggregates show a variety in photostability behaviour. The perylene fluorescence intensity has been enhanced by the presence of anthracene. Our basic approach for exciton probes is to search for one stable exciton intensity from the multiple emissions of the system. By using a binary mixture, a stable luminescence can be achieved by sacrificing one or all the other components in the system. We have been able to produce optically stable perylene/ anthracene, BASF dyes/anthracene, and uranyl/anthracene exciton probes, which are much more stable than their respective individual components. These mixtures have been employed in exciton probe fabrication, and the probes thus prepared are perylene emission enhanced and stabilized [21]. As expected, there is efficient energy transfer from anthracene to perylene in the binary system.

5   Characterization of molecular exciton probes

We have utilized several techniques to characterize the exciton light probes prepared by different methodologies. These techniques include scanning electron microscopy (SEM) technology, fluorescence microscopy, amplification tests, and optical and spectroscopic measurements. One of the biggest problems we have faced over the years has been the lack of a good characterization method for these exciton sources. However, the above techniques have provided useful information for guiding our efforts to improve the preparation of exciton probes.

5.1   *Optical microscopy of perylene nanocrystal*

To demonstrate the feasibility and usefulness of the idea of exciton sources and the methodologies developed to prepare such a source, we chose to work intensively with crystals of perylene. While this is not necessarily the best material available for the probe, it is easy to work with and its electronic and radiative properties have been extensively characterized [22]. In Figs 14.2 and 14.6, we showed two perylene crystal tips. These probes shine brightly at the tip. It should be noted that the spot sizes appear much larger on the photograph than the actual sizes, due to time integration or diffusively scattered light. For example, the submicron-sized crystal appears to be at least five times larger than it actually is. This is the result of the long exposure time needed in low-light photography through an optical microscope. We estimate that the real size of the tip crystal is much smaller than 0.1 μm.

5.2   *Exciton source light throughput*

To show that the active light source does indeed aid in guiding light through an aperture, the light throughput from both active and passive light sources was compared. An amplification factor was specifically defined [21] as the ratio of light throughput from an active light source over that from a similarly sized passive light source. This is a measure of the improvement achieved by an active light source in transmitting light compared with the same size passive light source. We have tested the amplification factor for anthracene tips, perylene tips, and other crystal tips. In Table 14.2 we show some of the results. For most of the 'good' active light sources, the photon output exceeds that of the passive sources (empty holes) by various factors. The highest amplification is about seven times, for perylene. The smaller the dimension of the micropipette the greater the amplification, although the absolute intensity of the signal from the tip will decrease. This has been observed for most of the tested exciton probes, although quantitative data were hard to obtain due to the difficulties in characterizing the precise profile of the micropipette. In addition, the crystals do not grow uniformly and the size and shape vary from micropipette to micropipette, with a corresponding change in the gain. The crystal even aids in transmitting light through micropipettes larger than half a wavelength, since much of the energy exiting the fibre tip is in higher order modes than TE_{11} and has correspondingly larger cutoff diameters [12,23]. Geometric factors may also work to increase the throughput in organic crystal-filled micropipettes even when the diameter is larger than the cutoff, since part of the light

Table 14.2 Amplification factors for exciton light probes.

Exciton probe	Probe size range	Amplification factor
Anthracene	$\sim 0.15\,\mu m$	3.2
1-Aminoanthracene	$\sim 0.2\,\mu m$	5.4
Perylene	$\sim 0.1\,\mu m$	6.8
DCM	$\sim 0.15\,\mu m$	2.9

incident on the inner wall, which otherwise would have been lost, will be re-emitted from the crystal coating the inner wall along the axis of the micropipette and will reach the aperture. It was only when the micropipette was broken at the tip to obtain a very large aperture (many microns) that we saw a decrease in transmission due to the crystal blocking the propagation of light. The light throughput from a passive source is usually measured with a short wavelength light (blue for perylene tip), while that for an active source is measured with a longer wavelength of light (orange for perylene tip). This makes the amplification factor smaller than it should be, considering the relationship between wavelength and refractive index of the medium. We also note that the amplification factor would be much larger if the initial time readings were used for calculations. This is especially true for the anthracene tip, since it loses intensity very quickly, due to photo-oxidation.

We have also demonstrated in this work that the amplification factor is not just due to the higher index of refraction for the crystal (i.e. decrease of wavelength inside the crystal). In Table 14.2 it is shown that perylene crystals have a higher amplification factor than anthracene. However, the large shift from the ultraviolet exciting light to the orange-coloured emission provides a wavelength increase of about two, compared with an empty micropipette. This just about cancels the index of refraction increase inside the perylene crystal, i.e. the effective wavelength λ is about the same for the empty and perylene-filled tips. Thus the intensity amplification factor cannot be attributed to a simple refractive index change but must be related to an excitonic effect. Despite increased exciton annihilation with higher power, there are crystal tips which outperform empty holes at the highest obtainable photon fluxes. In addition, we surmise that the actual size of the active light source is significantly smaller than that of the hole, in contrast to passive sources, due to the exciton–metal quenching [12].

5.3 *Power change experiments*

Optic tests were carried out for both empty and crystal tips by changing the excitation power in order to study the characteristics of these exciton probes. For a passive light source, optical linearity is preserved since light is just transmitted through a near-field aperture. But what we have observed from the exciton light source is somewhat more complicated. Figure 14.10 demonstrates the very different characteristics of empty tips, anthracene-filled tips, and perylene-filled tips. While the empty micropipette shows a practically linear increase to photon counts with incident power, this is no longer true for the crystal tips. The non-linear functional behaviour of the anthracene crystal is consistent with an exciton annihilation effect. The reason for the saturation behaviour of the perylene crystal is less obvious. Each curve describes a single micropipette, with

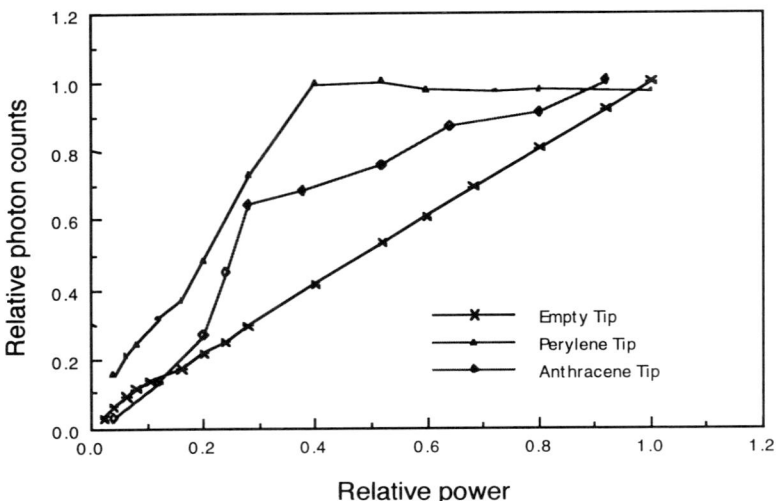

Figure 14.10 Optical linearity of different exciton and light probes.

a given tip size and shape; thus no comparison should be made among the different tips — they are shown here to demonstrate their different power characteristics and to prove that two of these tips are truly exciton light sources, even though physically we cannot characterize them well due to the quenching effects of the metal coating on the probe's outside wall.

5.4   Tip fluorescence spectrum

In previous studies, the relationship between crystal size and luminescence properties has not been understood well, even though there are reports indicating that the crystalline powders show different behaviour than the macroscopic crystals [41]. We have grown nearly strain-free crystallites at the tips of micropipettes. While the crystals are too small for observation through the best conventional optical microscopes, there is no problem in observing their fluorescence spectra. It is thus relatively simple to study the effects of the sizes of crystals on their optical spectra. Fluorescence spectra were taken from exciton probes with anthracene, perylene, DPA crystals or with dye-doped polymer, as well as with perylene/anthracene mixtures. Results show that there are peak shifts in the fluorescence spectra of nanometre-sized luminescent materials confined inside micropipettes from their corresponding bulk samples. Figure 14.11 shows a spectrum taken from an anthracene exciton light source: 355 nm was used to excite the tip crystal. There are peak shifts to the blue for the nanometre anthracene crystal, compared with bulk anthracene spectrum under similar experimental conditions [21]. These blue shifts are not very large. The shifts are possibly related to the higher ratio of surface area over volume for the nanometre crystals. The exciton states on the crystal surface usually have higher energy than those in the bulk. Thus our results are reasonable regarding the surface excitons in a nanometre crystal. Similarly, we have obtained optical spectra for perylene tips on fibre tips. We have used fibre tips to 'grow' perylene, tetracene, DPA, and fluorescein crystals from both organic solutions and doped polymer solutions. There are two things worth mentioning. First, sometimes

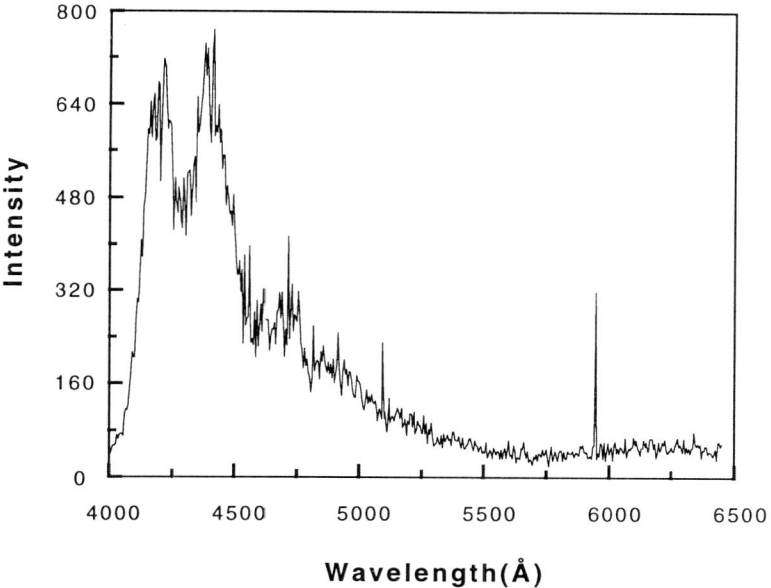

Figure 14.11 Fluorescence spectrum of anthracene exciton probe excited by 3550 Å at room temperature.

the perylene being picked up is in the form of isolated molecules. This is useful for the development of supertips where an isolated molecule is needed to act as an active centre. Second, the optical fibre tip end surface can hold organic materials. This provides the basic requirements for a nanometre light source.

5.5   *Photostability*

Photostability has been one of the most important considerations in selecting a luminescent material for the preparation of exciton light sources. We have searched widely for optically stable luminescent materials, ranging from inorganic to organic, from polymers to dendrimers, and from pure crystals to mixed crystals. The luminescent intensity of the probes is monitored in time. Different probes have different behaviours. For example, an anthracene tip gives an initial (for the first 1 min) fluorescence intensity 50 times higher than that in the period when its intensity is relatively stabilized. In our experiments, DPA and BASF dyes were much more stable than anthracene. Perylene has a moderate optical stability. Its fluorescence has been stabilized by doping with anthracene. We developed a new concept in protecting interesting species optically, which is to obtain a stable exciton light source by sacrificing other components in a mixture system. We used different ratios of two organic molecules to prepare a perylene solution in anthracene. The basic idea is to sacrifice one of the two crystals (lower in optical stability) to make the other more stable than when it is alone. We used perylene/anthracene, BASF dyes/anthracene and some other binary systems. This principle has been well demonstrated. The optical stability of perylene in a perylene/anthracene solution has been improved by at least 250% [21], compared to pure perylene.

6   Molecular exciton-based source–sample interaction modes

In traditional optical, X-ray, and electron microscopy, the source first produces photons or electrons and then the photons or electrons interact with the sample. Obviously, this is not the case for most SPMs [29]. But, in our application of the NSOM technique to biochemical and physical problems, it is apparent that the image contrast generated in the near-field is not always a simple convolution of the optical properties of the tip and sample. The effect of tip–sample interactions can often be seen in NSOM data. These interactions are a consequence of the close proximity of the probe and sample, a feature shared by all SPMs. Indeed, the basis of most SPM studies is to isolate the effect of a given tip–sample interaction and to exploit its dependence on some property of the sample to obtain a two-dimensional image. This is where NSOM differs from the other SPM methods, and is also the reason for its lack of built-in feedback. In an ideal NSOM, the intensity distribution at the exit of the NSOM tip is constant and there is no tip–sample interaction. The image obtained is then a linear convolution of the point spread function of the tip and the optical properties of the sample. As NSOM is applied to specific samples, it has become apparent that actual NSOM is not the ideal NSOM [33] mentioned above, but a hybrid of the ideal NSOM and various non-ideal image contrast mechanisms rooted in tip–sample interactions. In any case, theoretically, MEM should have its own built-in feedback mechanism based on the tip–sample energy transfer or quenching interactions. Here we list a few of the possible molecular exciton-based interactions.

6.1   *Excitation transfer interactions*

A molecular nanocrystal light source will transfer its energy to a chromophore molecule of a sample. Thus the light source becomes an energy donor and the sample molecule an energy acceptor. The result is a fluorescence emission typical of the sample. Superficially this may appear to be no different than the ordinary radiative process in which photons are first emitted from the source, then absorbed by the sample, causing it to fluoresce. However, in reality one has a non-radiative (Förster–Dexter energy transfer) process [22,42]. It may be much more efficient than the radiative process and it will exhibit very different quantitative behaviour. For instance, the dependence of fluorescence intensity on the distance from the probe is much steeper (fourth to sixth power), compared with a very weak distance dependence for the radiative process (depending on geometry). Also the light polarization dependence may differ for the two cases.

6.2   *Excitation quenching and transformation interactions*

A simple example consists of a molecular light source (e.g. anthracene) and a metallic sample (e.g. silver). At subwavelength ($<\lambda$) distances, the source's photon flux is modulated and even quenched by the sample — the Kuhn effect [48]. Again, this quenching effect can be considered a special case of the energy transfer mechanism, one without radiative re-emission. One example of this kind of quenching is the Kasha effect [43]. A 'heavy' atom (atomic mass >40) in the sample quenches the source emission. However, as a result the source may emit light at a longer wavelength (i.e. phosphorescence rather than fluorescence). This transformation effect is based on an intra-atomic spin–orbit coupling, where one atom (the heavy one) has a high degree of spin–orbit coupling ('relativistic effect') and the other atom (the light one) has very little of it on its own. The effect is 'internal' when both atoms belong to the same molecule and is 'external' when they do not. This effect is empirically observed spectroscopically, as a change in intensity or lifetime of an absorption or emission [22]. For instance, a molecule such as anthracene may have a fluorescence quantum efficiency of near unity because the first excited singlet state cannot transfer its energy to the lower lying first excited triplet state due to spin selection rules. However, upon 'induced' spin–orbit coupling such singlet–triplet transfer becomes possible both internally (with the aid of vibrational quanta that absorb the extra energy) and radiatively. Empirically one observes a 'quenching' of the fluorescence (both intensity and lifetime are reduced). These effects have been observed before only when both molecules are neighbours (or collide) inside the same phase, e.g. in liquid solution [44]. The interaction is extremely short range, e.g. 5 Å or less. Our experiment employed this effect at the interface of two distinct phases, i.e. the scanning supertip and the acceptor sample, where certain functional groups are chemically substituted with heavy atoms, such as Hg or I (see below).

6.3 *Single molecule sensitivity*

The goal of MEM is to achieve single molecule resolution with single molecule sensitivity. Elementary considerations [1,22] show that nearly a billion photons of the right frequency must hit a highly absorbing molecule (oscillator strength near unity) to cause a single excitation (on average). Assuming that this molecule is also an excellent light emitter (quantum efficiency near unity), and that the detector system requires about 100–1000 emitted photons s^{-1}, one needs about 10^{11}–10^{12} photons s^{-1} to emanate from the light source so as to detect a single molecule in fluorescence. This is now just possible with a 0.1 µm optical fibre. However, a problem still remains — filtering out the billion times stronger excitation light. This calls for very demanding spectral and/or time and/or other optical filtration.

An alternative method of single molecule excitation has been utilized in nature by photosynthesis [45]. An antenna made of hundreds of dye molecules absorbs the light and transmits the excitation to the desired 'active centre'. The transmission is done via excitation transfer (exciton transport), which is usually a multistep Förster energy transfer — from one antenna molecule to the next, and eventually from the nearest antenna molecule to the active centre (acceptor) molecule. An antenna of a million molecules needs only about 1000 photons to be excited (once). This excitation may be transmitted to the acceptor molecule with 90% efficiency, under favourable conditions [42]. Thus, the excitation sensitivity has been improved by five orders of magnitude. Furthermore, the problem of background discrimination (filtration of exciting light) has been reduced by a factor of a million. Even under less favourable antenna transfer conditions, the improvement is striking.

Our working scheme is given in Fig. 14.12. The antenna is a molecular crystallite or aggregate. One of its molecules ('active centre') is closest to the single sample molecule. This distance is 10–20 Å (much smaller than the Förster radius). The antenna (say, a million molecules) transfers excitations to the active centre, which in turn transfers its

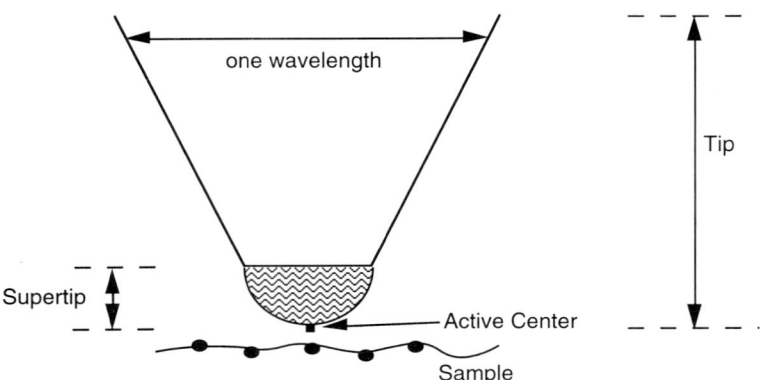

Figure 14.12 Schematic drawing of tip, supertip and active centre. A fibre optic tapered tip, crystallite antenna supertip and a single-molecule active centre. The sample shows acceptor molecules or moieties.

energy to the single sample molecule. Overall, only 10^4–10^6 photons are now required for a single excitation of the sample molecule, i.e. 10^3–10^5 times less than with direct excitation. The antenna tip is an exciton tip. First the excitons are trapped by the active centre. Then the exciton 'tunnels' from the tip's active centre to the sample molecule, causing the latter to emit a photon. Only about one exciton is needed for this single molecule excitation, however 10^4–10^6 photons were needed to produce that exciton. Thus the exciton approach, exemplifying a direct source–sample interaction, presents a significant improvement in single molecule sensitivity.

6.4 *Single molecule spatial resolution*

In STM, a single atom at the tip is the 'active centre'. This is not achieved by atomic precision fabrication but by 'natural selection'; the tip can be produced with ordinary scissors; the electrons find a defect site of atomic size [29]. If the same principle of defect site applies to the molecular exciton tip, then molecular spatial resolution will be achieved. The ultimate goal of biochemical analysis is single molecular localization both spatially and chemically. There are many reports on single molecular studies [46], but none has simultaneously achieved single molecule spatial and chemical resolution. STM has achieved single molecule resolution on well-defined structures [29], but only spatially, not chemically. Recent advances [1,16] achieved by NSOM have come closer, but are still far from real single molecule imaging. MEM holds the promise to achieve single molecule resolution both spatially and chemically [14].

7 Preliminary applications of exciton probes

Exciton probes are a new kind of scanning probe. They can be used both as a subwavelength light probe and as an exciton probe. There are many potential applications: we have explored a few of them. Exciton sources have been used in imaging, probe-to-sample energy transfer, probe-to-sample interfacial heavy atom effect, and chemical analysis experiments. Since the experimental procedures and conditions are different in each case, we will describe the conditions and results for each specific experiment.

7.1 *Exciton light source for imaging*

Perylene exciton light source tips have been used for scanning to obtain an image. We used a home-made NSOM set-up [47]. It consists of a 1″ long, 0.25″ outer diameter piezoelectric tube scanner which rasters the exciton tip in the plane transverse to the sample. The sample is supported by three stiff springs and is brought within the near field of the tip by adjusting three 80-pitch screws. There is no feedback in this system, and the NSOM scans are done at a constant height. For relatively small scan areas and flat samples, crashing can be avoided and high resolution can be obtained. The perylene tip is prepared by crystal deposition at a micropipette tip. As a demonstration, exciton light source-based NSOM images of a chromium grating replica are obtained with a resolution of about 500 Å. We have seen clear images of the chromium linear grating replica and its sharp edge. The periodicity of the grating is 0.46 μm and the line width is 0.1 μm. This imaging is comparable to the one we have obtained with an NSOM optical fibre tip. This shows that the exciton tip is capable of being used as a light probe even though there is no energy transfer mechanism but light retransmission. We have also used an exciton source tip for scanning condensed particles in the fluorescence mode of MEM. Similar results have been obtained. These results have shown the feasibility of using exciton probes as scanning optical probes in both transmission and fluorescence mode NSOM.

7.2 *Exciton sources for probe-to-sample energy transfer*

Probe-to-sample energy transfer is a distance-regulated energy transfer experiment. Distance-regulated energy transfer experiments were done before by using Langmuir–Blodgett (LB) films [48]. In our experiment, a multilayer LB film with the donor and the acceptor molecules separated by a few layers of optically inert materials, such as arachidic acid, was used as the sample. A light beam was shone onto the donor on the top surface of the LB film and luminescence from the acceptor was monitored. By varying the thickness of the arachidic acid, distance-regulated energy transfer was achieved. In the following experiment, we used exciton sources to monitor direct energy transfer from a nanoscopic point to a localized LB film and obtain distance-regulated energy transfer information from a much smaller area of sample.

Energy transfer experiments were conducted on dilute shallow solutions of rhodamine B, covered with lipid monolayers, and on monomolecular LB films [49] of arachidic acid : Di O (4 : 1). We prepared a series of exciton probes for energy transfer studies. Specifically, we used an anthracene or DPA exciton tip and a monolayer 1 : 100 (molar ratio) Di O-C18 : fatty acid LB film containing a dye molecule (Di O) at low concentration. These exciton probes were excited by UV laser lines transmitted by inserting an optical fibre tip into a micropipette. The experiment was performed on an inverted fluorescence microscope. The exciton probe was positioned on top of the LB film and scanned in the Z direction towards the sample. The fluorescence intensity of the dye inside the LB film was monitored by a photomultiplier tube attached to the microscope, and the intensity vs the separation was plotted to show the change in fluorescence intensity due to direct energy transfer. From a distance of approximately 0.5 μm, the exciton tip was moved closer and closer until physically touching the LB film, at which position zero separation is assumed. Then the tip was raised and the fluorescence intensity from the LB film monitored. Once consistent results were obtained at the same height, it was assumed that the tip was not broken.

When the tip–sample separation is approximately 0.5 μm, the DPA tip is a greenish blue light source. This emitted blue light is absorbed by the dye-containing film, resulting in a yellow–orange dye fluorescence. This absorption is fairly inefficient and is barely dependent on the Z distance (crystal tip to LB film), as the number of dye molecules in the beam is practically independent of Z ($Z < 0.5$ μm). A monolayer absorbs only a few per cent of the incident light even at its absorption maximum. However, once the DPA tip practically 'touches' the LB film (coming within the Förster radius of about 5 nm), the mechanism of energy (exciton) transfer dominates the LB film excitation process and much stronger yellow–orange fluorescence is observed. Up to 10 times fluorescence intensity enhancement has been observed when the probe physically touches the sample. Part of this enhancement could also be due to near-field light absorption [21]. The result of this experiment is shown in Fig. 14.13. If the background is subtracted properly, there may be an even larger enhancement in the fluorescence intensity for the acceptor molecules. Similar curves have been obtained with different exciton light sources. The large enhancement of acceptor fluorescence is attributed to non-radiative direct excitation (energy transfer), i.e. a direct source-to-sample coupling, which is highly sensitive to Z. In contrast, the long-range indirect coupling, via photons, depends only weakly on Z, and is trivially accounted for by considerations of geometrical optics. These results lend credence to our theoretical modelling of nanometre or subnanometre optical and exciton probes.

7.3   *Interfacial heavy atom effect*

The Kasha effect [43] is based on an intra-atomic spin–orbit coupling. These effects have never been observed before at the interface of two distinct phases. We investigated this effect at an interface by employing an exciton light source and a liquid sample. This provides us with information regarding the MEM operations based on quenching. Our experiment involved a solid film of perylene in contact with a liquid solution (water) containing heavy atoms (NaI). We observed significant fluorescence quenching effects for two situations: (i) the perylene film at the bottom of an optical fibre tip is scanned towards contact with the water solution (and back out again); (ii) the perylene is a thin solid film floating on the water solution. The perylene film experiment involved a floating solid film of perylene in contact with a liquid solution (water) containing heavy atoms of I^- (NaI). The perylene film is directly illuminated by an optical fibre that is scanned towards contact with the water solution. The second experiment was done with a perylene exciton light source. The experimental set-up was based on an inverted fluorescence microscope. The perylene tip was around 0.15 μm and was prepared from a perylene/benzene solution. The tip was driven down towards the solution containing the heavy ions, I^-. A well-controlled positioner enabled us to only let the tip touch the solution surface. The test solutions were changed from different concentrations of NaI solutions to NaCl solutions with corresponding concentrations. The light emission from the tip was monitored by a microscope objective. The results were compared by monitoring the perylene fluorescence intensities under different conditions.

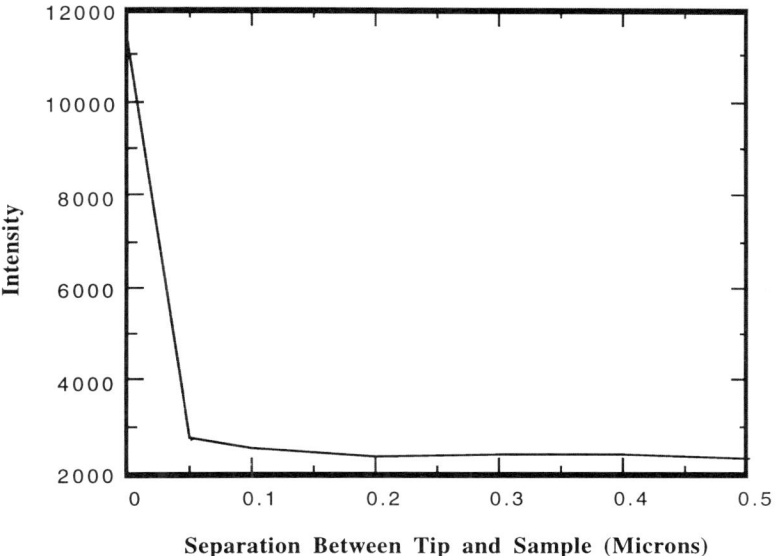

Figure 14.13 Preliminary energy transfer experiment by DPA crystal tip to scan 1 : 4 Di O/ arachidic acid Langmuir–Blodgett (LB) film.

We observed significant fluorescence quenching effects when the perylene thin solid film was floating on the water solution containing heavy atoms (NaI), shown in Fig. 14.14. The perylene film was underneath an optical fibre tip that was scanned towards contact with the water solution (and back out again). The fluorescence intensities of perylene were monitored on a macroscopic scale. It appears that perylene fluorescence is quenched by I^- (compared with Cl^-). This has been demonstrated several times with different thin films. Similar effects were observed in our exciton source experiment. As seen in Fig. 14.15, the quenching is also significant. We emphasize that the blank experiments, with NaCl water solutions, showed no quenching.

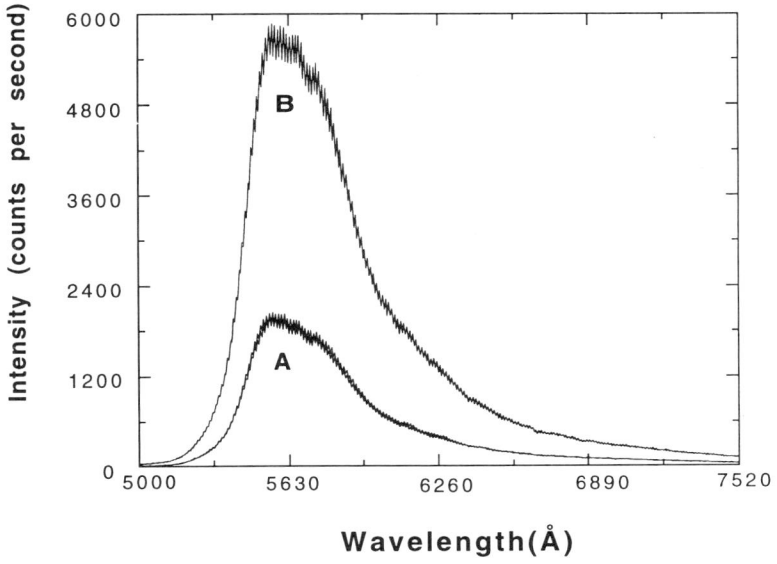

Figure 14.14 Emission spectra of perylene films interfaced with (A) NaI and (B) NaCl.

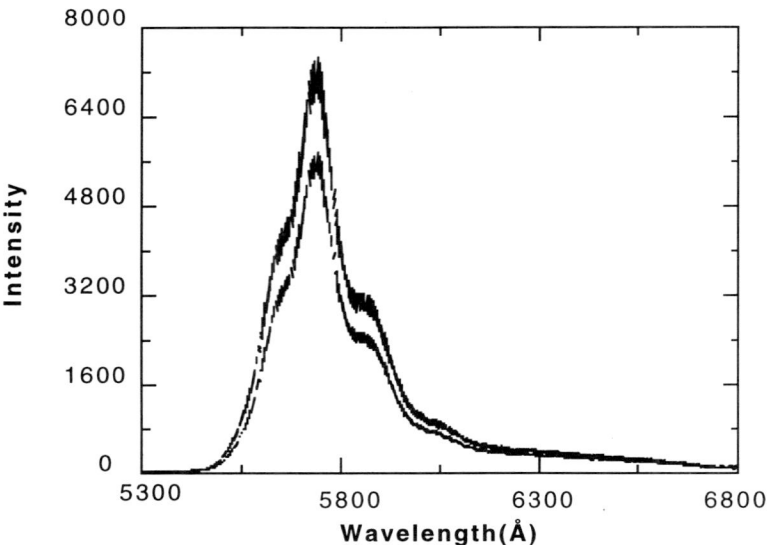

Figure 14.15 Perylene tip quenching experiment. Top: perylene crystal tip immersed in NaCl/water; bottom: perylene crystal tip immersed in NaI/water.

Furthermore, as soon as the optical fibre tip (with the perylene 'supertip') was pulled out of the water solution, the fluorescence was essentially restored to its old value. This eliminates the possibility that the perylene was dissolved, or chipped off, or reacted chemically with the iodide (or formed some permanent van der Waals complex with it). To the best of our knowledge, this is the first observation of an interfacial Kasha effect. This interfacial effect occurs in the near-field zone. Note that it is the presence of the quencher that causes the crystal surface to act as an exciton trap (but not necessarily the surface by itself).

7.4 Enhanced detection limit by exciton light sources

As mentioned above, in conventional absorption and excitation nearly a billion photons of the right frequency must hit a highly absorbing molecule to cause a single excitation [12]. With an extremely close exciton source, the excitation of a single molecule may need only one exciton. This opens the way for single molecule sensitivity and localization with an exciton source. It may also lead to new optical analysis techniques with much enhanced sensitivity. Here we illustrate the principle through the following example.

As shown in Fig. 14.16, there are two kinds of light sources with two different concentrations of samples. One of the optical light sources is a passive optical probe, such as an empty micropipette tip, while the other is an active light source, i.e. an exciton source such as an anthracene or DPA crystal light source. The samples are acceptor molecules which can be excited by both light sources. Suppose the sample and the probe are within Förster energy transfer distance and the fluorescence quantum efficiency of the sample molecule is unity. Then the detection limit will be dependent on the properties of the light sources and the concentrations of the samples. Suppose there are 10^6 photons per second (pps) from the incident light beam in both probes. In cases

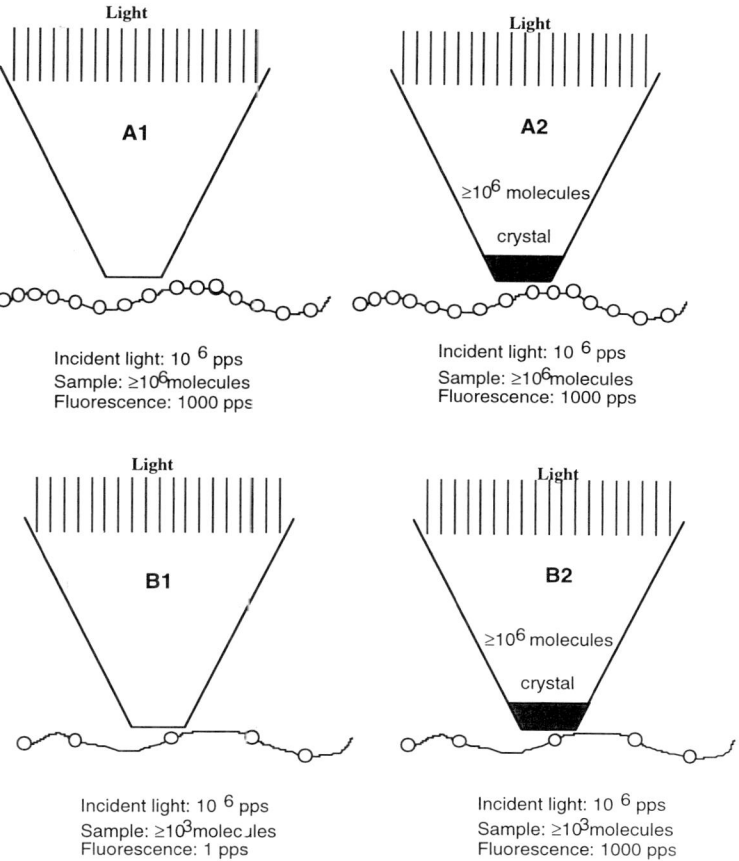

Figure 14.16 Schematic drawing of sensitivity enhancement with subwavelength exciton probe in biochemical analysis.

A1 and B1, all the photons will excite sample molecules directly, while in cases A2 and B2, photons are first absorbed by the crystals deposited at the nano-optical probe tip. As discussed above, excitons are produced and these excitons have their energy directly transferred to acceptor molecules in the sample if there are such acceptor molecules within Förster energy transfer distance.

Theoretically, the number of fluorescence photons produced from the sample in each situation can be estimated [21]. There are three assumptions. First, in far-field optics, about one billion photons are needed to excite one highly absorbent isolated molecule. Second, when a molecule is within the Förster radius of an exciton, the molecular excitation probability is on the order of unity. Third, the quantum yield of the sample molecule is assumed to be near unity. As shown in Fig. 14.16, for cases A1 and A2, there are large amounts of sample molecules, equal to the number of molecules in the crystal and thus there is no detection limit enhancement, and neither is there a need for enhancement since the analytes are in adequate supply. However, it is not the same for case B where we have a much lower concentration of sample molecules. The concentration is 1000 times less than that in case A, i.e. only 1000 molecules in the sample. The same calculation gives an enhancement of 1000 times for B2 over that of B1. This illustrates that by using an exciton source, one can have a 1000 times better detection limit in analysing very low concentration samples. It provides a novel approach for single molecule detection as well as for other fluorescence detection techniques aimed at extremely small or dilute samples [38].

The key for a successful demonstration of detection limit enhancement is that a large fraction of excitons produced in the exciton sources must be within Förster energy transfer distance to the sample molecules. There are several ways to construct such an exciton source [15,21]: the first is to have a thin layer of donor molecules, such as four layers of anthracene molecules, which is about 40 Å thick; the second is to have a 'funnel' effect [1], i.e. an active centre; the third is to let the crystal tip touch the acceptor molecules in the sample. In all three cases, the energy quanta will be transferred from the exciton source to the acceptor molecules in the sample.

We have done preliminary experiments to demonstrate the principle of detection limit enhancement by using exciton sources. Exciton probes, such as anthracene, DPA, and perylene crystal tips, were used as energy donors to transfer their energy to rhodamine 6G (R6G) in aqueous solutions of different concentrations, from 10^{-3} to 10^{-7} M. Laser light at 355 nm is used to excite both R6G and the active light sources. A large enhancement in detection limit has been observed. For example, a DPA crystal light source is scanned down towards a 10^{-6} M R6G aqueous solution. The R6G fluorescence intensity has been increased up to 15 times over a passive light source with the same photon output. It is thus clear that most of the light comes from excitons that are at the surface of the crystal.

We also used floating organic films for detection limit enhancement experiments. Anthracene crystal film was used as exciton source, while R6G was used as acceptor in aqueous solution. An anthracene crystal film was created on top of the 10^{-6} M R6G aqueous solution. The thin layer of anthracene was floated on the R6G aqueous solution by adding one drop of 0.025 M anthracene/benzene solution to the R6G aqueous solution. The donor/acceptor system is excited at 355 nm with an optical fibre. The fluorescence spectrum of R6G was recorded, shown in Fig. 14.17, and compared with that obtained from the same concentration of R6G aqueous solution by direct excitation with the same light-transmitting optical fibre. As shown in Table 14.3, there is a large enhancement of R6G fluorescence intensity. The actual enhancement value should be much larger — it is reduced by the non-linearity of the phototube.

There may be more than one factor responsible for the large enhancement in fluorescence intensity. One of the most obvious ones is the difference of excitation efficiency of R6G at 355 nm and at the anthracene fluorescence range. Anthracene has a wide fluorescence spectrum. We have taken an excitation spectrum of R6G. The difference of the excitation efficiency at 355 nm and at 450 nm is only 2.4 times, while Table 14.3 gives a fluorescence intensity difference of 24.8 times. This leaves a factor of at least 10 for the 'real' near-field (excitonic) enhancement. Thus it is fair to say that the enhancement in fluorescence intensity by exciton sources will result in a large enhancement in the detection limit in chemical analysis.

The principle of the detection limit enhancement is illustrated by our simple calculations and preliminary experiments, but the actual experiment is more complicated. We noticed that surface tension was a major difficulty in performing this experiment since there were a few factors which result in fluorescence enhancement

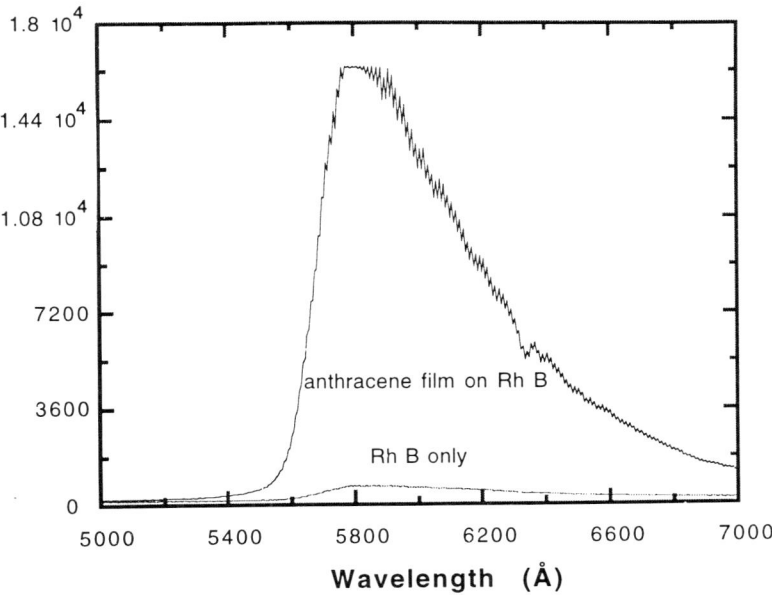

Figure 14.17 Spectra of rhodamine B in water. Top: with a thin film of anthracene on top; bottom: direct excitation by 355 nm laser line.

when the exciton light source is touching the liquid surface. For the film experiments, we used a thin layer R6G aqueous solution. The thickness of the solution is in the range of 500 μm, but the excitonic energy transfer distance from the anthracene film is less than 100 Å. Thus only a very small portion of the R6G molecules are within energy transfer range. This might be an important reason why only a 10-fold enhancement has been observed. It is easy to calculate that at the top 100 Å of the solution, the enhancement factor has to be about $10 \times 500/0.01 = 5 \times 10^5$. The selection of concentration of donor molecules is critical to demonstrate this effect experimentally. As shown in our calculation, high concentration will not result in any enhancement in fluorescence intensities. In addition, direct excitation of the sample from the incident light should be minimized, otherwise the background signal is very high and the difference in detection limit by these two optical probes will not be shown effectively.

Table 14.3 Fluorescence intensity enhancement of rhodamine 6G.

Excitation means	Samples	Integrated fluorescence intensity (arbitrary units)	Enhancement of fluorescence intensity
Optical fibre tip with 355 nm	Rhodamine 6G 10^{-6} M	74 500	
Optical fibre tip with 355 nm	One drop of 0.025 M anthracene floating on rhodamine 6G (10^{-6} M)	1 845 000	24.8

8   Excitonic supertip development

Even smaller probes are required and are in the making. In principle, an exciton source can be as small as a single molecule or atom. At the same time, its position and scanning have to be defined in space equally well to those of an STM tip (a randomly flying atom does not qualify). Existing designs for optical supertips [12,15] are based on the same principle as the green plant photosynthetic system. A submicrometre antenna collects the photons by absorption and transfers the excitation energy to a single active centre. From there the energy is either (i) radiated as a photon or (ii) transferred to the sample in an energy transfer process (Förster–Dexter). In either case the result is generally affected by the nearby sample molecule: (i) the radiated excitation may be effected, for example, by intramolecular spin–orbit coupling (Kasha effect); (ii) the energy transfer results in a fluorescence or phosphorescence typical of the sample molecule. In the latter case, only virtual photons are produced by the supertip; this gives an excitation transfer tip ('exciton tip') and only sample luminescence is detected.

Supertips were first introduced in AFM [50]. Since then they have been employed in a few other areas. We have prepared optical supertips for MEM. Supertip design considerations involve optical, excitonic, photochemical, and mechanical properties of the luminescent point source. There are many methods and reasons for making supertips. For example, the optical fibre tip can, in principle, be treated chemically, so as to produce specific supertips for a variety of purposes: (i) wavelength shifters, e.g. crystallites that fluoresce to the red of the tip emission; (ii) time 'extenders', as (i), utilizing prompt or delayed fluorescence or even phosphorescence; (iii) highly sensitive optochemical nanosensors; (iv) energy transfer supertips; (v) heavy atom sensors. There are many considerations in supertip development. An important aspect is an active centre. Taking photosysthesis as an example, a single light-absorbing molecule will only absorb about one out of 10^9 photons impinging on it [12]. A large photosynthetic antenna containing N absorbing chromophores will absorb about N times more photons. The N absorbed photons become N excitons and a large fraction of these make it to the reaction centre (the antenna is both a light collector and an exciton conductor). In MEM, the idea has been to create a miniscule photoactive centre, analogous to nature's reaction centre, relying on direct exciton transfer. Theoretically, for a given excited antenna with N excitons, one reaction centre will catch by itself nearly as large a fraction of the N excitons as 10 or 100 reaction centres, provided that energy transfer is fast enough. This is in contrast to direct photon absorption, for which the probability increases linearly with the number of absorbers (from one to 10 or to 100). The exciton trapping concept was demonstrated for molecular crystals with active centres called 'supertraps' on antenna islands (percolation clusters) [42]. In MEM this principle is highly important. The idea is to have a crystalline antenna that transports a large fraction of the excitations (excitons) to one or a few tip molecules. These molecules may act not only as 'supertraps' but also as the supertip's active centre. Such a supertip can become the ultimately smallest light source (size of a single molecule or chromophore). It can also serve as an 'exciton tip' from which the excitation energy jumps (tunnels, transfers) to an acceptor site on the sample. In such a way a small or dilute acceptor sample will be excited much more efficiently than by far-field (or even near-field) direct photon excitation (with a given primary photon source, such as a laser). Thus a supertip should be designed in such a way that an active centre will be able to perform the excitation energy jumps, and will have a plentiful supply of energy quanta from the supertip.

Supertips can double as NSOM and as MEM tips. The superresolution imaging of biologically interesting species relies on this dual function. The MEM operation itself involves different mechanisms: (i) active energy transfer (Förster–Dexter) where the supertip is the exciton (energy) donor and the sample is the acceptor; (ii) passive interaction (sensor mode), exemplified by the Kasha effect, where the sample either quenches the supertip's exciton or transforms it from 'singlet' to triplet'. This provides the basis for its various potential applications.

8.1 *Relation of tip, supertip and active centre*

MEM relies on quantum optics mechanisms, such as Förster energy transfer or the Kasha effect (external heavy atom effect). The interactions occur at the interface between the tip (its active centre) and the sample (the two are quantum mechanically coupled). For the best resolution, i.e. single molecule detection, this active centre consists of a single molecule, or molecular cluster, that does the imaging. This molecule is the energy donor site for the Förster energy transfer, or the spin–orbit interaction site for the Kasha effect. Figure 14.12 shows the relation between the tip, the supertip, and the active centre. It is quite similar to an antenna [45]. Here the antenna is a molecular crystallite of aggregate (such as anthracene, DPA). One of its molecules ('active centre') is closest to the single sample molecule, such as tetracene or perylene. The distance between them is 10–20 Å and is much smaller than the Förster radius. The antenna (about 10^6 molecules) transfers excitations to the active centre, which in turn transfers its energy to the single sample molecule. Overall, only 10^4–10^6 photons are now required for a single excitation of the sample molecule, i.e. 10^3–10^5 times less than with direct excitation. A completely analogous situation is found in scanning force microscopy (SFM) [51], where the active centre is the force contact site at the tip of the supertip. In MEM the active centre has to be optically excited repeatedly. The design of the tip system is thus geared towards the need of supplying the active centre with plenty of excitation quanta. The ideal tip for MEM is a supertip with an active centre of molecular size. We describe below several approaches for the developments of supertips which have been constructed on both micropipettes and optical fibre tips.

8.2 *Nanofabrication of supertips*

We have developed several methodologies in preparing supertips. They include crystal tip dipping, fibre tip dipping, controlled photonanofabrication, polymer matrix supertip preparation, and molecular engineering synthesis. All of these techniques are still in their primitive stages, and further development is needed. For example, a polymeric matrix attached to an optical fibre tip by spatially controlled photopolymerization is quite effective in making supertips. The polymer is a copolymer, consisting of acrylamide, N,N-methylenebis(acrylamide) (BIS) and appropriate dye monomer groups [18].

This polymer supertip acts as an antenna, even though less efficient compared with a molecular crystal. However, with the large photon flux emanating out of the tip this lower efficiency should suffice. At the very tip of this supertip the 'active centre' is attached by physical and chemical methods. The dye molecule on the top surface of the polymer is produced by photochemical reaction from a layer of precursor molecules deposited on the polymer tip by dipping into a solution. The highest probability for photochemical reaction is at the centre of the tip, thus making it likely that the first and only active molecule is produced at this centre. Thus supertips with relatively large physical size and extremely small active centre can be prepared [21].

8.3   *Nanocrystal designer probes*

Another approach employs exciton–conducting crystallites that absorb the light of the fibre optic tip, convert it into excitons, which ultimately produce photons again. We have successfully grown such crystals of perylene and DPA onto the fibre tip [21]. Alternatively, the crystallite is grown at the tip of a micropipette and the fibre optic tip is pushed deep into the pipette, very close to the crystal tip. The crystallite acts as an antenna that channels the excitons to an active centre which acts as an exciton trap [42]. This trap collects excitation from as far as 500–1000 Å.

The single impurity molecule (the 'supertrap') creates a host 'funnel' around it [22,42]. This funnel consists of host crystal molecules perturbed by the impurity ('guest') molecule. The closer the host molecule is to the trap, the lower its excitation energy. The molecules in the funnel act as exciton traps, catching the excitation from the host crystal and passing it deeper and deeper (in energy) to the deepest of them all, the supertrap. For MEM the guest molecule is deposited onto the surface of a molecular microcrystal and thus creates an energy funnel at its apex (active centre). The micropipette-based exciton tip is excited internally by a fibre optic tip. It gives as much intensity as with epi-illumination, i.e. external excitation method [35]. The crystal tips absorb the incident light, create excitons, and emit light again, or transfer energy to the active centre and emit a lower energy fluorescence. These tips are exciton supertips. By optimizing the selection of host-crystal guest-supertrap pairs, we have been able to prepare supertips with high intensity and good stability.

First, coated micropipettes are used to grow organic crystals and doped polymers. After very bright spots from the tip can be seen under a microscope, we dip the tip end surface into various concentrations of aqueous solutions with active centre molecules. The concentration is controlled so that predetermined numbers of molecules will stick to the luminescent material surface. For example, we dipped DPA crystal tips into rhodamine B aqueous solutions. A few rhodamine B molecules deposited onto the surface of the DPA crystal inside the micropipette tip. Similar procedures and results have been obtained for fibre tip dipping experiments. We have attached various molecular crystal supertips to the nanofabricated fibre optic tips. There are two different approaches: the first is to use bare tip [21], and the second is to use an exciton source prepared by near-field photochemical synthesis, such as a sensor tip. The dipping process is the same as described for a micropipette tip. The emission from the active centre is clearly observed in our experiments. For example, we have prepared a supertip with DPA and rhodamine B (RhG). The supertip was dipped into different concentra-

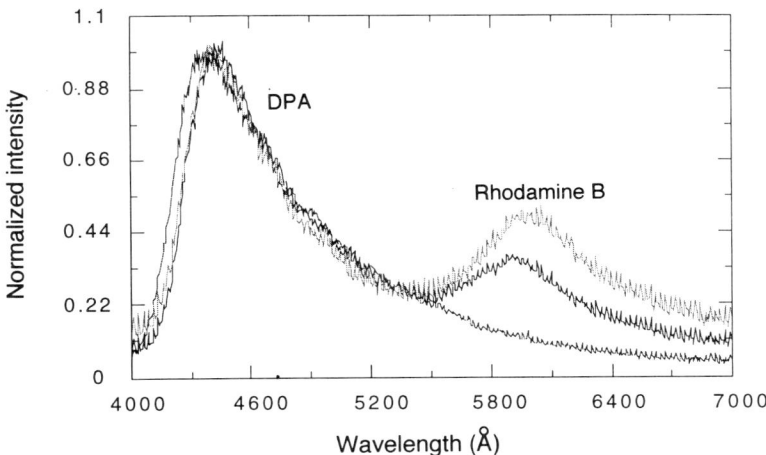

Figure 14.18 Fluorescence spectrum of active centres (rhodamine B) and supertip nanocrystal (diphenylanthracene).

tions of RhG aqueous solutions, and spectra of the supertip were taken with internal excitation by nanofabricated optical fibre tips. Figure 14.18 shows a group of spectra taken for DPA crystal tips dipped into RhG aqueous solutions. From the spectra, it is clear that even dipping into very diluted RhG solution, the emission of the tiny RhG active centre is still detectable in our low light collection efficiency optical multichannel analyser apparatus [19]. The result clearly demonstrates that supertips can be prepared by dipping nanometre crystal tips into solutions. We have estimated the number of RhG molecules on the supertip surface. Statistically, for the tip dipping volume (for a 0.5 μm tip), there are only about 80 RhG molecules for a concentration of 10^{-5} M RhG solution. Since the dipping time is short, only some of these 80 RhG molecules have been 'bonded' to the DPA crystal. The DPA crystal tip absorbs the incident light, creates excitons, and transfers its energy quanta to RhG molecules which emit light again.

8.4 Single dendrimer design

We have also designed a supertip made of a single symmetric macromolecule by using newly developed dendrimer supermolecules [27,28]. The so-called starburst phenylacetylene dendrimers include the largest so far synthesized structurally ordered molecule (D_{127}; see Fig. 14.19). These fractal, tree-like supermolecules have spatially localized eigenfunctions and, in particular, localized electronic states [58]. Furthermore, in the so-called SYNDROME family of dendrimers [28], simple theory leads to selectively lower excitation energies at the central locus of the ordered macromolecule, with energies increasing towards the rim. This is borne out experimentally by the vibronic spectra of the entire family of molecules (D_2 to D_{127}), with full internal consistency in the observed electronic energies (red shifts), vibrational quanta, Franck–Condon factors, overall transition moments, and picosecond spectral diffusion [58]. The architecture of this series of dendrimers is controlled by organic synthetic methods. For example, the overall shape of a 'D_{127}' molecule is bowl-like with a molecular size around 125 Å. It can thus act as both optical and force active centre. The large 'rim' is

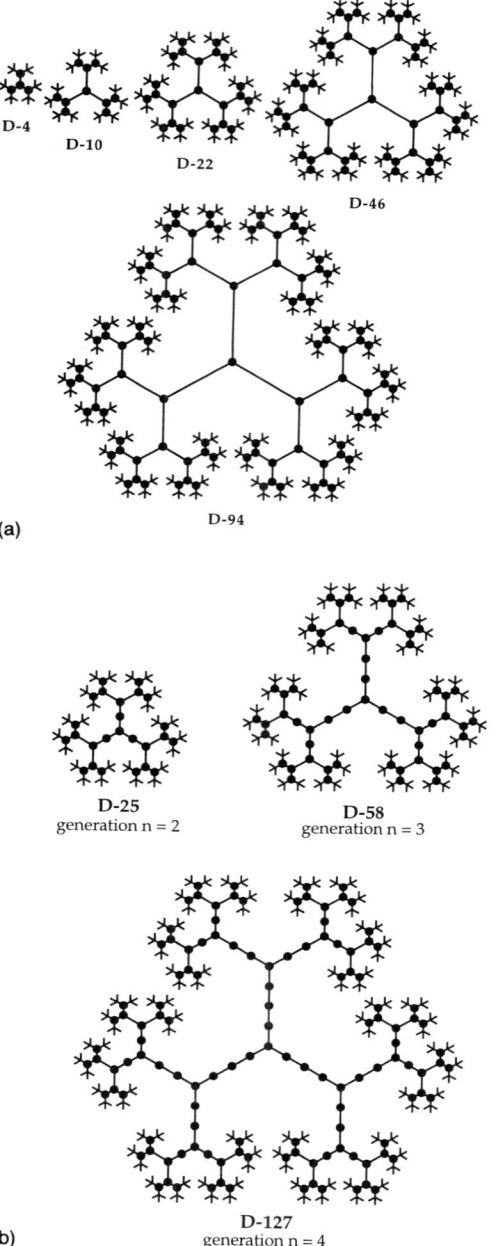

Figure 14.19 Dendrimers. (a) The phenylacetylene dendrimers by the convergent method. (b) The phenylacetylene dendrimers by the SYNDROME method [27,28].

bound to the tip by cumulative van der Waals bonding or covalent bonds D_{127} may be used in supertip preparation in two ways. The first is to make supertips in the range of 100 Å: in this case, D_{127} is an energy transfer acceptor. It traps most of the energy quanta from the bulk of the tip, as shown below. Thus D_{127} is the light-emitting active centre. The supertip prepared in this way follows this scheme:

tip → supertip → active centre → sample.

The second way for supertip preparation is to achieve about 10 Å resolution. To further demonstrate our 'energy funnel' model (see above) we synthesized partial dendrimeric wedges with (and without) an excitation acceptor, a perylene derivative pendant, at the locus. This molecule is an ordered supermolecule transducer of absorbed radiation (STAR) [52], in analogy to the primary excitation energy-collecting antenna of some natural photosynthetic systems. The key step in the nano-STAR synthesis is shown in Fig. 14.20, where the dendrimer fragment is represented in a simplified format such that each arene is indicated by a filled circle and acetylene linkages are shown as solid lines. The arene rings are either substituted at the 1,3,5-positions or the 1,4-positions as indicated by the 120° or 180° angles at each filled circle, respectively. Monodendron-iodide (**1**) reacted with 3-ethynylperylene (**2**) under standard cross-coupling conditions to give, after purification, a single high molecular weight product by size exclusion chromatography. The resulting solid is amorphous and can be cast into golden yellow films. As expected, the energy transfer from the large antenna ('tree canopy') to the small acceptor centre is indeed dramatic. The presence of the antenna (39 phenyl groups) increases the yield of the yellow perylenic emission by three orders of magnitude for a given excitation wavelength. Overall, such a photonic subwavelength nano-lens may play a role in developing molecular excitonics, including luminescent optical nanoprobes, scanning exciton tunnelling microscopy, and nanometre-scale fibre optic chemical and biochemical sensors. In order to synthesize such a STAR molecule, the nanoarchitecture process is modified to prepare a D_{127} molecule with a supertrap in the centre. A single such substituted group acts as a supertrap [42], collecting most of the excitation. It may act as an exciton donor, transferring excitation to an acceptor on the sample, and as the smallest possible light source. The supertrap in the nano-STAR has a size of about 10 Å. In such a supertip case, the supertrap centre has different optical and excitonic properties from its surrounding molecules. The centre can be used as a scanning tip which defines the scanning resolution. The supertip prepared in this way follows the following scheme:

tip → supertip → antenna → active centre → sample.

The design of the STAR molecule is not only for accelerated energy transfer but also for backward 'spill-over' of redundant energy in the pendant group. For instance, if a second excitation arrives while the first is still there, the two excitations are expected to 'fuse' [22] and the resulting higher energy excitation will transfer backward into the dendrimeric antenna state. There, having many more non-radiative energy decay channels, it will be reduced from an S_n to an S_1 excitation. Then the S_1 excitation will

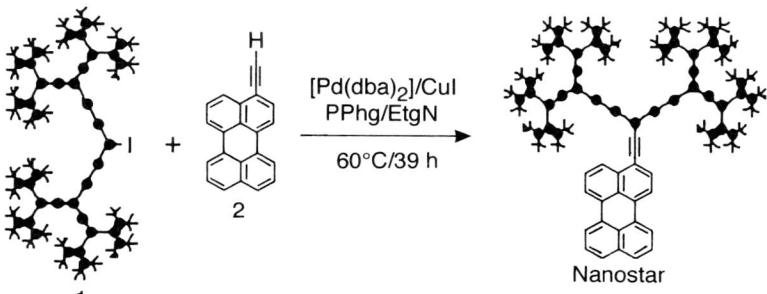

Figure 14.20 Nano-supermolecule transducer of absorbed radiation (STAR) synthesis.

flow again into the pendant 'supertrap'. The nano-STAR represents a new class of 'designer' molecules, tailor-made for single molecule light and exciton sources [1,2]. We note that its large size (125 Å for D_{127}), stability, and efficiency will allow it to be used as a 'supertip' for optical nanoprobes and nanosensors.

8.5   *Spectroscopic studies of D_{127} dendrimer*

We have studied the spectroscopic properties of D_{127} series, shown in Fig. 14.21. D_{127} is an excellent luminescent material for supertip development. D_{127} is optically very stable for supertip development. The fluorescence spectra of D_{127} at different time periods of continuous excitation with 350 nm have been obtained. It appears that the optical stability of D_{127} is much better than that of anthracene or perylene, and close to that of DPA (this is only a qualitative comparison). We calculated that the D_{127} fluorescence intensity was decreased 30% for the first 45 min, and 14% for the next 30 min under the above conditions, while for anthracene these amounts are 85% and 55% under similar conditions. We have also carried out D_{127} crystal growth experiments with micropipette and fibre tips at room temperature. For a concentration of 4.2 mg D_{127} in 15 ml THF, crystals of D_{127} are deposited at the micropipette tips. Under the fluorescence microscope, bright greenish light was observed. It is expected that D_{127} will be useful for exciton probe and supertip development.

9   Summary and discussion

A host of nanofabrication techniques have been developed to prepare subwavelength excitonic probes. Their sizes are as small as 200 Å. They have been constructed with the aid of exciton-transporting materials. More than 20 organic and inorganic materials and their doped polymers have been used in the preparation of exciton sources. Subwavelength exciton probes can be used as scanning, light-emitting, or exciton-generating donor tips of subwavelength dimensions. These exciton light sources are of both theoretical and practical interest. Design considerations involve optical, excitonic, photochemical, and other properties of the luminescent source. The principle of supertips for molecular exciton microscopy has been illustrated, and different methods for the preparation of supertips have been developed. The prelim-

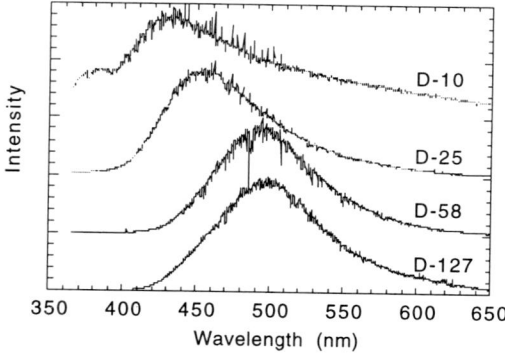

Figure 14.21 Fluorescence emission spectra of dendrimers. D10, D25, D58 and D127 solids irradiated by 355 nm UV light at room temperature.

inary applications of nanometre exciton sources have demonstrated their great potential in a wide variety of areas. Various subwavelength exciton probes have been used in probe-to-sample interaction experiments. When a luminescent point source is scanned over a sample, it senses a variety of perturbations such as quenching or external heavy atom effects. We have achieved a large enhancement in fluorescence when the probe and the sample are within energy range. Also, the first interfacial Kasha quenching effect has been demonstrated. Further, a 50 times enhancement of fluorescence detection sensitivity has been obtained using an exciton source. The exciton approach, using a direct source–sample interaction, presents a significant improvement in single molecule sensitivity. The large variety of nanoprobes should lead to new applications in NFO as well as in other microscopic analysis in a variety of sciences and technologies. We expect that exciton probes will be further developed, and will play an important role in the non-invasive detection and manipulation of single molecules. They will eventually lead to an optical microscopy technique with single molecule spatial resolution, nanosecond or picosecond temporal resolution, and single molecular biochemical sensitivity.

The principle of an exciton source has been demonstrated. The theoretical resolution limit of this kind of scanning probe microscopy is on the atomic or molecular scale. But the nanofabrication of exciton sources is still in its preliminary stage. There are many technical problems to be solved — from understanding the excitons in a nanoprobe to the control of photobleaching (a standard problem in fluorescence microscopy). However, the future looks bright. More controllable methods are needed to produce usable exciton sources. In recent years, there have been many new developments in molecular engineering. How to transfer the technologies in other fields to the production of small, intense, and optically stable exciton probes will be the key question for us to answer in the next few years.

The idea of using molecular excitons to prepare active light and exciton sources to improve the resolution of optical microscopy by orders of magnitude has been considered for over 7 years [4]. Theoretical results and limited experimental data were obtained. All these results lend credence to the theoretical modelling of MEM. However, there are many key questions that still need to be answered. What is the nature of the exciton scanning tip? What is the exciton concentration at the tip surface? Is it higher or lower than that in the bulk? How efficient is the process by which the excitons produced at the bulk crystal move towards the tip surface? What is the lowest limit for the source size? Does the 'funnel' principle work at the crystal tip? Does the imaging of an exciton light source involve photons, evanescent photons, or excitons? There are many other questions to be answered before any practical application of these exciton sources. We are facing some technical challenges in the following areas.

9.1 *Molecular engineering*

How can we prepare an ideal exciton tip in which there is a high concentration of excitons at the tip end surface? How can we prepare an exciton tip with molecular dimension? How can we prepare an optically stable scanning tip? All these problems are related to molecular engineering. There are two kinds of molecular engineering problems. The first has to do with the physical nature of the scanning tip, i.e. the smallness and the location of the active centre. The second molecular engineering problem is related to the electronic and optical nature of the probe. It deals with the question of how to prepare an optically stable scanning active centre with a high throughput of excitons. The position of the active centre is crucial for MEM imaging, especially when the tip–sample interaction creates the imaging. Recently, there has been great interest in controllable nanofabrication by molecular engineering. If the STM-controlled atom or molecule movement can be realized in the exciton light source preparation, then the location of any individual molecule is fully controllable. There should then be a way of creating an active centre at the top surface of an exciton tip. The location problem will also benefit from our efforts in studying and synthesizing dendrimers [21,52]. Traditional organic synthesis was used to prepare dendrimers by nanoarchitecture [27], and state-of-the-art photopolymerization was employed to produce nanometre-sized polymer tips [18] by controllable photonanofabrication. Also, work on an analogue photosynthesis antenna may help to design scanning exciton tips.

9.2 *Photostability of exciton probe*

The problem of photostability will need great progress in the second kind of molecular engineering. The new developments discussed above will help solve the photostability problem partially. As is well known, many organic and inorganic dyes are not optically stable enough to be used as a light- or exciton-generating source. We have used more than 20 luminescent materials to prepare exciton light probes, but photobleaching has always been a problem. Even though some tips may be used for scanning, the lifetime will be very short. Thus more stable luminescent materials are needed. Except for those mentioned above, there are a few possible sources for obtaining better luminescent materials: (i) organometallic compounds; (ii) rare earth metal ions and uranyl compounds; (iii) mixed solid solutions [21]; (iv) new dyes developed in other fields, such as for cellular biology and solar energy. Besides better optical materials, better protection of the active centre will prolong the lifetime of an exciton light source. For example, the removal of oxygen will greatly alleviate photo-oxidation of an anthracene exciton tip. Nevertheless, photostability will always be a problem in any fluorescence-related technology. Thus extra precautions will always be necessary. While there may be other problems such as image contrast, low light intensity detection, feedback mechanism, signal-to-noise ratio, or sample preparation, it is the tip preparation that has always been the main challenge in the development of any SPM. Technological advances in molecular engineering and in other fields are expected to speed up the development of MEM.

Acknowledgements. We thank our colleagues Jeff Moore, Zhong-You Shi, Michael Shortreed amd Zhifu Xu for their discussions and help. Financial support came from DOE grants DE-FG02-90ER61085 and DE-FG02-90ER60984. The excitonic light source research and development was supported by NSF grant DMR-9410709.

10 References

1 Kopelman R, Tan W. *Science* 1993; **262**: 1382.
2 Kopelman R, Tan W. In: Morris MD, ed. *Spectroscopic and Microscopic Imaging of the*

Chemical State. New York: Marcel Dekker, 1993: 227.

3 Tan, W *et al. ACS Symp Series* N. Akmal, ed. (in press).

4 Kopelman R, Lewis A, Lieberman K. In: Attwood D, Barton B. *X-Ray Microimaging for the Life Sciences.* CA: Lawrence Berkeley Laboratory, 1989: 166.

5 Abbè E. *Arch Mikroskop Anat* 1873; **9**: 413.

6 Lewis A, Isaacson M, Muray A, Harootunian A. *Ultramicroscopy* 1984; **13**: 227.

7 Pohl DW, Denk W, Lanz M. *Appl Phys Lett* 1984; **44**: 651.

8 Lieberman K, Harush S, Lewis A, Kopelman R. *Science* 1990; **247**: 59.

9 Betzig E, Trautman JK, Harris TD, Weiner JS, Kostelak RL. *Science* 1991; **251**: 1468.

10 Lewis A, Betzig E, Harootunian A, Isaacson M, Kratschmer E. In: Loew LM, ed. *Spectroscopic Membrane Probes*, Vol. II. CRC Press Cambridge, 1988: 81.

11 Kopelman R, Tan W, Birnbaum D. *J Lumin* 1994; **58**: 380.

12 Kopelman R, Tan W. *Appl Spectr Rev* 1994; **29**.

13 Betzig E, Trautman JK. *Science* 1992; **257**: 189.

14 Tan W *et al.* In: Masuhara H *et al. Microchemistry, Spectroscopy and Chemical in Small Domains.* Amsterdam: Elsevier Science, 1994: 301.

15 Tan W, Shi Z, Smith S, Kopelman R. Photonanofabrication and optical nanoprobes, *Mol Cryst Liq Cryst Sci Technol Sect. A* 1994; **252–253**: 535.

16 Betzig E, Chichester RJ. *Science* 1993; **262**: 1422.

17 Birnbaum D, Kook SK, Kopelman R. *J Phys Chem* 1993; **97**: 3091.

18 Tan W, Shi Z-Y, Smith S, Birnbaum D, Kopelman R. *Science* 1992; **258**: 778.

19 Tan W, Shi Z-Y, Kopelman R. *Anal Chem* 1992; **64**: 2985.

20 Zhao X, Kopelman R. *Ultramicroscopy,* 1995; **61**: 69.

21 Tan W. PhD thesis. University of Michigan, Ann Arbor, 1993.

22 Pope M, Swenberg E. *Electronic Processes in Organic Crystals.* New York: Oxford University Press, 1982.

23 McDonald A. *IEEE Trans Microwave Theory Tech MTT-20* 1972; **698**

24 Agranovich VM, Galanin MD. *Electronic Excitation Energy Transfer on Condensed Matter.* Amsterdam: North Holland, 1982.

25 Reddick RC, Warmack RJ, Chilcott DW, Sharp SL, Ferrell TL. *Rev Sci Instrum* 1990; **61**: 3669.

26 Binnig G, Rohrer H, Gerber Ch, Weibel E. *Phys Rev Lett* 1982; **49**: 57.

27 Xu Z, Moore JS. *Angew Chem* 1993; **32**: 1354.

28 Xu Z, Shi Z, Tan W, Kopelman R, Moore JS. *Polymer Preprints* 1993; **33**: 130.

29 Orr BG. In: Morris MD, ed. *Spectroscopic and Microscopic Imaging of the Chemical State.* New York: Marcel Dekker, 1993.

30 Nishimura H, Yamaoka T, Hattori K, Matsui A, Mizuno K. *J Phys Soc Japan* 1985; **54**: 4370.

31 Junter G. *Electrochemical Detection Techniques in the Applied Biosciences, Vol. 1: Analysis and Clinical Applications.* New York: Ellis Horwood, 1988.

32 Brown KT, Flaming DG. *Advanced Micropipette Techniques for Cell Physiology.* New York: Wiley, 1986.

33 Paesler, M and Moyer, P. *Neur-field optics: Theory, Instrumentation and Applications.* John Wiley & Sons. Inc. (1996).

34 Frank K, Becker MC. In: Nastuk WL, ed. *Physical Techniques in Biological Research*, Vol. V. New York: Academic Press, 1964: 22.

35 Lewis A, Lieberman K. *Nature* 1991; **354**: 214.

36 Kopelman R, Koo Y. *Israel J Chem* 1991; **31**: 147.

37 Wolfbeis OS, Weis L, Leiner MJP, Ziegler WE. *Anal Chem* 1988; **60**: 2028.

38 Tan W, Shi Z, Kopelman R. *Sensors and Actuators* 1995; **B28**: 157.

39 Van der Auweraer M, Verschuere B, Biesmans G, De Schryver FC, Willig F. *Langmuir* 1987; **3**: 992.

40 Vitukhnovskii AG, Sluch MI, Warren JG, Petty MC. *Chem Phys Lett* 1990; **173**: 425.

41 Jankowiak R, Kolinowski J, Konys M, Buchert J. *Chem Phys Lett* 1979; **65**: 549.

42 Francis AH, Kopelman R. In: Yen MW, Selzer PM, eds. *Topics in Applied Physics, Laser Spectroscopy of Solids*, 2nd edn. Berlin: Springer-Verlag, 1986; **49**: 241.

43 Kasha M. *J Chem Phys* 1952; **20**: 71.

44 Mataga N, Kubota T. *Molecular Interactions and Electronic Spectra*. New York: Marcel Dekker, 1970.

45 Fox MA, Jones WE, Watkins DM. *Chem Eng News* 1993; 38.

46 Barnes MD, Whitten WB, Ramsey JM. *Anal Chem* 1995; 418A.

47 Smith S, Monson E, Merritt G *et al. SPIE* 1993; **1858**: 81.

48 Kuhn H, Möbius D, Büchner H. In: Weissberger A, Rossiter BW, eds. *Physical Methods of Chemistry*, Part IIIB. New York: Wiley, 1972: 577.

49 Roberts G, ed. *Langmuir–Blodgett Films*. New York: Plenum Press, 1990.

50 Hansma, HG *et al. Science*, 1992; **256**: 1180.

51 Moers MHP, Tack RG, van Hulst NF, Bolger B. A combined near field optical and force microscope, *Scanning Microsc* 1993; **7**: 789.

52 Shortreed M, Shi Z, Tan W, Xu Z, Moore JS, Kopelmam R. *Science* 1995; (in press).

53 Avouris P. *Acc Chem Res* 1995; **28**: 95.

54 Born M, Wolf E. *Principles of Optics*. London: Pergamon Press, 1959.

55 Salling CT, Lagally MG. *Science* 1994; **265**: 502.

56 Tan W, Shi Z, Thorsrud BA, Harris C, Kopelman R. *SPIE* 1994; **2068**: 59.

57 Thomas RC. *Ion-sensitive Intracellular Microelectrodes*, New York: Academic Press, 1978.

58 Kopelman, R *et al.*, Phys Rev Cott, 1997; **78**: 1239.

15 Protein-based Three-dimensional Memories and Associative Processors

R.R. BIRGE, B. PARSONS, Q.W. SONG and J.R. TALLENT

Department of Chemistry and W.M. Keck Center for Molecular Electronics, Syracuse University, Syracuse, NY 13244, USA

1 Introduction

This chapter reviews the use of the light-transducing protein bacteriorhodopsin as the photoactive element in optoelectronic devices, holographic associative processors, and linear and non-linear three-dimensional optical memories. Biological molecules, used in their wild-type form or genetically and/or chemically modified, have inherent advantages as active components in optoelectronic devices [1]. Natural selection has already solved many of the key problems regarding the use of organic molecules in logic, switching, or data manipulative functions. Salient examples of biological molecules that have been investigated for optoelectronic applications include visual rhodopsin [1], bacteriorhodopsin [2–4], chloroplasts [5,6], photoactive yellow protein [7,8] and photosynthetic reaction centres [9]. This review will concentrate on optoelectronic applications of bacteriorhodopsin. Serendipity and natural selection have yielded a native protein with very useful properties for both linear and non-linear optical applications. Chromophore substitution, chemical modification, and genetic engineering are in active use to optimize the protein for specific applications. In this chapter, we will emphasize the use of the native protein in optoelectronic devices. However, the use of chemical and genetic methods to enhance the protein will be discussed briefly.

2 Bacteriorhodopsin

2.1 *The native protein*

Bacteriorhodopsin (bR; molecular weight ~26 000) is the light-harvesting protein contained within the purple membrane of *Halobacterium halobium* (also called *Halobacterium salinarium*) [10,11]. This organism thrives in marshes with salinity up to six times higher than that of sea water. The purple membrane consists of a semi-crystalline protein trimer in a phospholipid matrix (3 : 1 protein to lipid) which constitutes a specific functional site as a proton pump in the plasma membrane of the bacterial cell. The protein has seven trans-membrane α-helices which make up the protein's secondary structure. The bacterium synthesizes the purple membrane when dissolved oxygen concentrations become too low to sustain aerobic ATP production. The production of the purple membrane allows the organism to switch to photosynthesis as a means of energy production. The light-absorbing chromophore of bacteriorhodopsin is all-*trans* retinal (vitamin A aldehyde) (Fig. 15.1) and is connected to lysine via a protonated Schiff base linkage. The absorption of light by the chromophore initiates a complex photochemical cycle, of approximately 10 ms in length, characterized by a series of spectrally distinct thermal intermediates (see Fig. 15.2) [12]. The photocycle transports

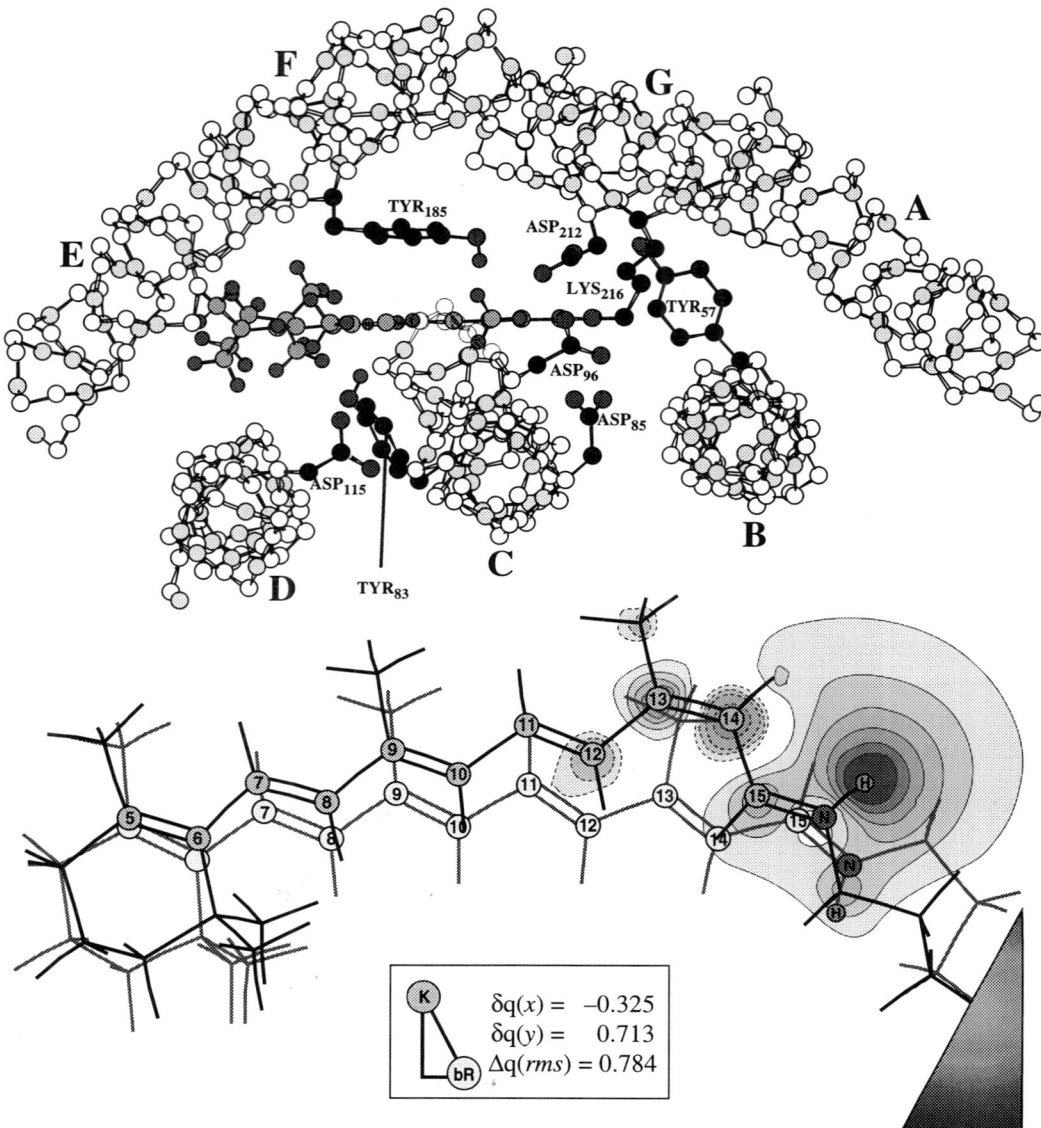

Figure 15.1 The chromophore binding site and the primary photochemical event in bacteriorhodopsin. The upper diagram shows membrane-spanning segments of the protein backbone along with the all-*trans* retinal chromophore and selected amino acids viewed from the cytoplasmic side (data from [4]). The protonated Schiff base chromophore is attached to lysine 216 (LYS$_{216}$) and the amino acids shown were selected based on their importance to the photochemical properties of the protein (ASP, aspartic acid; TYR, tyrosine). The bottom diagram shows a model of the primary photochemical event [bR (grey; underneath) → K (black; above)] and the shift in charge that is associated with the motion of the positively charged chromophore following 13-*trans* → 13-*cis* photoisomerization (Δq in Ångstroms). It is believed that the initial photoelectric signal is due primarily to the motion of the chromophore.

a proton from the intracellular to the extracellular side of the membrane. This light-induced proton pumping generates an electrochemical gradient which the bacterium uses to synthesize ATP in accordance with Mitchell's chemiosmotic model of energy transduction.

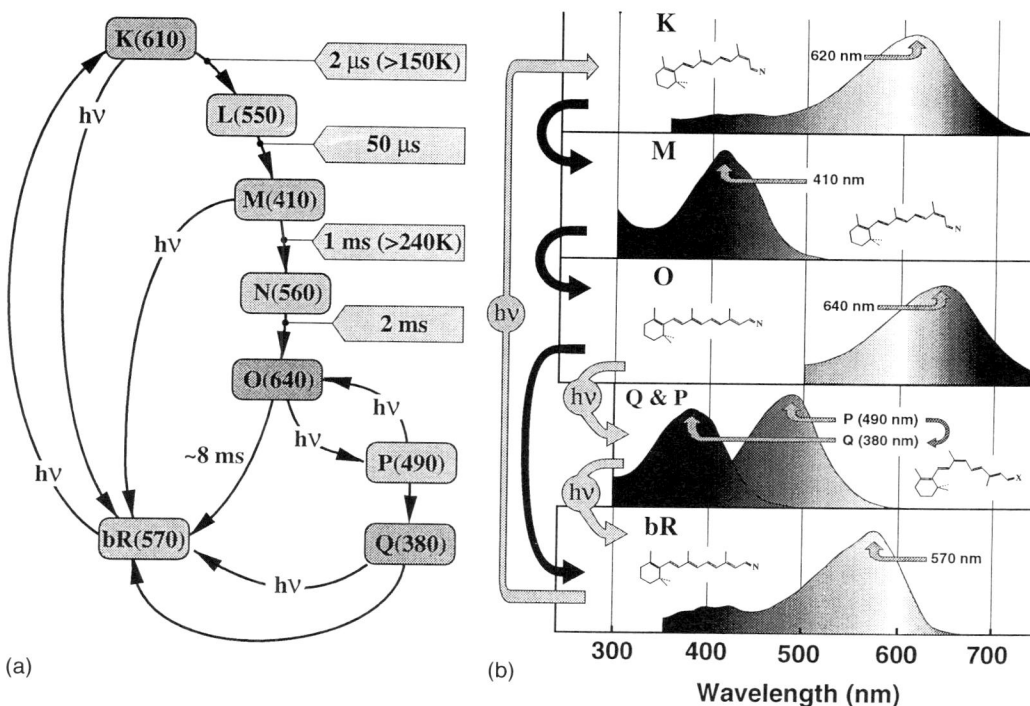

Figure 15.2 A simplified model of the light-adapted bacteriorhodopsin photocycle (a) and the electronic (one-photon) absorption spectra of selected intermediates in the photocycle (b). The height of the symbols in (a) is representative of the relative free energy of the intermediates, and the key photochemical transformations relevant to device applications are shown. Note that only selected intermediates are shown.

A robust protein resistant to both thermal and photochemical damage is required because this organism must function in the harsh salt marsh environment. Indeed, the protein can withstand temperatures as high as 140°C [13,14]. Photochemical stability is measured by the number of times the protein can be photochemically cycled between intermediates before denaturing. This cycling capability is called the 'cyclicity' and it exceeds 10^6 for bacteriorhodopsin, a value considerably higher than those observed in known synthetic photochromic materials. This high value is due to the protective features of the integral membrane protein which serves to isolate the chromophore from reactive oxygen, singlet oxygen, and free radicals. Thus, the common assumption that biological materials are too fragile to be used in optoelectronic devices does not apply to bacteriorhodopsin. The optoelectronic properties of bacteriorhodopsin have been reviewed in detail [1,2,15–22], and our discussion will be selective, with an emphasis on those properties most critical to device applications.

The absorption spectra of the key intermediates are shown in Fig. 15.2(b). The unique absorption spectrum of each intermediate is due to changes in the geometry of the chromophore as well as changes in the binding site. The chromophore, which is covalently bound to lysine 216, carries a net positive charge. The chromophore interacts strongly with neighbouring charged amino acids as well as a nearby positively charged divalent cation [23]. Through genetic engineering and/or chemical modification, the dispersive and electrostatic properties of the binding site can be modified to yield analogue proteins with different optoelectronic properties [2,24–32].

A near instantaneous ($<10^{-15}$ s) shift of electron density occurs with negative charge moving towards the nitrogen atom along the polyene chain upon absorption of a photon of light energy by the chromophore. This electron density shift interacts with negatively charged residues located nearby, which in turn potentiates a rotation around the $C_{13}=C_{14}$ double bond, thereby generating a 13-*cis* chromophore geometry (see Fig. 15.1). The primary photochemical event takes place in less than 500 fs [33–35]. The reason for the unusually high isomerization speed is due to a barrierless excited state potential surface [1], thus bacteriorhodopsin is the biological analogue of high electron mobility transistor (HEMT) devices [1,17].

Photoreversibility is another interesting feature of the chromophore *trans* → *cis* isomerization. Thus, irradiation of the protein with a wavelength within the absorption band of K results in the re-formation of the ground state. Many of the early proposed optical memories based on bacteriorhodopsin utilized the rapid and reversible photochemical switching between bR and K. However, the requirement that liquid nitrogen be used to arrest the photocycle in the K intermediate prevented commercialization of these devices (e.g. [36]). Finally, the isomerization of the protonated chromophore also induces a shift in positive charge perpendicular to the lipid membrane and generates a potential with an (instrumentation limited) 5 ps rise time [37].

From both a physiological and photonic perspective, the most significant photochemical intermediate is the blue-light-absorbing M intermediate. M forms ~50 μs after the absorption of a photon of light by bR and the rapid isomerization to the K state. Normally, M thermally reverts to the ground state with a time constant of about 10 ms. In this stage of the photocycle, the Schiff base proton of the chromophore is transferred to ASP_{85} which results in a chromophore exhibiting a highly blue-shifted absorption spectrum. Most importantly, bR can also be photochemically regenerated from M by the absorption of blue light. This property of a material, where a ground state photoinitiated reaction results in a relatively long-lived thermal intermediate which can also be photochemically driven back to the ground state, is called photochromism. The photochromic properties of bacteriorhodopsin are summarized below:

$$\text{bR (state 0) } (\lambda_{max} \simeq 570 \text{ nm}) \; \overset{\Phi_1 \sim 0.65}{\underset{\Phi_2 \sim 0.65}{\rightleftarrows}} \; \text{M (state 1) } (\lambda_{max} \simeq 410 \text{ nm}), \qquad \text{(Scheme 15.1)}$$

where the quantum yields of the forward reaction (bR to M) and reverse reaction (M to bR) are indicated by Φ_1 and Φ_2, respectively. The quantum yield is a measure of the probability that a reaction will take place after the absorption of a photon of light where a photochromic material possessing a quantum yield of unity and a comparatively long thermal intermediate lifetime is considered extremely light sensitive. One inherent advantage of bacteriorhodopsin as an optical recording medium is the high quantum efficiency with which it converts light into a state change. Complementing this property is the relative ease at which the thermal decay of M can be prolonged. The property where the M → bR thermal transition is highly susceptible to temperature, chemical environment, genetic modification, and chromophore substitution is exploited in many optical devices based on bacteriorhodopsin. However, it is difficult to modify the protein to generate an M state lifetime sufficiently long to provide for long-term data storage. An alternative form of the protein, however, offers a new photochromic equilibrium that exhibits two long-lived binary states. We discuss this altered form in the following section.

2.2 *The blue membrane*

At temperatures above $-30\,^\circ$C, M exhibits thermal decay back to bR. Consequently, purple membrane (PM)-based materials are, in general, suitable only for those applications in which light-induced refractive index changes need not be long lived. It is, however, possible to transform PM into a material with entirely different optical properties, potentially suitable in applications requiring very long-lived light-induced refractive index changes. The native protein binds approximately 4 mol Ca^{2+} and Mg^{2+} per mol protein at pH 6.0. Displacement or removal of these cations results in a reversible colour change from purple to blue, indicating the formation of a material referred to as the 'blue membrane' ($\lambda_{max} \sim 600$ nm; see Fig. 15.3) [38–43]. Cation displacement is accomplished by acidification of PM, resulting in the 'acid blue membrane' [42,44–46]. Cation removal, which produces the 'deionized blue membrane' [46], is accomplished by a number of methods. Among these are: repeated washings with EDTA/deionized water; treatment with cation exchange resin; treatment with high concentrations of NaCl or KCl with subsequent washings; treatment with 1-aminonaphthalene-3,6,8-trisulphonic acid (ANTSA); and electrolysis [38,41,46,47]. Although the acid form shows a greater tendency to aggregate, the above techniques appear to produce spectrally identical results [46]. The chromophore in blue membrane is present in both all-*trans* and 13-*cis* conformations [44–46]. The molar extinction coefficient of the blue membrane is $\sim 54\,760$ cm^{-1} M^{-1} at 603 nm, somewhat lower than for PM [48].

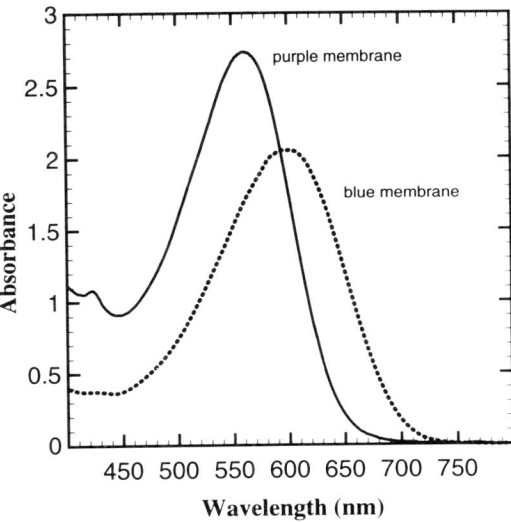

Figure 15.3 The absorption band of bacteriorhodopsin undergoes a red shift upon removal of the divalent cations ($\Delta\lambda = \sim 40$ nm). The resulting cation-free preparation is called the blue membrane, because the resulting red-shifted absorption spectrum yields a product that preferentially transmits light in the 400–500 nm region and appears blue to the eye. The above spectra were measured in dried polyvinyl alcohol films.

In addition to the colour change, blue membrane differs from PM in that it no longer exhibits a photocycle. Shown in Fig. 15.4 is the situation for the acid blue membrane as described by Fischer *et al.* [44]. Acid blue membrane, which forms reversibly below pH $\approx$ 3.2, will, under illumination with red light ($\lambda < 640$ nm), undergo reversible photoconversion to a species absorbing maximally at ~490 nm, with a molar extinction coefficient of ~44 500 cm^{-1} M^{-1} [48]. Like the purple and blue membranes, this species is sometimes referred to by its colour: the pink membrane [46,49]. It has been shown that in the pink membrane, the chromophore assumes a 9-*cis* conformation [44–46]. As was mentioned, this photoconversion is reversible — illumination of acid pink membrane with blue light results in conversion back to blue membrane [44–46]. These two species correspond to a photochromic pair of states, hereafter referred to as pink and blue. Other photoproducts of pink membrane are described as well: bR$_{450}$, which forms reversibly from pink membrane above pH ~ 6, and bR$_{380}$, a thermal decay product of bR$_{450}$ which has been shown to contain free 9-*cis* chromophore [44]. These species will not be discussed further with regard to device applications of the blue membrane. The pink $\leftrightarrow$ blue photochromic system is also observed in deionized blue membrane. Interestingly, the deionized pink membrane is stable in the presence of added cations [46].

The pink $\leftrightarrow$ blue photochromic pair differs significantly from bR $\leftrightarrow$ M in two respects. The first, which has already been alluded to, is that the lifetime of pink is very long, lasting weeks or months [50]. Secondly, the quantum efficiency of photoconversion for pink $\leftarrow$ blue is ~1.6×10^{-4}, roughly 1/4000 that of M $\leftarrow$ bR [48]. (The quantum yield of the reverse reaction, blue $\leftarrow$ pink, is about 50 times greater than the forward reaction, but is still much lower than the quantum yield of the M $\leftarrow$ bR reaction.) As a consequence, materials incorporating blue membrane as the optically active element can be expected to have much lower sensitivities than those based on purple membrane. In some applications, however, this is an advantage. When thin films of the blue membrane are used to store data by using highly focused laser irradiation,

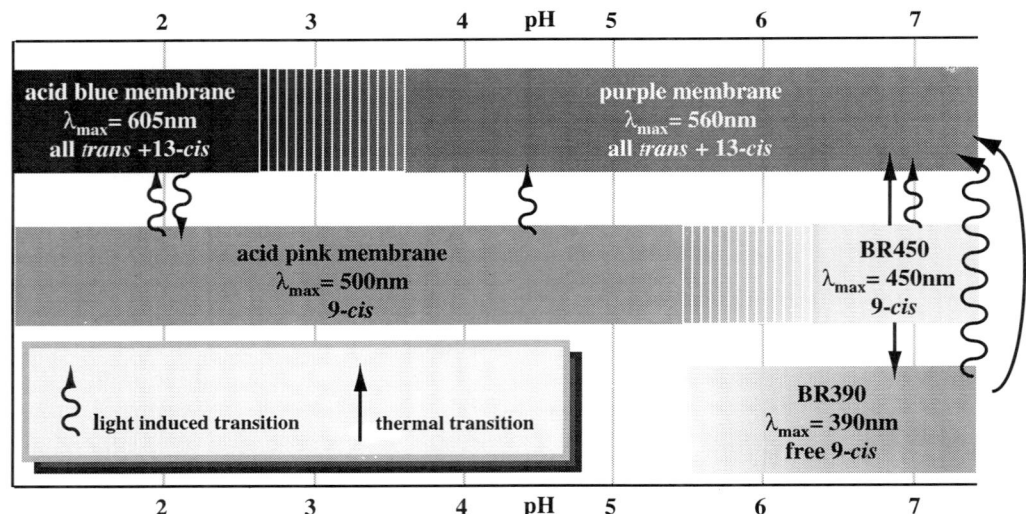

Figure 15.4 The photochemical and thermal interconversions among the family of long-lived intermediates of the acid blue membrane. (After Fischer *et al.* [44].)

the lower quantum yields are of minor consequence as standard diode lasers with milliwatt outputs provide adequate flux to provide interconversion. The lower quantum efficiencies provide stability to ambient light which is important for many applications, where light-tight environments can be provided for storage and operation, but are difficult to maintain during mounting and transfer.

2.3 *Holographic properties*

Thin films of bacteriorhodopsin fabricated by incorporating the protein into optically transparent polymers and polymer blends have shown good holographic performance and the capability of real optical processing [1,2,18,24,25,28–30,51–57]. The mechanism of diffracting light from a volume hologram produced in a photochromic material is important to understand as it is fundamental to many optical applications. We therefore provide a brief overview of this process.

The optical protocol used to record a plane wave hologram in a thin protein film is schematically shown in Fig. 15.5. Two laser beams derived from the same laser and of a wavelength (λ_w) absorbed by bR are overlapped at the plane of the film. Both beams are polarized perpendicular to the plane of incidence and make an angle ϕ_w with respect to the film normal. Due to the coherence properties of laser light, a three-dimensional interference pattern is imposed on the film. The resulting periodic spatial light intensity distribution is schematically shown in Fig. 15.6 and can be mathematically described in one dimension by using Eqn 15.1

$$I(x) = (I_1 + I_2)\,[1 + V\cos(2\pi x/P)], \tag{15.1}$$

where I_1 and I_2 represent the intensities of the individual beams, V is the contrast ratio

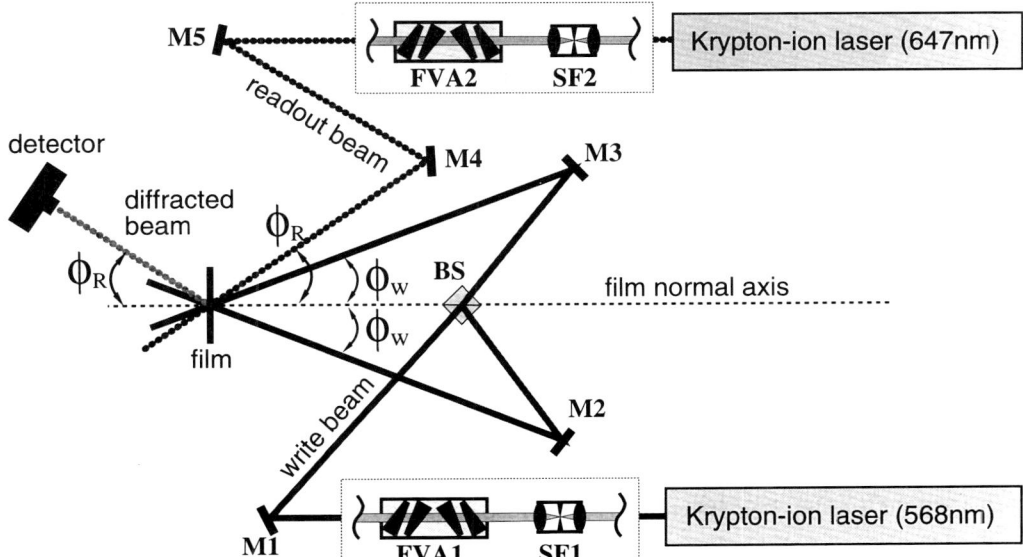

Figure 15.5 A schematic representation of the experimental set used in writing and reading a hologram. A bacteriorhodopsin hologram is written by overlapping two 568 nm beams derived from a krypton ion laser. The hologram is non-destructively read out at the Bragg angle using a probing beam not strongly absorbed by the protein (see text).

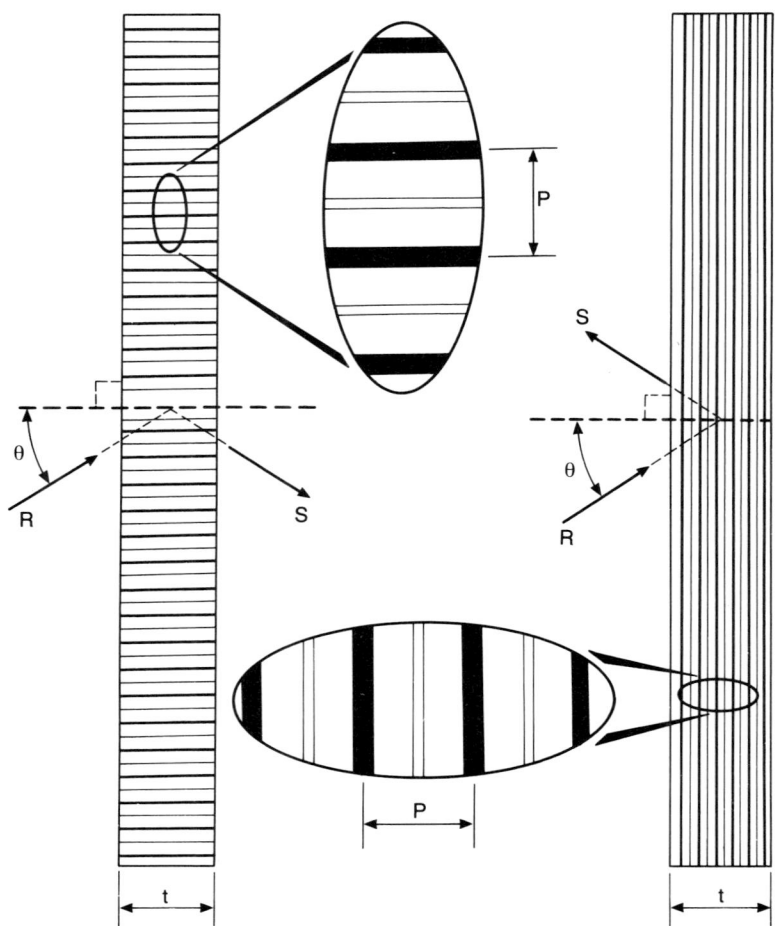

Figure 15.6 Volume transmission and reflection gratings and the key variables that define the properties of the gratings. The dark regions represent regions of high refractive index (or increased absorptivity) and the light regions represent regions of low refractive index (or decreased absorptivity), and the magnitude of the difference in refractive index (or absorptivity) determines in part the diffraction efficiency of the phase (or absorption) hologram.

of the interference pattern $[V = 2(I_1 I_2)^{1/2}/(I_1 + I_2)]$ and P is the fringe spacing of the grating (Fig. 15.6) given by $\lambda_w/(2 \sin \phi_w)$. The film records both the amplitude and phase information contained in the two incident beams as a periodic spatial concentration distribution of bR and M (since neither beam contains an object the hologram is 'structureless', compared with most eye-striking holograms). Thus, in places of constructive interference bR is driven to M and in regions of destructive interference no photochemistry is initiated. The spatial concentration distribution of bR and M can be more conveniently viewed as a spatial modulation of the material's absorption coefficient and index of refraction. These material properties, which are fundamental to the diffraction or reconstruction process in holography, are not independent in most photochromics and are related through the Kramers–Kronig transform [58,59]:

$$\Delta n(\lambda) = \frac{2.3026}{2\pi^2 t} \, \text{PV} \int_0^\infty \frac{A_M(\lambda') - A_{bR}(\lambda')\mathrm{d}\lambda'}{1 - \dfrac{\lambda'^2}{\lambda^2}} \tag{15.2}$$

where PV represents the principal value of the Cauchy integral, Δn is the wavelength-dependent change in refractive index, A_M and A_{bR} represent the absorbances of the ground state and thermal intermediate, respectively, t is the thickness of the hologram, and λ is the wavelength of the readout beam. The absorbance, $A(\lambda')$, is related to the absorption coefficient of a material, $\alpha(\lambda')$ (units of reciprocal length), by noting that:

$$\alpha(\lambda') = 2.3026 \frac{A(\lambda')}{t}. \tag{15.3}$$

In general, for a photochromic material to exhibit useful holographic properties it should possess a high quantum efficiency and a large shift in absorption maxima between photochromic states. The latter property generally results in a large photo-chemically induced change in refractive index. As we have seen in the previous section, the large blue shift in absorption maximum generated by deprotonation of the chromophore during the bR to M phototransformation makes bacteriorhodopsin an ideal optical recording material. Figure 15.7 shows a simulation of the refractive and diffractive properties of a thin film of bacteriorhodopsin based on application of the Kramers–Kronig transform and coupled wave theory (see below). It should be clear from the figure that the largest change in the refractive index is expected when the hologram is produced with a write wavelength that efficiently drives the bR to M photoconversion, and when readout wavelengths are used that yield non-destructive readout (not strongly absorbed by bR or M). The photodiffractive process can be analysed by using the coupled wave theory developed by Kogelnik [60]. The spatial modulations of the absorption coefficient and the index of refraction are described by the following truncated Fourier expansions:

$$\alpha(x) = \alpha_{avg} + \alpha_1 \cos(2\pi x/P), \tag{15.4}$$

$$n(x) = n_{avg} + n_1 \cos(2\pi x/P), \tag{15.5}$$

where P has been defined previously as the fringe spacing of the grating (Fig. 15.6), $\alpha(x)$ and $n(x)$ are the spatial-dependent values of the absorption constant and index of refraction, respectively, α_{avg} and n_{avg} are the average values of the absorption coefficient and the refractive index, respectively, and α_1 and n_1 represent the modulation amplitudes of the absorption coefficient (amplitude) and index of refraction. The latter parameters contribute to the total diffraction, the absorptive part through absorptive modulation of the light electric field amplitude and the refractive component through phase or optical path modulation of the light electric field amplitude. These parameters can be estimated through the use of Eqn 15.2 and taking into account the electric field description of the absorption coefficient described in Eqn 15.4.

The diffraction efficiency of a hologram is defined as the ratio of the diffracted light intensity I_D to the intensity of the reading beam I_0. As before, the diffraction process can have both an absorption and a phase component, and in the case of bacterio-rhodopsin, both contribute [61]:

$$\eta_{total} = \frac{I_D}{I_0} = \eta_{abs} + \eta_{phase} \tag{15.6}$$

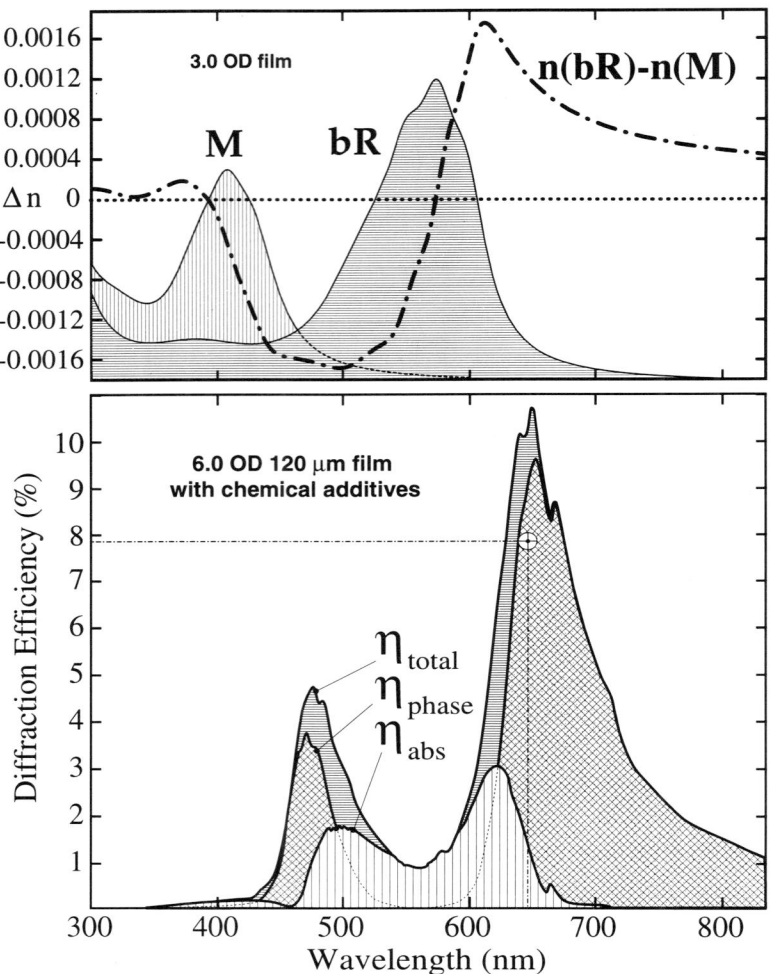

Figure 15.7 The change in refractive index associated with the bR → M photoisomerization for a 30 μm film of bacteriorhodopsin with an optical density (OD) of ~3 is shown as a function of wavelength in the upper panel. The refractive index change is expressed as the value for pure bR minus the value for pure M and is calculated by using the Kramers–Kronig transformation. The absorption spectra of bR and M are shown for reference. The diffraction efficiency associated with a 4 OD film for bR (100%) → bR (50%) + M (50%) photoconversion is shown in the lower panel and is calculated based on the observed absorption spectra by using the Kramers–Kronig relationship and Kogelnik approximation [60]. The dot at ~640 nm and ~8% diffraction efficiency shows an experimental result.

$$\eta_{abs} = \sinh^2 \left\{ \frac{\alpha_1(\lambda_R)t}{2 \cos \phi_R} \right\} D \tag{15.7}$$

$$\eta_{phase} = \sin^2 \left\{ \frac{\pi n_1(\lambda_R)t}{\lambda_R \, 2 \cos \phi_R} \right\} D \tag{15.8}$$

$$D = \exp \left\{ \frac{-\alpha_{avg}(\lambda_R)t}{\cos \phi_R} \right\} \tag{15.9}$$

where η_{total} is the total diffraction efficiency (1 = 100%), η_{abs} is the diffraction efficiency due to absorption, η_{phase} is the diffraction efficiency due to refraction, t is the thickness of the hologram (Fig. 15.6), λ_R is the wavelength of the read laser, ϕ_R is the angle of incidence of the read laser (Fig. 15.3), $\alpha_1(\lambda_R)$ is the modulation amplitude of the absorption coefficient at the read wavelength, $n_1(\lambda_R)$ is the modulation amplitude of the refractive index at the read wavelength, and $\alpha_{avg}(\lambda_R)$ is the average absorption coefficient of the hologram. Although the read angle ϕ_R is an experimentally adjustable variable, maximum efficiency is achieved by satisfying the Bragg condition:

$$\phi_R = \sin^{-1}\left\{ \frac{\lambda_R \sin(\phi_W)}{\lambda_W} \right\}, \tag{15.10}$$

where ϕ_W is the angle of the write beam relative to the hologram film normal (Fig. 15.5) and λ_W is the wavelength of the write beam. The D term defined in Eqn 15.9 which appears in Eqns 15.7 and 15.8 places a constraint on the maximum value of the absorptive component of the diffraction efficiency because the absorption modulation change, $\alpha_1(\lambda_R)$, that is required to generate diffraction, also contributes to the average absorption, $\alpha_{avg}(\lambda_R)$. The contribution of D limits the η_{abs} value to 3.7% or less. In contrast, η_{phase} is determined entirely by the change in refractive index, and values approaching 100% are possible. Thus, for applications requiring diffraction efficiencies exceeding 3.5%, phase holograms or mixed absorptive and phase holograms are preferred. This situation is found in the more traditional type of irreversible type of recording materials such as silver halide photographic films and dichromated gelatin. Figure 15.7 shows the results of the theoretically predicted diffraction efficiency of a 6 OD film of chemically enhanced bacteriorhodopsin as a function of varying readout wavelength. An experimental measurement is also shown indicating that the excellent diffraction efficiency that is predicted can, in fact, be experimentally realized. Holograms can be recorded in pure phase, pure absorption or mixed modes with recording wavelengths from 400 to 700 nm and readout from 400 to 850 nm. The recording sensitivity at ambient temperature is in the range 1–80 mJ cm^{-2}. An additional advantage of using this protein as an optical recording medium is its small size ($\sim$50 nm diameter) relative to the wavelength of light. This results in diffraction-limited performance ($>$5000 lines mm^{-1} for thin films). Diffraction efficiencies can also be improved by using genetically modified proteins [25], chromophore analogues, and chemical enhancement of the native protein [57] (for reviews see [1,2]).

2.3.1   HIGHER ORDER EFFECTS

The change in refractive index upon photoconversion of a bacteriorhodopsin film is responsible for the holographic properties of the medium. The change in refractive index due to absorption band shifts represents the dominant origin of the holographic properties. However, other more subtle effects can also contribute to the process. In the following paragraphs we will show that the intensity-dependent refractive index of a chemically enhanced bacteriorhodopsin film is composed of two components: the first arises from the shift in absorption band accompanying the bR to M phototransformation and follows the predictions made by the Kramers–Kronig transformation; the second is an additional modulation observed at high laser intensities and is likely the result of a thermal effect induced by laser heating.

The *Z*-scan method has been used to examine a bR film of thickness $\sim 150\,\mu m$ and $OD_{570} = 3$ prepared from a mixture containing poly(vinylalcohol) (PVA), bacterio-rhodopsin and the chemical compounds diaminopropane and guanidine hydrochloride [30,54,55]. Compared with a film containing only native bR at pH = 7, the chemical additives (i.e. chemical enhancement) increase the M-state lifetime by a factor of about 10^3 [61]. *Z*-scan modulation is a sensitive method used to measure the light-induced change of the refractive index in cubically non-linear materials [62]. The experimental set-up is shown in Fig. 15.8(a). Briefly, a focused Gaussian laser beam induces a spatially varying index profile in the bR film. The relationship between the total refractive index, *n*, and the illuminating intensity, *I*, is: $n = n_0 + n_2$, $I = n_0 + \Delta n$, where n_0 is the refractive index without illumination, n_2 is the second-order refractive index (or optical Kerr constant), and Δn is the total light-induced change in refractive index. When the film is translated into and out of the beam waist, the combination of the focusing lens and the induced lens modifies the beam divergence, and subsequently, the central intensity of the far field. The normalized intensity transmitted through the aperture is called the 'Z-scan curve'. The position of $Z = 0$ is assigned to be the focus point of the laser radiation. A pre-focus maximum followed by a post-focus minimum indicates that the induced lens and Δn are negative. Conversely, a pre-focus minimum followed by a post-focus maximum shows that the induced lens and Δn are positive. The change in refractive index, Δn, can be calculated through the *Z*-scan curve for any given incident intensity.

Figures 15.8(b) and 15.8(c) show the normalized aperture transmittance as a function of the film position for probing wavelengths of 633 and 476 nm, respectively. The aperture transmittance is defined as the ratio of the powers detected at D_1 and D_2. The normalization was performed by dividing the above transmittance by its value where the film is located at the focal point ($Z = 0$). In our experiment, the beam waist, ω_0, of the focused Gaussian beam is $50\,\mu m$ for 633 nm and $30\,\mu m$ for 476 nm, and the diffraction length of the beam is calculated by $z_0 = \pi\omega_0^2/\lambda$. The value of z_0 is calculated to be much larger than the thickness of bR film used. The film can therefore be considered as a thin sample. Δn can be calculated by noting that [62]:

$$\Delta n = \frac{\Delta P_{p-p}\lambda\alpha}{2\pi f(1 - e^{-\alpha h})},$$ (15.11)

where P_{p-p} is the maximum peak-to-valley value of the normalized *Z*-scan curve, α is the absorption coefficient, *h* is the thickness of the film, and $f = 0.406(1 - s)^{0.25}$ is an empirical constant. In this latter expression, *s* denotes the aperture transmittance which is defined as the ratio $P_1 : P_0$, where P_1 and P_0 are the powers detected by D_1 with and without the aperture, respectively. With this procedure, only the lensing effect is detected and the absorption effect of bR film is eliminated.

A useful parameter describing the intensity-dependent absorption properties of bR is the saturation intensity, $I_S(\lambda, \tau)$. This value is inversely proportional to the lifetime of the excited state (in this case the M state), and is conventionally defined as the intensity needed to produce an equal concentration of ground and excited state molecules:

$$I_S(\lambda, \tau) = \frac{hcN_a}{2.303\lambda\tau_M\{\Phi_{bR}\varepsilon_{bR}(\lambda) - \Phi_M\varepsilon_M(\lambda)\}},$$ (15.12)

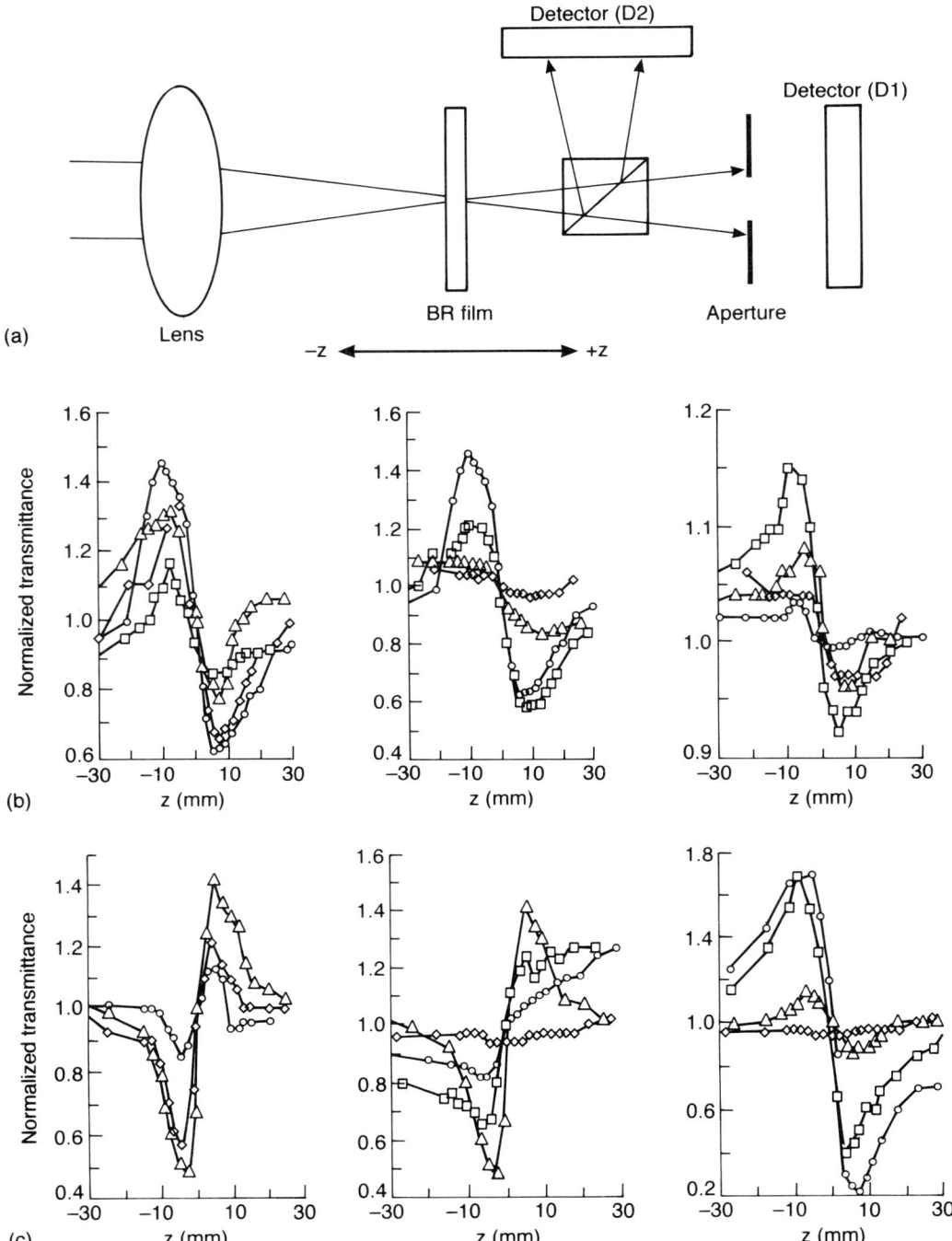

Figure 15.8 (a) Experimental set-up for Z-scan measurement of the bacteriorhodopsin films. (b) The results of the Z-scan measurements showing normalized aperture transmittance as a function of film position for 632.8 nm. (c) The results of the same measurement for 476 nm.

where h is Planck's constant, c is the speed of light, N_a is Avogadro's number, λ is the excitation beam wavelength, τ_M is the M-state lifetime, Φ_{bR} and Φ_M are the quantum yields of the forward and reverse reactions, respectively, and $\varepsilon_i(\lambda)$ are the corresponding molar extinction coefficients. Because the wavelengths used in the experiments were relatively far from the absorption peak of the M state, M-state absorption is very small and the back photoreaction can be neglected. The parameters are taken as: $\Phi_{bR} = 0.65$, $\varepsilon_{bR}(632) = 20\,000\ M^{-1}\ cm^{-1}$, $\varepsilon_{bR}(476) = 12\,000\ M^{-1}\ cm^{-1}$ and $\tau_M = 2.8\ s$ (1/e value). The value of τ_M was determined by completely bleaching the film with 570 nm light and subsequently monitoring the transmittance of a weak probe beam (570 nm) as a function of time. Substituting the above values into Eqn 15.12, the saturation intensities at 476 nm and 633 nm are calculated to be $\sim 3\ mW\ cm^{-2}$ and $\sim 4\ mW\ cm^{-2}$, respectively.

Using the saturation values defined above and calculating Δn from the data in Fig. 15.8(b) and 15.8(c) with Eqn 15.11, the following observations can be made.

1 At low illumination intensities near I_S, the Z-scan signature of refractive index change is positive at 476 nm and negative at 633 nm, and the magnitude of the refractive index change is proportional to the incident intensity. These results, which will be quantitatively presented later, are in good accord in both magnitude and sign with the predictions made from the Kramers–Kronig transformation [61].

2 At intermediate intensities ($I_S < I < 100 I_S$), Δn saturates and little or no lensing effect (focusing or self-defocusing) is observed.

3 At high intensities ($> 10^3 I_S$ for both 476 nm and 633 nm), the Z-scan signature of the refractive index change is negative for both wavelengths, and $|\Delta n|$ increases monotonically with intensity.

In order to exclude the possibility that the latter effect originated from the PVA polymer host, a control film containing no bacteriorhodopsin was examined and no lensing effect was observed in the intensity range investigated. This suggests that the refractive index change at high intensities is due to a thermal effect in the film. At high illumination intensities, a large number of molecules are pumped from the bR state to M state, resulting in a lower absorption coefficient for the bR state where the illumination wavelengths are located. Nonetheless, the larger number of incoming photons available now make the absolute amount of energy absorbed by the bR state to be much higher than that at low intensities. It is reasonable to argue that this strong laser heating expands the bR film resulting in a spatially dependent decrease in local density and n. This presumption is supported by the fact that the time response of the lensing process is slow (on the order of ms) and that the negative lensing effect is proportional to the illumination intensity. Because the molar absorptivity of bR at 633 nm is approximately one-half of that at 476 nm, the intensity required to show the thermal effect is greater at 633 nm.

The behaviour of the refractive index in different intensity ranges can be explained by using a simplified two-state photochromic model of the bR photocycle [26,61]. The two key states are bR and M which represent the ground state and longest lived thermal intermediate, respectively. The rate equation describing the time-dependent behaviour of the two states may be written as:

$$\frac{d[M]}{dt} = k_1[bR] - k_2[M] - k_t[M], \tag{15.13}$$

where [M] and [bR] are the molar concentrations of the M and bR state, and k_t is the thermal relaxation constant of the M state. It is defined by $k_t = 1/\tau_M = 0.357$ s^{-1}. Also in Eqn 15.13, k_1 and k_2 are the photochemical rate constants for bR $\rightarrow$ M and M $\rightarrow$ bR photoreactions, respectively. As discussed before, the absorption of the M state was relatively negligible in our experiments. Thus, we can neglect the back photoreaction from M to bR ($k_2 \approx 0$) and calculate k_1 as follows:

$$k_1(\lambda) = \frac{2.3026\Phi_{bR}\varepsilon_{bR}(\lambda)\lambda I(r)}{N_a hc}, \tag{15.14}$$

where $I(r)$ is the illumination intensity distribution. By using the data given earlier and for a given $I(r)$, Eqn 15.13 can be solved by noting that $[bR] + [M] = [bR]_0$, where $[bR]_0$ is the total molar concentration of the ground state. The refractive index change Δn_{linear}, due to the shift of absorption band for a given illumination intensity distribution $I(r)$, can be calculated by

$$\Delta n_{linear} = \frac{(n_0^2 + 2)^2}{6000 n_0} (R_{bR}\Delta[bR(I)] + R_M\Delta[M(I)]), \tag{15.15}$$

where R_{bR} and R_M are the molar refractions for each state, and $\Delta[bR]$ and $\Delta[M]$ indicate the change in molar concentration. They are related by: $\Delta[bR] = -\Delta[M]$ and $\Delta R = R_{bR} - R_M$. By substituting the solution of Eqn 15.13 into Eqn 15.15, one can solve Eqn 15.15 for a given illumination. When the average intensity, I_0, of the beam is near I_S, the spatial distribution of the normalized change in refractive index remains Gaussian, meaning Δn_{linear} is proportional to intensity. The distribution gradually becomes non-Gaussian at higher intensities and eventually approaches a rectangular function when I_0 is greater than $100 I_S$. In this region, Δn_{linear} is fully saturated and is constant throughout the illuminated area. As a consequence, the lensing effect vanishes, because it depends on the gradient of refractive index spatial profile. This model is consistent with the experimental results shown in Fig. 15.8.

In order to isolate the contribution of $\Delta n_{thermal}$ to the total change in refractive index, we biased the film to saturation by using a strong pump beam whose diameter was much larger in comparison to the Z-scan probe beam. In this case, Δn_{linear} is saturated (being a positive or negative constant depending on the wavelength) over the illuminated region, and the contribution of $\Delta n_{thermal}$ is superimposed on the refractive index modulation arising from nearly complete conversion of bR to M. At intensities much greater than I_S, the Z-scan signature of the refractive index change is negative for both wavelengths investigated. We attribute this self-defocusing effect to the gradient of the refractive index induced in the film by a thermal lens.

We can summarize that the intensity-dependent refractive index is composed of two effects. The Δn_{linear} component of the total refractive index change follows the prediction made by the Kramers–Kronig relation and arises from the shift in absorption band accompanying the bR to M phototransformation. This process can be clearly described by the two-state model. An additional modulation, observed only at high laser intensities, is the likely result of an induced thermal lens created by local heating in the bR sample. At intermediate intensities (where photochemistry becomes saturated), Δn is independent of the illuminating intensity. In general, this region should be avoided when bR is used in lensing or conventional holographic applications.

2.3.2   PHOTOREFRACTIVE PROPERTIES OF THE BLUE MEMBRANE

As discussed in Section 2.2 above, the blue membrane is the result of removal or displacement of cations from the purple membrane. In aqueous or hydrated environments, the blue membrane can be treated as a simple two-state photochromic material, the two states being blue ($\lambda_{max} \sim 600$ nm) and pink ($\lambda_{max} \sim 490$ nm) (see Fig. 15.9). As in the purple membrane, photoconversion of the blue membrane will result in a wavelength-dependent change in refractive index (Δn_λ). Quantitative measurements of the refractive index change accompanying the pink $\leftarrow$ blue transition have been performed by Gross *et al.* The results of this measurement, shown in Fig. 15.10, yield a value for Δn_{550} of $\sim 2 \times 10^{-4}$ [50,63]. This value, which was obtained for a 3.5 OD$_{603}$ hydrated polyacrylamide/blue membrane film, indicates yet another difference between blue membrane and purple membrane — photoconversion in the blue membrane will yield a smaller Δn than in the native protein. The reason for this difference is readily apparent from comparisons of the spectra shown in Figs 15.7 and 15.9. In a PM film, M $\leftarrow$ bR conversion results in a wavelength change of maximal absorption ($\Delta\lambda_{max}$) of ~ 147 nm. By comparison, photoconversion in a hydrated blue membrane film will result in a smaller $\Delta\lambda_{max}$ of ~ 110 nm. Application of the Kramers–Kronig transform (Eqn 15.2) yields the result that this smaller shift in absorption maximum equates to a smaller Δn. Accompanying the experimental data of Fig. 15.10 is the expected wavelength-dependent refractive index change predicted by the Kramers–Kronig transform.

A practical consequence of the smaller Δn obtainable from a blue membrane film is that such films will exhibit lower holographic efficiencies than those based upon purple membrane. Although diffraction efficiency data for hydrated blue membrane films are unavailable, such data are available for films incorporating protein from the ASP$_{85}$ $\rightarrow$ ASN$_{85}$ mutant, which is spectrally and photophysically similar to the blue membrane. Such films have been shown to exhibit $\sim 0.4\%$ diffraction efficiency per film OD [28]. While this efficiency is substantially lower than that achievable from purple membrane films, blue membrane films may nevertheless provide adequate diffraction efficiency for use in long-term holographic data storage applications.

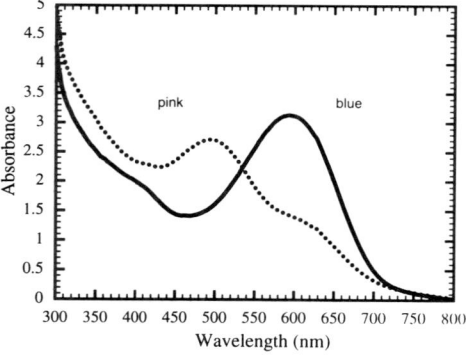

Figure 15.9 Visible absorption spectra of blue and pink states as they appear in a hydrated 3.5 OD$_{603}$ polyacrylamide/blue membrane film. Note the incomplete photoconversion due to the much higher quantum efficiency of the blue $\leftarrow$ pink back reaction. (After Gross *et al.* [63].)

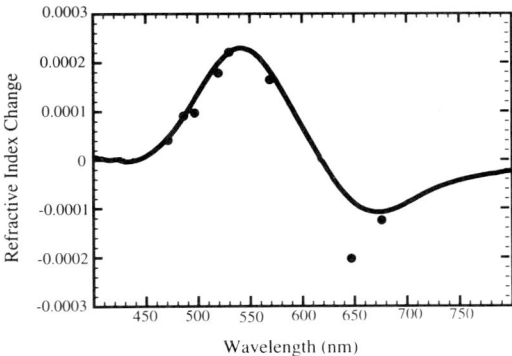

Figure 15.10 Refractive index changes accompanying blue → pink conversion in a 3.5 OD$_{603}$ polyacrylamide/blue membrane film. Experimental values of the refractive index changes (Δn) are shown for selected krypton ion laser wavelengths. The solid black line represents the theoretical wavelength-dependent refractive index change predicted by the Kramers–Kronig transformation of the spectra shown in Fig. 15.9. (After Gross *et al.* [63].)

3   Photochromic modulators and holographic optical processors

3.1   *Spatial light modulators*

Research in optical enginering during the past decade has demonstrated the unique capability of two-dimensional optical processing systems to perform complex mathematical processing functions such as pattern recognition, image processing, solution of partial differential and integral equations, linear algebra, and non-linear arithmetic [64,65]. Interest in exploring the architectures associated with optical processing is due to the speed and unique functionalities derived from the inherent parallel processing and interconnection capabilities of optical systems. Spatial light modulators (SLMs) are integral components in the majority of one-dimensional and two-dimensional optical processing environments. These devices modify the amplitude, intensity, phase, and/or polarization of a spatial light distribution as a function of an external electrical signal and/or the intensity of a secondary light distribution. The observation that a thin film of bacteriorhodopsin can act as a photochromic bistable optical device (either bR ↔ K or bR ↔ M photoreactions) or as a voltage-controlled bistable optical device (bR ↔ M photoreaction) suggests that it has significant potential as the active medium in SLMs [2,25,29,30,52,57,66–75]. Soviet scientists were the first to exploit this potential and deserve much of the credit for bringing bacteriorhodopsin to the attention of researchers working in optical engineering [24,51,53,76–78]. One of the most successful bacteriorhodopsin-based SLM devices was demonstrated by German researchers [26]. Their work exploits the bR ↔ M photoreaction of a mutant protein film in a Fourier optical architectural scheme that implements edge enhancement (spatial frequency filtering) on an input image [26]. More recently, similar results have been obtained by using chemically enhanced bacteriorhodopsin films [29,30,57]. The linear and non-linear optical properties of bacteriorhodopsin have also been exploited for application in optical neural architectures with great success [66–75].

3.2   *Holographic associative memories*

Associative memories operate in a fashion quite different from the serial memories that dominate current computer architectures. These memories take an input data block (or image), and independently of the central processor, 'scan' the entire memory for the data block that matches the input. In some implementations, the memory will find the closest match if it cannot find a perfect match. Finally, the memory will return the data block in memory that satisfies the matching criteria. Because the human brain operates in a neural, associative mode, many computer scientists believe that the implementation of large capacity associative memories will be required if we are to fully achieve artificial intelligence. Optical associative memories using Fourier transform holograms have significant potential for applications in optical computer architectures, optically coupled neural network computers, robotic vision hardware and generic pattern recognition systems. The ability to rapidly change the holographic reference patterns via a single optical input while maintaining both feedback and thresholding increases the utility of the associative memory, and in conjunction with solid-state hardware, opens up new possibilities for high speed pattern recognition architectures.

Bacteriorhodopsin's diverse range of photochromic responses and adequate holographic performance suggest its use as active elements in optical processors. One application currently under investigation is the use of bacteriorhodopsin thin films as the key holographic storage components in a real time optical associative memory [1,52,61]. Our optical design is shown in Fig. 15.11. This architecture employs both feedback and thresholding and is based on the closed-loop autoassociative design of Paek and Psaltis [79]. During the write operation, reference images stored in an electronically addressable spatial light modulator (ESLM) are optically fed into the loop by plane wave illumination (λ_w = 568 nm) from a krypton ion laser. The reference images are stored as Fourier transform holograms on thin polymer films containing bacteriorhodopsin (H1 and H2). For this real time application, no chemical additives are used to enhance the M lifetime. Accordingly, the hologram stores the reference image for approximately 10 ms before reverting to the ground state. During the readout operation, the input image (from transparencies or another ESLM) is read into the loop by using the optical imaging system shown in Fig. 15.8. Best results are obtained by illuminating the object by using plane wave illumination from a second krypton ion laser operating at a wavelength of 676.5 nm. The input image beam is passed through a microchannel plate spatial light modulator (MSLM) operating in threshold mode. Thereafter, the Fourier transformed product of the image reference is formed and retransformed at the plane of a pinhole array (PHA). The resulting correlation patterns are sampled by the pinholes (diameter ~500 µm) which are precisely aligned with the optical axis of the reference images. Light from the pinhole plane is retransformed and superimposed with the reference image stored on bacteriorhodopsin hologram 2 (H2). The resulting cross-correlation pattern represents the superposition of all images stored on the multiplexed holograms and is fed back through the microchannel plate spatial light modulator for another iteration. Thus, each image is weighted by the inner product between the pattern recorded on the MSLM from the previous iteration and itself. The output locks on to that image stored in the holograms which produces the largest correlation flux through its aligned pinhole.

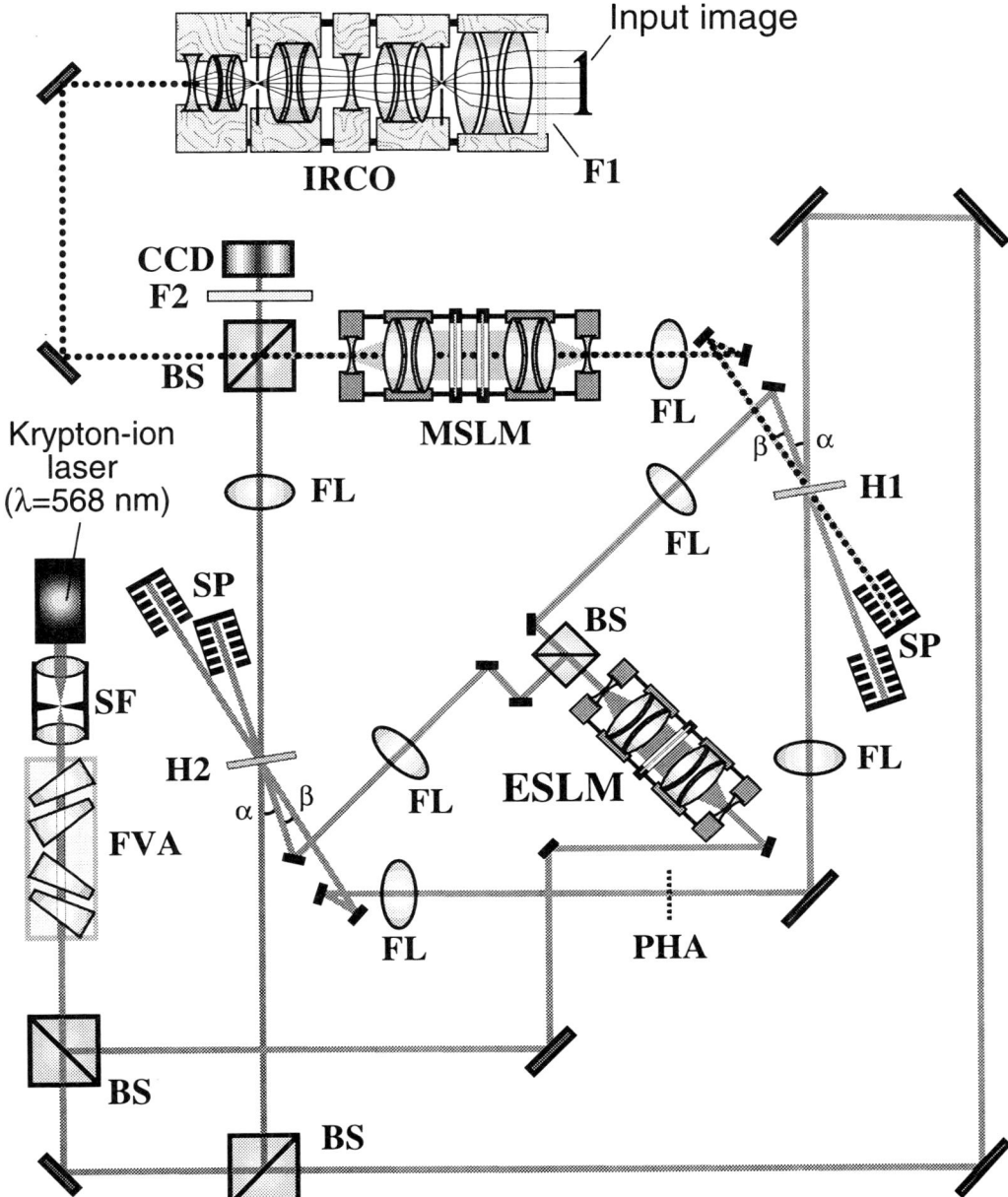

Figure 15.11 Schematic diagram of a Fourier transform holographic (FTH) associative memory with read/write FTH reference planes using thin polymer films of bacteriorhodopsin to provide real-time storage of the holograms (see text). The following symbols are used: BS, beam splitter; CCD, charge-coupled device two-dimensional array; CL, condensing lens; ESLM, electronically addressable spatial light modulator; FL, Fourier lens; FVA, Fresnel variable attenuator; F1, broadband filter for image; F2, interference filter with transmission maximum at laser wavelength, different from λ_{max} of F1; H1 and H2 holographic spatial light modulator (Fig. 15.7); IRCO, image reduction and condensing optics; MSLM, multichannel plate spatial light modulator; PHA, pin hole array; SF, spatial filter to select TEM_{00}.

The real time capability of the associative loop is made possible by using bacterio-rhodopsin films as the transient holographic medium. The high speed of phototrans-formation during write operation ($<50\,\mu s$) coupled with the quick relaxation time of the M state (~10 ms) allow for framing rates up to 100 frames s^{-1}. Slower or faster framing rates can be attained by simply altering the M lifetime with chemical additives and/or intermittently erasing the hologram with an external blue light source. The write and read wavelengths of the krypton ion lasers as well as the respective angle of incidence are chosen to optimize the diffraction efficiency of the bacteriorhodopsin holograms. During the process of sampling the correlation patterns, it is interesting to note that the inclusion of the pinholes destroys the shift invariance of the optical system. If the input pattern is shifted from its nominal position, the correlation peak shifts as well, and the correlation light flux will miss the pinhole. If the pinholes were removed, however, ghost holography would seriously impair image quality. The two apertures within the image reduction and collimation optics (IRCO) serve to provide correct registration, but the input image must still be properly centred to generate proper correlation. The problem of shift invariance represents one of the fundamental design issues that will have to be resolved for associative memories to reach their full potential. While there are a number of optical 'tricks' that can be used to counteract poor registration, the most easily implemented approach is to use the ESLM controller to scale and translate the reference images to maximize the correlation light flux as measured by the intensity of the image falling on the CCD output detector.

4   Volumetric memories

During the past decade, the speed of computer processors has increased between two and three orders of magnitude, with the largest increase occurring in small desktop workstations. This dramatic increase in processor capability has not been matched by a corresponding increase in data storage densities, which have increased by only one order of magnitude in both random access memory and hard disk technology [17]. We have now entered an era where a majority of computational algorithms are limited less by computer speed than by data storage, and the cost of a typical scientific workstation is determined in large part by the cost of random access memory and disk storage. One promising approach to increasing random access data storage is to use three-dimensional architectures. As we examine below, bacteriorhodopsin has properties which make it a useful material for both linear and non-linear volumetric architectures.

4.1   *Two-photon three-dimensional memories*

One new architecture that has received considerable attention during the past few years is based on the potential of using a two-photon absorption process to store data in three dimensions [17,19,80–82]. These memories read and write information by using two orthogonal laser beams to address an irradiated volume (1–200 μm^3) within a much larger volume of a non-linear photochromic material. A two-photon process is used to initiate the photochemistry, and this process involves the unusual capability of some molecules to capture two photons simultaneously to populate an energy level within the molecule with an energy equal to the sum of the energies of the two photons absorbed. Because the probability of a two-photon absorption process scales as the square of the intensity, photochemical activation is limited to a first approximation to regions within the irradiated volume. (Methods of preventing photochemistry outside of the irradiated volume from damaging data are described below.) The three-dimensional addressing capability derives from the ability to adjust the location of the irradiated volume in three dimensions. Two dimensional optical memories have a storage capacity that is limited to $\sim 1/\lambda^2$, where λ is the wavelength, which yields approximately 10^8 bits cm^{-2}. In contrast, three-dimensional memories can approach storage densities of $1/\lambda^3$, which yields densities of approximately 10^{12} bits cm^{-3}. In principle, a two-photon three-dimensional memory can store roughly three orders of magnitude more information in the same size enclosure relative to a two-dimensional optical disk memory. In practice, optical limitations and issues of reliability lower the above ratio to values closer to 100–300. Nevertheless, a two order of magnitude improvement in storage capacity is significant. Furthermore, the two-photon approach makes parallel addressing of data possible, which enhances data read/write speeds and system bandwidth.

In this section we examine the potential of this approach as well as the methods currently under study to implement both serial and parallel addressing. We will also demonstrate that bacteriorhodopsin has rather unique properties that are optimal for some two-photon volumetric architectures. Although a great deal has been learned during the past few years, much remains to be done before a viable commercial memory will be available. We will also examine the problems that remain to be solved.

4.1.1   THE TWO-PHOTON PROCESS

In order to understand how two-photon volumetric memories work, it is necessary to appreciate the nature of the two-photon absorption process. For the purposes of the present discussion, we will describe the process from a semiclassical perspective. While this approach lacks mathematical rigour, it preserves the relevant characteristics that relate to the performance of a potential memory material.

Two-photon absorption involves the ability of a molecule to absorb two photons (nearly) simultaneously and combine the energy of the two photons to access a stable excited state of the molecule [83–86]. The process is mediated by what is called a 'virtual state' of the molecule, a state of the molecule that has no classical analogue (see Fig. 15.12). The virtual state is created in response to the presence of the radiation field as it passes through the molecule, and it exists for an extremely short period of time (a few femtoseconds, the time it takes a photon to travel about 1 µm) [86]. A second photon must arrive before this virtual state decays (dephases), and for this reason, the probability of a two-photon absorption is proportional to the square of the light intensity. The lifetime of the virtual state is so brief that many refer to the process as 'simultaneous two-photon absorption'. There are two mechanisms that contribute to strong two-photon transitions. First, the molecule can have a low-lying strongly absorbing state that lies near the virtual level that is created by the initial photon. Heisenberg's uncertainty principle predicts that this state will have a probability of contributing to the virtual level with a lifetime that is roughly given by $h/(4\pi\Delta E)$, where h is Planck's constant and ΔE is the energy shift between the actual state and the virtual level. For example, a state separation of roughly 1 eV will support a virtual state lifetime of 1/3 of a femtosecond (0.3 fs = 3×10^{-16} s). Second, the molecule can have a large difference between the ground state and the final state dipole moments. This characteristic allows the two states to share participation in the formation of the virtual

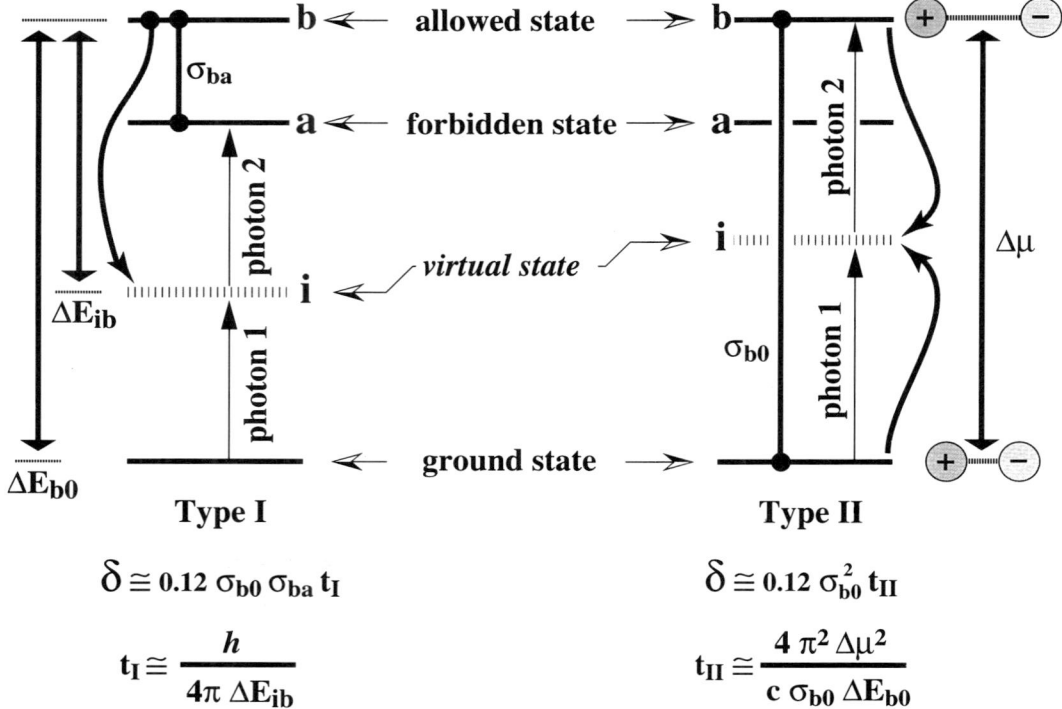

$$\delta \cong 0.12\ \sigma_{b0}\ \sigma_{ba}\ t_I \qquad\qquad \delta \cong 0.12\ \sigma_{b0}^2\ t_{II}$$

$$t_I \cong \frac{h}{4\pi\ \Delta E_{ib}} \qquad\qquad t_{II} \cong \frac{4\ \pi^2\ \Delta\mu^2}{c\ \sigma_{b0}\ \Delta E_{b0}}$$

Figure 15.12 A two-photon excitation process requires that the sum of the photon energies sum to the energy of an excited state of the molecule. A semi-classical description of this process predicts that the probability of two-photon absorption is proportional to the product of one-photon cross-sections (with values typically on the order of 1 Å^2 molecule^{-1}) and the lifetime of the virtual state, which is highly variable. In non-polar molecules a type I process occurs and only excited states that are optically forbidden via one-photon selection rules can be accessed via two-photon absorption. The lifetime of the virtual state, t_1, is described by Heisenberg's uncertainty principle equating energy (ΔE) and time (Δt) uncertainties [$\Delta E \Delta t \geqslant h/(4\pi)$]. The equation essentially predicts that an allowed state can participate in the formation of the virtual level for a time, t_1, roughly equal to $h/(4\pi)$ divided by the energy shift from the actual energy and the virtual state energy (ΔE on the figure). The type I process in large conjugated systems typically results in two-photon absorptivities on the order of 10 GM. In polar molecules, however, one-photon allowed states can also be two-photon active provided a change in dipole moment occurs upon excitation. In this case, the lifetime of the virtual level is proportional to the square of the dipole moment change, and for molecules with strongly allowed states with dipole moment changes on the order of 10 debye, two-photon absorptivities on the order of 100 GM are observed.

level, and can enhance the two-photon process by two or three orders of magnitude. (This latter characteristic is responsible for making the protein bacteriorhodopsin a successful two-photon memory medium.) These two types of two-photon processes are shown in Fig. 15.12.

The ability of a molecule to undergo two-photon excitation is linearly proportional to the two-photon absorptivity, δ, which has units of Göppert–Mayers or GMs (1 GM = 10^{-50} cm^4 s molecule^{-1} photon^{-1}). This definition was chosen because a 'typical' conjugated molecule will have a low-lying excited state with an absorptivity of approximately 1 GM [83–86]. The lowest lying excited singlet states of long-chain linear polyenes have two-photon absorptivities on the order of δ = 10 GM

[83,84,86,87]. The principal mode of enhancement is due to a type I process (Fig. 15.12). However, molecules which undergo large changes in dipole moment upon excitation can take advantage of type II processes (Fig. 15.12) and generate two-photon absorptivities exceeding 100 GM [23,83,87–90]. Of equal importance is the fact that the type II mechanism makes one-photon allowed states (state b in Fig. 15.2) also two-photon allowed. An efficient two-photon volumetric memory requires a material with a δ value at or above 100 GM. The protein bacteriorhodopsin is unusual in exhibiting a two-photon absorptivity of 290 GM, due to a large change in dipole moment upon excitation of the protein-bound chromophore ($\Delta\mu \approx 13.5$ debye) [23,89,91]. In fact, this protein exhibits the largest broad-band two-photon absorptivity reported for a molecule.

The two-photon induced photochromic behaviour is summarized in the scheme below:

$$\text{bR (state 0) } (\lambda_{max} \simeq 1140 \text{ nm}) \underset{h\omega^2;\ \Phi_2 \sim 0.65}{\overset{h\omega^2;\ \Phi_1 \sim 0.65}{\rightleftharpoons}} \text{M (state 1) } (\lambda_{max} \simeq 820 \text{ nm}). \quad \text{(Scheme 15.2)}$$

This scheme is identical to Scheme 15.1 except that the photochemistry is initiated by using a two-photon process and laser wavelengths twice those used to initiate normal (i.e. one-photon) photochemistry. We arbitrarily assign bR to binary state 0 and M to binary state 1. The above wavelengths are correct to only ± 40 nm, because the two-photon absorption maxima shift as a function of temperature and polymer matrix water content.

4.1.2 TWO-PHOTON VOLUMETRIC MEMORY ARCHITECTURES

There are many possible optical architectures for generating two-photon volumetric memories [80,81,92]. We will start with a discussion of serial designs and progress to more complex designs which provide for parallel read/write capability. Many investigators believe that the ability to implement parallel access represents one of the key attributes that differentiates the two-photon architecture from other volumetric memory schemes [81].

A basic two-photon memory design based on bacteriorhodopsin is shown in Fig. 15.13. The bacteriorhodopsin is contained in a cuvette and is oriented by using electric fields prior to polymerizing the polyacrylamide gel matrix. This orientation is required in order to use the photoelectric signal to monitor the state of the proteins occupying the irradiated volume. A write operation is carried out by firing simultaneously the two 1140 nm lasers (to write a 1) or the two 820 nm lasers (to write a 0). Contrary to one's preliminary impression, however, photochemistry also occurs along the laser axes, also due to two-photon processes. The resulting photochemical conversion process as a function of position is schematically shown in the top diagram of Fig. 15.14. To eliminate unwanted photochemistry along the laser axes, non-simultaneous firing of the lasers not used in the original write operation is carried out immediately following the write operation. The latter operation is known as 'photochemical cleaning'. and isolates the net photochemistry to the irradiated volume. The result is shown in the bottom diagram of Fig. 15.14. The position of the cuvette is controlled in three dimensions by using a series of actuators which independently drive

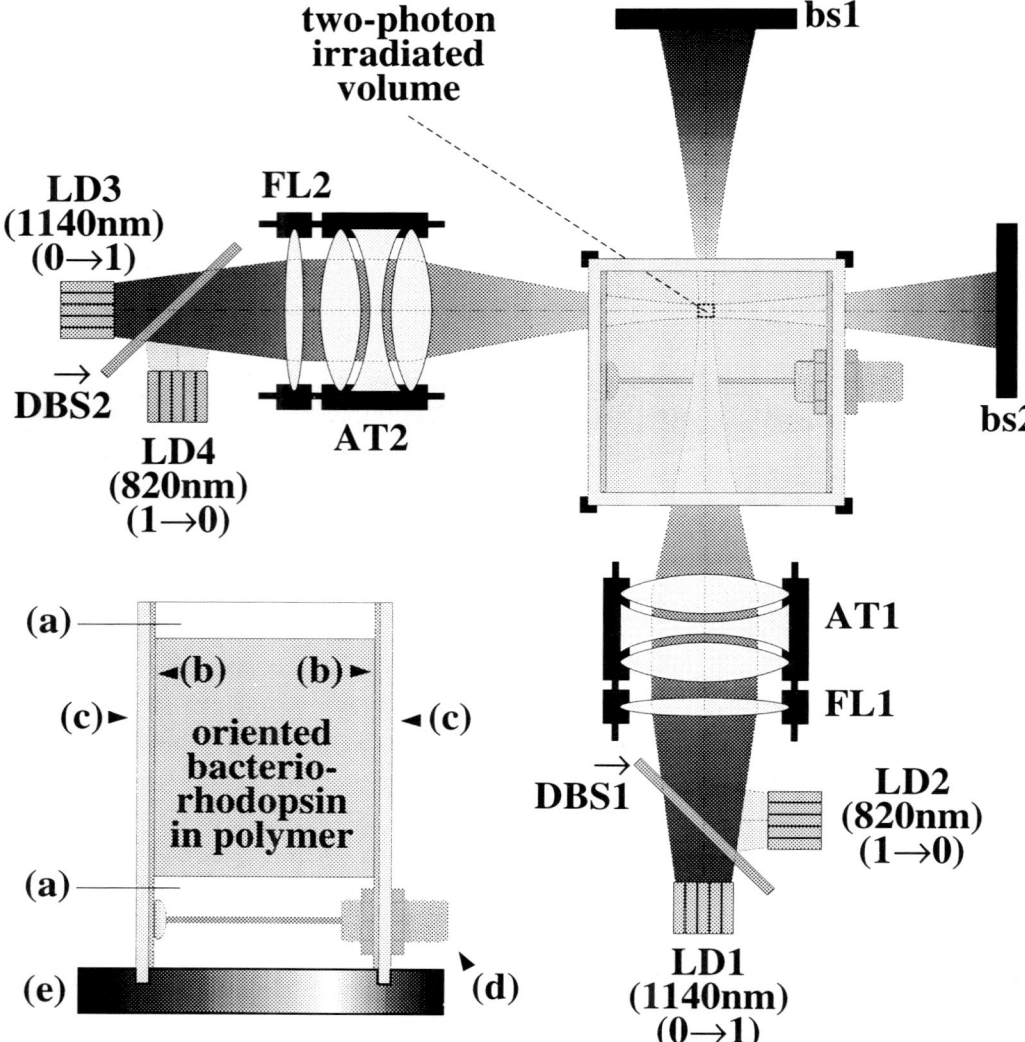

Figure 15.13 Schematic diagram of the principal optical components of a two-photon three-dimensional optical memory based on bacteriorhodopsin. The write operation involves the simultaneous activation of LD_1 and LD_3 ($0 \rightarrow 1$) or LD_2 and LD_4 ($1 \rightarrow 0$) to induce two-photon absorption within the irradiated volume and partially convert either bR to M ($0 \rightarrow 1$) or M to bR ($1 \rightarrow 0$). The write operation uses a 10 ns pulse and a pulse simultaneity of 1 ns. The protein is oriented within the cuvette by using an electric field prior to polymerization of the polyacrylamide gel. A polymer sealant is then used to maintain the correct polymer humidity. The SMA connector is attached to the indium–tin oxide conducting surfaces on opposing sides of the cuvette and is used to transfer the photoelectric signal to the external amplifiers and box-car integrators. Symbols and letter codes are as follows: (a) sealing polymer; (b) indium–tin oxide conductive coating; (c) BK7 optical glass; (d) SMA or OS50 connector; (e) peltier temperature-controlled base plate (0–20°C). AT, achromatic focusing triplet; bs, beam stop; DBS, dichroic beam splitter; LD, laser diode; FL, adjustable focusing lens.

the cuvette in the *x*, *y* or *z* direction. For slower speed maximum density applications, piezoelectric micrometers are used. For higher speed, lower density applications, voice-coil actuators are used.

A key requirement of the two-photon memory is to generate an irradiated volume

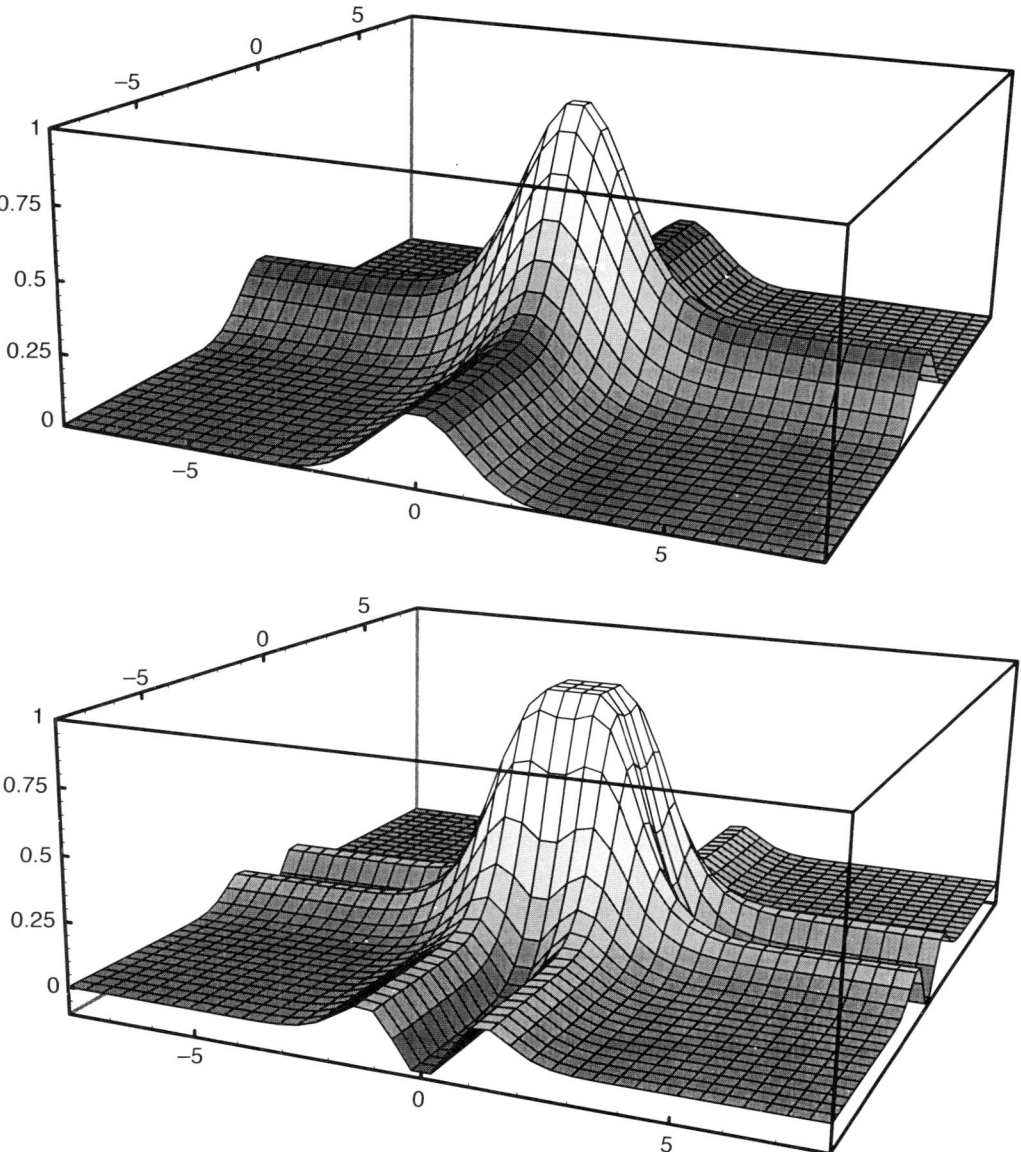

Figure 15.14 Computer simulation of the probability of two-photon-induced photochemistry ($P_\lambda^{2\omega}$) (vertical axis) as a function of location relative to the centre of the irradiated volume (ΔX_{focus} and ΔY_{focus}) in μm. The top contour plot shows the probability after two 1140 nm laser beams have been simultaneously directed along orthogonal axes crossing at the centre of the irradiated volume. The bottom contour plot shows the probability after two 820 nm 'cleaning pulses' have been independently directed along the same axes. The maximum conversion probability at $x = 0$, $y = 0$ is normalized to unity for both contour plots.

which is reproducible in terms of *xyz* location over lengths as large as 2 cm. In the present case, our memory cuvettes are typically ~1.6 cm in the *x* and *y* dimension and ~2 cm in the *z* direction (see Fig. 15.13). These dimensions are variable up to 2 cm on all sides, and can be as small as 1 cm on a side depending upon the desired storage capacity of the device. By using a set of fixed lasers and lenses, and moving the cuvette by using orthogonal translation stages, excellent reproducibility can be achieved

(± 1 μm for piezoelectric micropositioners, ± 3 μm for voice-coil actuators). Refractive inhomogeneities which develop within the protein–polymer cuvette as a function of write cycles adversely affect the ability to position the irradiated volume with reproducibility. This problem is due to the change in refractive index associated with the photochemical transformation (see Fig. 15.6). The problem is minimized by operating with a relatively large irradiated volume (~ 30 μm^3) and by limiting the photochemical transformation to 60 : 40 versus 40 : 60 in terms of relative bR : M percentages.

As one might anticipate, reading the data is much more difficult than writing the data. In the architecture shown in Fig. 15.13 a four laser read operation is used. By firing all four lasers simultaneously, the state of the irradiated volume can be probed by monitoring the differential photovoltage [17,93]. Careful adjustment of the relative intensities of the four lasers permits minimal disturbance of memory cells outside of the irradiated volume. A standard write operation is then performed to reset the memory cell to the correct state. This procedure serves to enhance data integrity by reducing the risk that multiple read/write cycles along the axis occupied by the interrogated memory cell corrupt the data in that memory cell. The total read process can be completed in ~ 50 ns, which means that the maximum serial data rate is ~ 20 Mbit s^{-1}. This data rate is decreased by the time necessary to move the cube to the next memory cell. This latency is determined by the extent to which the data to be read are contiguous and by the translation actuators (piezoelectric vs voice-coil). To maximize reliability, two additional bits are stored after each 8 bit byte to provide single bit error correction and double bit error detection. Additional data reliability can be provided by adding checksums at the end of each data block, but the checksum generation and verification process is handled by controller software and not by the two-photon three-dimensional memory control firmware.

Parthenopoulos and Rentzepis have proposed a creative approach that eliminates the photochemical cleaning that is employed in the above design [80]. These investigators use two different laser wavelengths to activate photochemistry and choose wavelengths that independently have a low probability of initiating two-photon processes. They also adjust the intensity of the two beams so that the laser with higher probability of inducing two-photon photochemistry is weak and the beam with the lower probability of inducing two-photon intensity is intense. The combination of these two characteristics dramatically decreases the amount of photochemistry outside of the irradiated volume. Unfortunately, it does not reduce the unwanted photochemistry to zero, and a large number of read/write operations can lead to data damage. Accordingly, most materials require some degree of photochemical cleaning. Materials with narrow absorption bands work best with the two-wavelength method while materials with broad absorption bands work best with single wavelength photochemical cleaning. Bacteriorhodopsin has broad absorption bands and thus photochemical cleaning is optimal for two-photon memories based on this material.

4.1.3   PARALLEL ARCHITECTURES

As noted previously, one of the key advantages of two-photon volumetric architectures derives from the potential of addressing data in parallel. There are numerous approaches to parallel addressing; they include the use of holographic lenses or phased arrays to focus light simultaneously onto multiple irradiated volumes, or by using two-dimensional spatial light modulators to create independently selected multiple beams [80,81,94]. The latter approach is shown schematically in Fig. 15.15. In this approach, the memory unit can be viewed as a multilayer storage system with each layer represented by a two-dimensional array of bits addressed independently by the information (or data) beam. The layer to be accessed is selected by turning on the appropriate addressing beam which illuminates one, and only one, two-dimensional array. This approach is called optical paging, and eliminates all moving parts. If we assume a high resolution spatial light modulator with 1024×1024 elements is employed, a total of more than 10^6 bits can be addressed in parallel. Thus, a relatively slow spatial light modulator (10 ms) can still provide a significant data bandwidth (~ 100 Mbit s^{-1}).

The architecture shown in Fig. 15.15 is fairly simple to understand in terms of the write operation. The read operation, however, is more complicated. If one is using a two-photon material that fluoresces, then the data can be read by using polarization selected imaging. Detailed designs based on this approach have been presented in the

Page Addressing SLM

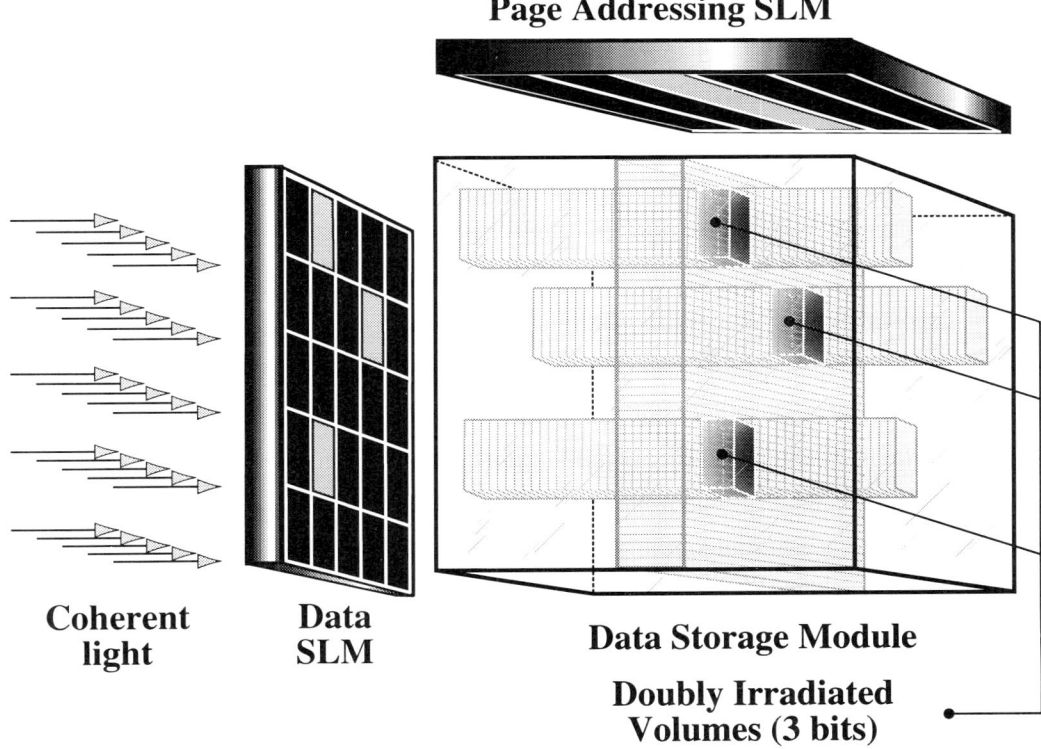

Coherent light **Data SLM** **Data Storage Module**

Doubly Irradiated Volumes (3 bits)

Figure 15.15 Parallel addressing of multiple data elements by using orthogonal spatial light modulators (SLMs) to create paging and data beams. A single page within the volumetric memory material is selected by the top SLM by irradiating an entire slice (page) of the memory with light. The data are imposed in parallel on the data array by the SLM at left which uses an active matrix liquid crystal array similar to those used in high resolution black and white notebook computer displays. Recent advances in two-dimensional laser arrays indicate that within the next 5–10 years monolithic data arrays will be available with sufficient resolution (1024×1024) to be useful in both non-linear (two-photon) and linear (branched-photocycle) volumetric memories.

literature and are promising [81]. If a non-fluorescent material is used, the states of the individual bits can be monitored by using phase shift detection by using an array detector to monitor fringe modulation before and after the addressing beam is activated.

4.1.4 PROBLEMS THAT REMAIN TO BE SOLVED

Much work remains to be done before a reliable (i.e. commercially viable) two-photon volumetric memory will be available. Many of the key optical components that are required are still in the developmental stage, and the prototypes that have been constructed all have reliability problems. There is one aspect of the problem that deserves special attention because it is inherent to all of the architectures described above. The key issue involves the inherent characteristic of all materials to change their refractive index when photochemically switched from one state to another when such a transformation changes the optical absorption spectrum of the material. The end result is that the process of writing data will produce small 'lenses' inside the material that will refract light and prevent tight focusing of the laser light onto the irradiated volume. Then the data bit appears to shift from the original location due to refraction of the light. The good news is that the effective change in refractive index is extremely small because light in the infrared region of the spectrum is used, and most materials do not exhibit large refractive index changes in the infrared region of the spectrum upon photochemical conversion. Nevertheless, the large number of refractive lenses generated when data are written produces an observable effect that invariably affects the ability to address microscopic irradiated volumes.

There are many approaches under study to eliminate or at least minimize the problem, and while all are interesting and worthy of discussion, we will limit our discussion to a few selected approaches. The first is trivial and involves the use of recording densities significantly below the diffraction limit so that even if the bit moves, it is large enough to find. This is not an optimal approach, however, because it essentially eliminates the dramatic increase in density that motivates investigation. The second is to use a method of encoding the data so that the integral refractive index change is minimized. This approach is under active study and simulations indicate that under a majority of situations, the problem can be minimized but not eliminated. A third approach is to use a combination of data encoding and orthogonal linear phased arrays that produce a tight focus by virtue of number rather than finesse. The theory behind the latter approach is that if enough lasers are involved, fluctuations in refractive index will be nulled out by the large number of sources.

4.2 *Branched-photocycle three-dimensional memories*

The simultaneous two-photon memory architecture has received a great deal of attention in the past few years, and because bacteriorhodopsin exhibits both high efficiency in capturing two photons [19] and a high yield of producing photoproduct after excitation, this material has been a popular memory medium. But more recent studies suggest that the branched-photocycle memory architecture may have comparable if not greater potential [21]. This sequential one-photon architecture completely eliminates unwanted photochemistry outside of the irradiated volume and provides for a particularly straightforward parallel architecture [95,96]

The memory is based on the use of the P and Q states for long-term data storage [12] (Fig. 15.12). The fact that these states can only be generated via a temporally separated pulse sequence provides a convenient method of storing data in three dimensions by using orthogonal laser excitation. The process is shown in Fig. 15.16 and is based on the following non-simultaneous sequence:

$$\text{bR (state 0)} \xrightarrow{\text{photon 1}} \text{K} \rightarrow \text{L} \rightarrow \text{M} \rightarrow \text{N} \rightarrow \text{O} \rightarrow \text{bR} \qquad \text{(paging)}$$

$$\text{O} \xrightarrow{\text{photon 2}} \text{P (state 1)} \rightarrow \text{Q (state 1}') \qquad \text{(write data)}$$

where K, L, M, N and O are all intermediates within the main photocycle, and P and Q are intermediates in the branching cycle (see Fig. 15.2). Note that the laser beam providing photon 2 is turned on after the laser beam providing photon 1 has been turned off. This sequential operation turns the protein into an optical AND gate which rigorously excludes photochemistry at locations outside of the doubly irradiated volume element [95,96].

A parallel write is accomplished by using an orthogonal optical architecture as shown in the upper right diagram in Fig. 15.16. Light activation is represented by using yellow lines to represent the presence and the directionality of the light beam and the vertical axis charts time relative to the firing of the paging laser. The timing is based on a memory curvette at ambient temperature (20–30°C). The paging beam with a wavelength of approximately 600 nm activates the photocycle of bacteriorhodopsin and after a few milliseconds the O intermediate reaches near maximum concentration. At this point, the data laser array is activated to photoselect the irradiated volume elements into which 1 bits are to be written. This process converts O to P in these, and only these locations within the memory cube. After many minutes (the decay time, τ_P, is highly dependent upon temperature and polymer matrix), the P state thermally decays to form the Q state. The latter is stable for months to years depending upon temperature and polymer matrix. We assign the bR state to binary state 0 and both P and Q to binary state 1. The entire write process is accomplished in ~10 ms, the time it takes the protein to complete the photocycle. If we use a 1024×1024 data array, we can write 1 048 576 data bits, or ~10.5 kB (one byte (B) is represented by eight data bits and two error correcting bits) within a 10 ms cycle. This represents an overall write data throughput of ten million characters per second (10 MB s^{-1}), which is comparable to slow semiconductor memory. By using more than one memory cell, data write times improve proportionally. An entire page of memory can be erased by using blue light (which converts both P and Q back to bR).

The read process takes advantage of the fact that light around 680 nm is absorbed by only two intermediates in the photocycle of light-adapted bacteriorhodopsin, the primary photoproduct K and the relatively long-lived O intermediate (see Fig. 15.2). A parallel read is accomplished by using a differential absorption process as shown in the right column of Fig. 15.16. The read sequence starts out in a fashion identical to that of the write process by activating the 600 nm paging beam. After 2 ms, the entire data array is turned on at a very low intensity (0.01% of the power used to write). A charge injection device (CID) array images the light passing through the data cube. Those

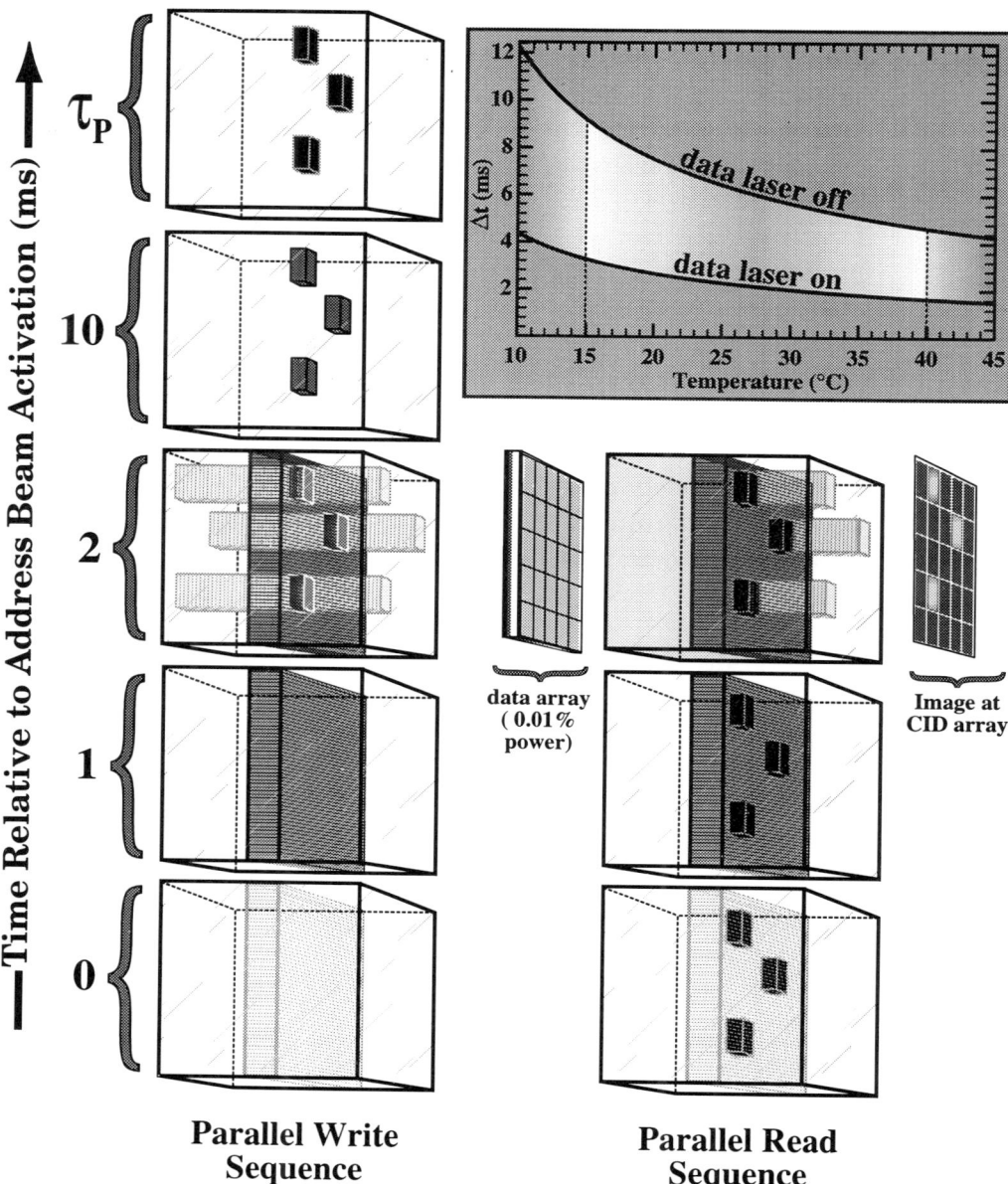

Figure 15.16 A schematic diagram of the parallel write and read sequences associated with a branched-photocycle three-dimensional memory based on bacteriorhodopsin. The basic optical architecture is shown in the insert at upper right where spatial light modulators (SLMs) are used to control the laser light. The write sequence is shown in the left column, and involves the following sequential process: (1) turn on 600 nm page beam to activate bR photocycle; (2) after 2 ms, activate appropriate 680 nm data beams to convert O to P. Note that time progresses from bottom to top. The read sequence is shown in the right column and involves the following sequential process: (1) turn on 600 nm page laser to activate bR photocycle; (2) after 2 ms, activate all data array elements at 0.01% power to read entire page as an image on the charge injection device (CID) array.

elements that are in the binary 1 state (P or Q intermediates) do not absorb the ~680 nm beams, but those volumetric elements that started out in the binary 0 state (bR) absorb the 680 nm light because these elements have cycled into the O state.

Noting that all of the volumetric elements outside of the paged area are restricted to the bR, P or Q states, the only significant absorption of the beam is associated with O states within the paged region. The CID detector array is therefore observing the differential absorptivity of the paged region, and the paged region alone. This is the key to the read operation, and it allows a reasonable signal-to-noise ratio even with thick (1–1.6 cm) memory media with $>10^3$ pages. Because the absorptivity of the O state within the paged region is more than 1000 times larger than the absorptivity of the remaining volume elements combined, a very weak beam can be used to generate a large differential signal. The read process is complete in $\sim$10 ms which gives a rate of 10 MB s^{-1} times the number of data cubes. One of the commercial requirements of volumetric memory systems is the need for highly homogeneous memory media. Three Space Shuttle flights have been carried out in collaboration with BioServe Space Technologies and NASA to investigate the potential of manufacturing bacteriorhodopsin memory cubes in microgravity, with promising results. Needless to say, if the data cubes require microgravity preparation, they will not be cost effective in comparison to other storage.

4.2.1 PROBLEMS THAT REMAIN TO BE SOLVED

The volumetric branched-photocycle memory is in the early prototyping stage, and as was the case for the two-photon 3D memory, reliability remains an important and as yet unresolved issue. The key problems are low conversion yield associated with the O $\rightarrow$ P photoreaction and refraction within the data cube requiring operation above diffraction-limited resolution. The former problem is being tackled by using two methods. First, organic cations are being added to the blue membrane to form an analogue bacteriorhodopsin which has an enhanced O state lifetime (factors of two to three are possible) [32]. We are also studying a number of genetically modified proteins which also increase the O state lifetime. These variants, kindly provided to us by Prof. Richard Needleman, provide similar improvement in O state lifetime, and we anticipate that a mutant can be found which enhances the O $\rightarrow$ P photoreaction quantum efficiency. Minimizing error due to refraction is being studied following two approaches. (We note that the origin of this probelm is analogous to that observed in the two-photon memory (Section 4.1.4).) Alternative methods of writing the data which forces the ratio of 1 vs 0 bits to approach unity regardless of data type is being investigated as a technique to minimize refractive index gradients from developing within the data cube. An alternative approach is to decrease the differential ratio of bR vs P/Q that represents the 0 vs 1 bits. While this lowers the signal-to-noise ratio of the read process, it enhances the optical registration of the bits by lowering the refractive index difference between 0 and 1 bits. The optimal trade-off of the many variables that define the characteristics of the memory has not yet been determined, and represents an active area of study.

Acknowledgements. The authors thank A. Aviram, D. Bloor, S.G. Boxer, C. Bräuchle, Z. Chen, A.B. Druzhko, T.G. Ebrey, M.A. El-Sayed, D. Govender, E. Greenbaum, R.B. Gross, J. Lanyi, R.A. Mathies, R.A. Needleman, D. Newns, N. Hampp, D. Oesterhelt, G.W. Rayfield, P.M. Rentzepis, H. Sasabe, W. Stoeckenius, J.A. Stuart, E.H.L. Tan,

A.R. Tanguay Jr., B.W. Vought, and M. Yamazaki for interesting and helpful discussions. The research in the authors' laboratory was sponsored in part by grants from the National Institutes of Health, Air Force Rome Laboratories, the W.M. Keck Foundation and the Corporate Affiliates of the W.M. Keck Center for Molecular Electronics.

5 References

1 Birge RR. *Annu Rev Phys Chem* 1990; **41**: 683.
2 Oesterhelt D, Bräuchle C, Hampp N. *Q Rev Biophys* 1991; **24**: 425.
3 Stuart JA, Birge RR. Characterization of the primary photochemical events in bacteriorhodopsin and rhodopsin. In: Lee AG, ed. *Biomembranes*, Vol. 2A. London: JAI Press, 1996: 33.
4 Henderson R, Baldwin JM, Ceska TA, Zemlin F, Beckmann E, Downing KH. *J Mol Biol* 1990; **213**: 899.
5 Greenbaum E. *J Phys Chem* 1992; **96**: 514.
6 Greenbaum E. *J Phys Chem* 1990; **94**: 6151.
7 Borgstahl GEO, Williams DR, Getzoff ED. *Biochemistry* 1995; **34**: 6278.
8 Hoff WD, Van Stokkum IHM, Van Ramesdonk HJ, *et al. Biophys J* 1994; **67**: 1691.
9 Boxer SG, Stocker J, Franzen S, Salafsky J. Re-engineering photosynthetic reaction centers. In: Aviram A, ed. *Molecular Electronics — Science and Technology*, Vol. 262. New York: AIPhys, 1992: 226.
10 Oesterhelt D, Stoeckenius W. *Nature (London), New Biol* 1971; **223**: 149.
11 Birge RR. *Biochim Biophys Acta* 1990; **1016**: 293.
12 Popp A, Wolperdinger M, Hampp N, Bräuchle C, Oesterhelt D. *Biophys J* 1993; **65**: 1449.
13 Shen Y, Safinya CR, Liang KS, Ruppert AF, Rothschild KJ. *Nature* 1993; **366**: 48.
14 Lukashev EP, Robertson B. *Bioelectrochem Bioenerget* 1995; **37**: 157.
15 Vsevolodov NN, Druzhko AB, Djukova TV. In: Hong FT, ed. *Molecular Electronics: Biosensors and Biocomputers*. New York: Plenum Press, 1989: 381.
16 Balashov SP, Litvin FF, Sineshchekov VA. *Physiochem Biol Rev* 1988; **8**: 1.
17 Birge RR. *IEEE Comp* 1992; **25**: 56.
18 Hampp N, Thoma R, Zeisel D, Bräuchle C. *Adv Chem* 1994; **240**: 511.
19 Birge RR. *Am Sci* 1994; **82**: 349.
20 Birge RR, Chen Z, Govender D *et al. CRC Handbook of Organic Photochemistry Photobiology* 1995; 1568.
21 Birge RR. *Sci Am* 1995; **272**: 90.
22 Birge RR, Gross RB. In: Petty MC, Bryce MR, Bloor D, eds. *Introduction to Molecular Electronics*. London: Edward Arnold, 1995: 315.
23 Stuart JA, Vought BW, Zhang CF, Birge RR. *Biospectroscopy* 1995; **1**: 9.
24 Bazhenov VY, Soskin MS, Taranenko VB, Vasnetsov MV. In: Arsenault HH, Szoplik T, Macukow B, eds. *Optical Processing and Computing*. New York: Academic, 1989: 103.
25 Hampp N, Bräuchle C, Oesterhelt D. *Biophys J* 1990; **58**: 83.
26 Thoma R, Hampp N, Bräuchle C, Oesterhelt D. *Opt Lett* 1991; **16**: 651.
27 Bräuchle C, Hampp N, Oesterhelt D. *Adv Mater* 1991; **3**: 420.
28 Hampp N, Popp A, Bräuchle C, Oesterhelt D. *J Phys Chem* 1992; **96**: 4679.
29 Song QW, Zhang C, Gross R, Birge RR. *Opt Lett* 1993; **18**: 775.
30 Song QW, Zhang C, Blumer R, Gross RB, Chen Z, Birge RR. *Opt Lett* 1993; **18**: 1373.
31 Druzhko AB, Chamorovsky SK. *BioSystems* 1995; **35**: 133.
32 Tan EHL, Govender DSK, Birge RR. *J Am Chem Soc* 1996; **118**: 2752.
33 Nuss MC, Zinth W, Kaiser W, Kolling E, Oesterhelt D. *Chem Phys Lett* 1985; **117**: 1.
34 Mathies RA, Lugtenburg J, Shank CV. In: Birge RR, Mantsch HH, eds. *Biomolecular Spectroscopy*, Vol. 1057. Bellingham, Washington: Int. Soc. Optical Engineering, 1989: 138.
35 Mathies RA, Brito Cruz CH, Pollard WT, Shank CV. *Science* 1988; **240**: 777.

36 Birge RR, Zhang CF, Lawrence AF. In: Hong F, ed. *Molecular Electronics*. New York: Plenum, 1989: 369.

37 Simmeth R, Rayfield GW. *Biophys J* 1990; **57**: 1099.

38 Chang C-H, Jonas R, Melchiore S, Govindhee R, Ebrey TG. *Biophys J* 1986; **49**: 731.

39 Jonas R, Ebrey TG. *Proc Natl Acad Sci USA* 1991; **88**: 149.

40 Zhang YN, Sweetman LL, Awad ES, El-Sayed MA. *Biophys J* 1992; **61**: 1201.

41 Chang C-H, Jonas R, Govindjee R, Ebrey TG. *Photochem Photobiol* 1988; **47**: 261.

42 Kimura Y, Ikegami A, Stoeckenius W. *Photochem Photobiol* 1984; **40**: 641.

43 Duñach M, Seigneuret M, Riguad J-L, Padròs E. *J Biol Chem* 1988; **263**: 17378.

44 Fischer UC, Towner P, Oesterheldt D. *Photochem Photobiol* 1981; **33**: 529.

45 Maeda A, Tatsuo I, Yoshizawa T. *Photochem Photobiol* 1981; **33**: 559.

46 Chang C-H, Liu SY, Jonas R, Govindjee R. *Biophys J* 1987; **52**: 617.

47 Liu SY, Ebrey TG. *Photochem Photobiol* 1987; **46**: 557.

48 Liu SY, Ebrey TG. *Photochem Photobiol* 1987; **46**: 263.

49 Pande C, Callender RH, Chang CH, Ebrey TG. *Biophys J* 1986; **50**: 545.

50 Gross RB. PhD thesis, Syracuse University, 1995.

51 Vsevolodov NN, Poltoratskii VA. *Sov Phys Tech Phys* 1985; **30**: 1235.

52 Birge RR, Fleitz PA, Gross RB *et al. Proc. IEEE EMBS* 1990; **12**: 1788.

53 Bazhenov VY, Soskin MS, Taranenko VB. *Sov Tech Phys Lett* 1987; **13**: 382.

54 Song QW, Zhang C, Gross RB, Birge RR. *Opt Commun* 1994; **112**: 296.

55 Song QW, Zhang C, Ku CY *et al. Opt Commun* 1995; **115**: 471.

56 Zhang C, Song QW, Ku CY, Gross RB, Birge RR. *Opt Lett* 1994; **19**: 1406.

57 Zhang YH, Song QW, Tseronis C, Birge RR. *Opt Lett* 1995; **20**: 2429.

58 Loudon R. *The Quantum Theory of Light*. Oxford: Clarendon, 1973.

59 Landau LD, Lifshitz EM. *Electrodynamics of Continuous Media*. New York: Pergamon, 1960.

60 Kogelnik H. *Bell Syst Tech J* 1969; **48**: 2909.

61 Gross RB, Izgi KC, Birge RR. *Proc SPIE* 1992; **1662**: 186.

62 Sheik-bahate M, Said AA, Stryland EW. *Opt Lett* 1989; **14**: 955.

63 Gross RB, Todorov AT, Birge RR. In: Lampropoulos GA, ed. *Applications of Photonic Technology*. New York: Plenum Press, 1995: 115.

64 Tanguay Jr AR. *Opt News* 1988; Feb: 23.

65 Tanguay Jr AR. *Opt Eng* 1985; **24**: 2.

66 Mobarry C, Lewis A. *Proc. SPIE* 1986; **700**: 304.

67 Werner O, Daisy R, Fisher B, Lewis A. *Opt Commun* 1992; **92**: 108.

68 Werner O, Fisher B, Lewis A. *Opt Lett* 1992; **17**: 241.

69 Werner O, Fisher B, Lewis A, Nebenzahl I. *Opt Lett* 1990; **15**: 1117.

70 Haronian D, Lewis A. *Appl Phys Lett* 1992; **61**: 2237.

71 Chen Z, Lewis A, Takei H, Nabenzahl I. *Appl Opt* 1991; **30**: 5188.

72 Chen Z, Takei H, Lewis A. *Proc Int Joint Conf Neural Networks* 1990; **II**: 803.

73 Takei H, Lewis A, Chen Z, Nebenzahl I. *Appl Opt* 1992; **30**: 500.

74 Haronian D, Lewis A. *Appl Opt* 1991; **30**: 597.

75 McMaster E, Lewis A. *Biochem Biophys Res Commun* 1988; **156**: 86.

76 Druzhko AB, Zharmukhamedov SK. In: Ivanitskiy GR, Vsevolodov NN, eds. *Photosensitive Biological Complexes and Optical Recording of Information*. Biological Research Center, Institute of Biological Physics, Pushchino: USSR Academy of Sciences, 1985: 119.

77 Ivanitskiy GR, Vsevolodov NN, eds. *Photosensitive Biological Complexes and Optical Recording of Information*. Biological Research Center, Institute of Biological Physics, Pushchino: USSR Academy of Sciences, 1985.

78 Savranskiy VV, Tkachenko NV, Chukharev VI. In: Ivanitskiy GR, Vsevolodov NN, eds. *Photosensitive Biological Complexes and Optical Recording of Information*. Biological Research Center, Institute of Biological Physics, Pushchino: USSR Academy of Sciences, 1985: 97.

79 Paek EG, Psaltis D. *Opt Eng* 1987; **26**: 428.

80 Parthenopoulos DA, Rentzepis PM. *Science* 1989; **245**: 843.

81 Hunter S, Kiamilev F, Esener S, Parthenopoulos DA, Rentzepis PM. *Appl Opt* 1990; **29**: 2058.

82 Lawrence AF, Birge RR. *Proc SPIE* 1992; **1773**: 401.

83 Birge RR. In: Kliger DS, ed. *Ultrasensitive Laser Spectroscopy*. New York: Academic Press, 1983: 109.

84 Birge RR. In: Sandorfy C, Theophanides T, eds. *Spectroscopy of Biological Molecules*, Boston: D. Reidel, 1984: 457.

85 McClain WM, Harris RA. In: Lim EC, ed. *Excited States*, Vol. 3. New York: Academic Press, 1977: 1.

86 Pierce BM, Birge RR. *Int J Quant Chem* 1986; **29**: 639.

87 Birge RR, Pierce BM, Murray LP. In: Sandorfy C, Theophanides T, eds. *Spectroscopy of Biological Molecules*. Boston: D. Reidel, 1984: 473.

88 Mortensen OS, Svendsen EN. *J Chem Phys* 1981; **74**: 3185.

89 Birge RR, Zhang CF. *J Chem Phys* 1990; **92**: 7178.

90 Birge RR, Murray LP, Pierce BM *et al. Proc. Natl Acad Sci USA* 1985; **82**: 4117.

91 Birge RR, Gross RB, Masthay MB *et al. Mol Cryst Liq Cryst Sci Technol Sec B Nonlinear Optics* 1992; **3**: 133.

92 Dvornikov AS, Rentzepis PM. *Adv Chem* 1994; **240**: 161.

93 Chen Z, Govender D, Gross R, Birge RR. *BioSystems* 1995; **35**: 145.

94 Marhic ME. *Opt Lett* 1991; **16**: 1272.

95 Birge RR, Govender DSK, Gross RB *et al. IEEE IEDM Tech Dig* 1994; **94**: 3.

96 Stuart JA, Tallent JR, Tan EHL, Birge RR. *Proc IEEE Nonvol Mem Tech (INVMTC)* 1996; **6**: 45.

Index

Note: page numbers in **bold** refer to tables and those in *italic* to figures.

Peter Anderson

19 ASHBROOK DRIVE

BELFAST

BT4 2FG

Tel. 01232 654223